AF594099

Wavelets: Theory, Algorithms, and Applications

Wavelet Analysis and Its Applications

The subject of wavelet analysis has recently drawn a great deal of attention from mathematical scientists in various disciplines. It is creating a common link between mathematicians, physicists, and electrical engineers. This book series will consist of both monographs and edited volumes on the theory and applications of this rapidly developing subject. Its objective is to meet the needs of academic, industrial, and governmental researchers, as well as to provide instructional material for teaching at both the undergraduate and graduate levels.

This is the fifth book of the series. It is a carefully edited volume that covers seven active research areas in this field. Multiresolution analysis, being the most powerful tool for constructing wavelets, is the first topic. It consists of at least four interesting points of view, ranging from non-stationary schemes to spectral theory. The second topic of discussion is wavelet transforms, where not only the geometric and algebraic aspects are studied, but periodic wavelet transforms are also investigated. Because spline functions are both local and refinable, they are particularly useful for representing wavelets. Spline-wavelets, as they are usually called, constitute the third topic of study in this volume. Polynomial splines and L-splines are discussed in the one-variable setting, and the four-directional box spline is used to construct bivariate frames. To analyze a signal adaptively, other localization schemes are required. The fourth topic of discussion includes Donoho's minimum entropy segmentation, Mallat's matching pursuit approach, and generalized Malvar wavelets. The property of self-similarity is common to both fractals and scaling functions, and hence wavelets. To link these two subjects, two interesting papers, constituting the fifth topic, are included. To the reader who is concerned with computations or applications, the last two topics are most important. A collection of five comprehensive articles on wavelet solution of PDE's, numerical computation, and algorithms, represents the sixth topic. Six contributions on applications constitute the seventh and concluding part of this volume. Topics included are: compression, analysis, and experimental study of turbulence; detection of seismic activities and mean-value jumps of a random process; as well as a comparison of four image compression methods: wavelet, wavelet packet, local cosine transform, and an adaptive best basis scheme. The series editor is most grateful to the authors of these 28 chapters for their outstanding contributions.

This is a volume in
WAVELET ANALYSIS AND ITS APPLICATIONS

Charles K. Chui, Series Editor
Texas A&M University, College Station, Texas

A list of titles in this series appears at the end of this volume.

Wavelets: Theory, Algorithms, and Applications

Edited by

Charles K. Chui

Department of Mathematics
Texas A&M University
College Station, Texas

Laura Montefusco

Departemento de Matematica
Università di Messina
Bologna, Italy

Luigia Puccio

Departemento de Matematica
Università di Messina
Messina, Italy

ACADEMIC PRESS, INC.

San Diego New York Boston
London Sydney Tokyo Toronto

This book is printed on acid-free paper. ♾

Academic Press, Inc.
A Division of Harcourt Brace & Company
525 B Street, Suite 1900, San Diego, California 92101-4495

United Kingdom Edition published by
Academic Press Limited
24-28 Oval Road, London NW1 7DX

Library of Congress Cataloging-in-Publication Data

Wavelets : theory, algorithms, and applications / edited by Charles K. Chui, Laura Montefusco, Luigia Puccio
p. cm. -- (Wavelet analysis and its applications : v. 5)
Includes index.
ISBN 0-12-174575-9
1. Wavelets. I. Chui, C. K. II. Montefusco, Laura. III. Puccio, Luigia. IV. Series.
QA403.3.W43 1994
515'2433--dc20 94-34457
CIP

PRINTED IN THE UNITED STATES OF AMERICA
94 95 96 97 98 99 MM 9 8 7 6 5 4 3 2 1

Contents

Part IV. Other Mathematical Tools for Time-Frequency Analysis

Part V. Wavelets and Fractals

Part VI. Numerical Methods and Algorithms

Part VII. Applications

Contributors

Numbers in parentheses indicate pages on which authors' contributions begin.

JEAN-LUC ATTIÉ (543), *Laboratoire d'Aérologie, Université Paul Sabatier, F-31077 Toulouse, France*

BRUNO BÉNECH (543), *Laboratoire d'Aérologie, Université Paul Sabatier, F-31077 Toulouse, France*

SILVIA BERTOLUZZA (425), *Instituto di Analisi Numerica, Consiglio Nazionale delle Ricerche, 27100 Pavia, Italy*

FRANCIS CASTANIÉ (573), *LEN7-GAPSE, ENSEEIHT, 31 071 Toulouse, France*

CHARLES K. CHUI (213), *Department of Mathematics, Center for Approximation Theory, Texas A&M University, College Station, Texas 77843*

ALBERT COHEN (3), *CEREMADE, Université Paris IX Dauphine, 75775 Paris, France*

GUY COURBEBAISSE (311), *ICPI-CPE-LTS, 69288 Lyon, France*

ECHEYDE CUBILLO (509), *Center for Theoretical Studies of Physical Systems, Clark Atlanta University, Atlanta, Georgia 30314*

STEPHAN DAHLKE (33), *Institut für Geometrie und Praktische Mathematik, RWTH Aachen, 52062 Aachen, Germany*

WOLFGANG DAHMEN (385), *Institut für Geometrie und Praktische Mathematik, RWTH Aachen, 52056 Aachen, Germany*

GEOFFREY DAVIS (271), *Computer Sciences Department, Courant Institute of Mathematical Sciences, New York University, New York, New York 10012*

LEONARDO DE ABREU SÀ (543), *Instituto Nacional de Pesquisas Espacials, Saõ Jose dos Campos, Brazil*

MIREILLE DEFRANCESCHI (495), *CEA/DSM/DRECAM/SRSIM, Ce - Saclay, F-91191 Gif-sur-Yvette, France*

ALINE DENJEAN (573), *LEN7-GAPSE, ENSEEIHT, 31 071 Toulouse, France*

DAVID L. DONOHO (233), *Department of Statistics, Stanford University, Stanford, California 94305*

AIMÉ DRUILHET (543), *Laboratoire d'Aérologie, Université Paul Sabatier, F-31077 Toulouse, France*

PIERRE DURAND (543), *Laboratoire d'Aérologie, Université Paul Sabatier, F-31077 Toulouse, France*

BERNARD ESCUDIÉ (311), *ICPI-CPE-LTS, 69288 Lyon, France*

MARIE FARGE (509), *Laboratoire de Metéorologie Dynamique du CNRS, Ecole Normale Supèrieure, 75231 Paris, France*

PATRICK FISCHER (495), *CEA / DSM / DRECAM / SRSIM, Ce - Saclay, F-91191 Gif-sur-Yvette, France*

CARL A. FRIEHE (533), *Department of Mechanical Engineering, University of California, Irvine, California 92717*

ERIC GOIRAND (509), *Mathematics Department, Washington University, St. Louis, Missouri 63130*

T. N. T. GOODMAN (53), *Department of Mathematical Sciences, University of Dundee, Dundee DD1 4HN, Scotland, United Kingdom*

PETER NIELS HELLER (13), *Aware, Inc., Cambridge, Massachusetts 02142*

MATTHIAS HOLSCHNEIDER (363), *CPT CNRS Luminy, F-13288 Marseille, France*

LONNIE HUDGINS (533), *Electronics Systems Division, Northrop Corporation, Hawthorne, California 90251*

STEPHANE JAFFARD (325), *CERMA, ENPC, La Courtine, 93167 Noisy le Grand, France; and CMLA, ENS Cachan, 94235 Cachan, France*

KURT JETTER (213), *FB Mathematik, Universität Duisburg, D-47048 Duisburg, Germany*

DMITRI KARAYANNAKIS (449), *Department of Physics, University of Crete, Iraklion, Crete, Greece*

JAROSLAV KAUTSKY (117), *School of Information Science and Technology, Flinders University and Cooperative Research Center for Sensor, Signal, and Information Processing (CSSIP), Adelaide, SA 5001, Australia*

LETIZIA LO PRESTI (561), *Department of Electronics, Politecnico, I-10129 Torino, Italy*

TOM LYCHE (197), *Institutt for Informatikk, University of Oslo, 0316 Oslo, Norway*

STÉPHANE MALLAT (271), *Computer Sciences Department, Courant Institute of Mathematical Sciences, New York University, New York, New York 10012*

MEINHARD E. MAYER (533), *Department of Physics, University of California, Irvine, California 92717*

CHARLES A. MICCHELLI (53), *IBM Research Division, T. J. Watson Research Center, Yorktown Heights, New York 10598*

LAURA BACCHELLI MONTEFUSCO (459), *Department of Mathematics, University of Bologna, 40127 Bologna, Italy*

GIOVANNI NALDI (425), *Dipartimento di Matematica, Università di Pavia, 27100 Pavia, Italy*

GABRIELLA OLMO (561), *Department of Electronics, Politecnico, I-10129 Torino, Italy*

MARK E. OXLEY (295), *Department of Mathematics and Statistics, Air Force Institute of Technology, Wright-Patterson AFB, Ohio 45433*

THIERRY PAUL (311), *CEREMADE, Université Paris IX Dauphine, 75775 Paris, France*

GERLIND PLONKA (137), *Fachbereich Mathematik, Universität Rostock, D-18051 Rostock, Germany*

SIEGFRIED PRÖSSDORF (385), *Institut für Geometrie und Praktische Mathematik, RWTH Aachen, 52056 Aachen, Germany*

JEAN CHRISTOPHE RAVEL (425), *Laboratoire d'Analyse Numérique, Université Pierre et Marie Curie, 75252 Paris, France*

SUSUMU SAKAKIBARA (179), *College of Science and Engineering, Iwaki Meisei University, Fukushima-ken 970, Japan*

REINHOLD SCHNEIDER (385), *Institut für Geometrie und Praktische Mathematik, RWTH Aachen, 52056 Aachen, Germany*

LARRY L. SCHUMAKER (197), *Department of Mathematics, Vanderbilt University, Nashville, Tennessee 37240*

GABRIELE STEIDL (155), *TH Darmstadt, Fachbereich Mathematik, 64289 Darmstadt, Germany*

JOACHIM STÖCKLER (213), *FB Mathematik, Universität Duisburg, D-47048 Duisburg, Germany*

BRUCE W. SUTER (295), *Department of Electrical and Computer Engineering, Air Force Institute of Technology, Wright-Patterson AFB, Ohio 45433*

MANFRED TASCHE (137), *Fachbereich Mathematik, Universität Rostock, D-18051 Rostock, Germany*

BRUNO TORRÉSANI (91), *CPT, CNRS-Luminy, 13288 Marseille, France*

RADKA TURCAJOVÁ (117), *School of Information Science and Technology, Flinders University and Cooperative Research Center for Sensor, Signal, and Information Processing (CSSIP), Adelaide, SA 5001, Australia*

RAYMOND O. WELLS, JR. (13), *Computational Mathematics Laboratory, Rice University, Houston, Texas 77251*

EVA WESFREID (509), *CEREMADE, Université Paris-Dauphine, 5775 Paris, France*

MLADEN VICTOR WICKERHAUSER (585), *Department of Mathematics, Washington University, St. Louis, Missouri 63130*

ZHIFENG ZHANG (271), *Computer Sciences Department, Courant Institute of Mathematical Sciences, New York University, New York, New York 10012*

Preface

Interest in wavelet analysis has been growing very rapidly in recent years. Advances in its theory, algorithms, and applications have greatly influenced the development of many disciplines of science and technology, including mathematics, physics, engineering, geosciences, meteorology, and will continue to influence other areas as well. However, while the rate of growth in each of these three directions of wavelet analysis has been phenomenal individually, the interaction among them is somewhat insufficient; and this creates an obstacle for the advancement of the field. Motivated by this need, we decided to invite several leading experts from each of the three areas and publish a well-balanced volume based on their contributions. With financial support from the Consiglio Nazionale delle Ricerche of Italy, we organized the International Conference on Wavelets, held in Taormina, Italy, during the period from October 14 to October 20, 1993. The conference brought together experts and professionals specializing in wavelet theories, algorithms, and applications, from four continents. This volume consists mainly of articles by the invited speakers and participants at the conference. Out of the 28 articles, 12 are invited papers by the main speakers, and 15 are selected from contributions by other participants. The author of the remaining paper, Donoho, was unable to accept our invitation to speak, but graciously agreed to contribute, which significantly strengthens the algorithmic area of this volume.

The seven topics of this volume serve as a reader's guide and are not intended to separate these somewhat closely related chapters into different areas. In addition, in order to avoid unnecessary discontinuities, chapters within each part are arranged according to subject matter, not alphabetically by contributor. Since the approach of multiresolution analysis has proved to be the most fruitful, not only for constructing wavelets, but also for algorithmic development, the first of the seven parts in this volume is devoted to multiresolution and multiscale analyses. Cohen allows the finite two-scale sequences to change from one iteration to another in order to construct compactly supported scaling functions that are infinitely differentiable. Heller and Wells discuss the spectral theory of multiresolution operators. In particular, they give certain asymptotic estimates of the Sobolev smoothness of rank 3 scaling functions for large genus. Dahlke generalizes the notion of multiresolution analysis to certain Riemannian manifolds. On the other hand, Goodman and Micchelli investigate various questions related to the important properties of orthogonality, compact support, finite bandwidth, symmetry, cardinal interpolation, and the two-scale relation that induces a multiresolution analysis.

The second part of this volume is devoted to the study of wavelet transforms from different points of view. Geometric aspects of continuous wavelet transforms are discussed by Torrésani. For example, a scheme for constructing wavelet decompositions of functions in homogeneous spaces is introduced with respect to certain group action. An approach based mainly on matrix algebra is described by Kautsky and Turcajová. In particular, the properties of banded block circulant matrices are used to study the conditions of orthogonality, biorthogonality, regularity, and symmetry of discrete wavelet transforms. Along this line, Plonka and Tasche discuss discrete wavelet transforms in the periodic setting. For example, they introduce a new approach to periodic wavelet analysis based on Fourier techniques and propose certain efficient algorithms for decompositions and reconstructions using FFT.

Spline functions are refinable by knot insertions and removals. In particular, cardinal B-splines are usually used to demonstrate the concept of multiresolution analysis. To give a firm mathematical setting, we begin the third part of this volume with a chapter by Steidl that studies a unified approach to cardinal, periodic, and discrete spline wavelets with any integer dilation factor $n \geqslant 2$. This is done by using Fourier analysis on locally compact albelian groups. We then specialize to polynomial spline wavelets with a contribution by Sakakibara that describes a method of implementation of the Chui–Wang spline wavelets for data smoothing. These polynomial spline wavelets are generalized by Lyche and Schumaker by using L-splines in the third chapter. They give an explicit construction of these compactly supported L-spline wavelets and apply their theory to develop multiresolution methods based on L-splines. On the other hand, Chui, Jetter, and Stöckler discuss the generalization to the higher dimensional setting by using box splines. In the bivariate case, a four-directional box spline is very attractive for surface modeling since it is most symmetric. However, since its integer translates do not constitute a Riesz basis, there is some difficulty in generating a stable wavelet basis with compact support. In this chapter, by using oversampling, the corresponding compactly supported semi-orthogonal wavelet with dilation matrix $A = \begin{pmatrix} 1 & 1 \\ 1 & -1 \end{pmatrix}$ is shown to generate a frame.

To analyze a signal adaptively, other localization schemes are sometimes needed. The fourth part of this volume is devoted to the description of other mathematical tools for time-frequency analysis. The minimum entropy segmentation approach is introduced by Donoho. In his chapter, he describes segmented multiresolution analyses of a bounded interval that lead to segmented wavelets that are well adapted to discontinuities, cusps, etc. This approach emphasizes the idea of average interpolation and leads to the adaptive wavelet transform of a signal of length n with only $O(n \log n)$ computations. In addition, Donoho describes an iterative approach, called segmentation pursuit, for edge identification. The adaptive method is described in the chapter by Davis, Mallat, and Zhang. A matching pursuit algorithm is introduced to compute a suboptimal expansion of a signal in a redundant dictionary of waveforms to best match the structures of the signal. In addition, they describe an algorithm for isolating the coherent structures of a signal and demonstrate with an application to pattern extraction in a noisy environment. Perhaps the best known adaptive wavelets are the so-called Malvar wavelets. Suter and Oxley give a complex-valued generalization and use these wavelets to get around the Balian-Low principle, obtaining something close to an exponential basis. The Wigner representation is another time–frequency localization tool. The final chapter in this part introduces an affine Wigner distribution and relates it with the affine Wigner operator. The authors, Courbebaisse, Escudié, and Paul, use a family of coherent states in the derivation of their expression and give a computational scheme.

Even before wavelets, fractals were already quite popular as a powerful image compression tool. The link between wavelets and fractals is self-similarity. The fifth part consists of two interesting contributions by two experts in this area. Jaffard investigates the mathematical validity of the multifractal formalism for functions, shows that it always yields an upper-bound for the spectrum of singularities, and demonstrates its validity by studying the Riemann function. Holschneider gives a new definition of fractal dimension in terms of certain small-scale behavior of the q-energy of wavelet transform. This is a generalization of the previous multifractal approaches and can be used to show that the correlation dimension of the spectral measure determines the long-time behavior of the time evolution generated by bounded self-adjoint operator acting on some Hilbert space.

We now come to numerical methods and algorithms. It is well-known that wavelets provide a hierarchical structure similar to the multi-level approach in numerical pseudo-differential equations (PDEs). The first two papers in this sixth part of the volume are concerned with this point of view. Dahmen, Prössdorf, and Schneider discuss the numerical solution of PDEs on smooth manifold. In particular, some relevant concepts and basic properties of a multiscale framework that facilitate an efficient and stable matrix compression for fast solutions of Galerkin-discretized linear systems are presented. The second chapter on numerical solutions is contributed by Bertoluzza, Naldi, and Ravel. They are concerned with boundary value problems on an interval and focus on the elliptic case. Both Galerkin and wavelet-collocation methods are described. The remaining chapters in this part are concerned with computations and algorithms. Karayannakis derives both upper and lower bound estimates for the value of the Franklin wavelet at its exceptional node 3/2, confirming a conjecture of Berkson. All calculations presented in this chapter can be carried out on an ordinary scientific pocket calculator. At the other end of the spectrum in calculations, Montefusco presents certain coarse-grained parallel algorithms for some basic problems of large-scale numerical linear algebra. It is shown that, starting from an efficient parallel wavelet packet matrix decomposition, these algorithms can reach good efficiency on distributed memory multiprocessors and require only few nearest neighbor communications. The last chapter in this part investigates the representation of the atomic Hartree-Fock equations in a wavelet basis by means of the Beylkin-Coifman-Rokhlin algorithm. The authors, Fischer and Defranceschi, demonstrate the effectiveness of their results by comparing with the Slater function, which provides the exact solution using the energy value as their criterion.

The seventh and final part of this volume is devoted to applications of wavelets to three important areas. The first three chapters are in the area of turbulence. Wickerhauser, Farge, Goirand, Wesfreid, and Cubillo compare the efficiency of two rank-reduction methods for representing the essential features of a two-dimensional turbulent vorticity field. The comparison is on efficiency of capturing the enstrophy or square-vorticity, the faithfulness to the power spectra, and the precision in resolving coherent structures. The second turbulence chapter by Mayer, Hudgins, and Friehe, presents a wavelet cospectral analysis of the boundary-layer flow in the atmosphere above ground. The cospectra show a bimodal structure similar to that previously found in shear-generated turbulence over ocean. The third chapter on turbulence deals with an experimental study of inhomogeneous turbulence in the lower-troposphere. The authors, Druilhet, Attié, de Abreu Sá, Durand, and Bénech, use the wavelet transform to analyze aircraft data for this purpose. Monitoring of seismic activities is the subject of the next contribution on wavelet applications. Olmo and Lo Presti demonstrate the effectiveness of the wavelet transform for estimation of arrival times of various seismic phases and for the capability as a de-noising process. The second detection application, presented by Denjean and Castanié, is on detections of mean-value jumps of a stationary random process. It surveys both the conventional method and the wavelet-based method. Perhaps the most well-known applications of wavelet techniques is to image compression. Wickerhauser gives a comprehensive comparison of these methods: wavelets, wavelet packets, local sines and cosines, and an adaptive best-basis method. Standard C algorithms for the local sine and cosine transforms are included in the last section of this interesting and useful chapter.

The collection of graphics featured in this volume is impressive. There are more than 135 figures throughout the volume; nine of these are in color, placed in the color section. These figures supply a visual dimension to the authors' texts and frequently illustrate the results of their accomplishments.

The reviewing and editing process, as well as producing the edited volume in a uniform style, requires an enormous amount of work. We are very grateful to the referees for their conscientious reviews and to the authors for their fine contributions. Many of the authors prepared their manuscripts in TEX and this helped a great deal in the editing process. Robin Campbell is to be thanked for typesetting some of the other papers. We are indebted to Margaret Chui for her most conscientious and diligent assistance in the editorial process. The fine quality of the volume owes much to her effort. We are thankful to Peter Renz and his editorial staff at Academic Press, Boston, for their helpful assistance and cooperation, and to N. Sivakumar for his valuable editorial comments. Finally, we would like to acknowledge the generous financial support of Consiglio Nazionale delle Ricerche of Italy.

College Station, Texas
Bologna, Italy
Messina, Italy
June, 1994

Charles K. Chui
Laura Montefusco
Luigia Puccio

Part I

Multiresolution and Multilevel Analyses

Non-stationary Multiscale Analysis

Albert Cohen

Abstract. Orthonormal bases of wavelets and wavelet packets yield linear, non-redundant time-scale and time-frequency decompositions for arbitrary functions and signals. Numerically, these decompositions are based on the iterative application of digital filter banks. Usually these filters are the same at every iteration. We show here the advantages of using filters that may vary from one iteration to the next one. An appropriate choice leads to C^∞ compactly supported wavelets and allows a better control of the time-frequency localization properties of wavelet packets. These results have been obtained jointly with N. Dyn of Tel-Aviv University and E. Séré of CEREMADE.

§1 Introduction

Wavelets and wavelet packets constitute useful tools for the decomposition of complicated functions into a small number of elementary waveforms that are localized both in time and frequency. We recall here the main ideas of their construction (a detailed review can be found in [11]).

One starts with the data of a multiresolution analysis, *i.e.*, a nested sequence of approximation subspaces

$$\{0\} \rightarrow \ldots V_{-2} \subset V_{-1} \subset V_0 \subset V_1 \subset V_2 \ldots \rightarrow L^2(\mathbb{R}) \qquad (1.1)$$

generated by a scaling function $\varphi \in V_0$ in the sense that for each $j \in \mathbf{Z}$ the family $\{\varphi_{j,k}\}_{k\in\mathbf{Z}} = \{2^{j/2}\varphi(2^j x - k)\}_{k\in\mathbf{Z}}$ is an orthonormal basis of V_j.

Such a function can either be expressed as the solution of a "dilation equation"

$$\varphi(x) = 2\sum_{n\in\mathbf{Z}} h_n \varphi(2x-n), \qquad (1.2)$$

or as the limit of a subdivision scheme, *i.e.*, iterative refinements of an initial Dirac sequence by convolutions with the coefficients h_n, since we have

$$\hat{\varphi}(\omega) = \prod_{k=1}^{+\infty} m_0(2^{-k}\omega), \qquad (1.3)$$

Wavelets: Theory, Algorithms, and Applications
Charles K. Chui, Laura Montefusco, and Luigia Puccio (eds.), pp. 3–12.

ISBN 0-12-174575-9

with $m_0(\omega) = \sum_n h_n e^{-in\omega}$ (for this reason the function φ is said to be "refinable" and Equation (1.2) is also called "refinement equation").

The sequence h_n and its Fourier series $m_0(\omega)$ need to satisfy specific constraints that are related to the properties of the function φ:

- If $\varphi \in L^1$, then $\sum_n h_n = m_0(0) = 1$.
- φ is compactly supported in $[p,q]$ if and only if $h_n = 0$ for $n < p$ or $n > q$.
- When the two previous constraints are satisfied, the orthonormality of $\{\varphi(x-k)\}_{k\in\mathbf{Z}}$ is ensured if and only if

$$|m_0(\omega)|^2 + |m_0(\omega+\pi)|^2 = 1 \tag{1.4}$$

and there exists a compact set K, congruent to $[-\pi,\pi]$ modulo 2π, such that $m_0(2^{-j}\omega) \neq 0$ for all $\omega \in K$ and $j > 0$ (see [2]).

From this framework, one derives an orthonormal wavelet by

$$\psi(x) = 2\sum_n g_n \varphi(2x-n), \tag{1.5}$$

with $g_n = (-1)^n h_{1-n}$. The family $\{\psi_{j,k}\}_{k\in\mathbf{Z}}$ is then an orthonormal basis of the orthogonal complement W_j of V_j into V_{j+1}. It characterizes the missing details between two successive levels of approximation. It follows that both $\{\psi_{j,k}\}_{j,k\in\mathbf{Z}}$ and $\{\varphi_{J,k}\}_{k\in\mathbf{Z}} \cup \{\psi_{j,k}\}_{j\geq J,k\in\mathbf{Z}}$ are orthonormal bases of $L^2(\mathbb{R})$.

The decomposition of V_1 into V_0 and W_0, as expressed by Equations (1.2) and (1.5), reveals a more general "splitting trick" : if $\{e_n\}_{n\in\mathbf{Z}}$ is an orthonormal basis of a Hilbert space H, and if h_n and g_n are coefficients that satisfy the previous constraints, then the sequences $u_n = \sqrt{2}\sum_k h_{k-2n}e_k$ and $v_n = \sqrt{2}\sum_k g_{k-2n}e_k$ are orthonormal bases of two orthogonal closed subspaces U and V such that $H = U \oplus V$.

Wavelet packets are obtained by using this trick to split the W_j spaces. More precisely, one can define a family $\{w_n\}_{n\in\mathbf{N}}$ by taking $w_0 = \varphi$, $w_1 = \psi$ and applying the following recursion:

$$w_{2n} = 2\sum_k h_k w_n(2x-k) \ \text{ and } \ w_{2n+1} = 2\sum_k g_k w_n(2x-k). \tag{1.6}$$

The families $\{w_n(x-m)\}_{m\in\mathbf{Z}} = \{w_{m,n}\}$ for $2^j \leq n < 2^{j+1}$ are the results of j splitting of the space W_j. Consequently, $\{w_{m,n}\}_{n\in\mathbf{N},m\in\mathbf{Z}}$ is an orthonormal basis of $L^2(\mathbb{R})$.

This is only a particular example of a wavelet packet basis. Many other bases can be obtained by splitting W_j more or less than j times. However, all the elements of these bases have the general form

$$w_{j,m,n}(x) = 2^{j/2} w_n(2^j x - m). \tag{1.7}$$

Such bases yield arbitrary decompositions of any discrete signal in the time-frequency plane, by means of fast algorithms. One assumes that the initial data $\{s_k\}_{k\in\mathbf{Z}}$ represents a function in V_j, *i.e.*, $f = \sum_k s_k w_{j,k,0}$. The splitting trick indicates that the coordinates of f in the basis $\{w_{m,n}\}_{m\in\mathbf{Z},0\leq n<2^j}$ can be obtained by j iterative convolutions of s_k with the discrete filters h_n and g_n, followed by a decimation of one sample out of two. Again, one has the possibility of splitting more or less than j times and use entropy tests to select the sparsest decomposition (see [8]). The computational cost is of order $CN\log N$, where N is the size of the initial data and C is the length of the filters h_n and g_n.

Wavelets and wavelet packets suffer, however, from two limitations:

- The regularity of the functions φ and ψ are bounded by the number of coefficients in the filter. In particular, there exists no C^∞ scaling function with compact support.
- The wavelet packets w_n suffer from a lack of frequency localization as n grows.

In the next section we analyze these two limitations in more details. Remarking that it is not necessary to use constantly the same pair of filters in the different splitting stages, since each of them is performed independently, we use this potential freedom in Section 3 to build wavelets that are both compactly supported and in C^∞ and in Section 4 to improve the properties of time-frequency localization of wavelet packets. The detailed proofs of these results, only sketched here, can be found in [5,6].

§2 Regularity and time-frequency localization in the standard case

Additional constraints can be put on $m_0(\omega)$ so that $\varphi(x)$ has some regularity. In particular, a family of trigonometric polynomials given by

$$m_{0,L}(\omega) = \Big(\frac{1+e^{i\omega}}{2}\Big)^L p_L(\omega), \quad L \in \mathbb{N}^*, \tag{2.1}$$

with

$$|p_L(\omega)|^2 = \sum_{j=0}^{L-1} \binom{L-1+j}{j} \sin^{2j}(\omega/2), \tag{2.2}$$

for all $L \leq 1$, has been constructed by Daubechies [9] in such a way that the smoothness of the associated scaling functions φ_L grows linearly with L, in the asymptotic sense. Numerous contributions have been made on the more general problem of evaluating the smoothness of the limit function generated by a subdivision scheme. (See [10,11,12,14,15,17], among others.)

A basic result is the following. If $\varphi \in C^r$ then necessarily

$$m_0(\omega) = \Big(\frac{1+e^{i\omega}}{2}\Big)^m p(\omega), \quad L \in \mathbb{N}^*, \tag{2.3}$$

with $m > r$. This means, in particular, that the number of non-zero h_n is bigger than r and that a scaling function cannot be both compactly supported and C^∞.

Let us now consider the wavelet packets w_n as defined in our introduction. These functions have a simple expression in the frequency domain. If we define $m_1(\omega) = \sum_n g_n e^{-in\omega} = e^{-i\omega} m_0(\omega + \pi)$, we can derive from Equations (1.3) and (1.6), for all $n < 2^j$,

$$\hat{w}_n(\omega) = [\prod_{k=1}^{n} m_{\varepsilon_k}(2^{-k}\omega)]\hat{w}_0(2^{-j}\omega) \tag{2.4}$$

with $n = \sum_{k=1}^{j} \varepsilon_k 2^{k-1}$. Two observations can be made from this formula:

- If $m_0(\omega)$ is a trigonometric polynomial of degree L, then all the functions w_n have a compact support of length L.
- In the case where $m_0(\omega) = \chi_{[-\pi/2,\pi/2]}(\omega)$, one has $\hat{w}_n(\omega) = \chi_{-J_n \cup J_n}(\omega)$ where the interval $J_n = [\pi g(n), \pi(g(n)+1)]$ is determined by the "Gray code" $g(n) = \sum_{k=1}^{j} \gamma_k 2^{k-1}$, $\gamma_j = \varepsilon_j$ and $\gamma_k = |\varepsilon_{k+1} - \varepsilon_k|$ for $1 \le k < j$.

The last observation suggests that if $m_0(\omega)$ is a trigonometric polynomial "close" to the ideal "cut-off filter" $\chi_{[-\pi/2,\pi/2]}(\omega)$, the function $w_{m,n}$ is localized both in space (around m) and in frequency (around $\pm\pi g(n)$), with a variance that does not depend on m and n in both domains. Unfortunately, this is not true. Several recent works (*e.g.*, [7,16]) have stated negative results concerning the frequency localization. For example, it was proved by Coifman, Meyer, and Wickerhauser [7] that the variances

$$\sigma_n = \min_{\omega_0 \ge 0} \int_0^{+\infty} |\hat{w}_n(\omega)|^2 |\omega - \omega_0|^2 d\omega \tag{2.5}$$

are not bounded in average, in the sense that $2^{-j}\sum_{n=0}^{2^j-1} \sigma_n$ goes to $+\infty$ as j increases. This type of result reveals the limitations of wavelet packets analysis in terms of time-frequency localization. The following result of Séré (1993) is even stronger.

Theorem 2.1. *Let w_n be the standard wavelet packets associated to finite filters. Then for any $M > 0$, we have*

$$\lim_{n\to\infty} \frac{1}{n} \operatorname{card}\{k \le n : \sigma_k > M\} = 1. \tag{2.6}$$

We now turn to the generalization in which one allows to use different filters at every splitting stage.

§3 Non-stationary wavelets

Let $\{m_0^k(\omega)\}_{k>0}$ be a family of trigonometric polynomials of degree $d(k)$ that satisfies the following properties:

$$|m_0^k(\omega)|^2 + |m_0^k(\omega+\pi)|^2 = 1, \tag{3.1}$$

$$m_0^k(0) = 1, \tag{3.2}$$

$$|\omega| \leq \pi/2 \Rightarrow m_0^k(\omega) \geq C > 0, \tag{3.3}$$

and

$$\sum_{k>0} 2^{-k} d(k) < +\infty. \tag{3.4}$$

(Condition (3.3) could be replaced by a weaker one: the existence of a compact set K congruous to $[-\pi, \pi]$ modulo 2π such that $m_0^k(2^{-l}\omega) \geq C > 0$ for all $k, l \geq 1$ and $\omega \in K$, see [2] for more details)

With these hypotheses the following results hold.

Theorem 3.1. *For all $j \leq 0$, the tempered distributions φ^j defined by*

$$\hat{\varphi}^j(\omega) = \prod_{k=1}^{+\infty} m_0^{k+j}(2^{-k}\omega) \tag{3.5}$$

are compactly supported L^2 functions. The families $\{\varphi^j(x-k)\}_{k\in\mathbf{Z}}$ are orthonormal. Moreover, the spaces V_j generated by $\{2^{j/2}\varphi^j(2^j x - k)\}_{k\in\mathbf{Z}}$ constitute a "half multiresolution analysis" in the sense that $V_j \subset V_{j+1}$ and the projection $P_j(f)$ of any L^2 function f onto V_j goes to f in L^2 as j goes to $+\infty$.

Theorem 3.2. *If $\{m_0^k(\omega)\} = \{m_{0,L(k)}(\omega)\}$ is a subset of the family given by Equations (2.1) and (2.2), and if $\lim_{k\to+\infty} L(k) = +\infty$, then the functions φ^j are smooth, i.e., contained in C^∞.*

The proofs of these results are straightforward consequences of the results in [5]. The particular case of Theorem 3.2 where $L(k) = k$ has also been considered by Berkolaiko and Novikov [1].

The proof of Theorem 3.1 goes in four steps:

- Using Bernstein inequality, one remarks that for a fixed ω, the sequence

$$s_k = |m_0^{k+j}(2^{-k}\omega) - 1| \leq C d(k) 2^{-k}|\omega| \tag{3.6}$$

  is absolutely summable and thus the infinite product in Equation (3.5) converges pointwise and in the sense of tempered distributions. By Equation (3.4), φ^j is a tempered distribution with compact support.
- As in the standard construction of wavelets (see [3]), one defines the approximants φ_n^j of φ^j by

$$\hat{\varphi}_n^j(\omega) = \prod_{k=1}^{n} m_0^{k+j}(2^{-k}\omega)\chi_{[-2^n\pi, 2^n\pi]}(\omega) \tag{3.7}$$

  and checks by recursion, using Equation (3.1), that $\{\varphi_n^j(x-k)\}_{k\in\mathbf{Z}}$ is an orthonormal system. By Fatou's lemma, it follows that φ^j is in $L^2(\mathbb{R})$.
- By Equation (3.3), we have

$$|\omega| \leq \pi \Rightarrow |\hat{\varphi}^j(\omega)| > C > 0 \tag{3.8}$$

and $|\hat{\varphi}_n^j(\omega)| \le C^{-1}|\hat{\varphi}^j(\omega)|$. By dominated convergence, it follows that $\{\varphi^j(x-k)\}_{k\in\mathbf{Z}}$ is an orthonormal system.

- By the definition of φ^j, it is clear that $V_j \subset V_{j+1}$. To check that $P_j(f)$ goes to f in L^2 for any f, it suffices to prove that $\|P_j(\chi_I)\|$ goes to $|I|$ for any finite interval I. This follows from $\int \varphi^j = 1$ for all j and from the decay of the support length of $\varphi^j(2^j x)$ to zero as j increases.

The proof of Theorem 3.2 is based on an important estimate on the trigonometric polynomials given by Equations (2.1) and (2.2).

Lemma 3.3. *The trigonometric polynomials constructed by Daubechies satisfy*

$$|m_{0,L}(\omega)| \le [m(\omega)]^{L-1} \tag{3.9}$$

with $m(\omega) = 1$ *if* $|\omega| \le \pi/2$ *and* $m(\omega) = |\sin(\omega)|$ *if* $\pi/2 \le |\omega| \le \pi$.

Proof: This is obvious for $|\omega| \le \pi/2$. For $\pi/2 \le |\omega| \le \pi$, we have $2\sin^2(\omega/2) \ge 1$ and consequently

$$\begin{aligned}
|p_L(\omega)|^2 &= \sum_{j=0}^{L-1} \binom{L-1+j}{j} \sin^{2j}(\omega/2) \\
&= \sum_{j=0}^{L-1} \binom{L-1+j}{j} [2\sin^2(\omega/2)]^j 2^{-j} \\
&\le [2\sin^2(\omega/2)]^{L-1} \sum_{j=0}^{L-1} \binom{L-1+j}{j} 2^{-j} \\
&= [2\sin^2(\omega/2)]^{L-1} |p_L(\pi/2)|^2 = [4\sin(\omega/2)]^{L-1}
\end{aligned}$$

(we have $|p_L(\pi/2)|^2 = 2^{L-1}$ by Equation (3.1)). It follows that

$$\begin{aligned}
|m_{0,L}(\omega)| &= \left(\cos^{2L}(\omega/2)|p_L(\omega)|^2\right)^{1/2} \\
&\le \left([4\cos^2(\omega/2)\sin^2(\omega/2)]^{L-1}\right)^{1/2} = |\sin(\omega)|^{L-1}. \quad \blacksquare
\end{aligned}$$

The proof of Theorem 3.2 goes then in two steps:

- By using a technique introduced by Daubechies [9], one checks that the function $P(\omega) = \prod_{k=1}^{+\infty} m(2^{-k}\omega)$ satisfies

$$|P(\omega)| \le C(1+|\omega|)^{-s}, \tag{3.10}$$

with $s > 0$ (it can be shown that the largest value for s is $1 - \frac{\log 3}{2\log 2}$).

- For a fixed j and for any $A > 0$, it is possible to find $j_A > j$ such that, for all $k > j_A$, $s(L(k)-1) \geq A$. It follows that φ^{j_A} satisfies

$$\begin{aligned}|\hat{\varphi}^{j_A}(\omega)| &= \prod_{k=1}^{+\infty} |m_{0,L(k+j_A)}(2^{-k}\omega)| \\ &\leq \prod_{k=1}^{+\infty} |m(2^{-k}\omega)|^{L(k+j_A)-1} \\ &\leq \prod_{k=1}^{+\infty} |m(2^{-k}\omega)|^{A/s} \\ &\leq |P(\omega)|^{A/s} \\ &\leq C_A(1+|\omega|)^{-A}\end{aligned}$$

Since $\hat{\varphi}^j(\omega) = \hat{\varphi}^{j_A}(2^{j-j_A}\omega)\prod_{k=1}^{j_A-j} m_{0,L(k+j)}(2^{-k}\omega)$, it follows that we have

$$|\hat{\varphi}^j(\omega)| \leq C'_A(1+|\omega|)^{-A}, \tag{3.11}$$

and since this holds for any $A > 0$, it follows that φ^j is a C^∞ function.

Finally let us mention that the half-multiresolution analysis $\{V_j\}_{j\geq 0}$ has the property of "spectral approximation." For all $f \in H^s$ and $0 \leq t \leq s$ one has

$$\|P_j f - f\|_{H^t} \leq C(s,t)2^{-j(s-t)}\|f\|_{H^s}\varepsilon(j,f) \tag{3.12}$$

where $\varepsilon(j,f) \in [0,1]$ and tends to 0 as j goes to $+\infty$ (see [5]).

§4 Non-stationary wavelet packets

The previous results allow us to construct "non-stationary wavelet packets" with the following definition.

Definition 4.1. *Let $\{m_0^k\}_{k>0}$ be a family of positive trigonometric polynomials that satisfies the set of Conditions* (3.1)-(3.4) *and define $m_1^k(\omega) = e^{-i\omega}\overline{m_0^k(\omega+\pi)}$. The non-stationary wavelet packets associated to $\{m_0^k\}_{k>0}$ are a sequence $\{w_n\}_{n\geq 0}$ of tempered distributions given by*

$$\hat{w}_n(\omega) = \prod_{k=1}^{+\infty} m_{\varepsilon_k}^k(2^{-k}\omega), \tag{4.1}$$

with $n = \sum_{k>0} 2^{k-1}\varepsilon_k$, $\varepsilon_k \in \{0,1\}$.

From Theorems 3.1 and 3.2, we can easily derive the following proposition.

Proposition 4.2. *The distributions $\{w_n\}_{n\geq 0}$ are compactly supported L^2 functions and the length of their support is independent of n.*

The family $w_{n,m}(x) = w_n(x-m)$, $n \geq 0$, $m \in \mathbf{Z}$, constitutes an orthonormal basis of $L^2(\mathbb{R})$.

If $\{m_0^k(\omega)\} = \{m_{0,L(k)}(\omega)\}$ is a subset of the family given by Equations (2.1) and (2.2) and if $\lim_{k\to+\infty} L(k) = +\infty$, then the functions w_n are infinitely smooth, i.e., contained in C^∞.

Proof: From the definition of w_n, we see that for $0 \leq n < 2^j$, the families $\{w_n(x-m)\}_{m\in\mathbf{Z}}$ are the result of j splittings of the space V_j. It follows that these functions are finite combinations of the functions $\varphi^j(2^j x - k)$. As a consequence, they are in L^2. They are also in C^∞ as soon as the hypotheses of Theorem 3.2 are satisfied.

From the denseness of the spaces V_j, it follows that the family $w_{n,m}(x) = w_n(x-m)$, $n \geq 0$, $m \in \mathbf{Z}$, constitutes an orthonormal basis of $L^2(\mathbb{R})$.

Finally, since m_1^k has the same degree as m_0^k, the support of w_n cannot be larger than $L = \sum_{k>0} 2^{-k} d(k) < +\infty$. ■

Remark. Since the generating functions of the V_j spaces are different at every scale, one can no longer use a recursion formula such as Equation (1.6) to derive the non-stationary wavelet packets .

However, fast algorithms can still be used to compute the non-stationary wavelet packet coefficients. The principle is exactly the same as in the standard case except that one uses different filters at every stage of the decomposition. Note that in the case where the degree $d(k)$ of m_0^k is increasing as k goes to $+\infty$, the decomposition starts at the finest scale with a long filter and ends at a coarser scale with a smaller filter. An opposite approach was used by Hess-Nielsen [13] to study the frequency spreading of wavelet packets. The choice of using longer filters for finer scales offers three important advantages:

- Since the functions φ^j are in C^∞, the reconstruction from the low frequency components will have a smooth aspect.
- In the algorithm, one needs to deal properly with the boundaries of the signal. Specific methods have been developed by Cohen, Daubechies, and Vial [4] to adapt the filtering process near the edges. These methods require, in particular, that the size of the filter is smaller than the size of the signal. This can only be achieved by using smaller filters in the coarse scales.
- Finally, these techniques can be used to combine time and frequency localization properties as shown by Theorem 4.3 below.

A positive result on frequency localization can be obtained with non-stationary wavelet packets, provided that the filterlength grows sufficiently fast. We assume, once again, that the hypotheses of Equations (3.1)-(3.4) are satisfied and that $\{m_0^k(\omega)\} = \{m_{0,L(k)}(\omega)\}$ is a subset of the family given by Equations (2.1) and (2.2).

Theorem 4.3. *Assume that* $L(k) \geq Ck^{3+r}$ *for some* $r > 0$. *Then, for any* $\varepsilon > 0$ *there is* $M_\varepsilon > 0$ *such that, for all* $n \geq 1$, *then*

$$\frac{1}{n} \text{ card } \{k \leq n \ / \ \tilde{\sigma}_k > M_\varepsilon \ \} \leq \varepsilon. \tag{4.2}$$

The proof of this result is detailed in [6]. It is rather technical, but essentially exploits the fact that, as $L(k)$ goes to $+\infty$, the function $m_{0,L(k)}(\omega)$ converges pointwise to 1 on $]-\pi/2, \pi/2[$, an immediate consequence of Lemma 3.3.

References

1. Berkolaiko, S. and I. Novikov, Sur des presque-ondelettes indéfiniment différentiables àsupport compact, *Dokl. Russ. Acal. Neuk.* **326** (1992), 935–938.
2. Cohen, A. Wavelets and digital signal processing, translated by R. Ryan, Chapman & Hall, 1994, to appear.
3. Cohen, A. and I. Daubechies, On the instability of arbitrary biorthogonal wavelet packets, *SIAM J. Math. Anal.* **24-5** (1993), 1340–1354.
4. Cohen, A. I. Daubechies and P. Vial, Wavelets and fast wavelet transforms on an interval, *Appl. and Comp. Harmonic Anal.* **1** (1993), 54–81.
5. Cohen, A. and N. Dyn, Nonstationary subdivision schemes and multiresolution analysis, CEREMADE, Université Paris-IX, Dauphine, 1994, preprint.
6. Cohen, A. and E. Séré, Time-frequency localization with nonstationary wavelet packets, CEREMADE, Université Paris-IX, Dauphine, 1994, preprint.
7. Coifman, R. R., Y. Meyer, and V. M. Wickerhauser, Size properties of wavelet packets, in *Wavelets and Their Applications*, G. Beylkin, R. R. Coifman, I. Daubechies, S. Mallat, Y. Meyer, L. A. Raphael, and M. B. Ruskai (eds.), Jones and Bartlett, 1992, 453–470.
8. Coifman, R. R., Y. Meyer, and V. M. Wickerhauser, Entropy-based algorithms for best basis selection, *IEEE Trans. on Inform. Theory* **38-2** (1992), 713–718.
9. Daubechies, I., Orthonormal bases of compactly supported wavelets, *Comm. in Pure and Appl. Math.* **41** (1988), 909–996.
10. Daubechies, I. and J. C. Lagarias, Two-scale difference equations I. Existence and global regularity of solutions, *SIAM J. Math. Anal.* **22-5** (1991), 1388–1410.
11. Daubechies, I. and J. C. Lagarias, Two-scale difference equations II. Local regularity, infinite products of matrices and fractals, *SIAM J. Math. Anal.* **23-4** (1992), 1031–1079.
12. Dyn, N. and D. Levin, Stationary and non-stationary binary subdivision schemes, in *Mathematical Methods in Computer-aided Geometric Design*, T. Lyche and L. Schumaker (eds.), Academic Press, 1992, 209–216.

13. Hess-Nielsen, N., Frequency localization of generalized wavelet packets, Wash. Univ., St. Louis., 1993, preprint.
14. Micchelli, C. A. and H. Prautzsch, Uniform refinements of curves, *Linear Algebra and Applications*, **114-115** (1989), 841–870.
15. Rioul, O., Simple regularity criteria for subdivision schemes, *SIAM J. Math. Anal.* **23-6** (1992), 1544–1576.
16. Séré, E., Localisation fréquentielle des paquets d'ondelettes, *Revista Matemática Iberoamericana*, 1993, to appear.
17. Villemoes, L., Sobolev regularity of wavelets and stability of iterated filter banks, in *Proc. Int. Conf. on Wavelets and Applications*, Y. Meyer and S. Roques (eds.), Frontiers, 1992.

Albert Cohen
CEREMADE-Université Paris IX Dauphine
75775 Paris Cedex 16, France
albert@ceremade.dauphine.fr

The Spectral Theory of Multiresolution Operators and Applications

Peter Niels Heller and Raymond O. Wells, Jr.

Abstract. For any wavelet matrix of rank m and genus g there is an associated wavelet system with a square integrable scaling function and $(m\text{-}1)$ wavelet functions; the fundamental scaling equation has mg coefficients. This paper concerns itself with a family of multiresolution operators associated with such a wavelet matrix and the corresponding wavelet system. These operators act on continuous even periodic functions and depend quadratically on the coefficients of the wavelet matrix; they have been studied for rank 2 wavelet systems by Lawton, Cohen, and Eirola in different contexts and under different names (transition operator, wavelet-Galerkin operator, etc.). Lawton and Cohen independently found that a rank 2 wavelet system yields an orthonormal basis if and only if 1 is a nondegenerate eigenvalue of the multiresolution operator. Eirola showed for the rank 2 case that the spectral radius of this operator could be used to explicitly compute the Sobolev smoothness of the scaling function (and hence the wavelet system), and that moreover, this spectral radius could be computed in terms of the spectral radius of a finite-dimensional operator. He found asymptotic formulas for this spectral radius and hence for the Sobolev smoothness as the genus (and number of coefficients in the scaling equation) grows large. This paper reviews the above results and outlines a generalization of Eirola's work to arbitrary rank wavelet systems. In particular we give asymptotic estimates of the Sobolev smoothness of rank 3 scaling functions for large genus by evaluating the kernel of the multiresolution operator at fixed points of an automorphism of the unit circle.

§1 Introduction

In this article we explore the notion of the *multiresolution operator*, its spectral theory, and applications. This operator (also called the transition operator) has appeared as a fundamental tool in several aspects of wavelet theory, such as the Lawton-Cohen theorem on wavelet orthonormal bases [2,18,19], and the work of Eirola [8] and others [1,3,4,5,23] on the Sobolev smoothness of wavelet scaling functions. After introducing the multiresolution operator and

Wavelets: Theory, Algorithms, and Applications
Charles K. Chui, Laura Montefusco, and Luigia Puccio (eds.), pp. 13–31.

ISBN 0-12-174575-9

observing its connection with the convolution and downsampling operations of multirate signal processing, we present a review of the work of Lawton, and that of Eirola. Throughout the paper we work in the setting of rank m wavelet systems (m is the integer dilation factor, not necessarily 2). This involves some generalization of previous work, and yields initial results on the differentiability of wavelet scaling functions for rank $m > 2$. In particular, we find that the minimal support rank 3 scaling functions with N vanishing wavelet moments have Sobolev smoothness which grows only logarithmically with N.

§2 Wavelets and multiresolution operators

We are considering here rank m wavelets of compact support [9,13], defined by a *rank m wavelet matrix* $a = (a_{r,k})$ which is an $m \times mg$ matrix satisfying the orthogonality condition

$$\sum_k a_{r,k} a_{r',k+ml} = m\delta_{r,r'}\,\delta_{l,0} \tag{2.1}$$

and the sum condition

$$\sum_k a_{r,k} = m\delta_{r,0}\ . \tag{2.2}$$

If we look at the Fourier transforms (frequency responses in signal processing terminology)

$$A_r(\omega) = \frac{1}{m}\sum_k a_{r,k}e^{ik\omega}$$

of the wavelet sequences (filters), the conditions in equations (2.1)-(2.2) can be restated as

$$\sum_{k,k'=0}^{m-1} A_r(\omega + 2\pi k/m)\,\overline{A_{r'}(\omega + 2\pi k'/m)} \equiv \frac{1}{m}\delta_{r,r'}\ , \tag{2.3}$$

and

$$A_r(0) = \delta_{r,0}\ . \tag{2.4}$$

Associated to the wavelet matrix is a system of scaling functions and wavelets; the rank m scaling function ϕ associated with the scaling sequence $a_{0,k}$ is the solution to the dilation equation

$$\phi(x) = \sum_k a_{0,k}\phi(mx - k)\ . \tag{2.5}$$

This dilation equation has a unique compactly supported solution in $L^2 \bigcap L^1$; its Fourier transform $\hat{\phi}$ satisfies

$$\hat{\phi}(\xi) = \prod_{j=1}^{\infty} A_0(\xi/m^j)\ .$$

The corresponding family of wavelets $\psi_{r,j,k}$ $(1 \leq r \leq m-1)$ is defined from ϕ by

$$\psi_r(x) = \sum_k a_{r,k}\phi(mx - k)$$

and

$$\psi_{r,j,k} = m^{-j/2}\psi_r(m^{-j}x - k) .$$

Lawton [17] proved that the collection $\{\psi_{r,j,k}(x)\}$ form a tight frame for $L^2(\mathbf{R})$.

We are now in a position to ask critical questions about the family of functions $\{\psi_{r,j,k}(x)\}$, such as "When is this tight frame an orthonormal basis?" and "How differentiable are the individual $\psi_{r,j,k}$?" Since the wavelets $\psi_{r,j,k}$ are linear combinations of dilates of the scaling function ϕ, these questions reduce to the study of ϕ: "When do the $\{\phi(x-k)\}$ form an orthonormal basis for the space $V_0 = \text{Span}\,\{\phi(x-k)\}$?" and "How differentiable can we make $\phi(x)$?"

It is in answering these questions that one is led to the (rank m) *multiresolution operator* associated with the scaling sequence $\{a_k\}$. (Because of the special role of the scaling sequence $a_{0,k}$, we will often abbreviate our notation and refer to the scaling sequence as $\{a_k\}$ and its Fourier transform as $A(\omega)$.) This operator is defined in the frequency domain as the map taking a function u on $[0, 2\pi)$ to the function

$$Tu\,(\omega) = \left|A\left(\frac{\omega}{m}\right)\right|^2 u\left(\frac{\omega}{m}\right) + \left|A\left(\frac{\omega+2\pi}{m}\right)\right|^2 u\left(\frac{\omega+2\pi}{m}\right) + \cdots$$
$$\cdots + \left|A\left(\frac{\omega+2\pi(m-1)}{m}\right)\right|^2 u\left(\frac{\omega+2\pi(m-1)}{m}\right) ,$$

where we have extended the domain of definition of u by its natural periodization. The multiresolution operator is defined in the time domain as the operator taking a sequence u_n to

$$(\check{T}u)_n = m\sum_{k,l} a_k\overline{a_l}u_{m,n+k-l}\ . \tag{2.6}$$

(The operator T has been referred to variously as the *transition operator* and the *wavelet-Galerkin operator* in the literature; we take the liberty of introducing the name *multiresolution operator* because it more clearly illustrates the central role of this operator in understanding wavelet bases and multiresolution analysis.) If the function $u(\omega)$ has the Fourier series expansion $\sum_n u_n e^{in\omega}$ then equation (2.6) expresses the relationship between the Fourier coefficients u_n of u and those of $\check{T}u$; *i.e.*, $(\check{T}u)_n$ is the n-th Fourier coefficient of Tu.

2.1 Multiresolution operators and multirate signal processing

The multiresolution operator has a natural meaning in the context of multirate signal processing: it represents convolution followed by downsampling – the two operations essential to moving between scales of a multiresolution analysis (hence the name multiresolution operator). In the time domain, we can interpret the multiresolution operator as performing convolution of a sequence with the autocorrelation $\overline{a} * a$, followed by downsampling by m (keeping every m-th sample). For further discussion of this from a signal processing perspective, see [6].

Given an arbitrary sequence $\{h_n\}$, the frequency domain representation of the map $u_n \rightarrow (h * u)_{mn}$ (also written $(h * u) \downarrow m$) is

$$u(\omega) \rightarrow H\left(\frac{\omega}{m}\right) u\left(\frac{\omega}{m}\right) + H\left(\frac{\omega + 2\pi}{m}\right) u\left(\frac{\omega + 2\pi}{m}\right) + \\ \ldots + H\left(\frac{\omega + 2\pi(m-1)}{m}\right) u\left(\frac{\omega + 2\pi(m-1)}{m}\right) . \tag{2.7}$$

We use the notation T_H for this map and refer to H as the *symbol* of the operator T_H, in analogy with the theory of pseudodifferential operators. If H is a trigonometric polynomial (*i.e.*, $\{h_n\}$ is a finite sequence), then T_H is a bounded linear operator on the Banach space $C[0, \pi]$ of continuous functions on $[0, \pi]$ with the sup norm. Since the multiresolution operator takes the form of equation (2.7) with $|A|^2$ playing the role of H, we can write $T_{|A|^2}$ for the multiresolution operator, and $\check{T}_{\overline{a}*a}$ for its time domain form.

2.2 Vanishing moments and the Daubechies coefficients

An early theorem on wavelets ([7], p. 153) states that if a wavelet $\psi_r(x)$ is N times continuously differentiable and has sufficient decay to be in L^1, then $\hat{\psi}_r(\xi)$ must vanish to order N at $\xi = 0$. In order for this to happen, A_r must vanish to order N at $\omega = 0$, which leads us to the following definition.

Theorem 2.1. *A rank m wavelet system is said to have degree N if any of the following equivalent conditions hold:*

(i) $(d/d\xi)^n \hat{\psi}_r(0) = 0$, *for* $n = 0, 1, \ldots, N-1$ *and* $r = 1, 2, \ldots, m-1$.
(ii) $(d/d\omega)^n A_r(0) = 0$, *for* $n = 0, 1, \ldots, N-1$ *and* $r = 1, 2, \ldots, m-1$.
(iii) $(d/d\omega)^n A_0\left(\frac{2\pi k}{m}\right) = 0$, *for* $n = 0, 1, \ldots, N-1$; $k = 1, 2, \ldots, m-1$.
(iv) $\sum_l (k + ml)^n a_{0,k+ml}$ *is independent of* k *for* $k = 0, 1, \ldots, m-1$ *and* $n = 0, 1, \ldots, N-1$.
(v) $A_0(\omega) = (1 + e^{i\omega} + \ldots + e^{i(m-1)\omega})^N Q(\omega)$ *for some trigonometric polynomial* $Q(\omega)$.

Further discussion of these issues can be found in [7,10,22]. We note that $1 + e^{i\omega} + \ldots + e^{i(m-1)\omega}$ is the Fourier transform of the Haar scaling sequence $\{1, 1, \ldots, 1\}$ of length m. This trigonometric polynomial and its square modulus play a critical role in the analysis of the multiresolution operator $T_{|A|^2}$, as we shall see below.

The main point is that we now have a computational condition to use in our attempt to find wavelet bases with a given degree of differentiability – since an N-times differentiable wavelet must have N vanishing moments, why not *impose* N vanishing moments and see how differentiable the resulting ϕ and ψ are? This was one of the ideas Daubechies used in the construction of orthonormal bases of compactly supported wavelets; she identified a minimal degree Q to define a wavelet system with N vanishing moments. This was generalized to rank m wavelets in [10,21]. In fact, characterizations of these minimal length wavelet systems with N vanishing moments invariably describe the autocorrelations $|A_0(\omega)|^2$ or $|Q(\omega)|^2$ and one can then perform a spectral factorization [20] to obtain A_0 or Q. For a wavelet system with N vanishing moments, it is convenient to write

$$P(\omega) = |A_0(\omega)|^2 = \left|\frac{1 + e^{i\omega} + \ldots + e^{i(m-1)\omega}}{m}\right|^{2N} R(\omega) \tag{2.8}$$

where we have factored the square modulus $P(\omega)$ into two parts, N powers of the modulus squared Haar polynomial (*à la* definition (v) above) and a term $R(\omega) = |Q(\omega)|^2$ designed to restore orthogonality to the overall symbol. With this new notation, T_P is the same as $T_{|A|^2}$, and henceforth we use T_P to denote the multiresolution operator associated with the scaling sequence $\{a_k\}$.

Lawton [19] used the multiresolution operator to give a time-domain characterization of the Daubechies scaling coefficients, which we generalize now to the rank m case. The autocorrelation of the rank m minimal length scaling coefficients of degree N is the solution to

$$\sum_{n=-N}^{N-1} (mn + r)^p (\overline{a} * a)_{mn+r} = \frac{1}{m}\, \delta_{p,0} \ , \tag{2.9}$$

where $0 \le p \le 2N - 1$, $1 \le r \le m - 1$, and

$$(\overline{a} * a)_{mn} = \delta_{n,0} \ ; \tag{2.10}$$

i.e., it solves $m - 1$ simultaneous Lagrange interpolation problems over the nonzero cosets of $\mathbf{Z}/m\mathbf{Z}$. The scaling sequence proper is then obtained from this autocorrelation by a spectral factorization. To connect these formulae with the multiresolution operator, let $u_r(n) = \delta_{n,r}$, for $r = 1, 2, \ldots, m - 1$. Then $v_r(n) = \check{T}_{\overline{a}*a} u_r(n) = (\overline{a} * a)_{mn+r}$ and equations (2.9)–(2.10) follow from (v) in Theorem 2.1. These expressions have an explicit solution given in [11] by a formula for the symbol $P(\omega)$:

$$P(\omega) = \frac{1}{m} + \frac{2}{m^2} \sum_{r=1}^{m-1} \sum_{n=-N}^{N-1} \prod_{\substack{k=-N \\ k \ne n}}^{N-1} \frac{mk + r}{m(k - n)} e^{i(mn+r)\omega} \ . \tag{2.11}$$

An earlier approach to finding the rank m Daubechies coefficients with N vanishing moments appears in [10,21]; there one obtained a formula for the minimal degree trigonometric polynomial $R(\omega)$:

$$R_N(x) = \sum_{n=0}^{N-1} \left\{ \frac{1}{n!} \left(\frac{d}{dx}\right)^n [H(x)]^{-N} \right\}_{x=1} (x - 1)^n \ , \tag{2.12}$$

where $x = \cos\omega$ and

$$H(x) = \left|\frac{1 + e^{i\omega} + \ldots + e^{i(m-1)\omega}}{m}\right|^2$$

is the modulus squared of the Haar trigonometric polynomial.

2.3 An invariant finite-dimensional subspace for T_P

Let $\{a_k\}$ be a scaling sequence and T_P be the associated multiresolution operator, where $P = |A|^2$ as before. The infinite-dimensional operator T_P has interesting properties when restricted to certain finite-dimensional subspaces, which we now develop after Daubechies [7]. Let the trigonometric polynomial P be given as in equation (2.11), expressed in the form

$$P(e^{i\omega}) = \sum_{j=-J}^{J} p_j e^{ij\omega} . \tag{2.13}$$

Moreover, define the $2J+1$-dimensional space of trigonometric polynomials

$$\mathcal{E}_J := \left\{ u(\omega) = \sum_{j=-J}^{J} u_j e^{ij\omega} \right\} ,$$

as well as the distinguished subspace

$$\mathcal{F}_{J,N} := \{ f \in \mathcal{E}_J : f(\omega) = (1 - \cos\omega)^N g(\omega) \} .$$

We now consider the action of the multiresolution operator T_P and the *reduced multiresolution operator* T_R defined by equation (2.7) with symbol R, the remainder in the factorization $P = H^N R$.

Lemma 2.2. *Suppose that the wavelet system associated with the scaling sequence $\{a_k\}$ has degree N (the first N wavelet moments vanish). Then*

(i) *The subspace $\mathcal{E}_J$ of $C[0,\pi]$ is an invariant subspace for the operator T_P, i.e.,*

$$T_P : \mathcal{E}_J \to \mathcal{E}_J .$$

The operator T_P acting on $\mathcal{E}_J$ has the matrix representation $t_{j,k} = p_{mj-k}$, or

$$T = \begin{bmatrix} 0 & 0 & 0 & \ldots & 0 & 0 & 0 \\ & \ddots & & & & & \\ 0 & 0 & 0 & \ldots & 0 & 0 & 0 \\ p_{J-m+2} & p_{J-m+3} & \cdots & p_J & 0 & \ldots & 0 \\ p_{J-2m+2} & p_{J-2m+3} & \cdots & p_{J-m} & \cdots & p_J & 0 \\ & & & \ddots & & & \\ 0 & \ldots & 0 & p_{J-m+2} & \cdots & p_{J-1} & p_J \\ 0 & 0 & 0 & \ldots & 0 & 0 & 0 \\ & & & & & \ddots & \\ 0 & 0 & 0 & \ldots & 0 & 0 & 0 \end{bmatrix}$$

with respect to the basis $\{e^{-iJ\omega}, \ldots, 1, \ldots, e^{iJ\omega}\}$.

(ii) $T_P|_{\mathcal{E}_J}$ *has eigenvalues* m^{-n}, $n = 0, 1, \ldots, , 2N-1$.

(iii) *Suppose* $f \in \mathcal{F}_{J,N}$, $f(\omega) = (1-\cos\omega)^N g(\omega)$; *then*

$$T_P f(\omega) = \frac{1}{m^{2N}}(1-\cos\omega)^N T_R\, g(\omega)\,. \tag{2.14}$$

Hence, $\mathcal{F}_{J,N}$ *is an invariant subspace for* T_P, *and moreover if* f *is an eigenvector for* T_P *in* $\mathcal{F}_{J,N}$ *with eigenvalue* μ, *then*

$$T_R g(\omega) = m^{2N}\mu g(\omega)\,,$$

so that g *is an eigenfunction for* T_R *with eigenvalue* $m^{2N}\mu$.

Remark. Part (iii) establishes a one-to-one correspondence between eigenfunctions of $T_P|_{\mathcal{F}_{J,N}}$ and eigenfunctions of $T_R|_{\mathcal{E}_{J-N}}$. This will enable us to reduce the study of the spectrum of T_P to that of T_R in the work to follow.

Proof: To prove (i), given $u \in \mathcal{E}_J$ and P as in equation (2.13), extend $T_P u$ to be an even function on $[-\pi, \pi]$, and expand it in a Fourier series on this interval. The sum over m terms in the definition of T_P offsets the dilation by $\frac{1}{m}$ so that no fractional powers of sine or cosine appear. The orthogonality of $\{\sin k\omega, \cos k\omega\}$ ensures that all but a finite number of coefficients in the Fourier expansion of $T_P u$ vanish, and we find that $T_P u \in \mathcal{E}_J$.

(ii) is strongly related to Theorem 2.1. As in that theorem, the factorization

$$P(\omega) = \left|\frac{1-e^{im\omega}}{m(1-e^{i\omega})}\right|^{2N} R(\omega) \tag{2.15}$$

implies that the partial moments

$$M_{k,n} = \sum_l (k+ml)^n p_{k+ml}$$

are equal to a number M_n independent of k, $k = 0, 1, \ldots, m-1$ for $n = 0, 1, \ldots, 2N-1$. Following [7,12], if we consider the polynomial sequence $e_n = (j^n)_{j=-J,\ldots,J} \in \mathcal{E}_J$, we find that

$$e_n T = m^{-(n+1)} \sum_{k=0}^{n} \binom{n}{k} M_{n-k} e_k\,.$$

Since $M_0 = 1$ with our normalizations, we see that each element e_n of $\mathcal{E}_J$ is a generalized left eigenvector of T_P with eigenvalue m^{-n} for $n = 0, 1, \ldots, 2N-1$.

(iii) follows by the factorization in equation (2.15) of P:

$$\begin{aligned} T_P f(\omega) &= T_P \left|1-e^{i\omega}\right|^{2N} g(\omega) \\ &= \left|\frac{1-e^{i\omega}}{m(1-e^{i\omega/m})}\right|^{2N} R(\frac{\omega}{m}) \left|1-e^{i\omega/m}\right|^{2N} g\left(\frac{\omega}{m}\right) + \cdots \\ &= \frac{1}{m^{2N}}\left|1-e^{i\omega}\right|^{2N} T_R\, g(\omega)\,. \end{aligned}$$

■

§3 The Lawton-Cohen-Gopinath Theorem on wavelet orthonormal bases

With the definition and elementary properties of the multiresolution operator in hand, we can now begin to apply it to problems of wavelet theory. The first question we asked was "When do the integer translates of ϕ, $\{\phi(x-k)\}$, form an orthonormal basis for the space V_0 defined as their span?" In other words, when does the sequence h_k defined by

$$h_k := \int \phi(x)\phi(x-k)dx$$

satisfy

$$h_k = \delta_{k,0} \ ?$$

By applying the scaling relation (2.5) to the definition of h_k, we find

$$\begin{aligned} h_k &= \sum_{j,l} a_j \overline{a}_l \int \phi(mx-j)\overline{\phi(mx-l-mk)}dx \\ &= \sum_{j,l} a_j \overline{a}_l h_{2k+l-j}. \end{aligned}$$

Simply put,

$$h = \check{T}_{\overline{a}*a} h \ .$$

However, since the scaling coefficients a_k satisfy the orthogonality relation

$$\sum_k a_k a_{k+ml} = \delta_{0,l} \, ,$$

then

$$\check{T}_{\overline{a}*a}\delta_{0,l} = \delta_{0,l} * (a * \overline{a}) \downarrow m = (a*\overline{a}) \downarrow m = \delta_{0,l} \, ,$$

or

$$\delta_{0,l} = \check{T}_{\overline{a}*a}\delta_{0,l} \ .$$

Thus both the (as yet undetermined) sequence $\{h_k\}$ and the Kronecker delta $\delta_{0,l}$ are eigensequences for $\check{T}_{\overline{a}*a}$ with eigenvalue 1. The fact that the scaling sequence $\{a_k\}$ satisfies the quadratic wavelet matrix condition (2.1) is *equivalent* to $\delta_{0,l}$ being an eigenvector of the multiresolution operator $\check{T}_{\overline{a}*a}$ with eigenvalue 1. This observation was made by Lawton [17], proving that the wavelet system $\{\psi_{r,j,k}\}$ is a tight frame if $\delta_{0,l}$ is such an eigenvector. Furthermore, Lawton showed that if $\delta_{0,l}$ is the only eigenvector for eigenvalue 1, then $\{\psi_{r,j,k}\}$ is an orthonormal wavelet basis. It is the content of a later theorem [2,18] that this is also a necessary condition for orthonormality.

Theorem 3.1. (Lawton-Cohen-Gopinath) *The set $\{\phi(x-k)\}$ form an orthonormal basis for V_0 (and, as a result, the set $\{\psi_{r,j,k}\}$ form an orthonormal basis for $L^2(\mathbf{R})$) if and only if 1 is a nondegenerate eigenvalue for the multiresolution operator $\check{T}_{\overline{a}*a}$ (and T_P as well).*

The lack of redundancy in the eigenvectors for eigenvalue 1 implies that there is no redundancy in the tight frame $\{\psi_{r,j,k}\}$. Observe that in the time domain we wish the unique eigensequence for $\check{T}_{\overline{a}*a}$ to be $\delta_{0,l}$, while in the frequency domain the unique eigenfunction of T_P to be the constant $u_0(\omega) \equiv 1$; in either case the eigenvalue is 1.

This theorem is not vacuous, as can be seen by the "degenerate Haar" scaling sequence

$$a^{dH} = \{1, 0, \ldots, 0, 1, 1, \ldots, 1\}$$

where m zeros follow the first 1 and then $m-1$ ones follow the string of zeros. The scaling function ϕ associated with a^{dH} is the characteristic function of the interval $\left[0, 2 + \frac{1}{m-1}\right]$, and $\phi(x)$ and $\phi(x-1)$ are clearly not orthogonal, although the corresponding wavelet system $\{\psi_{r,j,k}\}$ is a tight frame.

Cohen [2] has developed additional equivalent conditions for this theorem which involve "sets congruent to $[-\pi, \pi]$ mod 2π" and invariant cycles under the map $\omega \longrightarrow m\omega$ mod 2π. We do not elaborate on this here, other than to note that he also uses multiresolution operators in the Fourier domain in his characterization of orthonormality.

§4 Differentiability estimates via the spectral theory of the multiresolution operator

The multiresolution operator T_P arises in a second area of wavelet theory, the study of the differentiability of solutions to the scaling equation (2.5). In this section we present an overview of the work of Eirola [8], generalized to the rank m setting, and give our preliminary results for wavelet systems of rank $m = 3$. Detailed proofs and extensions of this work appear in [14]. Work on related problems appears in [1,3,4,5,23].

4.1 Sobolev estimates for the scaling function ϕ

We defined the scaling function ϕ as the unique solution in $L^2 \bigcap L^1$ of the dilation equation (2.5). Seeking to create smooth scaling functions, we identified a distinguished set of scaling sequences $\{a_k\}$, those of minimal length having degree N (*i.e.*, those whose wavelets have N vanishing moments). We now ask the question "Given such a scaling sequence, how differentiable is the corresponding ϕ?" We answer this by finding the maximal Sobolev space that ϕ belongs to. Given a real number s, the Sobolev space $\mathcal{H}^s$ is defined by

$$\mathcal{H}^s := \left\{ f \,:\, \|f\|^2_{\mathcal{H}^s} = \int_{\mathbf{R}} |\hat{f}(\xi)|^2 \left(1 + |\xi|^2\right)^s d\xi < \infty \right\}.$$

Recall that for $f : \mathbf{R} \to \mathbf{C}$, $f \in \mathcal{H}^s \Rightarrow f \in C^\alpha$, for $\alpha < s - \frac{1}{2}$. That is, if $s > n + \frac{1}{2}$, then f has n continuous derivatives. We define the *Sobolev smoothness* of a function $\phi \in L^2$ to be

$$s(\phi) := \sup \{ s : \phi \in \mathcal{H}^s \} .$$

Observe that $s(f) = s_0$ does not imply that $f \in \mathcal{H}^{s_0}$, only that $f \in \mathcal{H}^{s_0 - \epsilon}$ for all $\epsilon > 0$. Our goal in this section is, given a minimal length scaling sequence $\{a_k\}$, of degree N, to determine a formula relating the Sobolev smoothness $s(\phi)$ to the spectrum of the multiresolution operator T_P.

We obtain a solution to the scaling equation (2.5) by taking an initial guess $\phi_0(x) := \frac{sin(\pi x)}{\pi x}$ and then iterating:

$$\phi_{n+1} := \sum_k a_k \phi_n(mx - k) .$$

In the Fourier transform domain this equation becomes

$$\hat{\phi}_{n+1}(\xi) = A(\xi/m)\hat{\phi}_n(\xi/m) .$$

Letting χ_S be the characteristic function of a set S, we have $\widehat{\frac{sin(\pi x)}{\pi x}} \equiv \chi_{[-\pi,\pi]}$, and, moreover,

$$\hat{\phi}_{n+1}(\xi) = \prod_{j=1}^{n} A(\xi/m^j) \, \chi_{[-m^n\pi, m^n\pi]} .$$

Since the ϕ_n converge pointwise and in L^2 to ϕ, (proof in [7]), they will converge in $\mathcal{H}^s$ if there is a uniform bound on $\|\phi_n\|^2_{\mathcal{H}^s}$. The s-th Sobolev norm of ϕ_n is bounded by

$$\|\phi_n\|^2_{\mathcal{H}^s} \leq C \left(\|\hat{\phi}_n\|^2_{L^2} + \int_{\mathbf{R}} |\hat{\phi_n}(\xi)|^2 |\xi|^{2s} d\xi \right) , \tag{4.16}$$

and so we concentrate on bounding the right-hand term in equation (4.16).

4.1.1 A change of variable

We now perform a change of variable which relates the multiresolution operator T_P to the integral $\int_{\mathbf{R}} |\hat{\phi_n}(\xi)|^2 |\xi|^{2s} d\xi$ appearing in the right-hand side of inequality (4.16). Fixing the Sobolev exponent s, write

$$e_n(\xi) := |\hat{\phi}_n(\xi) - \hat{\phi}_{n-1}(\xi)|^2 |\xi|^{2s} .$$

If we can show that

$$\int_{\mathbf{R}} e_n(\xi) d\xi \leq C\epsilon^n \text{ for some } \epsilon < 1,$$

then

$$\int_{\mathbf{R}} |\hat{\phi}_n(\xi)|^2 |\xi|^{2s} d\xi \le C \sum_{k=1}^{n} \epsilon^k < C \frac{1}{1-\epsilon} < \infty$$

uniformly, and so $\phi(x) \in \mathcal{H}^s$ by inequality (4.16). We now estimate the integrals $\int_{\mathbf{R}} e_n(\xi) d\xi$ using the multiresolution operator T_P. Observe that $e_n(-\xi) = e_n(\xi)$ and

$$e_{n+1}(\xi) = \left| A\left(\frac{\xi}{m}\right)\right|^2 \left|\hat{\phi}_n\left(\frac{\xi}{m}\right) - \hat{\phi}_{n-1}\left(\frac{\xi}{m}\right)\right| |\xi|^{2s} = m^{2s} P\left(\frac{\xi}{m}\right) e_n\left(\frac{\xi}{m}\right) .$$

Since $\hat{\phi}_n(\xi) = 0$ for $|\xi| > m^{n-1}\pi$,

$$\int_{\mathbf{R}} e_n(\xi) d\xi = \int_{|\xi| \le m^{n-1}\pi} e_n(\xi) d\xi .$$

Now we use T_P to collapse this integral over a compact interval (that grows with n) into an integral over the fixed interval $[0, \pi)$. Define, for m even,

$$\begin{aligned}\kappa_1(\omega) = 2[\, & e_1(\omega) + e_1(\omega - 2\pi) + e_1(\omega + 2\pi) + e_1(\omega - 4\pi) + \\ & \cdots + e_1(\omega + \frac{m-2}{2} 2\pi) + e_1(\omega - \frac{m}{2} 2\pi)\,],\end{aligned}$$

and for m odd,

$$\begin{aligned}\kappa_1(\omega) = 2[\, & e_1(\omega) + e_1(\omega - 2\pi) + e_1(\omega + 2\pi) + e_1(\omega - 4\pi) + \\ & \ldots + e_1(\omega + \frac{m-3}{2} 2\pi)\,].\end{aligned}$$

Setting $\kappa_{n+1} := m^{2s} T_P \, \kappa_n$, we have the following lemma.

Lemma 4.1.

$$\int_{|\xi| \le m^n \pi} e_{n+1}(\xi) d\xi = \int_0^{\pi} \kappa_{n+1}(\omega) d\omega = \int_0^{\pi} m^{2sn} T_P^n \kappa_1(\omega) d\omega . \qquad (4.17)$$

The proof can be found in [14]. We see from Lemma 4.1 that the Sobolev smoothness of a scaling function ϕ defined by a scaling sequence $\{a_k\}$ is given by

$$s(\phi) = \sup \left\{ s : \int_0^{\pi} m^{2sn} T_P^n \kappa_1(\omega) d\omega \le C \epsilon^n \text{ for some } \epsilon < 1 \right\} . \qquad (4.18)$$

4.1.2 The study of the positive operator T_R

The Sobolev regularity of ϕ now depends solely on the family of integrals

$$\int_0^\pi T_P^n \kappa_1(\omega) d\omega \ .$$

As a first step in bounding these integrals, we reduce the study of iterates of the multiresolution operator T_P to the study of iterates of the *reduced* multiresolution operator T_R. It can be checked that $\kappa_1(\omega)$ vanishes to order $2N$ at $\omega = 0$, and

$$\kappa_1(\omega) = (1 - \cos\omega)^N v_1(\omega) \ ,$$

for some $v_1 \in \mathcal{C}[0, \pi]$. If $s > \frac{1}{2}$, then $v_1 \in \mathcal{C}^\infty_{[0,\pi]}$. Using relation (2.14),

$$\begin{aligned} \int_0^\pi T_P^n \kappa_1(\omega) d\omega &= \int_0^\pi m^{2Nn}(1 - \cos\omega)^N T_R^n \, v_1(\omega) \, d\omega \\ &\le c_N \, m^{2Nn} \int_0^\pi T_R^n v_1(\omega) d\omega \ . \end{aligned}$$

This is advantageous because $R(\omega)$ is a purely positive trigonometric polynomial, and so T_R is a positive operator on $\mathcal{C}[0, \pi]$. We will be able to bound the integral

$$\int_0^\pi T_R^n v_1(\omega) d\omega, \tag{4.19}$$

using the spectral radius ρ of the operator T_R on $\mathcal{C}[0, \pi]$.

We now examine this spectral radius, developing infinite-dimensional analogues of the Perron-Frobenius theorem. Recall that the spectral radius ρ of a linear operator T is defined as

$$\rho(T) = \lim_{n\to\infty} \left(\sup_{\|u\|=1} \|T^n u\|^{1/n} \right) \ .$$

We aim to compute $\rho(T_R)$ in terms of the spectral radius of the finite-dimensional operator $\widetilde{T}_R = T_R|_{\mathcal{E}_J}$. We will use the theory of positive operators as developed in [15]. A *cone* in $\mathcal{C}[0, \pi]$ is a closed subset which does not contain a one-dimensional subspace, and which contains halflines emanating from the origin. The subset K of $\mathcal{C}[0, \pi]$ consisting of nonnegative continuous functions on $[0, \pi]$ is such a cone. We also introduce the subset K^0 of K consisting of all elements $u \in K$ such that $u(\omega) > 0$ for all $\omega \in [0, \pi]$. With these preliminaries, the following theorem is proven in [8,14].

Theorem 4.2. *Let $R \in \mathcal{C}[0, \pi]$ be such that $R \in K^0 \cap \mathcal{E}_J$, then there exists a function $\tilde{u} \in K^0 \cap \mathcal{E}_J$ and a real number $\mu > 0$ such that:*

i) $\widetilde{T}_R \, \tilde{u} = \mu \tilde{u}$ *($\tilde{u}$ is an eigenvector of $\widetilde{T}_R$ with eigenvalue μ), and* $\rho(\widetilde{T}_R) = \mu$.
ii) $T_R \, \tilde{u} = \mu \tilde{u}$ *($\tilde{u}$ is an eigenvector of T_R with eigenvalue μ), and* $\rho(T_R) = \mu$.

iii) *If* $0 \neq u_0 \in K \cap \mathcal{C}^{\infty}_{[0,\pi]}$, *then there exists an* $n_0 > 0$ *such that* $T_R^{n_0} u_0 \in K^0$, *and*

$$\lim_{n\to\infty} \frac{T_R^n u_0}{\|T_R^n u_0\|} = \tilde{u} \ . \tag{4.20}$$

The limit in equation (4.20) is uniform for $\omega \in [0,\pi]$, *i.e., is a strong limit in the Banach space* $\mathcal{C}[0,\pi]$.

Applying this theorem with $R(\omega)$ as defined in equation (2.12), so that $J = N-1$, we obtain two critical corollaries.

Corollary 4.3. *If* μ *is the largest eigenvalue of* $T_R|_{\mathcal{E}_{N-1}}$ *and* u_0 *is an arbitrary nonnegative nonzero function* $\in \mathcal{C}^{\infty}_{[0,\pi]}$, *then given* $\epsilon > 0$, *there exist* c_ϵ, C_ϵ *such that*

$$c_\epsilon(\mu-\epsilon)^n \leq \int_0^\pi T_R^n u_0(\omega)d\omega \leq C_\epsilon(\mu+\epsilon)^n \ .$$

Applying this to integral (4.19), with $u_0 = v_1$, we find

$$s(\phi) = \sup\left\{s \ : \ m^{2s-2N}\mu < 1\right\} \ ,$$

yielding the following corollary.

Corollary 4.4. *If* μ *is the largest eigenvalue of* $T_R|_{\mathcal{E}_{N-1}}$ *and* $\mu < \frac{1}{m}$,

$$s(\phi) = N - \frac{\log\mu}{2\log m} \ . \tag{4.21}$$

Thus, we have obtained an explicit formula for the Sobolev smoothness of the scaling function ϕ in terms of the maximal eigenvalue of the associated reduced multiresolution operator T_R acting on the finite-dimensional space $\mathcal{E}_{N-1}$.

4.2 Results for $m = 2$, 3

In the cases $m = 2$ and $m = 3$, we are able to apply this machinery in two ways. First, given explicit formulas for the trigonometric polynomial $R(\omega)$, one may compute its eigenvalues on the subspace $\mathcal{E}_{N-1}$ and obtain the maximal Sobolev exponent from equation (4.21). Using the formula

$$R(\omega) = \sum_{n=0}^{N-1} \frac{(N-1+n)!}{(N-1)!\,n!}(1-\cos\omega)^n$$

in the $m = 2$ case (see [7] for a derivation) and the formula

$$R(\omega) = \sum_{n=0}^{N-1} \frac{(2N-1+n)!}{(2N-1)!\,n!}\left(\frac{2}{3}\right)^n (1-\cos\omega)^n \tag{4.22}$$

when $m = 3$ (obtained in [10]), we compute the Sobolev smoothness of minimal support scaling functions ϕ_N of degree N. (Warning! We have just switched notation and now use ϕ_N to refer to the degree N scaling function of minimal support, *not* the N-th iterative approximation to the scaling function ϕ.) The results are tabulated for $N = 1, 2, \ldots, 10$ in Table 4.1 and graphed for $N = 1, 2, \ldots, 25$ in Figure 4.1.

Table 4.1
Sobolev smoothness for minimal support scaling functions with N vanishing moments

N	m=2 Sobolev exponent	m=3 Sobolev exponent
1	0.5	0.5
2	1.0	0.9087
3	1.4150	1.1599
4	1.7756	1.2950
5	2.0968	1.3665
6	2.3884	1.4133
7	2.6587	1.4499
8	2.9147	1.4809
9	3.1617	1.5081
10	3.4027	1.5323

However, one can go much further. Eirola [8] used the notion of T_R as a positive operator mapping a particular cone K to itself to develop asymptotic estimates for the spectral radius of T_R as $N \to \infty$. For m odd, define the special cone

$$K^J = \left\{ u \in K^0 \cap \mathcal{E}_J \, : \, u(\omega) = \sum_{j=0}^{J} u_j (1 - \cos\omega)^j \, , \; u_j \geq 0 \right\} .$$

In [14], we prove the following lemma.

Lemma 4.5. *Given $R \in K^J$,*

$$T_R : K^J \to K^J \, .$$

Now define the dual cone K^{J^*} and adjoint operator T_R^*, and consider the evaluation functionals η_θ defined by

$$\langle \eta_\theta, u \rangle = u(\theta) \, .$$

One can prove the following lemma.

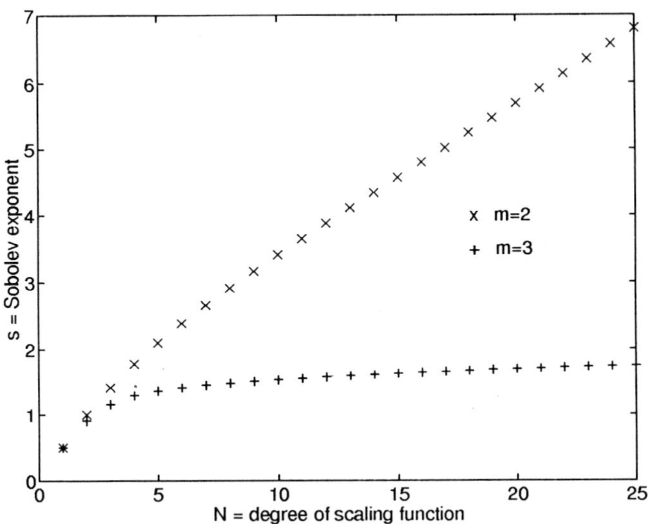

Figure 4.1. Sobolev exponents for minimal support degree N scaling functions $m = 2$ and $m = 3$, $N = 1, 2, \ldots, 25$.

Lemma 4.6. *If $R \in K^J$, and there exists $\lambda > 0$ and $\theta \in [0, \pi]$ such that*

$$T_R^* \eta_\theta - \lambda \eta_\theta \in K^{J^*} ,$$

then $\mu \geq \lambda$, and conversely, if

$$\lambda \eta_\theta - T_R^* \eta_\theta \in K^{J^*} ,$$

then $\mu \leq \lambda$.

In the case $m = 2$, Eirola applied this technique (using a different cone K^J) with $J = N - 1$, $\theta = \frac{2\pi}{3}$:

$$T_R^* \eta_{\frac{2\pi}{3}} = R\left(\frac{2\pi}{3}\right) \eta_{\frac{2\pi}{3}} + R\left(\frac{\pi}{3}\right) \eta_{\frac{\pi}{3}} .$$

Since $R\left(\frac{2\pi}{3}\right) >> R\left(\frac{\pi}{3}\right)$, this says that $\eta_{\frac{2\pi}{3}}$ is "almost" an eigenvector for T_R. Applying Lemma 4.6, it can be shown [8] that

$$R\left(\frac{2\pi}{3}\right) \leq \mu \leq R\left(\frac{2\pi}{3}\right)(1 + o(N)) . \tag{4.23}$$

Since we know

$$R\left(\frac{2\pi}{3}\right) = \sum_{n=0}^{N-1} \frac{(N-1+n)!}{(N-1)!\, n!} \left(\frac{3}{2}\right)^n ,$$

equation (4.23) yields an asymptotic value for the Sobolev exponent s:

$$s(\phi_N) \approx N - \frac{N \log 3}{2 \log 2} \approx .2075N \ .$$

This result has also been obtained by Cohen and Conze [3] and Volkmer [24], using different methods.

In the $m = 3$ case, we have applied the same cone-based estimates to obtain bounds on the maximal eigenvalue of T_R. This time we use η_π as the evaluation functional in Lemma 4.6, and find:

$$R(\pi) \leq \mu \leq R(\pi)\ (1 + o(N)) \ .$$

Computing

$$R(\pi) = \sum_{n=0}^{N-1} \frac{(2N-1+n)!}{(2N-1)!\, n!} \left(\frac{4}{3}\right)^n ,$$

the Sobolev differentiability of the $m = 3$ wavelets grows much more slowly with N:

$$s(\phi_N) \approx \frac{\log N}{4 \log 3} \ .$$

These results are developed in [14].

To show how differently the differentiability of these functions grows, in Figure 4.2 we juxtapose the graphs of ϕ_N for $m = 2, 3$ and $N = 2, 10$. While the two functions with $N = 2$ do not look that different, the $m = 2$, $N = 10$ function is clearly smoother than the example with $m = 3$, $N = 10$.

§5 Conclusion

We have introduced the multiresolution operator T_P associated with a wavelet scaling sequence $\{a_k\}$ and found the spectral theory of this operator to be critical for two important applications: deciding when a wavelet tight frame is actually an orthonormal basis, and in determining the Sobolev smoothness of a rank m scaling function of degree N (one whose first N wavelet moments vanish). Cohen [2] and Lawton [17,18] have shown that a scaling sequence $\{a_k\}$ leads to a wavelet orthonormal basis if and only if 1 is a nondegenerate eigenvalue of the associated multiresolution operator T_P.

In Section 4 we described the work of Eirola [8] and its generalization to the rank m case. We indicated that the minimal support degree N scaling function ϕ_N has Sobolev smoothness $s(\phi_N)$ which is linear in N for rank $m = 2$ and logarithmic in N for rank $m = 3$. It appears that, in general, $s(\phi_N)$ for these scaling functions will be linear in N for m even and logarithmic in N for m odd; this is still under investigation.

Are there families of rank m wavelet scaling functions ϕ_L defined by scaling sequences of length L which have Sobolev smoothness $s(\phi_L)$ which is linear in L for odd m? This question was raised by Daubechies [7] and has not

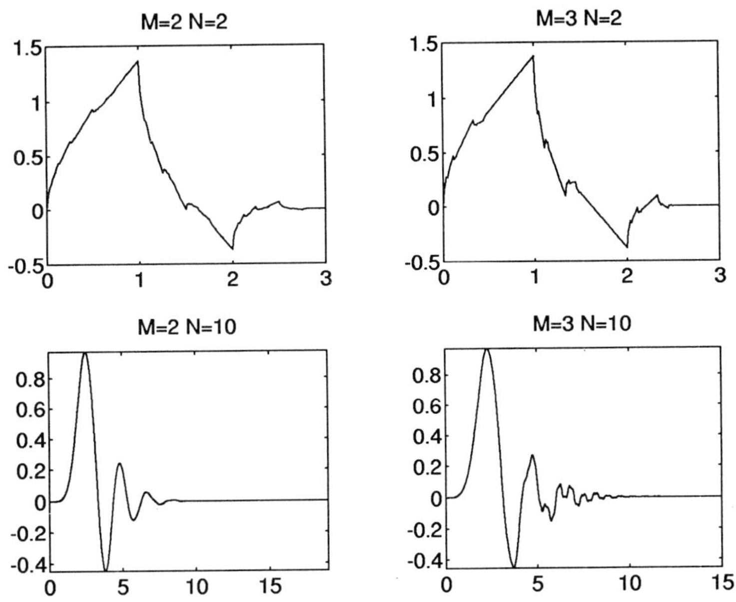

Figure 4.2. Scaling functions for $m = 2$, $N = 2$, $N = 10$ (top row) and $m = 3$, $N = 2$, $N = 10$ (bottom row).

been resolved, to our knowledge. Experimental results indicate the existence of a family of scaling sequences for rank $m = 3$ for which $s(\phi_L)$ depends algebraically on L (*e.g.*, $s(\phi_L) \approx L^{5/9}$), but these results are very preliminary. Further development of these ideas is reported in [14].

Acknowledgments. The authors would like to thank Gilbert Strang for pointing out the "coincidental" appearance of the multiresolution operator in the work of W. Lawton and T. Eirola. This research was supported in part by the Advanced Research Projects Agency of the Department of Defense and was monitored by the Air Force Office of Scientific Research under contract no. F49620-92-C-0054. The United States Government is authorized to reproduce and distribute reprints for governmental purposes notwithstanding any copyright notation hereon.

References

1. Cavaretta, A., W. Dahmen, and C. Micchelli, *Stationary Subdivision*, Mem. Amer. Math. Soc. **453**, 1991.

2. Cohen, A., Ondelettes, analyses multirésolutions et traitement numérique du signal, Ph.D. thésis, Université Paris IX, Dauphine, 1990.
3. Cohen, A. and J. P. Conze, Régularité des bases d'ondelettes et mesures ergodiques, *Rev. Math. Iberoamericana* **8** (1992), 351–365.
4. Cohen, A. and I. Daubechies, Non-separable bidimensional wavelet bases, *Rev. Math. Iberoamericana* **9** (1993).
5. Conze, J. -P. and A. Raugi, Fonctions harmonique pour un opérateur de transition et applications, *Bull. Soc. Math. France* **118** (1990), 273–310.
6. Crochiere, R. E. and L. R. Rabiner, *Fundamentals of Multirate Signal Processing*, Prentice-Hall, Englewood Cliffs, NJ, 1983.
7. Daubechies, I., *Ten Lectures on Wavelets*, CBMS Conference Series 61, SIAM, Philadelphia, 1992.
8. Eirola, T., Sobolev characterization of solutions of dilation equations, *SIAM J. Math. Anal.* **23** (1992), 1015–1030.
9. Gopinath, R. A. and C. S. Burrus, Wavelet transforms and filter banks, in *Wavelets: A Tutorial in Theory and Applications*, C. K. Chui (ed.), Academic Press, Boston, 1992.
10. Heller, P. N., Rank M wavelet matrices with N vanishing moments, *SIAM J. Matrix Anal.*, to appear.
11. Heller, P. N., A Lagrange interpolation approach to regular M-band wavelets, Aware Technical Report AD930623, Cambridge, MA, 1993.
12. Heller, P. N. and H. L. Resnikoff, Polynomials are generalized eigenfunctions of the wavelet transform, Aware Technical Report AD910912, Cambridge, MA, 1991.
13. Heller, P. N., H. L. Resnikoff, and R. O. Wells, Jr., Wavelet matrices and the representation of discrete functions, in *Wavelets: A Tutorial in Theory and Applications*, C. K. Chui (ed.), Academic Press, Boston, 1992.
14. Heller, P. N. and R. O. Wells, Jr., Sobolev regularity for rank m wavelet systems, in preparation.
15. Krasnoselskii, M. A., *Positive Solutions of Operator Equations*, Noordhoff, Groningen, 1964.
16. Krasnoselskii, M. A. et al., *Approximate Solution of Operator Equations*, Wolters-Noordhoff, Groningen, 1973.
17. Lawton, W., Tight frames of compactly supported affine wavelets, *J. Math. Phys.* **31** (1990), 1898–1901.
18. Lawton, W., Necessary and sufficient conditions for construction orthonormal wavelet bases, *J. Math. Phys.* **32** (1991), 57–61.
19. Lawton, W., Multiresolution properties of the wavelet-Galerkin operator, *J. Math. Phys.* **34** (1992).
20. Polya, G. and G. Szegö, *Problems and Theorems of Analysis*, Springer, New York, 1972.
21. Steffen, P., P. N. Heller, R. A. Gopinath, and C. S. Burrus, Theory of regular M-band wavelets, *IEEE Trans. on Signal Proc.* **41** (1993).
22. Strang, G., Wavelets and dilation equations, *SIAM Review* **31** (1989), 614–627.

23. Villemoes, L., Energy moments in time and frequency for two-scale difference equation solutions and wavelets, *SIAM J. Math. Anal.* **23** (1992), 1519–1543.
24. Volkmer, H., On the regularity of wavelets, *IEEE Trans. Inform. Theory* **38** (1992), 872–876.

Peter Niels Heller
Aware, Inc.
One Memorial Drive
Cambridge, MA 02142
heller@aware.com

Raymond O. Wells, Jr.
Computational Mathematics Laboratory
Rice University
Houston, TX 77251
wells@math.rice.edu

Multiresolution Analysis, Haar Bases and Wavelets on Riemannian Manifolds

Stephan Dahlke

Abstract. In this paper, the concept of a multiresolution analysis approximation of functions on $\mathbb{R}^n$ is generalized to specific Riemannian manifolds. An existence theorem based on generalized Haar functions is proved, and the associated Haar wavelets are constructed.

§1 Introduction

In general, a system of functions $\{\psi^i\}_{i=1,\ldots,N}$ is called a family of (mother) wavelets if the scaled and translated versions of $\{\psi^i\}_{i=1,\ldots,N}$ form an (orthonormal) basis of $L^2(\mathbb{R}^n)$. The most important tool for the construction of such a basis is the multiresolution approximation of functions introduced by Mallat [9] (see also [10]). In general, a multiresolution analysis consists of a nested sequence of closed subspaces of $L^2(\mathbb{R}^n)$ such that their union is dense in $L^2(\mathbb{R}^n)$ while their intersection is $\{0\}$. The space V_0 is assumed to be shift-invariant, *i.e.*,

$$f(\cdot) \in V_0 \iff f(\cdot - k) \in V_0\,, \quad k \in \mathbb{Z}^n\,,$$

and to be spanned by the integer translates of a single function ϕ, called the generator of the multiresolution analysis. Furthermore, let the space V_j be generated from V_{j-1} by a simple scaling of the argument, *i.e.*,

$$f \in V_{j-1} \iff f(2\cdot) \in V_j\ .$$

In this paper, we elaborate the conditions under which a multiresolution analysis, and therefore a wavelet basis, can be constructed on Riemannian manifolds. Our investigations were inspired by the observation that the construction of a multiresolution analysis on $L^2(\mathbb{R}^n)$ is essentially based on the fact that each

Wavelets: Theory, Algorithms, and Applications
Charles K. Chui, Laura Montefusco, and Luigia Puccio (eds.), pp. 33–52.

ISBN 0-12-174575-9

lattice point $k \in \mathbb{Z}^n$ induces a mapping of $\mathbb{R}^n$ with some natural properties and that the scaling process commutes with these mappings in a certain sense. Therefore, the construction of a multiresolution analysis can be reduced to the study of the diffeomorphism group of $\mathbb{R}^n$. According to this observation, one natural way to construct wavelets on Riemannian manifolds is to define them with the aid of certain diffeomorphisms associated with the manifold. This can be done as follows.

Let M be an oriented Riemannian manifold with Levi-Civita-connection and Riemannian volume dm. Let $\mathrm{Diff}^1(M)$ denote the group of C^1-diffeomorphisms of M. Suppose that there exists a (nontrivial) discrete countable subgroup K of isometries of $\mathrm{Diff}^1(M)$ which will be called a *lattice* in the sequel. The lattice K induces an action on M by

$$\begin{aligned} K \times M &\to M \\ (J, x) &\mapsto J(x) \,. \end{aligned} \tag{1.1}$$

This action can be used to define an equivalence relation on M by

$$x \sim y \Longleftrightarrow \text{there exists } J \in K \quad \text{such that } y = J(x) \,. \tag{1.2}$$

Let $M/_\sim$ denote the set of all cosets with respect to $\sim$. Equipped with the quotient topology, $M/_\sim$ is a well-defined topological space. Then the lattice is called *uniform* if $M/_\sim$ is compact.

To construct a multiresolution analysis on $\mathbb{R}^n$ one has to require that the lattice is invariant with respect to the scaling process. In our setting, this property can be interpreted as follows. We say that a lattice is *invariant* with respect to $A \in \mathrm{Diff}^1(M)$ if for each element J of K there exists a lattice point $\tilde{J}$ such that

$$A\left(J(x)\right) = \tilde{J}\left(A(x)\right), \tag{1.3}$$

which means that the mapping $\mathcal{A}$ defined by

$$\begin{aligned} \mathcal{A} : K &\to K \\ J &\mapsto A \circ J \circ A^{-1} \end{aligned} \tag{1.4}$$

is a well-defined group homomorphism of K into itself. We are now ready to define a multiresolution analysis on M.

Definition 1.1. *Let M be an oriented Riemannian manifold that admits a uniform lattice K. Suppose that there exists a diffeomorphism $A \in \mathrm{Diff}^1(M)$ such that K is invariant with respect to A. Let dm denote the Riemannian volume of M. Then a* multiresolution analysis *of $L^2(M, dm)$ is a sequence $\{V_j\}_{j \in \mathbb{Z}}$ of closed subspaces of $L^2(M, dm)$ such that*

$$\ldots \subset V_j \subset V_{j+1} \subset V_{j+2} \subset \ldots \tag{1.5}$$

$$V_j \ni f(\cdot) \iff f(A^{-j} \cdot) \in V_0, \tag{1.6}$$

$$\overline{\bigcup_{j=-\infty}^{\infty} V_j} = L^2(M, dm), \tag{1.7}$$

$$\bigcap_{j=-\infty}^{\infty} V_j = \{0\}, \tag{1.8}$$

$$V_0 \ni f(\cdot) \iff f(J(\cdot)) \in V_0, \quad J \in K\,. \tag{1.9}$$

There exists a function ϕ in V_0 called the generator of $\{V_j\}_{j \in \mathbb{Z}}$ such that

$$V_0 = \overline{\text{span}\{\phi(J(\cdot)),\ J \in K\}}$$

and

$$B_1 ||\lambda||^2_{l_2(K)} \leq \Big|\Big| \sum_{J \in K} \lambda_J\, \phi(J(\cdot)) \Big|\Big|_2^2 \leq B_2 ||\lambda||^2_{l_2(K)}, \quad B_1, B_2 \in \mathbb{R}^+\,. \tag{1.10}$$

This paper is organized as follows. In Section 2, we elaborate the conditions under which a multiresolution analysis in the sense of Definition 1.1 exists. We will see that generalized Haar functions play a fundamental role in this context. In Section 3, we give an outline on how such a generalized Haar function can be constructed. In Section 4, we study some examples and finally, in Section 5, we show the existence of Haar wavelets associated with the generalized multiresolution analysis presented here.

Since it is not attempted to present a complete study of multiresolution analysis on general manifolds, we confine ourselves here to some basic ideas and suggestions for further research.

This work is connected with some other investigations made before. Using a quite different approach, P. G. Lemarié has constructed a wavelet basis on certain Lie groups [8]. In [4], a multiresolution analysis on locally compact Abelian groups is given. There are also connections with the work of R. S. Strichartz concerning tilings on Lie groups [11].

§2 Construction of a multiresolution analysis

This section is devoted to the development of conditions on the manifold so that a multiresolution analysis in the sense of Definition 1.1 exists. In the sequel, let M always be an oriented Riemannian manifold with Levi–Civita connection and Riemannian volume dm that admits a uniform lattice K. The construction presented below is based on a generalized Haar function. To define such a function, we have to impose some conditions on the action (1.1).

We will always assume that this action is *effective* and satisfies the following natural assumptions:

(A) For all $x \in M$ there exists a neighbourhood U_x of x such that

$$J(U_x) \cap U_x = \emptyset$$

for all but finite many J in K;

(B) If x and y are not equivalent, then there exist neighbourhoods U_x, U_y of x and y, respectively, such that

$$J(U_x) \cap U_y = \emptyset$$

for all J in K; and

(C) The set of fixed points is nowhere dense in M.

It is a well-known fact that if the conditions (A), (B), and (C) are satisfied then there exists a *fundamental region* F for $M/_\sim$, *i.e.*, a complete set of representers that satisfies

$$\bigcup_{J \in K} J(F) \simeq M, \qquad J(F) \cap F \simeq \emptyset, \quad J \neq Id, \tag{2.1}$$

and which is the closure of its interior points. Without loss of generality, we will henceforth assume that the Riemannian volume is normalized in such a way that

$$m(F) = 1\,.$$

Furthermore, let us now assume that there exists a fundamental region Q which is *self-similar*, *i.e.*, Q satisfies (2.1) and a self-similarity relation

$$A(Q) = \bigcup_{i \in I} g_i(Q), \tag{2.2}$$

where $g_i \in K$ and I is a finite set. Then the characteristic function χ_Q is called a *generalized Haar function.* Based on this Haar function, one has the following.

Theorem 2.1. *Suppose that there exists a diffeomorphism A on M such that K is invariant with respect to A and that*

(i) *there exists a fundamental region Q for $M/_\sim$ such that*

$$A(Q) = \bigcup_{i \in I} g_i(Q)\,,$$

(ii) A^{-1} *is a contraction, i.e.,*

$$d\left(A^{-1}(x), A^{-1}(y)\right) \leq c_1\, d(x,y)\,,$$

where $c_1 < 1$, and d denotes the canonical metric induced by the Riemannian structure, and, let

$$\sup_{x \in M} \left|\det(T_x\, A^{-1})\right| \leq c_2 < 1\,.$$

Let V_0 be defined by

$$V_0 = \overline{\operatorname{span}\{\chi_Q\,(J(\cdot))\,,\ J \in K\}}. \tag{2.3}$$

Then the sequence $\{V_j\}_{j\in\mathbb{Z}}$ *of closed subspaces defined by*

$$V_j \ni f(\cdot) \iff f\left(A^{-j}(\cdot)\right) \in V_0 \tag{2.4}$$

is a multiresolution analysis of $L^2(M, dm)$.

Remark. A sufficient condition for (ii) is

$$\sup_{z\in M} ||T_z A^{-1}|| \le c_1 < 1 ,$$

for then we obtain

$$d\,(A^{-1}(x), A^{-1}(y)) \le \sup_{z\in M} ||T_z A^{-1}||\, d\,(x,y) \le c_1 d\,(x,y) ,$$

and the second statement of condition (ii) can also easily be checked.

Proof: We start the proof by showing (1.5). To this end, we first observe that χ_Q is a refinable function in the sense that

$$\chi_Q(x) = \sum_{i\in I} \chi_Q\left(g_i^{-1}\left(A(x)\right)\right) . \tag{2.5}$$

In fact, condition (i) implies that

$$\chi_Q(x) = \chi_{A^{-1}(A\,Q)}(x) = \chi_{A^{-1}\left(\bigcup\limits_{i\in I} g_i(Q)\right)}(x) = \begin{cases} 1 & \text{if } A(x) \in g_i(Q) \\ & \text{for some } i \in I \\ 0 & \text{elsewhere} \end{cases} .$$

But this means that

$$\chi_Q(x) = \sum_{i\in I} \chi_Q\left(g_i^{-1}(A\,x)\right)$$

holds. Since it is sufficient to check that $V_0 \subset V_1$, let

$$f(\cdot) = \sum_{J\in K} \lambda_J\, \chi_Q\,(J(\cdot)) ,$$

so that, employing Equations (2.5) and (1.3) we get

$$\begin{aligned} f(x) &= \sum_{J\in K} \lambda_J\, \chi_Q\,(J(x)) = \sum_{J\in K} \lambda_J \sum_{i\in I} \chi_Q\left(g_i^{-1}\left(A\,(J(x))\right)\right) \\ &= \sum_{J\in K}\sum_{i\in I} \lambda_J \chi_Q((g_i^{-1} \circ (A \circ J \circ A^{-1}) \circ A)(x)) \\ &= \sum_{J\in K}\sum_{i\in I} \lambda_J \chi_Q((g_i^{-1} \circ \mathcal{A}(J) \circ A)(x)) \\ &= \sum_{J\in K}\sum_{i\in I} \lambda_J \chi_Q((g_i^{-1} \circ \mathcal{A}(J))(A(x))) , \end{aligned}$$

which implies (1.5).

Condition (1.6) follows immediately from the definition. To show Equation (1.7), we first observe that M is a locally compact and second countable topological space, so that the continuous functions with compact support are dense in $L^2(M, dm)$ (see [2], Sections 43 and 44 for details). Let f be a continuous function with compact support. For an arbitrary point z in Q we define $f_j \in V_j$ by

$$f_j(x) := \sum_{J \in K} f\left(A^{-j}\left(J^{-1}(z)\right)\right) \chi_Q\left(J\left(A^j(x)\right)\right) . \tag{2.6}$$

Then it is easy to check that there exists a compact set $B \subset M$ such that $\operatorname{supp} f \subset B$ and $\operatorname{supp} f_j \subset B$ for all $j > 0$. Therefore, it is sufficient to check that f_j converges to f uniformly on B. Since Q is a fundamental region, the translated versions of χ_Q clearly form a partition of unity, *i.e.*,

$$1 \simeq \sum_{J \in K} \chi_Q\left(J(x)\right) . \tag{2.7}$$

Employing Equation (2.7), we get

$$\begin{aligned} \left|f(x) - f_j(x)\right| &= \left|f(x) - \sum_{J \in K} f\left((A^{-j} \circ J^{-1})(z)\right) \chi_Q\left(J\left(A^j(x)\right)\right)\right| \\ &= \left|\sum_{J \in K} \left(f(x) - f\left((A^{-j} \circ J^{-1})(z)\right)\right) \chi_Q\left(J\left(A^j(x)\right)\right)\right| \\ &\leq \sum_{J \in K} \left|\left(f(x) - f\left((A^{-j} \circ J^{-1})(z)\right)\right)\right| \chi_Q\left(J\left(A^j(x)\right)\right) . \end{aligned}$$

The sum on the right does not vanish if

$$x \in A^{-j}\left(J^{-1}(Q)\right)$$

holds. Then we have

$$\begin{aligned} d\left(x, A^{-j}\left(J^{-1}(z)\right)\right) &\leq \operatorname{diam}\left(A^{-j}\left(J^{-1} Q\right)\right) \\ &\leq c_1^j \operatorname{diam}\left(J^{-1}(Q)\right) \\ &\leq c_1^j \operatorname{diam}(Q) , \end{aligned}$$

where we used the facts that A^{-1} is a contraction and that J^{-1} is an isometry. Since the function f is compactly supported and continuous, it is uniformly continuous, so we can find, for a given $\epsilon > 0$, a number $\delta > 0$ such that

$$d\left(x, A^{-j}\left(J^{-1}(z)\right)\right) \leq \delta$$

implies

$$\left|f\left(A^{-j}\left(J^{-1}(z)\right)\right) - f(x)\right| \leq \epsilon .$$

Since $c_1 < 1$, there exists $j_0 \in \mathbb{N}$ such that

$$d\left(x, A^{-j}\left(J^{-1}(z)\right)\right) \leq \delta\,, \quad j \geq j_0\,,$$

so that we finally obtain

$$\begin{aligned} \left|f(x) - f_j(x)\right| &\leq \sum_{J \in K} \left|f(x) - f\left(\left(A^{-j} \circ J^{-1}\right)(z)\right)\right| \chi_Q\left(J\left(A^j(x)\right)\right) \\ &\leq \epsilon \sum_{J \in K} \chi_Q\left(J\left(A^j(x)\right)\right) \\ &= \epsilon\,, \end{aligned}$$

and (1.7) is proved.

Before we show (1.8) let us check (1.10). Since Q is a fundamental region, χ_Q has orthonormal translates in the sense that

$$\int_M \chi_Q(x)\, \chi_Q\left(J(x)\right)\, dm(x) = \begin{cases} 1 & \text{J} = \text{Id} \\ 0 & \text{otherwise} \end{cases}\,.$$

Therefore, (1.10) holds with $B_1 = B_2 = 1$. Employing (1.10), we are able to prove (1.8). Let

$$f \in \bigcap_{j=-\infty}^{\infty} V_j\,.$$

Then

$$f\left(A^j(\,\cdot\,)\right) \in V_0, \qquad \text{for all } j \in \mathbb{Z}\,.$$

Therefore, there exist sequences $\{\lambda_J^j\}_{J \in K}$ such that

$$f\left(A^j(\,\cdot\,)\right) = \sum_{J \in K} \lambda_J^j\, \chi_Q\left(J(\,\cdot\,)\right)\,.$$

Then we have

$$\begin{aligned} \left|f\left(A^j(\cdot)\right)\right| &= \left|\sum_{J \in K} \lambda_J^j\, \chi_Q\left(J(\,\cdot\,)\right)\right| \\ &\leq \sup_{J \in K} \left|\lambda_J^j\right| \sum_{J \in K} \left|\chi_Q\left(J(\,\cdot\,)\right)\right| \\ &\leq ||\lambda_J^j||_{l_2(K)}\,. \end{aligned}$$

Simple estimates yield

$$\begin{aligned}\left|f(x)\right| &= \left|f\left(A^j\left(A^{-j}(x)\right)\right)\right| \\ &\leq ||\lambda_J^j||_{l_2(K)} \\ &= \left|\left|\sum_{J\in K}\lambda_J^j\,\chi_Q\left(J(\cdot)\right)\right|\right|_2 \\ &= \left(\int_M \left|f\left(A^j(x)\right)\right|^2 dm(x)\right)^{\frac{1}{2}} \\ &= \left(\int_M \left|f(x)\right|^2 d\left(A^j\,m\right)(x)\right)^{\frac{1}{2}} .\end{aligned}$$

Employing condition (ii) we get

$$\begin{aligned}\left|f(x)\right| &= \left(\int_M \left|f(x)\right|^2 \left|\det\left(T_x\,A^{-j}\right)\right| dm(x)\right)^{\frac{1}{2}} \\ &\leq \left(\sup_{x\in M}\left|\det(T_x\,A^{-j})\right|\right)^{\frac{1}{2}} \left|\int_M \left|f(x)\right|^2 dm(x)\right|^{\frac{1}{2}} \\ &\leq c_2^{j/2}\,||f||_2 .\end{aligned}$$

If we let j tend to infinity, we obtain

$$f \equiv 0 ,$$

and (1.8) is proved. Finally, (1.9) follows from the definition, and so the theorem is proved. ■

It is natural to ask how a generator with higher smoothness can be constructed. On $\mathbb{R}^n$, this can be easily done by using a convolution. On a manifold, such a convolution is not well-defined. However, for a fixed function g, the convolution $f * g$ can be interpreted as an integral transformation with kernel

$$k(x,y) = g(x-y) .$$

Therefore, it seems natural to try to construct a kernel k on M such that the integral operator induced by k maps the refinable functions on M into themselves and acts as a smoothing procedure. Then a function in the range of the integral operator would be a candidate for a generator with higher smoothness. Concerning the conditions on the kernel, we observe the following facts.

Proposition 2.2. *Suppose that there exists a kernel $k(x,y)$ on M such that*

(i) *the mapping*
$$x \longmapsto k(x,\,\cdot)$$
is continuous in L^2-sense;

(ii) *$k(x,y)$ satisfies*
$$k\left(J(x), J(y)\right) = k(x,y)$$
for all $J \in K$, and
$$|\det(T_y A^{-1})|\, k\left(x, A^{-1}(y)\right) = \sum_{J\in K} a_J\, k\left(J\left(A(x), y\right)\right) \tag{2.8}$$
for some finitely supported sequence $\{a_J\}_{J\in K}$.

Let ϕ be a refinable function, i.e., there exists a finitely supported sequence $\{b_J\}_{J\in K}$ such that
$$\phi(x) = \sum_{J\in K} b_J\, \phi\left(J\left(A(x)\right)\right)\,. \tag{2.9}$$
Then the function $\tilde{\phi}$ defined by
$$\tilde{\phi}(x) := \int_M k(x,y)\,\phi(y)\,dm(y) \tag{2.10}$$
is refinable and continuous.

Proof: First we check that $\tilde{\phi}$ is continuous. Using Hölder's inequality, we get
$$\begin{aligned} \left|\tilde{\phi}(x) - \tilde{\phi}(z)\right| &\leq \int_M \left|k(x,y) - k(z,y)\right|\, \left|\phi(y)\right| dm(y) \\ &\leq \left(\int_M \left|k(x,y) - k(z,y)\right|^2 dm(y)\right)^{\frac{1}{2}} \left(\int_M \left|\phi(y)\right|^2 dm(y)\right)^{\frac{1}{2}} \\ &= ||k(x,\,\cdot) - k(z,\,\cdot)||_2\ \ ||\phi||_2\ , \end{aligned}$$
and the continuity of $\tilde{\phi}$ follows from condition (i). Now we show that $\tilde{\phi}$ is refinable. One has
$$\begin{aligned} \tilde{\phi}(x) &= \int_M k(x,y)\,\phi(y)\,dm(y) \\ &= \int_M k(x,y) \sum_{J\in K} b_J\,\phi\left(J\left(A(y)\right)\right)\, dm(y) \\ &= \sum_{J\in K} b_J \int_M k(x,y)\,\phi\left(J\left(A(y)\right)\right) dm(y) \\ &= \sum_{J\in K} b_J \int_M k\left(x, A^{-1}(y)\right)\ \phi\left(J(y)\right) \left|\det(T_y A^{-1})\right| dm(y)\,. \end{aligned}$$

Using condition (ii), we finally obtain

$$\begin{aligned}\tilde{\phi}(x) &= \sum_{J\in K} b_J \sum_{\tilde{J}\in K} a_{\tilde{J}} \int_M k\left(\tilde{J}(A(x)),y\right) \phi(J(y))\, dm(y) \\ &= \sum_{J\in K}\sum_{\tilde{J}\in K} b_J a_{\tilde{J}} \int_M k\left((J\circ\tilde{J})(A(x)),y\right)\phi(y)\, dm(y) \\ &= \sum_{J\in K}\sum_{\tilde{J}\in K} b_J a_{\tilde{J}} \tilde{\phi}\left((J\circ\tilde{J})(A(x))\right),\end{aligned}$$

and the proposition is proved. ■

Proposition 2.2 can be used to construct a generator with higher smoothness. However, for a given manifold, it is a hard problem to find a kernel that satisfies Equation (2.8), and moreover, it is not clear how the stability condition (1.10) can be checked for a generator constructed by Equation (2.10).

Remark. We have assumed that the action induced by the lattice K is effective. Let us now consider the case that the action is moreover *free*, *i.e.*, the mapping

$$\begin{aligned} H: \quad & K\times M \to M \\ & (J,x)\mapsto J(x)\end{aligned}$$

is one-to-one for each x in M. Furthermore, suppose that the mapping

$$\begin{aligned} \tilde{H}: \quad & K\times M \to M\times M \\ & (J,x)\mapsto (x,J(x))\end{aligned}$$

is *proper*, which means that if $B\subset M\times M$ is compact, then $\tilde{H}^{-1}(B)$ is compact. Under these conditions, it can be shown that $M/_{\sim}$ is a smooth compact manifold and the canonical projection $\pi:\ M\longrightarrow M/_{\sim}$ is a submersion (see [1], Chapter 4.1 for details). If we define the generalized periodization operator

$$[f](x) := \sum_{J\in K} f(J(x)) \tag{2.11}$$

and set

$$\phi_J^j(\cdot) := \left[\phi\left(J\left(A^j(\cdot)\right)\right)\right],$$

then the functions ϕ_J^j give rise to a nested sequence $\{V_j\}_{j\in\mathbf{N}_0}$ of subspaces of $L^2(M/_{\sim})$. So, a method to treat compact manifolds would be: given a compact manifold M, find a manifold $\tilde{M}$ that admits a lattice K such that $M\approx\tilde{M}/_{\sim}$ and apply the generalized periodization.

§3 Construction of a generalized Haar function

In this section, we present an algorithm for the construction of a generalized Haar function. Taking into account the self-similarity relation (2.2), it turns out that the construction of a Haar function is closely related to the problem of finding a generalized self-similar lattice tiling. On $I\!R^n$, this problem was intensively studied by Gröchenig and Madych [7] (see also [6]). Their construction can be generalized to our setting as follows. Recall that we assume that the mapping

$$\begin{aligned} \mathcal{A}: \quad & K \longrightarrow K \\ & J \longmapsto A \circ J \circ A^{-1} \end{aligned} \tag{3.1}$$

is a well-defined group homomorphism of K into itself. Then the quotient ${}^{K}/_{\mathcal{A}(K)}$ is well-defined. Let us suppose that ${}^{K}/_{\mathcal{A}(K)}$ is finite, and let $g_1, \dots, g_q$ be a complete enumeration of the cosets in ${}^{K}/_{\mathcal{A}(K)}$, *i.e.*,

$$K = \bigcup_{i \in I} g_i \circ \mathcal{A}(K), \qquad I = \{1, \dots q\}\,. \tag{3.2}$$

We want to find a fundamental region Q that satisfies

$$Q = \bigcup_{i \in I} A^{-1}\left(g_i(Q)\right)\,. \tag{3.3}$$

To this end, consider the mapping

$$\begin{aligned} \overline{A}: \quad & P_C(M) \longrightarrow P_C(M) \\ & \tilde{Q} \longmapsto \bigcup_{i \in I} A^{-1}\left(g_i(\tilde{Q})\right)\,, \end{aligned} \tag{3.4}$$

where $P_C(M)$ denotes the set of all compact subsets of M. With the metric

$$\rho(P, \tilde{Q}) = \max\left(r(P, \tilde{Q}), r(\tilde{Q}, P)\right)$$

where

$$r(P, \tilde{Q}) = \sup_{h \in P} \inf_{g \in \tilde{Q}} d(h, g)\,,$$

$P_C(M)$ is a complete metric space. To find a self-similar fundamental region, we have to check that the fix point iteration

$$Q_{n+1} = \overline{A}(Q_n) = \bigcup_{i \in I} A^{-1}\left(g_i(Q_n)\right)$$

converges to a compact set in the metric ρ.

Lemma 3.1. *If Q_0 is a compact set, then the sequence*

$$Q_{n+1} = \overline{A}(Q_n) = \bigcup_{i \in I} A^{-1}\left(g_i(Q_n)\right)$$

converges in the metric ρ to the set

$$Q := \{y \in M \mid y = \lim_{j \to \infty} (A^{-j} \circ \mathcal{A}^{j-1}(g_1) \circ \mathcal{A}^{j-2}(g_2) \circ \ldots \circ \mathcal{A}(g_{j-1}) \circ g_j)(z), z \in Q_0\}.$$

Remark. The set Q satisfies the self-similarity condition (2.2), since we have

$$\begin{aligned}
&\bigcup_{i \in I} A^{-1}\left(g_i(Q)\right) \\
&= \bigcup_{i \in I} \{ \lim_{j \to \infty} (A^{-1} \circ g_i \circ A^{-j} \circ \mathcal{A}^{j-1}(g_1) \circ \ldots \circ \mathcal{A}(g_{j-1}) \circ g_j)(z) \} \\
&= \bigcup_{i \in I} \{ \lim_{j \to \infty} (A^{-2} \circ A \circ g_i \circ A^{-1} \circ A^{-(j-1)} \circ \mathcal{A}^{j-1}(g_1) \circ \ldots \circ \mathcal{A}(g_{j-1}) \circ g_j)(z) \} \\
&= \bigcup_{i \in I} \{ \lim_{j \to \infty} (A^{-2} \circ \mathcal{A}(g_i) \circ A^{-(j-1)} \circ \mathcal{A}^{j-1}(g_1) \circ \ldots \circ \mathcal{A}(g_{j-1}) \circ g_j)(z) \} \\
&= \ldots = \bigcup_{i \in I} \{ \lim_{j \to \infty} (A^{-(j+1)} \circ \mathcal{A}^{j}(g_i) \circ \mathcal{A}^{j-1}(g_1) \circ \ldots \circ \mathcal{A}(g_{j-1}) \circ g_j)(z) \} \\
&= Q\,.
\end{aligned}$$

Proof of Lemma 3.1: The lemma can be proved by following the proof of Lemma 3 in [7]. Using the definition of $\mathcal{A}$, a simple calculation yields

$$\begin{aligned}
Q_n &= \bigcup_{i \in I} A^{-1}\left(g_i(Q_{n-1})\right) \\
&= \bigcup_{i \in I} A^{-1}\Big(g_i\Big(\bigcup_{k \in I} A^{-1}\left(g_k(Q_{n-2})\right)\Big)\Big) \\
&= \bigcup_{i \in I} \bigcup_{k \in I} (A^{-1} \circ g_i \circ A^{-1} \circ g_k)(Q_{n-2}) \\
&= \bigcup_{i \in I} \bigcup_{k \in I} (A^{-1} \circ A^{-1} \circ A \circ g_i \circ A^{-1} \circ g_k)(Q_{n-2}) \\
&= \bigcup_{i \in I} \bigcup_{k \in I} (A^{-2} \circ \mathcal{A}(g_i) \circ g_k)(Q_{n-2}) \\
&= \ldots = \bigcup_{(g_1, \ldots, g_n) \in I^n} (A^{-n} \circ \mathcal{A}^{n-1}(g_1) \circ \mathcal{A}^{n-2}(g_2) \circ \ldots \circ \mathcal{A}(g_{n-1}) \circ g_n)(Q_0).
\end{aligned}$$

Therefore, for an element $x_n \in Q_n$, there exists an element $x_0 \in Q_0$ such that

$$x_n = \left(A^{-n} \circ \mathcal{A}^{n-1}(g_1) \circ \ldots \circ \mathcal{A}(g_{n-1}) \circ g_n\right)(x_0)\,.$$

For $y_0 \in Q$, we define

$$\begin{aligned}
\tilde{y} &= (A^{-n} \circ \mathcal{A}^{n-1}(g_1) \circ \ldots \circ \mathcal{A}(g_{n-1}) \circ g_n)(y_0) \\
&= (A^{-n} \circ \mathcal{A}^{n-1}(g_1) \circ \ldots \circ \mathcal{A}(g_{n-1}) \circ g_n) \\
&\qquad \left(\lim_{j\to\infty} A^{-j} \circ \mathcal{A}^{j-1}(\tilde{g}_1) \circ \ldots \circ \tilde{g}_j(z)\right) \\
&= \lim_{j\to\infty} (A^{-n} \circ \mathcal{A}^{n-1}(g_1) \circ \ldots \circ \mathcal{A}(g_{n-1}) \circ g_n \circ A^{-j} \circ \mathcal{A}^{j-1}(\tilde{g}_1) \circ \ldots \circ \tilde{g}_j)(z) \,.
\end{aligned}$$

We want to check that $\tilde{y}$ is again a point in Q. Repeated insertion of the identity and employing the definition of $\mathcal{A}$ yields

$$\begin{aligned}
\tilde{y} &= \lim_{j\to\infty} (A^{-(n+1)} \circ A \circ \mathcal{A}^{n-1}(g_1) \circ A^{-1} \circ A \circ \mathcal{A}^{n-2}(g_2) \circ \ldots \circ \mathcal{A}(g_{n-1}) \\
&\qquad \circ A^{-1} \circ A \circ g_n \circ A^{-j} \circ \ldots \circ \tilde{g}_j)(z) \\
&= \lim_{j\to\infty} (A^{-(n+1)} \circ \mathcal{A}^{n}(g_1) \circ \mathcal{A}^{n-1}(g_2) \circ \ldots \circ \mathcal{A}(g_n) \\
&\qquad \circ A^{-(j-1)} \circ \mathcal{A}^{j-1}(\tilde{g}_1) \circ \ldots \circ \tilde{g}_j)\,(z) \\
&= \lim_{j\to\infty} (A^{-(n+j)} \circ \mathcal{A}^{n+j-1}(g_1) \circ \mathcal{A}^{n+j-2}(g_2) \circ \ldots \circ \mathcal{A}^{j-1}(\tilde{g}_1) \circ \ldots \circ \tilde{g}_j)\,(z).
\end{aligned}$$

From the definition of Q, it follows that $\tilde{y}$ is in fact an element in Q, and we can estimate

$$\begin{aligned}
\inf_{y\in Q} d(y, x_n) &\le \inf_{y_0\in Q} d\big((A^{-n} \circ \mathcal{A}^{n-1}(g_1) \circ \ldots \circ \mathcal{A}(g_{n-1}) \circ g_n)(y_0)\,, \\
&\qquad (A^{-n} \circ \mathcal{A}^{n-1}(g_1) \circ \ldots \circ \mathcal{A}(g_{n-1}) \circ g_n)(x_0)\big) \\
&\le c_1^n \inf_{y_0\in Q} d\big((\mathcal{A}^{n-1}(g_1) \circ \ldots \circ \mathcal{A}(g_{n-1}) \circ g_n)(y_0)\,, \\
&\qquad (\mathcal{A}^{n-1}(g_1) \circ \ldots \circ \mathcal{A}(g_{n-1}) \circ g_n)(x_0)\big) \\
&= c_1^n \inf_{y_0\in Q} d(y_0, x_0) \\
&\le c_1^n \sup_{x_0\in Q_0} \inf_{y_0\in Q} d(y_0, x_0) \\
&\le c_1^n \, r(Q_0, Q)\,,
\end{aligned}$$

where we once again used the facts that A^{-1} is a contraction and $\mathcal{A}^i(g_k)$ is an isometry. This yields

$$r(Q_n, Q) \le c_1^n \, r(Q_0, Q)\,.$$

Since $c_1 < 1$, it follows that $r(Q_n, Q)$ can be made arbitrarily small. Using similar arguments, it can be shown that the same holds for $r(Q, Q_n)$, and the lemma is proved. ■

The limit set Q can now be used to construct a multiresolution analysis.

Lemma 3.2. *Let Q denote the limit set constructed with the aid of Lemma 3.1. If*

$$|Q| = 1 ,$$

then χ_Q generates a multiresolution analysis of $L^2(M, dm)$.

Proof: We want to apply Theorem 2.1. According to the Remark from above, Q is a self-similar set. Therefore, it remains to check that Q provides a fundamental region for $M/_{\sim}$, *i.e.*,

(i) $\bigcup_{J \in K} J(Q) \simeq M$

(ii) $J(Q) \cap Q \simeq \emptyset, \qquad J \neq Id.$

First we show (i). Without loss of generality, let us assume that the set Q_0 is the canonical fundamental region F. Then it is easy to check that condition (i) holds for all

$$Q_n = \bar{A}^n(F) .$$

That means that for almost all x in M there exists an isometry J_n such that

$$J_n(x) \in Q_n .$$

Since all the Q_n's are contained in a fixed ball, there exists a subsequence $\{J_{n_i}\}_{i \in \mathbf{N}}$ such that

$$\lim_{i \to \infty} J_{n_i}(x) = y .$$

But then, x and y have to be equivalent, for otherwise, we have a contradiction to condition (B). Therefore, for some $\tilde{J} \in K$,

$$x = \lim_{i \to \infty} \tilde{J} \circ J_{n_i}(x) ,$$

which implies that $\{J_{n_i}\}_{i \in \mathbf{N}}$ contains a constant subsequence $J_{n_{i_k}} = J_N$, $k = 1, 2, \cdots$ because of (A). Since $J_N(x)$ is contained in Q_{n_i} for infinitely many indexes, it follows that $J_N(x)$ is in Q by virtue of the fact that Q_n converges to Q.

Now, let us attack the second part. Since Q satisfies (i), we have

$$\sum_{J \in K} \chi_Q(J(\cdot)) \geq 1 .$$

But since

$$1 = |Q| = \int_M \chi_Q(x) dm(x) = \int_F \sum_{J \in K} \chi_Q(J(x)) dm(x) ,$$

we obtain

$$\sum_{J \in K} \chi_Q(J(\cdot)) \simeq 1 ,$$

which implies (ii). ■

Remark. In this section, we always used the left cosets induced by the mapping $\mathcal{A}$, compare with Formula 3.2. Obviously, the right cosets can be treated similarly.

§4 Examples

In this section, we present some examples to illustrate the constructions of the previous sections.

Example 4.1. Clearly, the "classical" wavelet construction given by

$$M = \mathbb{R}^n, \quad K = \{I_k\}_{k \in \mathbf{Z}^n}, \quad I_k(x) := x + k \ ,$$

and an expanding integer scaling matrix is a special case of our construction. In [7], the self-similar tiles for several scaling matrices are studied.

Example 4.2. Our construction enables us to construct new wavelet bases on $\mathbb{R}^n$ by employing a different kind of isometries. If we use the canonical metric on $\mathbb{R}^n$, every isometry is of the form

$$J(x) = Bx + b$$

where

$$B \in O(n), \quad b \in \mathbf{Z}^n \ .$$

For example, on $\mathbb{R}^2$, let us define

$$J_{e,m,n}(x) := \begin{pmatrix} 1 & 0 \\ 0 & (-1)^e \end{pmatrix} \begin{pmatrix} x_1 \\ x_2 \end{pmatrix} + \begin{pmatrix} m \\ n \end{pmatrix}, \quad e \in \{0,1\},\ m,n \in \mathbf{Z} \ ,$$

and set

$$K := \{J_{e,m,n} | e \in \{0,1\},\ m,n, \in \mathbf{Z}\}.$$

This lattice is obviously uniform and conditions (A), (B), and (C) are satisfied. Any diagonal matrix with integer entries strictly larger than one leaves K invariant and satisfies condition (ii) in Theorem 2.1.
Let us choose

$$A = \begin{pmatrix} 2 & 0 \\ 0 & 2 \end{pmatrix}.$$

Then

$$\mathcal{A}(J_{e,m,n}) = A \circ J_{e,m,n} \circ A^{-1} = \begin{pmatrix} 1 & 0 \\ 0 & (-1)^e \end{pmatrix} \begin{pmatrix} \cdot \end{pmatrix} + \begin{pmatrix} 2m \\ 2n \end{pmatrix}.$$

Therefore, a complete set of representers of $K/_{\mathcal{A}(K)}$ is given by the four functions $J_{0,0,0}, J_{0,0,1}, J_{0,1,0}$, and $J_{0,1,1}$. A fundamental region is given by the rectangle

$$F = \{(x,y) \in \mathbb{R}^2 |\ 0 \le x < 1,.\ 0 \le y \le \frac{1}{2}\} \ .$$

It is easy to check that

$$A(F) \simeq J_{0,1,0}(F) \cup J_{1,0,1}(F) \cup J_{1,1,1}(F) \cup J_{0,0,0}(F) \ ,$$

and therefore F is self-similar with respect to $J_{0,1,0}, J_{1,0,1}, J_{1,1,1}$, and $J_{0,0,0}$.

We can also use discrete subgroups generated by nontrivial rotations. For instance, let us take the rotation about the angle $\frac{\pi}{4}$, *i.e.*, let us consider the matrix

$$J = \begin{pmatrix} \frac{1}{\sqrt{2}} & -\frac{1}{\sqrt{2}} \\ \frac{1}{\sqrt{2}} & \frac{1}{\sqrt{2}} \end{pmatrix} .$$

We fix a vector b in $\mathbb{R}^2$ and consider the translations

$$T_a^b := \sum_{i=0}^{7} a_i J^i(b) \ , \qquad a \in \mathbb{R}^8 .$$

Then a discrete subgroup K of $\mathrm{Diff}^1(\mathbb{R}^2)$ is given by

$$K := \{J^i(\cdot) + T_m^b \mid i = 0, \ldots, 7, \ m \in \mathbf{Z}^8\}.$$

Especially, if we choose

$$b = \begin{pmatrix} 1 \\ 0 \end{pmatrix} ,$$

then K can be described as

$$K = \{J^i(\cdot) + \begin{pmatrix} k \\ l \end{pmatrix} + \frac{1}{\sqrt{2}} \begin{pmatrix} m+n \\ m-n \end{pmatrix} |i = 0, \ldots, 7, \ k, l, m, n \in \mathbf{Z}\},$$

i.e., the translation component is given as a linear combination of the standard lattice and a scaled version of the quincunx grid. Once again, the conditions (A), (B), and (C) are satisfied, and any diagonal matrix with integer entries leaves K invariant. But there are in fact more possible choices for the expanding diffeomorphism. For instance, we can take a multiple of the rotation J itself, *i.e.*,

$$A(x) = \sqrt{2} J(x) = \begin{pmatrix} 1 & -1 \\ 1 & 1 \end{pmatrix} (x) .$$

For the standard group of isometries, the wavelets associated with this matrix were intensively studied (see [3] for details). In our case, we obtain

$$A\left(J^i(A^{-1}(x)) + \begin{pmatrix} k \\ l \end{pmatrix} + \frac{1}{\sqrt{2}} \begin{pmatrix} m+n \\ m-n \end{pmatrix} \right)$$

$$= (A \circ J^i \circ A^{-1})(x) + \begin{pmatrix} k-l \\ k+l \end{pmatrix} + \frac{1}{\sqrt{2}} \begin{pmatrix} 2n \\ 2m \end{pmatrix}$$

$$= J^i(x) + \begin{pmatrix} k-l \\ k+l \end{pmatrix} + \frac{1}{\sqrt{2}} \begin{pmatrix} (n+m)+(n-m) \\ (n+m)-(n-m) \end{pmatrix} .$$

To find a self-similar region, one has to use the iterative process described in Lemma 3.1.

Example 4.3. Let G be a Lie group that possesses a discrete subgroup H such that ${}^{G}/_{H}$ is compact. Let G be equipped with the canonical right-invariant connection. Then the lattice K is given by the set of right translations

$$K := \{J_n | J_n(g) = g \circ h, h \in H\}$$

induced by the discrete subgroup. If G admits a C^1 group automorphism A that satisfies

$$A(K) \subset K ,$$

then A satisfies (1.3) and can therefore be used as the scaling diffeomorphism. One example is given by the *Weyl–Heisenberg group* H_n. Given $(p, q, t) \in \mathbf{R}^{2n+1}$, we define the matrix

$$m(p,q,t) = \begin{pmatrix} 0 & p_1 & \dots & p_n & t \\ 0 & 0 & & 0 & q_1 \\ \vdots & \vdots & & & \vdots \\ 0 & 0 & \dots & 0 & q_n \\ 0 & 0 & \dots & 0 & 0 \end{pmatrix} .$$

Moreover, we define

$$M(p,q,t) = I + m(p,q,t) .$$

Then the set of all matrices $M(p,q,t)$ forms a well-defined Lie group, the (polarized) Weyl–Heisenberg group H_n with group law

$$M(p,q,t) \circ M(p',q',t') = M(p+p', q+q', t+t'+pq') .$$

A uniform lattice is given by the set

$$K := \{M(k,l,m) |\ k \in \mathbf{Z}^n, l \in \mathbf{Z}^n, m \in \mathbf{Z}\} .$$

Once again, (A), (B), and (C) are clearly satisfied. The automorphisms of H_n are classified (see [5] for details). It turns out that there exists a subgroup of dilations given by

$$\delta[r](M(p,q,t)) := M(rp, rq, r^2 t), \quad r > 0 .$$

For any r in $\mathbb{N}$ and larger than one $\delta[r]$ is an expanding automorphism that leaves K invariant.

§5 Wavelets

Finally, we construct a wavelet basis associated with the multiresolution analysis from Theorem 2.1. This wavelet basis can be interpreted as a natural generalization of the well-known Haar wavelet on $\mathbb{R}$. Following the investigations of Gröchenig and Madych [7], we have:

Theorem 5.1. *Let M be a Riemannian manifold such that the conditions of Theorem 2.1 are satisfied. Furthermore, let U be a unitary matrix whose first row contains constant elements. Then the functions $\{\psi_i\}_{i=2,\dots q}$ defined by*

$$\psi_i(x) := c^{-\frac{1}{2}} \sum_{j=1}^{q} u_{ij}\, \chi_Q\left(g_j(Ax)\right), \qquad c := \int_Q d\,(Am)(x) \tag{5.1}$$

satisfy

(i) $\langle \psi_i(J\,\cdot), \psi_{\tilde{i}}(\,\cdot\,)\rangle = \begin{cases} 1 & J = Id,\ i = \tilde{i} \\ 0 & \text{otherwise}, \end{cases}$

(ii) $V_1 = V_0 \oplus W_0, \qquad V_0 \perp W_0,$

where $W_0 = \overline{\operatorname{span}\{\psi_i\,(J(\,\cdot\,))\,,\quad i = 2, \dots, q\,,\quad J \in K\}}\,.$

Proof: First we show (i). The self-similarity relation (2.2) and definition (5.1) imply

$$\operatorname{supp} \psi_i \subset Q\,, \qquad i = 2, \dots, q\,,$$

and therefore

$$\langle \psi_i(J\,\cdot)\,,\, \psi_{\tilde{i}}\rangle = 0 \qquad \text{if} \quad J \neq Id\,.$$

Furthermore, one has

$$\begin{aligned}
\langle \psi_i(\,\cdot\,), \psi_{\tilde{i}}(\,\cdot\,)\rangle &= c^{-1} \sum_{j=1}^{q} \sum_{\tilde{j}=1}^{q} u_{ij}\, \overline{u}_{\tilde{i}\tilde{j}}\, \langle \chi_Q\left(g_j\left(A(\,\cdot\,)\right)\right), \chi_Q\left(g_{\tilde{j}}\left(A(\,\cdot\,)\right)\right) \rangle \\
&= c^{-1} \sum_{j=1}^{q} \sum_{\tilde{j}=1}^{q} u_{ij}\, \overline{u}_{\tilde{i}\tilde{j}} \int_M \chi_Q\left(g_j\left(A(x)\right)\right) \chi_Q\left(g_{\tilde{j}}\left(A(x)\right)\right) dm(x) \\
&= c^{-1} \sum_{j=1}^{q} \sum_{\tilde{j}=1}^{q} u_{ij}\, \overline{u}_{\tilde{i}\tilde{j}} \int_M \chi_Q\left(g_j(x)\right) \chi_Q\left(g_{\tilde{j}}(x)\right) d(Am)(x) \\
&= c^{-1} \sum_{j=1}^{q} \sum_{\tilde{j}=1}^{q} u_{ij}\, \overline{u}_{\tilde{i}\tilde{j}}\, \delta_{j\tilde{j}} \int_Q d(Am)(x) \\
&= \sum_{j=1}^{q} u_{ij}\, \overline{u}_{\tilde{i}j} \\
&= \delta_{i\tilde{i}}\,,
\end{aligned}$$

and (i) is shown.

Now we prove (ii) First we show that $V_0 \perp W_0$. We compute

$$
\begin{aligned}
\langle \phi(\cdot), \psi_i(\cdot) \rangle &= u_{11}^{-1}\, c^{-\frac{1}{2}} \sum_{j=1}^{q} \sum_{\tilde{j}=1}^{q} u_{1j}\, \overline{u}_{i\tilde{j}}\, \langle \chi_Q\left(g_j\left(A(\cdot)\right)\right), \chi_Q\left(g_{\tilde{j}}\left(A(\cdot)\right)\right) \rangle \\
&= u_{11}^{-1}\, c^{\frac{1}{2}} \sum_{j=1}^{q} u_{1j}\, \overline{u}_{i\,j} \\
&= 0\,.
\end{aligned}
$$

To see that $V_1 = V_0 \oplus W_0$, it is sufficient to check that for each $g_i, i \in I$, there exist sequences $\{a_J\}_{J\in K}$, $\{b_J^j\}_{J\in K}$, $j = 2, \ldots, q$ such that

$$
\phi\left(g_i\left(A(\cdot)\right)\right) = \sum_{J\in K} a_J\, \phi\left(J(\cdot)\right) + \sum_{j=2}^{q} \sum_{J\in K} b_J^j\, \psi_j\left(J(\cdot)\right)\ .
$$

We have

$$
\begin{aligned}
\chi_Q\left(g_i(A(\cdot))\right) &= \sum_{j=1}^{q} \delta_{ij}\, \chi_Q\left(g_j\left(A(\cdot)\right)\right) \\
&= \sum_{j=1}^{q} \sum_{k=1}^{q} \overline{u}_{ki}\, u_{kj}\, \chi_Q\left(g_j\left(A(\cdot)\right)\right) \\
&= \overline{u}_{1i} \sum_{j=1}^{q} u_{1j}\, \chi_Q\left(g_j\left(A(\cdot)\right)\right) + \sum_{k=2}^{q} \sum_{j=1}^{q} \overline{u}_{ki}\, u_{kj}\, \chi_Q\left(g_j\left(A(\cdot)\right)\right) \\
&= |u_{11}|^2 \sum_{j=1}^{q} \chi_Q\left(g_j\left(A(\cdot)\right)\right) + \sum_{k=2}^{q} \overline{u}_{ki} \sum_{j=1}^{q} u_{kj}\, \chi_Q\left(g_j\left(A(\cdot)\right)\right) \\
&= |u_{11}|^2\, \chi_Q(\cdot) + \sum_{k=2}^{q} c^{\frac{1}{2}}\, \overline{u}_{ki}\, \psi_k(\cdot)\,,
\end{aligned}
$$

and the theorem is proved. ■

Acknowledgments. The author is grateful to Angela Kunoth for many stimulating discussions and to Karl–Heinz Gröchenig for introducing him to the theory of tilings. Furthermore, he wants to thank Karsten Urban for some helpful remarks.

References

1. Abraham, R. and J. E. Marsden, *Foundations of Mechanics*, 2nd edition, Benjamin/Cummings Publ. Comp. Man., 1978.
2. Bauer, H., *Wahrscheinlichkeitstheorie und Grundzüge der Maßtheorie*, Walter de Gruyter, Berlin, New York, 1978.
3. Cohen, A. and I. Daubechies, Non–separable bidimensional wavelet bases, *Rev. Mat. Iberoamericana* **9** (1993), 51–137.
4. Dahlke, S., Multiresolution analysis and wavelets on locally compact Abelian groups, in *Curves and Surfaces II*, P. J. Laurent, A. Le Méhauté and L. L. Schumaker (eds.), Academic Press , Boston, 1994, to appear.
5. Folland, G. B., *Harmonic Analysis in Phase Space*, Princeton University Press, Princeton, New Jersey, 1989.
6. Gröchenig, K. and A. Haas, Self-similar lattice tilings, 1992, preprint.
7. Gröchenig, K. and W. R. Madych, Multiresolution analysis, Haar bases, and self-similar tilings of $\mathbb{R}^n$, *IEEE Trans. Inform. Theory* **38** (1992), 556–568.
8. Lemarié, P. G., Bases d'ondelettes sur les groupes de Lie stratifiés, *Bulletin de la Société Mathématique de France* **117** (1989), 211–232.
9. Mallat, S., Multiresolution approximation and wavelet orthonormal bases of L^2, *Trans. Amer. Math. Soc.* **315** (1989), 69–88.
10. Meyer, Y., *Ondelettes et Opérateurs I*, Hermann Editeurs des Sciences et des Arts, Paris, 1990.
11. Strichartz, R. S., Self-similarity on nilpotent Lie groups, 1992, preprint.

Stephan Dahlke
Institut für Geometrie und Praktische Mathematik
RWTH Aachen
Templergraben 55
52062 Aachen
Germany
dahlke@igpm.rwth-aachen.de

Orthonormal Cardinal Functions

T.N.T. Goodman and Charles A. Micchelli

Abstract. In this paper we investigate various questions relating to the construction of functions with one or more of the following properties: satisfying a refinement equation, orthonormal integer translates, compact support, band-limited, symmetric, or cardinal interpolatory.

§1 Introduction

The two functions $\phi(x) = \sin \pi x / \pi x$ and $\phi(x) = \chi_{[0,1)}(x)$ share three desirable properties. Both are cardinal functions, that is, they satisfy condition

$$\phi(j) = \delta_j := \begin{cases} 0, & j \notin \mathbf{Z}\backslash\{0\} \\ 1\ , & j = 0\ . \end{cases} \tag{1.1}$$

Both have orthonormal integer translates, that is,

$$\int_{\mathbf{R}} \phi(x)\phi(x-j)dx = \delta_j, \quad j \in \mathbf{Z} \tag{1.2}$$

and both are *refinable.* This last property means that ϕ satisfies a refinement equation,

$$\phi(x) = \sum_{j\in\mathbf{Z}} q_j \phi(2x-j), \quad x \in \mathbb{R}, \tag{1.3}$$

for some real constants q_j, $j \in \mathbf{Z}$. In particular, when both Equations (1.1) and (1.3) hold, Condition (1.1) implies, by setting $x = k/2$, $k \in \mathbf{Z}$, in Equation (1.3), that $q_k = \phi(k/2), k \in \mathbf{Z}$.

The validity of Equation (1.3) is obvious for the function $\phi(x) = \chi_{[0,1)}(x)$ while for the sinc function $\phi(x) = \sin \pi x/\pi x$ it follows from the cardinal series for functions in $L^2(\mathbb{R})$ which are band-limited to $[-\pi, \pi]$.

Wavelets: Theory, Algorithms, and Applications
Charles K. Chui, Laura Montefusco, and Luigia Puccio (eds.), pp. 53–88.

ISBN 0-12-174575-9

This paper investigates various questions related to constructing functions which have one or more of properties shown in Equations (1.1), (1.2), and (1.3) and are either symmetric or of compact support. Sections 2 and 3 deal with constructions using spline functions. Section 4 studies orthonormal refinable cardinal functions which are also studied in [8] from which we draw some of our motivation. The final section relates wavelets based on the sinc function with limits of spline wavelets as their degree tends to infinity.

§2 Local orthogonal spline projectors with a given accuracy

In [5], we constructed spline interpolation operators with a given polynomial accuracy which are local, that is, at every point the interpolant depends only on a finite number of the interpolated data. The basic idea in this construction was to add more knots than interpolation points, in fact twice as many, and use these extra degrees of freedom to identify a cardinal function of least support compatible with the requirement of prescribed accuracy. To achieve a truly local interpolation operator adding knots is unavoidable in general. In fact when both the knots and points of interpolation are infinite and equally spaced, the cardinal function when it exists has infinite support and is exponentially decaying, see [11]. In this section, we study similar problems of constructing *local* orthogonal spline operator with a prescribed accuracy.

We begin by recalling that the B-spline, M_n of degree n is defined as the $(n+1)$-fold convolution of the characteristic function of the interval $[0,1)$. Thus, the Fourier transform of M_n is given by

$$\hat{M}_n(x) := \int_{\mathbf{R}} e^{-ixt} M_n(t)dt = \left(\frac{1-e^{-ix}}{ix}\right)^{n+1}. \tag{2.1}$$

The integer translates of M_n, form a Riesz basis for all functions $f \in L^2(\mathbb{R})$ which are in $C^{n-1}(\mathbb{R})$ and are polynomials of degree at most n on each of the intervals $[j, j+1)$, $j \in \mathbf{Z}$. Thus, if we denote this space by V_0 we have

$$V_0 = \{[c, M_n] : \ c = (c_j : j \in \mathbf{Z}) \in \ell_2(\mathbf{Z})\}$$

where

$$[c, M_n](x) := \sum_{j \in Z} c_j M_n(x-j), \quad x \in \mathbb{R},$$

and there exist positive constants $\underline{\kappa}, \overline{\kappa} > 0$ such that

$$\underline{\kappa}\|c\|_{\ell_2} \leq \|[c, M_n]\|_{L^2} \leq \overline{\kappa}\|c\|_{\ell_2}.$$

We consider the larger space of all $f \in C^{n-1}(\mathbb{R}) \cap L^2(\mathbb{R})$ such that $f|_{[j/2,(j+1)/2]} \in \pi_n$, $j \in Z$ where $\pi_n :=$ the space of all polynomials of degree $\leq n$, viz.

$$V_1 = \{f(2\cdot) : \ f \in V_0\}.$$

Every $\phi \in V_1$ can be expressed in the form

$$\phi(x) = \sum_{j \in \mathbf{Z}} q_j M_n(2x - j) \tag{2.2}$$

for some (real) constants $q_j, j \in \mathbf{Z}$ where $q = (q_j)_{j \in \mathbf{Z}} \in \ell_2(\mathbf{Z})$. In this section, we are primarily interested in $\phi \in V_1$ that are of finite support, however, for later use we assume $q_j, j \in \mathbf{Z}$, decays exponentially. Hence every such ϕ likewise decays exponentially and determines the operator

$$(Tf)(x) = \sum_{j \in \mathbf{Z}} (f, \phi(\cdot - j))\phi(x - j), \quad x \in \mathbb{R} \tag{2.3}$$

where

$$(f, g) = \int_{\mathbf{R}} f(t)\overline{g}(t)dt.$$

When ϕ has orthonormal integer translates

$$(\phi, \phi(\cdot - j)) = \delta_j, \quad j \in \mathbf{Z} \tag{2.4}$$

T is a projection *i.e.*, $T^2 = T$ with range in the L_2-span of $\{\phi(\cdot - j) : j \in \mathbf{Z}\}$. We are interested in identifying $\phi \in V_1$ of least *finite* support satisfying Equation (2.4) and the accuracy constraint on T,

$$Tf = f, \quad f \in \pi_k, \tag{2.5}$$

for some k, $0 \leq k \leq n$.

Lemma 2.1. *Let ϕ in Equation (2.2) decay exponentially (equivalently, $q_j, j \in \mathbf{Z}$ decays exponentially). The Laurent series*

$$q(z) = \sum_{j \in \mathbf{Z}} q_j z^j$$

is analytic in some annulus $r < |z| < R$ for some $r < 1, R > 1$ and Equation (2.5) hold if and only if q satisfies

$$q(1) = \pm 2 \tag{2.6}$$

and

$$q^{(j)}(-1) = 0, \quad j = 0, 1, \ldots, k. \tag{2.7}$$

Proof: According to Equation (2.2), we have

$$\hat{\phi}(2x) = T(x)\hat{M}_n(x), \quad x \in \mathbb{R} \tag{2.8}$$

where

$$T(x) := \frac{1}{2}q(e^{-ix}).$$

Let us first assume that Equation (2.5) holds. In particular, for $k = 0$. This means that

$$1 = \hat{\phi}(0) \sum_{j \in \mathbf{Z}} \phi(x - j), \quad x \in \mathbf{R}$$

and so by the Poisson summation formula we get not only that $(\hat{\phi}(0))^2 = 1$ but also that $\hat{\phi}(2\pi j) = 0$, $j \in \mathbf{Z}\backslash\{0\}$. Now, according to Equation (2.8) specialized to $x = 0$ and $x = \pi$, we obtain, using the fact that $\hat{M}_n(2\pi j) = \delta_j$, $j \in \mathbf{Z}$ (see Equation (2.1)), that Equations (2.6)-(2.7) hold for $k = 0$.

Next, observe that for any polynomial p of degree m with leading coefficient one, there is a polynomial q of degree m such that

$$(p, \phi(\cdot - j)) = q(j), \quad j \in \mathbf{Z}$$

and q has leading coefficient $\hat{\phi}(0)$. Hence, we conclude that the map $\pi_k : p \to ((p, \phi(\cdot - j)), \ j \in \mathbf{Z}$ is invertible and hence Equation(2.5) implies that for *any* polynomial of degree ℓ, $\ell = 0, 1, \ldots, k$ with leading coefficient $\hat{\phi}(0)$,

$$\sum_{j \in \mathbf{Z}} q(x - j)\phi(x - j)$$

is a polynomial of degree m with leading coefficient one. It is well known that this is *equivalent* to the fact that

$$(\hat{\phi}(0))^2 = 1 \tag{2.9}$$

and

$$\hat{\phi}^{(\ell)}(2\pi j) = 0, \quad \ell = 0, 1, \ldots, k, \quad j \in \mathbf{Z}\backslash\{0\}. \tag{2.10}$$

Conversely, we claim that Equations (2.9)-(2.10) implies that Equation (2.5) is valid. The reason is that if $p \in \pi_k$ and ϕ satisfies Equations (2.4), (2.9)-(2.10), then not only is $Tp \in \pi_k$ but also

$$(Tp - p, \phi(\cdot - j)) = 0, \quad j \in \mathbf{Z}. \tag{2.11}$$

However, Equations (2.9)-(2.10) imply there is a $q \in \pi_k$ such that

$$Tp - p = \sum_{j \in \mathbf{Z}} q(j)\phi(\cdot - j), \tag{2.12}$$

and it follows that Equations (2.11)-(2.12) imply that $Tp = p$.

We will now show that Equations (2.9)-(2.10) is equivalent to Equations (2.7)-(2.8). To this end we note that for all $j \in \mathbf{Z}$, Leibniz's rule provides the equation

$$2^{\ell}\hat{\phi}^{(\ell)}(2\pi j) = \sum_{r=0}^{\ell} \binom{\ell}{r} T^{(\ell-r)}(\pi j)\hat{M}^{(r)}(\pi j). \tag{2.13}$$

If $j \in \mathbf{Z} \backslash \{0\}$ is even, then $\hat{M}_n^{(r)}(\pi j) = 0$, $r = 0, 1, \ldots, n$ (see Equation (2.1)). Then Equation (2.10) follows from Equation (2.13). If j is odd, the 2π-periodicity of T and Equations (2.7)-(2.8) imply that Equation (2.10) still holds. Conversely, if Equation (2.10) holds, then in particular, we have

$$0 = \sum_{r=0}^{\ell} \binom{\ell}{r} T^{(\ell-r)}(\pi) \hat{M}_n^{(r)}(\pi), \quad \ell = 0, 1, \ldots, k.$$

Since $\hat{M}_n(\pi) \neq 0$ we conclude that $T^{(j)}(\pi) = 0$, $j = 0, 1, \ldots, k$ and, therefore, Equation (2.7) follows. ■

There is a result, similar to Lemma 2.1, for the interpolation operator

$$(If)(x) = \sum_{j \in \mathbf{Z}} f(j)\phi(x - j)$$

which appears in [9]. We record it here for later use.

Lemma 2.2. *Let ϕ in Equation (2.2) decay exponentially (equivalently, $q_j, j \in \mathbf{Z}$, decays exponentially). Then the interpolation operator defined above satisfies*

$$If = f, \quad f \in \pi_k$$

for some $k, 0 \leq k \leq n$ if and only if

$$q(1) = 2$$

and

$$q^{(j)}(-1) = 0, \quad j = 0, \ldots, k.$$

Returning to our problem to construct a ϕ of the form of Equation (2.2) satisfying Equations (2.4)-(2.5), we see Lemma 2.1 implies that q must have the form

$$q(z) = 2^{-k-1}(1+z)^{k+1} d(z) \tag{2.14}$$

for some Laurent polynomial d. It should be understood here that when no demand for accuracy is made in Equation (2.5), $k = -1$ and so Equation (2.14) just means that $q = d$. Our next step is to express the condition of orthonormality, Equation (2.4) in terms of q itself. For this purpose we note that

$$\begin{aligned}(\phi, \phi(\cdot - j)) &= \sum_{\ell \in \mathbf{Z}} \sum_{m \in \mathbf{Z}} q_\ell \, q_m \int_{\mathbf{R}} M_n(2x - \ell) M_n(2x - 2j - m) dx \\ &= \frac{1}{2} \sum_{\ell \in \mathbf{Z}} \sum_{m \in \mathbf{Z}} q_\ell \, q_m a_{2j+m-\ell}\end{aligned}$$

where

$$a_j := \int_{\mathbf{R}} M_n(x)M_n(x-j)dx, \quad j \in \mathbf{Z}. \tag{2.15}$$

Thus, we see that Equation (2.4) is equivalent to the equation

$$\sum_{\ell \in \mathbf{Z}} \sum_{m \in \mathbf{Z}} q_\ell \, q_m a_{2j+m-\ell} = 2\delta_j, \quad j \in \mathbf{Z}. \tag{2.16}$$

To write this equation in a convenient form we let

$$u(z) := q(z)q(z^{-1})a(z);$$

in other words,

$$u_j = \sum_{\ell \in \mathbf{Z}} \sum_{m \in \mathbf{Z}} q_\ell q_m a_{2j+m-\ell}, \quad j \in \mathbf{Z}.$$

Therefore, Equation (2.16) means that $u(z) + u(-z) = 4$, or equivalently, in terms of q, we have

$$q(z)q(z^{-1})a(z) + q(-z)q(-z^{-1})a(-z) = 4, \quad z \in \mathbf{C}\backslash\{0\}. \tag{2.17}$$

Our objective is to solve this equation for q. For this purpose, we set

$$b(z) = 4^{-k-1}(1+z)^{k+1}(1+z^{-1})^{k+1}a(z), \quad z \in \mathbf{C}\backslash\{0\}. \tag{2.18}$$

We need to point out some properties of b. To this end, note that from Equation (2.15) we have $a_j = a_{-j}$, $j \in \mathbf{Z}$. Also, since $M_n(x) = 0$ for $x \notin (0, n+1)$ we see that $a_j = 0$ for $|j| > n$. Thus we have

$$a(z) = \sum_{-n}^{n} a_j z^j$$

and $a(e^{iw})$ is a cosine polynomial, viz.

$$a(e^{iw}) = a_0 + 2\sum_{j=1}^{n} a_j \cos jw. \tag{2.19}$$

It is known, (see Schoenberg [11]), that a has only *negative* zeros $\lambda_1, \ldots, \lambda_{2n}$ such that

$$\lambda_1\lambda_{2n} = \lambda_2\lambda_{2n-1} = \cdots = \lambda_n\lambda_{n-1} = 1. \tag{2.20}$$

In particular,

$$\lambda_j \neq -1, \quad j = 1, 2, \ldots, n. \tag{2.21}$$

This useful fact leads us to the following lemma.

Lemma 2.3. *Let*

$$c(z) = \sum_{-n}^{n} c_j z^j$$

have only zeros in $(-\infty, 0)$ *where* c_j *are real constants with* $c_j = c_{-j}, |j| \leq n$. *Then there is a polynomial* p *of degree* n *such that*

$$p(x) = c(e^{iw}), \quad x = \sin^2 w/2$$

and p *has only zeros in* $[1, \infty)$. *Moreover, if* $c(e^{iw}) \neq 0$, $|w| \leq \pi$ *then* $p(1) \neq 0$.

Proof: We express c in the form

$$c(z) = \sum_{-n}^{n} c_j z^j = \sum_{j=0}^{n} d_j (z + z^{-1})^j / 2^j$$

for some real constants $d_0, d_1, \ldots, d_n$ and define

$$p(z) = \sum_{j=0}^{n} d_j (1 - 2z)^j.$$

Then

$$c(z) = p(\frac{1}{2} - \frac{1}{4}(z + z^{-1})),$$

and therefore,

$$c(e^{iw}) = p(\frac{1}{2} - \frac{1}{2} \cos\ w) = p(\sin^2 w/2).$$

If $p(\tau) = 0$, then $\tau = \frac{1}{2} - \frac{1}{4}(z + z^{-1})$, where $z \in (-\infty, 0)$. Thus, $z + z^{-1} \leq -2$ and $\tau \geq 1$. Moreover, $\tau = 1$ if and only if $z = -1$. ■

Lemma 2.4. *Let* p *be a polynomial of degree* n *with all its zeros in* $[1, \infty)$ *having a positive leading coefficient. Then there exists a unique polynomial* q *with real coefficients of degree* $n - 1$ *such that*

$$q(x)p(x) + q(1 - x)p(1 - x) = 1. \tag{2.22}$$

Moreover, $(-1)^n q(x) > 0$ *for* $x \in (0, 1)$, *and if* R *is any polynomial which satisfies*

$$R(x)p(x) + R(1 - x)p(1 - x) = 1, \tag{2.23}$$

we have

$$R(x) = q(x) - p(1 - x)h(x), \tag{2.24}$$

where h *is any polynomial which is odd about* $1/2$.

Proof: Let $t_1 < \cdots < t_r$ be the distinct zeros of p in $[1,\infty)$ with multiplicities $m_1, \ldots, m_r$, respectively, with $m_1 + \cdots + m_r = n$. We define a polynomial q of degree at most $n-1$ by requiring that

$$q^{(j)}(1-t_\ell) = \frac{d^j}{dx}(p^{-1})(1-t_\ell), \quad j = 0,1,\ldots,m_\ell - 1, \quad \ell = 1,\ldots,r. \tag{2.25}$$

Then the polynomial

$$v(x) := q(x)p(x) + q(1-x)p(1-x),$$

is of degree at most $2n-1$ and

$$v^{(j)}(t_\ell) = v^{(j)}(1-t_\ell) = \delta_j, \quad j = 0,1,\ldots,m_\ell - 1, \quad \ell = 1,\ldots,r.$$

Hence Equation (2.22) holds for all $x \in \mathbb{C}$. Let $w := pq$ and observe that $w^{(j)}(t_\ell) = 0$, $j = 0,1,\ldots,m_\ell - 1$, $\ell = 1,\ldots,r$ and by Equation (2.25), $w^{(j)}(1-t_\ell) = \delta_j, j = 0,1,\ldots,m_\ell - 1$, $\ell = 1,\ldots,r$. Thus, by Rolle's theorem w' has $2n-2$ zeros outside of (0,1). Since w' has degree at most $2n-2$, we conclude that it is zero free in (0,1) and, thus, is strictly negative there. In particular, $w(x) > 0$ for $0 < x < 1$ and our hypothesis on p implies that $p(x)(-1)^n > 0$ for $0 < x < 1$. This proves the first claim. For the second, we let R be any polynomial satisfying Equation (2.23), then

$$(R(x) - q(x))p(x) + (R(1-x) - q(1-x))p(1-x) = 0. \tag{2.26}$$

Consequently, whenever $p(1-x) = 0$, x must be nonpositive and so $p(x) \neq 0$. Thus, Equation (2.26) implies $R(x) - q(x)$ is divisible by $p(1-x)$. ∎

Proposition 2.5. *Let c satisfy the hypothesis of Lemma 2.3 and suppose*

$$c(e^{iw}) \geq 0, \quad w \in \mathbb{R}. \tag{2.27}$$

Then there exists an algebraic polynomial d of degree $n-1$ with real coefficients such that

$$|d(e^{iw})|^2 c(e^{iw}) + |d(-e^{iw})|^2 c(-e^{iw}) = 4, \quad w \in \mathbb{R}. \tag{2.28}$$

Moreover, any other Laurent polynomial which satisfies

$$T(e^{iw})c(e^{iw}) + T(-e^{iw})c(-e^{iw}) = 4, \quad w \in \mathbb{R} \tag{2.29}$$

has the form

$$T(e^{iw}) = |d(e^{iw})|^2 - c(-e^{iw})h(\sin^2 w/2),$$

where h is an arbitrary algebraic polynomial which is odd about 1/2.

Before we prove this result, we note the special case considered by Daubechies [6]. She obtained, for any positive integer n, the identity

$$\left|\frac{1+e^{-iw}}{2}\right|^{2n} |S(e^{iw})|^2 + \left|\frac{1-e^{-iw}}{2}\right|^{2n} |S(-e^{iw})|^2 = 4, \quad w \in \mathbb{R},$$

where

$$|S(e^{iw})|^2 = 4\sum_{j=0}^{n-1} \binom{n+j-1}{j} (\sin w/2)^{2j}.$$

Note that the Laurent polynomial

$$c(z) = 2^{-2n}(1+z)^n(1+z^{-1})^n, \quad z \in \mathbb{C}\backslash\{0\}$$

satisfies the hypothesis of Proposition 2.5 and so the existence of a nonnegative Laurent polynomial satisfying the above equation is predicted by Proposition 2.5. This example is important for wavelet construction.

Proof: (Proposition 2.5) According to Lemma 2.3 there is an algebraic polynomial p of degree n such that

$$p(\sin^2 w/2) = c(e^{iw})$$

for $w \in \mathbb{R}$. Let q be an algebraic polynomial defined by the formula

$$q(\sin^2 w/2) = |d(e^{iw})|^2 = d(e^{iw})d(e^{-iw}). \tag{2.30}$$

Since $\sin^2(w+\pi)/2 = 1 - \sin^2 w/2$ Equation (2.28) takes the equivalent form

$$p(x)q(x) + p(1-x)q(1-x) = 4.$$

Moreover, by Lemma 2.3, p only has zeros in $[1,\infty)$, and according to our hypothesis in Equation (2.27), $p(0) \geq 0$. This means that the leading coefficient of p has sign $(-1)^n$. Thus, by Lemma 2.4, there is an algebraic polynomial q of degree $n-1$ which is positive on (0,1) that satisfies Equation (2.30). According to Lemma 2.3

$$T(w) := q(\sin^2 w/2)$$

is a nonnegative even trigonometric polynomial of degree at most $n-1$, which by Riesz's Lemma can be written in the form

$$T(w) = |d(e^{iw})|^2,$$

where d is an algebraic polynomial of degree at most $n-1$ with real coefficients. This proves the first part of the proposition. The remaining claims follow from Lemma 2.4. ■

This result leads us to the following theorem.

Theorem 2.6. *For every $k = -1, \ldots, n$ there exists a real sequence $q_0, q_1, \ldots,$ q_{n+k} such that*

$$\phi(x) = \sum_{j=0}^{n+k} q_j M(2x-j)$$

has orthonormal integer translates and the corresponding projector

$$(Tf)(x) = \sum_{j \in \mathbf{Z}} (f, \phi(\cdot - j))\phi(x - j)$$

is exact for all polynomials of degree k.

Note that ϕ has support in the interval $[0, n + \frac{k+1}{2}]$. As we shall see in the proof of this result, no function with support on $(0, \ell)$, with ℓ an integer $< n + \frac{k+1}{2}$, satisfies the conclusions of Theorem 2.6. We also remark that the function $\phi * \phi(-\cdot)$ is a spline of degree $2n+1$ with knots at $\frac{1}{2}\mathbf{Z}$ which is a cardinal function. This function was considered in [5].

Proof: According to Lemma 2.1 any sequence $(q_j)_{j \in \mathbf{Z}}$ of finite support with the desired properties must have an associated Laurent polynomial which can be factored in the form of Equation (2.14) where the remaining factor d satisfies Equation (2.28). The Laurent polynomial b satisfies the conditions on c in Proposition 2.5 with n replaced by $n+k$. Specializing this result to the present circumstances proves the result. ■

We conjecture the following extension of Theorem 2.6 is valid. Begin with an arbitrary biinfinite partition $\{x_i\}_{i \in \mathbf{Z}}$ of $\mathbb{R}$. That is, $x_i < x_{i+1}$ for all $i \in \mathbf{Z}$ and

$$\lim_{i \to \pm\infty} x_i = \pm\infty.$$

We embed this partition $\{x_i\}_{i \in \mathbf{Z}}$ in another partition $\{t_i\}_{i \subset n\mathbf{Z}}$ such that

$$t_{2i} = x_i, \quad i \in \mathbf{Z},$$

and let

$$\mathcal{S}_n = \{f : f|_{[t_i, t_{i+1})} \in \pi_n, i \in \mathbf{Z}, f \in C^{n-1}(\mathbb{R})\}.$$

Conjecture. For every $k = -1, \ldots, n$ and $j \in \mathbf{Z}$ there exist functions $\phi_j \in \mathcal{S}_n$ with support contained in $(t_{2j}, t_{2j+2n+k+1})$ such that

$$\int_{-\infty}^{\infty} \phi_j(x)\phi_\ell(x)dx = \delta_{j\ell}, j, \ell \in \mathbf{Z},$$

and the projection operator

$$(Tf)(x) = \sum_{j \in \mathbf{Z}} (f, \phi_j)\phi_j(x),$$

is exact for all polynomials of degree k.

§3 Orthonormal cardinal functions using cardinal splines

In this section, we study the problem of finding a function ϕ of the form of Equation (2.2) which not only has orthonormal integer translates, but also is a cardinal function in the sense that

$$\phi(j) = \delta_j, \quad j \in \mathbf{Z}. \tag{3.1}$$

Except in the case that $n = 1$, we shall see that our construction produces a cardinal spline which decays exponentially at $\pm\infty$. We begin our discussion with the linear case ($n = 1$). In this case, we choose ϕ to be supported on $[-1, 1]$ so that, in view of Equation (3.1) it must have the form

$$\phi(x) = \begin{cases} \text{-2a (x+1)}, & -1 \le x \le -1/2 \\ \text{1+2 x (1+a)}, & -1/2 \le x \le 0 \\ \text{1-2 x (1+b)}, & 0 \le x \le 1/2 \\ \text{2b (x-1)}, & 1/2 \le x \le 1 \\ 0\,, & |x| > 1 \end{cases}$$

for some constants a and b. The requirement that ϕ has integral one leads to the equation

$$2a^2 + 2b^2 - a - b = 4.$$

Since ϕ vanishes outside of $(-1, 1)$ the only additional condition on ϕ is that it be orthogonal to $\phi(\cdot - 1)$ which gives the equation

$$a + b - 4ab = 0.$$

There are four distinct solutions to these two equations. Two are given by

$$(a, b) = -\frac{1}{2}(1 + \sqrt{2}, 1 - \sqrt{2})$$
$$(a, b) = \frac{1}{2}(2 - \sqrt{2}, 2 + \sqrt{2}),$$

and the remaining solutions are obtained by interchanging a and b. Another direct computation shows that

$$\int_{\mathbf{R}} \phi(x)dx = 1 - \frac{1}{2}(1 + a + b)$$

and

$$\sum_{j \in \mathbf{Z}} \phi(x - j) = 1 - (1 - |2x - 1|)(1 + a + b), \quad 0 \le x \le 1.$$

Thus, the solutions for which $a + b = -1$ give an orthogonal projector T (see Equation (2.3)), which reproduces constants. We should also point out that T *never* reproduces linear functions since ϕ has a knot at $-1/2$ and has support on $[-1, 1]$.

We now turn our attention to the case $n \geq 2$. Obviously, Equation (3.1) means that

$$\sum_{\ell \in \mathbf{Z}} q_\ell M_n(2j-\ell) = \delta_j, \quad j \in \mathbf{Z} \tag{3.2}$$

and, as in Section 2, this is equivalent to the identity

$$q(z)b(z) + q(-z)b(-z) = 2,$$

valid for z in the unit circle $\mathcal{D} := \{z : |z| = 1\}$, where

$$b(z) = \sum_{j \in \mathbf{Z}} M_n(j) z^j. \tag{3.3}$$

Thus, appealing to Equation (2.17), we must solve both of the equations

$$|q(z)|^2 a(z) + |q(-z)|^2 a(-z) = 4, \quad z \in \mathcal{D} \tag{3.4}$$

and

$$q(z)b(z) + q(-z)b(-z) = 2, \quad z \in \mathcal{D}, \tag{3.5}$$

for q where

$$a(z) = \sum_{j \in \mathbf{Z}} (M_n(\cdot), M_n(\cdot - j)) z^j = z^{-n-1} \sum_{j \in \mathbf{Z}} M_{2n+1}(j) z^j. \tag{3.6}$$

It is appropriate to study solutions to Equations (3.4) and (3.5) such that $q = (q_j)_{j \in \mathbf{Z}} \in \ell_1(\mathbf{Z})$, (absolutely summable real sequences). Thus, in particular, $q(z)$ is continuous. We also only require below that both a and b are *arbitrary* Laurent polynomials with real coefficients. Later we will specialize our observations to the choice of Equations (3.3) and (3.6). Our first result gives necessary and sufficient conditions for Equations (3.4) and (3.5) to have a solution $q \in \ell_1(\mathbf{Z})$. To this end, we find the following Laurent polynomials useful,

$$\mu_{a,b}(z) = a(z)b(-z)b(-z^{-1}) + a(-z)b(z)b(z^{-1}), \quad z \in \mathbf{C}\backslash\{0\} \tag{3.7}$$

$$\delta_{a,b}(z) = \mu_{a,b}(z) - a(z)a(-z), \quad z \in \mathbf{C}\backslash\{0\}. \tag{3.8}$$

Lemma 3.1. *Let a, b be nontrivial Laurent polynomials with real coefficients such that*

$$a(z^{-1}) = a(z) > 0, \quad z \in \mathcal{D}.$$

Then Equations (3.4)-(3.5) have a solution if and only if $\delta_{a,b}$ is nonnegative on $\mathcal{D}$. Moreover, if these conditions are valid then there is a solution to Equations (3.4)-(3.5) such that the sequence $q = (q_j)_{j \in \mathbf{Z}}$ decays exponentially.

Proof: First, assume Equations (3.4) and (3.5) have a solution. Let $z \in \mathcal{D}$. We then solve Equation (3.5) for $q(-z)$ and substitute the result into Equation (3.4) to obtain

$$a(z)|b(-z)|^2|q(z)|^2 + a(-z)|2 - q(z)b(z)|^2 = 4|b(-z)|^2.$$

Upon simplification, this is easily seen to be equivalent to the equation

$$\mu_{a,b}(z)|q(z)|^2 - 2a(-z)(b(z)q(z) + \overline{b(z)q(z)}) = 4(|b(-z)|^2 - a(-z)). \quad (3.9)$$

Next, we multiply both sides of this equation by $\mu_{a,b}(z)$ and "complete the square" to obtain

$$|\mu_{a,b}(z)q(z) - 2a(-z)\overline{b(z)}|^2 = 4\mu_{a,b}(z)(|b(-z)|^2 - a(-z)) + 4|a(-z)b(z)|^2.$$

Simplifying the right hand side of this equation, we obtain

$$|\mu_{a,b}(z)q(z) - 2a(-z)\overline{b(z)}|^2 = 4|b(-z)|^2\delta_{a,b}(z), \quad z \in \mathcal{D}, \quad (3.10)$$

from which we conclude $\delta_{a,b}(z)$ is nonnegative on the unit circle.

To see that $\mu_{a,b}$ is positive on $\mathcal{D}$ we use definition in Equation (3.8) and our hypothesis on $a(z)$ to conclude that

$$\mu_{a,b}(z) \geq a(z)a(-z) > 0, \quad z \in \mathcal{D}.$$

Conversely, suppose $\delta_{a,b}(z) \geq 0$ for $z \in \mathcal{D}$. Since $\delta_{a,b}$ is an even function such that $\delta_{a,b}(z) = \overline{\delta_{a,b}(z)} = \delta_{a,b}(z^{-1})$, $z \in \mathcal{D}$, we may write it in the form

$$\delta_{a,b}(z) = \nu(z^2),$$

where ν is a Laurent polynomial with real coefficients which also satisfies $\nu(z) = \overline{\nu(z)} = \nu(z^{-1})$, $z \in \mathcal{D}$. Applying Riesz Lemma to ν, we are assured that there is an algebraic polynomial γ with real coefficients such that

$$\delta_{a,b}(z) = |\gamma(z^2)|^2, \quad z \in \mathcal{D}. \quad (3.11)$$

Consequently, from Equation (3.10) we see that the rational function

$$R(z) := \frac{\mu_{a,b}(z)q(z) - 2a(-z)b(z^{-1})}{2b(-z)\gamma(z^2)z}, \quad z \in \mathbb{C}\backslash\{0\}, \quad (3.12)$$

has the property that it is identically one in modulus on $\mathcal{D}$. We solve for $q(z)$ in Equation (3.12),

$$q(z) = \frac{2a(-z)b(z^{-1}) + 2R(z)zb(-z)\gamma(z^2)}{\mu_{a,b}(z)}, \quad z \in \mathbb{C}\backslash\{0\}. \quad (3.13)$$

Next, we check that q satisfies Equation (3.5). For this purpose, we observe for $z \in \mathcal{D}$ that

$$q(z)b(z)\mu_{a,b}(z) = 2a(-z)|b(z)|^2 + 2b(-z)b(z)\gamma(z^2)zR(z).$$

Now, replacing z by $-z$ and adding the resulting equations, we obtain

$$\mu_{a,b}(z)(q(z)b(z) + q(-z)b(-z)) = 2\mu_{a,b}(z) + 2b(-z)b(z)\gamma(z^2)z(R(z) - R(-z)).$$

We conclude that $R(z)$ must be an even function. With this additional requirement on R, we verify that

$$a(z)|q(z)|^2 = \frac{4a(-z)^2a(z)|b(z)|^2 + 4a(z)|b(-z)|^2\delta_{a,b}(z) + S(z)}{\mu_{a,b}^2(z)}$$

where S is an odd Laurent polynomial. Thus, we conclude that

$$a(z)|q(z)|^2 + a(-z)|q(-z)|^2 = \frac{4a(-z)a(z)\mu_{a,b}(z) + 4\mu_{a,b}(z)\delta_{a,b}(z)}{\mu_{a,b}^2(z)} = 4.$$

The above computation allows us to identify solutions to Equations (3.4) and (3.5). For instance, in Equation (3.13) we may choose R to be identically one. Then, using the fact that $\delta_{a,b}$ is nonnegative on $\mathcal{D}$ we conclude, as above, that $\mu_{a,b}$ is also positive on $\mathcal{D}$. Hence, $\mu_{a,b}^{-1}$ can be expanded in a trigonometric series on $\mathcal{D}$ which has coefficients which decay exponentially. ■

We shall now apply Lemma 3.1 to the B-spline in Equation (2.1). Recall that we already pointed out in Section 2 that $a(z)$ given by (3.6) is positive on $\mathcal{D}$ and so the hypothesis of Lemma 3.1 is satisfied. To prove the existence of an orthonormal cardinal function in V_1 which decays exponentially, we must show that $\delta_{a,b}(z)$ is nonnegative on $\mathcal{D}$. We begin by making some observations about $\mu_{a,b}(z)$.

Since

$$b(z) = \sum_{j=1}^{n} M_n(j)z^j = z\sum_{j=0}^{n-1} M_n(j+1)z^j,$$

we conclude that $|b(e^{i\theta})|^2$ is a trigonometric polynomial of degree $n-1$. Hence, it follows that $\mu_{a,b}(e^{i\theta})$ is a trigonometric polynomial of degree $2n-1$. Moreover, since

$$\overline{\mu_{a,b}(z)} = \mu_{a,b}(z) = \mu_{a,b}(z^{-1}) = \mu(-z),$$

for $z \in \mathcal{D}$, we conclude

$$\mu_{a,b}(z) = \nu(z^2),$$

where ν is a Laurent polynomial of degree $n-1$. For instance, when $n = 1$ and $z \neq 0$

$$a(z) = \frac{1}{6}(z^{-1} + 4 + z), \quad b(z) = z$$

and thus

$$\mu_{a,b}(z) = a(z) + a(-z) = \frac{4}{3}.$$

Similarly, for $n = 2$ and $z \neq 0$, we have

$$a(z) = \frac{1}{120}(z^{-2} + 26z^{-1} + 66 + 26z + z^2),$$
$$b(z) = \frac{1}{2}(1 + z),$$

from which it follows that

$$b(z)b(z^{-1}) = \frac{1}{4}(2 + z + z^{-1})$$

and

$$\mu_{a,b}(z) = -\frac{1}{30}(3z^{-2} - 10 + 3z^2).$$

We conjecture that $\mu_{a,b}(z) = \eta_{a,b}(-z^2)$ where $\eta_{a,b}$ has only simple negative zeros. We prove only the following proposition.

Proposition 3.2. *Let a and b be the Laurent polynomials defined by Equations (3.3) and (3.5), respectively. Then $\eta_{a,b}(z)$ has only positive coefficients.*

Proof: From what we already observed it follows that if

$$\phi(x) = \sum_{j\in\mathbf{Z}} q_j M_n(2x - j), \quad x \in \mathbb{R}, \tag{3.14}$$

and

$$v(z) := \sum_{j\in\mathbf{Z}} (\phi, \phi(\cdot - j))z^j, \quad z \in \mathcal{C}\backslash\{0\}, \tag{3.15}$$

then for $z \in \mathcal{D}$,

$$v(z^2) = \frac{1}{4}\{|q(z)|^2 a(z) + |q(-z)|^2 a(-z)\}.$$

We choose $q(z) = b(-z)$ above and obtain

$$4v(z^2) = \mu_{a,b}(z). \tag{3.16}$$

With this choice of q, Equation (3.14) becomes

$$\phi(x) = \sum_{j\in\mathbf{Z}} (-1)^j M_n(j) M_n(2x - j),$$

which has support in $[\frac{1}{2}, n + \frac{1}{2}]$. Note that for all $i \in \mathbf{Z}$,

$$\phi(i + \frac{1}{2}) = \sum_{j\in\mathbf{Z}} (-1)^j M_n(j) M_n(2i - j + 1) = 0.$$

Thus, ϕ has $n-1$ zeros in $(\frac{1}{2}, n+\frac{1}{2})$ and by the zero bound in [7], these are simple and are the only zeros of ϕ in this interval. Hence

$$(-1)^i(\phi, \phi(\cdot - i)) > 0, \quad -n+1 \le i \le n-1.$$

Hence, by Equation (3.15), $v(-z)$ has only positive coefficients. Since, by Equation (3.16), $\eta_{a,b}(z) = 4v(-z)$, the result follows. ■

Next, we turn our attention to $\delta_{a,b}$. As we pointed out above $|b(e^{i\theta})|^2$ is a trigonometric polynomial of degree $n-1$. Consequently, for $z = e^{i\theta}$, all of the factors $a(z)a(-z), a(z)|b(z)|^2, a(-z)|b(z)|^2$ are trigonometric polynomials of degree $2n$. Since, for $z \in \mathcal{D}$,

$$\overline{\delta_{a,b}(z)} = \delta_{a,b}(z) = \delta_{a,b}(z^{-1}) = \delta_{a,b}(-z),$$

we conclude that

$$\delta_{a,b}(z) = \alpha(z^2),$$

where α is a Laurent polynomial of degree n. We conjecture the following

(i) For n odd, α only has negative zeros distinct from -1. It would then follow that $\alpha(z) > 0$, for $z \in \mathcal{D}$.

(ii) For n even, α has only a double zero at $z = 1$ and the remaining are negative.

To support this conjecture we note that for $n = 1$,

$$\delta_{a,b}(z) = \frac{1}{36}(z^{-2} + 34 + z^2),$$

and for $n = 2$,

$$\begin{aligned}\delta_{a,b}(z) &= \frac{1}{14400}(-z^{-4} - 896z^{-2} + 1794 - 896z - z^4) \\ &= \frac{1}{14400}(z^2-1)(z^{-2}-1)(z^2 + 898 + z^{-2}),\end{aligned}$$

since

$$a(z)a(-z) = \frac{1}{14400}(z^{-4} - 544z^{-2} + 3006 - 544z^2 + z^4).$$

So far, in general, we can prove the following proposition.

Proposition 3.3. *Let a and b be given by Equations (3.3) and (3.6) and $n \ge 1$. Then*

$$\delta_{a,b}(z) \ge 0, \quad z \in \mathcal{D}.$$

Proof: Case 1. n is odd. In this case we use the Poisson summation formula and note that

$$a(e^{i\theta}) = \sum_{j\in\mathbf{Z}} |\hat{M}_n(\theta + 2\pi j)|^2$$

and

$$b(e^{i\theta}) = \sum_{j\in\mathbf{Z}} \hat{M}_n(\theta + 2\pi j).$$

According to Equation (2.1), we have $\hat{M}_n(\theta) \geq 0$, all $\theta \in \mathbb{R}$ and so

$$|b(e^{i\theta})|^2 \geq a(e^{i\theta}),$$

which implies, for $\theta \in \mathbb{R}$, that

$$\delta_{a,b}(e^{i\theta}) \geq a(e^{i\theta})a(-e^{i\theta}) > 0.$$

Case 2. n even. We set $m = n + 1$. This case requires more effort. Again, from the Poisson summation formula

$$\begin{aligned}\delta_{a,b}(e^{i\theta}) = &\\ (\frac{\sin\theta}{2})^{2m} &\left\{\sum_{k\in\mathbf{Z}} \frac{1}{(\frac{\theta}{2}+k\pi)^{2m}} \left[\sum_{k\in\mathbf{Z}} \frac{1}{(\frac{\theta}{2}+\frac{\pi}{2}+k\pi)^m}\right]^2\right.\\ &+\sum_{k\in\mathbf{Z}} \frac{1}{(\frac{\theta}{2}+\frac{\pi}{2}+k\pi)^{2m}} \left[\sum_{k\in\mathbf{Z}} \frac{1}{(\frac{\theta}{2}+k\pi)^m}\right]^2\\ &\left.-\sum_{k\in\mathbf{Z}} \frac{1}{(\frac{\theta}{2}+k\pi)^{2m}} \sum_{k\in\mathbf{Z}} \frac{1}{(\frac{\theta}{2}+\frac{\pi}{2}+k\pi)^{2m}}\right\}.\end{aligned}$$

Since $\delta_{a,b}(e^{i\theta}) = \delta_{a,b}(-e^{i\theta}) = \delta_{a,b}(e^{-i\theta})$, it is sufficient to consider the above expression when $0 \leq \theta \leq \pi/2$. Thus, putting $v = \theta/\pi$, we must show, for $v \in [0, \frac{1}{2}]$, that

$$\begin{aligned}&\sum_{k\in\mathbf{Z}} \frac{1}{(v+2k)^{2m}} \left\{\sum_{k\in\mathbf{Z}} \frac{1}{(v+2k+1)^m}\right\}^2\\ &+\sum_{k\in\mathbf{Z}} \frac{1}{(v+2k+1)^{2m}} \left\{\sum_{k\in\mathbf{Z}} \frac{1}{(v+2k)^m}\right\}^2\\ &-\sum_{k\in\mathbf{Z}} \frac{1}{(v+2k)^{2m}} \sum_{k\in\mathbf{Z}} \frac{1}{(v+2k+1)^{2m}} \geq 0. \qquad (3.17)\end{aligned}$$

We shall actually show, for $v \in [0, \frac{1}{2}]$, that

$$\left\{\sum_{k\in\mathbf{Z}} \frac{1}{(v+2k)^m}\right\}^2 \geq \left\{1-\left(\frac{v}{2-v}\right)^2\right\}^2 \sum_{k\in\mathbf{Z}} \frac{1}{(v+2k)^{2m}}, \qquad (3.18)$$

$$\left\{\sum_{k\in\mathbf{Z}} \frac{1}{(v+2k+1)^m}\right\}^2 \geq \frac{((1-v)^{-m}-(1+v)^{-m})^2}{(1-v)^{-2m}+(1+v)^{-2m}} \sum_{k\in\mathbf{Z}} \frac{1}{(v+2k+1)^{2m}}, \qquad (3.19)$$

and also

$$\left\{1-\left(\frac{v}{2-v}\right)^m\right\}^2+\frac{((1-v)^{-m}-(1+v)^{-m})^2}{(1-v)^{-2m}+(1+v)^{-2m}}\geq 1. \tag{3.20}$$

Then Equation (3.17) follows from Equations (3.18), (3.19), and (3.20).

First, to prove Equation (3.20), we note that

$$\frac{((1-v)^{-m}-(1+v)^{-m})^2}{(1-v)^{-2m}+(1+v)^{-2m}}=\frac{\left(1-\left(\frac{1-v}{1+v}\right)^m\right)^2}{1+\left(\frac{1-v}{1+v}\right)^{2m}}$$
$$\geq\frac{\left(1-\left(\frac{1-v}{1+v}\right)\right)^2}{1+\left(\frac{1-v}{1+v}\right)^2}=\frac{2v^2}{1+v^2}.$$

Thus, to show Equation (3.20), it is sufficient to show that

$$1-2\left(\frac{v}{2-v}\right)^m+\frac{2v^2}{1+v^2}\geq 1,$$

which is the same as the inequality

$$\frac{(2-v)^m}{v^{m-2}(1+v^2)}\geq 1.$$

Since $m\geq 3$, the left hand side of this inequality is decreasing in v and so it is sufficient to verify it holds for $v=\frac{1}{2}$, which is easily done.

We now prove Equation (3.18).

$$\left\{\sum_{k\in\mathbf{Z}}\frac{1}{(v+2k)^m}\right\}^2=\left\{\sum_{k=0}^{\infty}\left(\frac{1}{(v+2k)^m}-\frac{1}{(2k+2-v)^m}\right)\right\}^2$$
$$\geq\left(\frac{1}{v^m}-\frac{1}{(2-v)^m}\right)^2+2\left(\frac{1}{v^m}-\frac{1}{(2-v)^m}\right)\sum_{k=1}^{\infty}\left(\frac{1}{(v+2k)^m}-\frac{1}{(2k+2-v)^m}\right)$$
$$\geq\left(1-\left(\frac{v}{2-v}\right)^m\right)^2\left\{\frac{1}{v^{2m}}+\frac{2}{v^m}\sum_{k=1}^{\infty}\left(\frac{1}{(v+2k)^m}-\frac{1}{(2k+2-v)^m}\right)\right\}. \tag{3.21}$$

Moreover, for $k\geq 1$, $m\geq 3$, and $0\leq v\leq\frac{1}{2}$,

$$\frac{2}{v^m}\left(\frac{1}{(v+2k)^m}-\frac{1}{(2k+2-v)^m}\right)\geq\frac{4(1-v)m}{v^m(v+2k)(2k+2-v)^m}$$
$$>\frac{2}{v^{m-1}(2k+2-v)^{m-1}(2k-v)^2}>\frac{2}{(2k-v)^{2m}}$$
$$\geq\frac{1}{(2k+v)^{2m}}+\frac{1}{(2k-v)^{2m}}, \tag{3.22}$$

and so Equations (3.21) and (3.22) give Equation (3.18).

Finally, we prove Equation (3.19). For $0 < v \le \frac{1}{2}$,

$$\left\{\sum_{k\in\mathbf{Z}} \frac{1}{(v+2k+1)^m}\right\}^2 = \left\{\sum_{k=0}^{\infty}\left(\frac{1}{(2k+1-v)^m} - \frac{1}{(2k+1+v)^m}\right)\right\}^2$$

$$\ge \left(\frac{1}{(1-v)^m} - \frac{1}{(1+v)^m}\right)^2$$

$$+2\left(\frac{1}{(1-v)^m} - \frac{1}{(1+v)^m}\right)\sum_{k=1}^{\infty}\left(\frac{1}{(2k+1-v)^m} - \frac{1}{(2k+1+v)^m}\right)$$

$$\ge \frac{((1-v)^{-m}-(1+v)^{-m})^2}{(1-v)^{-2m}+(1+v)^{-2m}}\left\{\frac{1}{(1-m)^{2m}} + \frac{1}{(1+v)^{2m}} + \frac{2}{(1-v)^{2m}}\right.$$

$$\left.\frac{1}{(1-v)^{-m}-(1+v)^{-m}}\sum_{k=1}^{\infty}\left(\frac{1}{(2k+1-v)^m} - \frac{1}{(2k+1+v)^m}\right)\right\}. \quad (3.23)$$

Now for $k \ge 1$, there exists τ, $-v < \tau < v$, such that

$$\frac{2}{(1-v)^{2m}} \cdot \frac{1}{(1-v)^{-m}-(1+v)^{-m}}\left(\frac{1}{(2k+1-v)^m} - \frac{1}{(2k+1+v)^m}\right)$$

$$= \frac{2}{(1-v)^{2m}}\frac{(1+\tau)^{m+1}}{(2k+1+\tau)^{m+1}} > \frac{2}{(1-v)^{m-1}(2k+1-v)^{m+1}}$$

$$> \frac{2}{(2k+1-v)^{2m}} \ge \frac{1}{(v+2k+1)^{2m}} + \frac{1}{(v-2k-1)^{2m}}. \quad (3.24)$$

Then, Equations (3.23) and (3.24) give Equation (3.19). ■

We have thus established the existence, for all $n \ge 1$, of an orthonormal cardinal function ϕ in V_1. Recalling Theorem 2.6, it is natural to ask whether we can choose an orthonormal cardinal function ϕ so that the orthogonal projector of Equation (2.3) is exact for all polynomials of degree n. We note that by Lemmas 2.1 and 2.2, this is equivalent to requiring that the interpolation operator

$$(If)(x) = \sum_{j\in\mathbf{Z}} f(j)\phi(x-j) \quad (3.25)$$

is exact for all polynomials of degree n. We shall see that for $n = 2$, the operators are not exact even for polynomials of degree 1, and it seems likely that this is true for all even n. However, we conjecture that the result is true when n is odd. Below we show that ϕ can be chosen so that the operators are exact for constants for all n, and are exact for linear functions when n is odd.

Proposition 3.4. *For all $n \ge 1$, there is an orthonormal, cardinal function ϕ in V_1 such that*

$$\sum_{j\in\mathbf{Z}} \phi(\cdot - j) = \int_{\mathbf{R}} \phi(t)dt = 1.$$

Proof: By Lemmas 2.1 and 2.2, we must choose q so that

$$q(1) = 2, \quad q(-1) = 0. \tag{3.26}$$

From Equations (2.1), (3.3), and (3.6) it follows that $b(1) = a(1) = 1$. Specializing Equation (3.13) to $z = 1$ gives

$$q(1) = \frac{2a(-1) + 2R(1)b(-1)|b(-1)|}{a(-1) + b(-1)^2}. \tag{3.27}$$

If n is even, then $b(-1) = 0$ and so $q(1) = 2$. If n is odd, choose $R(1) =$ sgn $b(-1)$ and Equation (3.27) again gives $q(1) = 2$. Setting $z = -1$ in Equation (3.13) gives

$$q(-1) = \frac{2b(-1) - 2R(1)|b(-1)|}{a(-1) + b(-1)^2}$$

which vanishes if $R(1) =$ sgn $b(-1)$. Thus, Equation (3.26) is satisfied. ■

Proposition 3.5. *For odd n, there is an orthonormal, cardinal function ϕ in V_1 such that the orthogonal projection of Equation* (2.3) *and the interpolation operator of Equation* (3.25) *are exact for all linear functions.*

Proof: By Lemmas 2.1 and 2.2 we must choose q so that Equation (3.26) holds and $q'(-1) = 0$. As in the proof of Proposition 3.4, Equation (3.26) is satisfied if $R(1) =$ sgn $b(-1)$. By differentiating Equation (3.13) at $z = -1$, we see that we need to satisfy, in addition, that

$$R'(-1)\gamma(1) = -a'(1)b(-1) - b'(-1) + R(1)\gamma(1) + R(1)b'(1)\gamma(1) + 2R(1)\gamma'(1). \tag{3.28}$$

We claim that for any α in $\mathbb{R}$, we can choose R with $R(1) = 1$ and $R'(-1) = \alpha$ (and hence that we can choose R with $R(1) = -1$ and $R'(-1) = \alpha$). If $\alpha = 0$, take $R(z) = 1$. If $\alpha = -2$, take $R(z) = z^2$. Otherwise, take $R(z) = (az^2 + 1)/(z^2 + a)$, for some a in $\mathbb{R}$ with $|a| \neq 1$. Then, $R'(-1) = 2(1 - a)/(1 + a)$ and so $R'(-1) = \alpha$ if $a = (2 - \alpha)/(2 + \alpha)$.

Now for odd n, $\gamma(1) = |b(-1)| \neq 0$ and so we can choose $R'(-1)$ to satisfy Equation (3.28). ■

For even n, we have $\gamma(1) = |b(-1)| = 0$ and so, by Equation (3.28), the operators of Equations (2.3) and (3.25) are exact for all linear functions only if

$$b'(-1) = 2R(1)\gamma'(1) = \pm 2\gamma'(1).$$

A straightforward calculation shows that this condition is not satisfied for $n = 2$.

In general, for the operators of Equations (2.3) and (3.25) to be exact for all polynomials of degree k, we require Equation (3.26) and

$$q^{(j)}(-1) = 0, \quad j = 0, \ldots, k. \tag{3.29}$$

By Equation (3.9), a necessary condition for Equation (3.29) is that the polynomial $\eta(z) := b(z)b(z^{-1}) - a(z)$ satisfies

$$\eta^{(j)}(1) = 0, \quad j = 0, \ldots, k. \tag{3.30}$$

Since $\eta(z) = \eta(z^{-1})$, the multiplicity of the zero of η at 1 is even. We know that Equation (3.30) is always satisfied for $k = 0$ and hence for $k = 1$. We now show that for $n \geq 2$, Equation (3.30) is satisfied for $k = 2$ (and hence for $k = 3$).

By Marsden's identity [12], we have

$$\sum_{i \in \mathbf{Z}} iM_n(i) = \frac{n+1}{2}, \sum_{i \in \mathbf{Z}} (i - \frac{n+1}{2})^2 M_n(i) = \frac{n+1}{12}.$$

Then straightforward calculation shows that $b_n'(1) = \frac{1}{2}(n+1)$, $a_n'(1) = n+1$, $b_n''(1) = \frac{1}{4}(n+1)^2 - \frac{5}{12}(n+1)$, $a_n''(1) = (n+1)^2 - \frac{5}{6}(n+1)$, and hence $\eta''(1) = 0$.

§4 Orthonormal refinable cardinal functions

In this section, we change our emphasis away from cardinal spline functions and present some results concerning orthonormal refinable cardinal functions. Our results improve upon the conclusions presented in Lewis [8]. Some other results are given in [3,4] on characterizing refinable functions which are orthonormal and/or cardinal and also symmetric refinable cardinal functions with compact support. We begin with the following proposition.

Proposition 4.1. *Let $\phi \in L^2_{loc}(\mathbb{R})$ be of compact support such that*

$$\phi(j) = (\phi, \phi(\cdot - j)) = \delta_j, j \in \mathbf{Z}, \tag{4.1}$$

$\hat{\phi}(0) = 1$, and for some real sequence $q = (q_j)_{j \in \mathbf{Z}}$ we have

$$\phi(x) = \sum_{j \in \mathbf{Z}} q_j \phi(2x - j). \tag{4.2}$$

Then

$$\phi(x) = \chi_{[0,1)}(x), \qquad \text{a.e. } x \in \mathbb{R}.$$

Proof: First observe by the orthonormality condition and the refinement equation

$$q_j = 2 \int_{\mathbf{R}} \phi(x)\phi(2x - j)dx, \quad j \in \mathbf{Z}.$$

Thus

$$q(z) := \sum_{j \in \mathbf{Z}} q_j z^j$$

must be a Laurent *polynomial.* Moreover, by Equations (3.7) and (3.8) with M_n replaced by ϕ we have $\mu_{a,b} = 2$ and $\delta_{a,b} = 1$ since $a = b = 1$, and so Equation (3.10) gives

$$|q(z) - 1|^2 = 1 \tag{4.3}$$

for $|z| = 1$. This implies that $q(z) = 1 + z^N \rho$ where $\rho = \pm 1$. To see this we write $q - 1$ in the form

$$q(z) - 1 = z^N \{\rho_0 + \rho_1 z + \cdots + \rho_M z^M\}$$

where $\rho_0, \ldots, \rho_M$ are real coefficients with $\rho_0 \rho_M \neq 0$ and M, N are integers. Then

$$|q(z) - 1|^2 = \rho_M \rho_0 z^{-M} + \cdots + \rho_M \rho_0 z^M$$

and so either $M = 0$ or $\rho_M \rho_0 = 0$. Since we have made our choice to rule out the latter, we conclude that $M = 0$. Now, according to the requirement that

$$q(z) + q(-z) = 2 \tag{4.4}$$

we also know that N is odd. Integrating both sides of the refinement equation and using the fact that $\hat{\phi}(0) = 1$ we conclude $q(1) = 2$ and so $\rho = 1$. Thus we have

$$q(z) = 1 + z^N$$

and so using the refinement equation we get

$$\begin{aligned} \hat{\phi}(\theta) &= \frac{1 + e^{-iN\theta/2}}{2} \hat{\phi}\left(\frac{\theta}{2}\right) \\ &= \cdots = \frac{1}{2^k} \left(\frac{1 - e^{-iN\theta}}{1 - e^{-iN\theta/2^k}}\right) \hat{\phi}\left(\frac{\theta}{2^k}\right) \end{aligned}$$

and by letting $k \to \infty$, we get

$$\hat{\phi}(\theta) = \frac{1 - e^{-iN\theta}}{iN\theta} = \hat{\chi}_{[0,N)}(\theta).$$

This means that $\phi = \chi_{[0,N]}$ a.e. and by the orthonormality constraint of Equation (4.1) we conclude that $N = 1$. ■

In view of this negative result the question arises as to what can be said about the existence of a function ϕ satisfying all the conditions except for the demand that it be of compact support. Such functions, which decay exponentially fast, are constructed in Lewis [8]. Next, we shall characterize all orthonormal refinable cardinal functions which have a point of symmetry.

First, as a matter of review, we recall from [8] that there is *no* orthonormal refinable cardinal function ϕ which has a point x_0 a point of symmetry, *i.e.*, for all $x \in \mathbb{R}$ such that

$$\phi(x + x_0) = \phi(x_0 - x). \tag{4.5}$$

and so it seems that the additional technical conditions imposed by Lewis [8] rules out the important sinc function

$$\phi(x) = \frac{\sin \pi x}{\pi x}, \quad x \in \mathbb{R}$$

which clearly is symmetric about $x_0 = 0$ and is an orthonormal refinable cardinal function. To clarify these conditions and also to ultimately show the existence of *other* such functions, we dwell briefly on some properties of the sinc function. As pointed out in the introduction, it satisfies the refinement equation

$$\phi(x) = \sum_{j \in \mathbf{Z}} q_j \phi(2x - j), \quad x \in \mathbb{R} \tag{4.6}$$

where

$$q_j = \frac{\sin \pi j/2}{\pi j/2}, \quad j \in \mathbf{Z}.$$

The series on the right of Equation (4.6) is absolutely convergent since both $\phi(x) = 0(|x|^{-1})$, $x \to \pm\infty$, and $q_j = 0(|j|^{-1}), j \to \pm\infty$. Additionally, $\hat{\phi}(\omega) = \chi_{[-\pi,\pi]}(\omega), \omega \in \mathbb{R}$ and

$$Q(\theta) = \sum_{j \in \mathbf{Z}} q_j e^{ij\theta} = 1 + \frac{2e^{i\theta}}{\pi} \sum_{j \in \mathbf{Z}} \frac{e^{ij(2\theta+\pi)}}{2j+1},$$

so that

$$Q(\theta) = 2\chi_\Omega(\theta), \tag{4.7}$$

where

$$\Omega = \bigcup_{j \in \mathbf{Z}} \{[-\pi/2, \pi/2] + 2\pi j\}. \tag{4.8}$$

Consequently, Q given in Equation (4.6) for the sinc function is *not* continuous.

A reading of the proof in [8] suggests that whenever ϕ is *any* refinable function "sufficiently regular" in the sense that ϕ has enough derivatives or that $\hat{\phi}$ decays sufficiently rapidly at $\pm\infty$ which has a point of symmetry x_0 (see Equation (4.5)), and is refinable, *i.e.*, satisfies Equation (4.6) for some Q "sufficiently regular", it follows that $2x_0 \in \mathbf{Z}$. An argument, similar to the one used in Chui [2] in the study of *linear phase*, is used by Lewis to draw this conclusion. The rest of the argument in [8] says that x_0 must then be zero since $\phi(j) = \delta_j, j \in \mathbf{Z}$ and $\phi(0) = \phi(2x_0)$. Thus, Q is *real* and takes *only* the value zero or two. But then continuity rules out both. Unfortunately, there *are* indeed refinable functions with *arbitrary* points of symmetry. For example, choose any $a \in [\pi, 4\pi/3]$ and define

$$\hat{\phi}(\omega) = e^{-ix_0\omega}(a - |\omega|)_+, \quad \omega \in \mathbb{R}$$

where x_0 is *any* real number. A simple calculation shows that

$$\phi(x) = \frac{2\sin^2 \frac{1}{2}a(x_0 - x)}{\pi(x_0 - x)^2}, \quad x \in \mathbb{R}.$$

Clearly ϕ is symmetric about x_0 and it is easily seen that

$$\hat{\phi}(2\omega) = \frac{1}{2}Q(\omega)\hat{\phi}(\omega), \quad \omega \in \mathbb{R}$$

where Q is the 2π-periodic function such that

$$Q(\omega) = 2e^{-ix_0\omega}\frac{(a - 2|\omega|)_+}{a - |\omega|}, \quad |\omega| \leq \pi.$$

The function Q has a Fourier series

$$Q(\omega) = \sum_{j\in\mathbf{Z}} q_j e^{ij\omega},$$

where a simple calculation shows that $q_j = 0(j^{-2}), j \to \pm\infty$ and so $q = (q_j)_{j\in\mathbf{Z}} \in l_1(\mathbf{Z})$. Thus Q is certainly continuous and

$$\phi(x) = \sum_{j\in\mathbf{Z}} q_j \phi(2x - j), \quad x \in \mathbb{R}.$$

No amount of regularity on ϕ and Q obviates this situation as this example can easily be modified by replacing $\hat{\phi}$ above with

$$e^{-ix_0\omega}(a - |\omega|)_+^k,$$

where k is any positive integer, to achieve any desired regularity for Q and $\hat{\phi}$.

We prove the following result.

Proposition 4.2. *Let $\phi \in L^2(\mathbb{R})$ and Q a 2π-periodic measurable function be such that*

$$\hat{\phi}(\omega) = \frac{1}{2}Q(w/2)\hat{\phi}(w/2), \quad \text{a.e.}, \quad \omega \in \mathbb{R}. \tag{4.9}$$

If ϕ has x_0 as a point of symmetry and

$$\textit{meas } \{\omega : \hat{\phi}(2\omega)\hat{\phi}(2\pi + \omega) \neq 0\} > 0,$$

then $2x_0 \in \mathbf{Z}$.

Proof: The symmetry Relation (4.5) implies that

$$\hat{\phi}(\omega)e^{i\omega x_0} = \hat{\phi}(-\omega)e^{-i\omega x_0}, \quad \text{a.e. }, \ \omega \in \mathbb{R}. \tag{4.10}$$

Substitute Equation (4.9) into both sides of this equation and simplify the result to obtain the equation

$$\hat{\phi}(\omega)\{Q(\omega) - Q(-\omega)e^{-2ix_0\omega}\} = 0, \quad \text{a.e. }, \ \omega \in \mathbb{R}. \tag{4.11}$$

Now, replace ω by $\omega + 2\pi$ and use the 2π-periodicity of Q to obtain

$$\hat{\phi}(\omega + 2\pi)\{Q(\omega) - Q(-\omega)e^{-2ix_0(\omega+2\pi)}\} = 0, \quad \text{a.e. }, \ \omega \in \mathbb{R}. \tag{4.12}$$

If there is a point for which

$$Q(-\omega)\hat{\phi}(\omega)\hat{\phi}(\omega + 2\pi) \neq 0, \tag{4.13}$$

then Equations (4.11) and (4.12) imply that

$$e^{4\pi i x_0} = 1$$

from which we conclude that $2x_0 \in \mathbf{Z}$. In fact, Equation (4.13) is indeed true, since

$$Q(-\omega)\hat{\phi}(\omega)\hat{\phi}(\omega + 2\pi) = 2e^{2ix_0\omega}\hat{\phi}(2\omega)\hat{\phi}(\omega + 2\pi), \quad \text{a.e. }, \ \omega \in \mathbb{R}. \tag{4.14}$$

■

Corollary 4.3. *Let $\phi \in L^2(\mathbb{R})$ and Q a 2π-periodic measurable function satisfy Equation (4.9) and suppose that ϕ has x_0 as a point of symmetry. If $\hat{\phi}(w) \neq 0$ a.e. in a neighborhood of either 2π or $4\pi/3$, then $2x_0 \in \mathbf{Z}$.*

Proof: By Proposition 4.2 we need only consider the situation when

$$\hat{\phi}(2\omega)\hat{\phi}(2\pi + \omega) = 0, \ \text{a.e. }, \ \omega \in \mathbb{R}. \tag{4.15}$$

First suppose that $\hat{\phi}(\omega) \neq 0$ a.e. in a neighborhood of 2π. Then by Equation (4.15), $\hat{\phi}(\omega) = 0$ a.e. in a neighborhood of 0. By repeated application of Equation (4.9) we see that $\hat{\phi}$ vanishes a.e. on $\mathbb{R}$, which is a contradiction.

Next, suppose that $\hat{\phi}(\omega) \neq 0$ a.e. in a neighborhood of $4\pi/3$. Applying Equation (4.10) to Equation (4.15) and replacing ω by $-\omega$, we have

$$\hat{\phi}(2\omega)\hat{\phi}(2\pi - \omega) = 0, \ \text{a.e. }, \ \omega \in \mathbb{R}. \tag{4.16}$$

But for almost all ω close enough to $2\pi/3$, we have $\hat{\phi}(2\omega) \neq 0 \neq \phi(2\pi - \omega)$. which is again a contradiction. ■

The question arises as to whether or not there are *symmetric* functions other than the sinc function which are orthonormal refinable cardinal functions. Let

$$Q(\theta) = 2\chi_\Omega(\theta) \tag{4.17}$$

for some measurable set Ω such that $\Omega + 2\pi = \Omega$. We write Ω in the form

$$\Omega = \bigcup_{j\in\mathbf{Z}} (\Gamma + 2\pi j), \tag{4.18}$$

where

$$\Gamma := \Omega|_{[-\pi,\pi]}. \tag{4.19}$$

With this recipe for $Q(\theta)$ we try to find a corresponding refinable function ϕ.

Now, according to the refinement equation, $\hat{\phi}$ satisfies the equation

$$\hat{\phi}(\theta) = \hat{\phi}(\theta/2)\chi_\Omega(\theta/2).$$

This suggests the choice

$$\hat{\phi}(\theta) = \chi_\Delta(\theta) \tag{4.20}$$

where

$$\Delta = \bigcap_{n=1}^{\infty} 2^n\Omega.$$

Hence

$$\phi(x) = \frac{1}{2\pi}\int_\Delta e^{ix\omega}d\omega, \quad x \in \mathbb{R} \tag{4.21}$$

and to meet the requirements that ϕ be a cardinal function we need that

$$\sum_{j\in\mathbf{Z}} \chi_\Delta(\theta + 2\pi j) = 1, \text{ a.e. } \quad \theta \in \mathbb{R}, \tag{4.22}$$

and so the sets $\Delta + 2\pi j$, $j \in \mathbf{Z}$, should form a partition of $\mathbb{R}$ (in an almost everywhere sense). In particular, by integrating both sides of Equation (4.22) from $-\pi$ to π, this implies that meas $\Delta = 2\pi$.

To insure that ϕ is real and also Q has real Fourier coefficients we demand that Ω is symmetric about the origin, viz. $\Omega = -\Omega$. With this condition we have, for any $j \in \mathbf{Z}$,

$$\begin{aligned}(\phi, \phi(\cdot - j)) &= \frac{1}{2\pi}\int_{\mathbf{R}} e^{ij\omega}\hat{\phi}(\omega)\hat{\phi}(-\omega)d\omega \\ &= \frac{1}{2\pi}\int_{\mathbf{R}} e^{ij\omega}\chi_\Delta(\omega)d\omega \\ &= \phi(j) = \delta_j,\end{aligned}$$

by Equation (4.22). In summary we have proved the following proposition.

Proposition 4.4. *Given any symmetric set $\Omega \subseteq \mathbf{R}$ such that $\Omega + 2\pi = \Omega$ and the set*

$$\Delta = \bigcap_{n=1}^{\infty} 2^n \Omega$$

has integer translates $\Delta + 2\pi j$, $j \in \mathbf{Z}$ which partition $\mathbf{R}$. Then the function defined in Equation (4.21) is an orthonormal symmetric cardinal function which satisfies the refinement equation

$$\hat{\phi}(\omega) = \frac{1}{2} Q(\omega/2)\hat{\phi}(\omega/2), \quad \omega \in \mathbf{R} \tag{4.23}$$

where

$$Q(\omega) = 2\chi_{\Omega}(\omega), \quad \omega \in \mathbf{R}. \tag{4.24}$$

We remark when Δ and Ω consist of the union of a *finite* number of intervals then

$$\phi(\omega) = 0(|\omega|^{-1}), \quad \omega \to \pm\infty$$

and

$$Q(\omega) = \sum_{j \in \mathbf{Z}} q_j e^{ij\omega}, \quad \omega \in \mathbf{R}$$

where

$$q_j = Q(|j|^{-1}), \quad j \to \pm\infty.$$

Hence the refinement Equation (4.23) takes the form

$$\phi(x) = \sum_{j \in \mathbf{Z}} q_j \phi(2x - j), \quad x \in \mathbf{R}, \tag{4.25}$$

since the series on the right hand side of Equation (4.25) converges absolutely.

Below we give examples of sets which satisfy the demands of Proposition 4.4. Specifically, we define

$$\Gamma_a = \left[\frac{-\pi}{2} - a, \frac{-\pi}{2}\right] \cup \left[\frac{-\pi}{2} + a, \frac{\pi}{2} - a\right] \cup \left[\frac{\pi}{2}, \frac{\pi}{2} + a\right]$$

for any $a \in [0, \pi/6]$ and let Ω_a be the periodic extension of Γ_a, that is

$$\Omega_a = \bigcup_{j \in \mathbf{Z}} (\Gamma_a + 2\pi j).$$

Then certainly Ω is symmetric and $\Omega + 2\pi = \Omega$ even for $a \in [0, \pi/2]$. As for the remaining requirements, we note that for all $m = 1, 2, \ldots$

$$\bigcap_{n=1}^{m} 2^n \Omega_a = \bigcup_{j \in \mathbf{Z}} (2\Gamma_a + 2^{m+1} j\pi),$$

so that letting $m \to \infty$ gives

$$\Delta = 2\Gamma_a.$$

An easy computation shows that

$$\phi(x) = \frac{\sin \pi x}{\pi x}(2\cos 2ax - 1)$$

in this case.

There is a converse to the above result which we provide next.

The functions ϕ we deal with below are always assumed to be in $L^2(\mathbb{R}) \cap C(\mathbb{R})$ and have the property that the 2π-periodic function

$$|\hat{\phi}|(\omega) := \sum_{j \in \mathbf{Z}} |\hat{\phi}(\omega + 2\pi j)|, \quad \omega \in \mathbb{R}$$

is in $L^1(-\pi, \pi)$. Hence, it follows that $\hat{\phi} \in L^1(\mathbb{R})$ and so

$$\phi(x) = \frac{1}{2\pi} \int_{\mathbf{R}} e^{ix\omega} \hat{\phi}(\omega) d\omega, \quad x \in \mathbb{R}.$$

Thus, if in addition ϕ is an orthonormal cardinal function then

$$\sum_{j \in \mathbf{Z}} \hat{\phi}(\omega + 2\pi j) = \sum_{j \in \mathbf{Z}} |\hat{\phi}(\omega + 2\pi j)|^2 = 1, \text{ a.e. } \omega \in \mathbb{R}. \tag{4.26}$$

Theorem 4.5. *Let ϕ be an orthonormal cardinal function in $L^2(\mathbb{R}) \cap C(\mathbb{R})$ which is symmetric about some point x_0, has a Fourier transform $\hat{\phi}$ which is continuous and nonzero in some neighborhood of the origin and satisfies a refinement equation $\hat{\phi}(\omega) = \frac{1}{2}Q(\omega/2)\hat{\phi}(\omega/2)$, a.e. $\omega \in \mathbb{R}$ where Q is some 2π-periodic function in $L^\infty(-\pi, \pi)$ with $\overline{Q(\omega)} = Q(-\omega)$, a.e. $\omega \in \mathbb{R}$. Then $x_0 = 0, \hat{\phi}(0) = 1$ and there are sets, Ω and Δ, which satisfy the conditions of Proposition 4.4 such that*

$$\phi(x) = \frac{1}{2\pi} \int_{\Delta} e^{ix\omega} d\omega, \quad x \in \mathbb{R}.$$

Proof: The first part of the proof is to show that $x_0 = 0$. It suffices to show $2x_0 \in \mathbf{Z}$ because specializing the symmetry condition gives $1 = \phi(0) = \phi(2x_0)$. Hence we suppose $2x_0 \notin \mathbf{Z}$ and derive a contradiction. To this end, we substitute the refinement equation into Equation (4.26) and simplify to conclude that

$$Q(\omega) + Q(\omega + \pi) = 2, \text{ a.e. }, \omega \in \mathbb{R} \tag{4.27}$$

and

$$|Q(\omega)|^2 + |Q(\omega + \pi)|^2 = 4, \text{ a.e. }, \omega \in \mathbb{R}. \tag{4.28}$$

Moreover, since ϕ is symmetric about x_0, Equation (4.11) is valid,

$$\hat{\phi}(\omega)(Q(\omega) - Q(-\omega)e^{-2ix_0\omega}) = 0, \text{ a.e. } ,\omega \in \mathbb{R}. \tag{4.29}$$

As in the proof of Proposition 4.1, Equations (4.27) and (4.28) imply that

$$|Q(\omega) - 1|^2 = 1, \text{ a.e. } ,\omega \in \mathbb{R}. \tag{4.30}$$

If the second factor in Equation (4.29) vanishes, then the fact that $\overline{Q(\omega)} = Q(-\omega)$ implies that $Q(\omega) = \overline{Q(\omega)} = e^{-2ix_0\omega}$, from which it follows that

$$Q(\omega) = \lambda(1 + e^{-2ix_0\omega}),$$

for some $\lambda \in \mathbb{R}$, and Equation (4.30) becomes

$$2\lambda(\lambda - 1)(1 + \cos 2x_0\omega) = 0.$$

Thus, by Equation (4.29) we have for almost all ω either $\hat{\phi}(\omega) = 0$ or $\lambda = 0$ or 1 and hence

$$\hat{\phi}(\omega)Q(\omega)(Q(\omega) - 1 - e^{-2ix_0\omega}) = 0. \tag{4.31}$$

From Equation (4.26), we also get that for almost all $\omega \in \mathbb{R}$ there is a (least) integer $j(\omega)$ such that $\hat{\phi}(\omega + 2\pi j(\omega)) \neq 0$ and hence by Equation (4.31)

$$Q(\omega) = 0 \text{ or } Q(\omega) = 1 + e^{-2ix_0(w+2\pi j(\omega))}. \tag{4.32}$$

Suppose that the set $A := \{\omega \in [-\pi, \pi] : Q(\omega) = 0\}$ has positive measure. For almost all ω in A we have Equation (4.27) and so $Q(\omega + \pi) = 2$. By Equation (4.32), for almost all ω in A, we have $Q(\omega+\pi) = 1+e^{-2ix_0(\omega+\pi+2\pi j(\omega+\pi))}$ and so $e^{-2ix_0(\omega+\pi+2\pi j(\omega+\pi))} = 1$ and hence

$$x_0(\omega + \pi + 2\pi j(\omega + \pi)) = \pi\ell, \text{ some } \ell \in \mathbf{Z}. \tag{4.33}$$

Since Equation (4.33) holds for an uncountable number of ω in A, we can choose an integer j so that Equation (4.33) hold for an infinite number of ω in A with $j(\omega + \pi) = j$. Therefore, we can take a convergent sequence of distinct points ω_n in A so that for $n = 1, 2, \ldots$

$$x_0(\omega_n + \pi + 2\pi j) = \pi\ell(n), \text{ some } \ell(n) \in \mathbf{Z}.$$

So for all $m, n \geq 1$, $x_0(\omega_n - \omega_m) = \pi(\ell(n) - \ell(m))$ and, since (ω_n) is convergent, $\ell(n) = \ell(m)$ for all large enough n and m. Since $\omega_n \neq \omega_m$ for $n \neq m$, we have $x_0 = 0$, which is a contradiction. Thus A has measure zero and so $Q(\omega) \neq 0$ a.e.

Now let $S = \{\omega : \hat{\phi}(\omega) \neq 0\}$. By the refinement equation it follows that, $S = 2S$ up to a set of measure zero. By assumption, S contains an open interval containing 0. Thus $S = \mathbb{R}$ up to a set of measure zero and by Equation (4.31),

$$Q(\omega) = 1 + e^{-2ix_0\omega} \text{ a.e. } \omega \in \mathbb{R}.$$

Consequently, by Equation (4.27) we have $e^{2ix_0\omega} + e^{2ix_0(\omega+\pi)} = 0$ a.e. and so $e^{2\pi i x_0} = -1$ which contradicts the fact that $2x_0 \notin \mathbf{Z}$.

So far we have established that $x_0 = 0$. This means that ϕ is an even function. Next, we consider the quantity, $(\phi, \phi(2\cdot -j))$ for $j \in \mathbf{Z}$. Since ϕ is an even function, this is also even in j. Moreover, by the refinement equation

$$\begin{aligned}
2\int_{\mathbf{R}} &\phi(x)\phi(2x-j)dx \\
&= \frac{1}{2\pi}\int_{\mathbf{R}} \hat{\phi}(\omega)\overline{\hat{\phi}(\omega/2)}e^{ij\omega/2}d\omega \\
&= \frac{1}{2\pi}\int_{\mathbf{R}} |\hat{\phi}(\omega)|^2 Q(\omega)e^{ij\omega}d\omega \\
&= \frac{1}{2\pi}\int_{-\pi}^{\pi} Q(\omega)e^{ij\omega}d\omega
\end{aligned},$$

and, therefore, Q is real. Equation (4.30) implies $Q(\omega) = 0$ or $Q(\omega) = 2$ almost all $\omega \in \mathbb{R}$. Set

$$\Omega = \{\omega : Q(\omega) = 2\} \tag{4.34}$$

then $Q = 2\chi_\Omega$ and Ω is a symmetric set such that $\Omega + 2\pi = \Omega$. Finally, the refinement equation gives

$$\begin{aligned}
\hat{\phi}(\omega) &= \chi_\Omega(\omega/2)\hat{\phi}(\omega/2) \\
&= \cdots \\
&= \chi_{\Omega_m}(\omega)\hat{\phi}(\omega/2^m)
\end{aligned}$$

where

$$\Omega_m = \bigcap_{n=1}^{m} 2^n\Omega$$

and so

$$\hat{\phi}(\omega) = \hat{\phi}(0)\chi_\Delta(\omega), \text{ a.e. }, \omega \in \mathbb{R}.$$

Substituting this formula into both equations in Equation (4.26) gives $\hat{\phi}(0) = 1$ and that $\Delta + 2\pi j, j \in \mathbf{Z}$ partition $\mathbb{R}$. ■

Corollary 4.6. *Let ϕ, Q satisfy the conditions of Theorem 4.5. Then Q must be discontinuous.*

Proof: If, in fact, Q was continuous, then either $Q = 0$ or $Q = 2$. Either case is easily shown to be false. ■

This corollary is our interpretation of the remark made in [8] that there are no orthonormal refinable cardinal functions with a point of symmetry.

§5 Limits of wavelets

In the last section we demonstrated the special role that is played by a refinable function whose Fourier transform is the characteristic function of a set. In this section, we show how the special case of the sinc function can be obtained as the limit of certain judiciously chosen spline wavelets.

For any integer $n \geq 1$, we let V_n be the closed linear subspace of $L^2(\mathbb{R})$ given by $\{[c, M_{2n-1}] : c \in \ell_2(\mathbf{Z})\}$, $V_n^1 = \{f(2\cdot) : f \in V_n\}$, while W_n denotes the orthogonal complement of V_n in V_n^1. We let L_n denote the unique element of V_n satisfying

$$L_n(j) = \delta_j, \quad j \in \mathbf{Z}, \tag{5.1}$$

and define

$$\psi_n(x) = \sum_{j=-\infty}^{\infty} (-1)^{j-1} a_{j-1}^n L_n(2x - j), \tag{5.2}$$

where

$$a_j^n = \int_{-\infty}^{\infty} L_n(x) L_n(2x + j) dx. \tag{5.3}$$

The fact that L_n exists and decays exponentially appears in [11]. A special case of results from [10] implies that $\{\psi(\cdot - j) : j \in \mathbf{Z}\}$ forms a Riesz basis for W_n.

We shall consider convergence of ψ_n as $n \to \infty$. First we recall that

$$\hat{L}_n(w) = \left\{ \sum_{k=-\infty}^{\infty} \left(\frac{w}{w + 2\pi k} \right)^{2n} \right\}^{-1}, \tag{5.4}$$

from which it easily follows that

$$\begin{aligned} 0 \leq \hat{L}_n(w) &\leq 1, \\ 1 - \hat{L}_n(w) &\leq \left(\frac{w}{\pi}\right)^{2n} + o(1), \ |w| < \pi \\ \hat{L}_n(w) &\leq \left(\frac{\pi}{w}\right)^{2n}, \ |w| > \pi. \end{aligned} \tag{5.5}$$

To identify the limit of ψ_n as $n \to \infty$, we let ϕ be the sinc function and denote by V_∞ the space of entire functions of exponential type $\leq \pi$, equivalently, $V_\infty = \{[c, \phi] : c \in \ell_2(\mathbf{Z})\}$. Let $V_\infty^1 = \{f(2\cdot) : f \in V_\infty\}$, *i.e.*, the space of entire functions of exponential type $\leq 2\pi$, and let W denote the orthogonal complement of V_∞ in V_∞^1. Now, define ψ by

$$\psi(x) = \phi(2x - 1) - \frac{1}{2}\phi(x - \frac{1}{2}), \quad x \in \mathbb{R}. \tag{5.6}$$

Since $\hat{\phi} = \chi_{[-\pi,\pi]}$, we have

$$\begin{aligned} \hat{\psi}(w) &= \frac{1}{2} e^{-iw/2} \hat{\phi}\left(\frac{w}{2}\right) - \frac{1}{2} e^{-iw/2} \hat{\phi}(w) \\ &= \frac{1}{2} e^{-iw/2} \chi_{[-2\pi,-\pi] \cup [\pi,2\pi]}(w), \ a.e., \ w \in \mathbb{R}. \end{aligned} \tag{5.7}$$

Thus, a function f in $L^2(\mathbb{R})$ lies in W if and only if $\hat{f} = g\hat{\psi}$ for some g in $L^2\left([-2\pi, -\pi] \cup [\pi, 2\pi]\right)$ and $\|\hat{f}\|_2 = \|g\|_2$. Since any such function g can be extended to be 2π-periodic on $\mathbb{R}$, we see that $\{\psi(\cdot - j) : j \in \mathbf{Z}\}$ forms a Riesz basis for W. Now, by Equation (5.7)

$$\begin{aligned}\hat{\psi}(2w) &= \frac{1}{2}e^{-iw}\chi_{[-\pi,-\pi/2]\cup[\pi/2,\pi]}(w), \ a.e., \ w \in \mathbb{R} \\ &= \frac{1}{2}e^{-iw}a(w+\pi)\hat{\phi}(w), \ a.e., \ w \in \mathbb{R} \end{aligned} \tag{5.8}$$

where

$$a(w) = \begin{cases} 1, & -\frac{\pi}{2} + 2k\pi \le w \le \frac{\pi}{2} + 2k\pi, \quad k \in \mathbf{Z}\ , \\ 0, & \text{otherwise.} \end{cases}$$

Then

$$a(w) = \sum_{j=-\infty}^{\infty} a_j e^{ijw} \tag{5.9}$$

where, by the Parseval identity,

$$\begin{aligned} a_j &= \frac{1}{2\pi}\int_{-\pi/2}^{\pi/2} e^{-ijw}dw = \frac{1}{2\pi}\int_{-\infty}^{\infty} \hat{\phi}(2w)\hat{\phi}(w)e^{-ijw}dw \\ &= \frac{1}{4\pi}\int_{-\infty}^{\infty} \hat{\phi}(v)\hat{\phi}\left(\frac{v}{2}\right)e^{-ij/2}dv \\ &= \int_{-\infty}^{\infty} \phi(x)\phi(2x+j)dx. \end{aligned} \tag{5.10}$$

Therefore, by Equations (5.8) and (5.9), we have

$$\psi(x) = \sum_{j=-\infty}^{\infty} (-1)^{j-1}a_{j-1}\phi(2x-j), \tag{5.11}$$

where a_j is given by Equation (5.10).

Lemma 5.1. *For $n \ge 1$, $\hat{\psi}_n$ is uniformly bounded and* $\lim_{n\to\infty}\hat{\psi}_n(x) = \hat{\psi}(x)$, a.e. $x \in \mathbb{R}$.

Proof: Define

$$a_n(w) = \sum_{j=-\infty}^{\infty} a_j^n e^{ijw}, \quad w \in \mathbb{R}. \tag{5.12}$$

From Equation (5.3) and the exponential decay of L_n, we see that for each n, Equation (5.12) converges to give a continuous function a_n. From Equation (5.3), Parseval's identity gives

$$a_j^n = \frac{1}{2\pi}\int_{-\infty}^{\infty} \hat{L}_n(2w)\hat{L}_n(w)e^{ijw}dw.$$

Moreover, using the decay properties of $\hat{L}_n$, we can apply Poisson's summation formula to get

$$a_n(w) = \sum_{k=-\infty}^{\infty} \hat{L}_n(2w + 4k\pi)\hat{L}_n(w + 2k\pi), \quad w \in \mathbb{R}. \tag{5.13}$$

Now from Equation (5.2), we have

$$\hat{\psi}_n(w) = \frac{1}{2}e^{-iw/2}a_n\left(\frac{w}{2} + \pi\right)\hat{L}_n\left(\frac{w}{2}\right), \quad w \in \mathbb{R}, \tag{5.14}$$

and from Equations (5.5) and (5.13),

$$|\hat{\psi}_n(w)| \leq C|\hat{L}_n\left(\frac{w}{2}\right)|, \quad w \in \mathbb{R}, \tag{5.15}$$

for some C independent of w and n. In particular, $\hat{\psi}_n$ is uniformly bounded. Consequently, we conclude by Equations (5.5), (5.13), and (5.14) that

$$\begin{aligned} \lim_{n\to\infty} \hat{\psi}_n(w) &= \frac{1}{2}e^{-iw/2}\left\{\sum_{k=-\infty}^{\infty} \chi_{[-3\pi-4k\pi,-\pi-4k\pi]}(w)\chi_{[-4\pi-4k\pi,-4k\pi]}(w)\right\} \\ &\quad \times \chi_{[-2\pi,2\pi]}(w) \\ &= \frac{1}{2}e^{-iw/2}\chi_{[-2\pi,-\pi]\cup[\pi,2\pi]}(w) \\ &= \hat{\psi}(w), \quad \text{a.e.} \quad w \in \mathbb{R} \end{aligned}$$

by Equation (5.7). ■

We recall that any functions f_n in W_n and g in W_∞ can be represented as

$$\begin{aligned} f_n(x) &= \sum_{j=-\infty}^{\infty} b_j^n \psi_n(x - j), \quad \text{a.e. ,} \quad x \in \mathbb{R} \\ g_n(x) &= \sum_{j=-\infty}^{\infty} b_j \psi(x - j), \quad \text{a.e. ,} \quad x \in \mathbb{R}. \end{aligned}$$

for some $(b_j^n)_{-\infty}^{\infty}$ and $(b_j)_{-\infty}^{\infty}$ in $\ell_2(\mathbf{Z})$.

Proposition 5.2. *If $(b_j^n)_{-\infty}^{\infty} \to (b_j)_{-\infty}^{\infty}$ in $\ell_2(\mathbf{Z})$, then*

$$\lim_{n\to\infty} \sum_{j=-\infty}^{\infty} b_j^n \psi_n(\cdot - j) = \sum_{j=-\infty}^{\infty} b_j \psi(\cdot - j) \text{ in } L^2(\mathbb{R}).$$

Proof: Let $\Psi_n = \sum_{j=-\infty}^{\infty} b_j^n \psi_n(\cdot - j)$ and $\Psi = \sum_{j=-\infty}^{\infty} b_j \psi(\cdot - j)$. Then

$$\hat{\Psi}_n = b_n \hat{\psi}_n, \quad \hat{\Psi} = b\hat{\psi},$$

where

$$b_n(w) = \sum_{j=-\infty}^{\infty} b_j^n e^{-ijw}, \quad b(w) = \sum_{j=-\infty}^{\infty} b_j e^{-ijw}, \quad \text{a.e. } w \in \mathbb{R},$$

and so

$$\|\hat{\Psi}_n - \hat{\Psi}\|_2 \leq \|(b_n - b)\hat{\psi}_n\|_2 + \|b(\hat{\psi}_n - \hat{\psi})\|_2. \tag{5.16}$$

For the first term in the right of the inequality above we have

$$\begin{aligned}\|(b_n - b)\hat{\psi}_n\|_2^2 &= \sum_{k=-\infty}^{\infty} \int_{-\pi}^{\pi} |b_n(w) - b(w)|^2 |\hat{\psi}_n(w + 2k\pi)|^2 dw \\ &\leq \left\{ \sup_{w\in[-\pi,\pi]} \sum_{k=-\infty}^{\infty} |\hat{\psi}_n(w + 2k\pi)|^2 \right\} \|b_n - b\|_{L^2(-\pi,\pi)}^2. \end{aligned} \tag{5.17}$$

Moreover, by Equations (5.5) and (5.15), we see that $\sum_{k=-\infty}^{\infty} |\hat{\psi}_n(w + 2k\pi)|^2$ is uniformly bounded. Since $(b_j^n)_{-\infty}^{\infty} \to (b_j)_{-\infty}^{\infty}$ in $\ell_2(\mathbf{Z}), b_n$ converges to b in $L^2[-\pi, \pi]$ as $n \to \infty$ and by Equation (5.17)

$$\lim_{n\to\infty} \|(b_n - b)\hat{\psi}_n\|_2 = 0. \tag{5.18}$$

For the second term on the right of Equation (5.16), we observe that

$$\|b(\hat{\psi}_n - \hat{\psi})\|_2^2 = \sum_{k=-\infty}^{\infty} \int_{-\pi}^{\pi} |b(w)|^2 |\hat{\psi}_n(w + 2k\pi) - \hat{\psi}(w + 2k\pi)|^2 dw. \tag{5.19}$$

First we estimate the terms in the sum above for $k = -1, 0, 1$. To this end, we invoke Lemma 5.1 and the dominated convergence theorem to conclude that

$$\lim_{n\to\infty} \sum_{k=-1}^{1} |b(w)|^2 |\hat{\psi}_n(w + 2k\pi) - \hat{\psi}(w + 2k\pi)|^2 dw = 0. \tag{5.20}$$

For the remaining terms, we have $|k| \geq 2$. Thus, for $w \in [-\pi, \pi]$, Equation (5.7) gives

$$\begin{aligned}|\hat{\psi}_n(w + 2k\pi) - \hat{\psi}(w + 2k\pi)| &= |\hat{\psi}_n(w + 2k\pi)| \\ &\leq C(k - \frac{1}{2})^{-2n},\end{aligned}$$

by Equations (5.5) and (5.15). We conclude that

$$\sum_{|k|\geq 2} \int_{-\pi}^{\pi} |b(w)|^2 |\hat{\psi}_n(w + 2k\pi) - \hat{\psi}(w + 2k\pi)|^2 dw$$

$$\leq \|b\|^2_{L^2(-\pi,\pi)} C^2 \sum_{|k|\geq 2} (k - \frac{1}{2})^{-4n}$$

which goes to zero as $n \to \infty$. This observation together with Equations (5.19) and (5.20) shows that

$$\lim_{n\to\infty} \|b(\hat{\psi}_n - \hat{\psi})\|_2 = 0. \tag{5.21}$$

Hence, we have established that the right hand side Equation (5.16) goes to zero which proves that $\lim_{n\to\infty} \Psi_n = \Psi$ in $L^2(\mathbb{R})$. ■

Final Remark: We wish to thank Charles Chui for pointing out the reference [D.-X. Zhou, Construction of real-valued wavelets by symmetry] in which another example is given of a continuous refinable function with an arbitrary point of symmetry.

References

1. Cavaretta, A. S., W. Dahmen, and C. A. Micchelli, Stationary subdivision, *Memoirs of Am. Math. Soc.* **93** (1991), 453.
2. Chui, C. K., *An Introduction to Wavelets*, Academic Press, Boston, 1992.
3. Chui, C. K. and X. Shi, Wavelets and multiscale interpolation, in *Mathematical Methods in Computer Aided Geometric Design II*, T. Lyche, L. L. Schumaker (eds.), Academic Press 1992, 111–133.
4. Chui, C. K. and X. Shi Characterizations of fundamental scaling functions and wavelets, *Approx. Theory and Its Appl.* **9** (1993), 37–52.
5. Dahmen, W., T. N. T Goodman, and C. A. Micchelli, Compactly supported fundamental functions for spline interpolation, *Numer. Math.* **52** (1988), 639–664.
6. Daubechies, I., Orthonormal bases of compactly supported wavelets, *Comm. Pure and Appl. Math.* **41** (1988), 909–996.
7. Goodman, T. N. T., New bounds for the zeros of spline functions, *J. Approx. Theory* **76** (1994), 123–130.
8. Lewis, R. M., Cardinal interpolating multiresolutions, *J. Approx. Theory* **76** (1994), 177–202.
9. Micchelli, C. A., Banded matrices with banded inverses, *J. Comp. Appl. Math.* **41** (1992), 281–300.
10. Micchelli, C. A., Using the refinement equation for the construction of pre-wavelets, *Numer. Algorithms* **1** (1991), 75–116.
11. Schoenberg, I. J., *Cardinal Spline Functions*, Regional Conference Series in Applied Mathematics **12**, SIAM, Philadelphia, 1973.
12. Schumaker, L. L., *Spline Functions: Basic Theory*, Wiley, New York, 1981,
13. Xia, X.-G. and Z. Zhang, On sampling theorem, wavelets and wavelet transforms, *J. Comp. Appl. Math.*, 1993, preprint.

T. N. T. Goodman
University of Dundee
Department of Mathematical Sciences
Dundee DD1 4HN
Scotland, U.K.

Charles A. Micchelli
IBM Research Division
T.J. Watson Research Center
P.O. Box 218
Yorktown Heights, NY 10598
U.S.A.

Part II

Wavelet Transforms

Some Remarks on Wavelet Representations and Geometric Aspects

Bruno Torrésani

Abstract. We consider continuous wavelet decompositions, mainly from geometric and algebraic aspects. As examples we describe a scheme for construction of wavelet decompositions of functions on spaces that are homogeneous with respect to some group action. Restricting to discrete group actions we show how multiresolution structures can be directly derived from algebraic arguments. We finally describe approximate multiresolution structures adapted to cases in which no exact pyramidal algorithm can be associated with a given wavelet.

§1 Introduction

The main goal of this paper is to provide a description of (continuous) wavelet analysis and corresponding algorithms, mainly in terms of algebraic and geometric arguments. Today, sereval different aspects of wavelets have been developed, so that the word "wavelet" is now to be understood in a somewhat wider sense. Depending on the application of wavelets one has in mind, one may be more interested in some particular approaches to wavelets. For example, if one needs efficient and fast numerical computations, an appropriate algorithmic structure is required, which is more or less synonymous of multiresolution approach. If on the other hand one is more interested in signal *analysis*, one may be interested in considering instead redundant representations and emphasizing their symmetry properties.

This paper is devoted to the description of the latter aspect of wavelet decompositions. More precisely, we will essentially focus on the time-frequency, or higher-dimensional generalizations, representation theorems that can be obtained from covariance requirements. We will also show how the generic algorithmic (multiresolution) structure of wavelet analysis can be derived from group-theoretical arguments, as a consequence of some discrete covariance requirements on the wavelet transform, naturally leading to quadrature mirror

Wavelets: Theory, Algorithms, and Applications
Charles K. Chui, Laura Montefusco, and Luigia Puccio (eds.), pp. 91–115.

ISBN 0-12-174575-9

filters. Finally, we will describe the situation that occurs when no discrete filters are available, but the wavelet and scaling function are sufficiently localized in the Fourier space to ensure the existence of approximate discrete filters, yielding a good approximation of the wavelet transform.

The main sections of this paper have been written in such a way that they can be read independently, except the exception of the third section that contains examples related to the previous section.

§2 The group-theoretical picture

The group theoretical approach to time-frequency analysis was set first by A. Grossmann, J. Morlet, and T. Paul in [10]. The main interest of such an approach is that it emphasizes from the very beginning the covariance properties of the representation. Let us consider for instance the case of one-dimensional signal analysis. It is clear that at least for some purposes such as recognition, the representation has to be translation covariant; in other words the representation of a shifted signal must be a shifted copy of the representation of the signal. The main purpose of the group theoretical approach is to say: let us put the covariance requirement at the very beginning and derive the representation from it. By covariance we shall always mean covariance with respect to the action of a symmetry group, since we shall always need to consider products and inverse of transformations from the group.

2.1 Generalities

Let us begin from a measured space (X, m) and a covariance group G. We want to construct representations of functions in $L^2(X, m)$, covariant with respect to G. This implicitly means that G has a well-defined action on X, or in other words, X is *homogeneous* with respect to G, meaning that, $\forall x, x' \in X$, there exists $g \in G$ such that $x' = g \cdot x$. We will also require that m is quasi-invariant with respect to the action of G, *i.e.*, that the transformed measure $dm(g \cdot x)$ is absolutely continuous with respect to $dm(x)$. The existence of such quasi-invariant measures is a classical result of harmonic analysis on homogeneous spaces.

Then there is a canonical unitary representation of G on $L^2(X, m)$, given by:

$$\pi(g) \cdot f(x) = \sqrt{\frac{dm(g^{-1} \cdot x)}{dm(x)}} f(g^{-1} \cdot x). \tag{1}$$

Given such a set of data, and a "mother wavelet" $\psi \in L^2(X, m)$, there is a natural candidate for wavelet transform, namely the map

$$T : f \in L^2(X, m) \to T_f \in L^2(G, \mu) \tag{2}$$

$$T_f(g) = \langle f, \pi(g) \cdot \psi \rangle \tag{3}$$

that "tests" the signal $f(x)$ with all the elementary wavelets

$$\psi_{(g)}(x) = \sqrt{\frac{dm(g^{-1} \cdot x)}{dm(x)}} \psi(g^{-1} \cdot x). \tag{4}$$

The wavelet transform is a bounded function on G (as follows from the Cauchy-Schwarz inequality); moreover, since the representation is unitary, one also gets "for free" the covariance of the wavelet transform

$$T_{\pi(h) \cdot f}(g) = T_f(h^{-1}g). \tag{5}$$

Now arises the question of "invertibility" of the wavelet transform constructed in that way. For this it is convenient to introduce the following operator

$$\mathcal{A} : f \in L^2(X, m) \to \mathcal{A} \cdot f \tag{6}$$

$$\mathcal{A} \cdot f(x) = \int_G T_f(g) \psi_{(g)}(x) d\mu(g) \tag{7}$$

that "inverts" the wavelet transform. Here, μ is the (unique, up to a constant factor) left-invariant measure on G, *i.e.*, $d\mu(hg) = d\mu(g)$. This naturally leads to the following set of questions:

1. When is the group "sufficiently large" to characterize any $f \in L^2(X, m)$? In other words, when is the $\mathcal{A}$ operator invertible ?
2. When is the symmetry group "not too large?" In other words, when is the $\mathcal{A}$ operator bounded?
3. When is $\mathcal{A}$ a multiple of the identity (as happens for instance in the wavelet case), *i.e.*, T is a multiple of an isometry ?
4. What does G have to do with "time-frequency?"

2.2 Square-integrability

The answer to question 3 has been well-known for quite a long time. It deals with the theory of square-integrable representations.

Definition 1. *The representation π of G is square-integrable if*

1. *π is irreducible;*
2. *There exists a function $\psi \in L^2(X, m)$ such that the set of diagonal matrix elements $\varphi_\psi(g) = \langle \psi, \pi(g) \cdot \psi \rangle$ belongs to $L^2(G, \mu)$. ψ is said to be admissible.*

The main result for our purpose lies in the following orthonormality relations, proved by Duflo and Moore [6], Carey [4], and Grossmann, Morlet, and Paul [10]. Here is a weak form of the result:

Theorem 2. *Let π be a square-integrable unitary representation of the locally compact group G in $L^2(X, m)$, and let ψ be an admissible vector. Then the map*

$$T : f \in L^2(X, m) \to \frac{1}{\sqrt{c_\psi}} T_f \in L^2(G, \mu) \tag{8}$$

is an isometry, and can then be inverted by its adjoint.

In other words, any function $f \in L^2(X, m)$ can be decomposed as a superposition of modified copies of ψ, obtained by acting on it with the symmetry group G.

Remark. The result of Duflo and Moore is actually much stronger. They in particular show that the set of admissible vectors coincides with the domain of some densely defined linear operator K; in addition, K turns out to be a scalar whenever G is unimodular.

As an example, one may consider the natural action of the rotation group $SO(3)$ on the 2-dimensional sphere S^2. Such a representation is not square-integrable, because it is highly reducible. It splits into a countable infinity of invariant finite-dimensional representation spaces. Such a decomposition leads to the well-known spherical harmonics, and is not suitable for our purpose since in addition, it does not yield any natural notion of (local) frequency. However, the notion of frequency can be introduced into the game by considering an extension of the rotation group by local frequency translations, leading to the windowed Fourier transform on spheres. We shall come back in more details to this example in the next section, but for our current purpose, it is sufficient to say that such a group, the so-called Euclidian group, is "too large" for the representation space; in other words, that the reresentation is not square-integrable.

2.3 Going to quotient spaces

The case of spheres is particularly interesting because it allows one to really see where the departure from square-integrability comes from. Roughly speaking, it may be said that at any point, the corresponding "local Fourier space" can be identified with the (co)tangent space at this point, so that the space of all (local) frequency translations is isomorphic to $\mathbb{R}^n$. Then if one believes that the wavelet representations have something to do with "position-frequency" representations, one has to factor out the extra-dimension to go back to $\mathbb{R}^{n-1}$ (see, *e.g.*, [1]).

From a more general point of view, this means that one has to consider systems of vectors in the representation space, labeled by elements of some quotient space $P = G/H$ rather than the group itself. But since the representation π is defined on G and not on P (let us recall that P may be viewed as a set of equivalence classes of elements of G, two elements g and g' being equivalent if $g' = gh$ for some $h \in H$), one has to choose a representative element $\sigma(p) \in G$ for any $p \in P$. Then one can introduce the generalized coherent states (or wavelets): if $\psi \in L^2(X)$

$$\psi_{(p)} = \pi \circ \sigma(p) \cdot \psi, \tag{9}$$

and again the operator

$$\mathcal{A}_\sigma : f \in L^2(X) \to \int_P \langle f, \psi_{(p)} \rangle \psi_{(p)} d\eta(p) \tag{10}$$

where η is the unique up to a constant factor, quasi-invariant measure on P, and again ask the same questions as before, regarding the "size" of P and the properties of σ.

2.4 Phase-space

Among the possible candidates for the P space, there is a most natural one, obtained as follows. Let us recall that we started from a representation π of G. Differentiating π, *i.e.*, looking at the representatives of elements close to the identity of G, one obtains a representation $d\pi$ of the Lie algebra $\mathcal{G}$ of G, *i.e.*, the tangent space at the identity. It turns out that for a large class of groups, the representation $d\pi$ can be constructed from a linear form on $\mathcal{G}$, i.e. an element $F_\pi \in \mathcal{G}^*$, the dual space of $\mathcal{G}$.

Now, to this F_π element can be associated a family of vectors $F_\pi^h \in \mathcal{G}^*$:

$$F_\pi^h(X) = F_\pi(hXh^{-1}) \;\; \forall X \in \mathcal{G}. \tag{11}$$

By definition, the coadjoint orbit $\mathcal{O}_\pi$ associated with the representation π is the set of such F^h vectors, or in other words, the quotient of G by the subgroup of elements that leave F_π invariant.

$$\mathcal{O}_\pi \cong G/G_\pi, \tag{12}$$

where

$$G_\pi = \left\{h \in G, \;\; F^h = F\right\}. \tag{13}$$

It can be shown (see for example [12] for a review) that $\mathcal{O}_\pi$ inherits a structure of symplectic manifold. $\mathcal{O}_\pi$ is sometimes called the phase-space associated with π (in the geometrical sense), and is for us a natural candidate to study square-integrability of the representations.

Remark. It must be recalled here that the Lie groups and algebras under consideration are always finite-dimensional. The computations with the dual space are easy to do, since the dual can be identified with the space itself. This allows one to compute explicitely the phase-space associated with a given representation, provided that the representation can be generated from an element $F_\pi \in \mathcal{G}^*$. Moreover, the groups under consideration can in general be realized as simple matrices groups, so that explicit computations are much easier. For instance, the affine or "$ax+b$" group in dimension 1 can be realized as the group of matrices of the form

$$G_{aff} = \left\{ \begin{pmatrix} a & b \\ 0 & 1 \end{pmatrix}, b \in \mathbb{R}, a \in \mathbb{R}_+^* \right\}$$

while the Weyl-Heisenberg group is the group of matrices

$$G_{WH} = \left\{ \begin{pmatrix} 1 & q & t \\ 0 & 1 & p \\ 0 & 0 & 1 \end{pmatrix}, p, q \in \mathbb{R}, t \in [0, 2\pi] \right\}.$$

From such matrix realizations it is easy to compute the matrix realizations of the corresponding Lie algebras simply by differentiating at the origin C^1 curves of the form $M_x, x \in \mathbb{R}$, with $M_0 = e$.

§3 Examples

3.1 Square-integrable representations

The first examples are well-known, and yield the usual wavelet transforms. In such cases, the covariance group actually turns out to be isomorphic, up to a compact factor, to the geometric phase space of the representation.

This is the case of the so-called affine, or "$ax + b$ group", generated by translations and dilations. In such a case, the group law (see the matrix realization above) is the following

$$(b, a) \cdot (b', a') = (b + ab', aa'), \tag{14}$$

and it is well-known that the following representation, acting on the Hardy space $H^2(I\!R) = \{f \in L^2(I\!R), \hat{f}(\xi) = 0 \forall \xi \leq 0\}$

$$\pi(b, a) \cdot f(x) = \frac{1}{\sqrt{a}} f\left(\frac{x - b}{a}\right) := f_{(b,a)}(x) \tag{15}$$

is square-integrable. Then the direct application of Duflo-Moore's theorem yields the celebrated Calderón identity. If $\psi \in L^2(I\!R)$ is admissible, *i.e.*, such that $\int_0^\infty |\hat{\psi}(\xi)|^2 d\xi/\xi = 1$), then any $f \in H^2(I\!R)$ can be decomposed as

$$f = \int_0^\infty \int_{I\!R} \langle f, \psi_{(b,a)} \rangle \psi_{(b,a)} \frac{da}{a} \frac{db}{a}. \tag{16}$$

The extension to the $L^2(I\!R)$ context is trivially obtained under the additional assumption $\int_0^\infty |\hat{\psi}(-\xi)|^2 d\xi/\xi = 1$.

The set of admissible vectors is moreover isomorphic to the domain of a linear operator K, which is in this case a convolution operator, defined by the symbol $\sigma(\xi) = \xi^{-1/2}$.

The generalizations of affine wavelets to higher dimensions have also been considered by Murenzi in a specific case [15], and later on by Bernier and Taylor [3]. The key point is that to keep an irreducible representation, the scale parameter has to be replaced by matrices, *i.e.*, one has to consider subgroups of $GL(n)$. Let H be such a subgroup, and consider

$$G = H \times I\!R^n \tag{17}$$

with group multiplication given by

$$(x, h) \cdot (x', h') = (x + h \cdot x', hh'). \tag{18}$$

G can be realized as a matrix group as follows:

$$G = \left\{ \begin{pmatrix} h & b \\ 0 & 1 \end{pmatrix}, b \in I\!R^n, a \in H \right\},$$

where b is understood as a column vector, and 0 as a line vector of length n.

Then let $U = H \cdot x_0$ be an orbit of H in $\mathbb{R}^n$, and let $H^2_U(\mathbb{R}^2)$ be the subspace of $L^2(\mathbb{R}^n)$ consisting of functions whose Fourier transform is supported in U. This is a possible generalization of the classical complex Hardy space $H^2(\mathbb{R})$ to the multidimensional setting. Bernier and Taylor show that if the U-orbit is free, *i.e.*, if $h \cdot x_0 = x_0$ implies $h = e$, and open in $\mathbb{R}^n$, the restriction to $H^2_U(\mathbb{R}^2)$ of the natural representation

$$\pi(h,b) \cdot f(x) = \delta(h)^{-1/2} f\left(h^{-1} \cdot (x-b)\right) \tag{19}$$

is square-integrable. Here $\delta(h) = |\det h|$ is the Radon-Nikodym derivative $d(h \cdot x)/dx$. Notice that the assumptions on H force H to be n-dimensional. But it is very likely that the results can be generalized to weaker assumptions. For instance, for groups of the form $\mathbb{R}^*_+ \times SO(n) \times \mathbb{R}^n$, the representation given in Equation (19) is also square-integrable [15], although $dim(SO(n)) > n$ for $n > 2$.

This case is particularly interesting and has been studied in great details. It leads to rotation-covariant wavelet representations that were developed by R. Murenzi. More precisely, let us stick to the two-dimensional case. The plane $\mathbb{R}^2$ is clearly an open free orbit for $\mathbb{R}^*_+ \times SO(2)$. If one denotes by $r(\theta)$ the rotation matrix of angle θ in the plane, then

$$dm(r) = dm(\theta) = d\theta$$

is the rotation-invariant measure on $SO(2)$. One then has the following representation theorem for functions on the plane: if $\psi \in L^2(\mathbb{R}^2)$ is such that $\int_{\mathbb{R}^2} |\hat{\psi}(\xi)|^2 d\xi/||\xi||^2 = 1$, any $f \in L^2(\mathbb{R}^2)$ can be decomposed as

$$f = \int_{\mathbb{R}^2 \times \mathbb{R}^*_+ \times [0,2\pi]} \langle f, \psi_{(b,a,\theta)} \rangle \psi_{(b,a,\theta)} \frac{db}{a^2} \frac{da}{a} d\theta \tag{20}$$

with the obvious notations. It is important to notice that in such a case, the coefficients $T_f(b,a,\theta) = \langle f, \psi_{b,a,\theta} \rangle$ can be viewed as a position-frequency representation of $f(x)$, with b as position variable and $(1/a, \theta)$ as polar coordinates representation of frequency. This has been successfully used in image analysis (see, namely, [2] and [9]). In particular, it can be shown using standard approximation techniques [9] that if $f(x)$ can be modeled as

$$f(x) = A(x) \cos \varphi(x),$$

then $T_f(b,a,\theta)$ is locally maximum near the two sheets $a_r(b), \theta_r(b) \pm \pi$, where $a_r(b)^{-1}$ and $\theta_r(b)$ are polar coordinates of the vector $\nabla A(b)$, if $a_r(x)|\nabla A(x)| \ll |A(x)|$ and $|\nabla^2 \varphi(x)| \ll |(\nabla \varphi(x))^2/\varphi(x)|$. This is the two-dimensional generalization of the classical localization properties of the wavelet transform of asymptotic signals.

3.2 Phase-space representations

Now arises the question of the connection of such a formalism with phase-space, *i.e.*, position-frequency space. It turns out that geometry can yield a very natural notion of phase-space, as described before.

3.2.1 Time-frequency atoms

Consider the simple example of Gabor transform. It is well-known that Gabor transform is naturally associated with the so-called Weyl-Heisenberg group, a 3-dimensional group, *i.e.*, the group generated by time and frequency translations with group multiplication $(p,q,t)\cdot(p',q',t') = (p+p',q+q',t+t'+pq' \bmod 2\pi)$. (See the matrix realization in the previous section.) As a matter of fact, the Gabor transform of a function is in general not considered as a function on the group itself, but rather as a function of time and frequency, *i.e.*, as a function on the phase-space, considered as a quotient of the group. However, a direct computation of the phase-space, *i.e.*, the coadjoint orbit, shows that it is isomorphic to $\mathbb{R}^2 \cong \{(p,q)\}$, *i.e.*, the parameter set of the Gabor transform.

This remark can be generalized to the case of higher-dimensional groups. The simplest example is that of more general time-frequency atoms, generated by time-frequency translations and modulations. This amounts to introduce a larger group, generated by the above-mentioned transformations, called the affine, or inhomogeneous, Weyl-Heisenberg group.

$$G \cong \mathbb{R}^2 \times \mathbb{R}_+^* \times [0,2\pi] \tag{21}$$

with group multiplication

$$(p,q,a,t)\cdot(p',q',a',t') = (p+p'/a, q+aq', aa', t+t'+apq' \bmod 2\pi)). \tag{22}$$

It can be realized as the following matrix group:

$$G_{aWH} \cong \left\{ \begin{pmatrix} 1 & aq & t \\ 0 & a & p \\ 0 & 0 & 1 \end{pmatrix}, p,q \in \mathbb{R}, t \in [0,2\pi] \right\}.$$

The irreducible representations of such a group are well-known, and there is a natural one, acting on $L^2(\mathbb{R})$, given by

$$\pi(p,q,a,t)\cdot f(x) = \frac{1}{\sqrt{a}} e^{i(t+p(x-q))} f\left(\frac{x-q}{a}\right). \tag{23}$$

A simple calculation shows that π is *not* square-integrable. This simply comes from the fact that the considered group is "too big" for the representation space. However, it is also easy to compute the phase-space associated with such a representation (see [11]), and to show that it is isomorphic to the quotient space

$$\mathcal{O}_\pi \cong G/(I\!R_+^* \times [0, 2\pi]) \cong I\!R^2. \tag{24}$$

Let then $\sigma : \mathcal{O}_\pi \to G$ be a piecewise differentiable Borel function. It was shown in [11] that one can find σ functions such that the associated $\mathcal{A}_\sigma$ operator is a scalar. This is however not the case in general, but it is not very difficult to show that under some very weak assumptions, $\mathcal{A}_\sigma$ is bounded with bounded inverse, and then suitable for numerical applications.

Let us assume for instance that one is only interested in introducing a relationship between the frequency and the bandwidth of the analyzing functions. We consider sections $\sigma : \mathcal{O}_\pi \to G$ of the following form:

$$\sigma(q, p) = (q, p, \beta(p), 0). \tag{25}$$

Then Proposition 3 follows.

Proposition 3. *Let σ be a piecewise C^1 function: $\mathcal{O}_\pi \to G$ given as in Eq. (25). Then $\mathcal{A}_\sigma$ is a convolution operator and is a multiple of the identity if and only if β is of the form*

$$\beta(p) = \frac{1}{\alpha p + \alpha'} \tag{26}$$

for some constants α, α' non-simultaneously vanishing.

Proof: A direct computation shows that $\mathcal{A}_\sigma$ is indeed a convolution, defined by the multiplier

$$\gamma(\xi) = \int_{I\!R} \left| \hat{\psi}\left(\beta(p)(\xi - p)\right) \right|^2 |\beta(p)| dp.$$

The end of the proposition follows from the explicit computation of the Jacobian of the change of variable $p \to \beta(p)(\xi - p)$. ■

3.2.2 The case of the sphere

Let us now turn to more complex examples. As was discussed in the beginning of this paper, one is interested in building phase-space harmonic analysis on more general homogeneous spaces. Let us describe now a construction of windowed Fourier analysis on the simplest nontrivial homogeneous space, *i.e.*, the sphere S^2. Classical harmonic analysis on S^2 is based on the action of the rotation group $SO(3)$ on S^2, and leads to spherical harmonics. In order to introduce a notion of local frequency into the analysis, one needs to consider the (co)tangent spaces at any point of the sphere, in other words the cotangent bundle T^*S^2 of the sphere (see Figure 1).

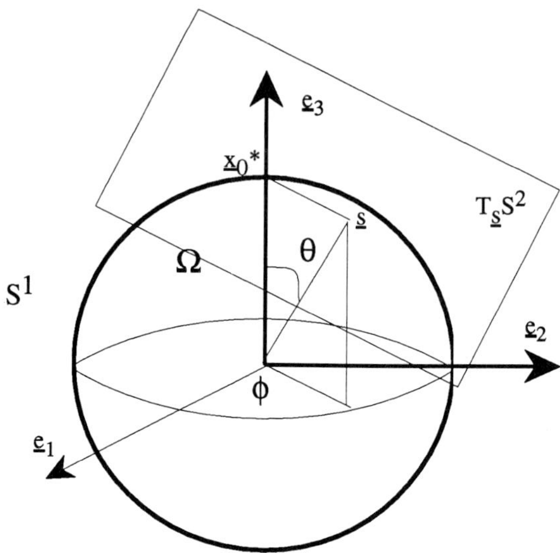

Figure 1. Windowed Fourier analysis on the sphere: geometry of the problem.

To build a windowed Fourier transform, one needs to introduce in addition frequency translations at any point, *i.e.*, translations in all the planes cotangent to the sphere. This then leads to the translations of $\mathbb{R}^3$, *i.e.*, $\mathbb{R}^3$ itself. Moreover, the action of rotations on frequency translations has to be nontrivial, since when one goes from one point of S^2 to another one by a rotation, one has to rotate the cotangent spaces accordingly.

Altogether, this leads to the so-called Euclidian group

$$E(3) \cong SO(3) \times \mathbb{R}^3, \tag{27}$$

with the multiplication

$$(r,p) \cdot (r',p') = (rr', p + r \cdot p'), \tag{28}$$

which is a particular case of the groups studied by Bernier and Taylor. It also admits the following matrix realization:

$$E(3) = \left\{ \begin{pmatrix} r & p \\ 0 & 1 \end{pmatrix}, p \in \mathbb{R}^n, r \in SO(3) \right\}. \tag{29}$$

The difference is that one is now no longer interested in its action on $L^2(\mathbb{R}^3)$ which leads to Murenzi's wavelets, but on the following representation, acting on $L^2(S^2)$ as follows:

$$\pi(r,p) \cdot f(s) = e^{ip \cdot s} f(r^{-1} \cdot s), \quad s \in S^2. \tag{30}$$

Such a representation is not square-integrable, as it can be directly checked. However, as stressed before, there is a natural candidate to restrict the representation in order to get a position-frequency representation theorem. Since computations on spheres are somewhat involved, it is simpler to consider the coset space $E(3)/\mathbb{R}$, which differs from the true phase-space, the cotangent bundle of S^2 only by a compact factor $SO(2)$ that does not change anything to square-integrability.

Before stating the representation theorem associated with the group action we considered, let us fix our notations. We shall denote by Ω the upper half-sphere, *i.e.*, the set $\{s = (x, y, z) \in \mathbb{R}^3,\ x^2 + y^2 + z^2 = 1 \text{ and } z > 0\}$ and by θ and φ the spherical coordinates of $s \in S^2$. Then we have [18].

Theorem 4. *Assume that $\psi \in L^1(S^2) \cap L^2(S^2)$ is such that $Supp(\psi) \subset \Omega$ and*

$$0 \neq c_\psi = 8\pi^3 \int_0^{2\pi} \int_{-\pi/2}^{\pi/2} \frac{\left|\psi(\theta, \varphi)\right|^2}{\cos\theta}\, d\theta d\varphi < \infty. \tag{31}$$

Then the map

$$f \in L^2(S^2) \to \frac{1}{\sqrt{c_\psi}} T_f \in L^2(G/\mathbb{R}) \tag{32}$$

is an isometry.

In other words, any function in $L^2(S^2)$ can be decomposed as

$$f = \int_{\mathbb{R}^2} \int_{SO(3)} \langle f, \psi_{(r,p)} \rangle\, \psi_{(r,p)}\, d^2p d\mu(r), \tag{33}$$

i.e., as a superposition of modulated and rotated copies of the window ψ. We are then in a situation very close to that of the "flat" windowed Fourier transform; the main modification is that the modulation is now perturbed by the curvature of the sphere.

Remark. The same analysis can also be carried out in any dimension, for instance in the one-dimensional case; the situation is even simpler since the $SO(2)$ rotation group is abelian and isomorphic to the circle itself. In such a case, one also gets a (periodic) time-frequency representation theorem; the building blocks are of the form

$$\psi_{(\alpha,p)}(\theta) = e^{ip\sin(\alpha+\theta)} \psi(\alpha + \theta), \tag{34}$$

where $p \in \mathbb{R}, \alpha, \theta \in [0, 2\pi]$, and $\psi(\theta)$ is a 2π periodic function, supported in an interval of length at most π. The perturbation of the modulation introduced by curvature is even clearer here. Moreover, it is effective only for large values of the angle, since for small α, $\sin\alpha \sim \alpha$. As an example, Figure 3 represents the modulus of the windowed Fourier transform on the circle of 125 milliseconds of the speech signal /one/.

Remark. Up to now, it seems that nothing is known about the problem of constructing "true wavelets" on spheres; otherwise stated, how to introduce dilations on spheres.

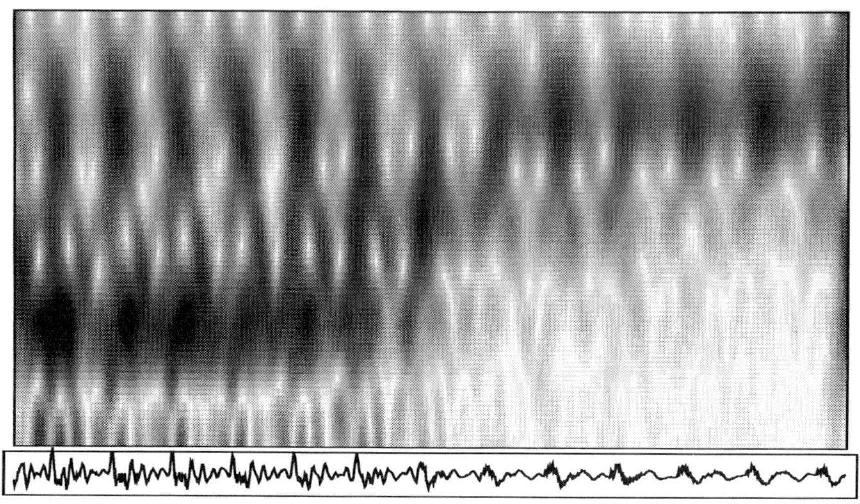

Figure 2. Windowed Fourier transform (modulus) on the circle of 125 ms of /one/.

§4 Algorithms from group theoretical considerations

Let us now go to more algorithmic aspects of wavelets. It is well-known that one of the reasons of the success of wavelet analysis is the existence of fast pyramidal algorithms for the computation of the wavelet transform. Such algorithms are of course adapted to discrete problems, and can actually be seen in the following way (this approach is very close to the so-called "algorithme à trous" [8]). Consider a field $\mathbb{F} = \mathbb{Z}$, or $\mathbb{Z}_p$ (the set of congruence classes modulo p, where p is a prime number), and the set of sequences on $\mathbb{F}$, that we will denote by $\ell^2(\mathbb{F})$. There exist well-defined translations and dilations on $\ell^2(\mathbb{F})$, according to:

$$T_b \cdot f(n) = f(n-b) \tag{35}$$

$$D_a \cdot f(n) = \begin{cases} f\left(\frac{n}{a}\right) & \text{if } a \text{ divides } n \\ 0 & \text{elsewhere .} \end{cases} \tag{36}$$

Then it can easily be verified that $\{T_b, D_a\}$ reproduce the affine multiplication law:

Lemma 5.

$$T_b D_a T_{b'} D_{a'} = T_{b+ab'} D_{aa'}. \tag{37}$$

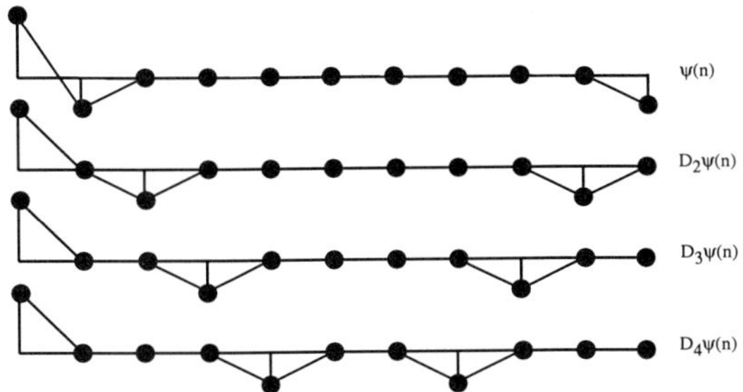

Figure 3. Discrete dilation "with holes."

Remark. When $F = \mathbb{Z}_p$ with p prime, a always divides n. Then, there exists a well-defined affine group on $\ell^2(\mathbb{Z}_p)$, and the operators T_b and D_a define a representation of such a group, which turns out to be square-integrable. The general result of Duflo and Moore can then be directly used, and yields a cyclic version of wavelet analysis of sequences, developed by K. Flornes and coworkers in [7].

Moreover, the D_a-dilation operators are suitable from algorithmic point of view, in the sense that they don't change the number of non-vanishing coefficients. But they have the "drawback of this advantage", in the sense that, for example, the dilate by 3 of the sequence $\{0, ..., 0, 1, 1, 1, 0, ...0\}$ is the sequence $\{0, ..., 0, 1, 0, 0, 1, 0, 0, 1, 0, ..., 0\}$, which is not convenient if we want to think to this sequence as some sampling of a function defined on the line or the circle, which is one of the goals of wavelet analysis. An example is shown in figure 2, where one sees the dilates of a sample discrete wavelet. Because of the presence of the intermediate zeroes, it is hard to interpret the discrete dilated wavelets as samplings (whatever it means...) of dilated copies of a function.

The effect of such a dilation "with holes" can be seen in Figure 4, which represents the cyclic (*i.e.*, on $\mathbb{Z}_p$) wavelet transform of a "Dirac sequence", *i.e.*, the set of cyclic convolutions products of the periodic sequence $(0, \cdots, 0, 1, 0, \cdots, 0)$ of length 97 with scaled copies $D_a \cdot \psi$ of a real-valued wavelet with 7 non-vanishing coefficients as represented at the bottom of the figure.

To avoid such effects, one needs a dilation operator allowing one:

- To think of sequences as discretization of functions, *i.e.*, mapping from $L^2(\mathbb{R})$ to $\ell^2(\mathbb{Z})$, for example.
- To think of the discrete dilation as the combination of a continuously defined dilation and the previous discretization.
- To keep the algebraic properties of translations and dilations, *i.e.*, to continue to satisfy Equation (37).

Such problem was already addressed in [8], and later on in [7].

Let us then introduce the following family of "pseudo-dilations"

$$\mathcal{D}_a = K_a D_a \tag{38}$$

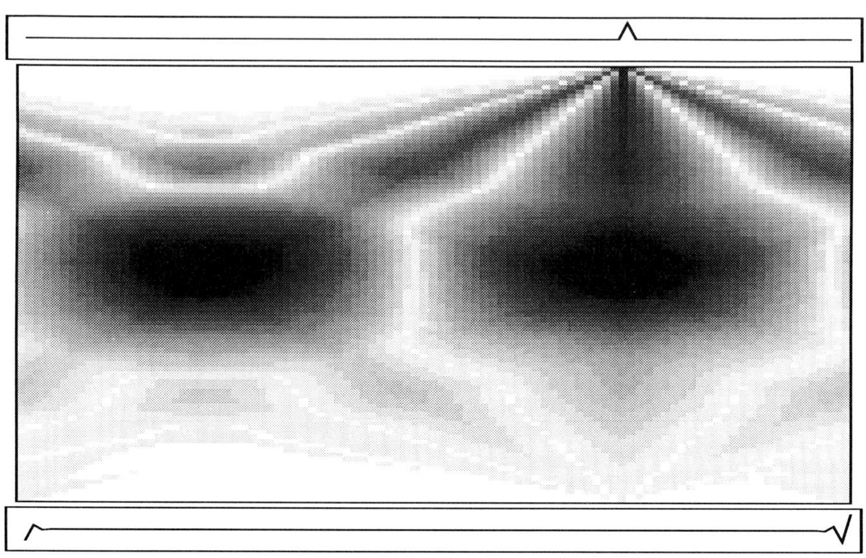

Figure 4. Cyclic wavelet transform of a "Dirac sequence" with the dilation "with holes."

where $\{K_a, a \in \mathbb{F}^*\}$ is a family of linear operators acting on $\ell^2(\mathbb{F})$.

Definition 6. *The family $\{\mathcal{D}_a, a \in \mathbb{F}^*\}$ is an admissible family of dilations if the operators $T_b\mathcal{D}_a$ satisfy the Equation* (37).

One then has the following result, the proof of which is easy and was given in [7].

Proposition 7. *The family $\{\mathcal{D}_a, a \in \mathbb{F}^*\}$ is an admissible family of dilations if and only if any $K_a, a \in \mathbb{F}^*$ is a convolution operator*

$$K_a \cdot f = F_a * f \tag{39}$$

and the discrete filters F_a satisfy the following compatibility conditions

$$F_{aa'} = F_a * (D_a \cdot F_{a'}) = F_{a'} * (D_{a'} \cdot F_a) \tag{40}$$

for any $a, a' \in \mathbb{F}$. Such families of filters will be called admissible.

Proof: The first argument is that the K_a operators have to be translation invariant, which force them to be convolution operators. In addition, one must have $\mathcal{D}_{aa'} = \mathcal{D}_a\mathcal{D}_{a'} = \mathcal{D}_{a'}\mathcal{D}_a$, which directly yields the admissibility relations. ■

To understand such conditions, let us assume that there exist admissible families of filters, and consider for instance the case $a = a' = 2$, and use the pseudo-dilation to compute corresponding wavelet coefficients. Let then $\psi \in \ell^2(\mathbb{F})$, and set as usual $\tilde{\psi}(n) = \psi(-n)^*$. The corresponding wavelet coefficients of $f \in \ell^2(\mathbb{F})$ read

$$T_f(\cdot, 4) = f * \left[\mathcal{D}_4 \cdot \tilde{\psi}\right] = f * K_4 * \left[D_4 \cdot \tilde{\psi}\right] = f * K_2 * (D_2 \cdot K_2) * \left[D_4 \cdot \tilde{\psi}\right]. \quad (41)$$

This basically amounts to say the following:

- The computation of wavelet coefficients at scale 2 can be obtained by a filtering (with $D_2 \cdot \tilde{\psi}$) of a "smoothed" copy of f, itself obtained by a smoothing with the filter F_2.
- The computation of wavelet coefficients at scale 4 can be obtained by a filtering (with $D_4 \cdot \tilde{\psi}$) of a "smoothed" copy of the previous smoothing of f, with the filter $D_2 \cdot F_2$.
- In any case, the number of multiplications is the same, since one always uses copies of the filters scaled with the D_a dilation operator.

Then, if one wants to compute discrete wavelet transform for scales that are only powers of 2, one immediately recovers the algorithmic structure of multiresolution analysis. Indeed, the K_2 filter is identical to the usual $H = \{h_n\}$ filter, and the samples of the wavelet $\tilde{\psi}(n)$ play the role of the $G = \{g_n\}$ high-pass filter.

From the point of view of the approach developed here, the next question is that of the existence of admissible families of filters. The existence is a well-known result in the case where one only considers scales that are powers of some integral number, but not in the case of general sequences of scales. It turns out that the situation is much simpler if there exists a global scaling function, *i.e.*, a function $\phi(x)$ such that in the Fourier space:

$$\hat{\phi}(q\xi) = \hat{F}_q(\xi)\hat{\phi}(\xi). \quad (42)$$

Then it is easy to see the following.

Proposition 8. *Let F_q be a family of discrete filters, labeled by the prime elements of $\mathbb{F}$, and such that there exists a function $\phi(x)$ fulfilling Equation (42) for all q. Let the filters $F_a(n), a \in \mathbb{F}^*$ be defined by*

$$\hat{F}_a(\xi) = \frac{\hat{\phi}(a\xi)}{\hat{\phi}(\xi)}. \quad (43)$$

Then the family $F_a, a \in \mathbb{F}^*$ is admissible.

Proof: Obviously, one has

$$\hat{F}_{aa'}(k) = \frac{\hat{\phi}(aa'k)}{\hat{\phi}(k)} = \frac{\hat{\phi}(aa'k)}{\hat{\phi}(ak)}\frac{\hat{\phi}(ak)}{\hat{\phi}(k)} = \hat{F}_{a'}(ak)\hat{F}_a(k) \quad (44)$$

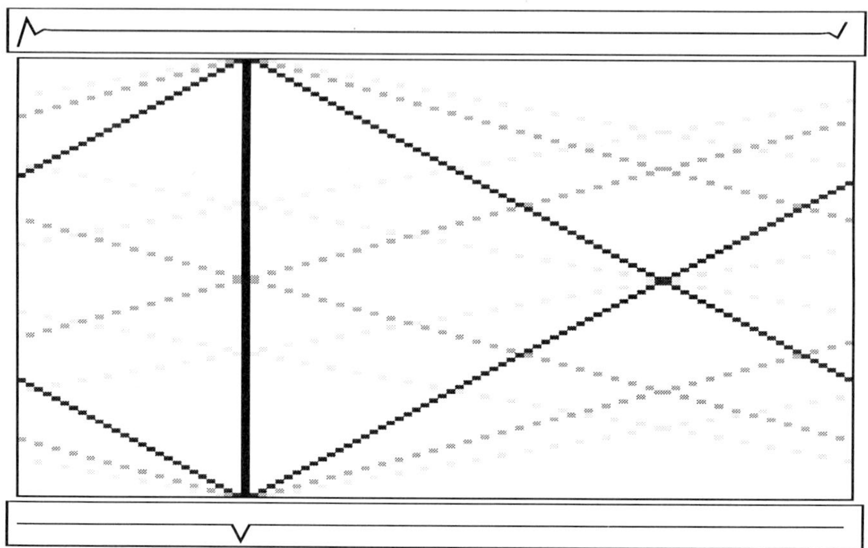

Figure 5. Cyclic wavelet transform of a "Dirac sequence" with the pseudodilation.

which is precisely the Fourier space version of the admissibility condition. ■

This reduces the problem to that of finding families of filters labeled by prime numbers. It turns out that some examples of filters are easy to find out, namely those associated with spline interpolation. More precisely, taking

$$\hat{F}_q(k) = \left(\frac{1 - e^{\frac{2i\pi qk}{p}}}{1 - e^{\frac{2i\pi k}{p}}} \right)^N , \tag{45}$$

we have exactly the B-spline filters, that satisfy Equation (42), in both the $\mathbb{F} = \mathbb{Z}$ and $\mathbb{F} = \mathbb{Z}_p$ cases (the $\mathbb{Z}_p$ case is more constrained, because of the periodicity; Equation (42) has to be fulfilled also when the product of scale factors is understood modulo p).

As an example, let us come back to the previously considered "Dirac sequence", the cyclic wavelet transform of which is now computed using a pseudo-dilation with cubic spline filters (see Figure 5). The cubic spline filter "fills the holes."

As a consequence of Equation (42) and since only the F_q filters with prime scale factors q are needed, one gets a pyramidal algorithm for the computation of the corresponding wavelet transform as follows. Given a scale factor $a \in \mathbb{F}^*$:

1. Decompose a into prime factors

$$a = q_1^{r_1} q_2^{r_2} \cdots q_n^{r_n}.$$

2. Use Equation (42) to derive the decomposition of F_a into products of scaled copies of the F_{q_i}s.

Remark. We have only described here the algorithms for the computation of wavelet transforms $\langle f, \psi_{(b,a)} \rangle$ of some discrete signal $f \in \ell^2(\mathbb{F})$. When $\mathbb{F} = \mathbb{Z}$, there does not exist a well-defined affine group acting on $\ell^2(\mathbb{Z})$, so that one cannot use the general approach of the first sections to get reconstruction formulas and algorithms. In the case $\mathbb{F} = \mathbb{Z}_p$, the general results can be used with the D_a dilations, yielding directly the usual reconstruction formulas. However, given a family of admissible dilations $\mathcal{D}_a, a \in \mathbb{Z}_p^*$, one has by construction a representation of the cyclic affine group, but it is in general not unitary. Then again the Duflo-Moore theorem cannot be used directly. It is shown in [7] how to generate a family of filters for the reconstruction algorithm.

§5 Approximate algorithms

We saw in the last section how the algorithmic structure of multiresolution analysis is, in some sense, implied by the use of dilations and the requirement of covariance with respect to dilations. The algorithms are based on the existence of a scaling function and a pair of 2π-periodic functions.

However, in many cases there does not exist such 2π-periodic filters. We shall see in this section that in some cases, it is still possible to use pyramidal algorithms if one accepts small errors in the computation of wavelet coefficients while still keeping exact reconstruction.

5.1 The one-dimensional case

We saw in the last section how the algorithmic structure of multiresolution analysis can be derived from purely algebraic arguments. Then, as is well-known, to get a pyramidal algorithm for the computation of wavelet coefficients, one must start from a pair (ϕ, ψ) of functions, with associated filters, denoted from now on $H = \{h_\ell\}$ and $G = \{g_\ell\}$ (with Fourier transforms denoted by $m_0(\xi)$ and $m_1(\xi)$, respectively), such that

$$\hat{\phi}(2\xi) = m_0(\xi)\hat{\phi}(\xi) \tag{46}$$

and

$$\hat{\psi}(2\xi) = m_1(\xi)\hat{\phi}(\xi). \tag{47}$$

Setting

$$S_j f(n) = 2^{-j} \int f(x) \phi\left(2^{-j}(x-n)\right)^* dx \tag{48}$$

and

$$T_j f(n) = 2^{-j} \int f(x) \psi \left(2^{-j}(x-n)\right)^* dx, \tag{49}$$

one directly gets the following pyramidal structure for the computation of the coefficients

$$S_j f(n) = \sum_k h_k^* S_{j-1} f(n - k2^{j-1}). \tag{50}$$

However, this is, in general, not the case in the sense that although it is in general easy to associate a scaling function $\phi(x)$ to a wavelet $\psi(x)$, there is no reason to expect the quotients $\hat{\phi}(2\xi)/\hat{\phi}(\xi)$ and $\hat{\psi}(2\xi)/\hat{\phi}(\xi)$ to exist as 2π-periodic functions.

Nevertheless, due to the good localization of the scaling function in the Fourier space, there are good reasons to expect that such quotients, or more precisely their product,with $\hat{\phi}(\xi)$, can be well-approximated by 2π-periodic functions. In other words, it may happen that there exists a pair of filters $\tilde{H}$ and $\tilde{G}$ such that the algorithm

$$\tilde{T}_j f(n) = \sum_k \tilde{g}_k^* \tilde{S}_{j-1} f(n - 2^{j-1}k) \tag{51}$$

$$\tilde{S}_j f(n) = \sum_k \tilde{h}_k^* \tilde{S}_{j-1} f(n - 2^{j-1}k) \tag{52}$$

yields an accurate result. This is expressed by the following theorem.

Theorem 9. *Let $\psi(x)$ and $\phi(x)$ be respectively a wavelet and a scaling function such that the collection of its integer translates of $\phi(x)$ forms an unconditional basis of their closed linear span, and denote by $T_j f(n)$ and $S_j f(n)$ respectively the discrete wavelet transform and the discrete scaling function transform of $f \in L^2(\mathbb{R})$. Then*

1. *There exists a unique pair of discrete filters $\tilde{H}$ and $\tilde{G}$ such that the distances $\mu_0 = ||\phi(x/2)/2 - \sum \tilde{h}_\ell^* \phi(x+\ell)||_2$ and $\mu_1 = ||\psi(x/2)/2 - \sum \tilde{g}_\ell^* \phi(x+\ell)||_2$ are minimal.*
2. *These approximate filters are given by*

$$\tilde{m}_0(\xi) = \frac{\sum_{k\in\mathbf{Z}} \hat{\phi}(2(\xi+2\pi k))\hat{\phi}(\xi+2\pi k)^*}{\sum_{k\in\mathbf{Z}} |\hat{\phi}(\xi+2\pi k)|^2} \tag{53}$$

$$\tilde{m}_1(\xi) = \frac{\sum_{k\in\mathbf{Z}} \hat{\psi}(2(\xi+2\pi k))\hat{\phi}(\xi+2\pi k)^*}{\sum_{k\in\mathbf{Z}} |\hat{\phi}(\xi+2\pi k)|^2}. \tag{54}$$

3. *The following estimates hold: for any $f \in L^2(\mathbb{R})$,*

$$||S_j f - \tilde{S}_j f||_{\ell^\infty} \le K_j \mu_0 ||f||_2 \tag{55}$$

and

$$||T_j f - \tilde{T}_j f||_{\ell^\infty} \le K_j' \mu_1 ||f||_2 \tag{56}$$

for some constants K_j and K'_j.

Proof: (sketched) The proof of existence and unicity of the $\tilde{m}_0$ and $\tilde{m}_1$ filters follows from classical arguments, and their explicit expressions are obtained by using, for example, the fact that $\phi(x/2)/2 - \sum \tilde{h}_\ell^* \phi(x+\ell)$ is perpendicular (in L^2) to $\phi(x-k)$ for all $k \in \mathbb{Z}$ (and the same for m_1). The proof of the estimates are also standard, involving basically the Poisson summation formula, and the Hölder and Cauchy-Schwarz inequalities, and can be found in [16], where the asymptotic behaviour of the K_j and K'_j constants is also discussed. ■

It is also worth noticing that since an explicit expression for the filters is available, one can easily relate the decay of the corresponding filter coefficients with the regularity of the scaling function in the Fourier space as follows (see [16] for a proof).

Proposition 10. *If both $\psi(\xi)$ and $\hat{\phi}(\xi)$ belong to the Sobolev space $H^m(\mathbb{R})$, then*

$$\tilde{h}_k \sim \tilde{g}_k \sim O(k^{-m}) \text{ as } k \to \infty. \tag{57}$$

Remark. It is also interesting to notice that if the scaling function and the wavelet have been chosen in such a way that

$$\hat{\psi}(2\xi) = \hat{\phi}(\xi) - \hat{\phi}(2\xi), \tag{58}$$

then one has the following "linear QMF condition"

$$\tilde{m}_0 + \tilde{m}_1 = 1, \tag{59}$$

ensuring simple reconstruction formulas:

$$\tilde{S}_{j_0} f(n) = \tilde{S}_J f(n) + \sum_{j=J}^{j_0 - 1} \tilde{T}_j f(n). \tag{60}$$

5.1.1 Examples: Gaussian-type wavelets

Let us consider for instance the case of wavelets canonically associated with Gaussian scaling functions. It is a well-known fact that in such a case, there is no multiresolution structure available that would lead to exact pyramidal algorithms. However, the approximate filters described by Theorem 9 turn out to be particularly efficient, due to the exponential localization of the scaling function in the Fourier space. Let, for example, the scaling function $\phi(x)$ be of the form

$$\phi_\alpha(x) = K_\alpha e^{x^2/\alpha} \tag{61}$$

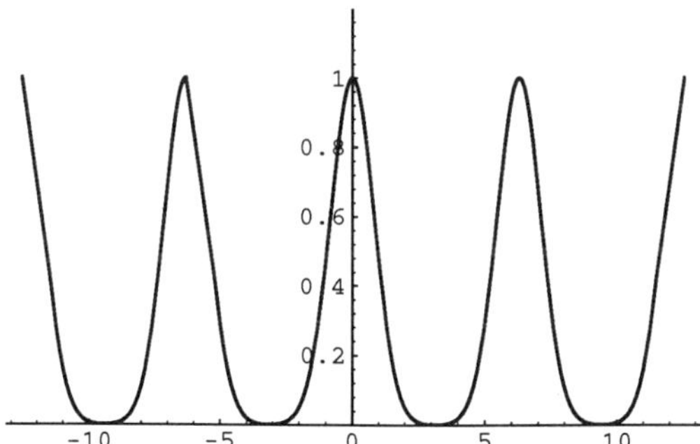

Figure 6. Approximate low-pass filter for Laplacian of Gaussian wavelet.

with K_α a constant such that $\int \phi_\alpha(x)dx = 1$, and a wavelet

$$\psi_\alpha(x) = 2\phi_\alpha(2x) - \phi_\alpha(x). \tag{62}$$

Then one obtains a pair of approximate filters such that $\tilde{m}_0 + \tilde{m}_1 = 1$.

The graph of $\tilde{m}_0(\xi)$ is shown in Figure 6. Filters coefficients are given in [16] and a sample Mathematica^{TM} program for the computation of the coefficients is available [19].

5.1.2 Variations on approximate filters

In many signal analysis applications, it is also desirable to introduce redundancy with respect to the scale parameter, or to use wavelets that are more concentrated in frequency. For instance, for speech processing, one typically uses bandwidthes of one third of an octave. The method developed in the previous section can also be adapted to such cases.

Let us first consider the first case, in which one is interested in getting a wavelet transform that is close to the continuous one, *i.e.*, in which one wants to keep redundancy between scales. The simplest way to proceed in such a case is to use the previous formulas which are all scale-covariant, and to run the algorithm starting from scaled copies of the scaling function.

More precisely, let us consider the case of a wavelet transform, with respect to wavelets of bandwidth approximatively one octave, and with N different scales (in geometric progression) per octave. Set $a_\lambda = a_{j,\mu} = 2^{\lambda/N}$, where $\lambda = Nj + \mu$. If $g(x)$ is an infinitesimal wavelet, normalized so that $c_g = 1$ for simplicity, set

$$\phi_\mu(x) = \int_{2^\mu}^{\infty} \frac{1}{a} g\left(\frac{x}{a}\right) \frac{da}{a} \tag{63}$$

and

$$\psi_\mu(x) = \int_{2^{\mu-1}}^{2^\mu} \frac{1}{a} g\left(\frac{x}{a}\right) \frac{da}{a} = 2\phi_\mu(2x) - \phi_\mu(x). \tag{64}$$

Then the same procedure as before will produce a family of low and high-pass filters, given by:

$$\tilde{m}_0^\mu(\xi) = \frac{\sum_{k\in\mathbf{Z}} \hat{\phi}_\mu(2(\xi+2\pi k))\hat{\phi}_\mu(\xi+2\pi k)^*}{\sum_{k\in\mathbf{Z}} |\hat{\phi}_\mu(\xi+2\pi k)|^2} \tag{65}$$

$$\tilde{m}_1^\mu(\xi) = \frac{\sum_{k\in\mathbf{Z}} \hat{\psi}_\mu(2(\xi+2\pi k))\hat{\phi}_\mu(\xi+2\pi k)^*}{\sum_{k\in\mathbf{Z}} |\hat{\phi}_\mu(\xi+2\pi k)|^2}. \tag{66}$$

Another possibility is to build a wavelet transform in which all sub-bands (*i.e.*, all scales) are independent. Then it is sufficient to work with only one scaling function and to introduce the following family of wavelets

$$w_\mu = \int_{2^{(\mu-1)/N}}^{2^{\mu/N}} \frac{1}{a} g\left(\frac{x}{a}\right) \frac{da}{a}. \tag{67}$$

One is then led to the following family of high-pass filters:

$$\tilde{m}_\mu(\xi) = \frac{\sum_{k\in\mathbf{Z}} \hat{w}_\mu(2(\xi+2\pi k))\hat{\phi}(\xi+2\pi k)^*}{\sum_{k\in\mathbf{Z}} |\hat{\phi}(\xi+2\pi k)|^2}, \tag{68}$$

and in such a case, one has the following property:

$$\tilde{m}_0 + \sum \tilde{m}_\mu = 1, \tag{69}$$

yielding the reconstruction formulas:

$$S_{j_0} f = S_J f + \sum_\mu \sum_{j=J}^{j_0-1} T_j^\mu f, \tag{70}$$

where one has set

$$T_j^\mu f(k) = \langle f, 2^{-j} w_\mu \left(2^{-j}(\cdot - k)\right)\rangle. \tag{71}$$

Of course, both schemes can be mixed together, for instance, to get the scale-redundant third of octaves wavelet representations that seem to be useful for speech processing.

5.2 Two-dimensional case

The method described in the previous section yields approximate filters in cases where there does not exist "true" 2π-periodic filters for a given pair (ϕ, ψ). However, this situation can be generalized to the two-dimensional case, where it is not known how to generate filters for radial scaling functions or to introduce angular localization in the Fourier space.

Let us consider the problem of constructing a pyramidal algorithm for 2-dimensional wavelets simply generated by integral translations and dilations by 2 of a unique mother wavelet. It is well-known that in such a case there does not exist any possible wavelet basis. The same discussion as before can be developed, *i.e.*, roughly speaking, computing the orthogonal projection of $\phi(x/2, y/2)$ and $\psi(x/2, y/2)$ on the closed linear span of the functions $\phi(x - k, y - \ell))$, leading to approximate filters of the form:

$$\tilde{m}_0(\xi,\zeta) = \frac{\sum_{k,\ell\in\mathbf{Z}} \hat{\phi}(2(\xi+2\pi k), 2(\zeta+2\pi\ell))\hat{\phi}(\xi+2\pi k, \zeta+2\pi\ell)^*}{\sum_{k,\ell\in\mathbf{Z}} |\hat{\phi}(\xi+2\pi k, \zeta+2\pi\ell)|^2} \tag{72}$$

$$\tilde{m}_1(\xi,\zeta) = \frac{\sum_{k,\ell\in\mathbf{Z}} \hat{\psi}(2(\xi+2\pi k), 2(\zeta+2\pi\ell))\hat{\phi}(\xi+2\pi k, \zeta+2\pi\ell)^*}{\sum_{k,\ell\in\mathbf{Z}} |\hat{\phi}(\xi+2\pi k, \zeta+2\pi\ell)|^2}. \tag{73}$$

If one is interested in cases for which

$$\hat{\psi}(2\xi, 2\zeta) = \hat{\phi}(\xi,\zeta) - \hat{\phi}(2\xi, 2\zeta), \tag{74}$$

one gets "decomposition-reconstruction" algorithms of the usual form

$$\tilde{S}_j f(m,n) = \sum_{k,\ell} \tilde{h}_{k,\ell} \tilde{S}_{j-1}(m - k2^{j-1}, n - \ell 2^{j-1}) \tag{75}$$

(and the same for the wavelet coefficients), and

$$\tilde{S}_{j_0} f(m,n) = \tilde{S}_J f(m,n) + \sum \tilde{T}_j f(m,n). \tag{76}$$

Figures 7 and 8 represent respectively the low- and high-pass filters for the two-dimensional Laplacian of Gaussian wavelets.

The problem of constructing discrete wavelet-like decompositions with angular selectivity in their frequency localization can be addressed in a similar way. For example, starting with a family of wavelets $\psi^\lambda(x)$ such that

$$\sum_\lambda \hat{\psi}^\lambda(2\xi, 2\zeta) = \hat{\phi}(\xi,\zeta) - \hat{\phi}(2\xi, 2\zeta), \tag{77}$$

one ends up with a family of filters $\tilde{m}_0(\xi,\zeta)$ with the same expression as before, and $\tilde{m}_1^\lambda$, simply given by

$$\tilde{m}_1^\lambda(\xi,\zeta) = \frac{\sum_{k,\ell\in\mathbf{Z}} \hat{\psi}^\lambda(2(\xi+2\pi k), 2(\zeta+2\pi\ell))\hat{\phi}(\xi+2\pi k, \zeta+2\pi\ell)^*}{\sum_{k,\ell\in\mathbf{Z}} |\hat{\phi}(\xi+2\pi k, \zeta+2\pi\ell)|^2}. \tag{78}$$

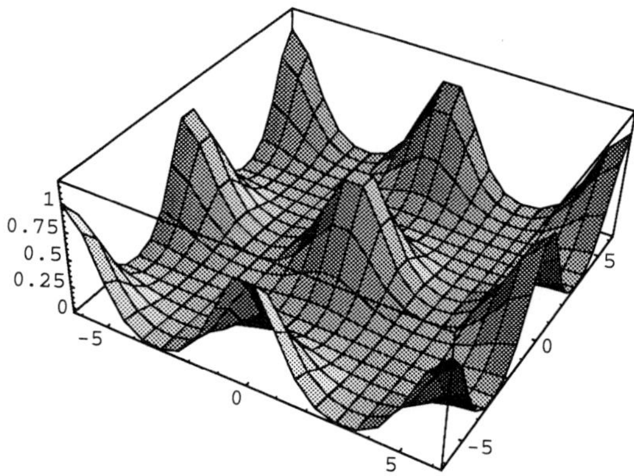

Figure 7. Approximate low-pass filter for 2D Laplacian of Gaussian wavelet.

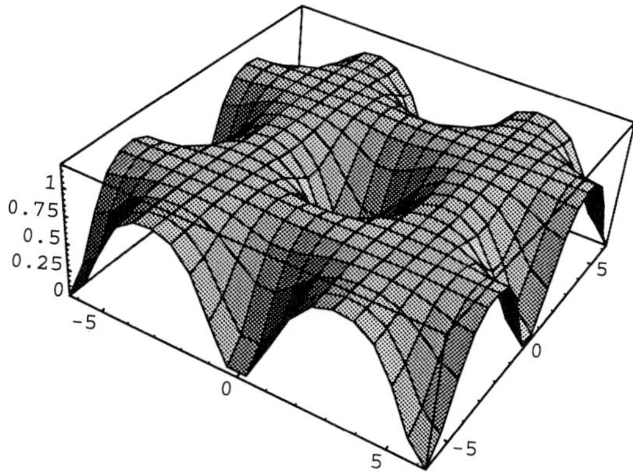

Figure 8. Approximate high-pass filter for 2D Laplacian of Gaussian (radial) wavelet.

Remark. It is important to point out that the methodology developed in this section only aims to produce pyramidal algorithms with exact reconstruction, with the corresponding discrete filters. Since the construction actually starts from a wavelet and a scaling function, it is also natural to expect the coefficients produced by the algorithm to be somewhat close to the coefficients of corresponding wavelet ans scaling function. This is quite natural in the 1D case, as we saw in the last section. In the 2D radial case, this is not necessarily the case, the algorithms are built on the square geometry imposed by the square lattice being not really adapted to the radial assumptions. Otherwise stated, the μ_0 and μ_1 coefficients may be larger than those in the 1D case. Along the same idea, it sounds natural to expect a better situation if one works with hexagonal lattice instead of the square one, the formulas we gave here are easily modified accordingly. Up to now we have actually no proof of this.

Acknowledgments. Many of the results described in this paper have been obtained in collaboration with M. Duval-Destin, K. Flornes, C. Gonnet, A. Grossmann, C. Kalisa, M. Holschneider, M. A. Muschietti. I would like to take the opportunity to thank them for the pleasure of collaboration. I would also like to thank J. P. Antoine, G. Beylkin, Y. Meyer and R. Murenzi for fruitful discussions. This work was supported in part by the late "GDR Ondelettes," CNRS.

References

1. Ali, S. T., J. P. Antoine, and J. P. Gazeau, Square-integrability of group representations on homogeneous spaces I, Reproducing triples and frames, *Ann. Inst. H. Poincaré* **55** (1991), 829–856; II, Generalized square-integrability and equivalent families of coherent states, *Ann. Inst. H. Poincaré* **55** (1991), 860–890.
2. Antoine, J. P., B. Carrette, R. Murenzi, and B. Piette, Image analysis with 2-dimensional wavelet transform, *Signal Processing* **31** (1993), 241–272.
3. Bernier, D. and K.F. Taylor, Wavelets from square-integrable representations, Dept .of Mathematics and Statistics, University of Saskatchewan, Canada, preprint.
4. Carey, A. L., Square-integrable representations of non-unimodular groups, *Bull. Austr. Math. Soc.* **15** (1976), 1.
5. De Bièvre, S., Coherent states over symplectic homogeneous spaces, *J. Math. Phys.* **30** (1989), 1401–1407.
6. Duflo, M. and C. C. Moore, On the regular representation of a nonunimodular locally compact group, *J. Funct. An.* **21** (1976) 209–243.
7. Flornes, K., A. Grossmann, M. Holschneider, and B. Torrésani, Wavelets on finite fields, *Appl. Comp. Harmonic Anal.* **1** (1994), to appear.
8. Holschneider, M., R. Kronland-Martinet, J. Morlet, and Ph. Tchamitchian, A real-time algorithm for signal analysis with the help of wavelet transform, in *Wavelets, Time-Frequency methods and Phase Space*, J. M. Combes, A. Grossmann, Ph. Tchamitchian (eds.), Marseille, IPTI, Springer, 1987, 286–297.
9. Gonnet, C. and B. Torrésani, Local frequency analysis with the help of two-dimensional wavelet transform, *Signal Processing* **38** (1994), to appear.
10. Grossmann, A., J. Morlet, and T. Paul, Transforms associated with square integrable group representations I, *J. Math. Phys.* **27** (1985), 2473. II, *Ann. Inst. H. Poincaré* **45** (1986), 293.
11. Kalisa, C. and B. Torrésani, N-dimensional affine Weyl-Heisenberg wavelets, *Ann. Inst. H. Poincaré, Phys. Théor.* **59** (1993), 201–236.
12. Kirillov, A., *Elements of representation theory*, Moscow.

13. Mallat, S., A theory for multiresolution signal decomposition: the wavelet representation, *IEEE Pattern Anal. and Machine Intell.* **11** (7) (1989), 674–693.
14. Meyer, Y., *Ondelettes et Opérateurs*, in two volumes, Hermann, Paris, 1990.
15. Murenzi, R., *Ondelettes multidimensionnelles et application à l'analyse d'image*, Ph. D. Thesis, Louvain-la-Neuve, Belgium.
16. Muschietti, M. A. and B. Torrésani, Pyramidal algorithms for Littlewood-Paley decompositions, *SIAM J. Math. An.* (1994), to appear.
17. Torrésani, B., Time-Frequency Representations: wavelet packets and optimal decomposition, *Ann. Inst. H. Poincaré, Phys. Théor.* **56** (1992), 1.
18. Torrésani, B., Phase-space decompositions; local Fourier analysis on spheres, CPT Marseille, submitted to Appl. and Comp. Harmonic Anal..
19. Torrésani, B., Approximate filters for LOG wavelets, Mathematica^{TM} program, available by anonymous ftp at cpt.univ-mrs.fr (internet number 139.124.7.5), directory /pub/preprints/94/wavelets , file LOG.ma.

Bruno Torrésani
CPT, CNRS-Luminy, Case 907
13288 Marseille Cedex 09, France.
torresan@marcptnx1.univ-mrs.fr

A Matrix Approach to Discrete Wavelets

Jaroslav Kautsky and Radka Turcajová

Abstract. Orthogonality, biorthogonality, regularity, and symmetry of discrete wavelet transforms are all studied via properties of banded block circulant matrices. It is thus shown that, in parallel with the standard approach of functional analysis, the basic aspects of wavelets can be derived using fairly elementary means of matrix algebra. In other words, matrices are everywhere, even where they are not.

§1 Introduction

In this paper we present an alternative approach to discrete wavelet transforms based on the methodology of linear algebra and matrix computations. Our aim is to show that they can be studied, characterized and constructed using the techniques of matrix calculus, often quite independently of the concepts involved in the continuous transforms.

In Section 2, we start with restating some basic results related to multiresolution analysis. We consider the case of a general integer dilation parameter (*i.e.*, possibly more than one wavelet function for one scaling function). While skipping most of the details we concentrate on those aspects of the projections from higher to lower resolution subspaces which reveal that the underlying linear operator is given by a matrix of a special structure, a block circulant matrix.

In Section 3, we discuss the orthogonal block circulant matrices that correspond to orthonormal wavelets with compact support. These matrices are generated by a finite block sequence (called a wavelet matrix), the length of which is independent of the size of the block circulant matrix. We show that under a special product all wavelet matrices can be factorized into elementary linear factors based on symmetric projectors.

The next section, Section 4, then deals with biorthogonal wavelets, which also have a group structure under an extension to the above product. While we successfully generalize the concept of linear factors (based on general projectors

Wavelets: Theory, Algorithms, and Applications
Charles K. Chui, Laura Montefusco, and Luigia Puccio (eds.), pp. 117–135.

ISBN 0-12-174575-9

here) we show that similar factorization results do not hold for biorthogonal matrices. However, we introduce a biorthogonal wavelet matrix of a special type that can be characterized by nilpotent block companion-like matrices and show that any biorthogonal wavelet matrix is a product of this special one and an orthogonal one.

In Section 5 we turn our attention to the regularity of the first row (or rows, in the biorthogonal case) and the vanishing of the discrete moments of all other rows of the wavelet matrix. It is a rather subtle fact that these two properties are, due to the orthogonality, equivalent. We prove this by direct algebraic means using, for higher regularity, the concept of "regularity vectors."

Then, in Section 6, we return to orthogonal wavelets and deal with the problem of completing the wavelet matrix when its first row is given. We present a simple algorithm for such a construction and show that it provides for all possible solutions. In fact, there are infinitely many solutions for values of the dilation parameter higher than 2.

Finally, in Section 7, we discuss the characterization of compactly supported symmetric (higher multiplicity) wavelets. We show there that the corresponding wavelet matrices, which have symmetric and antisymmetric rows, can be built up from simple factors based on centrosymmetric orthogonal matrices.

§2 Towards matrices

Let us consider an integer dilation parameter $m \geq 2$. A scaling function $\phi \in L^2(\mathbb{R})$ generating a multiresolution analysis [4,12] then satisfies a dilation equation of the form

$$\phi(x) = \sqrt{m} \sum_{k \in \mathbf{Z}} h_k \phi(mx - k) \tag{1}$$

and there are $m-1$ wavelet functions $\psi^{(s)}$, $s = 1, \ldots, m-1$, which are defined as some special linear combinations of the scaling function,

$$\psi^{(s)}(x) = \sqrt{m} \sum_{k \in \mathbf{Z}} g_k^{(s)} \phi(mx - k). \tag{2}$$

Let f be an arbitrary function from $L^2(\mathbb{R})$ and let us denote

$$f_{j,k}^{(0)} = \langle f, m^{j/2} \phi(m^j x - k) \rangle, \quad f_{j,k}^{(s)} = \langle f, m^{j/2} \psi^{(s)}(m^j x - k) \rangle, \; s = 1, \ldots, m-1.$$

Then

$$f_{j-1,l}^{(0)} = \sum_{k \in \mathbf{Z}} f_{j,k}^{(0)} h_{k-ml}, \qquad f_{j-1,l}^{(s)} = \sum_{k \in \mathbf{Z}} f_{j,k}^{(0)} g_{k-ml}^{(s)}, \; s = 1, \ldots, m-1.$$

Recursive applications of this formulae (with respect to j) are known as discrete (or fast) higher multiplicity wavelet transform. The transform is invertible,

$$f_{j,k}^{(0)} = \sum_{l \in \mathbf{Z}} f_{j-1,l}^{(0)} h_{k-ml} + \sum_{s=1}^{m-1} \sum_{l \in \mathbf{Z}} f_{j-1,l}^{(s)} g_{k-ml}^{(s)}.$$

In practice, some j_0 is usually chosen and a discrete signal is identified with the sequence $\{f_{j_0,k}^{(0)}\}_{k\in\mathbf{Z}}$. However, the finite signals have to be first extended to an infinite sequence. A standard solution is a periodic extension, which corresponds to the use of periodized wavelets, [4], and f being square integrable on an interval. When the sequence $\{f_{j,k}^{(0)}\}_{k\in\mathbf{Z}}$ is mN-periodic, the sequences $\{f_{j-1,k}^{(s)}\}_{k\in\mathbf{Z}}$, $s = 0, \ldots, m-1$ are N-periodic and for the data within one period we have

$$\boldsymbol{f}_{j-1} = C_N(A)\boldsymbol{f}_j^{(0)},$$

where

$$\boldsymbol{f}_j^{(0)} = \begin{pmatrix} f_{j,0}^{(0)} & f_{j,1}^{(0)} & \cdots & f_{j,mN-1}^{(0)} \end{pmatrix}^T,$$

$$\boldsymbol{f}_{j-1} = \begin{pmatrix} f_{j-1,0}^{(0)} & f_{j-1,0}^{(1)} \cdots & f_{j-1,0}^{(m-1)} & f_{j-1,1}^{(0)} & f_{j-1,1}^{(1)} & \cdots & f_{j-1,N-1}^{(m-1)} \end{pmatrix}^T$$

and $C_N(A)$ is the $mN \times mN$ block circulant matrix generated by the *wavelet matrix* $A = (\cdots \quad A_0 \quad A_1 \quad A_2 \quad \cdots)$,

$$A_n = \begin{pmatrix} h_{nm} & h_{nm+1} & \cdots & h_{nm+m-1} \\ g_{nm}^{(1)} & g_{nm+1}^{(1)} & \cdots & g_{nm+m-1}^{(1)} \\ \vdots & \vdots & \cdots & \vdots \\ g_{nm}^{(m-1)} & g_{nm+1}^{(m-1)} & \cdots & g_{nm+m-1}^{(m-1)} \end{pmatrix};$$

$$C_N(A) = \begin{pmatrix} \sum_k A_{kN} & \sum_k A_{1+kN} & \cdots & \sum_k A_{N-1+kN} \\ \sum_k A_{N-1+kN} & \sum_k A_{kN} & \cdots & \sum_k A_{N-2+kN} \\ \vdots & \vdots & \ddots & \vdots \\ \sum_k A_{1+kN} & \sum_k A_{2+kN} & \cdots & \sum_k A_{kN} \end{pmatrix}.$$

For the reconstruction we then have

$$\boldsymbol{f}_j^{(0)} = C_N(A)^T \boldsymbol{f}_{j-1},$$

which shows that an orthogonal transform is applied on each level of recursion. Indeed, the orthogonality of the integer translates of scaling and wavelet functions implies

$$\sum_{l\in\mathbf{Z}} A_l A_{k+l}^T = \delta_{0k} I, \qquad k \in \mathbf{Z}, \tag{3}$$

(δ_{0k} being the Kronecker delta), which guarantees that $C_N(A)$ is orthogonal independently of the value of N.

On the other hand, if a wavelet matrix A is given such that it contains only a finite number of nonzero blocks, satisfies the conditions above and the sum of the elements in the first row is $\sqrt{m}$, Equations (1) and (2) can be solved. The translates and dilates of the wavelet functions then form at least a tight frame; in almost all cases this frame turns out to be an orthonormal basis [4,11]. Thus the orthogonal block circulant matrices or the wavelet matrices which generate them can be used to study compactly supported higher multiplicity wavelets simply by means of linear algebra.

§3 Factorizations of compactly supported orthogonal wavelets

In this section our aim is to describe all matrices which can give rise to orthogonal wavelets with compact support. The wavelet matrix A must then contain only a finite number of nonzero blocks; without loss of generality we can suppose that $A_k = 0$ whenever $k < 0$ or $k \geq p$ for some p and write $A = (\, A_0 \quad A_1 \quad \cdots \quad A_{p-1}\,)$. The out-of-the-range blocks will be automatically considered to be zero. We will also identify the matrix A and its extension constructed from A by adding one or more $m \times m$ zero blocks to either side. We can thus add or discard zero blocks at either end and renumber blocks when desirable.

Let us first concentrate on the shifted orthogonality conditions (3). One can enforce the group structure to the set of the matrices conforming to Equations (3) using, for example, the following product.

Definition 1. *Let* $A = (\, A_0 \quad A_1 \quad \cdots \quad A_{p-1}\,)$, $B = (\, B_0 \quad B_1 \quad \cdots \quad B_{q-1}\,)$. *We define*

$$A \square B = (\, D_0 \quad D_1 \quad \cdots \quad D_{p+q-2}\,), \qquad D_j = \sum_k A_k B_{j-k}.$$

The product $\square$ is a polynomial-like product of sequences of square blocks. It is motivated by a simple fact that if orthogonal block circulant matrices are multiplied, the result is an orthogonal block circulant matrix; $C_N(A \square B) = C_N(A) C_N(B)$. Thus we see that the matrices conforming to Equation (3) indeed form a group under this product, the unit element being the identity matrix of order m and the element inverse to A having the form

$$A^{inv} = (\, A_{p-1}^T \quad A_{p-2}^T \quad \cdots \quad A_0^T\,). \tag{4}$$

The $\square$ product of p and q-block matrix comprises, generally, of $p + q - 1$ nonzero blocks. However, in some particular cases, the extreme blocks can vanish and the product is then shorter. For instance, let us choose an arbitrary symmetric projector P ($PP = P$, $P^T = P$) such that

$$PA_0^T = A_0^T, \quad PA_{p-1}^T = 0. \tag{5}$$

Such a projector always exists, because Conditions (3) imply $A_0 A_{p-1}^T = 0$; we can take, for example, $P_{p-1} = UU^T$, where U is a matrix the columns of which form an orthonormal basis for the range of A_0^T. The actual length of $C = A \square (I - P \quad P)$ is then only $p - 1$ blocks. If we apply this observation recursively, we obtain the following theorem.

Theorem 2. *Matrix* $A = (\, A_0 \quad A_1 \quad \cdots \quad A_{p-1}\,)$ *satisfies the shifted orthogonality Conditions* (3) *if and only if there exist symmetric projectors* P_j, $j = 1, \ldots, p-1$, *and an orthogonal matrix* H *such that*

$$A = H \square (\, P_1 \quad I - P_1\,) \square (\, P_2 \quad I - P_2\,) \square \cdots \square (\, P_{p-1} \quad I - P_{p-1}\,). \tag{6}$$

Let us now, for a while, turn our attention to the first factor in Equation (6), an orthogonal matrix H, which plays rather a special role in the factorization. If H_A denotes the *characteristic matrix* of A,

$$H_A = \sum_j A_j,$$

then $H_B \square H_F = H_{B\square F}$ for any B and F, and therefore, in Equation (6), $H = H_A$. This is very useful when we have in mind the other necessary condition for the existence of a wavelet frame, the one that the sum of the elements in the first row of A is to be equal to $\sqrt{m}$. Using the concept of the characteristic matrix we can rewrite it as

$$\boldsymbol{e}_1^T H_A \mathbf{1} = \sqrt{m};$$

here and hereafter $\boldsymbol{e}_j$ stands for the jth column of the identity matrix and $\mathbf{1}$ for the vector of ones, both of the appropriate size. If A satisfies the shifted orthogonality Conditions (3), the characteristic matrix H_A is orthogonal. The Cauchy inequality then gives us

$$\sqrt{m} = |\boldsymbol{e}_1^T H_A \mathbf{1}| \le \|H_A^T \boldsymbol{e}_1\| \, \|\mathbf{1}\| = \sqrt{m}.$$

Since the equality holds if and only if the vectors are linearly dependent, the first row of the characteristic matrix must be a multiple of $\mathbf{1}^T$, in particular

$$\boldsymbol{e}_1^T H_A = (1/\sqrt{m})\,\mathbf{1}^T.$$

Thus, if a wavelet matrix is to be constructed from the factors, the first one, a square orthogonal matrix, must have its first row equal to $(1/\sqrt{m})\,\mathbf{1}^T$.

Often one would like to parametrize the wavelet matrices. Then the question arises, whether the factorization in Theorem 2 is unique. A more thorough inspection shows that $C = A\square(\,I-P \quad P\,)$, where P is a symmetric projector, is shorter then A if and only if Conditions (5) hold. Thus, if $\operatorname{rank} A_0 + \operatorname{rank} A_{p-1} = m$ (A is *nondeficient*), the projector and consequently also the linear factor are determined uniquely. Furthermore C is then also nondeficient, which implies that the other factors are also unique.

If a unique factorization is required for *deficient* wavelet matrices (*i.e.*, when $\operatorname{rank} A_0 + \operatorname{rank} A_{p-1} < m$), one has to impose some additional conditions. For example, one can decide to choose in the process of factorizing always the projector with the lowest possible rank (which is unique). It is possible to show that such a choice leads to a sequence of projectors, in which

$$\operatorname{rank} P_{i-1} \ge \operatorname{rank} P_i, \tag{7}$$

$$\operatorname{rank} P_{i-1}P_i = \operatorname{rank} P_i, \quad i = 2,\ldots,p-1. \tag{8}$$

The nondeficient wavelet matrices are just a special case here, where the ranks of all projectors are equal.

Now, let a sequence of projectors satisfying Relations (7) and (8) and an orthogonal matrix H be given. Then, if one factorizes the matrix A defined by Equation (6), choosing the lowest rank projectors one obtains exactly the same sequence of projectors as the one he started with. Thus the following theorem holds.

Theorem 3. *There is a one to one correspondence between sequences of $p-1$ symmetric projectors satisfying Conditions* (7) *and* (8) *and matrices satisfying Conditions* (3) *comprising of p square blocks with a given characteristic matrix.*

There are several factorizations similar to the one we have described here. Their particular features make them more or less suitable for different applications. For example, Vaidyanathan (see, *e.g.*, [18]) describes a factorization where only the linear factors of the type $(P \quad I-P)$ where P is a symmetric projector *of rank one* are used. However, to have simpler basic building blocks one has to sacrifice one pleasant property—the number of factors is no more related in some straigthforward way to the length of the wavelet matrix and, generally, there are more than $p-1$ linear factors. Another way is to use some more elaborate products. While the Pollen product, which preserves the characteristic matrix (see, *e.g.*, [13,6]), presents basically only cosmetic changes (see [10]), the shift products from [17] can lead to completely different factorizations. The 0-shift product coincides with the $\square$ product presented here, but the nonzero shifts permit to use as the basic building blocks $m \times m$ orthogonal matrices. Furthermore, when the uniqueness of such a factorization is studied, the conditions analogous to Conditions (7) and (8) are simpler in a sense that they do not contain any constraints on relations between factors. Similar factorizations are again known from engineering circles; Vaidyanathan (see [18]) mentions one using the Householder reflections, unfortunately it has the same disadvantage as the factorization from the same source we mentioned earlier.

§4 Biorthogonal wavelet transforms

A biorthogonal wavelet transform of a discrete periodic signal can be also described as repeated multiplications by some block circulant matrices $C_N(A)$. However, unlike of the case of orthogonal wavelets, these matrices are not orthogonal here but only invertible for all N, the inverse being $C_N(\tilde{A})^T$ generated by some $\tilde{A} = (\cdots \tilde{A}_0 \quad \tilde{A}_1 \quad \tilde{A}_2 \quad \cdots)$. This is guaranteed by the shifted orthogonality conditions analogous to Equation (3),

$$\sum_{l \in \mathbf{Z}} A_l \tilde{A}_{l+k}^T = \delta_{0k} I, \qquad k \in \mathbf{Z}. \tag{9}$$

We will concentrate here on the case where all the basis functions are compactly supported, that is, both A and $\tilde{A}$ contain only a finite number of nonzero blocks. The necessary and sufficient conditions on A and $\tilde{A}$ then can be derived that secure that the two-scale equations analogous to Equations (1) and (2) have solutions and dual Riesz bases for $L^2(\mathbb{R})$ can be developed; for the case $m = 2$, see [1] or [2]. (By the way, matrices and their eigenvalues appear in the conditions presented there.) We will study here slightly different problems: What must A satisfy so as $\tilde{A}$ satisfying Conditions (9) contained only a finite number of nonzero blocks? How to construct such pairs? Can they be factorized in some way?

We can succesfully generalize the concepts from the previous section, but we have to be rather careful. When cancelling or adding zero square blocks or renumbering blocks, we have to do the same thing with both A and $\tilde{A}$, because the mutual position of the blocks of A and $\tilde{A}$ is important. Without loss of generality, we will suppose that $A_j = \tilde{A}_j = 0$, for $j < 0$, and we will describe the exact extent of nonzero blocks in the following way.

Definition 4. *We say that*

$$\mathcal{A} = \left\{ \begin{matrix} A \\ \tilde{A} \end{matrix} \right\}$$

is an (l,p,k,q)-biorthogonal pair (or a biorthogonal pair of type (l,p,k,q)) *if*

- $A_j = 0$ *for* $0 \le j < l$, $j \ge l+p$, $A_l \ne 0$ *and* $A_{l+p-1} \ne 0$,
- $\tilde{A}_j = 0$ *for* $0 \le j < k$, $j \ge k+q$, $\tilde{A}_k \ne 0$ *and* $\tilde{A}_{k+q-1} \ne 0$.

Due to our convention about cancelling zero extreme blocks, we can assume that at least one of A_0 and $\tilde{A}_0$ is nonzero, that is, $\min(l,k) = 0$. The total length of the pair, thus, is $r = \max(l+p, k+q)$ blocks.

It is not difficult to observe that if A and $\tilde{A}$, resp. B and $\tilde{B}$, form a biorthogonal pair, then $A \square B$ and $\tilde{A} \square \tilde{B}$ are also biorthogonal counterparts. Thus, we can extend the definition of the $\square$ product to biorthogonal pairs in the following way:

$$\left\{ \begin{matrix} A \\ \tilde{A} \end{matrix} \right\} \square \left\{ \begin{matrix} B \\ \tilde{B} \end{matrix} \right\} = \left\{ \begin{matrix} A \square B \\ \tilde{A} \square \tilde{B} \end{matrix} \right\}.$$

Similarly as in the case of orthogonal wavelets, the biorthogonal pairs form a group under this product. The unit element is the $(0,1,0,1)$-biorthogonal pair formed by two identity matrices; the element inverse to

$$\mathcal{A} = \left\{ \begin{matrix} (& A_0 & A_1 & \dots & A_r &) \\ (& \tilde{A}_0 & \tilde{A}_1 & \dots & \tilde{A}_r &) \end{matrix} \right\}$$

is then

$$\mathcal{A}^{inv} = \left\{ \begin{matrix} (& \tilde{A}_r^T & \tilde{A}_{r-1}^T & \dots & \tilde{A}_0^T &) \\ (& A_r^T & A_{r-1}^T & \dots & A_0^T &) \end{matrix} \right\}$$

(compare (4)). We can form an *orthogonal pair* from a matrix A that satisfies the shifted orthogonality conditions (3) by setting $\tilde{A} = A$. These orthogonal pairs then, naturally, form a subgroup of the group of biorthogonal pairs under the (extended) $\square$ product.

We can also generalize the concept of characteristic matrix. It follows from the shifted orthogonality conditions (9) that $H_A = \sum_j A_j$ is invertible and $H_A^{-1} = H_{\tilde{A}}^T = \sum_j \tilde{A}_j^T$. Thus, we can form the *characteristic pair* of $\mathcal{A}$,

$$\mathcal{H}_{\mathcal{A}} = \left\{ \begin{matrix} H_A \\ H_{\tilde{A}} \end{matrix} \right\} = \left\{ \begin{matrix} H_A \\ H_A^{-T} \end{matrix} \right\}.$$

The linear factors can be generalized as well. We simply replace an orthogonal matrix by a matrix which is merely invertible and use a general projector instead of a symmetric one. The resulting biorthogonal linear factor then has the form

$$\mathcal{A} = \left\{ \begin{matrix} H \\ H^{-T} \end{matrix} \right\} \square \left\{ \begin{matrix} (& P & I - P &) \\ (& P^T & I - P^T &) \end{matrix} \right\}$$

(H is nonsingular, P a projector).

It is not too difficult to prove that, in fact, any $(0,2,0,2)$-biorthogonal pair has the form of a linear factor. Unfortunately, the analogy with orthogonal pairs ends here. The biorthogonal linear factors can be used to construct longer biorthogonal pairs—a product of $r-1$ such factors is, generally, a biorthogonal pair of total length r. However, one can find biorthogonal pairs of total length r that can not be decomposed into a product of the characteristic pair and $r-1$ linear factors. To show this, it is enough to find a biorthogonal pair $\mathcal{A}$ such that $\mathcal{A}\square\mathcal{F}$ is at least as long as $\mathcal{A}$ for any biorthogonal linear factor $\mathcal{F}$. This happens, for example, when $\mathcal{R}(A_0^T) \cap \mathcal{R}(A_{r-1}^T) \neq \{0\}$ or $\mathcal{R}(\tilde{A}_0^T) \cap \mathcal{R}(\tilde{A}_{r-1}^T) \neq \{0\}$ ($\mathcal{R}$ denotes the range of the matrix, *i.e.*, the subspace generated by its columns). Here is an example of a biorthogonal pair with common directions in the extreme blocks of the upper matrix, which thus can not be factorized in the way described earlier.

$$\left\{ \begin{matrix} \frac{1}{2} \begin{pmatrix} 0 & 2 & 1 & 0 & 1 & -2 & -2 & 0 & 0 & 1 & 3 & 1 & 0 & 0 & -1 & 0 \\ 0 & 0 & -1 & 0 & 1 & 0 & 0 & 0 & 0 & 0 & 0 & 0 & 0 & 0 & 0 & 0 \\ 0 & 1 & 0 & 0 & 1 & 0 & 0 & 0 & 0 & -1 & -1 & -1 & 0 & 0 & 1 & 0 \\ 0 & 1 & 0 & 0 & 1 & -2 & -2 & 0 & 0 & 0 & 2 & 0 & 0 & 0 & 0 & 0 \end{pmatrix} \\ \\ \frac{1}{2} \begin{pmatrix} 0 & 2 & 0 & 0 & 0 & -2 & 2 & -1 & 2 & 1 & -1 & 2 & -1 & 0 & 0 & 0 \\ 0 & -2 & 0 & 0 & 4 & 2 & -2 & -1 & -2 & 1 & -1 & 2 & -1 & 0 & 0 & 0 \\ 0 & 2 & 0 & 0 & 0 & -2 & 2 & -1 & 2 & 1 & -1 & -2 & -1 & 0 & 0 & 0 \\ 0 & -2 & 0 & 0 & 0 & 2 & -2 & 3 & -2 & -3 & 3 & -2 & 3 & 0 & 0 & 0 \end{pmatrix} \end{matrix} \right\}$$

Since a simple complete characterization of biorthogonal pairs by factorization into biorthogonal linear factors is not possible, we shall adopt the "divide and conquer" strategy—we will factorize biorthogonal pairs into the product of an orthogonal pair and a biorthogonal pair in a special form.

Each biorthogonal pair $\mathcal{A}$ of type (l,p,k,q) has four extreme blocks A_l, A_{l+p-1}, $\tilde{A}_k$, and $\tilde{A}_{q+k-1}$. Generally, $A_l\tilde{A}_{q+k-1}^T = 0 = A_{l+p-1}\tilde{A}_k^T$. Following the same path as in the case of orthogonal wavelets, we can construct an *orthogonal* linear factor $\mathcal{F}$, based on a symmetric projector, exploiting, say, the second of these orthogonalities of the extreme blocks, which holds whenever $(l,p,k,q) \neq (0,p,p-1,q)$. This leads to vanishing of two extreme blocks in the product $\mathcal{A}\square\mathcal{F}$; the other two extreme blocks generally do not vanish but, in some special cases, may. Considering all such possibilities we arrive at the following result.

Lemma 5. *Let $\mathcal{A}$ be a biorthogonal pair of type $(l,p,k,q) \neq (0,p,p-1,q)$. Then there exists an orthogonal linear factor $\mathcal{F}$ such that $\mathcal{B} = \mathcal{A}\square\mathcal{F}$ is of the type $(\hat{l},\hat{p},\hat{k},\hat{q})$, where $\hat{q} \leq q$ and exactly one of the following three possibilities* occurs:

$$\begin{aligned} &\hat{l}=l, \quad \hat{p}=p, \quad \hat{k}=k+1 \quad or \\ &\hat{l}=l-1, \quad \hat{p}=p, \quad \hat{k}=k \quad or \\ &\hat{l}=l, \quad \hat{p}=p-1, \quad \hat{k}=k. \end{aligned} \tag{10}$$

The only case when we can not apply the previous lemma is when the type of the biorthogonal pair is $(0, p, p-1, q)$ for some p and q, because then $A_{p-1}\tilde{A}_{p-1}^T = I \neq 0$. We will call such a special biorthogonal pair a *biorthogonal atom* (or more precisely (p,q)-atom or atom of type (p,q)). When $A_{p-1} = \tilde{A}_{p-1} = I$, we will say that the atom is normalized. Obviously, if $\mathcal{A}$ is an atom then

$$\left\{ \begin{matrix} A_{p-1}^{-1} \\ A_{p-1}^T \end{matrix} \right\} \square \mathcal{A}, \qquad \mathcal{A} \square \left\{ \begin{matrix} A_{p-1}^{-1} \\ A_{p-1}^T \end{matrix} \right\}$$

are normalized atoms.

Note that in all three cases in Condition (10)

$$\hat{l} + \hat{p} - \hat{k} = l + p - k - 1.$$

Furthermore, type (l, p, k, q) equals $(0, p, p-1, q)$ if and only if $l + p - k = 1$. Thus, by repeated application of Lemma 5, we obtain the next theorem.

Theorem 6. *Every (l, p, k, q)-biorthogonal pair can be factorized into the product of a $(0, 1, 0, 1)$-biorthogonal pair, a normalized atom and $l + p - k - 1$ orthogonal linear factors.*

Note that if an atom is orthogonal, it is necessarily of type $(0, 1, 0, 1)$. Thus the atom really represents the nonorthogonality of the pair and the factorization above can be viewed as the decomposition of the biorthogonal pair into its orthogonal and nonorthogonal part. Let us have a look now, how a normalized atom can look like.

Suppose that $A = (\, A_0 \quad A_1 \quad \cdots \quad A_{p-2} \quad I\,)$ is given. The question is, when a biorthogonal counterpart of the form $\tilde{A} = (\, I \quad \tilde{A}_p \quad \cdots \quad \tilde{A}_{p+q-2}\,)$ exists.

The special shape of atom allows us to solve Equations (9) sequentially for $\tilde{A}_p$, $\tilde{A}_{p+1}$, ...; particularly, $\tilde{A}_n$ is given by a matrix difference equation

$$\tilde{A}_n^T = -(A_{p-2}\tilde{A}_{n-1}^T + A_{p-3}\tilde{A}_{n-2}^T + \cdots + A_0\tilde{A}_{n-p+1}^T),$$

$$\tilde{A}_{p-1} = I, \qquad \tilde{A}_{p-2} = \cdots = \tilde{A}_1 = 0.$$

If we denote $Y_n = (\, \tilde{A}_{n+p-1} \quad \tilde{A}_{n+p-2} \quad \ldots \quad \tilde{A}_{n+1}\,)^T$, the equation above can be rewritten as

$$Y_n = GY_{n-1}, \quad Y_0 = E_1,$$

where G is a block companion–like matrix,

$$G = \begin{pmatrix} -A_{p-2} & -A_{p-3} & -A_{p-4} & \cdots & -A_1 & -A_0 \\ I & 0 & 0 & \cdots & 0 & 0 \\ 0 & I & 0 & \cdots & 0 & 0 \\ \vdots & \vdots & \ddots & \ddots & \vdots & \vdots \\ 0 & 0 & 0 & \cdots & I & 0 \end{pmatrix}, \tag{11}$$

and $E_1 = (\, I \quad 0 \quad \cdots \quad 0\,)^T$. The general solution is

$$Y_n = G^n Y_0 = G^n E_1,$$

and it is not too difficult to prove that $\tilde{A}_j$ vanishes for all $j > p+q-2$ if and only if $G^{p+q-2} = 0 \neq G^{p+q-3}$. Thus, we have the following theorem.

Theorem 7. *Let $A = (\, A_0 \quad A_1 \quad \cdots \quad A_{p-2} \quad I\,)$. Then there exists $\tilde{A}$ such that the biorthogonal pair $\mathcal{A}$ is a normalized (p,q)-atom if and only if the matrix G defined by Equation* (11) *is nilpotent with index $p+2-2$.*

In the context of perfect reconstruction filter banks Vetterli has shown that for any A with a finite number of nonzero blocks there exists $\tilde{A}$ containing only a finite number of nonzero blocks and conforming to Equations (9) if and only if the determinant of $\sum_j A_j \lambda^j$ is a monomial in λ. The theorem we have just presented is consistent with this result, because, for G to be nilpotent, its characteristic polynomial must be equal to $\lambda^{m(p-1)}$, and we have

$$\det(\lambda I - G) = \det \left(\sum_{j=0}^{p-1} A_j \lambda^j \right) = \lambda^{m(p-1)}.$$

However, the information about the size of the atom (*i.e.*, the index of nilpotency of G) is lost.

Theorem 7 allows us to draw some conclusions about the possible types of atoms. Clearly, there are no (p,q)-atoms with $q > (m-1)p - m + 2$. For $p = 2$, $q = (m-1)p - m + 2 = m$ can be reached (A_0 being similar to the $m \times m$ Jordan block with eigenvalue 0). In the case of $p > 2$ this is not that clear because of the special structure of G. Using Lemma 5 and the above consideration, one can also derive bounds on the type of general biorthogonal pairs [8].

Atoms form a subgroup of the group of biorthogonal pairs under the product $\square$. For further characterization of atoms one can try to build them up, again, as products of simple factors, say $(2,q)$-atoms, which are quite easy to construct (A_0 must be nilpotent; $\tilde{A}_j^T = (-1)^{j-1} A_0^{j-1}$). Whether any atom can be factorized in this way remains an open question, but it seems that $(2,2)$-atoms only do not suffice (see [8]).

§5 Vanishing moments and regularity of wavelet matrices

One of the most important properties of wavelets is the vanishing of moments up to some degree. The vanishing of the moments of the wavelet functions is reflected by the vanishing of the discrete moments of all but the first row of A, respectively $\tilde{A}$. This is, then, equivalent to a certain property of the first row of the other matrix, usually called regularity (not to be confused with the regularity of the underlying functions; see [15] for the orthogonal case). Our aim here is to demonstrate that the mentioned equivalence can be shown purely by the means of linear algebra. Although we formulate the property of the first row in a way different from the usual, we will stick with the term regularity and show later how it relates to the standard form.

We denote $D = \operatorname{diag}\{0, 1, \ldots, m-1\}$ and $D_b = \operatorname{diag}\{0, 1, \ldots, mp-1\}$ the diagonal matrices with successive integer values in the diagonal and use them to define the discrete moments of A as follows.

Definition 8. *By $\boldsymbol{\mu}_s$ we denote the vector of degree s moments of the rows of A,*

$$\boldsymbol{\mu}_s = AD_b^s\mathbf{1}, \qquad s = 0, 1, \ldots .$$

We note that, the moments vector of degree 0, $\boldsymbol{\mu}_0$, equals to $H_A\mathbf{1}$, where H_A is the characteristic matrix of A. The vanishing of the 0th moment of all but the first row of A then means that $H_A\mathbf{1} = \beta_0\mathbf{e}_1$ for some constant β_0. Because the shifted orthogonality conditions (9) imply $H_{\tilde{A}}^T H_A = I$, this can be rewritten as

$$H_{\tilde{A}}^T\mathbf{e}_1 = (1/\beta_0)\mathbf{1}. \tag{12}$$

The above simple argument can be extended for higher degrees and higher regularity, although the technical details quickly become quite complicated. We introduce the following new concepts to simplify the presentation, starting with a generalization of characteristic matrices.

Definition 9. *By*

$$H_k = m^k \sum_{j=0}^{p-1} j^k A_j,$$

$k = 0, 1, \ldots$, we denote the generalized k-th characteristic matrix of A. Analogously, $\tilde{H}_k$ will stand for the k-th characteristic matrix of $\tilde{A}$.

These characteristic matrices are not biorthogonal, as $H_0 = H_A$ and $\tilde{H}_0 = H_{\tilde{A}}$, but the following property can be shown directly from their definition.

Lemma 10. *The biorthogonal wavelet A satisfies the shifted orthogonality conditions if and only if the generalized characteristic matrices satisfy*

$$\sum_{k=0}^{s}(-1)^k \binom{s}{k} H_{s-k}\tilde{H}_k^T = \delta_{s0}I, \quad s = 0, 1, \ldots .$$

As the jth diagonal block (of size m) of D_b equals $D + (j-1)mI$, we may express the discrete moments in terms of the generalized characteristic matrices:

$$\boldsymbol{\mu}_s = \sum_{t=0}^{s} \binom{s}{t} H_{s-t} D^t \mathbf{1}.$$

We also need the following vectors.

Definition 11. *Let* $\boldsymbol{\alpha} = (\alpha_0 \quad \alpha_1 \quad \alpha_2 \quad \dots)^T$ *be a vector of the appropriate length. We will call the vectors*

$$\boldsymbol{g}_s(\boldsymbol{\alpha}) = D^s \mathbf{1} - \sum_{t=1}^{s} (-1)^t \binom{s}{t} \alpha_{s-t} \tilde{H}_t^T \boldsymbol{e}_1 \ , \quad s = 0, 1, \dots \ ,$$

the regularity vectors of $\tilde{A}$.

Note that each $\boldsymbol{g}_s$ depends only on the first $s-1$ parameters in $\boldsymbol{\alpha}$. Particularly, $\boldsymbol{g}_0 = \mathbf{1}$ independently of $\boldsymbol{\alpha}$, and Equation (12) can be rewritten as

$$\boldsymbol{g}_0 = \beta_0 \tilde{H}_0^T \boldsymbol{e}_1 .$$

It is this relation which we will generalize to define the regularity.

Definition 12. *If, for some vector* $\boldsymbol{\alpha} = (\alpha_0 \quad \alpha_1 \quad \dots \quad \alpha_r)^T$,

$$\boldsymbol{g}_s(\boldsymbol{\alpha}) = \alpha_s \tilde{H}_0^T \boldsymbol{e}_1 \ , \quad s = 0, 1, \dots, r,$$

we say that $\tilde{A}$ *is* r*-regular.*

Observe that the regularity is the property of the first row only.

Theorem 13. *The moments of all but the first row of* A *up to degree* r *vanish, that is,*

$$\boldsymbol{\mu}_s = \beta_s \boldsymbol{e}_1 \ , \quad s = 0, 1, \dots, r,$$

for some $\boldsymbol{\beta} = (\beta_0 \quad \beta_1 \quad \dots \quad \beta_r)^T$ *if and only if*

$$\boldsymbol{g}_s(\boldsymbol{\beta}) = \beta_s \tilde{H}_0^T \boldsymbol{e}_1 \ , \quad s = 0, 1, \dots, r,$$

i.e., the first row of $\tilde{A}$ *is* r*-regular.*

We discussed the case $r = 0$ earlier. For $r > 0$ the statement can be proved by a fairly straightforward induction in which the following result is used.

Lemma 14. *Let* $r > 0$ *be an integer. If*

$$\boldsymbol{g}_s(\boldsymbol{\alpha}) = \alpha_s \tilde{H}_0^T \boldsymbol{e}_1, \qquad s = 0, 1, \dots, r-1,$$

for some vector $\boldsymbol{\alpha} = (\alpha_0 \quad \dots \quad \alpha_{r-1})^T$ *then*

$$\boldsymbol{\mu}_r = H_0 \boldsymbol{g}_r(\boldsymbol{\alpha}) \ .$$

Proof: One can observe that

$$\boldsymbol{\mu}_r - H_0 \boldsymbol{g}_r = Z \boldsymbol{e}_1 \ ,$$

where

$$Z = \sum_{s=0}^{r-1} \binom{r}{s} H_{r-s} \sum_{t=0}^{s} (-1)^{s-t} \binom{s}{t} \alpha_t \tilde{H}_{s-t}^T + H_0 \sum_{t=0}^{r-1} (-1)^{r-t} \alpha_t \tilde{H}_{r-t}^T \ .$$

Collecting terms at α_t and changing the summation order then gives

$$Z = \sum_{t=0}^{r} (-1)^{r-t} \binom{r}{t} \alpha_t \sum_{s=0}^{r-t} (-1)^s \binom{r-t}{s} H_s \tilde{H}_{r-t-s}^T - \alpha_r I$$

(for any α_r). From the orthogonality conditions of Lemma 10 it follows that $Z = 0$ as required. ∎

One of the classical ways to characterize the regularity of $\tilde{A}$ is to require that "the shifts of the first row span polynomials of a given degree". The next lemma highlights the connection to our approach.

Lemma 15. *Let $\boldsymbol{f}_{sk} = (D + kmI)^s \mathbf{1}$, $k \in \mathbf{Z}$, $s = 0, 1, 2, \ldots$. If $\tilde{A}$ is r-regular then, for $s = 0, 1, \ldots, r$,*

$$\boldsymbol{f}_{sk} = \sum_{j=0}^{p-1} \gamma_{s,k-j} \tilde{A}^t \boldsymbol{e}_1 ,$$

where

$$\gamma_{s,n} = \sum_{t=0}^{s} \binom{s}{t} (-mn)^t \alpha_{s-t} \ .$$

These are explicit formulae, in the terms of the constants $\boldsymbol{\alpha}$ from Definition 12, for the expansion of each monic power s of which $\boldsymbol{f}_{sk}$ is the kth block of length m. The proof is based on calculations similar to those described in the proof of Lemma 14 above.

We also want to mention that, for $m > 2$, some rows of A (or $\tilde{A}$) may have the number of vanishing moments larger. In fact, this can be achieved by a simple transformation (see [7] for details on the orthogonal case).

Theorem 16. *If a biorthogonal pair $\mathcal{A}$ has regularities r of $\tilde{A}$ and $\tilde{r}$ of A, then there exists an orthogonal matrix Q, $Q\boldsymbol{e}_1 = Q^T \boldsymbol{e}_1 = \boldsymbol{e}_1$, such that the biorthogonal pair*

$$\mathcal{B} = \left\{ \begin{matrix} B \\ \tilde{B} \end{matrix} \right\} = \left\{ \begin{matrix} Q^T A \\ Q^{-1} \tilde{A} \end{matrix} \right\}$$

has the same regularities as $\mathcal{A}$ and such that, for $2 \le k \le m$, $r + k - 1$ moments of the k-th row of B vanish.

The kth row of B $(2 \leq k \leq m)$ has $r+k-1$ vanishing moments if and only if the matrix of moments of B

$$\begin{aligned}(\, BD_b^r\mathbf{1} \quad BD_b^{r+1}\mathbf{1} \quad \ldots \quad BD_b^{r+m-1}\mathbf{1}\,) \\ = Q^T\,(\, AD_b^r\mathbf{1} \quad AD_b^{r+1}\mathbf{1} \quad \ldots \quad AD_b^{r+m-1}\mathbf{1}\,)\end{aligned}$$

is upper triangular. This can be achieved by choosing the transformation matrix Q to be an orthogonal matrix arising from the QR-decomposition of the corresponding matrix of the moments of A. Note that such a transformation will preserve the first rows of A and $\tilde{A}$ and hence the regularities r and $\tilde{r}$. Under some circumstances (matrix of moments of $\tilde{B}$ having an LU decomposition without pivoting), one can go further and increase the number of vanishing moments even in the other matrix. However, the transformation then is not orthogonal but only nonsingular and may be badly conditioned.

§6 Construction from the first row

Some properties of wavelets, as for example the minimal number of vanishing moments, are completely determined by the scaling function. Thus, it may make sense to construct the scaling function separately and then ask, how the wavelet functions can look like. This is equivalent to the construction of a wavelet matrix when its first row is given. The question, thus, is, when a vector can be completed to a matrix satisfying the shifted orthogonality conditions (3) and, if it can, how many such solutions exist.

For classical wavelets (multiplicity $m = 2$), the second row of the matrix can be chosen as the first one reversed with every other element multiplied by -1. This solution is unique up to a multiple by -1 and a shift by a multiple of 2. For general m, however, no such simple solution is known.

In [6], Heller, Resnikoff, and Wells proposed obtaining the rest of a matrix conforming to Conditions (3) from the first row and the characteristic matrix as a solution to a certain system of linear equations. They conjectured that their system was always nonsingular and suggested, for larger values of m, the existence of a finite number of different wavelet matrices for the same data. We presented an algorithm for obtaining all possible solutions to the problem based on the Pollen product factorization [10]. Later we derived algorithms also for the shift product factorizations [17]. The possibility to use a factorization was noticed also by Steffen *et al.* [15]. However, they used the Vaidyanathan's factorization and set the number of factors in the construction to be equal to the length (in blocks) of the first row and, thus, restricted themselves to a special class of solutions.

Here, we show an algorithm using the $\square$ product we have defined in Section 3. It is basically identical to the one from [10], because the Pollen product differs from the $\square$ product just by the addition of normalization of characteristic matrix.

It is clear that so as some vector $\boldsymbol{a}^T = (\,\boldsymbol{a}_0^T \quad \boldsymbol{a}_1^T \quad \ldots \quad \boldsymbol{a}_{p-1}^T\,)$, where $\boldsymbol{a}_j$ are m-vectors, could be completed to a matrix conforming to Conditions (3), it has to satisfy at least

$$\sum_{j=0}^{p-1-k} \boldsymbol{a}_j^T \boldsymbol{a}_{j+k} = \delta_{k0}, \quad k = 0, 1, \ldots, p-1. \tag{13}$$

We know that the matrix $A = (\, A_0 \quad A_1 \quad \cdots \quad A_{p-1}\,)$, which we want to construct, can be expressed as the $\square$ product of some matrix B comprising of only $p-1$ square blocks and a linear factor $(\, P \quad I-P\,)$ (P is a symmetric projector). If we know the first row of A and the projector P, we know also the first row of B. In particular, the first row of A is $\boldsymbol{a}^T = (\, \boldsymbol{a}_0^T \quad \boldsymbol{a}_1^T \quad \ldots \quad \boldsymbol{a}_{p-1}^T\,)$ if and only if the first row of B is

$$\boldsymbol{b}^T = (\, \boldsymbol{b}_0^T \quad \boldsymbol{b}_1^T \quad \ldots \quad \boldsymbol{b}_{p-2}^T\,), \quad \boldsymbol{b}_k = (I-P)\boldsymbol{a}_{k+1} + P\boldsymbol{a}_k.$$

In fact, instead of constructing A, we can construct B. This is the same problem as the one we started with, but B is one block shorter. We can repeat this, until we are required to construct just an orthogonal matrix (1-block matrix conforming to Conditions (3)) with a given first row.

The question is, of course, how to choose the projectors. We know, that for B to exist, the vector $\boldsymbol{b}$ must satisfy at least the shifted orthogonality conditions (13). This happens if and only if $\boldsymbol{a}^T$ satisfies these conditions and

$$P\boldsymbol{a}_0 = \boldsymbol{a}_0\,, \quad P\boldsymbol{a}_{p-1} = 0. \tag{14}$$

Thus, a projector satisfying conditions analogous to Conditions (14) must be chosen in each step of the recursion and it is always possible because the shifted orthogonality conditions (analogous to) (13) hold. Finally, in the last step, we need just to construct an orthogonal matrix, the first row of which is some normalized m-vector, which is possible as well. We thus see that the shifted orthogonality conditions (13) are not only necessary, but also sufficient.

We summarize the recursive process in the algorithm below. It follows from the discussion above and Theorem 2 that going through all possible choices in the algorithm, one obtains successively all solutions to the completion problem.

Algorithm 17. *Given: vector* $\boldsymbol{a}^T = (\, \boldsymbol{a}_0^T \quad \boldsymbol{a}_1^T \quad \ldots \quad \boldsymbol{a}_{p-1}^T\,)$ *satisfying the shifted orthogonality conditions* (13).

(a) *Set* $\boldsymbol{x}_k^{p-1} := \boldsymbol{a}_k, \quad k = 0, 1, \ldots, p-1.$

(b) *For* $j = p-1, p-2, \ldots, 1$ *do:*

   (i) *Choose a symmetric projector* P_j *such that* $P_j \boldsymbol{x}_0^j = \boldsymbol{x}_0^j$ and $P_j \boldsymbol{x}_j^j = 0$.

   (ii) *Set* $\boldsymbol{x}_k^{j-1} := P_j \boldsymbol{x}_k^j + (I - P_j)\boldsymbol{x}_{k+1}^j, \quad k = 0, 1 \ldots, j-1.$

(c) *Choose an orthogonal matrix* H *(which will be the characteristic matrix of* A*) such that* $H^T \boldsymbol{e}_1 = \boldsymbol{x}_0^0$.

(d) *Set* $A = H\square(\, P_1 \quad I - P_1\,)\square\cdots\square(\, P_{p-1} \quad I - P_{p-1}\,)$.

Finally, let us discuss the number of possible solutions to the problem for $m > 2$. The freedom in the choice of the projector P conforming to Conditions

(14), corresponds to the freedom in division of a space of dimension $m-2$, $m-1$ or m (depending on if the vectors $\boldsymbol{a}_0$ and $\boldsymbol{a}_{p-1}$ are zero or not) into two subspaces. Therefore, when $m=3$, for each projector, we may have only two orthogonal possibilities (we can choose the rank to be either 1 or 2) and, apart of the choice of the characteristic matrix, the number of the solutions may be finite. However, if $m \geq 4$, there are infinitely many possible choices for each projector. Thus, for any prescribed first row (conforming to Conditions (13)) and characteristic matrix, there are infinitely many possible solutions and, hence, the conjecture of Heller *et al.* about a finite number of solutions is obviously false.

§7 Symmetric orthogonal wavelets

In some applications, symmetry of wavelets is of importance. More precisely, one requires each of the sequences $\{h_k\}_{k\in\mathbf{Z}}$ and $\{g_k^{(s)}\}_{k\in\mathbf{Z}}$, $s=1,\ldots,m-1$, in Equations (1) and (2), respectively, to be symmetric or antisymmetric about some point. It is well-known that except for the trivial case of Haar basis, there are no classical (multiplicity $m=2$) compactly supported symmetric wavelets. Usually, when symmetry is required, biorthogonal wavelets are used, instead. Another possibility, however, is to use higher multiplicity wavelets.

We will consider here only the case when all the rows of the wavelet matrix $A = (\, A_0 \quad A_1 \quad \cdots \quad A_{p-1}\,)$ are symmetric or antisymmetric about the middle of the length of A. This means that $AJ = DA$ for some $m \times m$ diagonal matrix D with diagonal entries 1 or -1; J stands for a peridentity (exchange) matrix of the appropriate size. Using the square blocks we can rewrite this as

$$A_j = DA_{p-1-j}J, \quad j=0,\ldots,p-1. \tag{15}$$

The paraunitary filter banks with such impulse response matrices were studied already by Vetterli and Le Gall [19]. However, they restricted their attention to those having even number of rows, half of them symmetric, the other half antisymmetric. They proposed a way to construct them from orthogonal centrosymmetric matrices. However, the question of completeness of such a characterization remained open.

One can easily observe that the characteristic matrix of A conforming to Equations (15) must satisfy

$$H_A = DH_AJ$$

and therefore, whenever a row of A is symmetric (antisymmetric, resp.), so is the corresponding row of H_A. Thus, because for A satisfying the shifted orthogonality conditions (3), H_A is orthogonal, the assumption of Vetterli and Le Gall is for an even number of rows, in fact, necessary. For odd m, A must, then, have $(m+1)/2$ symmetric and $(m-1)/2$ antisymmetric rows. We have shown in Section 3 that, so as the tight frame (or orthogonal basis) could be developed, the first row of H_A must be equal to a multiple of $\mathbf{1}^T$. Since this vector is symmetric, the first row of A must be symmetric as well.

Another simple observation is that if A satisfies Conditions (15), the matrix $\hat{A} = H_A^T \square A$ is centrosymmetric, *i.e.*,

$$\hat{A}_j = J\hat{A}_{p-1-j}J, \quad j = 0, \dots, p-1.$$

On the other hand, if $\hat{A}$ is centrosymmetric and $H_A = DH_AJ$, then Conditions (15) hold for A. Thus, we see that instead of matrices with symmetric and antisymmetric rows, we can study centrosymmetric matrices. A pleasant fact is that the central symmetry is preserved by the product $\square$.

For a linear factor $(\,P \quad I-P\,)$ to be centrosymmetric, the symmetric projector P must satisfy $I - P = JPJ$. Suppose that m is even and let us express P as $P = UU^T$. Then $I - P = JPJ = JUJ(JUJ)^T$, which means that the matrix $(\,U \quad JUJ\,)$ (which is centrosymmetric) is orthogonal. Thus, the centrosymmetric linear factors can be easily produced from centrosymmetric orthogonal matrices and, actually, all of them can be constructed in this way.

Theorem 18. *A centrosymmetric matrix* $A = (\,A_0 \quad A_1 \quad \cdots \quad A_{p-1}\,)$ *with an even number of rows satisfies the shifted orthogonality conditions* (3) *if and only if there exist centrosymmetric linear factors* $(\,P_j \quad I - P_j\,)$, $j = 1, \dots, p-1$, *(*P_j *are symmetric projectors) and a centrosymmetric orthogonal matrix* H *such that*

$$A = H\square(\,P_1 \quad I - P_1\,)\square(\,P_2 \quad I - P_2\,)\square\cdots\square(\,P_{p-1} \quad I - P_{p-1}\,).$$

To prove the statement one can follow the lines of the proof of Theorem 2. In addition one just have to show that he really can choose the symmetric projector P conforming to Conditions (5) in such a way that $I - P = JPJ$. For the detailed proof, see [16].

A factorization into the $m/2$-shift product of centrosymmetric orthogonal matrices can also be derived [16]. Other similar factorizations can be found in [5] and [14].

When m, the number of rows, is odd, the situation is more complicated. The simple fact that $m/2$ is not an integer, causes a series of problems. For example, centrosymmetric linear factors do not exist;

$$A_0A_1^T = A_0JA_0^TJ = 0$$

implies rank $A_0 < m/2$, while

$$A_0A_1^T + A_0A_1^T = A_0A_0^T + JA_0A_0^TJ = I$$

requires rank $A_0 \geq m/2$. Nevertheless, a complete characterization similar to the one for m even is still possible.

If one takes two orthogonal centrosymmetric matrices with the same central column, let us say $(\,U_1 \quad \boldsymbol{v} \quad JU_1J\,)$ and $(\,U_2 \quad \boldsymbol{v} \quad JU_2J\,)$ ($\boldsymbol{v}$ is symmetric), then the $\square$ product of linear factors W_1 and W_2,

$$W_1 = (\,U_1U_1^T + \boldsymbol{v}\boldsymbol{v}^T \quad JU_1U_1^TJ\,) \qquad W_2 = (\,U_2U_2^T \quad JU_2U_2^TJ + \boldsymbol{v}\boldsymbol{v}^T\,),$$

is centrosymmetric. The following theorem then holds.

Theorem 19. *A centrosymmetric matrix* $A = (\, A_0 \quad A_1 \quad \cdots \quad A_{p-1}\,)$ *with an odd number of rows satisfies the shifted orthogonality conditions* (3) *if and only if*

$$A = H \square W_1 \square W_2 \square \cdots \square W_{p-1},$$

where, for $j = 1, 2, \ldots, (p-1)/2$,

$$\begin{aligned} W_{2j-1} &= (\, U_{2j-1}U_{2j-1}^T + \boldsymbol{v}_j\boldsymbol{v}_j^T \quad JU_{2j-1}U_{2j-1}^T J\,)\ , \\ W_{2j} &= (\, U_{2j}U_{2j}^T \quad JU_{2j-1}U_{2j-1}^T J + \boldsymbol{v}_j\boldsymbol{v}_j^T\,) \end{aligned}$$

and both $(\, U_{2j-1} \quad \boldsymbol{v}_j \quad JU_{2j-1}J\,)$ *and* $(\, U_{2j} \quad \boldsymbol{v}_j \quad JU_{2j}J\,)$ *are centrosymmetric orthogonal matrices.*

See [16] for the proof.

Note that if m is odd, the number of square blocks of A must be necessarily odd.

Finally, we would like to point out that, for m even, if one has a symmetric vector satisfying the shifted orthogonality conditions (13), one can always complete such a vector to a matrix conforming to Condition (3) with symmetric and antisymmetric rows. The reason is that if $\boldsymbol{x}^j$ in Algorithm 17 is symmetric, one can choose P_j such that $JP_jJ = I - P_j$ and $\boldsymbol{x}^{j-1}$ is then also symmetric. For m odd, the situation is slightly more complicated—the proposed symmetric vector must comprise of odd number of m-vectors. Then one can choose pairs of linear projectors as in Theorem 19, which guarantees $\boldsymbol{x}^j$ to be symmetric for every even j.

Acknowledgments. The second author is supported by the Australian Government as an Overseas Postgraduate Research Scholar.

References

1. Cohen, A., Biorthogonal wavelets, in *Wavelets- A Tutorial in Theory and Applications*, C. K. Chui (ed.), Academic Press, Boston, 1992, 123–152.
2. Cohen, A., I. Daubechies, and J. C. Feauveau, Biorthogonal bases of compactly supported wavelets, *Comm. Pure and Appl. Math.*, 1991, to appear.
3. Daubechies, I., Orthonormal bases of compactly supported wavelets, *Comm. Pure Appl. Math.* **41** (1988), 909–996.
4. Daubechies, I., *Ten Lectures on Wavelets*, SIAM, Philadelphia, 1992.
5. Gopinath, R. A. and C. S. Burrus, Unitary FIR filter banks and symmetry, Rice University, 1992, preprint, submitted to *IEEE Trans. on Circuits and Systems.*
6. Heller, P. N., H. L. Resnikoff, and R. O. Wells, Jr., Wavelet matrices and the representation of discrete functions, in *Wavelets– A Tutorial in Theory and Applications*, C. K. Chui (ed.), Academic Press, Boston, 1992, 15–50.
7. Kautsky, J., An algebraic construction of discrete wavelet transforms, *Applications of Mathematics* **38** (1993), 169–193.

8. Kautsky, J. and R. Turcajová, Discrete biorthogonal wavelet transforms as block circulant matrices, Flinders University, 1993, preprint, submitted to *Linear Algebra and its Applications*.
9. Kautsky, J. and R. Turcajová, Extremal characterization of regular wavelet matrices and an algorithm for their evaluation, in *Software and Algorithms of Numerical Analysis, Proceedings of 10th Summer School*, University of West Bohemia, Plzen, 1993.
10. Kautsky, J. and R. Turcajová, Pollen product factorization and construction of higher multiplicity wavelets, Flinders University, 1993, preprint, to appear in *Linear Algebra and its Applications*.
11. Lawton, W. M., Necessary and sufficient conditions for constructing orthonormal wavelet bases, *J. Math. Phys.* **32** (1991), 57–61.
12. Mallat, S., A theory for multiresolution signal decomposition: the wavelet representation, *IEEE Pattern Anal. and Machine Intell.* **11** (7) (1989), 674–693.
13. Pollen, D., Parametrization of compactly supported wavelets, Technical report, AWARE, Inc., 1989.
14. Soman, A. K., P. P. Vaidyanathan, and T. Q. Nguyen, Linear phase paraunitary filter banks: Theory, factorizations and applications, *IEEE Trans. on Signal Processing*, Special issue on wavelets, 1993, to appear.
15. Steffen, P., P. Heller, R. A. Gopinath, and C. S. Burrus, Theory of regular M-band wavelet bases, Rice University, 1992, preprint,
16. Turcajová, R., Factorizations and construction of linear phase paraunitary filter banks and higher multiplicity wavelets, Flinders University, 1993, preprint, submitted to *Numer. Algorithms*.
17. Turcajová, R. and J. Kautsky, Shift products and factorizations of wavelet matrices, Flinders University, 1993, preprint, submitted to *Numer. Algorithms*.
18. Vaidyanathan, P. P., *Multirate Systems and Filter Banks*, Signal Processing Series, Prentice Hall, Englewood Cliffs, 1992.
19. Vetterli, M. and D. Le Gall, Perfect reconstruction FIR filter banks: Some properties and factorizations, *IEEE Trans. ASSP* **37** (1989), 1057–1071.

Jaroslav Kautsky
School of Information Science and Technology, Flinders University, and Cooperative Research Centre for Sensor, Signal and Information Processing (CSSIP)
GPO Box 2100, Adelaide, SA 5001, Australia.
jarka@maths.flinders.edu.au

Radka Turcajová
School of Information Science and Technology, Flinders University, and Cooperative Research Centre for Sensor, Signal and Information Processing (CSSIP)
GPO Box 2100, Adelaide, SA 5001, Australia.
radka@maths.flinders.edu.au

A Unified Approach to Periodic Wavelets

Gerlind Plonka and Manfred Tasche

Abstract. We sketch a new approach to p–periodic wavelets for general periodic scaling functions. Our method is based on properties of periodic shift–invariant spaces and related bracket products. A special way to construct periodic wavelets is the periodization of a known cardinal multiresolution. Efficient decomposition and reconstruction algorithms using FFT–algorithms are proposed.

§1 Introduction

A theory of periodic wavelets is the basic tool for an investigation of periodic processes in signal processing and numerical analysis. One way to construct periodic wavelets is the periodization of known cardinal wavelets. Meyer [13] was the first to study such periodic multiresolutions (see also Daubechies [8], pp. 304–307). Further, Perrier and Basdevant [14] (see [20] for a different approach) investigated orthogonal periodic spline wavelets defined by periodization of Battle–Lemarié wavelets [1,10]. Recently, the authors [15] considered semiorthogonal periodic spline wavelets which can be obtained by periodization of Chui–Wang wavelets [4,6,7]. On the other hand, there are constructions of periodic wavelets which do not use this periodization technique. Various trigonometric wavelets were studied [5,17,18,19] without using the knowledge about the possible existence of corresponding cardinal multiresolutions.

The aim of this paper is to introduce a unified approach to periodic univariate wavelets and to the corresponding decomposition and reconstruction algorithms based on Fourier technique. It should be stressed that the presented theory does not depend on the cardinal approach, *i.e.*, it is not derived by periodization of a cardinal multiresolution. However, special periodic multiresolutions obtained by periodization are included.

It turns out that similar ideas used for the construction of cardinal wavelets also succeed in the periodic case. The basic tool of our method is the detailed analysis of p–periodic shift–invariant subspaces of the Hilbert space L_p^2 of all

Wavelets: Theory, Algorithms, and Applications
Charles K. Chui, Laura Montefusco, and Luigia Puccio (eds.), pp. 137–151.

ISBN 0-12-174575-9

p–periodic square integrable functions. For $j \in \mathbb{N}_0$, we put $h_j := p/d_j$ with $d_j := 2^j d$ $(d \in \mathbb{N})$. We are especially interested in h_j–shift–invariant spaces $S_j(\varphi_j)$ generated by $h_j\mathbf{Z}$–translations of one function $\varphi_j \in L_p^2$. Our research is influenced by [2,3,9]. In these papers, cardinal shift–invariant spaces in $L^2(\mathbb{R}^d)$ have systematically been studied, and the results have been applied to a new approach to wavelets on $\mathbb{R}^d$.

The main idea in studying cardinal multiresolutions and wavelets is to consider the corresponding problems in the Fourier transformed domain. In case of periodic multiresolutions, we will use the bijective mapping by the finite Fourier transform instead.

The outline of our paper is as follows. In Section 2, we consider p–periodic shift–invariant subspaces of L_p^2, which we describe by their finite Fourier transforms. The scalar product between functions of p–periodic h_j–shift–invariant spaces can be simplified to a finite sum by means of the so-called bracket product, which is closely related to the p–periodic autocorrelation symbol introduced in Section 3. This bracket product is convenient for the description of stable bases of $S_j(\varphi_j)$ $(j \in \mathbb{N}_0)$ as well as for the characterization of orthogonal shift–invariant spaces.

In Section 3, we define a p–periodic multiresolution of L_p^2 by a nested sequence of h_j–shift–invariant spaces $V_j := S_j(\varphi_j)$ $(j \in \mathbb{N}_0)$. The required conditions of a p–periodic multiresolution of L_p^2 and their consequences for φ_j are analyzed in some detail. In Section 4, we introduce the p–periodic wavelet space W_j $(j \in \mathbb{N}_0)$ as orthogonal complement of V_j in V_{j+1}. Periodic wavelets ψ_j are obtained by finding generators for the p–periodic h_j–shift–invariant space W_j. Using the two–scale symbol of φ_j and the periodic autocorrelation symbols of φ_j and φ_{j+1}, we characterize the possible wavelets ψ_j in Theorem 4.4. Further, the close connection between the Fourier transformed two–scale relations of φ_j and ψ_j $(j \in \mathbb{N}_0)$ and the two–scale (2,2)–matrices is discussed. With the assumption of the stability for the bases of V_j and W_j $(j \in \mathbb{N}_0)$, all two-scale symbol matrices are well–conditioned. Lastly, Section 5 is devoted to new efficient decomposition and reconstruction algorithms based on Fourier technique and two–scale symbol matrices. Our wavelet algorithms are very fast, numerically stable, and do not contain truncation errors.

§2 Periodic shift–invariant spaces

Let $p > 0$ and $d \in \mathbb{N}$ be fixed. Put $d_j := 2^j d$, $h_j := p/d_j$ $(j \in \mathbb{N}_0)$. Consider the Hilbert space L_p^2 of all p–periodic, square integrable functions $f : \mathbb{R} \to \mathbb{C}$ with the scalar product

$$\langle f, g \rangle := \frac{1}{p} \int_0^p f(t)\, \overline{g(t)}\, \mathrm{d}t, \quad f, g \in L_p^2 \,,$$

and the related norm $\|\cdot\|$. Introduce the finite Fourier transform of $f \in L_p^2$ by $\mathbf{c}(f) := (c_u(f))_{u=-\infty}^{\infty} \in l^2$ with

$$c_u(f) := \langle f, e^{2\pi i u \cdot /p} \rangle, \quad u \in \mathbf{Z}.$$

For $u \in \mathbf{Z}$ and $f \in L_p^2$, we have

$$c_u(f(\cdot - kh_j)) = \omega_j^{uk}\, c_u(f), \quad k = 0, \ldots, d_j - 1, \tag{2.1}$$

where $\omega_j := \exp(-2\pi i/d_j)$.

A linear subspace S of L_p^2 is called *h_j–shift–invariant,* if for each $f \in S$ all h_j–shifts $f(\cdot - kh_j)$ $(k = 0, \ldots, d_j - 1)$ are contained in S. The *h_j–shift–invariant subspace generated by* $\varphi \in L_p^2$ is defined by

$$S_j(\varphi) := \text{span}\, \{\varphi(\cdot - kh_j) : \; k = 0, \ldots, d_j - 1\}.$$

A useful characterization of the functions of $S_j(\varphi)$ can be given by their finite Fourier transforms.

Lemma 2.1. (see [16]) *Let $\varphi \in L_p^2$ and $j \in \mathbb{N}_0$ be given.*
(i) *We have $f \in S_j(\varphi)$ if and only if*

$$c_u(f) = \hat{a}_{j,u}(f)\, c_u(\varphi), \quad u \in \mathbf{Z},$$

where

$$\hat{a}_{j,u}(f) \in \mathbb{C}, \quad \hat{a}_{j,u}(f) = \hat{a}_{j,u+d_j}(f), \quad u \in \mathbf{Z}.$$

(ii) *Let $f \in S_j(\varphi)$. Then $S_j(f) = S_j(\varphi)$ if and only if*

$$\text{supp}\ \mathbf{c}(f) = \text{supp}\ \mathbf{c}(\varphi)$$

with the support supp $\mathbf{c}(f) := \{u \in \mathbf{Z} : \; c_u(f) \neq 0\}$ *of* $\mathbf{c}(f)$.

For a detailed analysis of periodic shift–invariant spaces we introduce the following notion. Let the *bracket product of level j* $(j \in \mathbb{N}_0)$ be defined for $\mathbf{a} := (a_u)_{u=-\infty}^{\infty}$, $\mathbf{b} := (b_u)_{u=-\infty}^{\infty} \in l^2$ by $[\mathbf{a}, \mathbf{b}]_j := ([\mathbf{a}, \mathbf{b}]_{j,k})_{k=0}^{d_j - 1}$, where

$$[\mathbf{a}, \mathbf{b}]_{j,k} := \sum_{u=-\infty}^{\infty} a_{k+ud_j}\, \overline{b_{k+ud_j}}, \quad k = 0, \ldots, d_j - 1.$$

Then, $[\mathbf{a}, \mathbf{a}]_{j,k} \geq 0$ $(k = 0, \ldots, d_j - 1)$ for $\mathbf{a} \in l^2$. Further, $[\mathbf{a}, \mathbf{a}]_{j,k} = 0$ $(k = 0, \ldots, d_j - 1)$ if and only if $\mathbf{a} = (0)_{u=-\infty}^{\infty}$. By Cauchy–Schwarz inequality, we have for $k = 0, \ldots, d_j - 1$,

$$|[\mathbf{a}, \mathbf{b}]_{j,k}|^2 \leq [\mathbf{a}, \mathbf{a}]_{j,k}\, [\mathbf{b}, \mathbf{b}]_{j,k} \leq \|\mathbf{a}\|_{l^2}^2\, \|\mathbf{b}\|_{l^2}^2 < \infty.$$

The bracket product will be an important tool for the description of shift–invariant spaces as well as for the characterization of their bases.

Lemma 2.2. *Let $\varphi,\ \psi \in L^2_p$ and $j \in \mathbb{N}_0$. Further let $f \in S_j(\varphi)$, $g \in S_j(\psi)$ with*

$$c_u(f) = \hat{a}_{j,u}(f)\, c_u(\varphi), \quad c_u(g) = \hat{b}_{j,u}(g)\, c_u(\psi), \quad u \in \mathbf{Z},$$

be given, where $\hat{a}_{j,u}(f)$, $\hat{b}_{j,u} \in \mathbf{C}$ possess the properties

$$\hat{a}_{j,u}(f) = \hat{a}_{j,u+d_j}(f), \quad \hat{b}_{j,u}(g) = \hat{b}_{j,u+d_j}(g), \quad u \in \mathbf{Z}.$$

Then we have

$$\langle f, g\rangle = \sum_{k=0}^{d_j-1} \hat{a}_{j,k}(f)\, \overline{\hat{b}_{j,k}(g)}\, [\mathbf{c}(\varphi), \mathbf{c}(\psi)]_{j,k}.$$

Proof: From the Parseval identity, it follows that

$$\begin{aligned}\langle f, g\rangle &= \sum_{u=-\infty}^{\infty} c_u(f)\, \overline{c_u(g)} \\ &= \sum_{k=0}^{d_j-1} \sum_{v=-\infty}^{\infty} c_{k+vd_j}(f)\, \overline{c_{k+vd_j}(g)} \\ &= \sum_{k=0}^{d_j-1} a_{j,k}(f)\, \overline{b_{j,k}(g)} \sum_{v=-\infty}^{\infty} c_{k+vd_j}(\varphi)\, \overline{c_{k+vd_j}(\psi)},\end{aligned}$$

and hence the assertion. ∎

As a consequence of Lemma 2.2 we obtain the following corollary.

Corollary 2.3. *Let $\varphi,\ \psi \in L^2_p$ and $j \in \mathbb{N}_0$ be given. Then we have*
(i)

$$\langle \varphi(\cdot - lh_j), \psi\rangle = \sum_{k=0}^{d_j-1} \omega_j^{kl}\, [\mathbf{c}(\varphi), \mathbf{c}(\psi)]_{j,k}, \quad l = 0, \ldots, d_j - 1. \tag{2.2}$$

(ii) $S_j(\varphi) \perp S_j(\psi)$ *if and only if*

$$[\mathbf{c}(\varphi), \mathbf{c}(\psi)]_{j,k} = 0, \quad k = 0, \ldots, d_j - 1.$$

For $\varphi \in L^2_p$, we consider the system $\mathcal{B}_j(\varphi) := \{\varphi(\cdot - lh_j) :\ l = 0, \ldots, d_j - 1\}$. By Equation (2.2), the corresponding Gramian matrix is circulant and reads as follows

$$(\langle \varphi(\cdot - lh_j), \varphi(\cdot - nh_j)\rangle)_{l,n=0}^{d_j-1} = \mathbf{F}_j\, (\mathrm{diag}\ [\mathbf{c}(\varphi), \mathbf{c}(\varphi)]_j)\, \overline{\mathbf{F}}_j \tag{2.3}$$

with the d_j–th Fourier matrix

$$\mathbf{F}_j := (\omega_j^{kl})_{k,l=0}^{d_j-1}\ .$$

Thus, we obtain the following lemma.

Lemma 2.4. *Let $\varphi \in L_p^2$ and $j \in \mathbb{N}_0$ be given.*
(i) *$\mathcal{B}_j(\varphi)$ is a basis of $S_j(\varphi)$ if and only if*

$$[\mathbf{c}(\varphi), \mathbf{c}(\varphi)]_{j,k} > 0, \quad k = 0, \ldots, d_j - 1. \tag{2.4}$$

(ii) *$\mathcal{B}_j(\varphi)$ is an orthonormal basis of $S_j(\varphi)$ if and only if*

$$d_j\,[\mathbf{c}(\varphi), \mathbf{c}(\varphi)]_{j,k} = 1, \quad k = 0, \ldots, d_j - 1.$$

(iii) *If φ satisfies Equation (2.4) and if $\varphi^\star \in L_p^2$ is defined by*

$$c_u(\varphi^\star) := d_j^{-1/2}\,[\mathbf{c}(\varphi), \mathbf{c}(\varphi)]_{j,u_j}^{-1/2}\, c_u(\varphi), \quad u \in \mathbf{Z};\ u_j := u \bmod d_j, \tag{2.5}$$

then $\mathcal{B}_j(\varphi^\star)$ is an orthonormal basis of $S_j(\varphi)$.

Proof:
1. By Equation (2.3), the Gramian matrix related to $\mathcal{B}_j(\varphi)$ is regular if and only if diag $[\mathbf{c}(\varphi), \mathbf{c}(\varphi)]_j$ is regular, *i.e.*, if Relation (2.4) is satisfied.
2. Note that

$$\mathbf{F}_j\,\overline{\mathbf{F}}_j = d_j \mathbf{I}_j$$

with the d_j-th identity matrix $\mathbf{I}_j$. Now, $\mathcal{B}_j(\varphi)$ is an orthonormal basis of $S_j(\varphi)$ if and only if the Gramian matrix of $\mathcal{B}_j(\varphi)$ is equal to $\mathbf{I}_j$. This is true if and only if (ii) holds.
3. Since

$$[\mathbf{c}(\varphi^\star), \mathbf{c}(\varphi^\star)]_{j,k} = d_j^{-1}, \quad k = 0, \ldots, d_j - 1,$$

the Gramian matrix of $\mathcal{B}_j(\varphi^\star)$ is equal to $\mathbf{I}_j$. Hence, $\mathcal{B}_j(\varphi^\star)$ is an orthonormal basis of $S_j(\varphi^\star)$. By Lemma 2.1 (ii) and by Equation (2.5), we have $S_j(\varphi^\star) = S_j(\varphi)$. ■

§3 Periodic multiresolution

For each $j \in \mathbb{N}_0$, we form h_j-shift-invariant subspaces $V_j := S_j(\varphi_j)$ generated by $\varphi_j \in L_p^2$. Put $\varphi_{j,k} := \varphi_j(\cdot - kh_j)$ $(k = 0, \ldots, d_j - 1)$. We say that $\{V_j\}_{j=0}^\infty$ forms a *p-periodic multiresolution*, if the following three conditions are satisfied (compare [12,11,3]):

(M1) $\quad V_j \subset V_{j+1}, \quad j \in \mathbb{N}_0.$

(M2) $\quad \operatorname{clos}\,(\bigcup_{j=0}^{\infty} V_j) = L_p^2.$

(M3) There exist positive constants α, β such that for all $j \in \mathbb{N}_0$ and for any $(a_{j,n})_{n=0}^{d_j-1} \in \mathbb{C}^{d_j}$,

$$\alpha \sum_{n=0}^{d_j-1} |a_{j,n}|^2 \le \| \sum_{n=0}^{d_j-1} a_{j,n}\, d_j^{1/2}\, \varphi_{j,n}\|^2 \le \beta \sum_{n=0}^{d_j-1} |a_{j,n}|^2.$$

By (M3), $\mathcal{B}_j(d_j^{1/2}\varphi_j)$ is a basis of V_j. Furthermore, if the condition (M3) is satisfied, then the system $\{\mathcal{B}_j(d_j^{1/2}\varphi_j) : j \in \mathbb{N}_0\}$ is called *L_p^2–stable.* The h_j–shift–invariant subspace V_j is called *sample space of level j.* A generating function $d_j^{1/2}\varphi_j$ of V_j is the *scaling function* or *generator* of V_j. If all systems $\mathcal{B}_j(d_j^{1/2}\varphi_j)$ are orthonormal bases of V_j ($j \in \mathbb{N}_0$), then we say that $d_j^{1/2}\varphi_j$ is an *orthonormal scaling function of level j.* In this case the constants in condition (M3) read $\alpha = \beta = 1$. Note that dim $V_j = d_j$. Concerning (M2) we observe the following theorem.

Theorem 3.1. (see [16]) *Let $\{V_j\}_{j=0}^{\infty}$ be a nested sequence of h_j–shift–invariant subspaces $V_j := S_j(\varphi_j)$ with $\varphi_j \in L_p^2$. Then we have*

$$\text{clos}\,(\bigcup_{j=0}^{\infty} V_j) = L_p^2$$

if and only if

$$\bigcup_{j=0}^{\infty} \text{supp}\ \mathbf{c}(\varphi_j) = \mathbf{Z}. \tag{3.1}$$

For (M3) the following equivalence is known.

Theorem 3.2. (see [16]) *The system $\{\mathcal{B}_j(d_j^{1/2}\varphi_j) : j \in \mathbb{N}_0\}$ is L_p^2–stable with positive constants α, β if and only if for $n = 0, \ldots, d_j - 1$ and for $j \in \mathbb{N}_0$,*

$$\alpha \leq d_j^2\,[\mathbf{c}(\varphi_j), \mathbf{c}(\varphi_j)]_{j,n} \leq \beta. \tag{3.2}$$

Further, a basis $\mathcal{B}_j(d_j^{1/2}\varphi_j)$ ($j \in \mathbb{N}_0$) is orthonormal if and only if

$$d_j^2\,[\mathbf{c}(\varphi), \mathbf{c}(\varphi)]_{j,n} = 1, \quad n = 0, \ldots, d_j - 1.$$

Remark. By dim $V_j < \infty$, we can find positive constants α_j, β_j satisfying Relation (3.2) in each level $j \in \mathbb{N}$ if Relation (2.4) is assumed. However, for L_p^2-stability, we need that

$$\alpha := \inf\{\alpha_j : j \in \mathbb{N}_0\} > 0, \quad \beta := \sup\{\beta_j : j \in \mathbb{N}_0\} < \infty,$$

i.e., Relation (3.2) sharpens Relation (2.4).

In the following, we assume that Condition (3.2) is satisfied. From (M1), it follows $\varphi_j \in V_{j+1}$, *i.e.*, there exist unique coefficients $\alpha_{j+1,k} \in \mathbb{C}$ ($k = 0, \ldots, d_{j+1} - 1$) such that

$$\varphi_j = \sum_{k=0}^{d_{j+1}-1} \alpha_{j+1,k}\,\varphi_{j+1,k}, \quad j \in \mathbb{N}_0.$$

This is the so-called *two-scale relation* or *refinement equation* of φ_j. Using Equation (2.1), we obtain the Fourier transformed two-scale relation of φ_j ,

$$c_u(\varphi_j) = 2\,A_{j+1}(\omega_{j+1}^u)\,c_u(\varphi_{j+1}), \quad u \in \mathbf{Z}, \tag{3.3}$$

with the *two-scale symbol* or *refinement mask* of φ_j

$$A_{j+1}(z) := \frac{1}{2} \sum_{k=0}^{d_{j+1}-1} \alpha_{j+1,k}\, z^k, \quad z \in \mathcal{T}_{j+1},$$

where $\mathcal{T}_j := \{\omega_j^n : n = 0, \ldots, d_j - 1\}$ denotes the set of all d_j–th complex roots of unity.

If a scaling function $d_j^{1/2}\,\varphi_j$ $(j \in \mathbb{N}_0)$ satisfying Condition (3.2) is given, then an orthonormal basis $\mathcal{B}_j(d_j^{1/2}\,\varphi_j^\star)$ $(j \in \mathbb{N}_0)$ can easily be obtained by the following orthogonalization trick. Let $\varphi_j^\star$ $(j \in \mathbb{N}_0)$ be defined by their Fourier coefficients

$$c_u(\varphi_j^\star) := \frac{1}{d_j\,([\mathbf{c}(\varphi_j), \mathbf{c}(\varphi_j)]_{j,u_j})^{1/2}}\, c_u(\varphi_j), \quad u \in \mathbf{Z},$$

where $u_j := u \bmod d_j$. Then from Lemma 2.4 (iii), it follows that $\mathcal{B}_j(d_j^{1/2}\,\varphi_j^\star)$ is an orthonormal basis of $V_j = S_j(\varphi_j)$ and the relation $d_j^2\,[\mathbf{c}(\varphi_j^\star), \mathbf{c}(\varphi_j^\star)]_{j,n} = 1$ $(n = 0, \ldots, d_j - 1)$ is obvious. Furthermore, the two-scale symbol $A_{j+1}^\star$ satisfying

$$c_u(\varphi_j^\star) = 2\,A_{j+1}^\star(\omega_{j+1}^u)\,c_u(\varphi_{j+1}^\star), \quad u \in \mathbf{Z},$$

is connected with A_{j+1} for $n = 0, \ldots, d_{j+1} - 1$ by

$$A_{j+1}^\star(\omega_{j+1}^n) := 2 \left(\frac{[\mathbf{c}(\varphi_{j+1}), \mathbf{c}(\varphi_{j+1})]_{j+1,n}}{[\mathbf{c}(\varphi_j), \mathbf{c}(\varphi_j)]_{j,n}} \right)^{1/2} A_{j+1}(\omega_{j+1}^n).$$

A different approach to the bracket product $[\mathbf{c}(\varphi_j), \mathbf{c}(\varphi_j)]_j$ can be described by the so-called *p-periodic autocorrelation symbol* of φ_j defined by

$$\Phi_j(z) := \sum_{l=0}^{d_j-1} \langle \varphi_{j,-l}, \varphi_j \rangle\, z^l, \quad z \in \mathcal{T}_j.$$

We observe a close connection between the bracket product $[\mathbf{c}(\varphi_j), \mathbf{c}(\varphi_j)]_j$, the two-scale symbol A_{j+1} and the p-periodic autocorrelation symbols Φ_j and Φ_{j+1}.

Lemma 3.3. *For $j \in \mathbb{N}_0$ and $k = 0, \ldots, d_j - 1$, we have*

$$\Phi_j(\omega_j^k) = d_j\,[\mathbf{c}(\varphi_j), \mathbf{c}(\varphi_j)]_{j,k}, \tag{3.4}$$

$$\Phi_j(z^2) = 2\,|A_{j+1}(z)|^2\,\Phi_{j+1}(z) + 2\,|A_{j+1}(-z)|^2\,\Phi_{j+1}(-z), \quad z \in \mathcal{T}_{j+1}. \tag{3.5}$$

The condition (M3) *is equivalent to*

$$0 < \alpha \le d_j\,\Phi_j(\omega_j^k) \le \beta < \infty, \quad j \in \mathbb{N}_0,\ k = 0, \ldots, d_j - 1. \tag{3.6}$$

Proof: Let $j \in \mathbb{N}_0$. By Equation (3.3), we obtain for $k = 0, \ldots, d_j - 1$,

$$\begin{aligned}[\mathbf{c}(\varphi_j), \mathbf{c}(\varphi_j)]_{j,k} &= \sum_{u=-\infty}^{\infty} |c_{k+ud_{j+1}}(\varphi_j)|^2 + \sum_{u=-\infty}^{\infty} |c_{k+d_j+ud_{j+1}}(\varphi_j)|^2 \\ &= 4\,|A_{j+1}(\omega_{j+1}^k)|^2\,[\mathbf{c}(\varphi_{j+1}), \mathbf{c}(\varphi_{j+1})]_{j+1,k} \\ &\quad + 4\,|A_{j+1}(-\omega_{j+1}^k)|^2[\mathbf{c}(\varphi_{j+1}), \mathbf{c}(\varphi_{j+1})]_{j+1,k+d_j}.\end{aligned}$$

By Equation (2.2), we have for $l = 0, \ldots d_j - 1$,

$$\langle \varphi_{j,-l}, \varphi_j \rangle = \sum_{n=0}^{d_j-1} \omega_j^{-nl}\,[\mathbf{c}(\varphi_j), \mathbf{c}(\varphi_j)]_{j,n},$$

and hence,

$$\Phi_j(z) = \sum_{l,n=0}^{d_j-1} \omega_j^{-nl}\,[\mathbf{c}(\varphi_j), \mathbf{c}(\varphi_j)]_{j,n}\,z^l, \quad z \in \mathcal{T}_j.$$

For $z = \omega_j^k$ ($k = 0, \ldots, d_j - 1$), this yields Equation (3.4) by

$$\sum_{l=0}^{d_j-1} \omega_j^{(k-n)l} = d_j\,\delta_{k,n}.$$

From Equation (3.4), Equation (3.5) follows immediately. ■

§4 Periodic wavelet spaces

We define the *p-periodic wavelet space* W_j *of level* j ($j \in \mathbb{N}_0$) as the orthogonal complement of V_j in V_{j+1}, *i.e.*,

$$W_j := V_{j+1} \ominus V_j, \quad j \in \mathbb{N}_0.$$

Then it follows dim $W_j = d_{j+1} - d_j = d_j$ and the orthogonal sum representation

$$V_{j+1} = V_j \oplus W_j, \quad j \in \mathbb{N}_0. \tag{4.1}$$

By definition, the wavelet spaces W_j ($j \in \mathbb{N}_0$) are mutually orthogonal. Note that $f \in V_{j+1}$ implies $f(\cdot - 2h_{j+1}) = f(\cdot - h_j) \in V_{j+1}$. It can be easily observed that W_j is h_j–shift–invariant.

By (M1)–(M2), we obtain the orthogonal sum decomposition

$$L_p^2 = V_0 \oplus \bigoplus_{j=0}^{\infty} W_j.$$

Assume that for each $j \in \mathbb{N}_0$, the h_j–shift–invariant subspace W_j is generated by a function $\psi_j \in W_j$, *i.e.*, $W_j = S_j(\psi_j)$. Further, we suppose that there exist positive constants γ, δ such that for all $j \in \mathbb{N}_0$ and for any $(b_{j,n})_{n=0}^{d_j-1} \in \mathbb{C}^{d_j}$,

$$\gamma \sum_{n=0}^{d_j-1} |b_{j,n}|^2 \le \| \sum_{n=0}^{d_j-1} b_{j,n}\, d_j^{1/2}\, \psi_{j,n} \|^2 \le \delta \sum_{n=0}^{d_j-1} |b_{j,n}|^2. \tag{4.2}$$

In other words, $\{\mathcal{B}_j(d_j^{1/2}\, \psi_j) : \ j \in \mathbb{N}_0\}$ is L_p^2–stable. Under these assumptions, $d_j^{1/2}\, \psi_j$ is called *p–periodic semi-orthogonal wavelet of level j* or *p–periodic prewavelet of level j*. If all $\mathcal{B}_j(d_j^{1/2}\, \psi_j)$ $(j \in \mathbb{N}_0)$ are orthonormal bases, then $d_j^{1/2}\psi_j$ $(j \in \mathbb{N}_0)$ is called *p–periodic orthonormal wavelet of level j*.

For orthonormal wavelets, Inequality (4.2) is automatically satisfied, since for $(b_{j,n})_{n=0}^{d_j-1} \in \mathbb{C}^{d_j}$,

$$\| \sum_{n=0}^{d_j-1} b_{j,n}\, d_j^{1/2}\, \psi_{j,n} \|^2 = \sum_{n=0}^{d_j-1} |b_{j,n}|^2,$$

i.e., $\gamma = \delta = 1$. From (M1) and Equation (4.1), it follows $\psi_j \in V_{j+1}$. Then there exist unique coefficients $\beta_{j+1,k} \in \mathbb{C}$ $(k = 0, \dots, d_{j+1} - 1)$ such that

$$\psi_j = \sum_{k=0}^{d_{j+1}-1} \beta_{j+1,k}\, \varphi_{j+1,k}.$$

This is the so–called *two–scale relation* or *refinement equation* of ψ_j. By Equation (2.1), we obtain the Fourier transformed two–scale relation of ψ_j

$$c_u(\psi_j) = 2B_{j+1}(\omega_{j+1}^u)\, c_u(\varphi_{j+1}), \quad u \in \mathbf{Z}, \tag{4.3}$$

with the *two–scale symbol* or *refinement mask* of ψ_j

$$B_{j+1}(z) := \frac{1}{2} \sum_{k=0}^{d_{j+1}-1} \beta_{j+1,k}\, z^k, \quad z \in \mathcal{T}_{j+1}.$$

Further, we introduce the *p–periodic autocorrelation symbol* of ψ_j by

$$\Psi_j(z) := \sum_{l=0}^{d_j-1} \langle \psi_{j,-l}, \psi_j \rangle\, z^l, \quad z \in \mathcal{T}_j.$$

Then we observe the following connection between the bracket product $[\mathbf{c}(\psi_j), \mathbf{c}(\psi_j)]_j$, the two–scale symbol B_{j+1}, and the p–periodic autocorrelation symbols Ψ_j and Ψ_{j+1}.

Lemma 4.1. *For $j \in \mathbb{N}_0$ and $k = 0, \ldots, d_j - 1$, we have*

$$\Psi_j(\omega_j^k) = d_j\,[\mathbf{c}(\psi_j), \mathbf{c}(\psi_j)]_{j,k},$$

$$\Psi_j(z^2) = 2\,|B_{j+1}(z)|^2\,\Phi_{j+1}(z) + 2\,|B_{j+1}(-z)|^2\,\Phi_{j+1}(-z), \qquad z \in \mathcal{T}_{j+1}. \tag{4.4}$$

Condition (4.2) *is equivalent to*

$$0 < \gamma \le d_j\,\Psi_j(\omega_j^k) \le \delta < \infty, \quad j \in \mathbb{N}_0,\ k = 0, \ldots, d_j - 1. \tag{4.5}$$

The proof is similar to that of Lemma 3.3 and is omitted here.

Now the following question is of interest: How can we choose the two–scale symbol B_{j+1} such that $S_j(\psi_j) \perp V_j$ and Conditions (4.2) or (4.5) are satisfied? One condition for B_{j+1} follows from the orthogonality of $S_j(\psi_j)$ and V_j.

Lemma 4.2. *For $j \in \mathbb{N}_0$, we have $S_j(\psi_j) \perp V_j$ if and only if for $z \in \mathcal{T}_{j+1}$*

$$A_{j+1}(z)\,\overline{B_{j+1}(z)}\,\Phi_{j+1}(z) + A_{j+1}(-z)\,\overline{B_{j+1}(-z)}\,\Phi_{j+1}(-z) = 0. \tag{4.6}$$

The proof is based on Corollary 2.3 (ii) and the two–scale Relations (3.3) and (4.3).

Now let us introduce the *two–scale symbol matrices of the j-th level* ($j \in \mathbb{N}_0$)

$$\mathbf{S}_{j+1}(z) := \begin{pmatrix} A_{j+1}(z) & B_{j+1}(z) \\ A_{j+1}(-z) & B_{j+1}(-z) \end{pmatrix}, \qquad z \in \mathcal{T}_{j+1}. \tag{4.7}$$

In the next section, these matrices and their inverses will play the main role for the decomposition and reconstruction algorithms. We investigate the invertibility of $\mathbf{S}_{j+1}(z)$.

Lemma 4.3. *Assume that Relations* (3.6) *and* (4.5) *are true. For $j \in \mathbb{N}_0$, the two–scale symbol matrices $\mathbf{S}_{j+1}(z)$ ($z \in \mathcal{T}_{j+1}$) are regular with*

$$\frac{2\sqrt{\alpha\,\gamma}}{\beta} \le |\det \mathbf{S}_{j+1}(z)| \le \frac{2\sqrt{\beta\,\delta}}{\alpha}. \tag{4.8}$$

Further, we have

$$\mathbf{S}_{j+1}(z)^{-1} = \\ \operatorname{diag}\,(\Phi_j(z^2)^{-1},\, \Psi_j(z^2)^{-1})^{\mathrm{T}}\,\overline{\mathbf{S}_{j+1}(z)}^{\mathrm{T}}\,\operatorname{diag}\,(\Phi_{j+1}(z),\, \Phi_{j+1}(-z))^{\mathrm{T}}. \tag{4.9}$$

Proof: Using Equations (3.5), (4.4), and (4.6), we obtain, for $z \in \mathcal{T}_{j+1}$,

$$\overline{\mathbf{S}_{j+1}(z)}^{\mathrm{T}}\,\operatorname{diag}\,(\Phi_{j+1}(z),\, \Phi_{j+1}(-z))^{\mathrm{T}}\,\mathbf{S}_{j+1}(z) = \operatorname{diag}\,(\Phi_j(z^2),\, \Psi_j(z^2))^{\mathrm{T}}. \tag{4.10}$$

By known properties of the determinant, this equation yields

$$\Phi_{j+1}(z)\,\Phi_{j+1}(-z)\,|\det \mathbf{S}_{j+1}(z)|^2 = \Phi_j(z^2)\,\Psi_j(z^2), \quad z \in \mathcal{T}_{j+1}.$$

Again, by Relations (3.6) and (4.5), all matrices $\mathbf{S}_{j+1}(z)$ ($z \in \mathcal{T}_{j+1}$) are regular with Inequality (4.8). From Equation (4.10), Equation (4.9) it follows directly. ■

Remark. The assertion of Lemma 4.3 emphasizes the importance of the L_p^2–stability of $\{\mathcal{B}_j(d_j^{1/2}\varphi_j) : j \in \mathbb{N}_0\}$ and $\{\mathcal{B}_j(d_j^{1/2}\psi_j) : j \in \mathbb{N}_0\}$. If the systems $\{\mathcal{B}_j(d_j^{1/2}\varphi_j) : j \in \mathbb{N}_0\}$ or $\{\mathcal{B}_j(d_j^{1/2}\psi_j) : j \in \mathbb{N}_0\}$ are not L_p^2–stable, then the two–scale symbol matrices are not well–conditioned such that the existence of numerically stable algorithms for decomposition and reconstruction is not ensured (see Section 5).

Now, with the help of Conditions (4.4)–(4.6), the two–scale symbol B_{j+1} can be described more exactly.

Theorem 4.4. (see [16]) *Assume that Relation* (3.6) *holds. For every* $j \in \mathbb{N}_0$, $B_{j+1} : \mathcal{T}_{j+1} \to \mathbb{C}$ *is a two–scale symbol of a p–periodic semiorthogonal wavelet* $d_j^{1/2}\,\psi_j \in L_p^2$ *satisfying Property* (4.5) *if and only if*

$$B_{j+1}(z) = \frac{\Phi_{j+1}(-z)\,\overline{A_{j+1}(-z)}}{z\,\Phi_j(z^2)}\,K_j(z^2), \quad z \in \mathcal{T}_{j+1},$$

where $K_j : \mathcal{T}_j \to \mathbb{C}$ *satisfies the condition*

$$0 < \mu \le |K_j(z)| \le \nu < \infty, \quad z \in \mathcal{T}_j,$$

with positive constants μ, ν.

In the case of orthonormal wavelets the corresponding two–scale symbol $B_{j+1}^{\star}$ even satisfies the following corollary.

Corollary 4.5. *Assume that* $\mathcal{B}_j(d_j^{1/2}\,\varphi_j^{\star})$ *are orthonormal bases of* V_j *and* $A_{j+1}^{\star}$ ($j \in \mathbb{N}_0$) *the corresponding two–scale symbols of* $\varphi_j^{\star}$. *Then for every* $j \in \mathbb{N}_0$, $B_{j+1}^{\star} : \mathcal{T}_{j+1} \to \mathbb{C}$ *is a two–scale symbol of a p–periodic orthonormal wavelet* $d_j^{1/2}\,\psi_j^{\star} \in L_p^2$ *if and only if* $B_{j+1}^{\star}$ *has the form*

$$B_{j+1}^{\star}(z) = \pm\, z^{2n-1}\overline{A_{j+1}^{\star}(-z)}, \quad z \in \mathcal{T}_{j+1},$$

where $n \in \{0, \ldots, d_j - 1\}$. *For the related two–scale matrices*

$$\mathbf{S}_{j+1}^{\star}(z) := \begin{pmatrix} A_{j+1}^{\star}(z) & B_{j+1}^{\star}(z) \\ A_{j+1}^{\star}(-z) & B_{j+1}^{\star}(-z) \end{pmatrix}, \quad z \in \mathcal{T}_{j+1},$$

we have

$$|\det \mathbf{S}^\star_{j+1}(z)| = |A^\star_{j+1}(z)|^2 + |A^\star_{j+1}(-z)|^2 = 1, \quad z \in \mathcal{T}_{j+1}.$$

The proof follows directly from Theorem 4.4, taking in consideration the relations

$$d_j\, \Phi^\star_j(\omega_j^k) = d_j^2\, [\mathbf{c}(\varphi^\star_j), \mathbf{c}(\varphi^\star_j)]_{j,k} = 1,$$
$$d_j\, \Psi^\star_j(\omega_j^k) = d_j^2\, [\mathbf{c}(\psi^\star_j), \mathbf{c}(\psi^\star_j)]_{j,k} = 1, \quad k = 0, \ldots, d_j - 1,$$

instead of Relations (3.6) and (4.5).

§5 Decomposition and reconstruction algorithms

In this section we shall derive efficient decomposition and reconstruction algorithms based on Fourier technique. In order to decompose a given function $f_{j+1} \in V_{j+1}$ ($j \in \mathbb{N}_0$) of the form

$$f_{j+1} = \sum_{l=0}^{d_{j+1}-1} a_{j+1,l}\, \varphi_{j+1,l}, \quad a_{j+1,l} \in \mathbb{C}, \tag{5.1}$$

uniquely determined functions $f_j \in V_j$ and $g_j \in W_j$ have to be found such that

$$f_{j+1} = f_j + g_j. \tag{5.2}$$

Assume that the coefficients $a_{j+1,l} \in \mathbb{C}$ ($l = 0, \ldots, d_{j+1} - 1$) of f_{j+1} or their DFT(d_{j+1}) data

$$\hat{a}_{j+1,k} := \sum_{l=0}^{d_{j+1}-1} a_{j+1,l}\, \omega_{j+1}^{kl}, \quad k = 0, \ldots, d_{j+1} - 1, \tag{5.3}$$

are known. The wanted functions $f_j \in V_j$ and $g_j \in W_j$ can uniquely be represented by

$$f_j = \sum_{n=0}^{d_j-1} a_{j,n}\, \varphi_{j,n}, \quad g_j = \sum_{n=0}^{d_j-1} b_{j,n}\, \psi_{j,n} \tag{5.4}$$

with unknown coefficients $a_{j,n}, b_{j,n} \in \mathbb{C}$.

In order to reconstruct $f_{j+1} \in V_{j+1}$ ($j \in \mathbb{N}_0$), we have to compute the sum in Equation (5.2) with given $f_j \in V_j$ and $g_j \in W_j$. Assume that $a_{j,n}, b_{j,n} \in \mathbb{C}$ ($n = 0, \ldots, d_j - 1$) in Equation (5.4) or their DFT(d_j) data

$$\hat{a}_{j,k} := \sum_{n=0}^{d_j-1} a_{j,n}\, \omega_j^{kn}, \quad \hat{b}_{j,k} := \sum_{n=0}^{d_j-1} b_{j,n}\, \omega_j^{kn}, \quad k = 0, \ldots, d_j - 1, \tag{5.5}$$

are known. The function $f_{j+1} \in V_{j+1}$ can be uniquely represented in the form of Equation (5.1) with unknown coefficients $a_{j+1,l}$ ($l = 0, \ldots, d_{j+1} - 1$). The decomposition and reconstruction algorithms are based on the following theorem.

Theorem 5.1. *Assume that*

$$\operatorname{supp} \boldsymbol{c}(\varphi_{j+1}) \supseteq \{-d_j, \ldots, d_j - 1\}, \quad j \in \mathbb{N}_0. \tag{5.6}$$

For $j \in \mathbb{N}_0$, let $f_{j+1} \in V_{j+1}$, $f_j \in V_j$ and $g_j \in W_j$ with Equations (5.1)–(5.5) be given. Then we have for $n = 0, \ldots, d_{j+1} - 1$,

$$\hat{a}_{j+1,n} = 2\,\hat{a}_{j,n}\,A_{j+1}(\omega_{j+1}^n) + 2\,\hat{b}_{j,n}\,B_{j+1}(\omega_{j+1}^n), \tag{5.7}$$

i.e., for $k = 0, \ldots, d_j - 1$,

$$\begin{pmatrix} \hat{a}_{j+1,k} \\ \hat{a}_{j+1,k+d_j} \end{pmatrix} = 2\,\boldsymbol{S}_{j+1}(\omega_{j+1}^k) \begin{pmatrix} \hat{a}_{j,k} \\ \hat{b}_{j,k} \end{pmatrix}. \tag{5.8}$$

Proof: From

$$c_u(f_{j+1}) = c_u(f_j) + c_u(g_j), \quad u \in \mathbf{Z},$$

it follows by Equations (5.1) and (5.4)

$$\sum_{l=0}^{d_{j+1}-1} a_{j+1,l}\, c_u(\varphi_{j+1,l}) = \sum_{n=0}^{d_j-1} (a_{j,n}\, c_u(\varphi_{j,n}) + b_{j,n}\, c_u(\psi_{j,n})), \quad u \in \mathbf{Z},$$

and hence by Equations (2.1), (5.3), and (5.5),

$$\hat{a}_{j+1,u}\, c_u(\varphi_{j+1}) = \hat{a}_{j,u}\, c_u(\varphi_j) + \hat{b}_{j,u}\, c_u(\psi_j), \quad u \in \mathbf{Z}.$$

Using the Fourier transformed two–scale Relations (3.3) and (4.3), we obtain

$$\hat{a}_{j+1,u}\, c_u(\varphi_{j+1}) = (2\,\hat{a}_{j,u}\, A_{j+1}(\omega_{j+1}^u) + 2\,\hat{b}_{j,u}\, B_{j+1}(\omega_{j+1}^u))\, c_u(\varphi_{j+1}), \quad u \in \mathbf{Z}.$$

Since the coefficients of $c_u(\varphi_{j+1})$ are d_{j+1}-periodic, we conclude from the Assumption (5.6) that Equation (5.7) holds. Thus, we have for $k = 0, \ldots, d_j - 1$,

$$\begin{aligned} \hat{a}_{j+1,k} &= 2\,\hat{a}_{j,k}\, A_{j+1}(\omega_{j+1}^k) + 2\,\hat{b}_{j,k}\, B_{j+1}(\omega_{j+1}^k), \\ \hat{a}_{j+1,k+d_j} &= 2\,\hat{a}_{j,k}\, A_{j+1}(-\omega_{j+1}^k) + 2\,\hat{b}_{j,k}\, B_{j+1}(-\omega_{j+1}^k), \end{aligned}$$

and Equation (5.8) follows. ■

Remark. From Inclusion (5.6), Equation (3.1) follows directly. In all theories on periodic wavelets known up to now (see [5,14,15,17,18]), Condition (5.6) is satisfied.

From Theorem 5.1, we obtain immediately the following algorithms.

Algorithm 5.2. (Decomposition Algorithm)
Input:
$j \in \mathbb{N}_0$, $d \in \mathbb{N}$ *(power of 2)*, $d_j := 2^j\, d$,
$\hat{a}_{j+1,k} \in \mathbb{C} \quad (k = 0, \ldots, d_{j+1} - 1)$.

1. Precompute $\boldsymbol{S}_{j+1}(\omega_{j+1}^k)^{-1}$ $(k = 0, \ldots, d_j - 1)$ *(given by Equation* (4.9)*) by* FFT.
2. Compute for $k = 0, \ldots, d_j - 1$

$$\begin{pmatrix} \hat{a}_{j,k} \\ \hat{b}_{j,k} \end{pmatrix} := \frac{1}{2}\, \boldsymbol{S}_{j+1}(\omega_{j+1}^k)^{-1} \begin{pmatrix} \hat{a}_{j+1,k} \\ \hat{a}_{j+1,k+d_j} \end{pmatrix}.$$

Output: $\hat{a}_{j,k}$, $\hat{b}_{j,k} \quad (k = 0, \ldots, d_j - 1)$.

Algorithm 5.3. (Reconstruction Algorithm)
Input:
$j \in \mathbb{N}_0$, $d \in \mathbb{N}$ *(power of 2)*, $d_j := 2^j\, d$,
$\hat{a}_{j,k}$, $\hat{b}_{j,k} \in \mathbb{C} \quad (k = 0, \ldots, d_j - 1)$.

1. Precompute $\boldsymbol{S}_{j+1}(\omega_{j+1}^k)$ $(k = 0, \ldots, d_j - 1)$ *(given by Equation* (4.7)*) by* FFT.
2. Compute (5.8) *for* $k = 0, \ldots, d_j - 1$.
Output: $\hat{a}_{j+1,k} \quad (k = 0, \ldots, d_{j+1} - 1)$.

Acknowledgments. This research was supported by Deutsche Forschungsgemeinschaft.

References

1. Battle, G., A block spin construction of ondelettes. Part 1: Lemarié functions, *Comm. Math. Phys.* **110** (1987), 601–615.
2. de Boor, C., R. A. DeVore, and A. Ron, Approximation from shift–invariant subspaces of $L_2(\mathbb{R}^d)$, *Trans. Amer. Math. Soc.* **341** (1994), 787–806.
3. de Boor, C., R. A. DeVore, and A. Ron, On the construction of multivariate (pre)wavelets, *Constr. Approx.* **9** (1993), 123–166.
4. Chui, C. K., *An Introduction to Wavelets*, Academic Press, Boston, 1992.
5. Chui, C. K. and H. N. Mhaskar, On trigonometric wavelets, *Constr. Approx.* **9** (1993), 167–190.
6. Chui, C. K. and J. Z. Wang, A general framework for compactly supported splines and wavelets, *J. Approx. Theory* **71** (1992), 263–304.
7. Chui, C. K. and J. Z. Wang, On compactly supported spline wavelets and a duality principle, *Trans. Amer. Math. Soc.* **330** (1992), 903–915.
8. Daubechies, I., *Ten Lectures on Wavelets*, SIAM, Philadelphia, 1992.
9. Jia, R. Q. and C. A. Micchelli, Using the refinement equation for the construction of pre-wavelets II: Powers of two, in *Curves and Surfaces*, P. J. Laurent, A. Le Méhauté, and L. L. Schumaker (eds.), Academic Press, New York, 1991, 209–246.

10. Lemarié, P. G., Ondelettes á localisation exponentielle, *J. Math. Pures Appl.* **67** (1988), 227–236.
11. Mallat, S. G., Multiresolution approximations and wavelet orthonormal bases of $L_2(\mathbb{R})$, *Trans. Amer. Math. Soc.* **315** (1989), 69–87.
12. Meyer, Y., *Ondelettes et Opérateurs* I: *Ondelettes*, Hermann, Paris, 1980.
13. Meyer, Y., Wavelets and operators, Rapport CEREMADE 8704, University of Paris–Dauphine, 1987.
14. Perrier, V. and C. Basdevant, Periodic wavelet analysis, a tool for inhomogeneous field investigation, theory and algorithms, *Rech. Aérospat.*, 1989–3, 53–67.
15. Plonka, G. and M. Tasche, On the computation of periodic spline wavelets, University of Rostock, 1993, preprint.
16. Plonka, G. and M. Tasche, Periodic wavelets, University of Rostock, 1993, preprint.
17. Prestin, J. and E. Quak, Trigonometric interpolation and wavelet decompositions, CAT Report #296, Texas A&M University, 1993.
18. Prestin, J. and E. Quak, A duality principle for trigonometric wavelets, in *Wavelets, Images, and Surface Fitting*, P. J. Laurent, A. Le Méhauté, and L. L. Schumaker (eds.), AKPeters, Boston, to appear.
19. Privalov, A. A., On an orthogonal trigonometric basis, *Mat. Sb.* **182** (1991), 384–394; English transl. in *Math. USSR–Sb.* **72** (1992), 363–372.
20. Tasche, M., Orthogonal periodic spline wavelets, in *Wavelets, Images, and Surface Fitting*, P. J. Laurent, A. Le Méhauté, and L. L. Schumaker (eds.), AKPeters, Boston, to appear.

Gerlind Plonka
Fachbereich Mathematik, Universität Rostock,
D–18051 Rostock, Germany
nfa016@nfa.uni-rostock.d400.de

Manfred Tasche
Fachbereich Mathematik, Universität Rostock,
D–18051 Rostock, Germany
tasche@mathematik.uni-rostock.d400.de

Part III

Spline Wavelets

Spline Wavelets over R, Z, R/NZ, and Z/NZ

Gabriele Steidl

Abstract. Using terms of Fourier analysis on locally compact abelian groups, we give a unified approach to cardinal, periodic, and discrete spline wavelets with arbitrary dilation factor $n \geq 2$. The concept is based on generalized cardinal B-splines defined with respect to towers of subgroups.

§1 Introduction

In this paper, we give a unified approach to spline wavelets over $\mathbb{R}, \mathbf{Z}, \mathbb{R}/N\mathbf{Z}$, and $\mathbf{Z}/N\mathbf{Z}$ in terms of Fourier analysis on these groups.

The basic idea can be sketched as follows: Let G be one of the classical groups of Fourier analysis $\mathbb{R}, \mathbf{Z}, \mathbb{R}/N\mathbf{Z}$, or $\mathbf{Z}/N\mathbf{Z}$. We consider towers of cyclic subgroups U_j of G

$$G \rhd \ldots \rhd U_{j-1} \rhd U_j \rhd U_{j+1} \rhd \cdots$$

with constant index $|U_j : U_{j+1}| = n \geq 2$ for all j. This index plays the role of the dilation factor. Following S. Dahlke [7], we introduce the generalized cardinal B-splines $\varphi_j = N_{m,j}$ with respect to special representation systems of U_j in G. These B-splines provide many useful properties known from the cardinal B-splines over $\mathbb{R}$. Especially, the Fourier transforms $\hat{\varphi}_j$ and $\hat{\varphi}_{j+1}$ are related by a two-scale symbol P and the bracket products $[\hat{\varphi}_j, \hat{\varphi}_j]_j$ can be described using I. J. Schoenberg's generalized Euler-Frobenius polynomials [16]. By V_j, we denote the vector space spanned by the U_j-translates of φ_j. Applying properties of the generalized Euler-Frobenius polynomials, we show that the U_j-translates of φ_j form a Riesz basis of V_j and estimate the Riesz bounds. The vector spaces V_j generate a multiresolution analysis of $L^2(G)$ which is stationary if G is not discrete.

Next, we construct $n-1$ spline wavelets ψ_{j+1}^s $(s = 1, \ldots, n-1)$ which generate the orthogonal complement W_{j+1} of V_{j+1} in V_j. For $n \geq 3$, these spline wavelets differ only by U_j-shifts such that we call them "shifted spline

Wavelets: Theory, Algorithms, and Applications
Charles K. Chui, Laura Montefusco, and Luigia Puccio (eds.), pp. 155–177.

ISBN 0-12-174575-9

wavelets." For $G = \mathbb{R}$ or $G = \mathbf{Z}$, our shifted spline wavelets are those with minimal support. Hence, we obtain the Chui-Wang wavelets [6] in the special case $n = 2$ and $G = \mathbb{R}$ and their N-periodizations considered in [14] in the corresponding periodic case $G = \mathbb{R}/N\mathbf{Z}$. Our constructions of discrete cardinal spline wavelets and of cardinal spline wavelets with dilation factor $n > 2$ are new. The functions ψ^s_{j+1} and φ_j are related by the two-scale symbols Q^s not depending on j, if $G = \mathbb{R}$, and by the two-scale symbols Q^s_j with $\lim_{j\to\infty} Q^s_j = Q^s$ if $G = \mathbf{Z}$. We specify the two-scale symbol matrix (see [1], Theorem 2.26 (iv); [14]) for our purposes and show the Riesz basis property of our spline wavelets by evaluating the determinant of the two-scale symbol matrix. For the dilation factor $n = 2$, we estimate the Riesz bounds.

Finally, we use the two-scale symbol matrix and their inverse for the decomposition and reconstruction of functions in $L^2(G)$.

The Appendix shortly summarizes the notations and theorems from the theory of Fourier analysis on locally compact abelian groups we want to apply. Among the rich literature on this topic we refer to [11,15].

§2 Generalized cardinal B-splines over classical LCA groups

Throughout this paper, let G be one of the classical LCA groups $\mathbb{R}, \mathbf{Z}, \mathbb{R}/N\mathbf{Z}$, $\mathbf{Z}/N\mathbf{Z}$ ($N := n^t$; $n, t \geq 2$) with Haar measure m_G normalized at the end of the Appendix. Let

$$U_j := \begin{cases} n^j\mathbf{Z} & G = \mathbb{R},\, \mathbf{Z}, \\ n^j\mathbf{Z}/N\mathbf{Z} & G = \mathbb{R}/N\mathbf{Z},\, \mathbf{Z}/N\mathbf{Z} \end{cases}$$

be the subgroup of G generated by n^j, *i.e.*, $U_j := \{kn^j : k \in I_j\}$,

$$I_j := \begin{cases} \mathbf{Z} & G = \mathbb{R},\, \mathbf{Z}, \\ \{0, \ldots, N/n^j - 1\} & G = \mathbb{R}/N\mathbf{Z},\, \mathbf{Z}/N\mathbf{Z}. \end{cases}$$

We consider the following towers of subgroups $G \triangleright \ldots \triangleright U_j \triangleright U_{j+1} \triangleright \ldots$:

$$\begin{array}{lllll} \mathbb{R} & \triangleright \ldots \triangleright n^j\mathbf{Z} & \triangleright\, n^{j+1}\mathbf{Z} & \triangleright \ldots \triangleright \ldots, & (j \in \mathbf{Z}), \\ \mathbf{Z} & \triangleright \ldots \triangleright n^j\mathbf{Z} & \triangleright\, n^{j+1}\mathbf{Z} & \triangleright \ldots \triangleright \ldots, & (j \in \mathbb{N}_0), \\ \mathbb{R}/N\mathbf{Z} & \triangleright \ldots \triangleright n^j\mathbf{Z}/N\mathbf{Z} & \triangleright\, n^{j+1}\mathbf{Z}/N\mathbf{Z} & \triangleright \ldots \triangleright N\mathbf{Z}/N\mathbf{Z}, & (j \in \mathbf{Z};\ j < t), \\ \mathbf{Z}/N\mathbf{Z} & \triangleright \ldots \triangleright n^j\mathbf{Z}/N\mathbf{Z} & \triangleright\, n^{j+1}\mathbf{Z}/N\mathbf{Z} & \triangleright \ldots \triangleright N\mathbf{Z}/N\mathbf{Z}, & (j \in \mathbb{N}_0;\ j < t). \end{array}$$

We refer to functions defined on $G = \mathbb{R}, \mathbf{Z}$ as non-periodic functions and to those defined on $G = \mathbb{R}/N\mathbf{Z}, \mathbf{Z}/N\mathbf{Z}$ as N-periodic functions. Similarly, we consider non-discrete functions in the cases $G = \mathbb{R}, \mathbb{R}/N\mathbf{Z}$ and discrete functions in the cases $G = \mathbf{Z}, \mathbf{Z}/N\mathbf{Z}$. Clearly, we can replace $G = \mathbf{Z}$ ($G = \mathbf{Z}/N\mathbf{Z}$) by the refined grid $G = n^r\mathbf{Z}$ ($G = n^r\mathbf{Z}/N\mathbf{Z}$) with $r < 0$. Further, for $G = \mathbb{R}/N\mathbf{Z}, \mathbf{Z}/N\mathbf{Z}$, it is possible to end with $\mathbf{Z}/N\mathbf{Z}$ instead of $N\mathbf{Z}/N\mathbf{Z} = \langle 0 \rangle$ which makes the restriction that $n|N$ superfluous (see [13,14]).

Let F_j be the following representation system of U_j in G:

$$F_j := \begin{cases} [0, n^j) & G = \mathbb{R},\ \mathbb{R}/N\mathbb{Z}, \\ \{0, \ldots, n^j - 1\} & G = \mathbb{Z},\ \mathbb{Z}/N\mathbb{Z}. \end{cases}$$

By χ_{F_j}, we denote the characteristic function of F_j. Following S. Dahlke [7], we define the *generalized cardinal B-spline of order m* with respect to G and F_j recursively by

$$N_{1,j} := \chi_{F_j}, \quad N_{m,j} := n^{-j}\left(N_{m-1,j} * N_{1,j}\right) \quad (m \geq 2), \tag{2.1}$$

where $*$ denotes the convolution on G. For $G = \mathbb{R}$, the generalized cardinal B-splines $N_{m,j}$ coincide with the box splines $n^j M_{[n^j,\ldots,n^j]}$ in [2], especially $N_{m,0}$ with the well-known cardinal B-spline $N_m = M_{[1,\ldots,1]}$. For $G = \mathbb{Z}$, we obtain discrete cardinal B-splines, which are related with the convenient discrete box splines $b^{\cdot}_{[\cdot]}$ in [2] by $N_{m,j}(k) = n^j b^{1/n^j}_{[1,\ldots,1]}(k/n^j)$. Our discrete B-splines can be simply described by

$$\sum_{k=0}^{mn^j-1} N_{m,j}(k) z^k = n^{-(m-1)j} \left(\frac{z^{n^j} - 1}{z - 1}\right)^m .$$

Note that for $n \geq 3$ other choices of representation systems F_j may be interesting since the corresponding generalized B-splines define multiresolution analyses different from those in Section 4 (see [9]).

By Equation (2.1), it follows that

$$\operatorname{supp} N_{m,j} := \begin{cases} [0, mn^j) & G = \mathbb{R}, \\ \{0, \ldots, m(n^j - 1)\} & G = \mathbb{Z}\ , \end{cases} \tag{2.2}$$

and that the sum of the U_j-translates of $N_{m,j}$ is a *partition of unity, i.e.*,

$$\sum_{k \in I_j} N_{m,j}(x + kn^j) = 1. \tag{2.3}$$

Further, it is easy to see that

$$N_{m,j+1}(nx) = \begin{cases} N_{m,j}(x) & G = \mathbb{R}, \\ \sum_{r=0}^{n-1} N_{m,j}(x - rN/n) & G = \mathbb{R}/N\mathbb{Z} \end{cases} \tag{2.4}$$

which is not true for the discrete groups $G = \mathbb{Z}$, $\mathbb{Z}/N\mathbb{Z}$.

By the following lemma, the N-periodic generalized cardinal B-splines coincide with the N-periodization of the corresponding non-periodic generalized cardinal B-splines.

Lemma 2.1. *Let $N^I_{m,j}$ be the generalized cardinal B-spline over $G = \mathbb{R}$ or $G = \mathbf{Z}$ and let $N^{II}_{m,j}$ be the corresponding generalized cardinal B-spline over $G/N\mathbf{Z}$. Then*

$$N^{II}_{m,j}(x) = \sum_{k\in Z} N^I_{m,j}(x+kN) \quad (j \le t).$$

Proof: Let

$$g(x) := \sum_{k\in Z} N^I_{m,j}(x+kN) \ \in C(G/N\mathbf{Z}).$$

Then it holds by Theorem A.6 (i)

$$\hat{g}(\gamma) = \hat{N}^I_{m,j}(\gamma) \quad (\gamma \in \widehat{G/N\mathbf{Z}}).$$

On the other hand, we conclude by Equation (2.1),

$$\hat{N}^{II}_{m,j}(\gamma) = n^{-j(m-1)}\Big(\int_{F_j} \bar{\gamma}(x)\, \mathrm{d}m_{G/NZ}(x)\Big)^m = \hat{N}^I_{m,j}(\gamma). \ \blacksquare \tag{2.5}$$

By the translation property and the convolution property of Fourier transform, we obtain the following.

Theorem 2.2. *It holds*

$$N_{m,j+1}(x) = \sum_{k\in Z} p_k\, N_{m,j}(x - kn^j) \quad (x \in G), \tag{2.6}$$

$$\hat{N}_{m,j+1}(\gamma) = n\, P(\bar{\gamma}(n^j))\, \hat{N}_{m,j}(\gamma) \quad (\gamma \in \hat{G}) \tag{2.7}$$

with the two-scale symbol P of $N_{m,j}$ defined by

$$P(z) = n^{-1} \sum_{k\in Z} p_k\, z^k := n^{-m}\left(\frac{z^n - 1}{z-1}\right)^m. \tag{2.8}$$

If $G = \mathbb{R}/N\mathbf{Z}$ or $G = \mathbf{Z}/N\mathbf{Z}$, then $kn^j = kn^j \bmod N$ $(k \in \mathbf{Z})$ and the two-scale relation in Equation (2.6) can be replaced by

$$N_{m,j+1}(x) = \sum_{k\in I_j} \Big(\sum_{l\in Z} p_{lN/n^j+k}\Big)\, N_{m,j}(x - kn^j).$$

§3 Generalized Euler-Frobenius polynomials

Let

$$M_{2m,j}(y) := n^{-j} \int_G N_{m,j}(x) N_{m,j}(x+y)\, \mathrm{d}x = N_{2m,j}(y + mn^j - \delta m) \quad (3.1)$$

with

$$\delta := \begin{cases} 0 & G = \mathbb{R},\ \mathbb{R}/N\mathbf{Z}, \\ 1 & G = \mathbf{Z},\ \mathbf{Z}/N\mathbf{Z} \end{cases}$$

denote the autocorrelation of $N_{m,j}$.

Lemma 3.1. *Let $\Lambda_j \cong \widehat{G/U_j}$ denote the annihilator of U_j in $\hat{G}$. Then*

$$n^{-2j} \sum_{\alpha \in \Lambda_j} |\hat{N}_{m,j}(\alpha\gamma)|^2 = \sum_{k \in I_j} M_{2m,j}(kn^j)\, \bar{\gamma}(kn^j) \quad (\gamma \in \hat{U}_j). \qquad (3.2)$$

Proof: By the symmetry property and the convolution property of the Fourier transform, we obtain

$$\hat{M}_{2m,j}(\gamma) = n^{-j} |\hat{N}_{m,j}(\gamma)|^2 \quad (\gamma \in \hat{G}).$$

Thus,

$$\sum_{\alpha \in \Lambda_j} \hat{M}_{2m,j}(\alpha\gamma) = n^{-j} \sum_{\alpha \in \Lambda_j} |\hat{N}_{m,j}(\alpha\gamma)|^2 \quad (\gamma \in \hat{U}_j).$$

Recall that Λ_j is discrete and that every $\alpha \in \Lambda_j$ is assigned the measure n^{-j} (see Appendix). Then the duality relation in Theorem A.5 and the Poisson summation formula in Theorem A.6 (ii) with respect to $\hat{G}$ and $\hat{U}_j$ yield

$$\begin{aligned} n^{-j} \sum_{\alpha \in \Lambda_j} \hat{M}_{2m,j}(\alpha\gamma) &= \sum_{k \in I_j} (\hat{M}_{2m})\hat{}\,(kn^j)\, \gamma(kn^j) \\ &= \sum_{k \in I_j} M_{2m}(-kn^j)\, \gamma(kn^j) \quad (\gamma \in \hat{U}_j). \quad \blacksquare \end{aligned}$$

For $G = \mathbb{R}$, the lefthand side of Equation (3.2) is also referred as *(j-th) bracket product*

$$[\hat{N}_{m,j}, \hat{N}_{m,j}]_j := \sum_{\alpha \in \Lambda_j} |\hat{N}_{m,j}(\alpha\gamma)|^2 \quad (\gamma \in \hat{U}_j)$$

and plays a crucial rule in the consideration of shift-invariant spaces [1]. Fortunately, the bracket product of $\hat{N}_{m,j}$ with itself can be described in terms of the (Laurent) polynomial

$$\Phi_{2m,j}(z) := \sum_{k \in I_j} M_{2m,j}(kn^j)\, z^k. \qquad (3.3)$$

Corollary 3.2. *The polynomials $\Phi_{2m,j}$ satisfy*

$$n^{-2j} \sum_{\alpha \in \Lambda_j} |\hat{N}_{m,j}(\alpha\gamma)|^2 = \Phi_{2m,j}(\bar{\gamma}(n^j)) \quad (\gamma \in \hat{U}_j).$$

Moreover, the polynomials $\Phi_{2m,j}$ have the following useful properties: By $N_{2m,j}(kn^j + mn^j - \delta m) = N_{2m,j}(-kn^j + mn^j - \delta m)$ $(k \in I_j)$, it holds

$$\Phi_{2m,j}(z) = \Phi_{2m,j}(z^{-1}) \tag{3.4}$$

such that $\Phi_{2m,j}(\bar{\gamma}(n^j)) = \Phi_{2m,j}(\gamma(n^j))$. By Equation (2.5), we get

$$\Phi^{II}_{2m,j}(\gamma(n^j)) = \Phi^{I}_{2m,j}(\gamma(n^j)) \quad (\gamma \in n^j\widehat{\mathbf{Z}/N\mathbf{Z}};\ j \le t), \tag{3.5}$$

where the notation $\Phi^I_{2m,j}$ is used for $\Phi_{2m,j}$ in the non-periodic case $G = \mathbb{R}, \mathbf{Z}$, and the notation $\Phi^{II}_{2m,j}$ in the corresponding periodic case $G/N\mathbf{Z}$. The property of Equation (3.5) restricts later considerations to the non-periodic case.

In the case $G = \mathbb{R}$, we obtain by Equations (2.4) and (3.1) that the polynomials $\Phi_{2m,j}$ coincide with the $2m$*-th generalized Euler-Frobenius (Laurent) polynomials*

$$\Phi_{2m}(z) := \sum_{k=-(m-1)}^{m-1} N_{2m}(k+m)\, z^k$$

introduced by I. J. Schoenberg [16], *i.e.*, $\Phi_{2m,j}(z) = \Phi_{2m}(z)$ for all $j \in \mathbf{Z}$.

In the discrete case $G = \mathbf{Z}$, the polynomials $\Phi_{2m,j}$ indeed depend on the level j. However, since

$$\lim_{j \to \infty} N_{m,j}(kn^j) = N_m(k)$$

(see [2, p. 138]), they are related with the generalized Euler-Frobenius polynomials by

$$\lim_{j \to \infty} \Phi_{2m,j} = \Phi_{2m}. \tag{3.6}$$

To estimate the Riesz bounds in the next two sections, we need the following three lemmata.

Lemma 3.3. (i) *Let*

$$\sigma_m(z) := \sum_{k \in Z} \left(\frac{z}{z+k}\right)^{m+1} \quad (m \in \mathbb{N}_0;\ z \notin \mathbf{Z}).$$

Then

$$\sigma_m(z) = \left(\frac{\pi z}{\sin(\pi z)}\right)^{m+1} U_{m-1}(\cos(\pi z)) \quad (m \ge 1;\ z \notin \mathbf{Z}),$$

where U_m is defined by

$$U_0 = 1,\ U_m(z) = z\,U_{m-1}(z) + \frac{1-z^2}{m+1}\,U'_{m-1}(z) \quad (m \geq 1).$$

(ii) *The polynomial U_m has the following properties :*

$$U_{2m}(x) = a \prod_{j=1}^{m} (x^2 + a_j) \quad (a_m > \ldots > a_1 > 0;\ a > 0), \tag{3.7}$$

$$U_{2m+1}(x) = b\,x \prod_{j=1}^{m} (x^2 + b_j) \quad (b_m > \ldots > b_1 > 0;\ b > 0),$$

$$U_{2m-2}(\cos\frac{v}{2}) = \Phi_{2m}(e^{iv}) \quad (v \in [0,\pi)). \tag{3.8}$$

The proof is given in [17].

Lemma 3.4. *Let $\Phi_{2m,j}$ be defined with respect to $G = \mathbf{Z}$. Then it holds*

$$\Phi_{2m}(-1) \leq \Phi_{2m}(e^{iv}) \leq \Phi_{2m,j}(e^{iv}) \leq \Phi_{2m}(1) \quad (v \in [0, 2\pi);\ j \geq 0)$$

with

$$\Phi_{2m}(1) = 1,\ \Phi_{2m}(-1) = \frac{2^{2m}(2^{2m}-1)}{(2m)!}\,|B_{2m}|.$$

Here B_{2m} denotes the $2m$-th Bernoulli number.

Proof: By Equation (3.4), we obtain

$$\begin{aligned} &\Phi_{2m}(-1) \leq \Phi_{2m}(e^{iv}) \leq \Phi_{2m}(1), \\ &\Phi_{2m,j}(-1) \leq \Phi_{2m,j}(e^{iv}) \leq \Phi_{2m,j}(1) \qquad (v \in [0, 2\pi)) \end{aligned}$$

and by Equations (3.1), (3.3), and (2.3)

$$\Phi_{2m,j}(1) = \Phi_{2m}(1) = 1.$$

In the following, let $v \in (0, 2\pi)$. Then we have by Corollary 3.2 that

$$\begin{aligned} \Phi_{2m}(e^{iv}) &= \sum_{k \in Z} |\hat{N}_m(v + 2\pi k)|^2 \\ &= |1 - e^{-iv}|^{2m} \sum_{k \in Z} (v + 2\pi k)^{-2m} \end{aligned}$$

which implies for $v := \pi$

$$\Phi_{2m}(-1) = \left(\frac{2}{\pi}\right)^{2m} \sum_{k \in Z} (2k+1)^{-2m} = \frac{2^{2m}(2^{2m}-1)}{(2m)!}\,|B_{2m}|.$$

By Lemma 3.3 (i), we continue with

$$\Phi_{2m}(e^{iv}) = |1-e^{iv}|^{2m} \sum_{l=0}^{n^j-1} \sum_{k\in Z} (v+2\pi l+2\pi k n^j)^{-2m}$$
$$= \frac{|1-e^{iv}|^{2m}}{(2n^j)^{2m}} \sum_{l=0}^{n^j-1} \left(\sin \frac{v+2\pi l}{2n^j}\right)^{-2m} U_{2m-2}(\cos \frac{v+2\pi l}{2n^j}).$$

By Equation (3.8), it holds $0 < U_{2m-2}(\cos \frac{v+2\pi l}{2n^j}) \le 1$. Thus

$$\Phi_{2m}(e^{iv}) \le \frac{|1-e^{iv}|^{2m}}{(2n^j)^{2m}} \sum_{l=0}^{n^j-1} \left(\sin \frac{v+2\pi l}{2n^j}\right)^{-2m} \quad (v \in (0,\pi)). \tag{3.9}$$

On the other hand, we get by Corollary 3.2

$$\Phi_{2m,j}(e^{iv}) = n^{-2j} \sum_{l=0}^{n^j-1} |\hat{N}_{m,j}\left(\frac{v+2\pi l}{n^j}\right)|^2 \quad (v \in (0,2\pi);\ j \ge 1).$$

Now it follows from Equations (2.7) and (2.8) that

$$\Phi_{2m,j}(e^{iv}) = n^{-2j} \sum_{l=0}^{n^j-1} |n^j P(e^{-i(v+2\pi l)/n}) \dots P(e^{-i(v+2\pi l)/n^j}) \hat{N}_{m,0}\left(\frac{v+2\pi l}{n^j}\right)|^2$$
$$= \sum_{l=0}^{n^j-1} \left| n^{-j} \frac{e^{-i(v+2\pi l)}-1}{e^{-i(v+2\pi l)/n}-1} \cdots \frac{e^{-i(v+2\pi l)/n^{j-1}}-1}{e^{-i(v+2\pi l)/n^j}-1} \right|^{2m}$$
$$= \frac{|1-e^{-iv}|^{2m}}{(2n^j)^{2m}} \sum_{l=0}^{n^j-1} \left(\sin \frac{v+2\pi l}{2n^j}\right)^{-2m}.$$

The comparison with Equation (3.9) completes the proof. ■

Lemma 3.5. *Let* $\Psi_{2m}(z) := \Phi_{2m}(z)\Phi_{2m}(-z)\Phi_{2m}(z^2)$. *Then*

$$\min\{\Psi_{2m}(z) : |z| = 1\} = \Psi(i) = \Phi_{2m}(i)^2 \Phi_{2m}(-1).$$

Proof: Setting $z := e^{iv}$ $(v \in [0,2\pi))$ in Ψ_{2m}, we obtain

$$\Psi_{2m}(e^{iv}) = \Phi_{2m}(e^{iv})\Phi_{2m}(-e^{iv})\Phi_{2m}(e^{2iv}).$$

For $v \in \{0,\pi\}$, we have

$$\Psi_{2m}(1) = \Psi_{2m}(-1) = \Phi_{2m}(-1) \ge \Phi_{2m}(-1)\Phi_{2m}(i)^2.$$

If $v \in (0, \pi)$, then by Equation (3.8),

$$\Psi_{2m}(e^{iv}) = U_{2m-2}(\sin \frac{v}{2})\, U_{2m-2}(\cos \frac{v}{2})\, U_{2m-2}(\cos^2 \frac{v}{2} - \sin^2 \frac{v}{2}) \quad (v \in (0,\pi)).$$

Since $\Psi_2 = 1$, we restrict our attention to the case $m \geq 2$. Using Equation (3.7), we obtain

$$\Psi_{2m}(e^{iv}) = a^3 \prod_{j=1}^{m-1} ((\sin^2 \frac{v}{2}+a_j)(\cos^2 \frac{v}{2}+a_j)((\cos^2 \frac{v}{2}-\sin^2 \frac{v}{2})^2+a_j)) \quad (m \geq 2)$$

and substituting $u := \cos^2(v/2) - (1/2)$, we have

$$\begin{aligned}\Psi_{2m}(e^{iv}) =& a^3 \prod_{j=1}^{m-1} ((\frac{1}{2} - u + a_j)(\frac{1}{2} + u + a_j)(4u^2 + a_j)) \quad (u \in (-\frac{1}{2}, \frac{1}{2})), \\ =& a^3 \prod_{j=1}^{m-1} \Big(-\Big(2u^2 - \frac{4a_j^2 + 3a_j + 1}{4}\Big)^2 + \\ &+ \Big(\frac{4a_j^2 + 3a_j + 1}{4}\Big)^2 + a_j(a_j + \frac{1}{2})^2 \Big). \end{aligned} \tag{3.10}$$

The above product is minimal if and only if $u = 0$, *i.e.*, if and only if $v = \pi/2$.
■

By Equation (3.10), the corresponding value

$$\max\{\Psi_{2m}(z) : |z| = 1\}$$

depends on the numbers a_j $(j = 1, \ldots, m-1)$ if $m \geq 3$. Hence, its determination is only numerically possible.

Finally, we list a property of $\Phi_{m,j}$ which is a direct consequence of Corollary 3.2. We will need this property in the proof of Theorem 5.3.

Lemma 3.6. *Let $G = \mathbb{R}$ or $G = \mathbf{Z}$. If $|z| = 1$, then*

$$\sum_{r=0}^{n-1} |P(zw_n^r)|^2\, \Phi_{2m,j}(zw_n^r) = \Phi_{2m,j+1}(z^n) \quad (w_n := e^{-2\pi i/n}).$$

§4 Multiresolution analysis

In the sequel, we fix m. We set

$$\varphi_j := N_{m,j},$$
$$\varphi_{j,k} := \varphi_j(x - kn^j) \quad (x \in G),$$

and introduce the vector space V_j by

$$V_j := \operatorname{clos}_{L^2(G)}\{\operatorname{span}\{n^{-j/2}\varphi_{j,k} : k \in I_j\}\}. \tag{4.1}$$

Note that the vector spaces V_j have finite dimension

$$\dim V_j = N/n^j. \tag{4.2}$$

in the periodic case $G = \mathbb{R}/N\mathbf{Z}$, $\mathbf{Z}/N\mathbf{Z}$. Clearly, for each $f_j \in V_j$, the shifts $f_j(x - kn^j)$ $(k \in I_j)$ are also in V_j which means that V_j is U_j-translation invariant. The sequence of vector spaces $\{V_j\}$ forms a multiresolution analysis of $L^2(G)$, *i.e.*,

(i) $V_j \supset V_{j+1}$,

(ii) $\operatorname{clos}_{L^2(R)}\{\bigcup_{j\in Z} V_j\} = L^2(\mathbb{R})$ if $G = \mathbb{R}$,

$\operatorname{clos}_{L^2(R/NZ)}\{\bigcup_{j\in Z,\, j\le t} V_j\} = L^2(\mathbb{R}/N\mathbf{Z})$ if $G = \mathbb{R}/N\mathbf{Z}$,

$\operatorname{clos}_{L^2(Z)}\{\bigcup_{j\in N_0} V_j\} = L^2(\mathbf{Z})$ if $G = \mathbf{Z}$,

$\operatorname{clos}_{L^2(Z/NZ)}\{\bigcup_{j\in N_0,\, j\le t} V_j\} = L^2(\mathbf{Z}/N\mathbf{Z})$ if $G = \mathbf{Z}/N\mathbf{Z}$,

(iii) $\bigcap_{j\in Z} V_j = \{0\}$ if $G = \mathbb{R}$,

$\bigcap_{j\in Z,\, j\le t} V_j = V_t = \operatorname{span}\{\chi_{R/NZ}\}$ if $G = \mathbb{R}/N\mathbf{Z}$,

$\bigcap_{j\in N_0} V_j = \{0\}$ if $G = \mathbf{Z}$,

$\bigcap_{j\in N_0,\, j\le t} V_j = V_t = \operatorname{span}\{\chi_{Z/NZ}\}$ if $G = \mathbf{Z}/N\mathbf{Z}$.

By Equation (2.6), we obtain that $V_j \supset V_{j+1}$. The property (ii) holds for the discrete groups by the definition of V_0 and for the non-discrete groups by arguments used in [1, Theorem 4.3; 3]. Finally, (iii) follows by the definition of V_t in the periodic case and by [1, Corollary 4.14; 10] in the non-periodic case. Note that the proof of (iii) for $G = \mathbb{R}$ in [10] uses the Riesz basis property $\varphi_{j,k}$ $(k \in I_j)$ which we will show for our general setting in the following theorem. Then we can prove (iii) for $G = \mathbf{Z}$ in the same way as in [10].

Theorem 4.1. *The set $\{n^{-j/2}\varphi_{j,k} : k \in I_j\}$ forms a Riesz basis of V_j, i.e., there exist constants $0 < A \le A' < \infty$ such that*

$$A\,\|\lambda\|^2_{L^2(U_j)} \le \|\sum_{k\in I_j} \lambda_k n^{-j/2}\varphi_{j,k}\|^2_{L^2(G)} \le A'\,\|\lambda\|^2_{L^2(U_j)} \tag{4.3}$$

for all $\lambda \in L^2(U_j)$ *with* $\lambda(kn^j) := \lambda_k$ $(k \in I_j)$. *The Riesz bounds*

$$A = A_m := \frac{2^{2m}(2^{2m}-1)}{(2m)!}\,|B_{2m}|,\ A' = A'_m := 1$$

are the best possible constants in Equation (4.3).

Proof: We show that Equation (4.3) is equivalent to

$$A \le n^{-2j} \sum_{\alpha \in \Lambda_j} |\varphi_j(\alpha\gamma)|^2 \le A' \quad (\gamma \in \hat{U}_j).$$

Then the assertion follows immediately by Corollary 3.2, Equations (3.5) and (3.6), and Lemma 3.4. By the Parseval identity and the translation property of the Fourier transform, we obtain

$$\|\sum_{k \in I_j} \lambda_k\, n^{-j/2} \varphi_{j,k}\|^2_{L^2(G)} = n^{-j} \int_{\hat{G}} |\hat{\lambda}(\gamma)\hat{\varphi}_j(\gamma)|^2 \mathrm{d}m_{\hat{G}}(\gamma)$$

with $\hat{\lambda}(\gamma) := \sum_{k \in I_j} \lambda_k \bar{\gamma}(kn^j)$. Since $\alpha(kn^j) = 1$ for all $\alpha \in \Lambda_j$ and $\hat{G}/\Lambda_j \cong \hat{U}_j$, we conclude further that

$$\int_{\hat{G}} |\hat{\lambda}(\gamma)\hat{\varphi}_j(\gamma)|^2\, \mathrm{d}m_{\hat{G}}(\gamma) = \int_{\hat{U}_j} |\hat{\lambda}(\gamma)|^2\, n^{-j} \sum_{\alpha \in \Lambda_j} |\hat{\varphi}_j(\alpha\gamma)|^2\, \mathrm{d}m_{\hat{U}_j}(\gamma).$$

Moreover, we have by the Parseval identity

$$\|\lambda\|^2_{L^2(U_j)} = \|\hat{\lambda}\|^2_{L^2(\hat{U}_j)} = \int_{\hat{U}_j} |\hat{\lambda}(\gamma)|^2\, \mathrm{d}m_{\hat{U}_j}(\gamma).$$

Hence, Equation (4.3) is equivalent to

$$A\|\hat{\lambda}\|^2_{L^2(\hat{U}_j)} \le \int_{\hat{U}_j} |\hat{\lambda}(\gamma)|^2\, n^{-2j} \sum_{\alpha \in \Lambda_j} |\hat{\varphi}_j(\alpha\gamma)|^2\, \mathrm{d}m_{\hat{U}_j}(\gamma) \le A'\|\hat{\lambda}\|^2_{L^2(\hat{U}_j)},$$

and we are finished. ■

By Equation (2.4), the sequence $\{V_j\}$ is a stationary multiresolution analysis if $G = \mathbb{R}$, $\mathbb{R}/N\mathbf{Z}$, *i.e.*,

$$f(x) \in V_{j+1} \Longleftrightarrow f(nx) \in V_j \quad (x \in G).$$

§5 Shifted spline wavelets

Our next aim consists in constructing $n-1$ functions ψ_{j+1}^s $(s=1,\ldots,n-1)$ whose U_{j+1}-translates span the orthogonal complement W_{j+1} of V_{j+1} in V_j.

For $n \geq 3$, one can establish additional assumptions on the relations between the functions ψ_{j+1}^s. For $G = \mathbb{R}$ and the dilation factor $n = 3$, Chui and Lian [5] have constructed wavelets ψ_{j+1}^s $(s = 1,2)$ which are symmetric and antisymmetric, respectively, and which satisfy $\psi_{j+1}^1 \perp \psi_{j+1}^2$. Another approach to orthogonal wavelets with arbitrary dilation factor $n \geq 2$ $(n \in \mathbf{Z})$ was given by Welland and Lundberg [19]. In this paper, we simply assume that

$$\psi_{j+1}^s(x) = \psi_{j+1}^1(x-(s-1)n^j) \quad (s=1,\ldots,n-1), \tag{5.1}$$

i.e., the functions ψ_{j+1}^s are U_j-shifted versions of an initial function ψ_{j+1}^1. Furthermore, in the non-periodic case, our functions should have minimal support among all functions with the property in Equation (5.1).

By Equation (5.1), it suffices to determine the initial function ψ_{j+1}^1. We introduce the polynomial

$$Q_j(z) := n^{-1} \sum_{k\in Z} q_k\, z^k = z^{m(n-1)}\,(1-z)^m\, \frac{R_j(z^n)}{\Phi_{2m,j}(z)} \tag{5.2}$$

with

$$R_j(z^n) := \prod_{r=0}^{n-1} \Phi_{2m,j}(zw_n^r).$$

For $z \in \mathbb{C}$ with $|z| = 1$, this is equivalent to

$$Q_j(z) = \Big(\frac{1-z^n}{n}\Big)^m \frac{1}{\overline{P(z)}}\, \frac{R_j(z^n)}{\Phi_{2m,j}(z)}. \tag{5.3}$$

If $G = \mathbb{R}$, then the polynomial Q_j is independent of j, *i.e.*,

$$Q_j(z) = Q(z) := z^{m(n-1)}\,(1-z)^m\, \frac{R(z^n)}{\Phi_{2m}(z)}, \quad R(z) := \prod_{r=0}^{n-1} \Phi_{2m}(zw_n^r).$$

This is not the case for $G = \mathbf{Z}$, but we have by Equation (3.6) that

$$\lim_{j\to\infty} Q_j = Q.$$

Now we define the *two-scale symbol* Q_j^s of ψ_{j+1}^s by

$$Q_j^s(z) := n^{-1} \sum_{k\in Z} q_k^s\, z^k = z^s\, Q_j(z) \quad (s=1,\ldots,n-1), \tag{5.4}$$

i.e., $q_k^s = q_{k-s}$ and

$$\psi_{j+1}^s(x) = \sum_{k \in Z} q_k^s \, \varphi_{j,k}(x) \quad (x \in G),$$
$$\hat{\psi}_{j+1}^s(\gamma) = n \, Q_j^s(\bar{\gamma}(n^j)) \hat{\varphi}_j(\gamma) \quad (\gamma \in \hat{G};\ s = 1, \ldots, n-1). \tag{5.5}$$

Obviously, ψ_{j+1}^s $(s = 2, \ldots, n-1)$ are only U_j-shifted versions of ψ_{j+1}^1. We refer to ψ_{j+1}^s $(s = 1, \ldots, n-1)$ as $(j+1)$-*th spline wavelets* although the notation "wavelet" is not justified up to now.

Note that a similar construction idea for multivariate shifted wavelets with dilation factor $n = 2$ can be found in [1, Theorem 7.11].

For $G = \mathbb{R}$ and $n = 2$, the two-scale symbol

$$Q^1(z) = z \, Q(z) = z^{m+1} \, (1-z)^m \, \Phi_{2m}(-z)$$

coincides (except of the shift by z^{-2} and the scaling factor $(-1)^{m-1}2^{-m}$) with those of the Chui-Wang wavelet [6]. Clearly, for $G = \mathbb{R}$, we get stationary wavelets, *i.e.*,

$$\psi_{j+1}^s(nx) = \psi_j^s(x) \quad (x \in \mathbb{R}).$$

This is not the case for $G = \mathbf{Z}$. Again, by

$$Q_j^{II}(\bar{\gamma}(n^j)) = Q_j^I(\bar{\gamma}(n^j)) \quad (\gamma \in n^j \widehat{\mathbf{Z}/N\mathbf{Z}};\ j < t), \tag{5.6}$$

where Q_j^I is used in the non-periodic case $G = \mathbb{R}, \mathbf{Z}$ and Q_j^{II} in the corresponding periodic case, it is possible to restrict many considerations to the non-periodic case. Especially, Equation (5.6) implies the following.

Lemma 5.1. *Let $\psi_{j+1}^{s,I}$ be the spline wavelet over $G = \mathbb{R}$ or $G = \mathbf{Z}$ and let $\psi_{j+1}^{s,II}$ be the corresponding spline wavelet over $G/N\mathbf{Z}$. Then*

$$\psi_{j+1}^{s,II}(x) = \sum_{k \in Z} \psi_{j+1}^{s,I}(x + kN) \quad (j < t).$$

To justify the notation "wavelets" for our functions ψ_{j+1}^s, we show that

$$W_{j+1} := \operatorname{clos}_{L^2(G)}\{\operatorname{span}\{\psi_{j+1,k}^s : s = 1, \ldots, n-1;\ k \in I_{j+1}\}\} \tag{5.7}$$

is the orthogonal complement of V_{j+1} in V_j and that the functions $\psi_{j+1,k}^s$ $(s = 1, \ldots, n-1; k \in I_{j+1})$ provide a Riesz basis of W_{j+1}.

Let us begin with the following theorem.

Theorem 5.2. *Let V_{j+1} and W_{j+1} be given by Equations* (4.1) *and* (5.7), *respectively. Then*

$$V_{j+1} \perp W_{j+1}.$$

Proof: By the Parseval identity and the two-scale relations in Equations (2.7) and (5.5), we obtain

$$\langle \psi_{j+1}^s, \varphi_{j+1,k}\rangle_{L^2(G)} = n^2 \int_{\hat G} \bar\gamma(sn^j - kn^{j+1}) Q_j(\bar\gamma(n^j)) P(\gamma(n^j)) |\hat\varphi_j(\gamma)|^2 \, dm_{\hat G}(\gamma)$$

and further by splitting the integral and applying Corollary 3.2

$$\begin{aligned}\langle \psi_{j+1}^s, \varphi_{j+1,k}\rangle_{L^2(G)} =& n^{j+2} \int_{\hat U_j} \bar\gamma(sn^j - kn^{j+1})\, Q_j(\bar\gamma(n^j)) \\ & \times P(\gamma(n^j))\, \Phi_{2m,j}(\bar\gamma(n^j)) dm_{\hat U_j}(\gamma).\end{aligned}$$

By $U_j/U_{j+1} \cong \mathbf{Z}/n\mathbf{Z}$, the annihilator Λ_{j+1}^j of U_{j+1} in $\hat U_j$ consists of β_r ($r = 0, \ldots, n-1$) with $\beta_r(n^j) = \bar w_n^r$. By $\hat U_j/\Lambda_{j+1}^j \cong \hat U_{j+1}$, we conclude

$$\begin{aligned}\langle \psi_{j+1}^s, \varphi_{j+1,k}\rangle_{L^2(G)} =& n^{j+1} \int_{\hat U_{j+1}} \sum_{r=0}^{n-1} \bar\gamma\bar\beta_r(sn^j - kn^{j+1}) Q_j(\bar\gamma\bar\beta_r(n^j)) \\ & \times P(\gamma\beta_r(n^j))\, \Phi_{2m,j}(\bar\gamma\bar\beta_r(n^j))\, dm_{\hat U_{j+1}}(\gamma) \\ =& n^{j+1} \int_{\hat U_{j+1}} \bar\gamma\bar\beta_r(sn^j - kn^{j+1}) A(z)\, dm_{\hat U_{j+1}}(\gamma)\end{aligned}$$

with $z := \bar\gamma(n^j)$ and

$$A(z) := \sum_{r=0}^{n-1} w_n^{rs} Q_j(zw_n^r)\, \overline{P(zw_n^r)}\, \Phi_{2m,j}(zw_n^r).$$

By Equation (5.3) and the orthogonality relation of characters, we get finally

$$A(z) = \Big(\frac{1-z^n}{n}\Big)^m R_j(z^n) \sum_{r=0}^{n-1} w_n^{rs} = 0 \quad (s = 1, \ldots, n-1). \quad \blacksquare$$

To verify $V_j = V_{j+1} \oplus W_{j+1}$, it remains to prove that $V_j = V_{j+1} + W_{j+1}$. In the periodic case $G = \mathbf{R}/N\mathbf{Z}, \mathbf{Z}/N\mathbf{Z}$, the latter follows immediately by the dimensions in Equation (4.2) of the vector spaces. We decompose the polynomials P and Q_j^s as follows:

$$P(z) := \sum_{r=0}^{n-1} z^r P_r(z^n), \quad Q_j^s(z) := \sum_{r=0}^{n-1} z^r Q_{j,r}^s(z^n)$$

with

$$P_r(z) := n^{-1} \sum_{k\in Z} p_{nk+r}\, z^k, \,. \tag{5.8}$$

$$Q_{j,r}^s(z) := n^{-1} \sum_{k\in Z} q_{nk+r}^s\, z^k \quad (r = 0, \ldots, n-1). \tag{5.9}$$

To unify the notation, we set

$$\psi_{j+1}^{0} := \varphi_{j+1}, \;\; Q_j^0 := P, \;\; Q_{j,r}^0 := P_r,$$

and rewrite the two-scale relations in Equations (2.7) and (5.5) as

$$\begin{aligned}\hat{\psi}_{j+1}^{s}(\gamma) &= n\, Q_j^s(\bar{\gamma}(n^j))\, \hat{\varphi}_j(\gamma)\\ &= n \sum_{r=0}^{n-1} \bar{\gamma}(rn^j)\, Q_{j,r}^s(\bar{\gamma}(n^{j+1}))\, \hat{\varphi}_j(\gamma)\\ &= n \sum_{r=0}^{n-1} Q_{j,r}^s(\bar{\gamma}(n^{j+1}))\, \hat{\varphi}_{j,r}(\gamma) \quad (s = 0, \ldots, n-1).\end{aligned}$$

This is equivalent to

$$\left(\hat{\psi}_{j+1}^{s}(\gamma)\right)_{s=0}^{n-1} = n\, \mathbf{S}_j(\bar{\gamma}(n^{j+1}))\left(\hat{\varphi}_{j,r}(\gamma)\right)_{r=0}^{n-1} \quad (\gamma \in \hat{G}) \tag{5.10}$$

with the *two-scale symbol matrix*

$$\mathbf{S}_j(z) := \left(Q_{j,r}^s(z)\right)_{s,r=0}^{n-1}.$$

By Theorem 4.1, the set $\{\varphi_{j,r}(x - kn^{j+1}) : r = 0, \ldots, n-1; k \in I_{j+1}\}$ provides a Riesz basis of V_j. Then it is well-known [1, Theorem 2.26] that $\{\psi_{j+1,k}^s : s = 0, \ldots, n-1; k \in I_{j+1}\}$ is a Riesz basis of V_j if and only if

$$|\det \mathbf{S}_j(z)| > 0 \tag{5.11}$$

for all $z \in \mathbb{C}$ with $|z| = 1$. Since $\{\psi_{j+1,k}^0 : k \in I_{j+1}\}$ is a Riesz basis of V_{j+1} and $V_{j+1} \perp W_{j+1}$, this implies that $\{\psi_{j+1,k}^s : s = 1, \ldots, n-1; k \in I_{j+1}\}$ is a Riesz basis of W_{j+1}.

To prove Equation (5.12), we use that

$$Q_j^s(zw_n^l) = \sum_{r=0}^{n-1} z^r\, w_n^{rl}\, Q_{j,r}^s(z^n) \quad (l = 0, \ldots, n-1)$$

which yields

$$\mathbf{S}_j(z^n)\, \mathrm{diag}(z^r)_{r=0}^{n-1}\, \mathbf{F}_n = \mathbf{M}_j(z) \tag{5.12}$$

with

$$\mathbf{M}_j(z) := \left(Q_j^s(zw_n^l)\right)_{s,l=0}^{n-1}$$

and with the *n-th Fourier matrix* $\mathbf{F}_n := (w_n^{rl})_{r,l=0}^{n-1}$. For $G = \mathbb{R}$ and $n = 2$, the matrix $\mathbf{M}_j$ coincides with

$$\mathbf{M}(z) = \begin{pmatrix} P(z) & P(-z) \\ Q(z) & Q(-z) \end{pmatrix}$$

considered in [3, p.141].

We use the special structure of $\mathbf{M}_j(z)$ to prove the following theorem.

Theorem 5.3. *For all $z \in \mathbb{C}$ with $|z| = 1$, it holds*

$$\det \mathbf{S}_j(z) = n^{m-1}\,(-z)^{m(n-1)}\,R_j(z)^{n-2}\,\Phi_{2m,j+1}(z) > 0.$$

Proof: By Equation (5.4), the matrix $\mathbf{M}_j$ has the special structure

$$\mathbf{M}_j(z) = \operatorname{diag}\,(z^r)_{r=0}^{n-1} \begin{pmatrix} P(z)P(zw_n)\ldots P(zw_n^{n-1}) \\ \hline \tilde{\mathbf{F}}_n \operatorname{diag}(Q_j(zw_n^r))_{r=0}^{n-1} \end{pmatrix},$$

where $\tilde{\mathbf{F}}_n$ denotes the $(n-1,n)$-matrix which follows from $\mathbf{F}_n$ by deleting the first row. Now we obtain by Equation (5.12) and by $\mathbf{F}_n^{-1} = n^{-1}\bar{\mathbf{F}}_n$ that

$$\det \mathbf{S}_j(z^n) = (\det \mathbf{M}_j(z))\,(\det \mathbf{F}_n)^{-1}\,z^{-n(n-1)} = n^{-1}\prod_{r=0}^{n-1} Q_j(zw_n^r)\sum_{l=0}^{n-1}\frac{P(zw_n^l)}{Q_j(zw_n^l)}$$

and by Equation (5.3) that

$$\begin{aligned}\det \mathbf{S}_j(z^n) =& n^{-1}\Big(\frac{1-z^n}{n}\Big)^{m(n-1)} R_j(z^n)^{n-2}\,\Big(\prod_{r=0}^{n-1}\overline{P(zw_n^r)}\Big)^{-1} \\ &\cdot \sum_{l=0}^{n-1} |P(zw_n^l)|^2 \Phi_{2m,j}(zw_n^l).\end{aligned}$$

Applying Lemma 3.6 and Equation (2.8), we finish by

$$\det \mathbf{S}_j(z^n) = (-1)^{m(n-1)}\,n^{m-1}\,z^{n(n-1)m}\,R_j(z^n)^{n-2}\,\Phi_{2m,j+1}(z^n). \quad \blacksquare$$

Thus $\{\psi_{j+1,k}^s : s = 1,\ldots,n-1; k \in I_{j+1}\}$ is a Riesz basis of W_{j+1}. Moreover, we conclude that $V_j = V_{j+1} \oplus W_{j+1}$.

For $n = 2$, we can estimate the best possible Riesz bounds using Lemma 3.5 and similar techniques as in the proof of Theorem 4.1. More precisely, we have proved in [18] the following theorem.

Theorem 5.4. *Let $n = 2$. The set $\{2^{-(j+1)/2}\psi_{j+1,k} : k \in I_{j+1}\}$ forms a Riesz basis of W_{j+1}, i.e., there exist constants $0 < D \le D' < \infty$ such that*

$$D\,\|\lambda\|^2_{L^2(U_{j+1})} \le \|\sum_{k\in I_{j+1}} \lambda_k\,2^{-(j+1)/2}\psi_{j+1,k}\|^2_{L^2(G)} \le D'\,\|\lambda\|^2_{L^2(U_{j+1})} \quad (5.13)$$

for all $\lambda \in L^2(U_{j+1})$ with $\lambda(k\,2^{j+1}) := \lambda_k$ $(k \in I_{j+1})$. The Riesz bounds

$$D = D_m := 2^{2m}\Psi(i),$$

$$D' = D'_m := \begin{cases} 2^{2m} \max\{\Psi(z) : |z| = 1\} & G = \mathbf{R}, \mathbf{R}/N\mathbf{Z}, \\ 2^{2m} & G = \mathbf{Z}, \mathbf{Z}/N\mathbf{Z}, \end{cases}$$

are the best possible constants in Equation (5.13).

Finally, we remark that in the non-periodic case a completely similar approach as in [6] shows that our wavelets are those with minimal support among all wavelets with the shift property in Equation (5.1). By Equations (2.2), (3.1), (3.3), and (5.2), we check that

$$m_G(\operatorname{supp} \psi_{j+1}) = \begin{cases} 2n^j(n(m-1)+1) & G = \mathbf{R}, \\ 2n^j(n(m-1)+1) - m + 1 & G = \mathbf{Z}, \end{cases}$$

supposed that $j \geq \log_n m$ if $G = \mathbf{Z}$.

§6 Decomposition and reconstruction of functions

Assume that $f \in L^2(G)$ can be approximated by $f_j \in V_j$ with sufficiently small error $||f - f_j||_{L^2(G)}$. Let

$$f_j = \sum_{k \in I_j} c_{j,k}\, \varphi_{j,k}, \tag{6.1}$$

with $(c_{j,k})_{k \in I_j} \in L^2(U_j)$. By $V_j = V_{j+1} \oplus W_{j+1}$, the function f_j can be uniquely decomposed as

$$f_j = f_{j+1} + g_{j+1}, \tag{6.2}$$

where

$$f_{j+1} = \sum_{k \in I_{j+1}} c^0_{j+1,k}\, \psi^0_{j+1,k} \in V_{j+1}, \tag{6.3}$$

$$g_{j+1} = \sum_{s=1}^{n-1} \sum_{k \in I_{j+1}} c^s_{j+1,k}\, \psi^s_{j+1,k} \in W_{j+1} \tag{6.4}$$

with $(c^s_{j+1,k})_{k \in I_{j+1}} \in L^2(U_{j+1})$ $(s = 0, \ldots, n-1)$.

We are interested in computing $(c_{j,k})_{k \in I_j}$ from given $(c^s_{j+1,k})_{k \in I_{j+1}}$ $(s = 0, \ldots, n-1)$ and conversely. The first relation will be needed for the decomposition of functions and the second for its reconstruction.

Applying the Fourier transform in Equations (6.1), (6.3), and (6.4), we obtain

$$\hat{f}_j(\gamma) = \sum_{r=0}^{n-1} \hat{c}_{j,r}(\bar{\gamma}(n^{j+1}))\, \hat{\varphi}_{j,r}(\gamma),$$

$$\hat{f}_{j+1}(\gamma) = \hat{c}^0_{j+1}(\bar{\gamma}(n^{j+1}))\, \hat{\psi}^0_{j+1}(\gamma),$$

$$\hat{g}_{j+1}(\gamma) = \sum_{s=1}^{n-1} \hat{c}^s_{j+1}(\bar{\gamma}(n^{j+1}))\, \hat{\psi}^s_{j+1}(\gamma) \quad (\gamma \in \hat{G})$$

with

$$\hat{c}_{j,r}(\bar{\gamma}(n^{j+1})) := \sum_{k \in I_{j+1}} c_{j,nk+r}\, \bar{\gamma}(kn^{j+1}) \quad (r = 0, \ldots, n-1), \tag{6.5}$$

$$\hat{c}^s_{j+1}(\bar{\gamma}(n^{j+1})) := \sum_{k \in I_{j+1}} c^s_{j+1,k}\, \bar{\gamma}(kn^{j+1}) \quad (s = 0, \ldots, n-1) \tag{6.6}$$

and by Equation (6.2)

$$\sum_{r=0}^{n-1} \hat{c}_{j,r}(\bar{\gamma}(n^{j+1}))\, \hat{\varphi}_{j,r}(\gamma) = \sum_{s=0}^{n-1} \hat{c}^s_{j+1}(\bar{\gamma}(n^{j+1}))\, \hat{\psi}^s_{j+1}(\gamma) \quad (\gamma \in \hat{G}).$$

The last equation together with Equation (5.10) and Theorem 5.3 implies the following theorem.

Theorem 6.1. *For $z := \bar{\gamma}(n^{j+1})$ $(\gamma \in \hat{G})$, it holds*

$$(\hat{c}_{j,r}(z))_{r=0}^{n-1} = n\, (\mathbf{S}_j(z))^{\mathrm{T}}\, (\hat{c}^s_{j+1}(z))_{s=0}^{n-1},$$
$$(\hat{c}^s_{j+1}(z))_{s=0}^{n-1} = n^{-1}\, ((\mathbf{S}_j(z))^{\mathrm{T}})^{-1}\, (\hat{c}_{j,r}(z))_{r=0}^{n-1}.$$

Let

$$\Big(H^s_{j,r}(z)\Big)_{r,s=0}^{n-1} := \mathbf{S}_j(z)^{-1} = (\det \mathbf{S}_j(z))^{-1} \Big((-1)^{r+s} S_j(z)_{sr}\Big)_{r,s=0}^{n-1}, \tag{6.7}$$

where $S_j(z)_{sr}$ denotes the determinant of the matrix which follows from $\mathbf{S}_j(z)$ by cancelling the s-th row and the r-th column. Let

$$H^s_{j,r}(z) := n \sum_{k \in I_{j+1}} h^s_{nk-r}\, z^k \quad (r, s = 0, \ldots, n-1). \tag{6.8}$$

In the non-periodic case, the sequences $(h^s_{nk-r})_{k \in Z}$ are not finite, but by Theorem 5.3, they are rapidly decreasing, *i.e.*, $\lim_{k \to \infty} |k|^\mu\, |h^s_{nk-r}| = 0$ for all $\mu > 0$.

By Equations (6.5)-(6.8), (5.8), and (5.9), Theorem 6.1 can be rewritten as the following corollary.

Corollary 6.2. *It holds*

$$c_{j,k} = \sum_{l \in I_{j+1}} \Big(c^0_{j+1,l}\, p_{k-nl} + \sum_{s=1}^{n-1} c^s_{j+1,l}\, q_{k-nl-s}\Big) \quad (k \in I_j),$$

$$c^s_{j+1,l} = \sum_{r=0}^{n-1} \sum_{k \in I_{j+1}} c_{j,nk+r}\, h^s_{n(l-k)-r} \quad (l \in I_{j+1};\, s = 0, \ldots, n-1).$$

Theorem 6.1 and Corollary 6.2 give rise to efficient decomposition and reconstruction algorithms which are straightforward generalizations of the algorithms in [3,14]. Note that especially the algorithms in [14] apply Fast Fourier Transform (FFT) techniques.

Appendix: Locally compact abelian groups

Let G be an LCA group with addition as group operation and 0 as identity element. The abbreviation LCA will be used for "locally compact abelian". There exists a non-negative regular measure m_G on G, the so-called *Haar measure of* G which is translation-invariant and not identically 0. The Haar measure is unique up to multiplicative constants. Let $L^p(G)$ denote the set of all measurable complex-valued functions on G with finite norm

$$\|f\|_{L^p(G)} := \{\int_G |f(x)|^p \, dm_G(x)\}^{1/p}.$$

We identify functions which differ only on a set of Haar measure 0. Then $L^p(G)$ becomes a Banach space for $1 \leq p < \infty$. Especially, $L^2(G)$ is a Hilbert space with inner product

$$\langle f, g\rangle_{L^2(G)} := \int_G f(x)\overline{g(x)} \, dm_G(x) \quad (y \in G).$$

The *convolution* of f, g is defined by

$$(f * g)(y) := \int_G f(x) \, g(y-x) \, dm_G(x),$$

provided that $\int_G |f(x) \, g(y-x)| \, dm_G(x) < \infty$. For $f, g \in L^1(G)$ it holds that $f*g \in L^1(G)$ and that $\|f*g\|_1 \leq \|f\|_1 \|g\|_1$. With convolution as multiplication, $L^1(G)$ becomes a commutative Banach algebra over $\mathbb{C}$. Let $C(G)$ denote the set of all bounded continuous complex-valued functions on G and $C_0(G)$ the set of all bounded continuous complex-valued functions which vanish at infinity. With

$$\|f\|_\infty := \sup_{x\in G} |f(x)|,$$

$C_0(G)$ becomes a Banach space and, with multiplication defined by the ordinary multiplication of functions, a commutative Banach algebra. A continuous homomorphism γ of G into the multiplicative group of all complex numbers of absolute value 1 is called a *character of* G. With the group operation defined by

$$(\gamma_1\gamma_2)(x) := \gamma_1(x) \, \gamma_2(x) \quad (x \in G),$$

all characters of G form an abelian group, the *character group* $\hat{G}$ of G. If G is compact, then $\hat{G}$ is discrete and conversely. For all $f \in L^1(G)$ the function $\hat{f}$ defined on $\hat{G}$ by

$$\hat{f}(\gamma) := \int_G f(x)\, \bar{\gamma}(x)\, \mathrm{d}m_G(x) \quad (\gamma \in \hat{G})$$

is called the *Fourier transform of* f. The set of all functions $\hat{f}$ so obtained will be denoted by $A(\hat{G})$. Then $A(\hat{G})$ is a dense subalgebra of $C_0(\hat{G})$. If we give $\hat{G}$ the weak topology induced by $A(\hat{G})$, it becomes an LCA group.

Theorem A.1. *The map* $F: \ L^1(G) \to A(\hat{G})$ *with* $f \mapsto \hat{f}$ *is an algebra isomorphism.*

As direct consequence, we obtain the following corollary.

Corollary A.2. *Let* $f, g \in L^1(G)$. *Then it holds*

$$\begin{aligned} \hat{h}(\gamma) &= \bar{\gamma}(g)\hat{f}(\gamma) && (h(x) := f(x-g)) && \textit{(translation property)},\\ \hat{h}(\gamma) &= \overline{f}\hat{\ }(\gamma) && (h(x) := \bar{f}(-x)) && \textit{(symmetry property)},\\ (f * g)\hat{\ }(\gamma) &= \hat{f}(\gamma)\hat{g}(\gamma) && (\gamma \in \hat{G}) && \textit{(convolution property)}. \end{aligned}$$

Let Π be the class of positive definite functions on G and let $[L^1 \cap \Pi]$ be the vector space generated by $L^1 \cap \Pi$. Note that $[L^1 \cap \Pi]$ is dense in $L^1(G)$ and in $L^2(G)$.

Theorem A.3. (**Inversion Theorem**) *If* $f \in [L^1 \cap \Pi]$, *then* $\hat{f} \in L^1(\hat{G})$ *and*

$$f(x) = \int_{\hat{G}} \hat{f}(\gamma)\, \gamma(x)\, \mathrm{d}m_{\hat{G}}(\gamma)$$

for almost all x, *where* $m_{\hat{G}}$ *is the Haar measure of* $\hat{G}$ *suitably normalized.*

By the Plancherel theorem, the Fourier transform can be extended to an isomorphism of $L^2(G)$ onto $L^2(\hat{G})$.

Theorem A.4. (**Plancherel Theorem**) *The Fourier transform restricted to* $[L^1 \cap \Pi]$ *is an isometry onto a dense linear subspace of* $L^2(\hat{G})$. *It may be extended, in a unique manner, to an isometry of* $L^2(G)$ *onto* $L^2(\hat{G})$. *There holds the Parseval identity*

$$\langle f, g\rangle_{L^2(G)} = \langle \hat{f}, \hat{g}\rangle_{L^2(\hat{G})}.$$

There exist the following duality relations between G and $\hat{G}$.

Theorem A.5. *Let G be an LCA group and let U be a closed subgroup of G. Then*

$$\Lambda_U := \{\alpha \in \hat{G} : \ \alpha(u) = 1 \ \text{ for all } \ u \in U\}$$

is a closed subgroup of $\hat{G}$, called annihilator of U in $\hat{G}$. It holds

$$\Lambda_U \cong \widehat{G/U}, \quad \hat{G}/\Lambda_U \cong \hat{U}.$$

Especially, every LCA group G is the character group of its character group.

Finally, we want to apply the following theorem.

Theorem A.6. *Let G be an LCA group and let U be a closed subgroup of G. Suppose that $m_G, m_U, m_{G/U}$ are Haar measures of the indicated groups. Let $\xi = \xi(y)$ be the coset of U which contains $y \in G$. Let $f \in [L^1 \cap \Pi]$ such that*

$$g(\xi(y)) := \int_U f(x+y) \, \mathrm{d}m_U(x)$$

is a continuous function on G/U. As convenient, we write y instead of $\xi(y)$. Suppose that the Haar measures on G, U, and G/U are adjusted so that

$$\int_G f(x) \, \mathrm{d}m_G(x) = \int_{G/U} \int_U f(x+y) \, \mathrm{d}m_U(x) \, \mathrm{d}m_{G/U}(y).$$

Then it holds

$$\text{(i) } \hat{g}(\alpha) = \hat{f}(\alpha) \quad (\alpha \in \widehat{G/U}),$$

$$\text{(ii) } g(y) = \int_{\widehat{G/U}} \hat{g}(\alpha) \, \alpha(y) \, \mathrm{d}m_{\widehat{G/U}}(\alpha) \quad (y \in G/U).$$

We refer to the last equation as the Poisson summation formula although this notation is often used only in the special case $y = 0$.

In this paper, we consider the classical groups $\mathbb{R}, \mathbb{R}/N\mathbf{Z}, \mathbf{Z}$, and $\mathbf{Z}/N\mathbf{Z}$, where we normalize the Haar measure of G and its character group $\hat{G}$ with respect to the Theorems A.3 and A.6 as follows:

1. Let $G = \mathbb{R}$ be the additive group of real numbers with the natural topology of the real line and with Lebesgue measure as Haar measure. Then

$$\hat{\mathbb{R}} := \{\gamma_y(x) := e^{ixy} : \ y \in \mathbb{R}\}.$$

Under the correspondence $\gamma_y \leftrightarrow y$, it holds

$$\hat{\mathbb{R}} \cong \mathbb{R}.$$

It is convenient to identify γ_y with y taking into account that $\gamma_x\gamma_y$ corresponds to $x+y$ $(x,y\in\mathbb{R})$. Then the Fourier transform and its inverse are given by

$$\hat{f}(y) = \int_{\mathbf{R}} f(x)\, e^{-ixy}\, \mathrm{d}x \qquad (y \in \mathbb{R}),$$
$$f(x) = \frac{1}{2\pi}\int_{\mathbf{R}} \hat{f}(y)\, e^{ixy}\, \mathrm{d}y \qquad (x \in \mathbb{R}).$$

2. Let $G = \mathbb{R}/N\mathbf{Z}$ $(N \in \mathbb{R})$ be the additive group of reals modulo N with Lebesgue measure as Haar measure. Here G is compact such that $\hat{G}$ is discrete. It holds

$$\widehat{\mathbb{R}/N\mathbf{Z}} = \{\gamma_{2\pi k/N}(x) := e^{2\pi ikx/N} \;:\; k \in \mathbf{Z}\} \cong (2\pi/N)\mathbf{Z},$$
$$\hat{f}(2\pi k/N) = \int_0^N f(x)\, e^{-2\pi ikx/N}\, \mathrm{d}x \qquad (k \in \mathbf{Z}),$$
$$f(x) = \frac{1}{N}\sum_{k\in Z} \hat{f}(2\pi k/N)e^{2\pi ikx/N} \qquad (x \in [0,N)).$$

3. Let $G = N\mathbf{Z}$ with the discrete topology and with Haar measure $m_G(g) = 1$ for all $g \in G$. Since G is discrete, $\hat{G}$ is compact. It holds

$$\widehat{N\mathbf{Z}} = \{\gamma_v(Nk) := e^{ivNk} \;:\; v \in [0, 2\pi/N)\} \cong \mathbb{R}/(2\pi/N)\mathbf{Z},$$
$$\hat{f}(v) = \sum_{k\in Z} f(Nk)e^{-ivNk} \qquad (v \in [0,2\pi/N)),$$
$$f(Nk) = \frac{N}{2\pi}\int_0^{2\pi/N} \hat{f}(v)\, e^{ivNk}\, \mathrm{d}v \quad (k \in \mathbf{Z}).$$

4. Let $G = \mathbf{Z}/N\mathbf{Z}$ be the cyclic group of integers modulo N with the discrete topology and Haar measure $m_G(g) = 1$ for all $g \in G$. It holds

$$\widehat{\mathbf{Z}/N\mathbf{Z}} = \{\gamma_{2\pi k/N}(j) := e^{2\pi ijk/N} = w_N^{-jk} \;:\; k = 0,\ldots,N-1\} \cong \mathbf{Z}/N\mathbf{Z},$$
$$\hat{f}(2\pi k/N) = \sum_{j=0}^{N-1} f(j)\, w_N^{jk} \qquad (k = 0,\ldots,N-1),$$
$$f(j) = \frac{1}{N}\sum_{k=0}^{N-1} \hat{f}(2\pi k/N)\, w_N^{-jk} \qquad (j = 0,\ldots,N-1).$$

References

1. de Boor, C., R. A. DeVore, and A. Ron, On the construction of multivariate (pre)wavelets, *Constr. Approx.* **9** (1993), 167–190.

2. de Boor, C., K. Höllig, and S. Riemenschneider, *Box Splines*, Springer, New York, 1993.
3. Chui, C. K., *An Introduction to Wavelets*, Academic Press, Boston, 1992.
4. Chui, C. K., Wavelets and spline interpolation, in *Advances in Numerical Analysis II*, W. Light (ed.), Clarendon Press, Oxford, 1992.
5. Chui, C. K. and J. A. Lian, Construction of compactly supported symmetric and antisymmetric orthonormal wavelets, CAT-Report, 1993.
6. Chui, C. K. and J. Z. Wang, On compactly supported spline wavelets and a duality principle, *Trans. Amer. Math. Soc.* **330** (1992), 903–915.
7. Dahlke, S., Multiresolution analysis on locally compact abelain groups, RWTH Aachen, 1992, preprint.
8. Daubechies, I., *Ten Lectures on Wavelets*, SIAM, Philadelphia, 1992.
9. Gröchening, K. and W. R. Madych, Multiresolution analysis, Haar basis and self-semilar tilings of $\mathbb{R}^n$, *IEEE Trans. Inform. Theory* **38** (1992), 556–568.
10. Jia, R. Q. and C. A. Micchelli, Using the refinement equations for the construction of pre-wavelets II: Powers of two, in *Curves and Surfaces*, P. J. Laurent, A. Le Méhauté, and L. L. Schumaker (eds.), Academic Press, New York, 1991.
11. Loomis, L. H., *An Introduction to Abstract Harmonic Analysis*, D. van Nostrand Company, Princeton, 1953.
12. Meyer, Y., *Ondelettes et Opérateurs I*, Hermann, Paris, 1990.
13. Perrier, V. and C. Basdevant, La decomposition en ondelettes periodiques, un outil pour l'analyse de champs inhomogenes, theorie et algorithmes, *Rech. Aérospat.* **3** (1989), 53–67.
14. Plonka, G. and M. Tasche, Periodic spline wavelets, Univ. Rostock, 1993, preprint.
15. Rudin, W., *Fourier Analysis on Groups*, J. Wiley & Sons, New York, 1979.
16. Schoenberg, I. J., Cardinal interpolation and spline functions, *J. Approx. Theory* **2** (1969), 167–206.
17. Schoenberg, I. J., Contributions to the problem of approximation of equidistant data by analytic functions. Part B: On the problem of osculatory interpolation. A second class of analytic approximation formulae, *Quart. Appl. Math.* **IV** (1946), 112–141.
18. Steidl, G., Spline wavelets on cyclic groups, TH Darmstadt, 1993, preprint.
19. Welland G. V. and M. Lundberg, Construction of compact p-wavelets, *Constr. Approx.* **9** (1993), 347–370.

Gabriele Steidl
TH Darmstadt, Fachbereich Mathematik
Schlossgartenstrasse 7
64289 Darmstadt, Germany
steidl@mathematik.th-darmstadt.de

A Practice of Data Smoothing by B-spline Wavelets

Susumu Sakakibara

Abstract. We describe how to use the multiresolution analysis generated by the compactly supported cardinal B-spline wavelets for data smoothing. The parameters needed for the implementation of the algorithm are determined for the case of cubic B-splines. Truncation is needed for the interpolation and decomposition, and the reasonable orders of truncation are discussed. An example of applications is also presented.

§1 Introduction

Time series data obtained from a measurement often represent a smooth function of time. In recovering the smooth function from the noisy data, the hierarchical structure of the multiresolution analysis (MRA) [5,4] generated by wavelets provides us with a viable framework. By expanding a noisy function in terms of wavelets, the smooth function can be identified as its larger scale wavelet components.

The orthogonal wavelets of Daubechies [3] have been dominantly used in most of the wavelet analyses. Although orthogonality offers advantage in theoretical arguments, their values may not be computed in a straightforward way. On the other hand, the compactly supported cardinal B-spline wavelets constructed by Chui and Wang [2,1] have several desirable properties in applications. In this paper, we pursue the effectiveness of the cardinal B-spline wavelets in data smoothing.

The advantages of the compactly supported B-spline wavelets are as follows:

- Easy interpolation algorithms are available to find the representation of time series data in terms of the scaling functions.
- The scaling functions and wavelets have good regularity with small support.

Wavelets: Theory, Algorithms, and Applications
Charles K. Chui, Laura Montefusco, and Luigia Puccio (eds.), pp. 179–196.

ISBN 0-12-174575-9

- The total positivity of the scaling functions (do not vanish within their supports) is a property particularly desirable for data smoothing.
- The scaling functions and wavelets are either symmetric or antisymmetric with respect to their center, so that the distortion of the data will be minimal. In addition, they are explicitly written as a linear combination of truncated powers, so that their values at an arbitrary point may be readily obtained.
- The algorithm can be constructed using only the convolution of discrete data of real numbers, and hence it is fast.

Since the B-spline wavelets and scaling functions are not orthogonal to their respective integer translates, truncation is needed in the interpolation and decomposition algorithms. In this paper, we examine the truncation order with which the method gives the practical value. We also compare two alternative methods in interpolating a set of discrete data points.

We begin with a review of the Chui-Wang theory [1] in Section 2. In Section 3, we specialize in the cubic cardinal B-splines, and determine the parameters and the truncation order. An example of applications is given in Section 4. Finally, we conclude in Section 5.

§2 Compactly supported B-spline wavelets

2.1 The two-scale relation and wavelets

Let $\phi \in L^2(\mathbb{R})$ be a scaling function which satisfies the two-scale relation

$$\phi(x) = \sum_{k\in\mathbf{Z}} p_k\, \phi(2x-k). \tag{2.1}$$

We assume that the two-scale sequence $\{p_k\}$, $k \in \mathbf{Z}$, is finite, so that ϕ has a compact support. The scaling function ϕ generates a multiresolution analysis (MRA), $\cdots \subset V_{-1} \subset V_0 \subset V_1 \subset \cdots$, where V_j is the space of all functions of the form

$$f_j(x) = \sum_{k\in\mathbf{Z}} c_k^{(j)}\, \phi(2^j x-k). \tag{2.2}$$

In terms of the two-scale symbol of the two-scale sequence

$$P(z) = \frac{1}{2}\sum_{k\in\mathbf{Z}} p_k\, z^k, \tag{2.3}$$

the Fourier transform of the two-scale relation in Equation (2.1) may be written as

$$\hat{\phi}(\omega) = P(e^{-i\omega/2})\hat{\phi}\Big(\frac{\omega}{2}\Big). \tag{2.4}$$

We consider the autocorrelation function $F_\phi(x) = \int_{-\infty}^{\infty} \phi(x+y)\overline{\phi(y)}dy$, where $\overline{\phi(y)}$ denotes the complex conjugate of $\phi(y)$. The generalized Euler-Frobenius Laurent polynomial is defined by

$$E_\phi(z) = \sum_{k\in\mathbf{Z}} F_\phi(k)\, z^k. \tag{2.5}$$

Note that the Fourier transform $\hat{F}_\phi(\omega)$ of $F_\phi(x)$ is equal to $|\hat{\phi}(\omega)|^2$. By applying the Poisson summation formula to the sum $\sum_{k=-\infty}^{\infty} |\hat{\phi}(x+2\pi k)|^2$, we obtain the following identity, which plays the key role in the construction below.

$$P(z)\overline{P(z)}\frac{E_\phi(z)}{E_\phi(z^2)} + P(-z)\overline{P(-z)}\frac{E_\phi(-z)}{E_\phi(z^2)} = 1, \quad z = e^{-i\omega/2}. \tag{2.6}$$

We now define the symbols

$$\begin{aligned} G(z) &= \frac{1}{2}\sum_{k\in\mathbf{Z}} g_k\, z^k = \frac{E_\phi(z)}{E_\phi(z^2)}\overline{P(z)} \\ Q(z) &= \frac{1}{2}\sum_{k\in\mathbf{Z}} q_k\, z^k = -z^{2m-1}E_\phi(-z)\overline{P(-z)} \\ H(z) &= \frac{1}{2}\sum_{k\in\mathbf{Z}} h_k\, z^k = -z^{-2m+1}\frac{P(-z)}{E_\phi(z^2)}, \quad z = e^{-i\omega}, \end{aligned} \tag{2.7}$$

where $\overline{P(z)} = \frac{1}{2}\sum_k \overline{p}_k z^{-k}$, $z = e^{-i\omega}$, and $m \in \mathbf{Z}$ is arbitrary, but later identified as the order of B-spline. The wavelet is then defined by

$$\psi(x) = \sum_{k\in\mathbf{Z}} q_k\, \phi(2x-k). \tag{2.8}$$

2.2 The decomposition algorithm

Let W_j be the space of all functions of the form

$$g_j(x) = \sum_{k\in\mathbf{Z}} d_k^{(j)}\psi(2^j x - k). \tag{2.9}$$

Then W_j is the complementary subspace of V_j, *i.e.*, $V_j = V_{j-1}\dot{+}W_{j-1}$. Thus, $f_j(x)$ in Equation (2.2) can be uniquely decomposed into a sum of the form

$$f_j(x) = f_{j-1}(x) + g_{j-1}(x). \tag{2.10}$$

The decomposition in Equation (2.10) is determined by the sequences $\{g_k\}$ and $\{h_k\}$. In fact, using Equation (2.7), the identity in Equation (2.6) may be rewritten in several different forms, from which, when combined with Equations (2.1) and (2.8), one can show the decomposition formula [1]

$$\phi(2x-l)=\frac{1}{2}\sum_k\Big[g_{2k-l}\phi(x-k)+h_{2k-l}\psi(x-k)\Big],\quad l\in\mathbf{Z}.$$

When this is applied to Equation (2.10), where f_j and g_j are given by Equations (2.2) and (2.9), we have the decomposition algorithm

$$\begin{aligned}c_k^{(j-1)}&=\sum_{l\in\mathbf{Z}}\frac{1}{2}\,g_{2k-l}\,c_l^{(j)}\\ d_k^{(j-1)}&=\sum_{l\in\mathbf{Z}}\frac{1}{2}\,h_{2k-l}\,c_l^{(j)}.\end{aligned}\tag{2.11}$$

2.3 The interpolatory graphical display algorithm

We apply the two-scale relation in Equation (2.1) to Equation (2.2) repeatedly, and find that

$$f_j(x)=\sum_{l\in\mathbf{Z}}c_l^{(j+k)}\phi(2^{j+k}x-l),\quad k\ge 1.\tag{2.12}$$

The coefficients $c_l^{(j+k)}$, $k\ge 1$, are obtained recursively by applying

$$c_l^{(j+i)}=\sum_{k\in\mathbf{Z}}p_{l-2k}\,c_k^{(j+i-1)},\quad i\ge 1.\tag{2.13}$$

To find the similar formula for a function $g_j\in W_j$, we apply the two-scale relation in Equation (2.8) to Equation (2.9), and then apply Equation (2.1) repeatedly, and find that

$$g_j(x)=\sum_{l\in\mathbf{Z}}d_l^{(j+k)}\phi(2^{j+k}x-l).\tag{2.14}$$

The coefficients $d_l^{(j+k)}$, $k\ge 1$, are obtained by

$$d_l^{(j+1)}=\sum_{k\in\mathbf{Z}}q_{l-2k}\,d_k^{(j)},\quad d_l^{(j+i)}=\sum_{m\in\mathbf{Z}}p_{l-2m}\,d_m^{(j+i-1)},\quad i\ge 2.\tag{2.15}$$

With Equations (2.12) and (2.14), the values $f_j(n/2^{j+k})$ and $g_j(n/2^{j+k})$ can be obtained from $\phi(n)$, $n\in\mathbf{Z}$. In particular, by choosing $c_l^{(0)}=\delta_{l,0}$ or $d_l^{(0)}=\delta_{l,0}$, we obtain $\phi(n/2^{j+k})$ or $\psi(n/2^{j+k})$, respectively, in terms of $\phi(n)$, $n\in\mathbf{Z}$. The Formulas (2.12)–(2.15) define the interpolatory graphical display algorithm.

2.4 The cardinal B-splines

Let $N_1(x)$ be the characteristic function of the interval $[0,1)$, *i.e.*,

$$N_1(x) = \begin{cases} 1; & 0 \le x < 1, \\ 0; & \text{otherwise.} \end{cases} \tag{2.16}$$

The support of $N_1(x)$ is $[0,1]$. The cardinal B-spline of order m with knot sequence $\mathbf{Z}$ may be obtained by the recursion relation

$$N_m(x) = \frac{x}{m-1} N_{m-1}(x) + \frac{m-x}{m-1} N_{m-1}(x-1). \tag{2.17}$$

Since Equation (2.17) increases the support by one, we have $\operatorname{supp} N_m = [0,m]$.

It can be shown that N_m is the m-fold convolution of N_1, and hence

$$\hat{N}_m(\omega) = \left(\frac{1-e^{-i\omega}}{i\omega}\right)^m. \tag{2.18}$$

Using this formula, it is easy to show that $\phi = N_m$ satisfies the two-scale relation in Equation (2.1). In fact, applying Equation (2.18) to Equation (2.4), we can show in a straightforward fashion that

$$p_k = \frac{1}{2^{m-1}} \binom{m}{k}, \quad k = 0, \ldots, m \tag{2.19}$$

and $p_k = 0$, $k \in \mathbf{Z}$, otherwise. Also using Equation (2.18), we can compute the generalized Euler-Frobenius Laurent Polynomial (2.5) for $\phi = N_m$,

$$E_{N_m}(z) = \sum_{k=-m+1}^{m-1} N_{2m}(m+k) z^k = \frac{1}{(2m-1)!\, z^{m-1}} E_{2m-1}(z), \tag{2.20}$$

where we have introduced the Euler-Frobenius polynomial $E_{2m-1}(z)$ of order $2m-1$. Once $\{p_k\}$, $k \in \mathbf{Z}$, is found, Equations (2.7) and (2.8) allow us to define the wavelet ψ in Equation (2.8) associated with the scaling function $\phi = N_m$, where

$$q_k = \frac{(-1)^k}{2^{m-1}} \sum_{l=0}^{m} \binom{m}{l} N_{2m}(k+1-l), \quad k = 0, \ldots, 3m-2, \tag{2.21}$$

and $q_k = 0$, $k \in \mathbf{Z}$, otherwise. Since $\operatorname{supp} N_m = [0,m]$, we have $\operatorname{supp} \psi = [0, 2m-1]$.

2.5 The fundamental spline

The fundamental spline of order m is defined by

$$L_m(x) = \sum_{k\in\mathbf{Z}} \beta_k^{(m)} N_m\Big(x + \frac{m}{2} - k\Big), \tag{2.22}$$

which has the interpolation property $L_m(k) = \delta_{k,0}$, $k \in \mathbf{Z}$. This is used to find the interpolation of a data sequence. In fact,

$$\begin{aligned} f_0(x) = \sum_{k\in\mathbf{Z}} f_k L_m(x-k) &= \sum_{k\in\mathbf{Z}} c_k^{(0)} \phi(x-k), \\ c_k^{(0)} &= \sum_{l\in\mathbf{Z}} \beta_{k+2-l}^{(m)} f_l \end{aligned} \tag{2.23}$$

interpolates the data sequence $\{f_k\}$ at the integer knots $k \in \mathbf{Z}$.

The sequence $\{\beta_k^{(m)}\}$, $k \in \mathbf{Z}$, in Equation (2.22) is given by

$$B_m(z) = \sum_k \beta_k^{(m)} z^k = \Big[\sum_k N_m\Big(\frac{m}{2} + k\Big) z^k\Big]^{-1}. \tag{2.24}$$

For $m \geq 3$, the Laurent series of $B_m(z)$ does not terminate, and hence $L_m(x)$ is not compactly supported.

§3 The cubic cardinal B-spline wavelet analysis

3.1 The scaling function and wavelet

In this paper, we specialize splines of order $m = 4$, and consider the cubic cardinal B-spline N_4 as our scaling function $\phi(x) = N_4(x)$. Its values at the integer knots are

$$\phi(1) = \frac{1}{6}, \quad \phi(2) = \frac{2}{3}, \quad \phi(3) = \frac{1}{6}, \tag{3.1}$$

and $\phi(n) = 0$, $n \in \mathbf{Z}$, otherwise. The two-scale symbol $P(z)$ reduces to

$$P(z) = \Big(\frac{1+z}{2}\Big)^4. \tag{3.2}$$

By using the Euler-Frobenius Laurent polynomial in Equation (2.20) (see also Equation (3.4) below), the other symbols in Equation (2.7) can also be obtained. The sequences $\{p_k\}$ and $\{q_k\}$ consist of five and ten nonzero rational numbers, respectively, which are listed in Table 1. We can use the interpolatory graphical display algorithm of Subsection 2.3 to plot $\phi(x)$ and $\psi(x)$, as shown in Figure 1.

Table 1. The values of the two-scale sequences.

k	0	1	2	3	4	5
p_k	$\frac{1}{8}$	$\frac{1}{2}$	$\frac{3}{4}$	$\frac{1}{2}$	$\frac{1}{8}$	
q_k	$\frac{1}{40320}$	$-\frac{31}{10080}$	$\frac{559}{13440}$	$-\frac{247}{1260}$	$\frac{9241}{20160}$	$-\frac{337}{560}$
	$q_{5+l} = q_{5-l}$, $l = 1, \ldots, 5$. All other p_k and q_k are zero.					

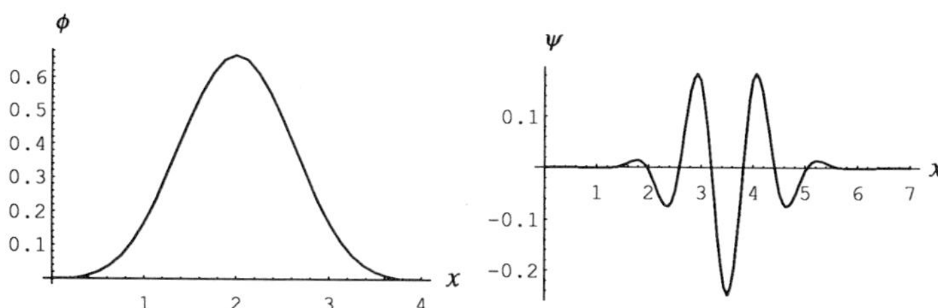

Figure 1. The scaling function $\phi(x) = N_4(x)$ (left) and the wavelet $\psi(x)$ (right).

3.2 Interpolation by the fundamental spline

The sequence $\{\beta_k^{(4)}\}$ in Equation (2.22) can be determined by Equation (2.24) with $m = 4$. Since

$$\frac{1}{B_4(z)} = \sum_{k \in \mathbf{Z}} N_4(2+k)\, z^k = \frac{E_3(z)}{6z} = \frac{1}{6} z^{-1} + \frac{2}{3} + \frac{1}{6} z$$

is symmetric with respect to the exchange $z \leftrightarrow z^{-1}$, if a is one of the zeros of $E_3(z)$, then the other zero is $1/a$. Let a be the one such that $|a| < 1$, namely

$$a = \sqrt{3} - 2 \approx -0.268.$$

Then we can write $B_4(z)$ in the following form

$$\begin{aligned} B_4(z) &= \frac{6z}{(z-a)(z-a^{-1})} = \frac{6}{a - a^{-1}} \left[\frac{1}{1-az} + \frac{az^{-1}}{1-az^{-1}} \right] \\ &= \sqrt{3} \sum_{k=-\infty}^{\infty} a^{|k|}\, z^k, \end{aligned}$$

from which we find

$$\beta_k^{(4)} = \sqrt{3}\, a^{|k|}.$$

In practice, we must truncate the series in Equation (2.22), and, therefore, define

$$L_4^{(n)}(x) = \sum_{k=-n}^{n} \beta_k^{(4)} N_4(x+2-k). \tag{3.3}$$

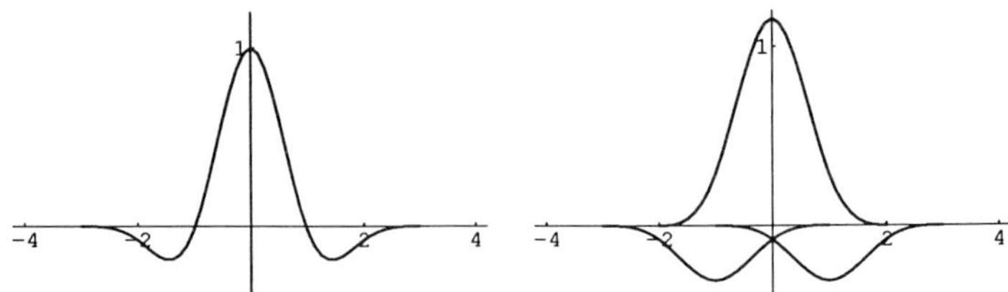

Figure 2. The plot of $L_4^{(1)}(x)$ (left), which consists of three pieces (right).

Although the sequence $\{\beta_k^{(4)}\}$ decreases exponentially, $|a| \approx 0.268$ is not very small so that the decay does not appear very fast. However, the truncated version in Equation (3.3) becomes very accurate at $n = 5$, as we shall see. Recall that $L_4(k) = 0$, $k \in \mathbf{Z}$, $k \neq 0$. The truncated version $L_4^{(1)}(x)$, plotted in Figure 2 (left), consists of three pieces, as shown in Figure 2 (right). The added pieces on both sides tend to make $L_4^{(1)}(\pm 1)$ closer to 0, at the expense of inducing small nonzero value at $n = \pm 2$. Adding more and more pieces by increasing the order n makes the decay faster, as shown in Figure 3. The values of $L_4^{(n)}(m)$, $0 \leq m \leq n+1$, are summarized in Table 2. Note that $|L_4^{(n)}(m)|$ is the largest at $m = \pm(n+1)$.

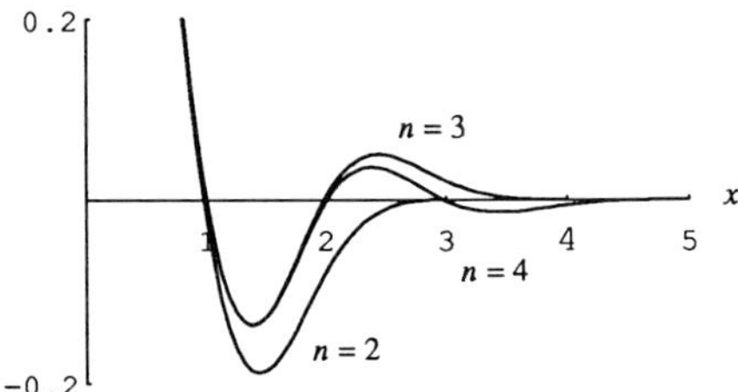

Figure 3. The decay of $L_4^{(n)}(x)$ for some lower values of n.

In order to determine an appropriate value n, we consider the data $f_k = 1$, $k = 0, \ldots, 18$, and $f_k = 0$, otherwise, and apply the interpolation algorithm in Equation (2.23) with $L_4^{(n)}(x)$, $n = 3, 4, 5$. In Table 2, we also list the values of $f_{\mathrm{O}}(9)$ for the three cases, which is to be compared with the original data $f_9 = 1$.

Figure 4 shows the interpolating function $f_{\mathrm{O}}(x)$ together with the original data points, for the cases of $n = 3$ and 5. Although the difference $|f_{\mathrm{O}}(9) - f_9|$ is numerically small, there is a noticeable difference in the plot for $n = 3$. We see that $n = 5$ is a fairly good approximation.

Table 2. The values of $L_4^{(n)}(m)$, $0 \le m \le n$, and $f_0(9)$.

	$L_4^{(n)}(0)$	$L_4^{(n)}(m)$, $0 < m < n$	$L_4^{(n)}(n)$	$L_4^{(n)}(n+1)$	$f_0(9)$
n	1	0	$\frac{2-\sqrt{3}}{2\sqrt{3}}a^n$	$\frac{2}{\sqrt{3}}a^n$	
3				−0.022214	0.985917
4				+0.005952	1.003774
5				−0.001595	0.998989
6				+0.000427	

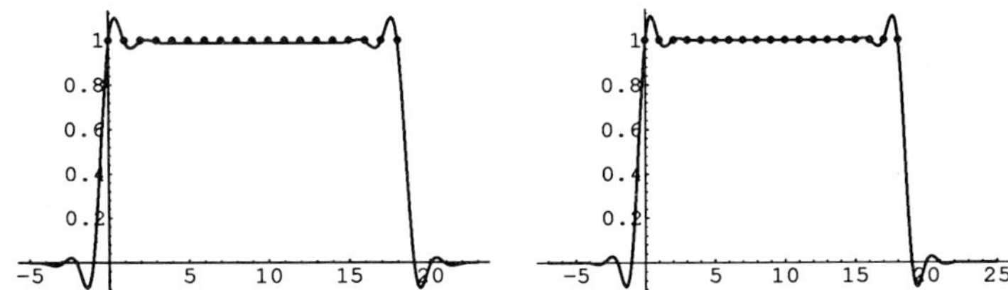

Figure 4. $f_0(x)$ obtained by using $L_4^{(3)}(x)$ (left), and $L_4^{(5)}(x)$ (right).

3.3 The decomposition algorithm

When the symbols $G(z)$ and $H(z)$ in Equation (2.7) are expanded in powers of z, we encounter the series $z^3/E_7(z) = \sum_{n\in\mathbf{Z}} \alpha_n z^n$ which we truncate as $\sum_{n=-N}^{N} \alpha_n z^n$. Since

$$\frac{E_7(z)}{z^3} = z^3 + 120z^2 + 1191z + 2416 + 1191z^{-1} + 120z^{-2} + z^{-3} \tag{3.4}$$

is symmetric with respect to the exchange $z \leftrightarrow z^{-1}$, it can be expressed as a cubic polynomial of $z + z^{-1}$. So the equation $E_7(z)/z^3 = 0$ reduces to $r^3 - 903r + 10332 = 0$ with $r = \frac{1}{2}(z + z^{-1}) + 20$. The three solutions can be found by Cardano's formula.

$$r_k = 2\rho^{1/3}\cos\left(\frac{\varphi + 2\pi k}{3}\right), \quad k = 0, 1, 2$$

$$\rho = 301\sqrt{301}, \quad \tan\varphi = \frac{7\sqrt{11905}}{-5166}, \quad \frac{\pi}{2} < \varphi < \pi.$$

The zeros of $E_7(z)$ are then given by $\{a, b, c\} = r_i - 20 + \sqrt{(r_i - 20)^2 - 1}$, $i = 0, 2, 1$, and $1/a$, $1/b$, $1/c$. The numerical values are summarized in Table 5. Thus, we can write

$$\begin{aligned}\frac{z^3}{E_7(z)} &= \frac{z^3}{(z-a)(z-a^{-1})(z-b)(z-b^{-1})(z-c)(z-c^{-1})} \\ &= C_a\left[\frac{1}{1-az} + \frac{az^{-1}}{1-az^{-1}}\right] + \ldots,\end{aligned}$$

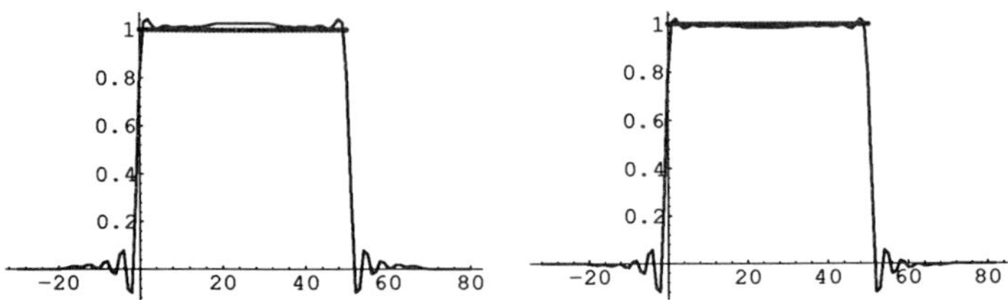

Figure 5. The smoothed function $f_{-1}(x)$ obtained by the decomposition with $N = 8$ (left), and 9 (right).

Table 3. The once-smoothed function at its center.

N	7	8	9
$f_{-1}(25)$	0.94622	1.02724	0.983866

where ... denotes two terms obtained from the first term by replacing a with b and c, respectively. The constant C_a is given by

$$C_a = \frac{bc}{(a - a^{-1})(b - a^{-1})(c - a^{-1})(b - a)(c - a)}$$

and C_b, C_c are obtained from C_a by cyclic permutations of $\{a, b, c\}$. The values of these parameters are summarized in Table 5. Thus, we find

$$\alpha_k = C_a\, a^{|k|} + C_b\, b^{|k|} + C_c\, c^{|k|}.$$

In the truncated form, the decomposition sequences are given by

$$\begin{aligned} G(z) &= \sum_{k=-2N-7}^{2N+3} \frac{1}{2}\, g_k = \frac{E_7(z)}{z^3} \left(\frac{1 + z^{-1}}{2}\right)^4 \sum_{k=-N}^{N} \alpha_k\, z^{2k} \\ H(z) &= \sum_{k=-2N-7}^{2N-3} \frac{1}{2}\, h_k = -\frac{7!}{z^7} \left(\frac{1 - z}{2}\right)^4 \sum_{k=-N}^{N} \alpha_k\, z^{2k}. \end{aligned} \tag{3.5}$$

In order to determine the appropriate value of N, we consider 51 data points $f_k = 1$, $k = 0, \ldots, 50$, and $f_k = 0$, otherwise. Let $f_0(x)$ be the interpolating function which is obtained by the algorithm using the fundamental spline $L_4^{(5)}(x)$. The decomposition algorithm in Equation (2.11) with Equation (3.5) yields the decomposition $f_0(x) = f_{-1}(x) + g_{-1}(x)$. Near the center of its support, the function $f_0(x)$ is flat, and, hence, we expect that $g_{-1}(x) = 0$ and $f_{-1}(x) = 1$ around $x = 25$. The values of $f_{-1}(25)$ are listed for some values of N in Table 3.

In Figure 5, the function $f_{-1}(x)$ is plotted together with the data points, for $N = 8$, and 9. So, with $N = 9$ the approximation is fairly good. The values of the decomposition sequences $\{g_k\}$ and $\{h_k\}$ for $N = 9$ are listed in Table 6.

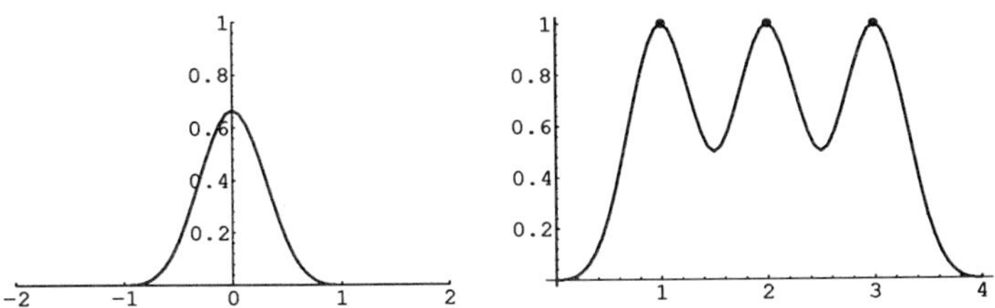

Figure 6. The plot of $\phi(2x+2)$ (left) and $\frac{3}{2}\sum_{k=1}^{3}\phi(2x-2k)$ (right).

3.4 The interpolation by the quasi-interpolation operator

There is an alternative method to find the interpolating function that fits the discrete data. This is the quasi-interpolation operator of Chui [1], which we now consider.

The basic idea is to note that $\phi(2x+2) = N_4(2x+2)$ has the support $[-1,1]$, and that $\phi(2n+2) = \frac{2}{3}\delta_{n,0}$, $n \in \mathbf{Z}$, as shown in Figure 6 (left). So, one might consider the sum

$$\frac{3}{2}\sum_{k\in\mathbf{Z}} f_k\,\phi(2x+2-2k)$$

as the interpolation of a data sequence $\{f_k\}$, $k \in \mathbf{Z}$. However, when it is applied to the flat data, $f_1 = f_2 = f_3 = 1$, and $f_k = 0$, otherwise, this sum produces dips between integer points, as shown in Figure 6 (right). This defect can be remedied by adding several integer translates of $\phi(2x)$. This is the quasi-interpolation operator P_4 of Chui [1], which is defined by

$$(P_4\,f)(x) = \sum_l \sum_k v_{l-2k}^{(4)} f(k)\,\phi(2x+2-l), \tag{3.6}$$

where the convolution sequence $\{v_k^{(4)}\}$ are listed in Table 4.

In Figure 7 (right), $(P_4\,f)(x)$ for the case of a single point $f(l) = \delta_{l,0}$ is plotted. This function is quite similar to the fundamental spline $L_4(x)$ (left), but it has a compact support. Also, the convolution sequence consists of rational numbers, as given in Table 4, which provides an advantage for not producing roundoff errors.

As an example, we consider the 19 data points, $f_k = 1$, $k = 0, \ldots, 18$, $f_k = 0$, otherwise, the same example as in Subsection 3.2. The interpolating function $f_1(x)$ is plotted in Figure 8 (left), which shows the excellent interpolation property. In fact, the value of $f_1(x)$ at its center $x = 9$ is exactly equal to unity.

However, it should be noted that this function $f_1(x)$ is in the space V_1, but not in V_0, in contrast to the case of fundamental spline. When this function

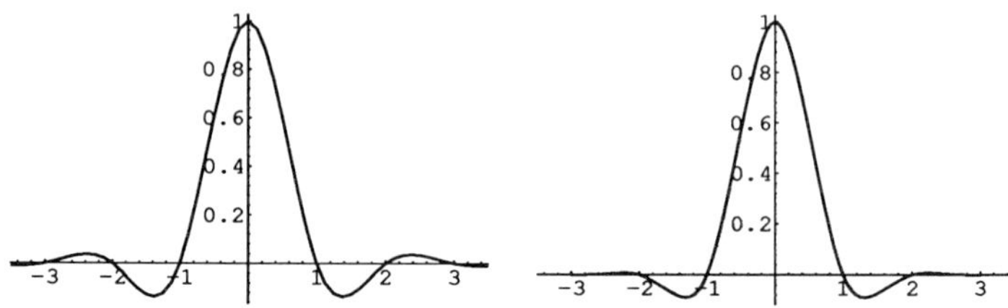

Figure 7. The fundamental spline (left) and the quasi-interpolator (right).

Table 4. The quasi-interpolation sequence.

k	0	± 1	± 2	± 3	± 4	otherwise
$v_k^{(4)}$	$\frac{29}{24}$	$\frac{7}{12}$	$-\frac{1}{8}$	$-\frac{1}{12}$	$\frac{1}{48}$	0

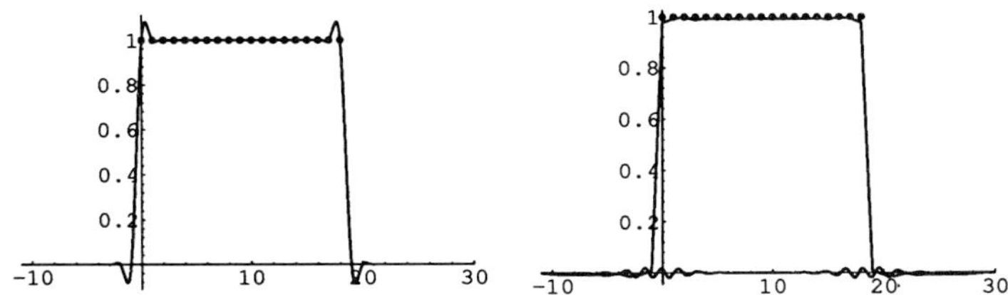

Figure 8. The function $f_1(x)$ found by the quasi-interpolator (left), and $f_0(x)$ and $g_0(x)$ obtained by the decomposition (right).

is decomposed as $f_1(x) = f_0(x) + g_0(x)$, we get the interpolating function $f_0 \in V_0$. The functions $f_0(x)$ and $g_0(x)$ are plotted in Figure 8 (right). When the decomposition algorithm of Subsection 2.2 with the truncation order $N = 9$ is used, we find that $f_0(9) = 0.993965$. The error $|f_0(9) - f_9| \approx 0.006$ in this case is slightly larger than $|f_0(9) - f_9| \approx 0.004$ in the case of the fundamental spline interpolation with $n = 4$ (see Table 2).

To study the advantage or disadvantage of the quasi-interpolation operator P_4, let us consider another slightly less trivial example with the data

$$f_k = \sin\frac{2\pi k}{40} + \frac{1}{5}\sin\frac{2\pi k}{3}, \quad k = 0, \ldots, 39$$

and $f_k = 0$, otherwise. Applying the quasi-interpolation operator P_4, we get $f_1(x)$, which is then decomposed into the sum of $f_0(x)$ and $g_0(x)$. Figure 9 (left) shows these functions obtained by the decomposition. The wavelet component

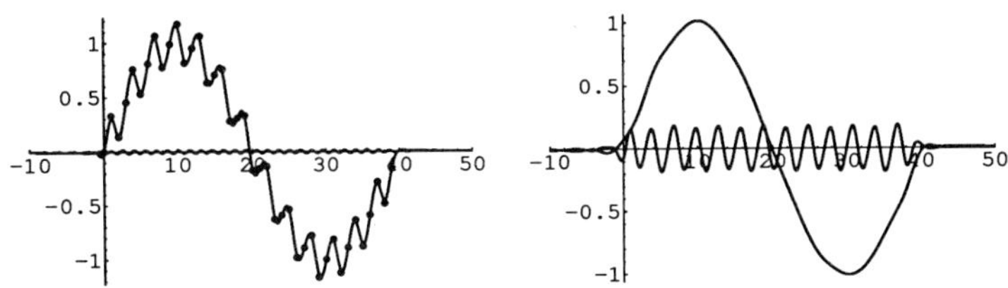

Figure 9. The interpolating function $f_0(x)$ and the wavelet components $g_0(x)$ obtained by using quasi-interpolation operator and decomposition, together with the data points (left). $f_{-1}(x)$ and $g_{-1}(x)$ obtained by decomposition of $f_0(x)$ (right).

$g_0(x)$ at this level is an artifact of our scheme, and it is small as it should be. When $f_0(x)$ is further decomposed into $f_{-1}(x) + g_{-1}(x)$, the property of the original data will be revealed, as in Figure 9 (right).

The fundamental spline interpolation will directly generate $f_0(x)$. Therefore, to extract the information of the data by the decomposition, one needs a single decomposition operation. The result is shown in Figure 10, which is similar to the plots in Figure 9.

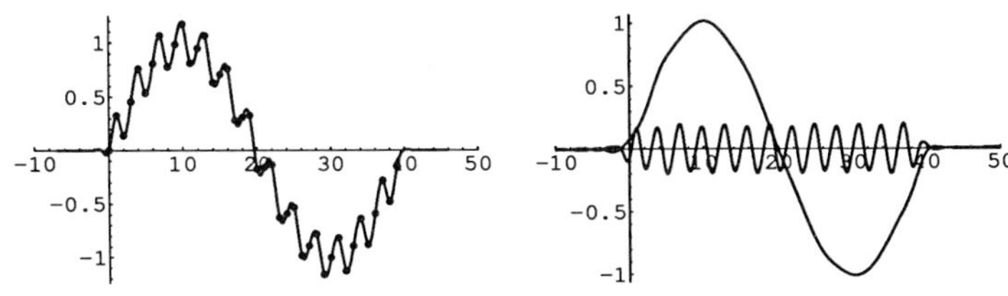

Figure 10. The interpolation by the fundamental spline, together with the data points (left), and $f_{-1}(x)$, $g_{-1}(x)$ obtained by the subsequent decomposition (right).

§4 An application to mechanical vibration

Here we apply the algorithm to study the property of data, obtained from measurements on a mechanical system. The fundamental spline $L_4^{(5)}$ is used for the interpolation, and the decomposition sequences with the truncation order $N = 9$ are used.

The mechanical system vibrates in response to a simulated seismic force with the duration of about 30 seconds. There is a dry friction which damps the vibration in a complicated manner. The data consists of 6320 numbers, f_k, $k = 0, \ldots, 6319$, shown in Figure 11, which represent the force applied to the system, measured at the time interval of 0.005 second.

By the interpolation algorithm with $L_4^{(5)}(x)$, the interpolating function $f_0(x)$ is obtained. By applying the decomposition algorithm of Equation (2.11), we get the smoothed signal $f_{-1}(x)$ and $g_{-1}(x)$. Further decomposition of $f_{-1}(x)$ yields $f_{-2}(x)$ and $g_{-2}(x)$. Figure 12 shows $f_0(x)$ (top), the wavelet component $g_{-1}(x)$ (middle), and the smoothed signal $f_{-1}(x)$ (bottom) in the range $[0, 500]$. The first 300 data seem to represent the background noise, which remains in $f_{-1}(x)$. There is also a modulation noise which is separated out into $g_{-1}(x)$.

Figure 13 shows the interpolating function $f_0(x)$ (top), its wavelet component $g_{-1}(x)$ and the smoothed signal $f_{-1}(x)$ (middle), the wavelet component $g_{-2}(x)$ and the smoothed signal $f_{-2}(x)$ (bottom), in the range $[1701, 1800]$.

The second data set, also of 6320 data points, represents the response displacement of the system. Figure 14 shows the original data (top) and the twice smoothed $f_{-2}(x)$ (bottom). Figure 15 shows the interpolating function $f_0(x)$ (top), its wavelet component $g_{-1}(x)$ and the smoothed signal $f_{-1}(x)$ (middle), and the smoothed signal $f_{-2}(x)$ (bottom), in the range $[2401, 3200]$.

Due to the dry friction which damps the vibration, the response of the system is not simple as in the linear oscillation. In particular, there will be no motion when the force is sufficiently small. The response displacement shows the characteristic features of the vibration with dry friction, consisting of successive stick and slip motion. By removing the noise contaminated in the original signal, such characteristic behavior seems to become evident.

§5 Conclusion

The scaling functions and wavelets based on the cardinal B-splines are not orthogonal to their integer translates. This requires the use of duals when a given function is to be expanded in terms of the scaling functions or wavelets. However, when the data is given as a sequence of numbers, as in the case of experimental data, there are very accurate interpolation methods available. We have shown that the use of the scaling functions and wavelets based on the cubic B-splines, together with the fundamental spline $L_4^{(5)}$ and the decomposition series truncated at the order 9, is an effective, efficient method of data smoothing.

The algorithm has been implemented in *Mathematica* [6]. Arithmetic with rational numbers is performed exactly, so that the chance of introducing round-off errors is minimal. The calculations presented here have been carried out on a personal computer. The *Mathematica* package developed by the author, '`SplineWavelet.m`',which is used in the analysis of Section 4, is available from *MathSource* at Wolfram Research, Inc.

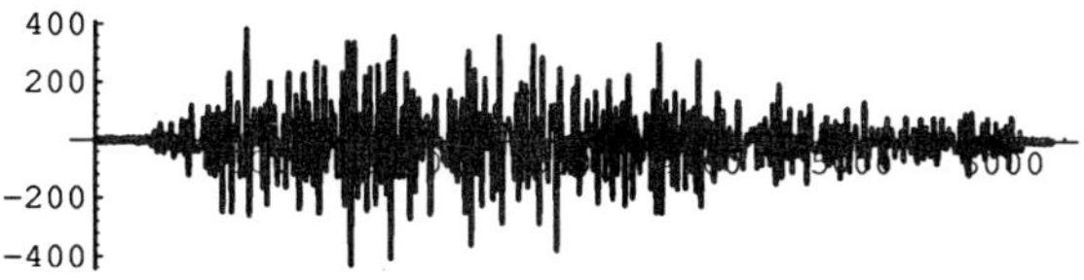

Figure 11. The input data.

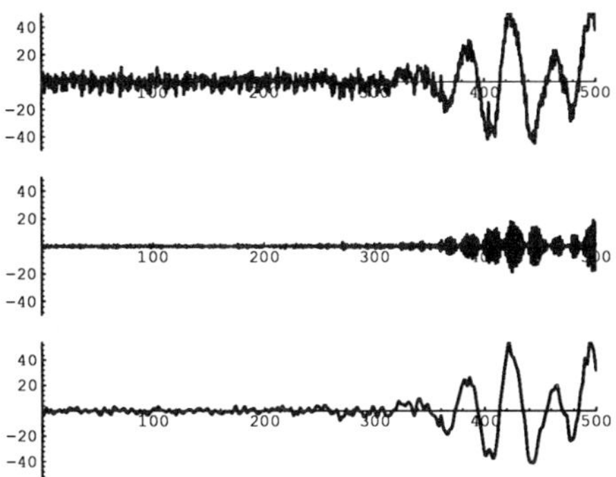

Figure 12. $f_0(x)$ for the input (top) and its wavelet component $g_{-1}(x)$ (middle) and smoothed signal $f_{-1}(x)$ (bottom) in the range $[0, 500]$.

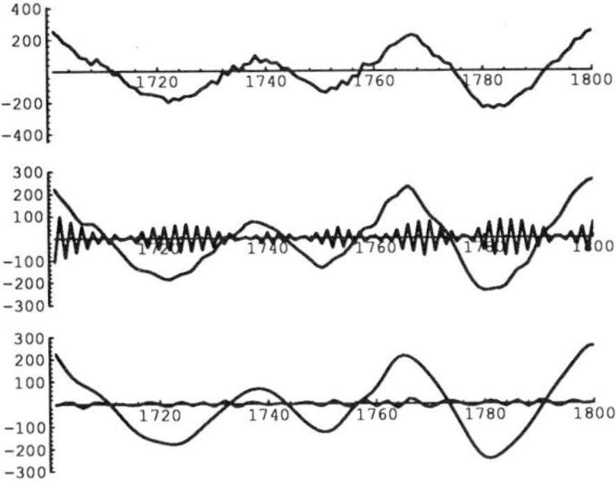

Figure 13. $f_0(x)$ for the input (top), its wavelet component $g_{-1}(x)$ and smoothed signal $f_{-1}(x)$ (middle), and $g_{-2}(x)$ as well as $f_{-2}(x)$ (bottom), in the range $[1701, 1800]$.

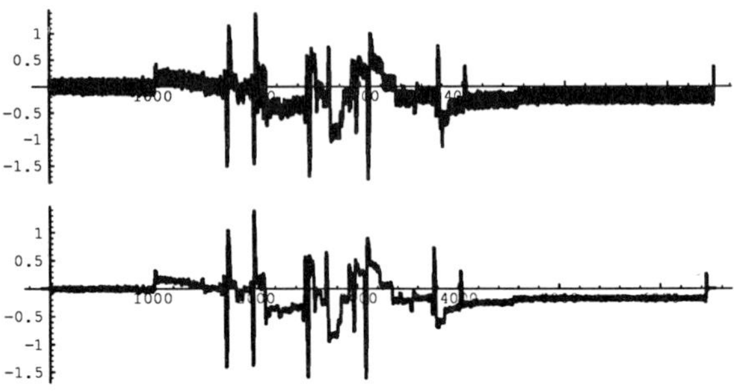

Figure 14. The response signal (top) and $f_{-2}(x)$ (bottom) which is obtained by applying decomposition twice.

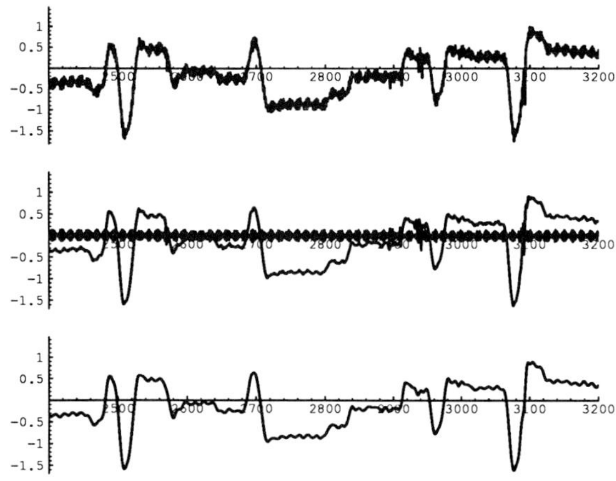

Figure 15. The interpolating function $f_0(x)$ (top) and its wavelet component $g_{-1}(x)$ and smoothed signal $f_{-1}(x)$ (middle), and the twice-smoothed $f_{-2}(x)$ (bottom) in the range [2401, 3200].

Acknowledgments. The author would like to thank the organizing committee of the International Conference on Wavelets, Taormina, October 14–20, 1993, Prof. Charles Chui, in particular, for discussions as well as his warm hospitality. He would also like to thank Prof. Akira Sakurai of Tokyo Denki University for encouragement. He has enjoyed discussions with Mei Kobayashi of IBM corporation, and Nobuyuki Shimizu of Iwaki Meisei University has provided him with the data of his experiments.

Table 5. Numerical values of the parameters in Subsection 3.3.

ρ	5222.15482344	a	-0.535280430796	C_a	0.001193707143
φ	2.994810383657	b	-0.122554615192	C_b	-0.000209488156
		c	-0.009148694810	C_c	0.000000847059

Table 6. The values of the decomposition sequences.

k	$\frac{1}{2}g_k$	$\frac{1}{2}h_k$
-25	$-0.00000026915966297914523 93$	$+0.00135656470141489200 6$
-24	$-0.0000333757982094140096 8$	$-0.0054262588056595680 26$
-23	$-0.0004508779167287551934$	$+0.0056050842486416321$
-22	$-0.0020650860533655137 2$	$+0.0047109570337313117 26$
-21	$-0.0041322888065089145 53$	$-0.0091147451381498069 03$
-20	$-0.0026729418933561988 54$	$-0.0088008398386956031 19$
-19	$+0.0027452464050414515 41$	$+0.0170280294661341211 3$
-18	$+0.0028661140621359558 59$	$+0.0164409446865050500 9$
-17	$-0.0055799832840118699 08$	$-0.0318118113183492517 2$
-16	$-0.0053879298185509605 39$	$-0.0307097008714741750 8$
-15	$+0.0104240521865301267 1$	$+0.0594333883903593729 9$
-14	$+0.0100667475195376882 8$	$+0.0573309522542995152 9$
-13	$-0.0194732693559731063 8$	-0.1110584407107915363
-12	$-0.0188156866209032657 9$	-0.1067758033725942632
-11	$+0.0363735869888239028 2$	$+0.2076908383800019567$
-10	$+0.0352261016744787495 5$	$+0.1967942773044243825$
-9	$-0.0679036084985637087 2$	-0.3897455807997501814
-8	$-0.0664208373870100612 7$	-0.3457708907749591328
-7	$+0.1264574463559396853$	$+0.7420976984778135588$
-6	$+0.1290835712180778118$	$+0.4684225966330769047$
-5	-0.2329246261335044913	$-1.4753945198919583 59$
-4	-0.2822118708111977771	$+0.4684225966330769047$
-3	$+0.4006808254672730 81$	$+0.7420976984778135588$
-2	$+0.8931628563140348035$	-0.3457708907749591328
$k \geq -1$	$\frac{1}{2}g_k = \frac{1}{2}g_{-4-k}$	$\frac{1}{2}h_k = \frac{1}{2}h_{-10-k}$

References

1. Chui, C. K., *An Introduction to Wavelets*, Academic Press, 1992.
2. Chui, C. K. and J. Z. Wang, On compactly supported spline wavelets and a duality principle, *Trans. Amer. Math. Soc.* **330** (1992), 903–915.
3. Daubechies, I., Orthonormal bases of compactly supported wavelets, *Comm. Pure and Appl. Math.* **41** (1988), 909–996.
4. Mallat, S., Multiresolution approximation and wavelet orthonormal bases of $L^2(\mathbb{R})$, *Trans. Amer. Math. Soc.* **315** (1989), 69–87.
5. Meyer, Y. Ondelettes et fonctions splines, *Séminaire EDP*, École Polytechnique, Paris, December 1986.

6. S. Wolfram, *Mathematica: A System for Doing Mathematics by Computer*, Second edition, Addison-Wesley, 1992.

Susumu Sakakibara
Iwaki Meisei University
College of Science and Engineering
Iwaki-shi, Fukushima-ken, 970, Japan
susumu@iwakimu.ac.jp

L-Spline Wavelets

Tom Lyche and Larry L. Schumaker

Abstract. We explicitly construct compactly-supported wavelets associated with L-spline spaces. We then apply the theory to develop multiresolution methods based on L-splines.

§1 Introduction

Given a nested sequence of finite dimensional linear spaces $\mathcal{S}_0 \subset \mathcal{S}_1 \subset \mathcal{S}_2 \cdots$ in L_2, it is an important problem to find bases for the orthogonal complements W_j of $\mathcal{S}_j$ in $\mathcal{S}_{j+1}$, *i.e.*, such that $\mathcal{S}_{j+1} = W_j \oplus \mathcal{S}_j$. Such basis functions are called *(pre) wavelets*, see *e.g.*, [3,5]. When the $\mathcal{S}_j$ are nested spaces of polynomial splines on dyadic equally spaced simple knots, it turns out that all of the wavelets are translates and dilations of a single *mother wavelet* ψ. A simple and elegant formula for ψ was found by Chui & Wang, see [3].

The case of polynomial splines with arbitrary simple knots and where a fixed number of simple knots are inserted between each pair of knots to go from one spline space to the next was treated in [2]. The case of uniform multiple knots was discussed in [6]. Lyche & Mørken [10] showed how to construct spline wavelets of minimal support for arbitrary knot sequences. The purpose of this paper is to extend the results of [10] to spaces $\mathcal{S}_j$ of L-splines on general knot sequences (see Section 2 below for a definition). Part of our motivation is to provide a framework for developing wavelets associated with trigonometric splines (see Section 7), which have important applications to doing a multiresolution analysis on the sphere.

The paper is organized as follows. In Section 2 we review the basic theory of L-splines, while in Section 3 we discuss associated locally supported basis functions (B-splines) and show that they form a Riesz basis for the L-spline space. In Section 4 we explicitly construct L-spline wavelets and show that they have properties analogous to the well-known polynomial spline wavelets. The important case where the splines are translation invariant is examined in

Wavelets: Theory, Algorithms, and Applications
Charles K. Chui, Laura Montefusco, and Luigia Puccio (eds.), pp. 197–212.

ISBN 0-12-174575-9

Section 5; in this case all wavelets *at a given level* are translates of one mother wavelet which we find explicitly. In Section 6 we introduce a kind of *multiresolution* framework for our L-splines and L-spline wavelets, and show how to derive the corresponding analogs of the *two-scale relation* and *decomposition* and *reconstruction relations*. We conclude the paper by briefly discussing some examples, including trigonometric and hyperbolic splines.

§2 L-splines

Let $I = [a, b]$ be a closed subinterval of the real line $\mathbb{R}$. Given functions

$$a_j \in C^j[a, b], \qquad j = 0, \ldots, m-1, \tag{2.1}$$

let

$$L = D^m + \sum_{j=0}^{m-1} a_j(x) D^j \tag{2.2}$$

be the corresponding linear differential operator of order m. Let

$$N_L = \{f \in L_1^m[a, b] : Lf(x) = 0, \; a \le x \le b\} \tag{2.3}$$

be the *null space* of L, where $L_1^m[a, b]$ is the usual Sobolev space of functions whose m-th derivatives lie in $L_1[a, b]$.

It is well-known from the theory of ordinary differential equations that N_L is an m-dimensional linear subspace of $C^m[a, b]$. Any set of functions $u_1, \ldots, u_m$ which span N_L is called a *fundamental solution set* for L.

We are now ready to recall the definition of L-splines. We follow the notation and development in [12]. Let

$$\Delta = \{a = x_0 < x_1 < \cdots < x_k < x_{k+1} = b\}$$

be a partition of I into $k+1$ subintervals. Let $\mathcal{M} = (m_1, \ldots, m_k)$ be a vector of integers satisfying $1 \le m_i \le m$, $i = 1, \ldots, k$. Then the associated *space of L-splines* is defined by

$$\begin{aligned} \mathcal{S} :=\mathcal{S}(N_L; \mathcal{M}; \Delta) = \{s : s|_{(x_i, x_{i+1})} \in N_L, \quad i = 0, \ldots, k, \text{ and} \\ D_-^{j-1} s(x_i) = D_+^{j-1} s(x_i), \quad j = 1, \ldots, m - m_i, \quad i = 1, \ldots, k\}. \end{aligned} \tag{2.4}$$

It is well-known that

$$\dim \mathcal{S}(N_L; \mathcal{M}; \Delta) = n := m + \sum_{i=1}^{k} m_i.$$

§3 A basis of locally supported splines

In this section we review the construction of locally supported basis splines. Given a partition Δ as above, let

$$t = \{t_1 \le t_2 \le \cdots \le t_{n+m}\} \tag{3.1}$$

where

$$a = t_1 = \cdots = t_m, \qquad t_{n+1} = \cdots = t_{n+m} = b,$$

and where

$$\{t_{m+1} \le \cdots \le t_n\}$$

is the set obtained by repeating each x_i a total of m_i times, $i = 1, \ldots, k$. This is called the *extended knot sequence* corresponding to Δ. For a fixed differential operator L, we denote the associated spline space defined in Equation (2.4) by $\mathcal{S}_t$.

To construct a local basis for $\mathcal{S}_t$, we use the fact (see Theorem 10.5 of [12]) that for a given differential operator L as in Equation (2.2), there exists $H > 0$ such that for any subinterval $J \subseteq I$ of length less than H, the null space N_L is an *Extended Complete Tchebycheff* system on J. From now on we suppose that

$$t_{i+m} - t_i < H, \qquad i = 1, \ldots, n. \tag{3.2}$$

Fix $1 \le i \le n$, and choose an interval J_i of length less than H which contains $[t_i, t_{i+m}]$. Then there exists an ECT-system $U_{m,i} = \{u_{j,i}\}_{j=1}^m$ which spans N_L on J_i. We may suppose that this system is in *canonical form, i.e.*,

$$\begin{aligned} u_{1,i}(x) =& w_{1,i}(x) \\ u_{2,i}(x) =& w_{1,i}(x) \int_{t_i}^{x} w_{2,i}(\xi_2) d\xi_2 \\ & \cdots \\ u_{m,i}(x) =& w_{1,i}(x) \int_{t_i}^{x} w_{2,i}(\xi_2) \int_{t_i}^{\xi_2} \cdots \int_{t_i}^{\xi_{m-1}} w_{m,i}(\xi_m) d\xi_m \ldots d\xi_2, \end{aligned} \tag{3.3}$$

where $w_{j,i} \in C^{m-j+1}[J_i]$ are positive weight functions for $j = 1, \ldots, m$. We shall also have need for the *dual canonical ECT-system* consisting of the functions

$$\begin{aligned} u^*_{1,i}(x) =& 1 \\ u^*_{2,i}(x) =& \int_{t_i}^{x} w_{m,i}(\xi_m) d\xi_m \\ & \cdots \\ u^*_{m,i}(x) =& \int_{t_i}^{x} w_{m,i}(\xi_m) \int_{t_i}^{\xi_m} \cdots \int_{t_i}^{\xi_3} w_{2,i}(\xi_2) d\xi_2 \ldots d\xi_m. \end{aligned} \tag{3.4}$$

The functions $U^*_m := \{u^*_{1,i}, \ldots, u^*_{m,i}\}$ span the null space of the adjoint operator L^* corresponding to L.

Next, we introduce a local Green's function associated with L and the interval J_i:

$$g_{m,i}(x;y) = \begin{cases} h_{m,i}(x;y), & x \geq y, \\ 0, & \text{otherwise}, \end{cases} \tag{3.5}$$

where

$$\begin{aligned} h_{m,i}(x;y) =& w_{1,i}(x) \int_y^x w_{2,i}(\xi_2) \int_y^{\xi_2} \cdots \int_y^{\xi_{m-1}} w_{m,i}(\xi_m) d\xi_m \ldots d\xi_2 \\ =& \sum_{j=1}^{m} (-1)^{m-j} u_{j,i}(x) u^*_{m-j+1,i}(y). \end{aligned} \tag{3.6}$$

Given a set of points $t_1 \leq \cdots \leq t_m$ in J_i and a sufficiently smooth function f defined on J_i, the associated *generalized divided difference* of f over the t_j's is defined by

$$[t_1, \cdots, t_m]_{U^*_{m,i}} f = \frac{D\begin{pmatrix} t_1, & \cdots & t_{m-1}, & t_m \\ u^*_{1,i} & \cdots & u^*_{m-1,i} & f \end{pmatrix}}{D\begin{pmatrix} t_1, & \cdots & t_{m-1}, & t_m \\ u^*_{1,i} & \cdots & u^*_{m-1,i} & u^*_{m,i} \end{pmatrix}}. \tag{3.7}$$

Here the D's stand for the determinants of the matrices formed from the functions in the bottom row evaluated at the points in the top row (see [12]).

We now define a B-spline associated with the knots $t_i, \ldots, t_{i+m}$ by

$$Q_i(x) = \lambda[t_i, \ldots, t_{i+m}]\, g_{m,i}(x;y) \tag{3.8}$$

where λ operates on the y variable, and is defined by

$$\lambda[t_i, \ldots, t_{i+m}] = (-1)^m \big([t_{i+1}, \cdots, t_{m+i}]_{U^*_{m,i}} - [t_i, \cdots, t_{m+i-1}]_{U^*_{m,i}}\big).$$

Although they appear to be different, as shown in [9] these are the same as the N_i defined in Equation (9.54) in [12]. They are in fact Tchebycheffian B-splines, and have many of the properties of the classical polynomial B-splines [12]. For example:

- Q_i has support on $[t_i, t_{i+m}]$;
- $0 \leq Q_i(x) \leq w_{1,i}(x)$ for $x \in (t_i, t_{i+m})$;
- the $\{Q_i\}_{i=1}^n$ form a basis for $\mathcal{S}_t$;
- the Q_i provide a representation for the generalized divided difference Equation (3.7), *i.e.*, for any function f with $L^* f \in L_1[t_i, t_{i+m}]$,

$$\lambda[t_i, \ldots, t_{i+m}] f = \int_{t_i}^{t_{i+m}} Q_i(x) L^* f(x) dx. \tag{3.9}$$

For later use, we state the following lemma.

Lemma 3.1. *Let $\{Q_i\}_{i=1}^n$ be the set of B-splines defined above, and let $1 \le \nu_1 < \cdots < \nu_p \le n$ and $\xi_1 \le \cdots \le \xi_p$ be prescribed. Then for any $1 \le p \le n$,*

$$D\begin{pmatrix} \xi_1, & \cdots & ,\xi_p \\ Q_{\nu_1}, & \cdots & ,Q_{\nu_p} \end{pmatrix} \ge 0,$$

and strict positivity holds if and only if

$$\xi_i \in (t_{\nu_i}, t_{\nu_i+m}) \cup \{x : D_+^{d_i} Q_{\nu_i}(x) \neq 0\}, \quad i = 1, \ldots, p, \tag{3.10}$$

where

$$d_i = \max\{j : \xi_i = \cdots = \xi_{i-j}\}.$$

For proofs of this result for polynomial splines and for Tchebycheffian splines for distinct ξ_i's, see Theorems 4.65 and 9.34 in [12]. For polynomial splines, the general case is proved in [11] using total positivity properties of a certain discrete spline collocation matrix. The techniques in [11] can also be used to prove Lemma 3.1.

Lemma 3.1 implies (see [8]) that

$$S^-\Big(\sum_{j=1}^{p} c_j Q_{\nu_j}(x)\Big) \le S^-(c_1, \ldots, c_p), \tag{3.11}$$

for any coefficients c_j, where S^- stands for the number of strong sign changes.

B-splines can be *normalized* in various ways. For our applications we want them to form a *stable basis* for L_2 in the sense that the ℓ_2 norm of the coefficient vector of a spline is of a comparable size to the L_2 norm of the spline itself. Let

$$B_i(x) = \frac{Q_i(x)}{(t_{i+m} - t_i)^{1/2}}, \quad i = 1, \ldots, n. \tag{3.12}$$

Lemma 3.2. *There exist positive constants A, B depending only on L such that*

$$A\|c\|_2 \le \|s\|_2 \le B\|c\|_2, \tag{3.13}$$

for all

$$s(x) = \sum_{j=1}^{n} c_j B_j(x), \tag{3.14}$$

with $c = (c_1, \ldots, c_n)$. Here $\|\cdot\|_2$ denotes the usual ℓ_2 norm for vectors and the usual L_2 norm for functions.

Proof: By a slight modification of the proof of Theorem 9.26 in [12], there exist linear functionals λ_i defined on $\mathcal{S}$ and a constant C depending only on L such that

$$\lambda_i Q_j = \delta_{i,j}, \qquad \text{all } i, j, \tag{3.15}$$

and

$$|\lambda_i f| \leq C(t_{i+m} - t_i)^{-1/2} \|f\|_{L_2[t_i, t_{i+m}]}, \qquad i = 1, \ldots, n.$$

Then

$$\sum_{i=1}^{n} |c_i|^2 = \sum_{i=1}^{n} (t_{i+m} - t_i)|\lambda_i s|^2 \leq C^2 \sum_{i=1}^{n} \|s\|^2_{L_2[t_i, t_{i+m}]} \leq mC^2 \|s\|_2^2.$$

This establishes the first Inequality (3.13) with $A = 1/\sqrt{m}C$.

For the second inequality, we have

$$\begin{aligned}
\int_a^b |s(x)|^2 dx &= \sum_{j=m}^{n} \int_{t_j}^{t_{j+1}} | \sum_{i=j-m+1}^{j} c_i B_i(x)|^2 dx \\
&\leq m \sum_{j=m}^{n} \int_{t_j}^{t_{j+1}} \sum_{i=j-m+1}^{j} |c_i B_i(x)|^2 dx \\
&\leq m \sum_{i=1}^{n} |c_i|^2 \|B_i\|_2^2 \\
&\leq C \sum_{i=1}^{n} |c_i|^2,
\end{aligned}$$

where $C = mW^2$, and $W = \max_{1 \leq i \leq n} \|w_{1,i}\|_\infty$. ■

§4 L-spline wavelets

Let

$$\tau = \{\tau_i\}_{i=1}^{N}, \qquad t = \{t_i\}_{i=1}^{N+M}$$

be two extended knot sequences corresponding to a closed interval $I = [a, b]$. We assume that the analogue of assumption of Equation (3.2) holds for τ, and that τ is a subsequence of t, *i.e.*, t is a *refinement* of τ. We express this as $\tau \subset t$. Then there exists a subsequence $s = \{s_1 \leq \cdots \leq s_M\}$ of t such that $t = \tau \cup s$. We can think of the s_i as *new knots*. Note that since $\tau \subset t$ assumption of Equation (3.2) also holds for t.

We are interested in the spline spaces $\mathcal{S}_\tau$ and $\mathcal{S}_t$. Since $\mathcal{S}_\tau$ is a subspace of $\mathcal{S}_t$, we can examine the orthogonal complement W_τ of $\mathcal{S}_\tau$ in $\mathcal{S}_t$. W_τ is the M-dimensional linear subspace of $\mathcal{S}_t$ such that

$$\mathcal{S}_t = W_\tau \oplus \mathcal{S}_\tau. \tag{4.1}$$

Our aim is to construct a basis for W_τ consisting of minimally supported functions w_i, $i = 1, \ldots, M$. We shall refer to these basis functions as *L-wavelets*.

Before proceeding, we introduce some new notation. Suppose that B_1, ..., B_n are the B-splines which span the space $\mathcal{S}_t$. Following [10], if a spline $f \in \mathcal{S}_t$ has support on the interval $[t_\ell, t_r]$ (so that it can be written as a linear combination of the B-splines $B_\ell, \ldots, B_{r-m}$), we use the abbreviation $[\ell : r]$ for its support set. A spline $f \in W_\tau$ has *minimal support* provided that if g is any other nonzero spline with support $[u : v]$ satisfying $\ell \le u < v \le r$, then $u = \ell$ and $v = r$.

Following [10], we define the *right* and *left* multiplicities of a knot $t_i \in t$ by

$$\rho_t(i) = \max\{j : t_{i+j-1} = t_i\} \tag{4.2}$$

$$\lambda_t(i) = \max\{j : t_{i-j+1} = t_i\}. \tag{4.3}$$

Throughout the remainder of this paper we use # to stand for the cardinality of a set. We now show that for each s_i in t but not in τ, there exists a corresponding L-wavelet.

Proposition 4.1. *Let s_i be one of the knots in t but not in τ, and let ℓ be the largest integer such that*

$$\#\{j : s_j \in (t_\ell, s_i] \text{ and } j < i\} + \rho_t(\ell) = m,$$

and let r be the smallest integer such that

$$\#\{j : s_j \in [s_i, t_r) \text{ and } j > i\} + \lambda_t(r) = m.$$

Then there exists an L-wavelet ψ_i with support $[\ell : r]$.

Proof: Suppose that the τ_j's contained in (t_ℓ, t_r) are $\tau_{\mu+1}, \ldots, \tau_{\mu+p}$. The selection of ℓ and r assures that $r - \ell = p + 2m$. We can interpret this as saying that $[t_\ell, t_r]$ contains m knots to the left of s_i which are either new knots or are equal to t_ℓ, and m knots to the right of s_i which are either new knots or are equal to t_r. The B-splines in $\mathcal{S}_t$ with support on $[t_\ell, t_r]$ are precisely

$$\gamma_j := B_{\ell+j-1}, \qquad j = 1, \ldots, q, \tag{4.4}$$

where $q = r - \ell - m + 1 = p + m + 1$. The B-splines in $\mathcal{S}_\tau$ whose supports intersect (t_ℓ, t_r) are precisely

$$\phi_j := \widetilde{B}_{\mu+j-m}, \qquad j = 1, \ldots, q-1, \tag{4.5}$$

where $\widetilde{B}_j$ are the B-splines in the space $\mathcal{S}_\tau$. Now let

$$\psi_i(x) = \det \begin{pmatrix} (\phi_1, \gamma_1) & \cdots & (\phi_1, \gamma_{q-1}) & (\phi_1, \gamma_q) \\ (\phi_2, \gamma_1) & \cdots & (\phi_2, \gamma_{q-1}) & (\phi_2, \gamma_q) \\ \vdots & \ddots & \vdots & \vdots \\ (\phi_{q-1}, \gamma_1) & \cdots & (\phi_{q-1}, \gamma_{q-1}) & (\phi_{q-1}, \gamma_q) \\ \gamma_1(x) & \cdots & \gamma_{q-1}(x) & \gamma_q(x) \end{pmatrix}. \tag{4.6}$$

Proposition 4.2 below implies that ψ_i is nonzero.

The entries in Equation (4.6) are the usual L_2 inner-products on $[a,b]$. Expanding out this determinant, we have

$$\psi_i(x) = \sum_{j=1}^{q} c_{i,j}\gamma_j(x). \tag{4.7}$$

This is clearly a spline in $\mathcal{S}_t$, and is orthogonal to each of the $\phi_1,\ldots,\phi_{q-1}$. Since these are the only basis elements in $\mathcal{S}_\tau$ whose supports intersect $[t_\ell, t_r]$, it follows that ψ_i is orthogonal to all of $\mathcal{S}_\tau$, and hence is an element of the orthogonal complement W_τ. The fact that ψ_i has minimal support follows from the choice of ℓ and r. ■

The following proposition shows that the L-wavelet defined in Equation (4.6) is nontrivial, and that the coefficients in Equation (4.7) oscillate in sign.

Proposition 4.2. *The coefficients $c_{i,j}$ in the L-wavelet ψ_i defined in Equations (4.6)–(4.7) oscillate strictly in sign. Moreover, ψ_i is nonzero on (t_ℓ, t_r) except for at most $q-1$ points where it changes sign.*

Proof: Expanding out Equation (4.6), it is clear that

$$c_{i,j} = (-1)^{q-i} d_j, \qquad d_j := \det G_j,$$

where

$$G_j := G\begin{pmatrix} \phi_1, & \cdots & & & \cdots & \phi_{q-1} \\ \gamma_1, & \cdots & \gamma_{j-1}, & \gamma_{j+1}, & \cdots & \gamma_q \end{pmatrix}, \tag{4.8}$$

for $j = 1,\ldots,q$, where G stands for the Gram matrix formed from the functions listed in the first and second rows. To prove our claim, we need to show that $d_j > 0$ for $j = 1,\ldots,q$.

By a basic composition formula (see [7, pp. 16–17]),

$$d_j = \int_\Omega D\begin{pmatrix} y_1, & \ldots, & y_{q-1} \\ \phi_1, & \ldots, & \phi_{q-1} \end{pmatrix} D\begin{pmatrix} y_1, & \ldots, & & & \cdots & y_{q-1} \\ \gamma_1, & \ldots, & \gamma_{j-1}, & \gamma_{j+1}, & \cdots & \gamma_q \end{pmatrix}, \tag{4.9}$$

where $\Omega = \{(y_1,\ldots,y_{q-1}) : a \le y_1 < \cdots < y_{q-1} \le b\}$. This immediately implies $d_j \ge 0$. Since the B-splines are continuous functions, d_j will be positive if the integrand is positive at some point in the interior of Ω. But by Lemma 3.1, this happens if the diagonals of the matrices in Equation (4.9) are positive for some point in Ω, *i.e.*, if and only if

$$\begin{aligned} \operatorname{supp_O}\phi_k \cap \operatorname{supp_O}\gamma_k &\ne \emptyset, \qquad \text{for } k = 1, 2,\ldots, j-1, \\ \operatorname{supp_O}\phi_k \cap \operatorname{supp_O}\gamma_{k+1} &\ne \emptyset, \qquad \text{for } k = j, j+1, \ldots, q-1, \end{aligned} \tag{4.10}$$

where $\operatorname{supp_O}$ denotes the interior of the support set. For the purpose of this proof, let us relabel the knots so that $\mu = m$ and $\ell = 1$, which also gives $r - m =$

$q = m+p+1$. Then $\text{supp}_0\,\phi_k = (\tau_k, \tau_{k+m})$ and $\text{supp}_0\,\gamma_k = (t_k, t_{k+m})$, and the interior τ-knots in $(t_\ell, t_r) = (t_1, t_{q+m})$ are $\tau_{m+1}, \ldots, \tau_{q-1}$. By construction we know that $\text{supp}_0\,\phi_1 \cap \text{supp}_0\,\gamma_1 \neq \emptyset$ since $\tau_m \leq t_1 < \tau_{m+1}$. Likewise, we have $\text{supp}_0\,\phi_{q-1} \cap \text{supp}_0\,\gamma_q \neq \emptyset$ since $\tau_{m+p} < t_r \leq \tau_{m+p+1}$ or $\tau_{q-1} < t_{q+m} \leq \tau_q$.

Suppose now that $\text{supp}_0\,\phi_\sigma \cap \text{supp}_0\,\gamma_\sigma = \emptyset$ for some $\sigma > 1$. Since Equation (4.10) holds for $k = 1$ and τ is a subsequence of t, we must then have $t_{p+m} \leq \tau_p$ for $p \geq \sigma$ and in particular $t_{q+m-1} \leq \tau_{q-1}$. Together with $\tau_{q-1} < t_{q+m}$, this means that

$$t_{q+m-1} \leq \tau_{q-1} < t_{q+m} = t_r.$$

But by Lemma 4.3 below, this contradicts the minimality of the support of ψ_i. A similar contradiction is obtained if $\text{supp}_0\,\phi_k \cap \text{supp}_0\,\gamma_{k+1} = \emptyset$ for some k. This shows that $d_j > 0$ for all j, and therefore that the $c_{i,j}$'s oscillate strictly in sign.

Since the B-spline coefficients of ψ_i oscillate strictly in sign, ψ_i cannot vanish identically on any subinterval of (t_ℓ, t_r). Moreover, Equation (3.11) implies ψ_i can have at most $q-1$ sign changes, and they have to occur at isolated points. ∎

Lemma 4.3. *Let ξ be a knot in τ with multiplicity u, and let $[\ell : r]$ be the support of an L-wavelet with $t_\ell < \xi < t_r$. Then ℓ satisfies*

$$\#\{j : s_j \in (t_\ell, \xi)\} + \rho_t(\ell) \geq u+1,$$

and r satisfies

$$\#\{j : s_j \in (\xi, t_r)\} + \lambda_t(r) \geq u+1.$$

Proof: Suppose to the contrary that there is an L-wavelet ψ with support $[\ell : r]$ and that $t_\ell < \xi$, for which

$$\#\{j : s_j \in (t_\ell, \xi)\} + \rho_t(\ell) = v \leq u.$$

Since $r - \ell - p = 2m$, if we do not count the τ-knots in (t_ℓ, t_r) then the sequence $(t_j)_{j=\ell}^{r}$ should contain exactly $2m+1$ knots. In our case we have $v \leq u \leq m$ knots to the left of ξ, at most m knots at ξ, and therefore at least one knot to the right of ξ so that $t_r > \xi$.

To complete the proof, we define w by

$$w = \#\{j : \tau_j \in (t_\ell, \xi)\},$$

and set $\ell' = \ell + u + w$. Let V_0 be the span of the splines γ_j defined in Equation (4.4), and let V_1 be the span of the ϕ_j in Equation (4.5). The space of splines V_1' with support $[\ell' : r]$ has dimension $\dim V_1' = \dim V_1 - u - w$, while the number of B-splines in $\mathcal{S}_\tau$ with support intersecting (t_ℓ', t_r) is $\dim V_0 - u - w$, so that $\dim V_0' = \dim V_0 - u - w$. This means that we have $\dim V_1' = \dim V_0' + 1$ so that we can construct an L-wavelet on (possibly a subinterval of) $[\ell' : r]$. The function ψ therefore cannot be an L-wavelet. ∎

The following theorem is the main result of this section.

Theorem 4.4. *Suppose that t, τ, and s are as above, and let $\psi_1, \ldots, \psi_M$ be the L-wavelets constructed in Proposition 4.1. Then the set $\{\psi_i\}_{i=1}^M$ forms a basis for the orthogonal complement W_τ of $\mathcal{S}_\tau$ in $\mathcal{S}_t$.*

Proof: Suppose the support of ψ_i is $[\ell_i : r_i]$, $i = 1, \ldots, M$. We note that $\ell_{i-1} < \ell_i$ and $r_{i-1} < r_i$, for $i = 2, \ldots, M$. To show that the ψ_i are linearly independent, suppose that

$$f = \sum_{i=1}^{M} d_i \psi_i \equiv 0. \tag{4.11}$$

Consider $x = t_{\ell_1}$, and let $\rho = \rho_t(\ell_1)$. Then since $\ell_1 < \ell_2 < \cdots < \ell_M$, using the fact that the $c_{i,j}$'s in Equation (4.7) are nonzero, it is easy to see that $D_+^{m-\rho}\psi_1(x) \neq 0$ while $D_+^{m-\rho}\psi_i(x) = 0$ for $i > 1$. Applying $D_+^{m-\rho}$ to f and evaluating at x, we conclude that $d_1 = 0$. This argument can be repeated to show that each of the other d_i is zero, and we have shown that the splines $\psi_1, \ldots, \psi_M$ are linearly independent. Since $\dim W = M$, it follows that they form a basis. ■

We now give an alternate representation for L-wavelets spanning the orthogonal complement of $\tau \subset t$. Let $\tilde{t}$ be the extended knot sequence obtained from t by adding t_1 and t_{m+n} each a total of m times. Suppose that $\mathcal{S}_{\tilde{t}}^{2m}$ is the spline space associated with the knots $\tilde{t}$ and the null space of the operator LL^*. Let $\{N_i\}$ be the B-splines in $\mathcal{S}_{\tilde{t}}^{2m}$.

Theorem 4.5. *Let ψ be the L-wavelet with support $[\ell : r]$. Suppose that $\tau_{\mu+1}, \ldots, \tau_{\mu+p}$ are the knots of τ lying in (t_ℓ, t_r). Then $\psi(x) = L^*\theta(x)$, where $\theta \in \mathcal{S}_{\tilde{t}}^{2m}$ is given by*

$$\theta(x) = \gamma D\begin{pmatrix} x, & \tau_{\mu+1}, & \cdots & \tau_{\mu+p} \\ N_\ell, & N_{\ell+1}, & \cdots & N_{\ell+p} \end{pmatrix}. \tag{4.12}$$

Here γ is a constant and $\ell + p = r - 2m$.

Proof: Let

$$\theta(x) = u^*(x) + \int_a^b (-1)^m g_m(y; x)\psi(y)dy,$$

where $(-1)^m g_m(y;x)$ is the Green's function for L^* and $u^* \in N_{L^*}$ is arbitrary for the moment. Then $\psi = L^*\theta$. Since ψ is a spline which belongs to the null space N_L of L, we conclude that θ belongs to $\mathcal{S}_{\tilde{t}}^{2m}$. We claim that $\lambda[\tau_i, \ldots, \tau_{i+m}]\theta = 0$ for $i = 1, \ldots, N$. To show this we fix i and observe that the restriction to $[\tau_i, \tau_{i+m}]^2$ of the global Green's function $g_m(x; y))$ for L must, by uniqueness of the Green's function, reduce to the τ-analogue of the local Green's function $g_{m,i}(x; y)$ given by Equation (3.5). Therefore, when $x \in [\tau_i, \tau_{i+m}]$ we can use $g_{m,i}$ instead of g_m when evaluating $\theta(x)$. The fact that $\lambda[\tau_i, \ldots, \tau_{i+m}]\theta = 0$ now follows from Equation (3.9). Choosing u^* so that θ vanishes on $\tau_1, \ldots, \tau_m$, we conclude that it vanishes on τ. But then the support of θ must be $[t_l, t_r]$. It follows that it is a linear combination of the B-splines $N_l, \ldots, N_{r-2m}$ in $\mathcal{S}_t^{2m}$ that vanishes on $\tau_{\mu+1}, \ldots, \tau_{\mu+p}$. We conclude that Equation (4.12) holds.

We still have to show that the determinant is not identically zero. It suffices to show that one of the coefficients in the expansion of the determinant is nonzero, e.g.,

$$D\begin{pmatrix} \tau_{\mu+1}, \ldots, \tau_{\mu+p} \\ N_{\ell+1}, \ldots, N_{\ell+p} \end{pmatrix} \neq 0.$$

By the analog of Lemma 3.1, we see that this holds if $\tau_{\mu+j} \in \operatorname{supp} N_{\ell+j}$ for $j = 1, \ldots, p$. It is therefore sufficient to show that $t_{\ell+1} < \tau_{\mu+1}$ and $\tau_{\mu+p} < t_{\ell+p+2m}$. The last inequality is trivial since $r = \ell + p + 2m$. For the first one, note that $t_\ell < \tau_{\mu+1}$, so that the only case that can cause problems is $t_\ell < t_{\ell+1} = \tau_{\mu+1}$. But this is impossible by Lemma 4.3. ∎

§5 The translation invariant case

In this section we assume that the differential operator L has constant coefficients, and so the corresponding null space is translation invariant. Our aim is to show that with uniformly spaced knots, there is a simple formula for the L-wavelets which generalizes the explicit formula for polynomial spline wavelets given in [4].

It is easy to show that with uniformly spaced knots, the B-splines associated with L are all translates of a single B-spline. In particular, suppose $t_i = ih/2$ and $\tau_i = ih$ for all i, where $h > 0$ is given. Then the B-splines in $\mathcal{S}_\tau$ are given by

$$\widetilde{B}_i(x) = N_h(x - ih),$$

and the B-splines in $\mathcal{S}_t$ are given by

$$B_i(x) = N_{h/2}(x - ih/2),$$

for all i, where N_h is the B-spline with knots $\{0, h, \ldots, mh\}$.

It suffices to give a formula for one wavelet, since it follows immediately from Equation (4.6) that all wavelets are translates of a single wavelet.

Theorem 5.1. *Let ψ be the L-wavelet defined in Proposition 4.1 corresponding to the knot $s = (m - 1/2)h$. Then ψ has support on $[0, (2m-1)h]$, and is given by*

$$\psi = \sum_{j=1}^{3m-1} (-1)^{j-1} w_j B_{j-1}, \tag{5.1}$$

where

$$w_j = \int_{\mathbf{R}} \widetilde{B}_0(x) B_{2m-j}(x)\, dx. \tag{5.2}$$

Proof: Taking $s = (m - 1/2)h$ in Proposition 4.1, we find $\ell = 0$ and $r = 4m-2$. Thus the L-wavelet has support $[\ell h/2, rh/2] = [0, (2m-1)h]$. To prove Equation (5.1) it is enough to show that $(\psi, \tilde{B}_{i-m}) = 0$ for $i = 1, 2, \ldots, 3m-2$. (These are the B-splines on τ with support overlapping the support of ψ.) By the translation invariance, we have

$$(B_{j-1}, \tilde{B}_{i-m}) = w_{2i+1-j}.$$

It follows that

$$\rho_i := (\psi, \tilde{B}_{i-m}) = \sum_{j=1}^{3m-1} (-1)^{j-1} w_j (B_{j-1}, \tilde{B}_{i-m}) = \sum_{j=1}^{3m-1} (-1)^{j-1} w_j w_{2i+1-j}.$$

Note that $w_j = 0$ for $j \leq 0$ and $j \geq 3m$. To show that $\rho_i = 0$ for $i = 1, 2, \ldots, 3m-2$ we consider two cases.

Suppose first $2i \leq 3m - 1$. We find

$$\rho_i = \sum_{j=1}^{2i} (-1)^{j-1} w_j w_{2i+1-j}.$$

Splitting this sum into two,

$$\rho_i = \sum_{j=1}^{i} (-1)^{j-1} w_j w_{2i+1-j} + \sum_{j=i+1}^{2i} (-1)^{j-1} w_j w_{2i+1-j},$$

and replacing j by $2i+1-j$ in one of the sums, we see that the two sums have the same value, but with opposite signs. It follows that $\rho_i = 0$.

Suppose then $2i > 3m - 1$. In this case

$$\rho_i = \sum_{j=2i+2-3m}^{3m-1} (-1)^{j-1} w_j w_{2i+1-j},$$

and the number of terms is even. We write

$$\rho_i = \sum_{j=2i+2-3m}^{i} (-1)^{j-1} w_j w_{2i+1-j} + \sum_{j=i+1}^{3m-1} (-1)^{j-1} w_j w_{2i+1-j}.$$

Again replacing j by $2i + 1 - j$ in one of the sums we see that the two sums have the same value, but with opposite signs. It follows that $\rho_i = 0$ also in this case. ■

A theory of wavelets associated with translation invariant spaces has recently been developed by [1]. Using Fourier transform methods, they also get formulae for the wavelets in this case.

§6 A multiresolution framework

In this section we show how to construct a nested sequence of L-splines which resembles a classical multiresolution analysis, see [3,5]. Let L be an m-th order linear differential operator as in Equation (2.2), and suppose that

$$\tau_0 \subset \tau_1 \subset \cdots$$

is a nested sequence of extended partitions of the interval I. Let $\mathcal{S}_j$ be the corresponding spaces of L-splines of dimension n_j. Suppose that for all j, $\{Q_{i,j}\}_{i=1}^{n_j}$ are the B-splines forming a basis for $\mathcal{S}_j$ defined in Section 3, and that $\{B_{i,j}\}_{i=1}^{n_j}$ are obtained by normalizing as in Equation (3.12).

Theorem 6.1. *The spaces $\mathcal{S}_j$, $j \geq 0$ form a kind of multiresolution analysis in the sense that*

1) $\mathcal{S}_0 \subset \mathcal{S}_1 \subset \mathcal{S}_2 \cdots$;
2) $\cup \mathcal{S}_j$ *is dense in* $L_2[a,b]$;
3) $\{B_{i,j}\}_{i=1}^{n_j}$ *is a Riesz basis for* $\mathcal{S}_j$, *i.e., the stability condition of Lemma 3.2 holds;*
4) *there exist L-spline wavelets* $\{\psi_{i,j}\}_{i=1}^{n_{j+1}-n_j}$ *which span the orthogonal complements* W_j *of* $\mathcal{S}_j$ *in* $\mathcal{S}_{j+1}$ *defined by*

$$\mathcal{S}_{j+1} = W_j \oplus \mathcal{S}_j,$$

for all j.

Proof: Properties 1), 3), and 4) are clear from the above discussion. Property 2) follows from well-known results about the approximation power of L-splines, see Theorem 10.25 in [12]. ■

As shown above, the wavelets $\psi_{i,j}$ in Theorem 6.1 are finite linear combinations of the B-splines $B_{i,j}$:

$$\psi_{i,j} = \sum_{\nu=\ell_i}^{r_i - m} q_{i,j,\nu} B_{\nu,j}, \tag{6.1}$$

where ℓ_i and r_i are as in Proposition 4.1.

In this setting, the classical *two-scale relation* does not necessarily hold. In its place we have the following formula for writing B-splines in $\mathcal{S}_j$ as finite combinations of the B-splines in the next finer space $\mathcal{S}_{j+1}$:

$$B_{i,j} = \sum_{\nu=\alpha_{i,j}}^{\beta_{i,j}} p_{i,j,\nu} B_{\nu,j+1}, \tag{6.2}$$

where $\alpha_{i,j}$ and $\beta_{i,j}$ are such that $B_{\nu,j+1}$ for $\nu = \alpha_{i,j}, \ldots, \beta_{i,j}$ have support contained in the support of $B_{i,j}$, see [9].

Clearly, we also have the *decomposition relation*

$$B_{i,j+1} = \sum_{\nu} g_{i,j,\nu} B_{\nu,j} + \sum_{\nu} h_{i,j,\nu} \psi_{\nu,j}. \tag{6.3}$$

In general, the g's and h's can be found by computing the L_2 projections of $B_{i,j+1}$ onto $\mathcal{S}_j$ and W_j, respectively.

In the translation invariant case with uniformly spaced knots, the coefficients $p_{i,j}$ and $q_{i,j}$ appearing in Equations (6.1) and (6.2) are independent of i.

§7 Examples

One of the most interesting examples of L-splines is provided by the choice

$$L = \begin{cases} (D^2 + (r-\frac{1}{2})^2)\cdots(D^2+(\frac{1}{2})^2), & m = 2r \\ (D^2 + r^2)\cdots(D^2+1)D, & m = 2r+1. \end{cases} \tag{7.1}$$

In this case the null space N_L of L is spanned by the functions

$$\begin{cases} \text{span}\,\{\cos(x/2), \sin(x/2), \ldots, \cos((r-\frac{1}{2})x), \sin((r-\frac{1}{2})x)\}, & m = 2r, \\ \text{span}\,\{1, \cos(x), \sin(x), \ldots, \cos(rx), \sin(rx)\}, & m = 2r+1. \end{cases}$$

The resulting splines are called *trigonometric splines*, and have been heavily studied in the approximation theory literature (see *e.g.*, [8] and references therein). For the sake of the reader not familiar with trigonometric splines, in Figure 1 we show a typical "linear" B-spline corresponding to the case $m = 2$ with uniform knots with spacing 2.5. In Figure 2 we show the wavelet with uniformly spaced knots with spacing 2.5.

Because of space limitations, we do not further discuss trigonometric spline wavelets here. Details along with applications to multi-resolution analysis on the sphere will be presented in a separate paper.

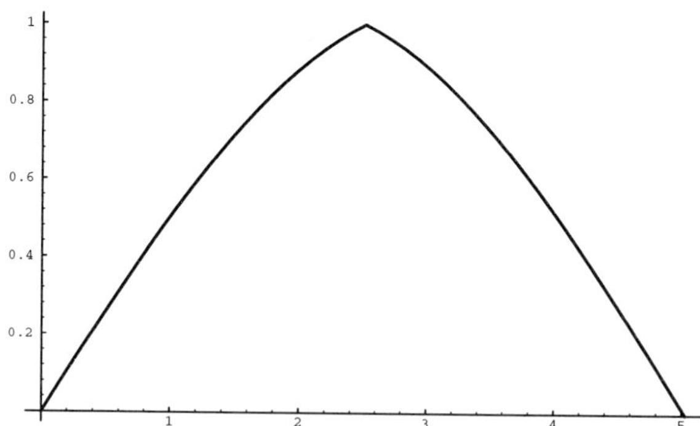

Figure 1. A trigonometric B-spline ($m = 2$).

Another interesting example is provided by the *hyperbolic splines*. These arise by replacing the plus signs in the above definition of L by minus signs. Then the null spaces are spanned by hyperbolic sine and cosine functions.

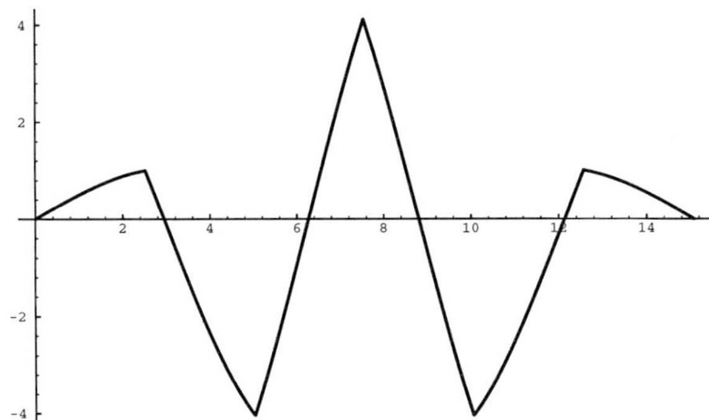

Figure 2. The corresponding trigonometric spline wavelet.

Acknowledgments. The first author was supported in part by the Research Council of Norway through project STP28402 and SINTEF SI. The second author was partially supported by NSF Grant 9208413.

References

1. de Boor, C., R. DeVore, and A. Ron, On the construction of multivariate (pre)wavelets, *Constr. Approx.* **9** (1993), 123–166.
2. Buhmann, M. and C. A. Micchelli, Spline pre-wavelets for non-uniform knots, *Numer. Math.* **61** (1992), 455–474.
3. Chui, C. K., *An Introduction to Wavelets*, Academic Press, Boston, 1992.
4. Chui, C. K. and J. Z. Wang, On compactly supported spline wavelets and a duality principle, *Trans. Amer. Math. Soc.* **330** (1992), 903–915.
5. Daubechies, I., *Ten Lectures on Wavelets*, CBMS Lecture Notes 61, SIAM, Philadelphia, 1992.
6. Goodman, T. N. T and S. L. Lee, Wavelets of multiplicity r, *Trans. Amer. Math. Soc.*, to appear.
7. Karlin, S., *Total Positivity, Vol. I*, Stanford University Press, Stanford, California, 1968.
8. Koch, P.-E., T. Lyche, M. Neamtu, and L. L. Schumaker, Control curves and knot insertion for trigonometric splines, preprint.
9. Lyche, T., A recurrence relation for Chebyshevian B-splines, *Constr. Approx.* **1** (1985), 155–173.
10. Lyche, T. and K. Mørken, Spline-wavelets of minimal support, in *Numerical Methods in Approximation Theory*, D. Braess, L. L. Schumaker (eds.), Birkhäuser, Basel, 1992, 177–194.
11. Mørken, K., Total positivity of the discrete spline collocation matrix II, *J. Approx. Theory*, to appear.
12. Schumaker, L. L., *Spline Functions: Basic Theory*, Wiley, New York, 1981.

T. Lyche
Institutt for Informatikk, University of Oslo
P.O.Box 1080, Blindern
0316 Oslo, Norway
tom@ifi.uio.no

Larry L. Schumaker
Dept. of Mathematics
Vanderbilt University
Nashville, TN 37240
s@mar.cas.vanderbilt.edu

Wavelets and Frames on the Four-Directional Mesh

Charles K. Chui, Kurt Jetter, and Joachim Stöckler

Abstract. Although the integer translates of a box spline on the four-directional mesh do not constitute a Riesz basis, yet this box spline generates an orthonormal basis of the closed linear subspace of $L^2(\mathbb{R}^2)$ that it spans. This paper is devoted to the study of the multiresolution analysis $\{V_j\}$, with dilation matrix $A = \begin{pmatrix} 1 & 1 \\ 1 & -1 \end{pmatrix}$, and its orthogonal complementary subspaces W_j, generated by any four-directional box spline. We will show that the corresponding compactly supported semi-orthogonal wavelet ψ does not satisfy the Riesz condition either, and that the orthonormalization $\psi^\perp$ of ψ may not be in $L^1(\mathbb{R}^2)$. The main result, however, is that by oversampling by $\mathbf{Z}^2 + \left(\frac{1}{2}, \frac{1}{2}\right)$, the compactly supported wavelet ψ generates a frame of $L^2(\mathbb{R}^2)$.

§1 Introduction

A box spline with some (multi-) integer direction set X in $\mathbb{R}^d$, $d > 1$, is perhaps the most natural generalization of the univariate cardinal (polynomial) B-splines (see [1,3]). In particular, when a three-directional mesh of $\mathbb{R}^2$ is generated by such an X, each corresponding box spline satisfies the Riesz condition, and hence, generates a multiresolution analysis (MRA) of $L^2 := L^2(\mathbb{R}^2)$. By following the univariate recipe as described in [4, pp. 182-184], it is therefore not very difficult to construct compactly supported semi-orthogonal and stable wavelets corresponding to any three-directional box spline, as carried out in [6,9]. On the other hand, when the direction set X generates a four-directional mesh of $\mathbb{R}^2$, the box spline $\phi := B_X$, with direction set X, no longer satisfies the Riesz condition. This lack of stability, however, does not prevent the validity of the standard orthonormalization procedure applied to ϕ, yielding an orthonormal basis $\phi^\perp(\cdot - \alpha)$ of

$$V_0 := \operatorname{clos}_{L^2} \operatorname{span}\{\phi(\cdot - \alpha) \colon \alpha \in \mathbf{Z}^2\}.$$

Wavelets: Theory, Algorithms, and Applications
Charles K. Chui, Laura Montefusco, and Luigia Puccio (eds.), pp. 213–230.

ISBN 0-12-174575-9

Hence, with the matrix

$$A = \begin{pmatrix} 1 & 1 \\ 1 & -1 \end{pmatrix}$$

as the dilation matrix, we have an MRA $\{V_j\}$, $j \in \mathbf{Z}$, of L^2, where

$$V_j := \{f\colon\ f(A^{-j}\cdot) \in V_0\}.$$

Also, since $|\det A| = 2$, a straightforward generalization of the univariate recipe [4, pp. 182-184] yields a compactly supported wavelet ψ (with "minimum support"), that generates the orthogonal complementary subspaces W_j, namely: $V_{j+1} = V_j \oplus W_j$, $V_j \perp W_j$, $j \in \mathbf{Z}$, and

$$\begin{cases} W_j := \{g\colon\ g(A^{-j}\cdot) \in W_0\}, \\ W_0 = \operatorname{clos}_{L^2} \operatorname{span}\{\psi(\cdot - \alpha)\colon\ \alpha \in \mathbf{Z}^2\}. \end{cases}$$

Of course, just as ϕ itself, ψ does not satisfy the Riesz condition. Also, just as ϕ, the standard orthonormalization process can be applied to ψ to generate $\psi^\perp \in L^2$, for which

$$\{\psi^\perp(\cdot - \alpha)\colon\ \alpha \in \mathbf{Z}^2\}$$

is an orthonormal basis of W_0. However, $\psi^\perp$ has very poor decay property, and may not even be in $L^1(\mathbb{R}^2)$, in general. Therefore, in spite of its instability, it is still more tempting to use the compactly supported ψ instead of $\psi^\perp$ in applications.

In order to make use of the unstable ψ, we may appeal to frame stability. In Section 4 of this paper, we will see that the notion of oversampling as introduced in [5] can be applied here, even to the unstable ψ, to produce a (stable) frame of L^2, as follows. In addition to the lattice points $\mathbf{Z}^2$ for the shifts in W_0, we consider the extra lattice points $\left(\frac{1}{2}, \frac{1}{2}\right)^T + \mathbf{Z}^2$, yielding

$$\mathcal{B} := \mathbf{Z}^2 \cup \left(\begin{pmatrix} \frac{1}{2} \\ \frac{1}{2} \end{pmatrix} + \mathbf{Z}^2 \right).$$

Let us now consider the standard notation

$$\psi_{j;\alpha} := |\det A|^{j/2} \psi(A^j \cdot - \alpha).$$

Then the main result to be discussed in Section 4 is that the family

$$\{\psi_{j,\alpha}\colon\ j \in \mathbf{Z}, \alpha \in \mathcal{B}\}$$

constitutes a frame of L^2.

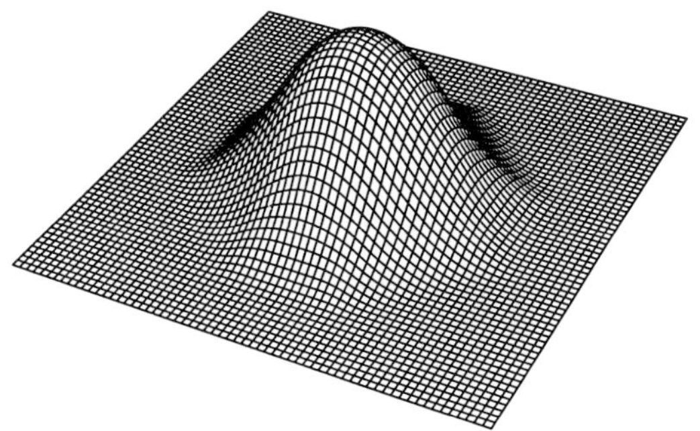

Figure 1. The Zwart-Powell element B_{1111}.

§2 Four-directional box splines as scaling functions

There are several equivalent ways to define box splines of d variables (see [1, Chapter 1] and [3, pp.16-17]). However, perhaps the most appropriate definition for our purpose is the one in terms of the Fourier transform as follows. Given a directional set in terms of a matrix

$$X = (\zeta_1, \zeta_2, \ldots, \zeta_N) \in \mathbf{Z}^{d \times N}$$

with rank $(X) = d$, the corresponding *forward box spline* $B_X \in L^2(\mathbb{R}^d)$ is the inverse Fourier transform of

$$\widehat{B}_X(\xi) = \int_{\mathbf{R}^d} B_X(x)\; e^{-ix\cdot\xi} dx = \prod_{j=1}^{N} \frac{1 - e^{-i\zeta_j \cdot \xi}}{i\zeta_j \cdot \xi}$$

(with $\zeta_j \cdot \xi$ denoting the scalar product). It is well-known that B_X is a compactly supported piecewise polynomial function, and the global smoothness of B_X can be easily controlled by a judicious choice of the directional vectors $\zeta_1, \ldots, \zeta_N$.

While box splines in general were successfully used for approximation of multivariate functions, we are interested only in the special case of bivariate ($d = 2$) *four-directional box splines*. Here, the directional matrix is of type

$$X = (\underbrace{e_1, \ldots, e_1}_{k}, \underbrace{e_2, \ldots, e_2}_{l}, \underbrace{e_1 + e_2, \ldots, e_1 + e_2}_{m}, \underbrace{e_1 - e_2, \ldots, e_1 - e_2}_{n}) \tag{2.1}$$

for $k, l, m, n \in \mathbb{N}$ with $e_1 = \binom{1}{0}$ and $e_2 = \binom{0}{1}$, the two canonical unit vectors in $\mathbb{R}^2$; hence $B_X = B_{klmn}$ has the Fourier transform

$$\widehat{B}_{klmn}(\xi_1, \xi_2) = \tag{2.2}$$

$$\left(\frac{1 - e^{-i\xi_1}}{i\xi_1}\right)^k \left(\frac{1 - e^{-i\xi_2}}{i\xi_2}\right)^l \left(\frac{1 - e^{-i(\xi_1+\xi_2)}}{i(\xi_1 + \xi_2)}\right)^m \left(\frac{1 - e^{-i(\xi_1-\xi_2)}}{i(\xi_1 - \xi_2)}\right)^n.$$

More specifically, the case $k = l = m = n = 1$ leads to the famous *Zwart-Powell element* B_{1111} which is a piecewise quadratic C^1-function supported in the octagon $\{\lambda_1 e_1 + \lambda_2 e_2 + \lambda_3(e_1 + e_2) + \lambda_4(e_1 - e_2)\colon\ \lambda_i \in [0, 1], i = 1, \ldots, 4\}$ (see Figure 1).

We consider the (principal) shift-invariant space

$$S_\phi := \operatorname{clos}_{L^2} \operatorname{span}\{\phi(\cdot - \alpha)\colon \alpha \in \mathbf{Z}^2\} \tag{2.3}$$

spanned by the $\mathbf{Z}^2$-translates of $\phi = B_{klmn}$. Since $\phi \in L^2$, we have the following useful characterization in terms of the Fourier transform:

Theorem 2.1. ([2]). *$f \in S_\phi$ if and only if $f \in L^2$ and $\widehat{f} = \widehat{\phi}\,\omega$ with ω a 2π-periodic function.*

In order to describe the density of scaled versions of S_ϕ, we start with the autocorrelation function

$$\Phi := \phi * \phi^* \tag{2.4}$$

of ϕ (which is a centered four-directional box spline); here $\phi^*(x) = \overline{\phi(-x)}$ denotes the involution of ϕ and hence $\widehat{\Phi} = |\widehat{\phi}|^2$. Its symbol is given by

$$\widetilde{\Phi}(\xi) = \sum_{\alpha\in\mathbf{Z}^2} |\widehat{\phi}(\xi + 2\pi\alpha)|^2 = \sum_{\alpha\in\mathbf{Z}^2} \Phi(\alpha) e^{-i\alpha\cdot\xi}. \tag{2.5}$$

If we put

$$\widehat{\Lambda}_\Phi := \widehat{\Phi}/\widetilde{\Phi} \tag{2.6}$$

then we have the following result.

Theorem 2.2. ([2]). *If $P_\phi : L^2 \to S_\phi$ denotes the orthogonal projector onto the shift-invariant space generated by $\phi = B_{klmn}$, then*

$$||f - P_\phi f||_2^2 = \frac{1}{(2\pi)^2} \int_C (1 - \widehat{\Lambda}_\Phi(\xi)) |\widehat{f}(\xi)|^2 d\xi, \qquad f \in L^2, \tag{2.7}$$

provided that supp $\widehat{f} \subset C := [-\pi, +\pi]^2$.

On the other hand, it is not too hard to see that for $\phi = B_{klmn}$ there is a $p \in \mathbb{N}$ such that

$$|1 - \widehat{\Lambda}_\Phi(\xi)| \le \text{const}|\xi|^{2p} \quad \text{for} \quad \xi \in C\,,$$

so that Equation (2.7) implies that

$$||f - P_\phi(f)||_2 \le \text{const } |f|_{2,p} \tag{2.8}$$

with $|f|_{2,p}$ denoting the Sobolev seminorm of order p. In particular, Theorem 2.2 yields the following density result.

Theorem 2.3. *Let $A \in \mathbb{R}^{2\times 2}$ be a dilation matrix, i.e., a matrix with eigenvalues λ_1, λ_2 where $|\lambda_1| \geq |\lambda_2| > 1$. Let $V_0 := S_\phi$ with $\phi = B_{klmn}$, and for $j \in \mathbf{Z}$, let V_j be defined by*

$$f \in V_j \iff f(A^{-j}\cdot) \in V_0. \tag{2.9}$$

If $P_j : L^2 \to V_j$ denotes the orthogonal projection onto V_j, then

$$\lim_{j\to\infty} P_j f = f, \qquad f \in L^2.$$

A particularly interesting consequence (for wavelet decompositions based on Mallat's multiresolution analysis) of Theorem 2.3 follows, if the spaces V_j are nested, namely:

$$\ldots \subset V_{-1} \subset V_0 \subset V_1 \subset \ldots$$

According to Equation (2.9), the inclusion requirement is that $\phi(A^{-1}\cdot) \in S_\phi$, or equivalently, by Theorem 2.1,

$$|\det A|\, \widehat{\phi}(A^T\xi) = \widehat{\phi}(\xi)\, \omega(\xi) \tag{2.10}$$

with a 2π-periodic function ω. We consider two cases.

The usual dyadic scaling refers to $A = \left(\begin{smallmatrix} 2 & 0 \\ 0 & 2 \end{smallmatrix}\right)$. It is easy to see from Equation (2.2) that for this case the scaling equation holds with the 2π-periodic function ω given by the trigonometric polynomial

$$\omega(\xi_1, \xi_2) = \tag{2.11}$$
$$4\left(\frac{1+e^{-i\xi_1}}{2}\right)^k \left(\frac{1+e^{-i\xi_2}}{2}\right)^l \left(\frac{1+e^{-i(\xi_1+\xi_2)}}{2}\right)^m \left(\frac{1+e^{-i(\xi_1-\xi_2)}}{2}\right)^n.$$

This is well-known and often used, *e.g.*, in the construction of wavelets from three-directional box splines (see [6,9,10]). The disadvantage here is that in order to write the orthogonal decomposition

$$V_1 = V_0 \oplus W_0^{(1)} \oplus W_0^{(2)} \oplus W_0^{(3)},$$

we need $|\det A| - 1 = 3$ wavelet spaces $W_0^{(i)}$, $i = 1, 2, 3$, each spanned by integer translates of a function $\psi^{(i)} \in V_1$.

For simplicity, it seems to be more attractive to try a matrix $A \in \mathbf{Z}^{2\times 2}$ with $|\det A| = 2$ as in [7]. Among the examples discussed there, the matrix

$$A = \begin{pmatrix} 1 & 1 \\ 1 & -1 \end{pmatrix}, \quad \text{so that} \quad A^2 = \begin{pmatrix} 2 & 0 \\ 0 & 2 \end{pmatrix}, \tag{2.12}$$

is most appropriate to the study of four-directional box splines, since $Ae_1 = e_1 + e_2$, $Ae_2 = e_1 - e_2$, $A(e_1 + e_2) = 2e_1$, and $A(e_1 - e_2) = 2e_2$ which is the basis of the following scaling result.

Lemma 2.4. *If $\phi = B_{klmn}$ with $k = m$ and $l = n$, and $A = \begin{pmatrix} 1 & 1 \\ 1 & -1 \end{pmatrix}$, then*

$$2\,\widehat{\phi}(A\xi) = \widehat{\phi}(\xi)\,P(\xi) \tag{2.13}$$

with

$$P(\xi) = 2\left(\frac{1+e^{-i\xi_1}}{2}\right)^k \left(\frac{1+e^{-i\xi_2}}{2}\right)^l. \tag{2.14}$$

Moreover, $4\,\widehat{\phi}(2\xi) = \widehat{\phi}(\xi)\,\omega(\xi)$ with $\omega(\xi) = P(\xi)\,P(A\xi)$.

Theorem 2.5. *Let $\phi = B_{klmn}$ with $k = m$ and $l = n$, and $A = \begin{pmatrix} 1 & 1 \\ 1 & -1 \end{pmatrix}$; and set $V_0 := S_\phi$ and*

$$V_j := \{f : f(A^{-j}\cdot) \in V_0\}. \tag{2.15}$$

Then $\{V_j\}_{j\in\mathbf{Z}}$ defines an A-scaled multiresolution analysis of L^2, i.e.,

(a) *$\{V_j\}_{j\in\mathbf{Z}}$ is a nested sequence*

$$\ldots \subset V_{-1} \subset V_0 \subset V_1 \subset \ldots$$

of closed subspaces of L^2.

(b) *$\bigcup_{j\in\mathbf{Z}} V_j$ is dense in L^2, and $\bigcap_{j\in\mathbf{Z}} V_j = \{0\}$.*

(c) *V_j as defined by Equation (2.15) is $A^{-j}\mathbf{Z}^2$- shift-invariant; i.e.,*

$$f \in V_j \Longrightarrow f(\cdot - A^{-j}\alpha) \in V_j \quad \text{for} \quad \alpha \in \mathbf{Z}^2.$$

(d) *There is a function $\phi^\perp \in S_\phi$ such that $\phi^\perp(\cdot - \alpha), \alpha \in \mathbf{Z}^2$, constitutes a complete orthonormal system of S_ϕ.*

In particular, $S_\phi = S_{\phi^\perp}$.

Proof: (a) follows from the definition in Equation (2.3) and Lemma 2.4, and (c) is trivial since S_ϕ is $\mathbf{Z}^2$-shift-invariant. To prove (d), we first apply the usual orthonormalization process

$$\widehat{\phi^\perp} = \frac{\widehat{\phi}}{\sqrt{\widetilde{\Phi}}} \tag{2.16}$$

to obtain $\phi^\perp$, where $\widetilde{\Phi}$ is given in Equation (2.5). Here, however, we have to face the fact that the symbol $\widetilde{\Phi}$ has zeros, so that the usual arguments need to be extended. We may apply Theorems 2 and 3 of [8] (which apply here, provided $k = l = m = n$) in order to see that $\widetilde{\Phi}(\xi) = 0$ if and only if $\xi = \binom{\pi}{\pi} + 2\pi\alpha$, $\alpha \in \mathbf{Z}^2$, and hence, $|\widehat{\phi^\perp}|^2 = \widehat{\Phi}/\widetilde{\Phi} \in L^1(\mathbb{R}^2)$. Or we may directly conclude from

$$\begin{aligned}\int_{\mathbb{R}^2} |\widehat{\phi^\perp}(\xi)|^2 d\xi &= \int_{\mathbb{R}^2} \frac{\widehat{\Phi}(\xi)}{\widetilde{\Phi}(\xi)} d\xi \\ &= \sum_{\gamma\in\mathbf{Z}^2} \int_C \frac{\widehat{\Phi}(\xi + 2\pi\gamma)}{\widetilde{\Phi}(\xi)} d\xi = (2\pi)^2\end{aligned}$$

that $\phi^{\perp} \in L^2$. Therefore, $S_\phi = S_{\phi^\perp}$, and the orthogonality of the translates $\phi^{\perp}(\cdot - \alpha)$, $\alpha \in \mathbf{Z}^2$, is clear.

It remains to show that (b) holds. The first part follows from Theorem 2.3, and the second part of (b) is known to follow from the remaining conditions. For completeness, we include the following argument. That is, we will verify the equivalent statement

$$\lim_{j \to -\infty} P_j f = 0 \quad \text{for} \quad f \in L^2. \tag{2.17}$$

According to an apparent density argument (and since $\|P_j\| = 1$ for all j), we may assume that $\widehat{f}$ is continuous with compact support in $\mathbb{R}^2 \setminus \{0\}$.

Now since $P_0 f = \sum_{\alpha \in \mathbf{Z}^2} \langle f, \phi^{\perp}_{0,\alpha} \rangle \phi^{\perp}_{0,\alpha}$ with $\phi^{\perp}_{0,\alpha} = \phi^{\perp}(\cdot - \alpha)$, we have

$$(\widehat{P_0 f})(\xi) = \widehat{\phi}^{\perp}(\xi)\, \omega_f(\xi)$$

where

$$\begin{aligned} \omega_f(\xi) &= \sum_{\alpha \in \mathbf{Z}^2} \langle f, \phi^{\perp}_{0,\alpha} \rangle e^{-i\alpha \cdot \xi} \\ &= \sum_{\alpha \in \mathbf{Z}^2} \widehat{f}(\xi + 2\pi\alpha)\, \overline{\widehat{\phi}^{\perp}(\xi + 2\pi\alpha)}. \end{aligned}$$

Hence, by the A-scaling property, we get

$$P_j f = (P_0 f_{A^{-j}})(A^j \cdot) \quad \text{with} \quad f_{A^{-j}} = f(A^{-j} \cdot), \tag{2.18}$$

so that, since $A^T = A$, it follows that

$$(\widehat{P_j f})(\xi) = 2^{-j}\, \widehat{\phi}^{\perp}(A^{-j}\xi)\, \omega_{f_{A^{-j}}}(A^{-j}\xi).$$

By the orthogonality of $P_j f$ and $f - P_j f$, we arrive at

$$\begin{aligned} \|P_j f\|_2^2 &= \frac{1}{(2\pi)^2} \langle \widehat{P_j f}, \widehat{f} \rangle \\ &= \frac{1}{(2\pi)^2} \int_{\mathbf{R}^2} \left\{ \sum_{\alpha \in \mathbf{Z}^2} \widehat{f}(\xi + 2\pi A^j \alpha)\, \overline{\widehat{\phi}^{\perp}(A^{-j}\xi + 2\pi\alpha)} \right\} \times \\ &\qquad \times \widehat{\phi}^{\perp}(A^{-j}\xi)\, \overline{\widehat{f}(\xi)}\, d\xi \,. \end{aligned}$$

Also, from the Cauchy-Schwarz inequality for the expression in the braces and the identity

$$\sum_{\alpha \in \mathbf{Z}^2} |\widehat{\phi}^{\perp}(\xi + 2\pi\alpha)|^2 \equiv 1,$$

we see that

$$\|P_j f\|_2^2 \le \frac{1}{(2\pi)^2} \int_{\mathbf{R}^2} \left\{ \sum_{\alpha \in \mathbf{Z}^2} |\widehat{f}(\xi + 2\pi A^j \alpha)|^2 \right\}^{\frac{1}{2}} |\widehat{\phi}^{\perp}(A^{-j}\xi)|\, |\widehat{f}(\xi)|\, d\xi \,. \tag{2.19}$$

Now from the assumption on $\widehat{f}$ we observe that $\sum_{\alpha\in\mathbf{Z}^2}|\widehat{f}(\xi+2\pi\alpha)|^2$ is a continuous and 2π-periodic function, hence is bounded by a constant, say, c_f^2. From this, for any $j\in\mathbf{Z}$ with $j\le 0$, we have

$$\sum_{\alpha\in\mathbf{Z}^2}|\widehat{f}(\xi+2\pi A^j\alpha)|^2\le 2^{-j}c_f^2\ . \tag{2.20}$$

Take any $\varepsilon>0$ so that $\widehat{f}(\xi)=0$ for $|\xi|\le\varepsilon$. Then combining Equations (2.19) and (2.20), we have

$$\begin{aligned}\|P_j(f)\|_2^2 &\le \frac{1}{(2\pi)^2}\,c_f\int_{|\xi|\ge\varepsilon}2^{-j/2}\,|\widehat{\phi}^{\perp}(A^{-j}\xi)|\,|\widehat{f}(\xi)|\,d\xi\\ &\le \frac{1}{(2\pi)^2}\,c_f\Big\{\int_{|\xi|\ge\varepsilon}|\widehat{f}(\xi)|^2d\xi\int_{|\xi|\ge\varepsilon}2^{-j}\,|\widehat{\phi}^{\perp}(A^{-j}\xi)|^2\,d\xi\Big\}^{1/2}\\ &= c_f\,\|f\|_2\Big\{\frac{1}{(2\pi)^2}\int_{|\xi|\ge\varepsilon}2^{-j}\,|\widehat{\phi}^{\perp}(A^{-j}\xi)|^2\,d\xi\Big\}^{1/2}\\ &\to 0 \text{ as } j\to-\infty,\end{aligned}$$

since $\widehat{\phi}^{\perp}\in L^2$. This completes the proof of the theorem. ■

Remark. A similar analysis could be given when the matrix (2.12) is replaced by $A=\begin{pmatrix}1&-1\\1&1\end{pmatrix}$, but we do not elaborate on this.

§3 Four-directional wavelets

With $\phi=B_{klkl}$ and $A=\begin{pmatrix}1&1\\1&-1\end{pmatrix}$ we now extend the construction of semi-orthogonal (s.o.) wavelets (also called pre-wavelets) as given, *e.g.*, in [4; pp. 15, 75, 182-184] to the case of A-scaled spaces. With

$$|\det A|\;\widehat{\phi}(A^T\xi)=P(\xi)\;\widehat{\phi}(\xi) \tag{3.1}$$

as in Lemma 2.4, we define ψ according to

$$|\det A|\;\widehat{\psi}(A^T\xi)=Q(\xi)\;\widehat{\phi}(\xi) \tag{3.2}$$

with

$$\begin{aligned}Q(\xi):&=e^{i\xi\cdot e_1}\;\overline{P(\xi+2\pi A^{-T}e_1)}\;\widetilde{\Phi}(\xi+2\pi A^{-T}e_1)\\ &=e^{i\xi_1}\;\overline{P(\xi+\pi(e_1+e_2))}\;\widetilde{\Phi}(\xi+\pi(e_1+e_2))\end{aligned} \tag{3.3}$$

and $\widetilde{\Phi}$ as in Equation (2.5). Here, we write $A^{-T}:=(A^{-1})^T$ (of course, with our specific choice of $A=A^T$ we find $A^{-T}=A^{-1}=\frac{1}{2}A$, but we want to give the general formulas for other choices of A).

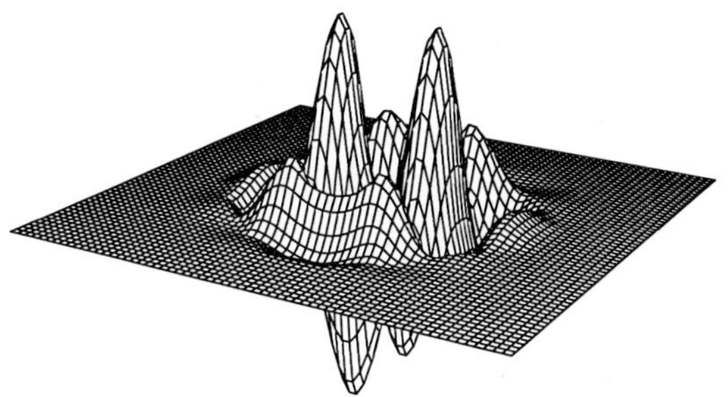

Figure 2. The function ψ in case $\phi = B_{1111}$.

Lemma 3.1.

(a) *With ψ as defined in Equation (3.2) and $\widetilde{\Phi}$ given by Equation (2.5), it follows that*

$$\begin{aligned}\widetilde{\Psi}(\xi) &:= \sum_{\alpha\in\mathbf{Z}^2} |\widehat{\psi}(\xi + 2\pi\alpha)|^2 \\ &= \widetilde{\Phi}(\xi)\ \widetilde{\Phi}(A^{-T}\xi)\ \widetilde{\Phi}(A^{-T}(\xi + 2\pi e_1))\ .\end{aligned} \tag{3.4}$$

(b) *In addition, for*

$$\widehat{\psi}^{\perp} := \widehat{\psi}/\sqrt{\widetilde{\Psi}}, \tag{3.5}$$

then $S_\psi = S_{\psi^\perp}$, and $\{\psi^\perp(\cdot - \alpha)\colon\ \alpha \in \mathbf{Z}^2\}$ is a complete orthonormal system of S_ψ.

Proof: With $\mathcal{A} = A^T\mathbf{Z}^2$, and hence $\mathbf{Z}^2 = \mathcal{A} \cup (e_1 + \mathcal{A})$, we see that

$$\begin{aligned}\widetilde{\Phi}(A^T\xi) &= \sum_{\alpha\in\mathbf{Z}^2} |\widehat{\phi}(A^T\xi + 2\pi\alpha)|^2 \\ &= \sum_{\alpha\in\mathcal{A}} \{\ |\widehat{\phi}(A^T\xi + 2\pi\alpha)|^2 + |\widehat{\phi}(A^T\xi + 2\pi\alpha + 2\pi e_1)|^2\ \} \\ &= \sum_{\alpha\in\mathbf{Z}^2} \{\ |\widehat{\phi}(A^T(\xi + 2\pi\alpha))|^2 + |\widehat{\phi}(A^T(\xi + 2\pi\alpha + 2\pi A^{-T}e_1))|^2\ \}.\end{aligned}$$

By inserting the scaling equation, this becomes

$$|\det A|^2\ \widetilde{\Phi}(A^T\xi) = |P(\xi)|^2\ \widetilde{\Phi}(\xi) + |P(\xi + 2\pi A^{-T}e_1)|^2\ \widetilde{\Phi}(\xi + 2\pi A^{-T}e_1). \tag{3.6}$$

Similarly, from Equation (3.2) we have

$$|\det A|^2\ \widetilde{\Psi}(A^T\xi) = |Q(\xi)|^2\ \widetilde{\Phi}(\xi) + |Q(\xi + 2\pi A^{-T}e_1)|^2\ \widetilde{\Phi}(\xi + 2\pi A^{-T}e_1), \tag{3.7}$$

where it follows from Equation (3.3) that

$$|Q(\xi)| = |P(\xi + 2\pi A^{-T}e_1)|\ \widetilde{\Phi}(\xi + 2\pi A^{-T}e_1)$$

and, since $2A^{-T}e_1 = Ae_1 = e_1 + e_2 \in \mathbf{Z}^2$, that

$$|Q(\xi + 2\pi A^{-T}e_1)| = |P(\xi)|\ \widetilde{\Phi}(\xi).$$

Inserting this into Equation (3.7) and using Equation (3.6), we arrive at

$$|\det A|^2\ \widetilde{\Psi}(A^T\xi) = |\det A|^2\ \widetilde{\Phi}(A^T\xi)\ \widetilde{\Phi}(\xi)\ \widetilde{\Phi}(\xi + 2\pi A^{-T}e_1).$$

This verifies (a).

In order to see (b), we note that $\widetilde{\Phi}$, and hence $\widetilde{\Psi}$, vanishes only on a set of measure zero. Therefore, by Theorem 2.1, we have $S_\psi = S_{\psi^\perp}$. Finally, the orthonormality of the translates $\psi^\perp(\cdot - \alpha)$, $\alpha \in \mathbf{Z}^2$, is an immediate consequence of the identity

$$\sum_{\alpha\in\mathbf{Z}^2} |\widehat{\psi}^\perp(\xi + 2\pi\alpha)|^2 \equiv 1. \quad \blacksquare$$

Remark. The construction of $\psi^\perp$ could be effected in the following equivalent way. We start with the scaling equation for $\phi^\perp$ in Equation (2.16), namely:

$$|\det A|\ \widehat{\phi}^\perp(A^T\xi) = P^\perp(\xi)\ \widehat{\phi}^\perp(\xi),$$

where $P^\perp = P\sqrt{\widetilde{\Phi}/\widetilde{\Phi}(A^T\cdot)}$, and put

$$|\det A|\ \widehat{\eta}(A^T\xi) = Q^\perp(\xi)\ \widehat{\phi}^\perp(\xi)$$

with

$$Q^\perp(\xi) := e^{i\xi\cdot e_1}\ \overline{P^\perp(\xi + 2\pi A^{-T}e_1)}\ . \tag{3.8}$$

Then we have

$$\begin{aligned} Q^\perp(\xi) &= e^{i\xi\cdot e_1}\ \overline{P(\xi + 2\pi A^{-T}e_1)}\ \sqrt{\frac{\widetilde{\Phi}(\xi + 2\pi A^{-T}e_1)}{\widetilde{\Phi}(A^T\xi)}} \\ &= Q(\xi)\ \frac{1}{\sqrt{\widetilde{\Phi}(A^T\xi)\ \widetilde{\Phi}(\xi + 2\pi A^{-T}e_1)}} \\ &= Q(\xi)\ \sqrt{\frac{\widetilde{\Phi}(\xi)}{\widetilde{\Psi}(A^T\xi)}} \end{aligned}$$

and

$$\frac{\widehat{\psi}^{\perp}(A^T\xi)}{\widehat{\eta}(A^T\xi)} = \frac{\widehat{\psi}(A^T\xi)}{\widehat{\eta}(A^T\xi)} \frac{1}{\sqrt{\widetilde{\Psi}(A^T\xi)}}$$

$$= \frac{Q(\xi)\ \widehat{\phi}(\xi)}{Q^{\perp}(\xi)\ \widehat{\phi}^{\perp}(\xi)\ \sqrt{\widetilde{\Psi}(A^T\xi)}}$$

$$= \frac{\widehat{\phi}(\xi)}{\sqrt{\widetilde{\Phi}(\xi)}\ \widehat{\phi}^{\perp}(\xi)} = 1$$

by Equation (2.16). This shows that $\eta = \psi^{\perp}$.

Theorem 3.2. *With the notations given in Theorem 2.5, the orthogonal decomposition*

$$V_1 = V_0 \oplus W_0$$

holds with $W_0 = S_{\psi} = S_{\psi^{\perp}}$, *where* ψ *is defined as in Equation (3.2).*

Proof: Applying

$$f \in V_1 \Longleftrightarrow f(A^{-1}\cdot) \in V_0$$

together with Theorem 2.1, we see that $f \in V_1$ if and only if $f \in L^2$ and $\widehat{f}(A^T\xi) = \widehat{\phi}(\xi)\ \omega(\xi)$ with a 2π-periodic function ω. This shows that W_0 is a subspace of V_1.

To see that V_0 and W_0 are orthogonal, it is equivalent to showing that

$$\phi^{\perp}(\cdot - \alpha) \perp \psi^{\perp}(\cdot - \beta) \quad \text{for all} \quad \alpha, \beta \in \mathbf{Z}^2 .$$

Now with $C = [-\pi, +\pi]^2$, we have

$$\begin{aligned}
&\langle\, \phi^{\perp}(\cdot - \alpha), \psi^{\perp}(\cdot - \beta)\, \rangle \\
&= \frac{1}{(2\pi)^2} \int_{\mathbf{R}^2} e^{-i(\alpha-\beta)\cdot\xi}\ \widehat{\phi}^{\perp}(\xi)\ \overline{\widehat{\psi}^{\perp}(\xi)}\ d\xi \\
&= \frac{1}{(2\pi)^2} \int_C e^{-i(\alpha-\beta)\cdot\xi} \sum_{\gamma\in\mathbf{Z}^2} \widehat{\phi}^{\perp}(\xi + 2\pi\gamma)\ \overline{\widehat{\psi}^{\perp}(\xi + 2\pi\gamma)}\ d\xi \\
&= \frac{1}{(2\pi)^2} \frac{1}{|\det A|^2} \int_C e^{-i(\alpha-\beta)\cdot\xi} \sum_{\gamma\in\mathbf{Z}^2} \{\ P^{\perp}(A^{-T}(\xi + 2\pi\gamma)) \times \\
&\qquad \times\ \overline{Q^{\perp}(A^{-T}(\xi + 2\pi\gamma))}\ |\widehat{\phi}^{\perp}(A^{-T}(\xi + 2\pi\gamma))|^2\ \}\ d\xi\ .
\end{aligned}$$

Writing $\mathbf{Z}^2 = \mathcal{A} \cup (e_1 + \mathcal{A})$ with $\mathcal{A} = A^T\mathbf{Z}^2$, and using the fact that

$$\sum_{\gamma\in\mathbf{Z}^2} |\widehat{\phi}^{\perp}(\eta + 2\pi\gamma)|^2 \equiv 1\ ,$$

we arrive at

$$\begin{aligned}
&\langle\, \phi^{\perp}(\cdot - \alpha), \psi^{\perp}(\cdot - \beta)\, \rangle \\
&= \frac{1}{(2\pi)^2} \frac{1}{|\det A|^2} \int_C e^{-i(\alpha-\beta)\cdot\xi} \,\{\, P^{\perp}(A^{-T}\xi)\, \overline{Q^{\perp}(A^{-T}\xi)} + \\
&\qquad + P^{\perp}(A^{-T}\xi - 2\pi A^{-T}e_1)\, \overline{Q^{\perp}(A^{-T}\xi - 2\pi A^{-T}e_1)}\,\}\, d\xi = 0
\end{aligned}$$

since the expression in the braces vanishes by Equation (3.8).

To complete the proof, we show that the orthogonal projection onto $V_0 \oplus W_0$ is the identity on V_1. Here, we may use the fact that

$$\{\phi^{\perp}(\cdot - \alpha), \psi^{\perp}(\cdot - \alpha) : \alpha \in \mathbf{Z}^2\}$$

is an orthonormal basis of $V_0 \oplus W_0$, and that

$$\{|\det A|^{1/2} \phi^{\perp}(A \cdot -\alpha) : \alpha \in \mathbf{Z}^2\}$$

is an orthonormal basis of V_1. Hence, it suffices to show that, for each $\alpha \in \mathbf{Z}^2$,

$$\begin{aligned}
|\det A|^{-1} &= \langle\, \phi^{\perp}(A \cdot -\alpha), \phi^{\perp}(A \cdot -\alpha)\, \rangle \\
&= \sum_{\gamma \in \mathbf{Z}^2} \{|\langle\, \phi^{\perp}(A \cdot -\alpha), \phi^{\perp}(\cdot - \gamma)\, \rangle|^2 + |\langle\, \phi^{\perp}(A \cdot -\alpha), \psi^{\perp}(\cdot - \gamma)\, \rangle|^2\}\ .
\end{aligned}$$

This can be verified by applying the same techniques as above (Parseval's identity, periodization, and using the partitioning $\mathbf{Z}^2 = \mathcal{A} \cup (e_1 + \mathcal{A})$), where this time, we apply the identity

$$|\det A|^2 \equiv |P^{\perp}(\xi)|^2 + |Q^{\perp}(\xi)|^2. \quad \blacksquare$$

Remark. The essential identities that are basic in this proof are the following relations for $P^{\perp}$ and $Q^{\perp}$:

$$P^{\perp}(\xi)\, \overline{Q^{\perp}(\xi)} + P^{\perp}(\xi + 2\pi A^{-T}e_1)\, \overline{Q^{\perp}(\xi + 2\pi A^{-T}e_1)} = 0 \tag{3.9a}$$

and

$$|P^{\perp}(\xi)|^2 + |Q^{\perp}(\xi)|^2 = |\det A|^2\ . \tag{3.9b}$$

These follow from the specific choice of $Q^{\perp}$ as described in the foregoing remark, viz.

$$Q^{\perp}(\xi) := e^{i\xi\cdot e_1}\, \overline{P^{\perp}(\xi + 2\pi A^{-T}e_1)} \tag{3.9c}$$

with

$$P^{\perp}(\xi) := P(\xi)\, \sqrt{\widetilde{\Phi}(\xi)/\widetilde{\Phi}(A^T\xi)}\ , \tag{3.9d}$$

and Equation (3.6). (For the definition of P and $\widetilde{\Phi}$ we refer to Lemma 2.4 and Equation (2.5), respectively.)

As a simple consequence of this construction, we obtain the following result which will be needed in the sequel.

Lemma 3.3. *Let $\mathcal{A} = A\mathbf{Z}^2$ and $C = [-\pi, +\pi]^2$. Then*
(a) *$P^\perp$ and $Q^\perp$ are 2π-periodic functions that are continuous on $\mathbf{R}^2 \backslash \pi(e_1 + \mathcal{A})$;*
(b) *$|P^\perp|$ and $|Q^\perp|$ are even functions;*
(c) *$\operatorname{ess\,sup} |P^\perp| = \operatorname{ess\,sup} |Q^\perp| = |\det A| = 2$; and*
(d) *$\operatorname{ess\,inf}\{|P^\perp(\xi)| \,:\, \xi \in \frac{1}{2}C\} = \operatorname{ess\,inf}\{|Q^\perp(\xi)| \,:\, \xi \in \binom{\pi}{\pi} + \frac{1}{2}C\} =: m_0 > 0$.*

Proof: (a) is clear from Equations (3.9c) and (3.9d) since P and $\widetilde{\Phi}$ are trigonometric polynomials and (as already mentioned in the proof of Theorem 2.5) $\widetilde{\Phi}(A^T\xi) = 0$ if and only if $A^T\xi \in \binom{\pi}{\pi} + 2\pi\mathbf{Z}^2$; *i.e.*, $\xi \in \pi(e_1 + \mathcal{A})$. (b) holds since $|P|$ and $\widetilde{\Phi}$ are even, and

$$\begin{aligned} |Q^\perp(-\xi)| &= |P^\perp(-\xi + \pi(e_1 + e_2))| = |P^\perp(-\xi - \pi(e_1 + e_2))| \\ &= |P^\perp(\xi + \pi(e_1 + e_2))| = |Q^\perp(\xi)| \; . \end{aligned}$$

(c) follows from Equation (3.9b), since $P^\perp$ vanishes at $\xi = \binom{\pi}{\pi}$ which is a point of continuity, and (d) holds since $\frac{1}{2}C$ and $\binom{\pi}{\pi} + \frac{1}{2}C$ do not contain a zero of $P^\perp$ or $Q^\perp$, respectively. ■

We are now ready to state our main result of this section. Starting with the space

$$W_0 = S_\psi = S_{\psi^\perp} \tag{3.10}$$

we define the corresponding A-scaled spaces $W_j, j \in \mathbf{Z}$, by

$$f \in W_j \iff f(A^{-j}\cdot) \in W_0 \; . \tag{3.11}$$

Since $\{\psi^\perp(\cdot - \alpha)\colon \alpha \in \mathbf{Z}^2\}$ is an orthonormal basis for W_0, we see that

$$\psi^\perp_{j,\alpha} := |\det A|^{j/2}\, \psi^\perp(A^j \cdot - \alpha) \;, \quad \alpha \in \mathbf{Z}^2 \;, \tag{3.12}$$

is an orthonormal basis for W_j. Since $V_1 = V_0 \oplus W_0$, we have the corresponding orthogonal decompositions

$$V_{j+1} = V_j \oplus W_j \;, \quad j \in \mathbf{Z} \; . \tag{3.13}$$

Moreover, if $P_j : L^2 \longrightarrow V_j$ denotes the orthogonal projection onto V_j, we see that

$$Q_j := P_{j+1} - P_j \tag{3.14}$$

is the orthogonal projection onto W_j. Finally, since $f = \lim_{j\to+\infty} P_j f$ and $\lim_{j\to-\infty} P_j f = 0$, we have the orthogonal decomposition

$$f = \sum_{j\in\mathbf{Z}} Q_j f \tag{3.15}$$

for every $f \in L^2$. In conclusion, we have proved the following.

Theorem 3.4. *Let $\phi = B_{klkl}$, and consider the A-scaled multiresolution analysis $\{V_j\}_{j\in\mathbf{Z}}$ of L^2. Let ψ be defined as in Equation (3.2) and $\psi^\perp$ as in Lemma 3.1. Then with W_j and $\psi^\perp_{j,\alpha}$ as in Equations (3.10) – (3.12),*

(a) *the spaces $W_j, j \in \mathbf{Z}$, form an orthogonal decomposition of L^2; and*
(b) *the functions $\psi^\perp_{j,\alpha}, \alpha \in \mathbf{Z}^2$, form an orthonormal basis of $W_j, j \in \mathbf{Z}$.*

In other words, ψ is an s.o. wavelet, and $\psi^\perp$ is an o.n. wavelet for L^2.

§4 Four-directional frames

In Section 3 we have constructed two different bases for W_0 (and hence, by A-scaling the corresponding bases for L^2). The first one,

$$\psi(\cdot - \alpha)\ , \quad \alpha \in \mathbf{Z}^2\ , \tag{4.1}$$

is compactly supported, but not stable (*i.e.*, not a Riesz basis); one can show that this is not even a frame of W_0. The second one,

$$\psi^{\perp}(\cdot - \alpha)\ , \quad \alpha \in \mathbf{Z}^2\ , \tag{4.2}$$

is orthonormal, hence trivially stable, but not compactly supported. In fact, the decay property of $\psi^{\perp}$ is so poor that (for example in case $\phi = B_{1111}$) it is not even integrable. Therefore, at first sight it seems to be hopeless to use these bases for any numerical calculations.

The situation is not that bad since, after all, we can construct a compactly supported frame by using twice as many shifts as compared with Equation (4.1). We use the following notation:

$$A = \begin{pmatrix} 1 & 1 \\ 1 & -1 \end{pmatrix} = A^T, \quad B := A^{-1} = A^{-T}\ . \tag{4.3}$$

Hence, for the sublattice $\mathcal{A} = A\mathbf{Z}^2$ of $\mathbf{Z}^2$, we have the dual lattice

$$\mathcal{B} := B\mathbf{Z}^2 = \mathbf{Z}^2 \cup (\begin{pmatrix} 1/2 \\ 1/2 \end{pmatrix} + \mathbf{Z}^2)\ , \tag{4.4}$$

from which we take the shifts.

Theorem 4.1. *Let $\phi = B_{klkl}$, and let ψ be the corresponding s.o. wavelet as given by Equation* (3.2). *Then*

$$\psi_{j,\alpha} := |\det A|^{j/2}\ \psi(A^j \cdot - \alpha)\ , \quad j \in \mathbf{Z}\ , \quad \alpha \in \mathcal{B}\ ,$$

constitutes a frame of L^2, i.e., there exist constants $0 < m \leq M < \infty$ such that, for any $f \in L^2$,

$$m\ \|f\|_2^2 \leq \sum_{j\in\mathbf{Z}} \sum_{\alpha\in\mathcal{B}} |\langle f, \psi_{j,\alpha} \rangle|^2 \leq M\ \|f\|_2^2\ .$$

Since the proof of this theorem is somewhat complicated, we only give a very brief sketch and delay a more detailed argument to a forthcoming paper. First, we decompose $f \in L^2$ according to the orthogonal wavelet spaces,

$$f = \sum_{j\in\mathbf{Z}} f_j\ , \quad f_j \in W_j\ , \tag{4.5}$$

and observe that we can represent each f_j by

$$\widehat{f_j}(\xi) = |\det B|^{j/2}\ \tau_j(B^j\xi)\ \widehat{\psi}^{\perp}(B^j\xi)\ , \quad j \in \mathbf{Z}\ , \tag{4.6}$$

with certain 2π-periodic functions τ_j; hence

$$\|f\|_2^2 = \sum_{j\in\mathbf{Z}} \|\tau_j\|_{L^2(C)}^2\ . \tag{4.7}$$

It is not very hard to see that $\langle\ f_l, \psi_{j,\alpha}\ \rangle\ =\ 0$ for the following cases: (i) $j < l$ and $\alpha \in \mathcal{B}$, (ii) $j > l$ and $\alpha \in \mathbf{Z}^2$. From this one derives, after some straightforward calculations, the following.

Lemma 4.2. *For any $j \in \mathbf{Z}$ and $f \in L^2$,*

$$\sum_{\alpha\in\mathbf{Z}^2} |\langle\ f, \psi_{j,\alpha}\ \rangle|^2 = \|\sqrt{\widetilde{\Psi}}\ \tau_j\|_{L^2(C)}^2 \tag{4.8a}$$

and

$$\sum_{\alpha\in\mathcal{B}\backslash\mathbf{Z}^2} |\langle\ f, \psi_{j,\alpha}\ \rangle|^2 = \|S_j\|_{L^2(C)}^2\ ; \tag{4.8b}$$

here, $\widetilde{\Psi}$ is given as in Lemma 3.1, and S_j is the 2π-periodic function defined by

$$S_j(\xi) := e^{ie_1\cdot B\xi}\ \sqrt{\widetilde{\Psi}(\xi)}\ \sum_{\nu\geq 0} \tau_{j-\nu}(B^{-\nu}\xi)\ T_\nu(\xi)$$

with

$$T_\nu(\xi) := |\det A|^{\nu/2}\ \sum_{\beta\in\mathbf{Z}^2} (-1)^{|\beta|}\ \widehat{\psi}^{\perp}(B^{-\nu}(\xi + 2\pi\beta))\ \overline{\widehat{\psi}^{\perp}(\xi + 2\pi\beta)}\ .$$

The next observation is that the functions T_ν in this lemma can be rewritten in terms of $P^{\perp}$ and $Q^{\perp}$ as follows:

$$T_0(\xi) = |\det B|^2\ \{\ |Q^{\perp}(B\xi)|^2 - |Q^{\perp}(B\xi + \tbinom{\pi}{\pi})|^2\ \} \tag{4.9a}$$

and, for $\nu > 0$,

$$T_\nu(\xi) = |\det B|^{1+\nu/2}\ \overline{Q^{\perp}(B\xi)}\ Q^{\perp}(B^{1-\nu}\xi)\ \prod_{\lambda=-1}^{\nu-2} P^{\perp}(B^{-\lambda}\xi)\ . \tag{4.9b}$$

This shows that T_0 and T_ν, $\nu \geq 3$, vanish at $Z = \{\tbinom{\pi}{0}, \tbinom{0}{\pi}\}$ and allows us to find, for any $\varepsilon > 0$, a δ such that the balls $B_\delta\tbinom{\pi}{0}$ and $B_\delta\tbinom{0}{\pi}$, with radius δ

and centers at $\binom{\pi}{0}$ and $\binom{0}{\pi}$, respectively, are disjoint with the zero set of $\widetilde{\Psi}$ and that by setting $U := B_\delta\binom{\pi}{0} \cup B_\delta\binom{0}{\pi}$, we have

$$\operatorname*{ess\,sup}_{\xi\in U} |T_0(\xi)| + \sum_{\nu\geq 3} \operatorname*{ess\,sup}_{\xi\in U} |T_\nu(\xi)| < \varepsilon \,. \tag{4.10}$$

Now we separate the sum that defines S_j according to

$$\begin{aligned} S_j^{(1)}(\xi) &:= \tau_{j-1}(B^{-1}\xi)\; T_1(\xi) \;+\; \tau_{j-2}(B^{-2}\xi)\; T_2(\xi)\;, \\ S_j^{(2)}(\xi) &:= \tau_j(\xi)\; T_0(\xi) \;+\; \sum_{\nu\geq 3} \tau_{j-\nu}(B^{-\nu}\xi)\; T_\nu(\xi)\;, \end{aligned} \tag{4.11}$$

so that

$$|S_j(\xi)|^2 \geq \widetilde{\Psi}(\xi)\; \{\; |S_j^{(1)}(\xi)|^2 - 2\; |S_j^{(1)}(\xi) S_j^{(2)}(\xi)|\; \}\,.$$

From this and applying the Cauchy-Schwarz inequality, we obtain the estimate

$$\sum_{j\in\mathbf{Z}} \|S_j\|^2_{L^2(C)} \geq \frac{1}{(2\pi)^2} \sum_{j\in\mathbf{Z}} \int_U |S_j(\xi)|^2\, d\xi \geq \sigma_1^2 - 2\sigma_1\sigma_2 \tag{4.12}$$

with

$$\sigma_\nu := \left(\frac{1}{(2\pi)^2} \sum_{j\in\mathbf{Z}} \int_U \widetilde{\Psi}(\xi) |S_j^{(\nu)}(\xi)|^2 d\xi \right)^{\frac{1}{2}}, \quad \nu = 1, 2. \tag{4.13}$$

A detailed analysis based on Equations (4.9) and (4.10) leads to the bounds

$$\sigma_2^2 \leq \varepsilon^2 \sup_{\xi\in U} \widetilde{\Psi}(\xi) \|f\|_2^2, \quad \text{and} \tag{4.14a}$$

$$\begin{aligned} \sigma_1^2 &\geq \frac{c_1(\varepsilon)}{(2\pi)^2} \sum_{j\in\mathbf{Z}} \int_{B_\delta\binom{\pi}{0}} \{|\det B^{-1}|\; |\tau_{j-1}(B^{-1}\xi)|^2 + |\det B^{-2}|\; |\tau_{j-2}(B^{-2}\xi)|^2\} d\xi \\ &= \frac{c_1(\varepsilon)}{(2\pi)^2} \sum_{j\in\mathbf{Z}} \int_{\mathcal{M}} |\tau_j(\xi)|^2, \end{aligned} \tag{4.14b}$$

where $c_1(\varepsilon)$ is independent of f and $c_1(\varepsilon) \geq c_1(\varepsilon_0) > 0$ for all $\varepsilon \leq \varepsilon_0$. The set $\mathcal{M} = \mathcal{M}_\varepsilon$ in Equation (4.14b) is defined by

$$\mathcal{M} \subset C,\; \mathcal{M} \equiv B^{-1}(B_\delta\tbinom{\pi}{0}) \cup B^{-2}(B_\delta\tbinom{\pi}{0})\ (\text{mod } C),$$

so that $\mathcal{M}$ is a neighborhood of the zero set of $\widetilde{\Psi}$ in $C = [-\pi, +\pi]^2$. The crucial step in order to achieve Equation (4.14b) is to separate the terms τ_{j-1} and τ_{j-2} in the definition of $S_j^{(1)}$.

The rest of the proof for the lower frame bound is then completed by the distinction between the following two cases.

If the terms in Equation (4.13) satisfy $\sigma_2 \le \frac{\sigma_1}{4}$, then

$$\sum_{j\in\mathbf{Z}} \|S_j\|^2_{L^2(C)} \ge \frac{\sigma_1^2}{2}$$

by Equation (4.12). From Equations (4.14b) and (4.7), and applying Lemma 4.2, we obtain

$$\sum_{j\in\mathbf{Z}}\sum_{\alpha\in\mathcal{B}} |\langle f, \psi_{j,\alpha}\rangle|^2 \ge c_2(\varepsilon)\|f\|_2^2, \tag{4.15}$$

where $0 < c_2(\varepsilon) := \min\{\frac{c_1(\varepsilon)}{2}, \inf_{\xi\in C\setminus\mathcal{M}} \widetilde{\Psi}(\xi)\}$. In case $\sigma_1 < 4\sigma_2$, however, the inequalities in Equations (4.14a) and (4.14b) imply

$$\frac{1}{(2\pi)^2}\sum_{j\in\mathbf{Z}}\int_{\mathcal{M}} |\tau_j(\xi)|^2 d\xi < \frac{16\varepsilon^2 \sup_{\xi\in U}\widetilde{\Psi}(\xi)}{c_1(\varepsilon)}\|f\|_2^2.$$

This gives, by Lemma 4.2 and Equation (4.7),

$$\sum_{j\in\mathbf{Z}}\sum_{\alpha\in\mathcal{B}} |\langle f, \psi_{j,\alpha}\rangle|^2 \ge \frac{c_2(\varepsilon)}{(2\pi)^2}\sum_{j\in\mathbf{Z}}\int_{C\setminus\mathcal{M}} |\tau_j(\xi)|^2 d\xi > m\|f\|_2^2,$$

with the lower frame bound given by

$$m := c_2(\varepsilon)\left(1 - \frac{16\varepsilon^2 \sup_{\xi\in U}\widetilde{\Psi}(\xi)}{c_1(\varepsilon)}\right). \tag{4.16}$$

The factor m in Equation (4.16) is independent of f and is positive for all $0 < \varepsilon \le \varepsilon_0$ with sufficiently small positive ε_0.

The proof of the existence of an upper frame bound is simpler. For more details we refer the reader to our forthcoming paper.

Acknowledgments. This research was initiated when the last two authors spent a short research visit at Texas A&M University in March of 1993. We greatfully acknowledge partial support by NATO Grant # CRG 900 158.

References

1. de Boor, C., K. Höllig and S. D. Riemenschneider, *Box Splines*, Springer-Verlag, New York, 1993.
2. de Boor, C., R. DeVore, and A. Ron, Approximation from shift-invariant subspaces of $L^2(\mathbb{R}^d)$, *Trans. Amer. Math. Soc.*, to appear.
3. Chui, C. K., *Multivariate Splines*, CBMS Series in Applied. Math. #54, SIAM Publ., Philadelphia, 1988.
4. Chui, C. K., *An Introduction to Wavelets*, Academic Press, Boston, 1992.

5. Chui, C. K. and X. L. Shi, Bessel sequences and affine frames, *Appl. and Comp. Harmonic Anal. (ACHA)* **1** (1993), 29–49.
6. Chui, C. K., J. Stöckler and J. D. Ward, Compactly supported box spline wavelets, *Approx. Theory and Appl.* **6** (1993), 77–100.
7. Cohen, A. and I. Daubechies, Non-separable bidimensional wavelet bases, *Rivista Mathemática Iberoamericana* **9** (1993), 51-137.
8. Jetter, K. and J. Stöckler, Algorithms for cardinal interpolation using box splines and radial basis functions, *Numer. Math.* **60** (1991), 97–114.
9. Riemenschneider, S. D. and Z. W. Shen, Wavelets and pre-wavelets in low dimensions, *J. Approx. Theory* **71** (1992), 18–38.
10. Shen, Z. W., Non-tensor product wavelet packets in $L^2(\mathbb{R}^s)$, CMS-TSR 93-10, University of Wisconsin, Madison.

Charles K. Chui
Center for Approximation Theory
Department of Mathematics
Texas A&M University
College Station, TX 77843, USA
cchui@tamu.edu

Kurt Jetter
FB Mathematik
Universität Duisburg
D-47048 Duisburg, Germany
hn277je@unidui.uni-duisburg.de

Joachim Stöckler
FB Mathematik
Universität Duisburg
D-47048 Duisburg, Germany
hn277st@math.uni-duisburg.de

Part IV

Other Mathematical Tools for Time-Frequency Analysis

On Minimum Entropy Segmentation

David L. Donoho

Abstract. We describe segmented multiresolution analyses of $[0,1]$. Such multiresolution analyses lead to segmented wavelet bases which are adapted to discontinuities, cusps, etc., at a given location $\tau \in [0,1]$. Our approach emphasizes the idea of *average-interpolation* – synthesizing a smooth function on the line having prescribed boxcar averages. This particular approach leads to methods with *subpixel resolution* and to wavelet transforms with the advantage that, for a signal of length n, all n pixel-level segmented wavelet transforms can be computed simultaneously in a total time and space which are both $O(n \log(n))$.

We consider the search for a segmented wavelet basis which, among all such segmented bases, minimizes the "entropy" of the resulting coefficients. Fast access to all segmentations enables fast search for a best segmentation. When the "entropy" is Stein's Unbiased Risk Estimate, one obtains a new method of edge-preserving de-noising.

When the "entropy" is the ℓ^2-energy, one obtains a new multi-resolution edge detector, which works not only for step discontinuities but also for cusp and higher-order discontinuities, and in a near-optimal fashion in the presence of noise.

We describe an iterative approach, *Segmentation Pursuit*, for identifying edges by the fast segmentation algorithm and removing them from the data.

§1 Introduction

1.1 Improved de-noising

Several recent papers (see [25] and references therein) have shown that wavelet methods can be used to de-noise data of various kinds, obtaining a level of theoretical performance not approached by pre-existing methods. In general, these methods have the following character: first, one takes the empirical wavelet transform of the noisy data; next one subjects the coefficients to a simple coordinatewise nonlinearity, applying specially-chosen thresholding to wavelet

Wavelets: Theory, Algorithms, and Applications
Charles K. Chui, Laura Montefusco, and Luigia Puccio (eds.), pp. 233–269.

ISBN 0-12-174575-9

coefficients; finally one inverts the empirical wavelet transform, obtaining de-noised coefficients.

While the theoretical benefits of this approach are now well-established, and the actual reconstructions obtained by wavelet de-noising methods have seemed to us quite good, particularly in comparison with pre-existing methods, we have received comments from users and others that indicate some improvement is to be desired. These comments include

1. *Gibbs Phenomenon.* In the neighborhood of strong jump discontinuities, wavelet shrinkage methods often exhibit alternating overshoot. Although the phenomenon is much more localized than in the case of Fourier series, it would be desirable to improve further.
2. *Peak Shrinkage.* In the analysis of data such as NMR spectra, there is a tendency of wavelet shrinkage to "pull down" strong peaks, thereby distorting amplitudes. It would be desirable to reduce this tendency.
3. *Edge Erosion.* In the analysis of certain edge data, there is a tendency of wavelet shrinkage to erode weak edges, reducing the sharpness of transitions. It would be desirable to reduce or avoid this tendency.
4. *Inter-Scale Correlations.* The theory underlying the optimality of wavelet shrinkage techniques shows quite clearly that from a minimax-theoretic point of view, the wavelet transform plays the role of a de-correlating transform, mapping the object into a space where the different coordinates (wavelet coefficients) have no information about each other, and therefore scalar processing (*e.g.*, coordinatewise thresholding) cannot essentially be improved upon. On the other hand, in "real world" objects, containing edges, one can expect to see nonzero wavelet coefficients at the same locations across several scales. Therefore, in real objects, the information that a certain wavelet coefficient is large leads to the presumption that similarly located coefficients at other scales be large. One expects that by exploiting such correlations, the accuracy of reconstruction might be improved over what simple thresholding offers. It would be desirable to develop a method to exploit such inter-level coefficient correlations and improve on ordinary wavelet shrinkage.

All of these problems seem to call for improving on wavelet de-noising, in a second pass, to clean up the "mess" left by the presence of singularities in 1 and 2-dimensional data. For example, if a user complains of Peak Shrinkage and asks for an improvement, there ought to be a simple, automatic procedure to correct it.

1.2 Segmented MRA and best segmentation

In this paper, we develop an approach to these problems based on the concept of *segmented multi-resolution analysis* for the interval $[0, 1]$, which allows a kind of wavelet decomposition and reconstruction adapted to the presence of segmentation. The functions in a segmented MRA need not be continuous across a certain point τ internal to the interval $[0, 1]$. We develop in Section

2 a special MRA, based on a biorthogonal system of wavelets called average-interpolating wavelets by Donoho [19]; these derive from smooth wavelets which are biorthogonal to Haar wavelets, and they lead to fast algorithms for computing the segmented wavelet transform.

The idea is that if an object contains a sharp separation between one "phase" and another, we can develop an MRA adapted to that two-phase structure, and a corresponding wavelet transform, so as to avoid the presence of nonzero wavelet coefficients associated with the inter-phase transition.

Of course, the application of such a segmented MRA depends heavily on information about the location τ of the inter-phase boundary. In empirical work, this is not generally available *a-priori*. Hence, in order to exploit segmented MRA's, we must have some way to *infer* τ from data.

In Section 3 we approach this problem from a *best-basis* point of view; compare Coifman and Wickerhauser [10]. Section 2 makes available to us a collection of *segmented* wavelet expansions

$$f \sim \sum_{j,k} \alpha^t_{j,k} \psi^{(t)}_{j,k} ;$$

each expansion determined by a segmentation point t. We seek among all these representations of f for one with minimal representation cost; we call the optimizing basis a best basis for f, and label the optimizing segmentation point $\hat{\tau}$. To measure representation cost we depart slightly from Coifman and Wickerhauser [10], who were analyzing noiseless data and selecting best bases in a different collection of bases; they proposed the use of a $-\theta^2 \log(\theta^2)$ entropy as a measure of the cost of a representation. For measuring the cost of a representation in the noiseless cases we consider here other entropies, including θ^2, $|\theta|$, and $|\theta|^{1/2}$. In dealing with noisy data, we adapt ideas of Donoho and Johnstone [24,25], and suggest the use of a certain Stein Unbiased Estimate of Risk as an entropy measure. This measures the quality of a given segmented-wavelet basis in which to de-noise the data. We test the functioning of the SURE Best-basis paradigm on some simple examples.

The practicality of any best-basis method depends on the existence of a fast search algorithm for searching through a collection of bases. In Section 4, we describe a fast algorithm for obtaining the optimizer $\hat{\tau}$ when the entropy measure is an additive measure of information. This algorithm searches through all n pixel-level segmentations and identifies the optimum one in order $n \log(n)$ space and time.

When we infer τ from noisy data, we are actually identifying edges. In fact, using the θ^2-entropy leads to a new multiresolution edge locator which adapts to the type of edge (step edge; cusp; etc.); we hope to show elsewhere that this has near-optimal properties in locating such types of edges in noise. Section 5 describes in heuristic terms some properties of this locator.

How can one handle the presence of multiple segmentation points? In Section 6 we propose an iterative method, *segmentation pursuit*, based on iteratively identifying the best current segmentation point and "stripping away" the singularity at the identified position. We illustrate by a computational example.

This is not a "math paper." There are no theorems here, only computational experiments, motivated by our work in other "math papers." We aim here only to develop a collection of fast computational tools which may be employed as a "vacuum cleaner" to remove structure from residuals in ordinary de-noising caused by the presence of singularities in the object to be recovered. The algorithms described here are all available for use in MATLAB and may be obtained by anonymous FTP to playfair.stanford.edu.

1.3 Credit where credit is due

The idea to somehow adapt "wavelets ideas" to the presence of singularities and discontinuities is not new. See the nonlinear multi-scale edge reconstruction ideas of Mallat and Zhong [30] and Mallat and Froment [29].

The need to use some sort of wavelet transform adapted to the presence of edges has been mentioned by Jawerth of the University of South Carolina at several conference presentations in 1992 and 1993. After the work reported here was done, the author learned that Deng, Jawerth, Peters, and Sweldens [14] have independently and somewhat earlier come up with fast algorithms for computing all pixel-level segmented 1-d transforms. Their method is based on "breaking" the dataset into pieces and computing boundary-adjusted transforms of the left and right pieces, while ours is based on segmented refinement schemes. The underlying logic is somewhat different, while the resulting algorithms are similar. Their approach is ultimately more general; ours does have its own merits – a solid derivation from refinement principles and a special role in sub-pixel resolution.

§2 1-d Segmented wavelet transforms

In this section we briefly describe a method for constructing 1-d segmented multi-resolution analyses.

2.1 Refinement by average-interpolation

Suppose we have an array $(a_{j,k})_{k=-\infty}^{\infty}$ which represents averages of a function f on dyadic intervals $I_{j,k} = [k/2^j, (k+1)/2^j]$. We may synthesize mock-averages at finer scales by the following procedure (see Figure 2.1). Let D be an *even* integer greater than 0.

Step 1. At each site k, find the polynomial $\pi_{j,k}$ of degree D which generates the same averages in the neighborhood $(a_{j,k'}, k' = k - D/2, \dots, k + D/2)$, *i.e.*,

$$Ave_{j,k'}\pi_{j,k} = a_{j,k'}, \qquad k' = k - D/2, \dots, k + D/2.$$

As the polynomial has $D + 1$ coefficients and there are $D + 1$ constraints to satisfy, the polynomial is uniquely determined.

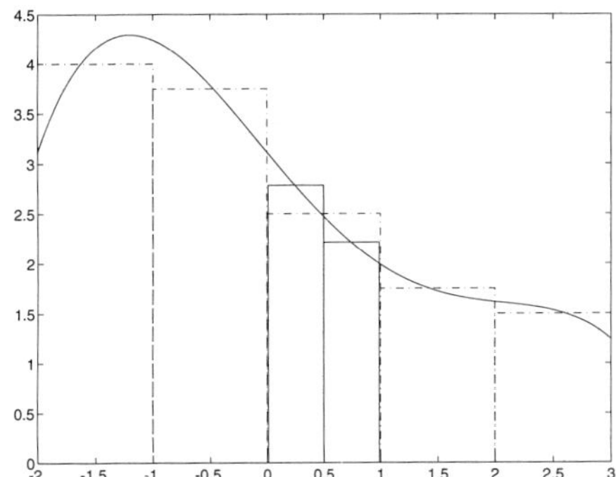

Figure 2.1. Averages, polynomial fit, averages imputed to finer scale.

Step 2. Define the mock-averages at the next finer scale as averages of that polynomial. On the left half of the sub-interval, we obtain

$$a_{j+1,2k} = Ave_{j+1,2k}\pi_{j,k} \ ,$$

while on the right half,

$$a_{j+1,2k+1} = Ave_{j+1,2k+1}\pi_{j,k}.$$

Step 3. After having synthesized all the $a_{j+1,k}$'s, set $j := j+1$ and goto Step 1.

This refinement scheme, which is analogous to the interpolating refinement scheme of Deslauriers-Dubuc [13,26], is discussed at length in [19]. The main point is that it defines a sequence of refinements which in some sense converge: the function $A_{j',D}(t) = \sum_k a_{j',k} 1_{[k/2^{j'},(k+1)/2^{j'}]}(t)$ converges to a continuous limit $A_D(t)$ on the line, which has the averages $a_{j,k}$ at scale 2^{-j}. In fact, the limit has C^R regularity, where $R = R(D)$ increases with D.

The above describes average-interpolation on the line. On the interval $[0,1]$, we have only averages $(a_{j,k})_{k=0}^{2^j-1}$. At the heart of the interval, refinement can proceed exactly as above; at the edges we redefine the set of neighboring intervals in Step 1 to refer only to the $D+1$ nearest intervals fitting inside the interval $[0,1]$.

Return again to the case of functions on the line.

2.2 Average-interpolating multiresolutions

The vector space V_j of functions obtainable by refining sequences $(a_{j,k})_k$ has an alternate description. Refining the Kronecker sequence $a_{0,k} = \delta_{k,0}$ yields fundamental functions $\phi = \phi_D$. These functions and their integer translations and dyadic dilations $\phi_{j,k}(t) = 2^{j/2}\phi(2^j t - k)$ generate the spaces $V_j = \{f : f = \sum_k \beta_{j,k}\phi_{j,k}(t)\}$. The parameters are identical to rescaled averages:

$\beta_{j,k} = 2^{-j/2}a_{j,k}$. The V_j make up a multiresolution analysis which is, therefore, biorthogonal to the usual Haar MRA. The operations of calculating the averages $(a_{j,k})_k$ of f at scale 2^{-j} and then refining those averages to produce a limit function $\tilde{f}$; the linear operator implicitly defined by $\tilde{f} = P_j f$ acts as the identity on V_j and is, therefore, a nonorthogonal projection. Because the average interpolation scheme is exact on polynomials of degree D, we have

$$P_j \pi = \pi \ ,$$

whenever π is a polynomial of degree D.

Given the averages at a scale $j+1$, we of course know the averages at scale j, because $a_{j,k} = (a_{j+1,2k} + a_{j+1,2k+1})/2$. The vector space W_j obtained by refining sequences $(a_{j+1,k})_k$ where $a_{j+1,2k} = -a_{j+1,2k+1}$, consists entirely of functions whose coarser-scale averages are zero; it is in fact the difference space $W_j = V_{j+1} - V_j$. This space has an alternate description. Refining the Kronecker sequence $a_{0,k} = (\delta_{k,1} - \delta_{k,0})/\sqrt{2}$ yields Mother wavelets $\psi = \psi_D$. These functions and their integer translations and dyadic dilations $\psi_{j,k}(t) = 2^{j/2}\psi(2^j t - k)$ generate the difference spaces: $W_j = \{f : f = \sum_k \alpha_{j,k}\psi_{j,k}(t)\}$. Let $h(t)$ be the Haar function $h(t) = 1_{(1/2,1]} - 1_{(0,1/2]}$, and $h_{j,k}(t) = 2^{j/2}h(2^j t - k)$. The parameters $\alpha_{j,k}$ of an object $f \in W_j$ are identical to Haar coefficients of f: $\alpha_{j,k} = 2^{-j/2}(a_{j+1,2k+1} - a_{j+1,2k})/2 = \int f h_{j,k}$. The difference space W_j is, therefore, biorthogonal to the usual Haar detail spaces. Now consider the operator which, given the averages of f at scale 2^{-j-1}, calculates the averages one scale coarser, refines those coarser averages, producing mock averages $(\hat{a}_{j+1,k})$; then forms the residuals $r_{j+1,k} = (a_{j+1,k} - \hat{a}_{j+1,2k})$ and average-interpolates that residual sequence. This produces an element of W_j, which we denote $Q_j f$; Q_j is a non-orthogonal projection on W_j.

This multiresolution system arose before, without the average-interpolation interpretation, in [8], where it was called a system biorthogonal to spline of degree 0; see the discussion in [19].

Corresponding to these schemes on the line are boundary-corrected multiresolutions on the interval $[0,1]$. These are built from a boundary-corrected refinement scheme for the interval. This scheme has spaces $V_j^{[]}$ and $W_j^{[]}$, with projectors $P_j^{[]}$ and $Q_j^{[]}$. These retain key properties from the line, such as biorthogonality with respect to the Haar system, and the polynomial exactness $P_j^{[]}\pi = \pi$, valid whenever π is a polynomial of degree D on $[0,1]$ There are 2^j basis elements of $V_j^{[]}$, obtained by refinement of appropriately normalized Kronecker sequences and also 2^j elements of $W_j^{[]}$. We call these functions $\phi_{j,k}$ and $\psi_{j,k}$, respectively. See Figure 2.2. Fix j_0 so that $2^{j_0} > 2(D+2)$. Then every function in $L^2[0,1]$ has an expansion

$$f = \sum_{k=0}^{2^{j_0}-1} \beta_{j_0,k}\phi_{j_0,k} + \sum_{j \geq j_0} \sum_{k=0}^{2^j-1} \alpha_{j,k}\psi_{j,k}$$

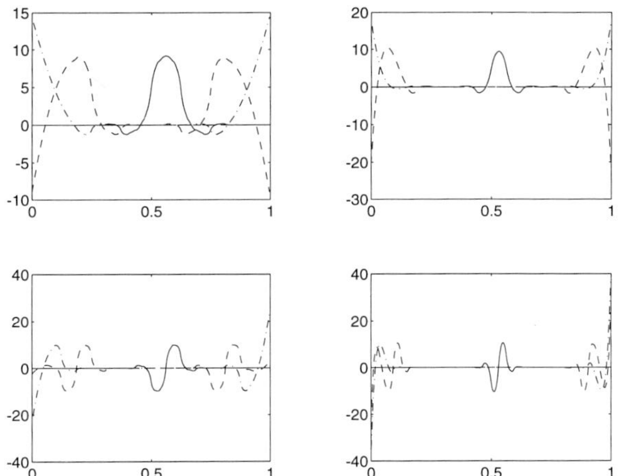

Figure 2.2. Interval wavelets: (top left) $\phi_{3,k}, D = 2$; (top right) $\phi_{4,k}$, $D = 4$; (bottom left) $\psi_{3,k}, D = 2$; (bottom right) $\psi_{4,k}, D = 4$.

unconditionally convergent in L^2 norm.

Here, the coefficients are computed by

$$\beta_{j,k} = \int_0^1 \chi_{j,k}(t)(f - P_j^{[]} f)(t)dt,$$

where $\chi_{j,k} = 2^{j/2} 1_{(2^{-j}k, 2^{-j}(k+1)]}$ is a normalized boxcar and

$$\alpha_{j,k} = \int_0^1 h_{j,k}(t)(f - P_j^{[]} f)(t)dt.$$

The mapping from f to its coefficients $((\beta_{j_0,k})_k, (\alpha_{j_0,k})_k, (\alpha_{j_0+1,k})_k, \ldots))$ is the unsegmented wavelet transform we are interested in.

This transform has the property that its basis elements are C^R regular, with an $R = R(D)$, and the $\alpha_{j,k}$ coefficients all vanish for polynomials of degree D, that the $\alpha_{j,k}$ coefficients of a function in C^r, $0 < r < R$, are of order $2^{-j(r+1/2)}$; see [19] for more information.

Moreover, the transform has associated with it a fast algorithm. Given the boxcar integrals $\beta_{j_1,k}$ at a fine scale $2^{-j_1} \ll 2^{-j_0}$, coefficients $(\beta_{j_0,k})$ and $(\alpha_{j,k})$ at all coarser levels $j_0 \leq j < j_1$ can be computed in order n time, where $n = 2^{j_1}$.

2.3 Segmented refinement

Consider now the following segmented refinement procedure, with segmentation point τ. We assume that the segmentation point is in the heart of the interval, so that $D/2^j < \tau < (2^j - D/2^j)$. Given a sequence of averages $a_{j,k}$,

$0 \leq k < 2^j$, as in Figure 2.3, we synthesis mock averages at finer scales by the following procedure:

Step 1. At each site k which is more than $D/2$ sites away from the boundaries 0 and 1 and more than $D/2$ sites away from the segmentation point τ, use the earlier procedure to find the polynomial $\pi_{j,k}$ of degree D which generates the same averages in the neighborhood $(a_{j,k'}, k' = k - D/2, \ldots, k + D/2)$.

Step 2. At each site k which is at most $D/2$ sites away from the boundaries 0 and 1 find the polynomial $\pi_{j,k}$ of degree D which generates the same averages in the neighborhood $(a_{j,k'})_{k' \in N(k)}$, where $N(k)$ consists of the $D+1$ nearest neighbors of k.

Step 3. At each site k which is at most $D/2$ sites away from the segmentation point τ, we distinguish two cases.

Step 3a. If $\tau \notin [2^{-j}k, 2^{-j}(k+1)]$, then find the polynomial $\pi_{j,k}$ of degree D which generates the same averages in the neighborhood $(a_{j,k'})_{k' \in N(k)}$, where $N(k)$ consists of the $D+1$ nearest neighbors of k which are all on the same side of the segmentation point as k.

Step 3b. If $\tau \in [2^{-j}k, 2^{-j}(k+1)]$, then fit, by constrained least squares, left and right polynomials $\pi^L_{j,k}$ $\pi^R_{j,k}$ of degree D to the block averages in the neighborhoods on the right and left of the segmentation point, respectively; the constraint is that the piecewise polynomial $\pi^\tau_{j,k}$ which is π^L on the left of τ and π^R on the right of τ should have an average equal to the average $a_{j,k}$.

Step 4. In Steps 1, 2, and 3a, define the mock-averages at the next finer scale as averages of the polynomial. On the left half of the sub-interval we get

$$a^\tau_{j+1,2k} = Ave_{j+1,2k}\pi_{j,k},$$

while on the right half,

$$a^\tau_{j+1,2k+1} = Ave_{j+1,2k+1}\pi_{j,k}.$$

In Step 3b, the procedures are the same, only using the piecewise polynomial $\pi^\tau_{j,k}$.

Step 5. After having synthesized all the $a^\tau_{j+1,k}$'s, set $j := j+1$ and goto Step 1.

This segmented average-interpolating resolution has a variety of properties; a key one being that *if π^τ is a piecewise polynomial of degree D, with one knot, at τ, then the refinement process recovers π^τ exactly.*

To see the possible benefit of this procedure, consider a piecewise linear function, with jump discontinuity at $\tau = \lfloor .37 \cdot 256 \rfloor / 256$. Figure 2.4a depicts the boxcar averages at scale $j = 8$; Figure 2.4b depicts the same averages at scale $j = 4$. Refinement from the coarse data at scale $j = 4$ by the usual nonsegmented method produces Figure 2.4d; the attempted refinement misses the fine-scale structure entirely; in fact, the transition in the synthesized fine-scale data retains the same slope as in the coarse scale data. In contrast, Figure 2.4c illustrates the result of segmented refinement; this has perfectly reconstructed the fine-scale data, despite being derived from much coarser-scale data.

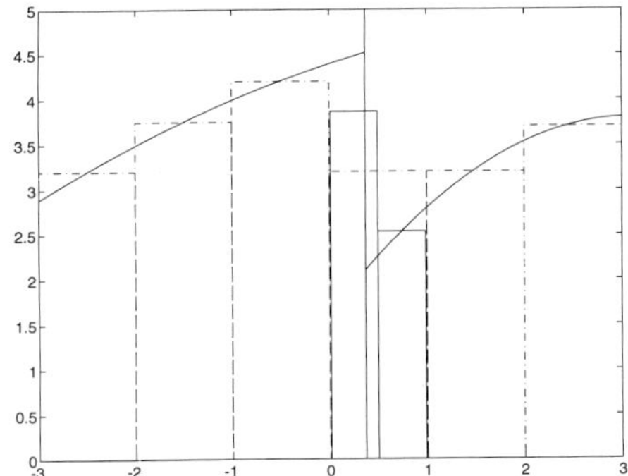

Figure 2.3. Averages, left & right polynomials, segmented imputations to finer scale.

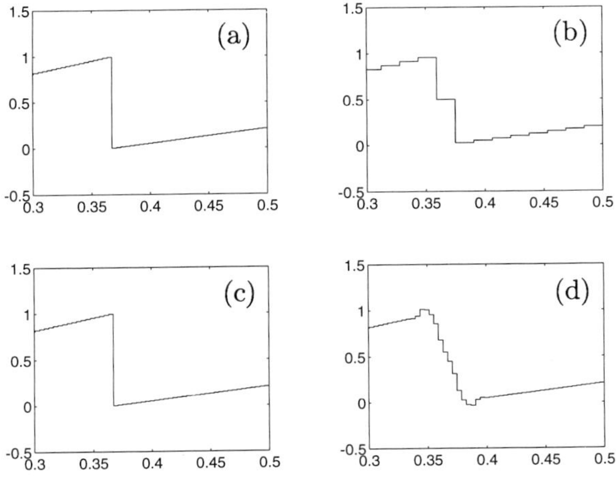

Figure 2.4. (a) (Original) Ramp; (b) (Coarsened) Ramp; (c) Segmented Refinement; (d) Ordinary Refinement.

2.4 Fast segmented transforms

Combining results of the last two sections allows us to define fast segmented transforms, as follows. Given a segmentation point τ at the "Heart" of the interval, and a j_0 satisfying $D/2^{j_0} < \tau < (2^{j_0} - D)/2^{j_0})$, we transform f to coefficients

$$\beta_{j_0,k} = \int_0^1 \chi_{j_0,k}(t) f dt, \qquad k = 0, \ldots, 2^{j_0} - 1,$$

and

$$\alpha_{j,k}^{\tau} = \int_0^1 h_{j,k}(t)(f - P_j^{\tau} f)(t) dt.$$

Given integrals at a fine scale $\beta_{j_1,k}$, $j_1 \gg j_0$, we can calculate all needed coefficients at coarser scales $j_0 \le j < j_1$ in order 2^{j_1} time. Given $(\beta_{j+1,k})_k$ we calculate $\beta_{j,k} = (\beta_{j+1,2k} + \beta_{j+1,2k+1})/\sqrt{2}$ exactly as in the fast algorithm for the Haar transform. We then calculate $(\alpha^{\tau}_{j,k})$ as follows: refine the coarser sequence $(\beta_{j,k})$, getting $(\hat{\beta}^{\tau}_{j+1,k})_k$, (this takes $O(2^j)$ time), and form the residuals $r^{\tau}_{j+1,k} = \beta_{j+1,k} - \hat{\beta}^{\tau}_{j+1,k}$. These residuals obey $r^{\tau}_{j+1,2k} = -r^{\tau}_{j+1,2k+1}$ and so

$$\alpha^{\tau}_{j,k} = (r^{\tau}_{j+1,2k+1} - r^{\tau}_{j+1,2k})/\sqrt{2}.$$

As all computations at level j can be performed in a time proportional to 2^j, this algorithm for computing at all levels $j_0 \le j < j_1$ is order 2^{j_1}.

2.5 Examples of segmented transforms

Now assume that we have been given a segmentation point τ and n boxcar averages, $(a_{j_1,k})_k$, of f, where $n = 2^{j_1}$. We have seen that we can calculate the first n wavelet coefficients of f in order n time. (Incidentally, the reconstruction of boxcar averages from those wavelet coefficients is also order n).

We now give some examples of this fast transform in action. Figure 2.5 presents four objects: *Ramp, Cusp, Noise*, and *HeaviSine*. They all are segmented at the point $\tau = .3696$. Data consist of boxcar averages at resolution $j = 11$. We use $D = 2$ and $j_0 = 4$ below.

Figure 2.6 presents the traditional wavelet coefficients $(\alpha_{j,k})$ of these objects. In panels a,b, and d, the presence of the singularity is clearly signaled by the significant wavelet coefficients in the vicinity of τ. Similar information is contained in Figure 2.7, which presents the multi-resolution displays $P_{j_0}f$ and $Q_j f$ for these objects.

Figure 2.8 presents the segmented wavelet coefficients $(\alpha^{\tau}_{j,k})$ of these objects. In Panels a, and b, they are all essentially zero; in Panel d they are essentially zero after two resolution levels. The segmented coefficients of object *Noise* scarcely differ from the ordinary nonsegmented ones. Figure 2.9 portrays the same information in the form of multi-resolution displays $P^{\tau}_{j_0}f$ and $Q^{\tau}_j f$ for these objects. The discontinuity in the MRA's is visible.

2.6 Application areas

We now indicate three possible applications of segmented wavelet transforms.

2.6.1 Data compression

It is evident on comparing Figures 2.6 and 2.8 that coefficients which are significant in the ordinary wavelet expansion become zero in the segmented wavelet expansion. This is a consequence of the fact that if π^{τ} is a piecewise polynomial of degree D, with breakpoint at τ, then

$$\alpha^{\tau}_{j,k}(\pi^{\tau}) = 0, \qquad j \ge j_0, \quad k = 0, \ldots, 2^j - 1.$$

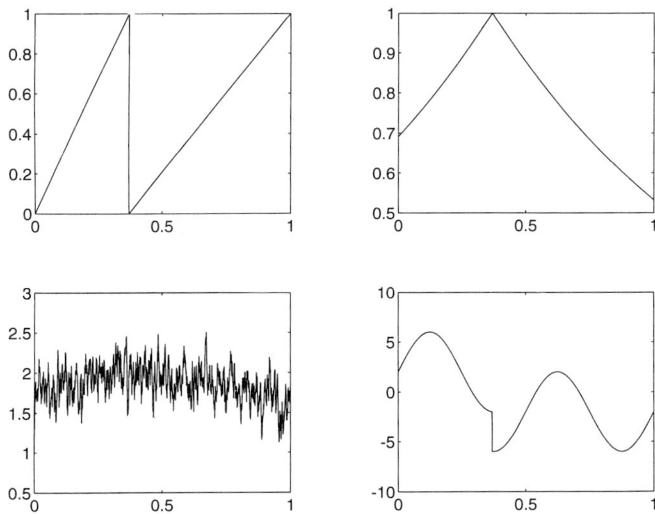

Figure 2.5. (top left) Ramp; (top right) Cusp; (bottom left) Noise; (bottom right) HeaviSine.

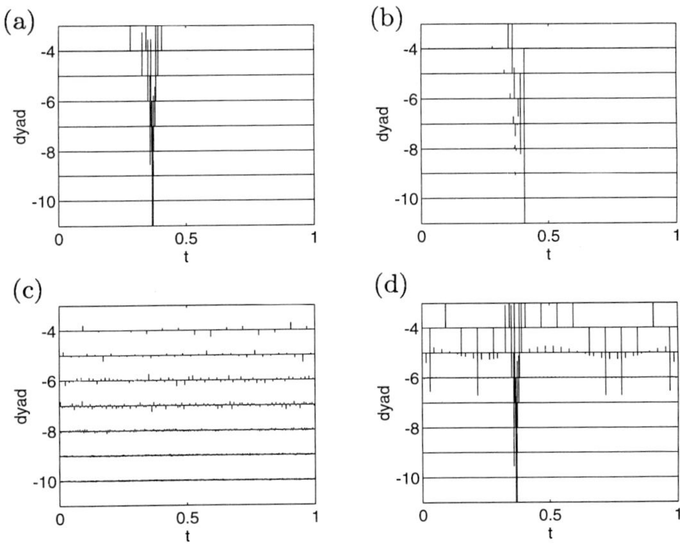

Figure 2.6. (a) WT (Ramp); (b) WT (Cusp); (c) WT (Noise); (d) WT (HeaviSine).

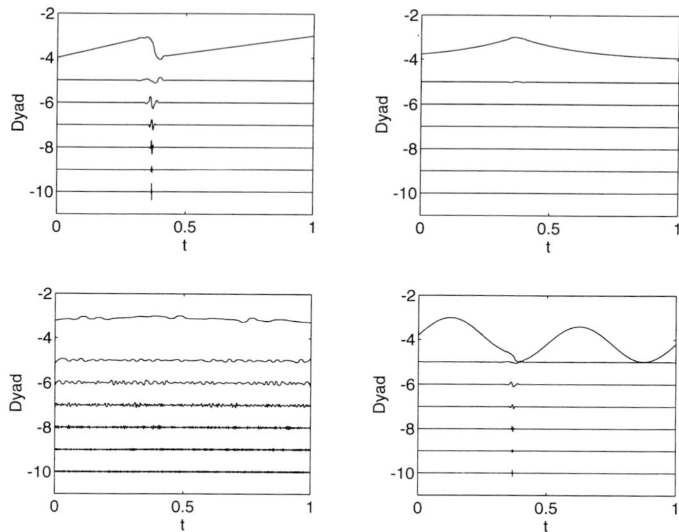

Figure 2.7. (top left) MRA (Ramp); (top right) MRA (Cusp); (bottom left) MRA (Noise); (bottom right) MRA (HeaviSine).

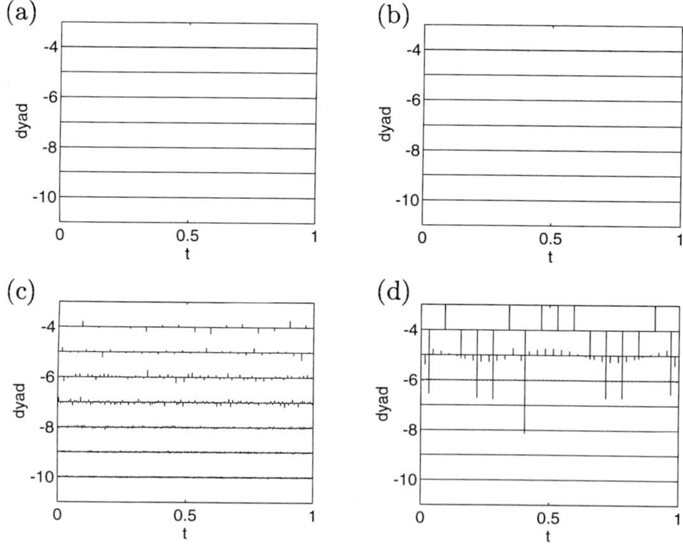

Figure 2.8. (a) SWT (Ramp); (b) SWT (Cusp); (c) SWT (Noise); (d) SWT (HeaviSine).

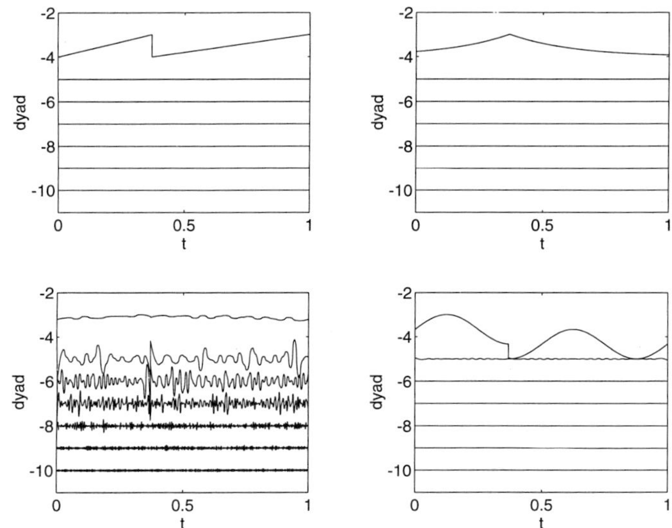

Figure 2.9. (top left) SMRA (Ramp); (top right) SMRA (Cusp); (bottom left) SMRA (Noise); (bottom right) SMRA (HeaviSine).

Ramp is piecewise linear; *Cusp* and *HeaviSine* are piecewise analytic, and so well approximated by polynomials; hence the segmented wavelet transform really should be sparser than the ordinary one.

To quantify sparsity, we use the following approach. Suppose that $\theta = T(f)$ is a transform of f into a sequence space; we measure sparsity in the transform domain as follows. Let $|\theta|_{(i)}$ denote the i-th from largest coefficient in θ, so that $|\theta|_{(0)} = \max_k |\theta_k|$, and define the compression number

$$c_m = \sum_{i>m} |\theta|_{(i)}^2.$$

The compression numbers measure how well we can approximate the vector θ by a vector with only m nonzero entries. If c_m tends to zero rapidly with m, then there are very few big coefficients in θ.

Figure 2.10 portrays the compression numbers c_m for the two different transforms of each of the objects. In each case, the dashed line is the ordinary transform and the solid line is the properly segmented transform. Evidently, segmented compression is much better than ordinary compression in cases a, b, and d, while it performs about the same as ordinary compression for object *Noise*.

2.6.2 Subpixel resolution

As indicated in Figure 2.4, a properly segmented multiresolution operator $P_j^\mathcal{T}$ has the ability to reconstruct jumps much more precisely than the usual $\Omega(2^{-j})$

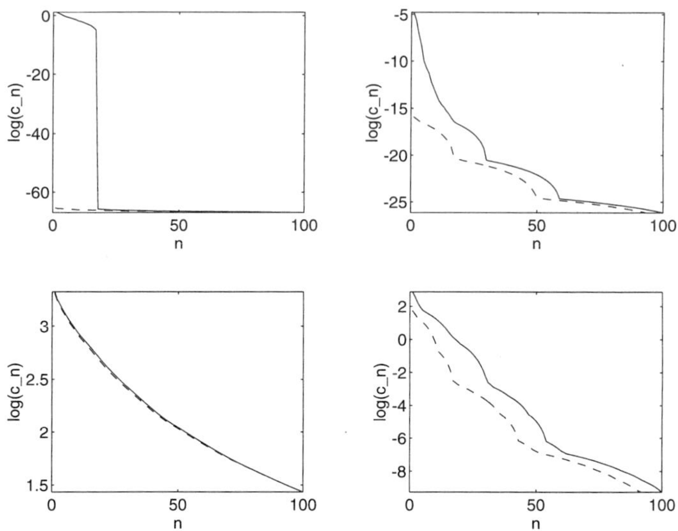

Figure 2.10. (top left) Ramp; (top right) Cusp; (bottom left) Noise; (bottom right) HeaviSine.

resolution of unsegmented operators. In principle this can even continue at scales finer than the data gathering, so that if we know the segmentation point τ with extreme precision, we can reconstruct at a resolution which is just as accurate as our knowledge of τ, and much more accurately than our measurement model seems naively to permit.

2.6.3 De-noising

As indicated in the introduction, one of our main interests in the present topic is in improving the behavior of wavelet shrinkage de-noising. To illustrate how this can be done using segmented transforms, we present in Figure 2.11 a noisy version of the *Ramp* object, its segmented wavelet transform, a thresholded version of the transform, and the reconstruction which was obtained by inverting the transform. The result displays a clean break, with no messy Gibbs phenomena, nor any appreciable shrinkage of the jump.

In contrast, Figure 2.12 gives a side-by-side comparison of the segmented recovery with the usual nonsegmented method described in [25], using a periodized wavelet transform. The difference is pronounced; the nonsegmented method is plagued by Gibbs artifacts.

We also present, in Figure 2.13, a noisy version of the *Cusp* object, its segmented wavelet transform, a thresholded version of the transform, and the reconstruction which was obtained by inverting the transform. The result displays a clean cusp in the correct location and amplitude.

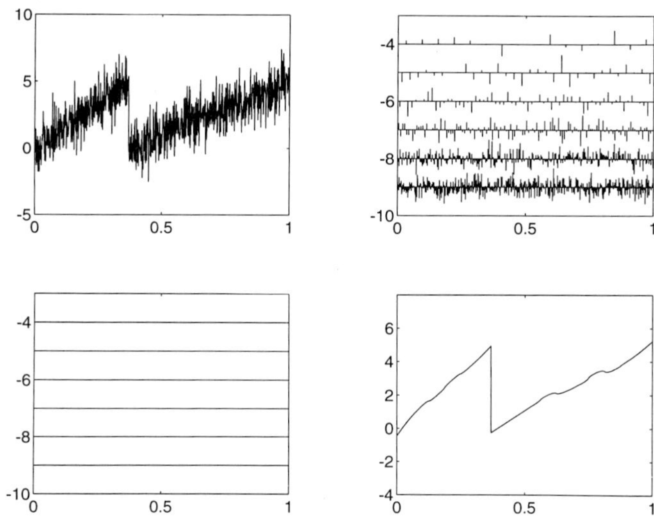

Figure 2.11. (top left) Noisy Ramp; (top right) SWT (Noisy Ramp); (bottom left) Threshold (SWT (Noisy Ramp)); (bottom right) Seg-DeNoise Ramp.

Figure 2.14 gives a superposed comparison of the segmented recovery with the usual nonsegmented method described in [25], using a periodized wavelet transform. The difference is not very pronounced; the nonsegmented method is plagued by downward shrinkage of peak amplitudes.

For fairness, it must be emphasized at this point that we are using *ideally* segmented transforms, in which the exact value τ of the break is known and employed. Such exact knowledge would not be available in most situations.

2.7 Variations

Before leaving the topic of 1-dimensional segmented MRA's, it is worth remarking that many variations on these ideas are possible.

First, the specific average-interpolating refinement scheme we are studying is not the only one we could have used. Here, we have obtained a polynomial by interpolating averages at blocks in a neighborhood of a point. We could equally well have fit, by constrained least squares, the averages at blocks in a larger neighborhood, with the constraint that the polynomial in question had an average that agreed exactly with the block average to be refined. This would give a method with perhaps more numerical stability, at the expense of longer filters.

Second, it is not necessary to use average-interpolating wavelets; for example, rather than modelling the system on biorthogonality with respect to the

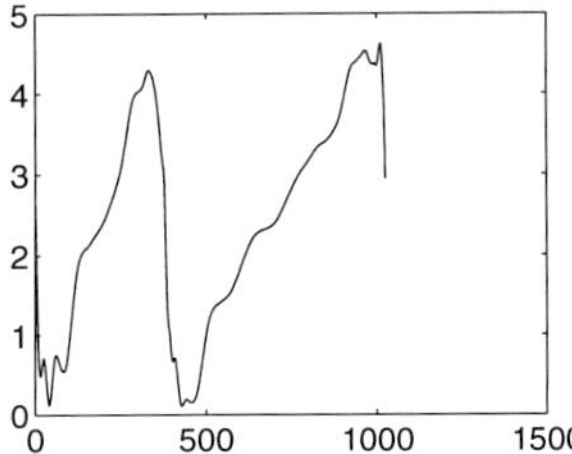

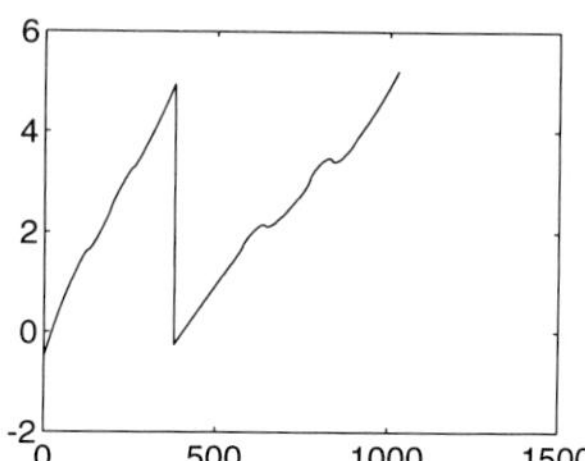

Figure 2.12. (left) Ordinary DeNoise[y Ramp]; (right) Seg-DeNoise[y Ramp].

Haar system, one could have used higher-order spline systems – at the expense of greater complexity in certain refinement calculations.

Third, the system considered here is only biorthogonal. To be maximally consistent with the best-basis notions we will present below, it would be very interesting to consider segmented orthogonal wavelet expansions. The ideas of Andersson, Hall, Jawerth, and Peters [1] may be useful in this connection.

§3 Adapting by minimum entropy

There is one obvious objection to the direction we have been headed. This would argue that while segmented transforms may be attractive in the ideal case where the appropriate point of segmentation is known exactly, one never knows this point in advance, and so the concept of segmented transform is of uncertain usefulness.

In this section, we will investigate the idea of selecting, adaptively, from data, an appropriate segmentation. Let $\mathcal{E} = \mathcal{E}(\theta)$ be an *entropy*, which, in our terms, means merely a functional which is small for sparse vectors containing very few nonzero components and which is large for vectors containing very many nonzero components all of the same size. An example is the ℓ^1 entropy

$$\mathcal{E}^1(\theta) = \sum_i |\theta_i|.$$

Other examples will be given below. We use the convention that if x denotes a vector of dyadic length $n = 2^{j_1}$ containing block averages at scale

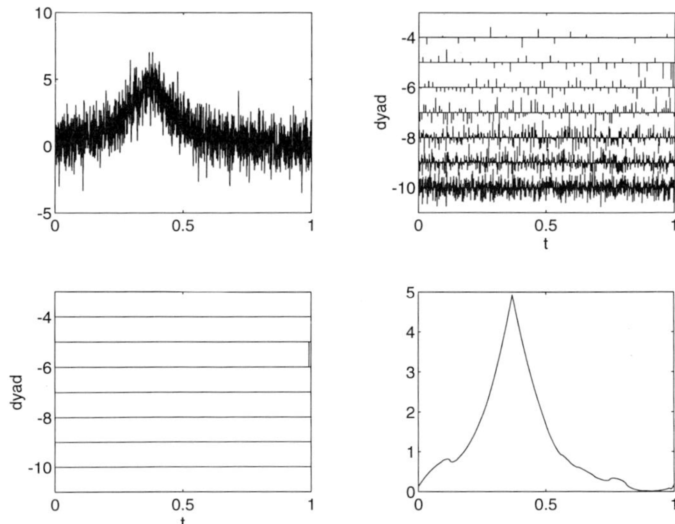

Figure 2.13. (top left) Noisy Cusp; (top right) SWT (Noisy Cusp); (bottom left) Threshold (SWT (Noisy Cusp)); (bottom right) Seg-DeNoise [cusp].

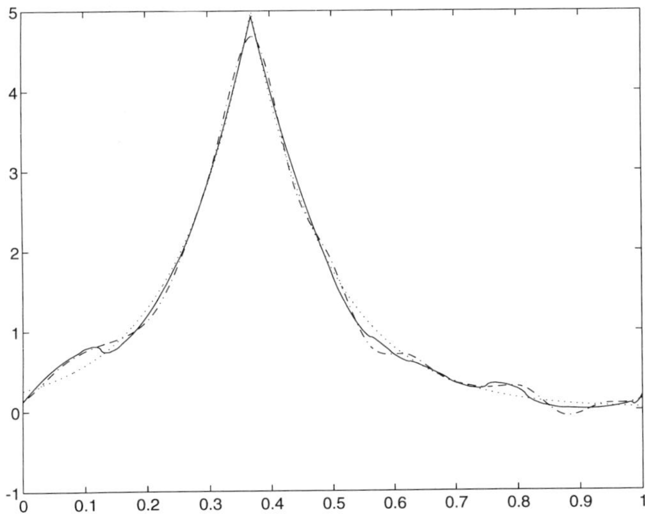

Figure 2.14. Object Cusp $(\cdots)$ Unsegmented $(-\cdots)$ and Segmented $(-)$ DeNoising.

2^{-j_1}, then $W_n^t x$ denotes the segmented wavelet coefficients $((\beta_{j_0,k})_k, (\alpha_{j,k}^t)_k)$ obtained with segmentation point t. The *Minimum Entropy Segmentation* principle (MES) is to select, from among all possible segmented bases for representing x, that basis which gives the coefficients with smallest entropy in the wavelet coefficient domain:

$$\hat{\tau} = \arg\min_{t \in [0,1]} \mathcal{E}(W_n^t x).$$

One ought to choose the entropy $\mathcal{E}$ so that the resulting segmentation reflects the task at hand. Consequently, there will be different implementations of this principle, depending on whether one's goal is data compression or de-noising.

3.1 Data compression

Coifman and Wickerhauser [10] in now-classic work, have proposed a method of best-basis selection which, translated into the present framework, goes as follows. First, given wavelet coefficients $W_n^t x$, define $p_{j,k} = (\alpha_{j,k}^t)^2$. Then $p_{j,k} \geq 0$.

Second, one defines the Coifman-Wickerhauser entropy (C-W entropy) by

$$\mathcal{E}^{CW}(\theta) = -\sum_{j,k} p_{j,k} \log(p_{j,k})$$

(We differ here from Coifman-Wickerhauser in two ways. First, they were working with orthogonal transformations, and it was, therefore, natural to normalize the object to unit ℓ^2-norm 1. We do *not* adopt this convention here. Second, our sum does not include the $(\beta_{j_0,k})$ terms, which, anyway, are the same regardless of segmentation point t.)

To the original C-W entropy we add the following general family of entropies $\mathcal{E}^\alpha$, $\alpha \in [0, 2]$,

$$\mathcal{E}^\alpha(\theta) = \sum_{j,k} p_{j,k}^{\alpha/2}.$$

We are particularly interested in the ℓ^1 entropy $\mathcal{E}^1$, the $\ell^{1/2}$ entropy $\mathcal{E}^{1/2}$, and the ℓ^2 entropy $\mathcal{E}^2$.

All of these measures are measures of anti-sparsity. The limit, as $\alpha \to 0$, is simply the numerosity,

$$\mathcal{E}^0 = \#\{(j,k) : \alpha_{j,k}^t \neq 0\}.$$

At the other limit, as $\alpha \uparrow 2$, we can obtain the C-W entropy:

$$\frac{d}{d\alpha}\mathcal{E}^\alpha(\theta)|_{\alpha \to 2} = \mathcal{E}^{CW}(\theta).$$

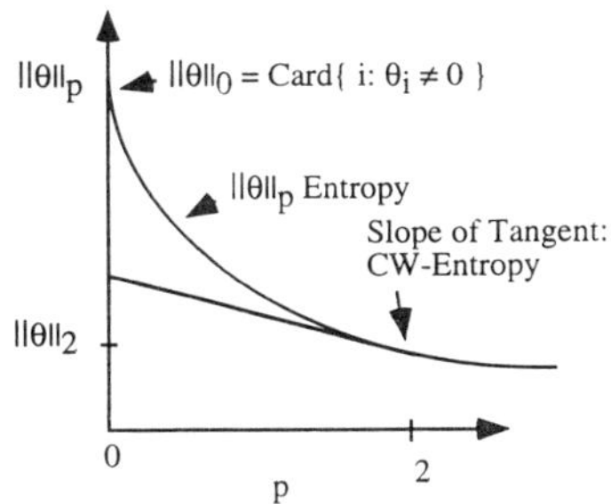

Figure 3.1. Relationships of ℓP entropies.

Hence, the C-W entropy is the tangent on a curve which measures sparsity; the other entropies are simply points on that curve. See Figure 3.1.

Note that in the original best-basis setting, one was considering a choice among orthogonal bases, and the ℓ^2 entropy would not vary among bases; but here, one is considering a choice among nonorthogonal bases, and the ℓ^2 entropy is more reasonable.

We now compare the use of these entropies in choice of segmentation. First we consider object *Ramp*; here n = 2048 and the correct segmentation is at $\tau = 757/2048$. Figure 3.2 shows the unsegmented wavelet transform and three segmentations at pixel boundaries 756, 757, and 758. Because the object is piecewise linear, at the correct segmentation 757, the wavelet coefficients $\alpha_{j,k}$ all vanish. The figure shows that at nearby segmentations, the segmented wavelet transform entropy is intermediate between an unsegmented one and the appropriately segmented one. Figure 3.3 shows entropy profiles for pixel-level segmentations running from 749 to 765. At the correct segmentation 757, all of the entropies vanish. As we move away from the correct segmentation, the entropy more or less increases, but is more well-behaved for the 2, 1- and 1/2 entropies than for the C-W entropy, which seems erratic.

Next we consider the object *Cusp*; here $n = 2048$ and the correct segmentation is again $\tau = 757/2048$. Figure 3.4 shows the unsegmented wavelet transform and three segmentations at pixel boundaries 756, 757, and 758. Because the object is piecewise analytic, at the correct segmentation 757, the wavelet coefficients $\alpha_{j,k}$ nearly vanish. The figure shows that at nearby segmentations, the segmented wavelet transform entropy is intermediate between an unsegmented one and the appropriately segmented one. Figure 3.5 shows entropy profiles for pixel-level segmentations running from 749 to 765. At the correct segmentation 757, all of the entropies vanish. As we move away from the correct segmentation, the entropy more or less increases, but is again more well-behaved for the 2, 1- and 1/2 entropies than for the C-W entropy, which has rather a broad minimum, and fails to point to a unique optimum.

In both examples, the 1/2-entropy indicates a sharper preference for a specific segmentation than the other entropies.

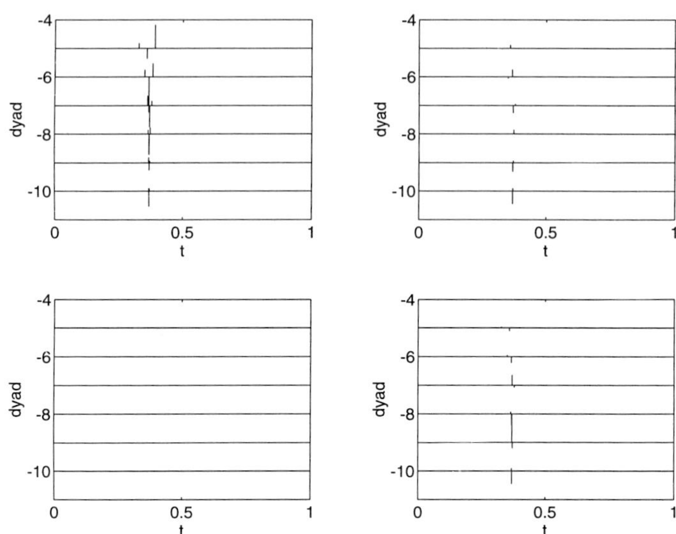

Figure 3.2. (top left) Ordinary Transform of Ramp; (top right) Transform Segmented at $t = 756/2048$; (bottom left) Transform Segmented at $t = 757/2048$; (bottom right) Transform Segmented at $t = 758/2048$.

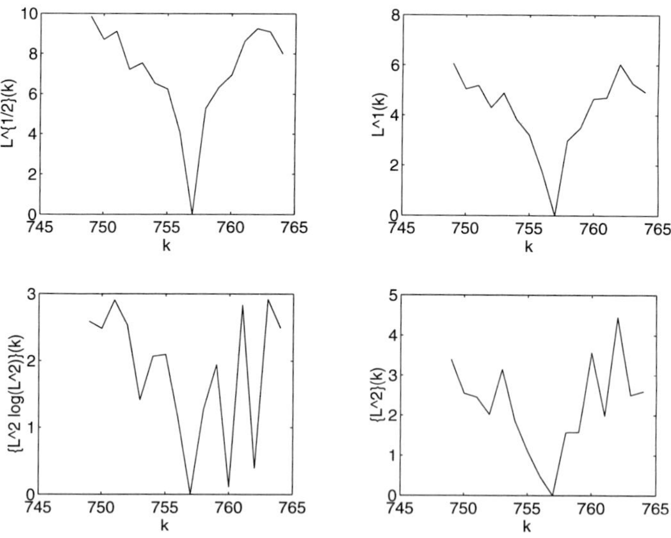

Figure 3.3. Entropies for Ramp Object, $t = k/n$.

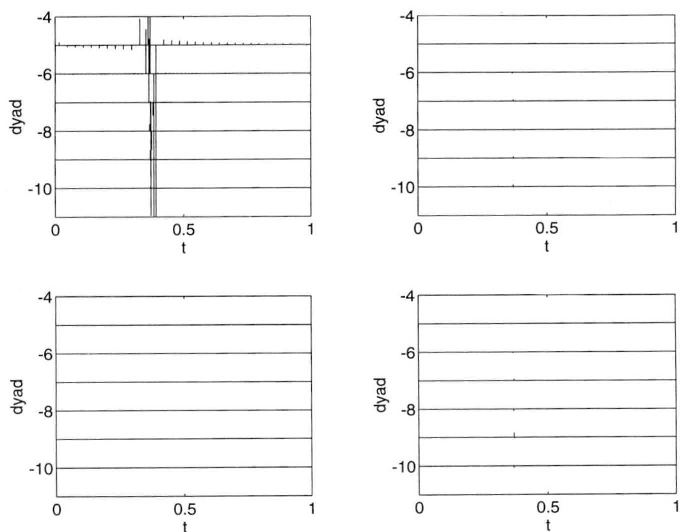

Figure 3.4. (top left) Ordinary Transform of Cusp^4; (top right) Transform Segmented at $t = 756/2048$; (bottom left) Transform Segmented at $t = 757/2048$; (bottom right) Transform Segmented at $t = 758/2048$.

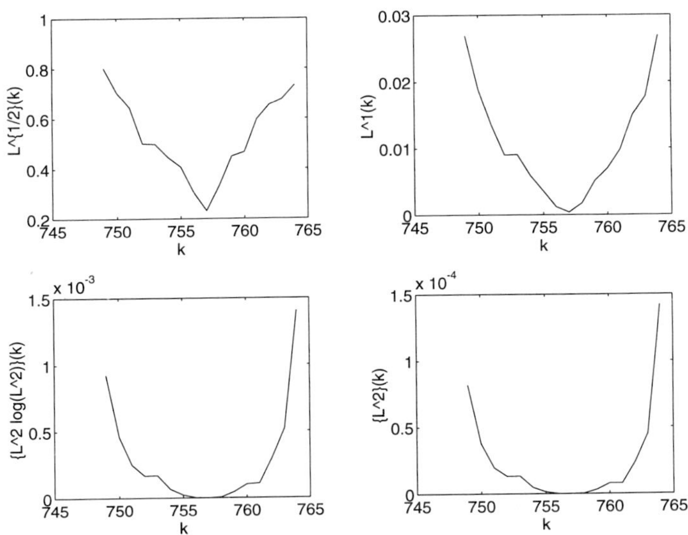

Figure 3.5. Entropies for Cusp^4 Object, $t = k/n$.

3.2 De-noising

We now consider adaptive choice of basis in the presence of noise. With $n = 2^{j_1}$, we suppose that we have noisy block-averages

$$d_k = Ave\{f|I_{j_1,k}\} + \epsilon \cdot z_k, \qquad k = 0, \ldots, n-1$$

where the z_k are a Gaussian white noise. We process these data as if they were noiseless block averages, obtaining coarse-scale empirical block averages

$$v_{j_0,k} = \beta_{j_0,k} + \epsilon \cdot \xi_{j_0,k}, \qquad k = 0, \ldots, 2^{j_0} - 1,$$

and empirical wavelet coefficients

$$w^t_{j,k} = \alpha^t_{j,k} + \epsilon \zeta^t_{j,k}, \qquad k = 0, \ldots, 2^j - 1.$$

We act provisionally as if the ξ's and ζ's were independent and constant variance 1, which they are not, owing to the lack of orthogonality of the transforms.

We consider the problem of recovering the vector of coefficients $\theta^t = \left((\beta_{j_0,k})_k, (\alpha^t_{j_0,k})_k, \ldots,\right)$, and we group the noisy empirical wavelet coefficients together into a vector $y^t = \left((v_{j_0,k})_k, w^t_{j,k}, \ldots,\right)$.

Initially, consider the following ideal problem (compare Donoho and Johnstone [21]). We have available an oracle which furnishes optimal weights (w_i) for use in a diagonal linear estimator $\hat{\theta}^t = (w_i y^t_i)_i$; these weights being optimal in the sense that they minimize the mean squared error

$$E \sum (w_i y_i - \theta_i)^2.$$

In reality, such an oracle and such optimal weights are never available to us. The risk of such an ideal procedure is within a factor of 2 of the following proxy:

$$\mathcal{R}(\hat{\theta}^t) = \sum_i \min((\theta^t_i)^2, \epsilon^2).$$

In terms of the compression number introduced earlier, we have

$$\mathcal{R}(\hat{\theta}^t) = c_{N(\epsilon)} + \epsilon^2 N(\epsilon),$$

where $N(\epsilon) = \#\{i : |\theta_i| > \epsilon\}$. As this "ideal risk" is, therefore, large for dense vectors containing lots of entries and small for sparse vectors containing only a few nonzero coefficients, it is an entropy.

Figure 3.6, top left panel displays the behavior of this ideal risk measure in segmenting the *Ramp* object; top right panel displays the behavior in segmenting the *Cusp* object. In both cases the minimum risk segmentation is at

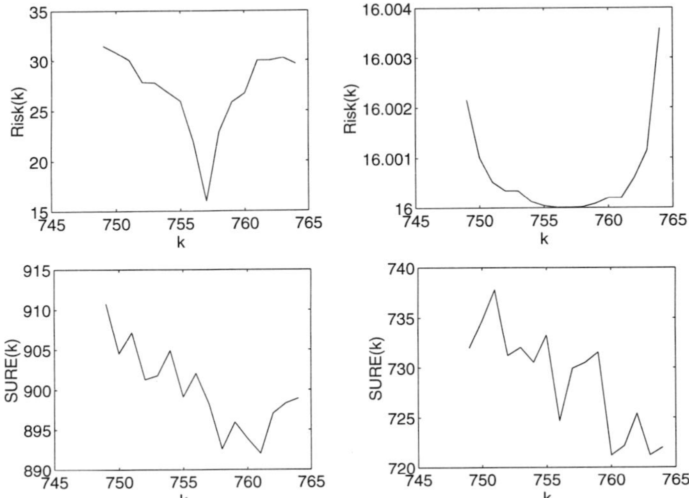

Figure 3.6. (top left) Ramp Object, Ideal Risk; (top right) Cusp Object; Ideal Risk; (bottom left) SURE Measure, $t = k/n$, Ideal $= 757$; (bottom right) SURE Measure, $t = k/n$, Ideal $= 757$.

the natural location. There is a sharper preference for the minimum in the case of the *Ramp* object, which is connected with the object's sharper discontinuity.

Now consider the behavior of a "real" de-noising procedure, as follows. Setting a threshold $\lambda = \sqrt{2\log(n)}\epsilon$ we apply soft thresholds, getting estimates

$$\hat{\theta}_i = \eta_\lambda(y_i), i = 1, \ldots, n.$$

(Again we are acting provisionally as if the y_i all have the same variance ϵ^2). As in [24,25], we can estimate the risk of this estimator using Stein's Unbiased Estimate of Risk for this nonlinear estimator:

$$SURE(y) = \epsilon^2 \cdot \left(n - 2\sum_i 1_{|y_i|<\lambda} \right) + \sum_i \min(y_i^2, \lambda^2).$$

This risk measure is smaller for sparse vectors and larger for dense vectors, and so represents a kind of entropy.

Figure 3.6, bottom left panel displays the behavior of this empirical risk measure in segmenting the noisy *Ramp* object; bottom right panel displays the behavior in segmenting the noisy *Cusp* object. In both cases the SURE profile is much noisier than the Risk profile (as expected); and the minimum SURE segmentation is near, but not exactly at, the natural location. There is again sharper preference for the minimum in the case of the *Ramp* object, which is connected with the object's sharper discontinuity.

Apparently, even in the presence of noise, one can adaptively select a transform which preserves the structure of strong discontinuities.

§4 Fast computation of all segmentations

A further objection to the direction we have been headed is computational. The segmented wavelet transform is an order n operation; to calculate all n pixel-level segmentations, therefore, seems naively to require an $O(n^2)$ procedure; this is unsuitable for many applications.

Fortunately, there is a method for calculating all n pixel-level segmentations in order $n \log(n)$ time and space. The method is based on the following observation. In performing the segmented refinement, only those blocks within a distance $D/2$ of the block containing the segmentation point are affected by the segmentation. That is to say, the resulting values are the same as they would be in a nonsegmented refinement based on the average-interpolating wavelets of Subsection 2.1. We, therefore, propose to calculate, for each pixel-level segmentation $t = i/n, i = 0, \ldots, n-1$, only those specific coefficients which differ from the unsegmented transform.

Label these coefficients

$$\nu_{j,l}(t), j = j_0, \ldots, j_1 - 1, l = -D/2, \ldots, D/2,$$

where the subscript j again indicates resolution level and the subscript l indicates offset from the block $\lfloor t2^j \rfloor$ containing the segmentation point.

Suppose these coefficients are all available for a given segmentation point t, and that we also have available the unsegmented transform. By copying values from the array of ν into the appropriate locations of the array of unsegmented wavelet coefficients, we obtain the segmented wavelet coefficients.

Each one of these coefficients depends in a linear fashion on a fixed number of block averages at scale $j+1$ in a neighborhood of a given block. Therefore each coefficient can be computed from scratch in order $C \cdot D$ work.

There are order $log_2(n) \cdot D$ $\nu's$ attached to a given t; therefore the calculation of all the $\nu's$ attached to that t, starting from scratch, is an order $D^2 \log_2(n)$ operation.

It follows that we can evaluate, in sequence, all n pixel-level segmented transforms by the following approach.

Step 1. Compute the unsegmented AI wavelet transform. Make an extra copy of the transform array.

Step 2. For $i = 0, \ldots, n-1$ do:

Step 2.a. Calculate the ν-coefficients for $t = i/n$.

Step 2.b. Copy them into the unsegmented array at the appropriate positions (these depend on t).

Step 2.c. Evaluate the entropy of the resulting array.

Step 2.d. Restore the unsegmented array with the original unsegmented wavelet coefficients.

Step 3. Using the best i arising in Step 2, perform Steps 2.a and 2.b once more at that i.

The cost of this procedure, excepting the evaluations of entropy, is order $n \cdot \log(n) \cdot D^2$.

We now point out how to rapidly minimize the entropy functional. Let θ^0 denote the coordinates of the unsegmented wavelet transform. Define the differential entropy

$$\delta\mathcal{E}(t) = \mathcal{E}(\theta^t) - \mathcal{E}(\theta^0).$$

The minimum of the entropy $\mathcal{E}(\theta^t)$ will be at the same value of t as the minimum of the differential entropy, so it is sufficient to minimize differential entropy.

Now we remark that all the entropy functionals we have discussed are coordinatewise sums; in addition, most of the coefficients of θ^0 and θ^t agree. Therefore most terms in the entropy difference $\mathcal{E}(\theta^t) - \mathcal{E}(\theta^0)$ disappear, and only those coefficients which are potentially different need be considered. The differential entropy $\delta\mathcal{E}(t)$ is, up to a quantity which does not depend on t, simply a functional of the ν coefficients, and of the unsegmented wavelet coefficients that they replace. Call the coefficients being replaced

$$\mu_{j,l}(t), j = j_0, \ldots, j_1 - 1, l = -D/2, \ldots, D/2.$$

(Of course, these are all present in the single n-element array of wavelet coefficients, so that one does not actually store the μ's; it is convenient to have a notation referring to them). One can, therefore, simply evaluate the *entropy of the ν-coefficients*, subtract the entropy of the μ-coefficients, and minimize this difference as a function of t.

We, therefore, have the following streamlined algorithm, which requires less time and space.

Step 0. Calculate the ordinary unsegmented transform.

Step 1. For $i = 0, \ldots, n-1$ do:

Step 2.a. Calculate the ν-coefficients for $t = i/n$.

Step 2.b. Evaluate the entropy difference between those ν coefficients, and the corresponding μ coefficients from the unsegmented transform.

Step 3. Using the best i arising in Step 2, calculate the segmented wavelet transform for $t = i/n$.

The complexity of the unsegmented transform is again $O(n)$, and the whole algorithm is order $n \cdot \log_2(n) \cdot D^2$.

Remark.

1. With the exception of offsets $l = 0$, each $\nu_{j,l}$ is actually the result of a filtering operation – simple convolution – applied to the block averages. Thus $\nu_{j,l}$ is constant in blocks of size $2^{(j_1-j)}$. By exploiting this remark, all the $\nu_{j,l}(t)$ for $l \neq 0$ can be computed simultaneously in order $O(n)$ time (rather than $n\log(n)$).

2. The vector $\nu(t)$ is, in fact, a kind of multiresolution filter bank, with $\approx (D+1)\log_2(n)$ outputs at each "time" i. Therefore, *we are searching for an optimal segmentation by applying multiresolution filters, and evaluating the entropy of the output, searching for a minimum entropy output.* Applying this idea, we present, in Figure 4.1, a display of the entire filter bank output for object *Ramp*.

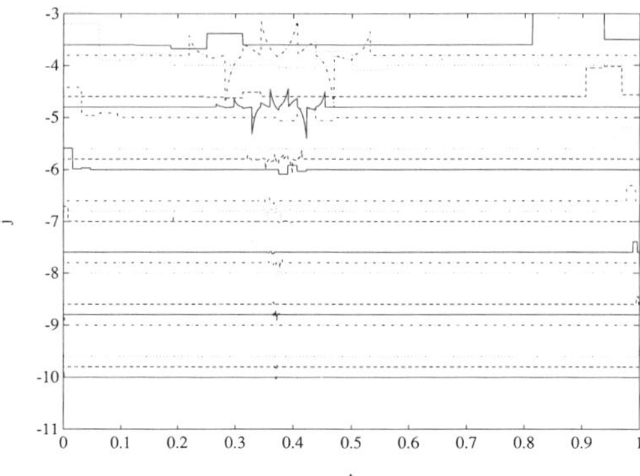

Figure 4.1. Object Ramp; Multiresolution filter Bank $\nu_{j,l}(t)$.

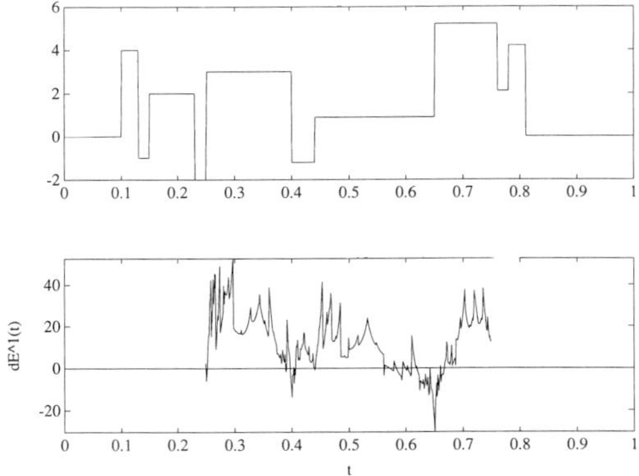

Figure 4.2. (top) Object Blocks; (bottom) L^1 Differential Entropy Profile, Object Blocks.

3. Although we started from the point of view of looking for a single segmentation point, we can instead look for several. By displaying the whole differential entropy profile $\delta\mathcal{E}(t)$ as a function of t, we may perhaps identify several points of segmentation. To illustrate this fact, we display in Figure 4.2, top panel, the object *Blocks*, which has many edges, and in bottom panel, the corresponding differential entropy profile.

4. Although the method we are discussing is order $n\log(n)$, the constants involved are worse than, say, the constants involved in the Fast Fourier transform, as order $n\log(n)$ inversions of a $2D+2$ by $2D+2$ matrix must be made. (The matrices are all the same except for two rows, however, so systematic use of the Sherman-Morrison-Woodbury formulas could improve the constants over naive inversion.)

5. Most fundamentally, the current ideas completely change our opinion

of what makes sense in treating noisy data. We began with the prejudice that using SURE to select bases was the most sensible approach. This was based on the favorable experience of Donoho and Johnstone [24,25] in using SURE to adaptively select parameters of a de-noising procedure. In particular, we would not have considered the use of an ℓ^2 entropy, because the noise in such entropy would seemingly swamp any signal. But from the present point of view, it is clear that minimizing the simple ℓ^2 entropy criterion involves in an essential way only quadratic forms in order $\log_2(n)$ noise variables. Therefore, the noise variance in such a criterion grows with n in a moderate way, and selecting a segmentation by minimizing the ℓ^2 entropy is no longer ruled out.

§5 MES as an edge locator

The observation just made suggests a new multiresolution filter bank edge locator: *find the t minimizing the ℓ^2-entropy of the segmented empirical transform.* There is, of course, a massive literature on edge detection and location; it is interesting to compare this *Minimum Energy Segmentation* method with existing approaches.

1. The segmented wavelet approach allows a definition of edges which is very broad, including not only step discontinuities, but also cusps, and, if high polynomial degrees D are employed in wavelet construction, discontinuities in higher derivatives. Moreover, the segmented wavelet approach allows considerable variety in behavior near the edge – a jump discontinuity needn't be a simple Heaviside; it could also be a jump with different slopes on the two sides of the jump. In contrast, many existing schemes depend on a specific shape of discontinuity (*e.g.*, Heaviside); they may miss more subtle effects and may produce biased locations if the simple models they assume do not fit.

2. The segmented wavelet approach is based on a multiresolution filter, whereas the traditional approaches are mono-resolution – based on filtering at one fixed scale. It is evident that if one filters at one fixed scale, a variety of tuning parameters need to be specified; if these are misspecified, the results will be poor. The segmented wavelet approach based on the ℓ^2 entropy has no tuning parameters.

3. The segmented wavelet approach seems to be near-optimal from the point of view of statistical theory. Roughly speaking, when the object contains a step discontinuity, the Minimum Energy criterion seems to give a localization of the edge at a rate approaching the $O_P(1/n)$ which statistical theory says is optimal; when the object contains a cusp discontinuity, the Minimum-Energy criterion seems to give a localization of the edge of accuracy $O_P(1/\sqrt{n})$, which statistical theory says is again optimal, and so on for other discontinuity types.

This is not the place for an extended analysis or proof of the asymptotic properties of this edge locator. However, we do offer a simple heuristic analysis which may be persuasive, and with work can be refined into a rigorous analysis. For simplicity, we argue below as if the various coefficient functionals had equal, unit, norms. A rigorous argument would allow for the fact that they do not.

Suppose that we have a function like *Ramp* with jump discontinuity at τ, and polynomial behavior on both sides (polynomial of degree D). Consider the segmented wavelet coefficients $\nu_{j,l}(t)$ for $t \neq \tau$. We say that a non-central coefficient ($l \neq 0$) is "contaminated" if the formula producing it involves using data from blocks containing the segmentation point, or on the opposite side.

For "uncontaminated" coefficients with $\ell \neq 0$, the magnitude of the coefficient $\nu_{j,l}(t)$ is $O(2^{-j(D+1/2)})$. For "contaminated" coefficients, the magnitude of the coefficient, unless a happy accident intervenes, is $O(2^{-j/2})$.

The ℓ^2 entropy is an additive measure, so we may partition the risk measure by resolution level:

$$\delta\mathcal{E}^2(t) = \sum_j \left(Q_j^\nu(t) - Q_j^\mu(t)\right),$$

with component at level j

$$Q_j^\nu = \sum_l \nu_{j,l}^2(t); \qquad Q_j^\mu = \sum_l \mu_{j,l}^2(t).$$

Roughly speaking therefore, Q_j^ν is of order

$$Q_j^\nu \approx \#\{\text{uncontaminated } \nu_{j,l}\} \cdot 2^{-j(2D+1)} + \#\{\text{contaminated } \nu_{j,l}\} 2^{-j}.$$

Now, again roughly speaking

$$\#\{\text{contaminated } \nu_{j,l}\} \approx \min(2^j|t-\tau|, D/2+1)$$

and

$$\#\{\text{uncontaminated } \nu_{j,l}\} \approx D - \#\{\text{contaminated } \nu_{j,l}\}.$$

We conclude that each Q_j^ν has a "well" of order 2^{-j} wide, and a depth which is of order 2^{-j} also. Combining over all scales, we get for expected behavior that $\delta\mathcal{E}^2(t)$ has a well with sides behaving like $\asymp |t-\tau|$, for $|t-\tau| \geq 1/n$.

In case the underlying function has a cusp, "contaminated" coefficients are of size $2^{-j(3/2)}$. Repeating the above analysis, we get an expected behavior that $\mathcal{E}^2(t)$ has a well with sides behaving like $\asymp |t-\tau|^2$, for $|t-\tau| \geq 1/n$.

In general, for a discontinuity in the m-th derivative, $\mathcal{E}^2(t)$ has a well with sides behaving like $\asymp |t-t|^{m+1}$, for $|t-\tau| \geq 1/n$.

Let us now consider the noise in the objective function. Define the noise process $Z(t) = \delta\mathcal{E}^2(y^t) - E\{\delta\mathcal{E}^2(y^t)\}$. This is a continuous zero-mean stochastic process, at each t a diagonal quadratic form in $O(\log_2(n))$ random variables, each one a Gaussian with variance σ^2/n. Hence $Z(t)$ has tail probabilities bounded by a double exponential distribution with variance parameter $C(\sigma^2 \log_2(n)/n)^2$. We ignore, in this heuristic treatment, the issue of the non-centrality parameters of various quadratic forms.

Now for the minimum to occur at a certain, fixed t, it is necessary that the noise in $Z(t) - Z(\tau)$ exceed the drift $Q(t) - Q(\tau)$. The chance that the noise is smaller than some multiple of $\sigma^2 \log_2(n)/n$ is overwhelming. Therefore, a given t has a non-negligible chance to be better than τ only if $Q(t) - Q(\tau)$ is smaller than some multiple of $\sigma^2 \log_2(n)/n$. This suggests that

$$Q(\hat{\tau}) - Q(\tau) = O_P(\sigma^2 \log_2(n)/n).$$

(Rigorous proof of such a relation of course requires the use of techniques from the theory of empirical processes.) Combining these relations with the fact that, in the presence of a jump discontinuity, $Q(\hat{\tau}) - Q(\tau) \asymp |\hat{\tau} - \tau|$, and that, in the presence of a cusp, $Q(\hat{\tau}) - Q(\tau) \asymp |\hat{\tau} - \tau|^2$, and one gets the following heuristic predictions.

First, the rate of convergence of the minimizer at a simple discontinuity with polynomial behavior on either side is predicted to be

$$\hat{\tau} - \tau = O_P(\sigma^2 \log_2(n)/n).$$

Second, the rate of convergence of the minimizer at a simple cusp with polynomial behavior on each side is predicted to be

$$\hat{\tau} - \tau = O_P(\sigma \sqrt{\log_2(n)/n}).$$

Third, the rate of convergence of the minimizer at an m-th order discontinuity, with polynomial behavior on each side, is predicted to be

$$\hat{\tau} - \tau = O_P((\sigma^2 \log_2(n)/n)^{1/(m+1)}),$$

provided $D > m$.

We know from asymptotic decision theory that this behavior is essentially the best one may expect. It is possible that if we knew that the discontinuity were of a certain type, say a ramp, we could invent a method which converges at a slightly faster rate – avoids the logarithm terms. But the new method makes no assumptions whatever, and achieves a near-optimal rate for the given type of singularity without advance knowledge of the type of singularity. We conjecture (based on related experience in [22]) that if one wants to adapt to an *unknown* type of singularity, the logarithm terms can not be avoided.

We carried out a small simulation experiment to assess the performance of the estimator. In the simulation, we attempted to locate the segmentation point for objects *Ramp* and *Cusp* at various signal-to-noise ratios (SNR) and sample sizes.

The simulations show that the estimator had an all-or-nothing character. At sufficiently high signal-to-noise ratio, the methods give accuracy at the pixel level, while at signal-to-noise ratio below some critical threshold, the methods fail completely. There was very little evidence of continuous or gradual degradation in the estimator's quality with decreasing SNR. We did not succeed in identifying a heuristic formula which would predict the SNR at which the pixel-level resolution degraded completely.

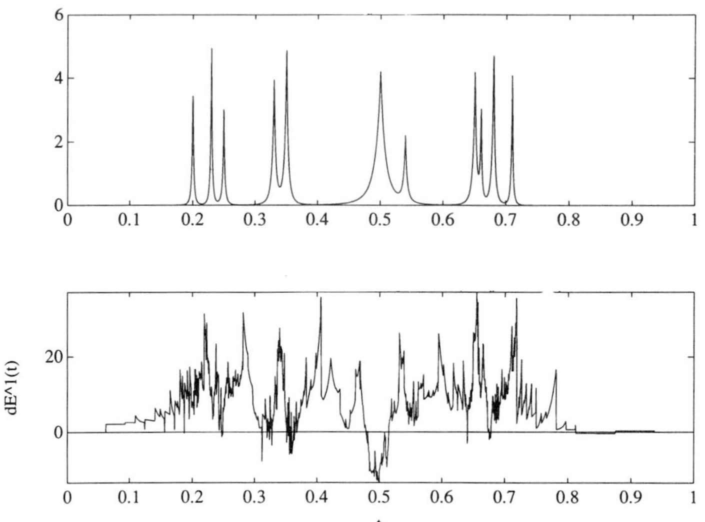

Figure 6.1. (top) Object Bumps, (bottom) L^1 Differential Entropy Profile, Oject Bumps.

§6 Multi-segmented analysis

Figure 4.2 shows that, if one evaluates the differential entropy profile on an object with several discontinuities, the profile will exhibit several local minimizers. This suggests that tools for finding a single best segmentation might profitably be employed in the case of multiple-segmented objects.

The key issue in such an undertaking is that some kind of sequential unmasking is necessary. Figure 6.1 shows an object, *Bumps*, together with its differential entropy profile. Evidently, not all the bumps result in visible local minima of the entropy profile. Figure 6.2 displays its segmentation-coefficients $\nu_{j,l}$.

6.1 Sharp- and flat-components

Any function $f_j \in V_j^\tau$ is, in principle, smooth except at τ. Hence we can decompose the function into "potentially singular" and "certainly smooth" parts solely by location. If $K_j(\tau)$ denotes those indices k where $\text{clos}(\text{supp}(\phi_{j,k}))$ contains τ, then for an $f_j \in V_j$ we may write

$$f_j = f_j^{\#,\tau} + f_j^{\flat,\tau}$$

with "potentially singular" (sharp) part

$$f_j^{\#,\tau} = \sum_{k \in K_j(\tau)} \beta_{j,k}^\tau \phi_{j,k},$$

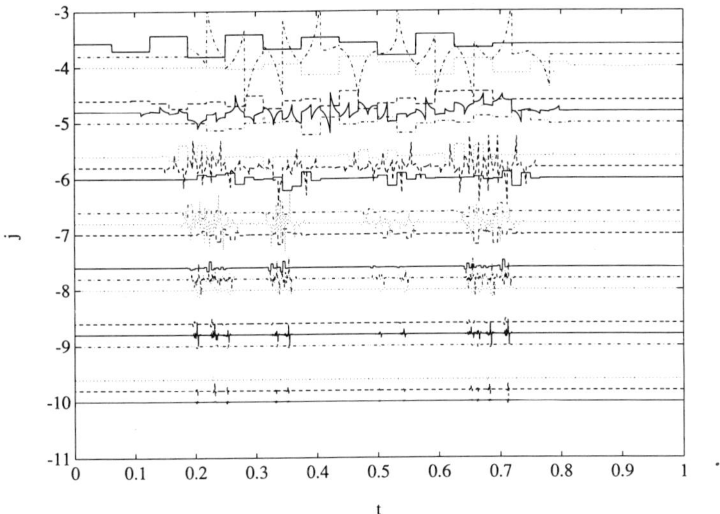

Figure 6.2. Object Bumps, Multiresolution Filter Bank $\nu_{j,l}(t)$.

and "certainly smooth" (flat) part

$$f_j^{\flat,\tau} = \sum_{k \notin K_j(\tau)} \beta_{j,k}^{\tau} \phi_{j,k}.$$

We note also that the mapping $f \to f_j^{\#,\tau}$ is a nonorthogonal projection, and similarly for $f \to f_j^{\flat,\tau}$

We can do the same sort of thing for a function d in W_j, getting nonorthogonal projection operators $S_j^{\sharp,\tau} d$ and $S_j^{\flat,\tau} d$.

If we now write $f = f_{j_0} + \sum_{j_0 \le j < J} Q_j^t f$, we can decompose each individual term into sharp- and flat- components, producing

$$f = f^{\#,\tau} + f^{\flat,\tau},$$

where

$$f^{\#,\tau} = f_{j_0}^{\#,\tau} + \sum_{j_0 \le j < J} S_j^{\sharp,\tau} Q_j f.$$

The function $f^{\#,\tau}$ is potentially singular at τ and of compact support; the complementary function $f^{\flat,\tau}$ is zero at τ and also in a vicinity of width $\asymp 2^{-J}$.

To illustrate these ideas, we present in Figure 6.3 the corresponding functions $f^{\#,\tau}$ for object *Bumps*, where τ runs through the points t_i underlying the construction of the *Bumps* object.

We may think of the functions $f^{\#,\tau}$ as representing the "part of" f "explained by" any singularity at τ.

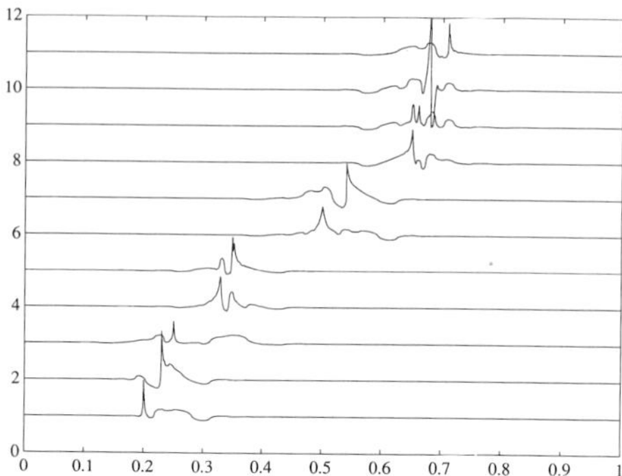

Figure 6.3. Segmentation-Compenents of Object Bumps.

6.2 Segmentation pursuit

The ability to identify the part of a function "explained by" a singularity at a fixed point suggests a sort of iterative cleaning operation, analogous to Friedman and Stuetzle's Projection Pursuit in statistics and Mallat and Zhong's Matching Pursuit in signal analysis.

Step 1. Set $r := f$ and $i := 1$.

Step 2. Identify a point of likely segmentation via

$$t_i := \arg\min_t \mathcal{E}(W^t r).$$

Step 3. Calculate $f^{\sharp,t_i}$, the component of r "explained by" the segmentation.

Step 4. Remove this component

$$r := r - f^{\sharp,t_i}.$$

Step 5. Unless satisfied, set $i := i + 1$ and go to Step 2.

We call this "segmentation pursuit;" it is in formal analogy with "projection pursuit" [27] and "matching pursuit" [31].

Figure 6.4 gives the result of applying segmentation pursuit to object *Bumps*; shown are the functions $f^{\#,t_i}$ extracted in the first ten iterations of the procedure. Several issues deserve comment. First, that while some of the functions extracted are indeed sharp peaks, corresponding to sharp peaks in the original object, some of the extracted objects are rather "dull" and not exactly what one expects. The reason is that the points of segmentation do not always correspond to actual singularities; the extracted components in those cases are smooth rather than peaked. Second, the method appears, in general, to leave "peaky" residuals even when peaks are being successfully extracted; this is caused by the influence of peak shapes which differ from piecewise polynomial. Figure 6.5 displays the residual vector at several stages.

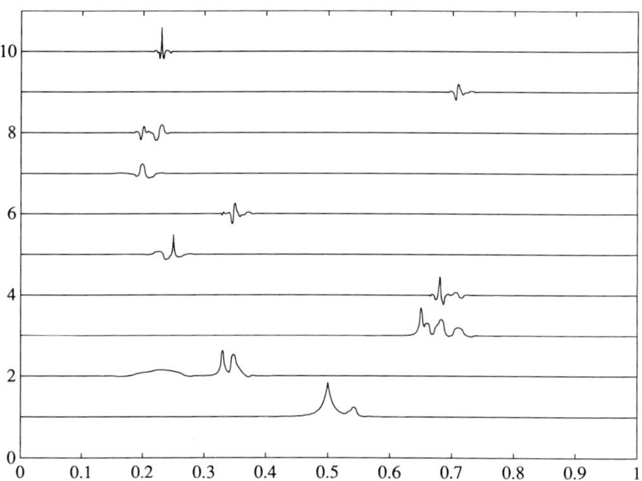

Figure 6.4. First Ten Segmentation-Components Extracted by Pursuit.

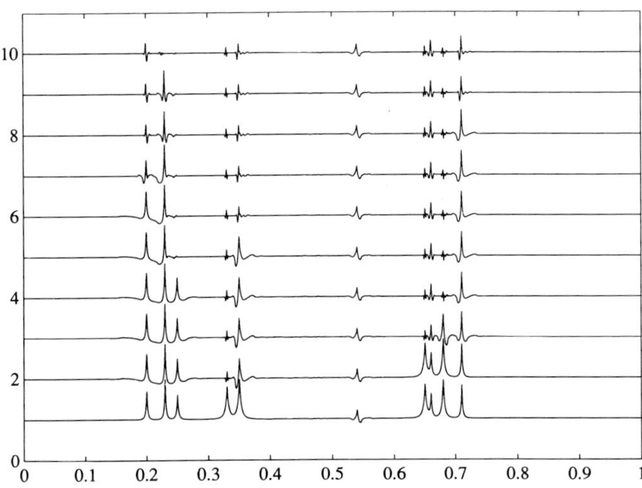

Figure 6.5. First Ten Residual Components Extracted by Pursuit.

Based on experience with projection pursuit [27], a variety of simple modifications to the above should also be useful, and should address the two objectionable features just seen. One example is "backfitting," where, after adding a new term into the equation, we cycle through all previous terms; on each cycle we add back to r the term under consideration, and then we locate and extract a singular component all over.

Step 1. Set $r := f$ and $i := 1$.
Identify a point of likely segmentation via

$$t_i := \arg\min_t \mathcal{E}(W^t r).$$

Step 2. Calculate $f^{\sharp,t_i}$, the component of r "explained by" the segmentation.

Step 3. Remove this component

$$r := r - f^{\sharp,t_i}.$$

Step 4. For $j := 1$ to $i-1$, set $r := r + f^{\sharp,t_j}$, and perform the analog of Steps 2 and 3, extracting an "improved" $f^{\sharp,t_j}$.

Step 5. Unless satisfied, set $i := i+1$ and go to Step 2.

The idea is that we can, thereby, adjust the locations of segmentations to allow for improved segmentation after neighboring peaks are unmasked.

Efficient implementation of this idea requires the implementation of an efficient updating scheme for the all-segmentations algorithm. When extracting a sharp-component of f, the $\nu_{j,l}$ coefficients of the new residual r differ from the previous coefficients only in order $O(\log(n))$ positions. If we develop a method to efficiently update just those coefficients, the intrinsic computational complexity of this iterative scheme can be made quite small. We have not yet implemented this scheme.

6.3 A vacuum cleaner

The computational toolkit we have assembled consists of several dozen procedures, expressed as MATLAB m-files.

The principal application we have in mind for this toolkit at the present time is the one indicated in the introduction: improving on ordinary wavelet de-noising by adapting to the presence of a few singularities.

1. Apply several iterations of Segmentation pursuit to remove edges.
2. Apply standard wavelet de-noising to the edge-less object.
3. Apply segmented wavelet de-noising to each segmentation-component.
4. Superpose the results.

Armed with a fast all-segmentations algorithm and a fast updating method for extracting sharp-components, such ideas are practical and bear further study.

§7 Discussion

This article describes computational experiments applying the persuasive best-basis heuristic to the segmentation problem. It does not try to prove anything, but to develop the implications of the wavelet/best-basis formalism. It also represents a report on the development of a considerable body of software, which the reader may wish to obtain for experimental purposes.

The author sees two principal issues raised by this work.

1. There are many existing edge-detection schemes. A number of these are based on wavelets themselves [32,33]. The method we have discussed here is different, in that it tries to be true to the *internal* logic of the wavelets/best basis paradigm.

2. Fidelity to the internal logic of wavelets puts us in a kind of straight jacket. We are limited in this paper to certain methods and attitudes; this makes some questions, like how to handle fully two-d segmentation problems, seem very difficult.

The author also wishes he had the time to repeat these experiments using more stable refinement schemes. This would be his first priority for further work. For a discussion of experiments in 2-d based on these ideas, see the Technical report on which this article is based.

Acknowledgments. Supported by NSF DMS 92-09130 and by ONR N00014-92-0066 (Statistical Sciences, Inc.). The author's interest in this topic was stimulated by good-natured complaints against wavelet de-noising lodged by Jeff Scargle and Stéphane Mallat. The author thanks Iain Johnstone for suggestions leading to the Minimum Energy Segmentation edge detector, and Wim Sweldens for informing us of the priority of Deng, Jawerth, Peters, and Sweldens in introducing fast pixel-level segmentations.

References

1. Andersson, L., N. Hall, B. Jawerth and G. Peters, Wavelets on closed subsets of the real line. in *Recent Advances in Wavelet Analysis*, L. L. Schumaker and G. Webb (eds.), Academic Press, 1993, 1–62.
2. Antonini, M., M. Barlaud, P. Mathieu, and I. Daubechies, Image coding using wavelet transforms, *IEEE Proc. ASSP*, 1991, to appear.
3. Beylkin, G., R. Coifman, and V. Rokhlin, Fast wavelet transforms and numerical algorithms, *Comm. Pure Appl. Math.* **43** (1991), 141–183.
4. Cavaretta, A. S., W. Dahmen, and C. A. Micchelli, Stationary subdivision, *Mem. Amer. Math. Soc.* **453** (1991).
5. Chui, C. K., *An Introduction to Wavelets*, Academic Press, Boston, 1992.
6. Chui, C. and E. Quak, Wavelets on a bounded interval, *Numerical Methods of Approximation Theory*, D. Braess and L. L. Schumaker (eds.), Birkhauser Verlag, Basel, 1992, 1–24.
7. Cohen, A., Thèse, Université Paris IX - Dauphine, 1990.

8. Cohen, A., I. Daubechies, and J. C. Feauveau, Biorthogonal bases of compactly supported wavelets, *Comm. Pure Appl. Math.* **45** (1990), 485–560.
9. Cohen, A., I. Daubechies, B. Jawerth, and P. Vial, Multiresolution analysis, wavelets, and fast algorithms on an interval, *Comptes Rendus Acad. Sci. Paris* (A), 1992, to appear.
10. Coifman, R. R. and M. V. Wickerhauser, Entropy-based algorithms for best-basis selection, *IEEE Trans. Inform. Theory* **38** (1992), 713–718.
11. Daubechies, I., *Ten Lectures on Wavelets*, Philadelphia, SIAM, 1992.
12. Daubechies, I., Two recent results: Wavelet bases for the interval and biorthogonal wavelets diagonalizing the derivative operator, in *Recent Advances in Wavelet Analysis*, L. L. Schumaker and G. Webb (eds.), Academic Press, 1993, 237–259.
13. Deslauriers, G. and S. Dubuc, Symmetric iterative interpolation processes, *Constr. Approx.* **5** (1989), 49–68.
14. Deng, B., B. Jawerth, G. Peters and W. Sweldens, Wavelet probing for compression-based segmentation, in *Proc. SPIE Symp. Math. Imaging: Wavelet Applications in Signal and Image Processing*, Proceedings of SPIE Conference, San Diego, July 1993.
15. DeVore, R. A., B. Jawerth and B. J. Lucier, Image compression through wavelet transform coding, *IEEE Trans. Inform. Theory.* **38** (2) (1992), 719–746.
16. DeVore, R. A. and B. J. Lucier, Fast wavelet techniques for near-optimal image processing, *Proc. IEEE Mil. Commun. Conf.*, IEEE Communications Soc., NY, October 1992.
17. Donoho, D. L., De-noising via soft-thresholding, *IEEE Trans. Inform. Theory*, 1992, to appear.
18. Donoho, D. L., Wavelet Shrinkage and W.V.D.: a ten-minute tour, in *Progress in Wavelet Analysis and Applications*, Y. Meyer and S. Roques (eds.), Editions Frontières: Gif-sur-Yvette, 1993, 109–128.
19. Donoho, D. L., Smooth wavelet decompositions with blocky coefficient kernels, in *Recent Advances in Wavelet Analysis*, L. L. Schumaker and G. Webb, (eds.), Academic Press, Boston, 1993.
20. Donoho, D. L., Unconditional bases are optimal bases for data compression and for statistical estimation, *Appl. and Comp. Harmonic Anal.* **1** (1993), 100–115.
21. Donoho, D. L. amd I. M. Johnstone, Ideal spatial adaptation via wavelet shrinkage, *Biometrika*, 1992, to appear.
22. Donoho, D. L. amd I. M. Johnstone, Neo-classical minimax theorems, thresholding, and adaptation, Technical Report, Department of Statistics, Stanford University, 1992.
23. Donoho, D. L. and I. M. Johnstone, Minimax estimation by wavelet shrinkage, *Ann. Stat.*, 1992, to appear.
24. Donoho, D. L. and I. M. Johnstone, Adapting to unknown smoothness via wavelet shrinkage, *J. Am. Stat. Assn.*, 1993, to appear.

25. Donoho, D. L., I. M. Johnstone, Keryacharian and D. Picard, Wavelet shrinkage: asymptopia? *J. Roy. Stat. Soc.* ser (**B**), 1993, to appear.
26. Dubuc, S., Interpolation through an iterative scheme, *J. Math. Anal. and Appl.* **114** (1986) 185–204.
27. Friedman, J. H. and W. Stuetzle, Projection pursuit regression, *J. Am. Stat. Assn.* **76** (1981) 817–823.
28. Lucier, B. J., Wavelets and image compression, in *Mathematical Methods in CAGD and Image Processing*, T. Lyche and L. L. Schumaker (eds.), Academic Press, Boston, 1992, 1–10.
29. Mallat, S. and J. Froment, Second-generation compact coding from wavelet edges, in *Wavelets: A Tutorial in Theory and Applications*, C.K. Chui (ed.), Academic Press, Boston, 1992.
30. Mallat, S. and S. Zhong, Wavelet transform maxima and multiscale edges, in *Wavelets and Their Applications*, G. Beylkin, R. R. Coifman, I. Daubechies, Y. Meyer, L. Raphael, and M. B. Ruskai (eds.), Academic Press, Boston, 1992.
31. Mallat, S. and S. Zhong, Matching pursuits with time-frequency dictionaries, in *IEEE Trans. Signal Proc.* **41** (1993), 3397–3415.
32. Moreau, E., G. Charbassier, and J. C. Lassau, Détection et localisation d'un obstacle grâce à l'utilisation d'une transformée en ondelettes, in *Progress in Wavelet Analysis and Applications*, Y. Meyer and S. Roques (eds.), Editions Frontières: Paris, 1993.
33. Serir, A. and B. Sansal, Détecteurs de contours optimaux basés sur la transformation en ondelettes, in *Progress in Wavelet Analysis and Applications*, Y. Meyer and S. Roques (eds.), Editions Frontières, Paris, 1993.

David L. Donoho
Department of Statistics
Stanford University
Stanford, CA 94305
donoho@playfair.stanford.edu

Adaptive Time-Frequency Approximations with Matching Pursuits

Geoffrey Davis, Stéphane Mallat, and Zhifeng Zhang

Abstract. Computing the optimal expansion of a signal in a redundant dictionary of waveforms is an NP-complete problem. We introduce a greedy algorithm called a matching pursuit which computes a sub-optimal expansion. The dictionary waveforms which best match a signal's structures are chosen iteratively. An orthogonalized version of the matching pursuit is also developed. Matching pursuits are general procedures for computing adaptive signal representations. With a dictionary of Gabor functions, a matching pursuit defines an adaptive time-frequency transform. We derive a signal energy distribution in the time-frequency plane which does not contain interference terms, unlike the Wigner and Cohen class distributions. A matching pursuit is a chaotic map whose asymptotic properties are studied. We describe an algorithm which isolates the coherent structures of a signal and show an application to pattern extraction from noisy signals.

§1 Introduction

In this paper we focus on the problem of approximating functions using linear combinations of a small number waveforms. To obtain a compact expansion of a function which contains complex structures, we must adapt our expansion to the various components of the function. Examples of such a linear expansions include triangular mesh approximations to surfaces, used in scientific computing applications. These mesh expansions can be adapted to obtain low approximation error with a small number of triangles by varying the size and shape of the triangles according to the approximated surface's local properties. Adaptive linear expansions can be used to extract information from signals. We obtain an adaptive time-frequency decomposition of a signal by expanding the signal into a sum of waveforms whose localizations in time and frequency match those of the different signal structures. Such adaptive time-frequency representations are important in signal processing applications such as speech analysis.

Wavelets: Theory, Algorithms, and Applications
Charles K. Chui, Laura Montefusco, and Luigia Puccio (eds.), pp. 271–293.

ISBN 0-12-174575-9

The waveforms which we use for our expansions are drawn from a large and redundant collection, called a dictionary. In section 2 we examine the computational complexity of optimally approximating a function with a linear combination of vectors from a dictionary. We prove that in a finite dimensional space, computing the optimal solution is an NP-complete problem, which motivates the use of sub-optimal greedy algorithms. We introduce the matching pursuit algorithm, a greedy algorithm which computes function expansions by iteratively selecting dictionary vectors which best correlate to signal structures. An orthogonal version of the matching pursuit algorithm is also described and compared with the non-orthogonal algorithm. An application of matching pursuits to finding adaptive time-frequency decompositions is explained in section 6 A signal is decomposed into waveforms selected from a dictionary of time-frequency atoms, a collection of dilations, translations, and modulations of a single window function. We construct a time-frequency energy distribution by summing the Wigner distributions of the selected time-frequency atoms. Unlike the Wigner distribution or Cohen's class distributions, this energy distribution does not include interference terms and thus provides a clear picture of the time-frequency plane.

Matching pursuits have chaotic properties which are analyzed for particular dictionaries. As the number of iterations of a matching pursuit increases, the approximation error converges to the realization of a noise process whose energy is uniformly spread across all the dictionary vectors. From this asymptotic behavior, we derive an algorithm that selects coherent signal structures from noisy signals. We describe an application to noise removal in speech recordings.

§2 Optimal adaptive approximations in dictionaries

We expand functions from a Hilbert space $\mathbf{H}$ into linear combinations of vectors from a large collection $\mathcal{D} = (g_\gamma(t))_{\gamma\in\Gamma}$, with $\|g_\gamma\| = 1$, called a dictionary. The dictionary is constructed so that linear combinations of dictionary vectors are dense in $\mathbf{H}$. The smallest possible dictionary is a basis of $\mathbf{H}$, but in practice a dictionary is a very redundant set. The redundancy gives an increased degree of freedom in constructing function expansions, and this freedom is used in order to obtain improved convergence properties of the expansions.

For signal processing applications, we study the properties of dictionaries composed of waveforms that are well concentrated both in time and frequency. Our signal space is $\mathbf{L}^2(\mathbf{R})$ and we construct such a dictionary by scaling, translating and modulating a single window function $g(t) \in \mathbf{L}^2(\mathbf{R})$. We suppose that $g(t)$ is real and centered at 0. We also impose that $\|g\| = 1$, that the integral of $g(t)$ is non-zero and $g(0) \neq 0$. For any scale $s > 0$, frequency modulation ξ and translation u, we denote $\gamma = (s, u, \xi)$ and define

$$g_\gamma(t) = \frac{1}{\sqrt{s}} g\left(\frac{t-u}{s}\right) e^{i\xi t}. \tag{1}$$

The index γ is an element of the set $\mathbf{\Gamma} = \mathbf{R}^+ \times \mathbf{R^2}$. The factor $\frac{1}{\sqrt{s}}$ normalizes to 1 the norm of $g_\gamma(t)$. The function $g_\gamma(t)$ is centered at the abscissa u and its energy is concentrated in a neighborhood of u, whose size is proportional to s. Let $\hat{g}(\omega)$ be the Fourier transform of $g(t)$. Equation (1) yields

$$\hat{g}_\gamma(\omega) = \sqrt{s}\hat{g}(s(\omega - \xi))e^{-i(\omega-\xi)u}. \tag{2}$$

Since $|\hat{g}(\omega)|$ is even, $|\hat{g}_\gamma(\omega)|$ is centered at the frequency $\omega = \xi$. Its energy is concentrated in a neighborhood of ξ, whose size is proportional to $1/s$. The dictionary of time-frequency atoms $\mathcal{D} = (g_\gamma(t))_{\gamma\in\mathbf{\Gamma}}$ is a very redundant set of functions in $\mathbf{L}^2(\mathbf{R})$ that includes window Fourier frames and wavelet frames [3]. Instead of expanding the signal on such a frame that is chosen a priori, we want to choose within $\mathcal{D}$ the time-frequency atoms that are best adapted to expand a given $f(t)$. A first issue is to define a notion of "optimal" approximation within a given dictionary.

Definition 1. *Let $\epsilon > 0$. An ϵ-approximation of $f \in \mathbf{H}$ is a linear expansion of dictionary vectors*

$$\tilde{f} = \sum_{0\leq n<M} \beta_n g_{\gamma_n},$$

for which

$$\|\tilde{f} - f\| < \epsilon. \tag{3}$$

$\tilde{f}$ is an optimal ϵ-approximation if M is the minimum integer for which (3) is satisfied.

To determine the computational complexity of obtaining these optimal solutions, we consider a space $\mathbf{H}$ of finite dimension and dictionaries $\mathcal{D}$ of size $O(N^k)$ for some $k > 0$. We encode all quantities with $O(N^p)$ bits for some positive p. We say that an algorithm solves the optimal ϵ-approximation problem if, for any given $f \in \mathbf{H}$, any $\mathcal{D}$ of size $O(N^k)$, and any $\epsilon > 0$, we can find an optimal ϵ-approximation for f. The following result shows that this problem is computationally intractable.

Theorem 1. *The optimal ϵ-approximation problem is NP-hard.*

In large dimensional spaces, it is therefore not feasible to compute optimal ϵ-approximations. Theorem 1 is proved by showing that any instance of the Exact Cover by 3-Sets problem [6] can be transformed in polynomial time into an optimal ϵ-approximation problem. Thus, an algorithm which solves the ϵ-approximation problem can solve the NP-complete Exact Cover by 3-Sets problem. The intractability of the approximation problem is due to the coupling between terms in the function expansions when the dictionary elements are not orthogonal. We map the overlapping sets in the Exact Cover by 3-Sets problem to a set of coupled, non-orthogonal dictionary vectors. When the dictionary vectors are orthogonal, this coupling-induced complexity vanishes. We can solve the problem in $O(N \log N)$ by sorting the inner products

$\{|\langle f, g_\gamma \rangle|^2\}_{\gamma \in \Gamma}$ and finding the minimum M for which the sum of the M largest terms satisfies $\|f\|^2 - \sum_{i=1}^{M} |\langle f, g_{\gamma_i} \rangle|^2 < \epsilon$.

In addition to the computational complexity, an important issue is that this optimization criteria can lead to numerically unstable expansions. We can construct examples where the l^2 norm of the expansion coefficients is arbitrarily larger than $||f||^2$, *i.e.*,

$$\sum_{0 \leq n < M} |\beta_n|^2 >> ||f||^2. \tag{4}$$

This instability can be avoided by imposing the constraint that the approximation must satisfy

$$\sum_{0 \leq n < M} |\beta_n|^2 \leq K||f||^2, \tag{5}$$

for some fixed $K \geq 1$. One can prove that there always exist such ϵ-approximations.

When ϵ is modified, the dictionary vectors that appear in the optimal approximation can change completely, which prevents us from computing optimal approximations using progressive refinement. This instability of the optimal approximations, together with the computational intractibility of computing them, leads us to use greedy sub-optimal algorithms that progressively refine the functional approximation by choosing appropriate dictionary vectors.

§3 Matching pursuit

Let $f \in \mathbf{H}$. We want to compute a linear expansion of f over a set of vectors selected from $\mathcal{D}$ which best matches the inner structures of f. A matching pursuit is a greedy algorithm which successively approximates f with orthogonal projections onto elements of $\mathcal{D}$. Let $g_{\gamma_0} \in \mathcal{D}$. The vector f can be decomposed into

$$f = \langle f, g_{\gamma_0} \rangle g_{\gamma_0} + Rf, \tag{6}$$

where Rf is the residual vector after approximating f in the direction of g_{γ_0}. Clearly g_{γ_0} is orthogonal to Rf, hence

$$\|f\|^2 = |\langle f, g_{\gamma_0} \rangle|^2 + \|Rf\|^2. \tag{7}$$

To minimize $\|Rf\|$, we must choose $g_{\gamma_0} \in \mathcal{D}$ such that $|\langle f, g_{\gamma_0} \rangle|$ is maximal. In some cases, it is only possible to find a vector g_{γ_0} that is close to the maximum in the sense that

$$|\langle f, g_{\gamma_0} \rangle| \geq \alpha \sup_{\gamma \in \Gamma} |\langle f, g_\gamma \rangle|, \tag{8}$$

where $\alpha \in (0, 1]$ is an optimality factor.

We sub-decompose the residue Rf by projecting it onto the vector of $\mathcal{D}$ that best matches Rf, as was done for f. This projection of Rf generates a second residue, $R^2 f$, which we again decompose to obtain a third residue, and so on.

Let us explain by induction how the matching pursuit is carried further. Let $R^0 f = f$. We suppose that we have computed the nth order residue $R^n f$, for $n \geq 0$. We choose with the choice function C an element $g_{\gamma_n} \in \mathcal{D}$ which closely matches the residue $R^n f$

$$|\langle R^n f, g_{\gamma_n}\rangle| \geq \alpha \sup_{\gamma \in \Gamma} |\langle R^n f, g_\gamma \rangle|. \tag{9}$$

The residue $R^n f$ is sub-decomposed into

$$R^n f = \langle R^n f, g_{\gamma_n}\rangle g_{\gamma_n} + R^{n+1} f, \tag{10}$$

which defines the residue at the order n+1. Since $R^{n+1} f$ is orthogonal to g_{γ_n}, we have

$$\|R^n f\|^2 = |\langle R^n f, g_{\gamma_n}\rangle|^2 + \|R^{n+1} f\|^2. \tag{11}$$

Let us carry this decomposition up to the order m. We decompose f into the telescoping sum

$$f = \sum_{n=0}^{m-1} (R^n f - R^{n+1} f) + R^m f. \tag{12}$$

Equation (10) yields

$$f = \sum_{n=0}^{m-1} \langle R^n f, g_{\gamma_n}\rangle g_{\gamma_n} + R^m f. \tag{13}$$

Similarly, we write $\|f\|^2$ as a telescoping sum

$$\|f\|^2 = \sum_{n=0}^{m-1} (\|R^n f\|^2 - \|R^{n+1} f\|^2) + \|R^m f\|^2 \tag{14}$$

which we combine with Equation (11) to obtain an energy conservation equation

$$\|f\|^2 = \sum_{n=0}^{m-1} |\langle R^n f, g_{\gamma_n}\rangle|^2 + \|R^m f\|^2. \tag{15}$$

Thus, the original vector f is decomposed into a sum of dictionary elements which are chosen to best match its residues. Although this decomposition is non-linear, we maintain an energy conservation as though it were a linear, orthogonal decomposition. An important issue is to understand the behavior of the residue $R^m f$ when m increases. By transposing a result proved by Jones [9] for projection pursuit algorithms [5], one can prove [10] that the matching pursuit algorithm converges, even in infinite dimensional spaces.

Theorem 2. *Let $f \in \mathbf{H}$. The residue $R^m f$ defined by the induction Equation (10) satisfies*

$$\lim_{m \to +\infty} \|R^m f\| = 0. \tag{16}$$

Hence

$$f = \sum_{n=0}^{+\infty} \langle R^n f, g_{\gamma_n} \rangle g_{\gamma_n}, \tag{17}$$

and

$$\|f\|^2 = \sum_{n=0}^{+\infty} |\langle R^n f, g_{\gamma_n} \rangle|^2. \tag{18}$$

When $\mathbf{H}$ is of finite dimension, $||R^m f||$ decays exponentially to zero.

This theorem proves that any vector f is characterized by the double sequence $(\langle R^n f, g_{\gamma_n} \rangle, \gamma_n)_{n \in \mathbf{N}}$, called a structure book, which specifies the expansion coefficients and the index of each chosen vector within the dictionary.

§4 Back-projection and orthogonal pursuit

After m iterations, a matching pursuit decomposes a signal f into

$$f = \sum_{n=0}^{m-1} \langle R^n f, g_{\gamma_n} \rangle g_{\gamma_n} + R^m f. \tag{19}$$

If we stop the algorithm at this stage and only record the partial structure book $(\langle R^n f, g_{\gamma_n} \rangle, \gamma_n)_{0 \leq n < m}$, the summation of Equation (19) recovers an approximation of f with error $R^m f$. However, this sum is not the linear expansion of the vectors $(g_{\gamma_n})_{0 \leq n < m}$ which best approximates f. Let $\mathbf{V}_m$ be the space generated by $(g_{\gamma_n})_{0 \leq n < m}$ and $\mathbf{P}_{\mathbf{V}_m}$ be the orthogonal projector onto $\mathbf{V}_m$. For any $f \in \mathbf{H}$, $\mathbf{P}_{\mathbf{V}_m} f$ is the closest vector to f that can be written as linear expansion of the m vectors $(g_{\gamma_n})_{0 \leq n < m}$. We derive from Equation (19) that

$$\mathbf{P}_{\mathbf{V}_m} f = \sum_{n=0}^{m-1} \langle R^n f, g_{\gamma_n} \rangle g_{\gamma_n} + \mathbf{P}_{\mathbf{V}_m} R^m f. \tag{20}$$

If the family of vectors $(g_{\gamma_n})_{0 \leq n < m}$ is not orthogonal, which is generally the case, then $\mathbf{P}_{\mathbf{V}_m} R^m f \neq 0$. The computation of

$$\mathbf{P}_{\mathbf{V}_m} R^m f = \sum_{n=0}^{m-1} x_n g_{\gamma_n}, \tag{21}$$

is called a back-projection. Instead of storing the inner products $\langle R^n f, g_{\gamma_n} \rangle$ in the structure book, we store $\langle R^n f, g_{\gamma_n} \rangle + x_n$ in order to recover $\mathbf{P}_{\mathbf{V}_m} f$ with Equation (20). In this case, the approximation error

$$\mathbf{P}_{\mathbf{W}_m} f = f - \mathbf{P}_{\mathbf{V}_m} f \tag{22}$$

is the orthogonal projection of f on the space $\mathbf{W}_m$, the orthogonal complement of $\mathbf{V}_m$ in $\mathbf{H}$. The calculation of the coefficients $(x_n)_{0\leq n<m}$ requires that we solve the following linear system. For any g_{γ_k}, $0 \leq k < m$,

$$\langle \mathbf{P}_{\mathbf{V}_m} R^m f, g_{\gamma_k}\rangle = \langle R^m f, g_{\gamma_k}\rangle = \sum_{n=0}^{m-1} x_n \langle g_{\gamma_n}, g_{\gamma_k}\rangle. \tag{23}$$

Let us denote $X = (x_n)_{0\leq n<m}$ and $Y = (\langle R^m f, g_{\gamma_k}\rangle)_{0\leq k<m}$. Let $G = (\langle g_{\gamma_n}, g_{\gamma_k}\rangle)_{0\leq k<m, 0\leq n<m}$ be the Gram matrix of the family of selected vectors. The linear system of Equations (23) can be written $Y = GX$. A solution of this system is computed efficiently with a conjugate gradient algorithm [10]. If $\mathbf{H}$ is of finite dimension N, there are many classes of dictionaries for which any collection of N distinct dictionary vectors is a basis of $\mathbf{H}$. This is the case for the Gabor dictionary used for time-frequency decompositions. Hence, after selecting N different vectors with a matching pursuit, the back-projection reduces to 0 the remaining residue.

Instead of recovering the orthogonal projection $\mathbf{P}_{\mathbf{V}_m} f$ at the end of the matching pursuit, one can modify the pursuit algorithm by computing this orthogonal projection when selecting each new vector of the dictionary. It is more efficient to orthogonalize the family of selected vectors with a Gram-Schmidt algorithm than to perform a back projection. This type of algorithm was first introduced for control applications [1] and also studied independently from this work by Pati et al. [11]. It has the advantage of providing better approximations than the matching pursuit algorithm, but it requires much more computation and can introduce numerical instabilities into the expansions. We describe by induction this orthogonal pursuit.

For $n = 0$, we set $R^0 f = f$. Like in a matching pursuit, we define a supremum factor α, with $0 < \alpha \leq 1$, and choose $g_{\gamma_0} \in \mathcal{D}$ which satisfies

$$|\langle f, g_{\gamma_0}\rangle| \geq \alpha \sup_{\gamma\in\mathbf{\Gamma}} |\langle f, g_\gamma\rangle| \ . \tag{24}$$

The space $\mathbf{V}_1$ is generated by the single vector g_{γ_0}. The first vector u_0 of the Gram-Schmidt basis is g_{γ_0}. The next residue is defined by

$$Rf = f - \mathbf{P}_{\mathbf{V}_1} f = f - \langle f, g_{\gamma_0}\rangle g_{\gamma_0} \ . \tag{25}$$

Let us explain by induction how to compute the orthogonal residue $R^{n+1}f$ from $R^n f$. We suppose that we have already selected n vectors $(g_{\gamma_p})_{0\leq p<n}$ that are linearly independent and that we computed the corresponding Gram-Schmidt orthogonal basis $(u_p)_{0\leq p<n}$. Both $(g_{\gamma_p})_{0\leq p<n}$ and $(u_p)_{0\leq p<n}$ are bases of the space $\mathbf{V}_n$ and

$$R^n f = f - \mathbf{P}_{\mathbf{V}_n} f \ . \tag{26}$$

We choose a vector $g_{\gamma_n} \in \mathcal{D}$ which satisfies

$$|\langle R^n f, g_{\gamma_n}\rangle| \geq \alpha \sup_{\gamma\in\mathbf{\Gamma}} |\langle R^n f, g_\gamma\rangle|. \tag{27}$$

If $\langle R^n f, g_{\gamma_n}\rangle \neq 0$, then the vector g_{γ_n} cannot belong to the space $\mathbf{V}_n$ since $R^n f$ is orthogonal to $\mathbf{V}_n$. Hence, the vectors $(g_{\gamma_p})_{0\leq p\leq n}$ are linearly independent. The next vector u_n of the Gram-Schmidt basis is obtained by subtracting from g_{γ_n} its projection on the space $\mathbf{V}_n$

$$u_n = g_{\gamma_n} - \sum_{p=0}^{n-1} \frac{\langle g_{\gamma_n}, u_p\rangle}{||u_p||^2} u_p \ . \tag{28}$$

The family $(u_p)_{0\leq p\leq n}$ is an orthogonal basis of $\mathbf{V}_{n+1}$. The residue $R^{n+1}f$ is defined by

$$R^{n+1}f = f - \mathbf{P}_{\mathbf{V}_{n+1}} f = f - \sum_{p=0}^{n} \frac{\langle f, u_p\rangle}{||u_p||^2} u_p \ . \tag{29}$$

This can also be rewritten

$$R^{n+1}f = R^n f - \frac{\langle R^n f, u_n\rangle}{||u_n||^2} u_n \ . \tag{30}$$

Since $R^n f$ is orthogonal to the vectors $(g_{\gamma_p})_{0\leq p<n}$, Equation (28) implies that $\langle R^n f, u_n\rangle = \langle R^n f, g_{\gamma_n}\rangle$ and, thus,

$$R^{n+1}f = R^n f - \frac{\langle R^n f, g_{\gamma_n}\rangle}{||u_n||^2} u_n \ . \tag{31}$$

This equation is similar to the residue updating Equation (10) of a matching pursuit, but instead of subtracting a vector in the direction of g_{γ_n}, we subtract a component in a direction orthogonal to all vectors previously selected. Since $R^{n+1}f$ and u_n are orthogonal,

$$||R^{n+1}f||^2 = ||R^n f||^2 - \frac{|\langle R^n f, g_{\gamma_n}\rangle|^2}{||u_n||^2} \ . \tag{32}$$

An orthogonal pursuit guarantees that the selected vectors $(g_{\gamma_n})_{0\leq n\leq m}$ are linearly independent, and computes the best possible approximation of f from these vectors. Since $R^0 f = f$, we derive from Equations (31) and (32) that for any $m > 0$

$$f = \sum_{0\leq n<m} \frac{\langle R^n f, g_{\gamma_n}\rangle}{||u_n||^2} u_n + R^m f \ , \tag{33}$$

and

$$\|f\|^2 = \sum_{0\leq n<m} \frac{|\langle R^n f, g_{\gamma_n}\rangle|^2}{||u_n||^2} + \|R^m f\|^2 \ . \tag{34}$$

The derivations are similar to those for Equations (13) and (14). The next theorem is similar to Theorem 2 and guarantees the convergence of the orthogonal pursuit [4].

Theorem 3. *Let $f \in \mathbf{H}$. Let N be the dimension of $\mathbf{H}$ (N may be infinite). The orthogonal matching pursuit converges in $M \leq N$ iterations (M may be infinite if N is infinite). The residue $\mathbf{R^n f}$ defined inductively by Equation (31) satisfies*

$$\lim_{n \to M} \|R^n f\| = 0, \tag{35}$$

$$f = \sum_{0 \leq n < M} \frac{\langle R^n f, g_{\gamma_n} \rangle}{||u_n||^2} u_n \ , \tag{36}$$

and

$$\|f\|^2 = \sum_{0 \leq n < M} \frac{|\langle R^n f, g_{\gamma_n} \rangle|^2}{||u_n||^2} \ . \tag{37}$$

If $\mathbf{H}$ is of finite dimension, the orthogonal pursuit converges within a finite number of iterations.

Our primary objective is not to expand f over $(u_n)_{0 \leq n < M}$ but rather over $(g_{\gamma_n})_{0 \leq n < M}$. We want coefficients $(\beta_n)_{0 \leq n < M}$ such that

$$f = \sum_{0 \leq n < M} \beta_n g_{\gamma_n}. \tag{38}$$

Since $u_n \in \mathbf{V}_n$ and $(g_{\gamma_p})_{0 \leq p \leq n}$ is a basis of $\mathbf{V}_n$, we can decompose u_n into

$$u_n = \sum_{p=0}^{n} b_{p,n} g_{\gamma_p}. \tag{39}$$

The coefficients $b_{p,n}$ can be calculated while computing the orthogonal matching pursuit, as explained in Section 5. Inserting Equation (39) into Equation (36) yields

$$f = \sum_{0 \leq n < M} \frac{\langle R^n f, g_{\gamma_n} \rangle}{||u_n||^2} \sum_{p=0}^{n} b_{p,n} g_{\gamma_p} \ . \tag{40}$$

One could naively try to rearrange the terms of this double summation to obtain

$$f = \sum_{0 \leq p < M} g_{\gamma_p} \sum_{p \leq n < M} b_{p,n} \frac{\langle R^n f, g_{\gamma_n} \rangle}{||u_n||^2} \ . \tag{41}$$

However, when $M = +\infty$, the infinite sum over n that defines each coefficient β_p may not converge. Such a situation arises when the family $(g_{\gamma_n})_{0 \leq n < M}$ is not a Riesz basis of the closed space $\mathbf{V}_M$ that it generates. For such a case, we cannot obtain an expansion of the form of Equation (38) from the orthogonal matching pursuit. If the signal space $\mathbf{H}$ has a finite dimension N, then M is finite, so we can always invert the two sums of Equation (40) to obtain Equation (41). The basis $(g_{\gamma_n})_{0 \leq n < M}$ may, however, be very badly conditioned, in which case we can have numerical instabilities:

$$\sum_{0 \leq n < M} |\beta_n|^2 >> ||f||^2 \ . \tag{42}$$

The residues of orthogonal matching pursuits in general decrease faster than the residues of non-orthogonal matching pursuits. However, this orthogonal procedure can yield unstable expansions and requires much more numerical computation because of the Gram-Schmidt orthogonalization. The implementation and computational complexity of these two algorithms is compared in the next section.

§5 Numerical implementations of matching pursuits

We describe fast implementations of non-orthogonal and orthogonal matching pursuits in finite dimensional spaces, and compare their performance. Software implementing matching pursuits for time-frequency dictionaries is available through anonymous ftp at the address cs.nyu.edu . Instructions are in the file README of the directory /pub/wave/software.

At stage n of the pursuit, we suppose that the inner products $(\langle R^n f, g_\gamma\rangle)_{\gamma\in\mathbf{\Gamma}}$ have already been computed. We choose the optimality factor $\alpha = 1$, so that the first step is to find g_{γ_n} such that

$$|\langle R^n f, g_{\gamma_n}\rangle| = \sup_{\gamma\in\mathbf{\Gamma}} |\langle R^n f, g_\gamma\rangle|.$$

If all the inner products are stored in an open hash table, finding this supremum requires on average $O(1)$ operations. Otherwise, one needs to search across the whole set of inner products.

For a non-orthogonal matching pursuit, once the vector g_{γ_n} is selected, we compute the inner product of the new residue $R^{n+1}f$ with all $g_\gamma \in \mathcal{D}$, with an updating formula derived from Equation (10)

$$\langle R^{n+1} f, g_\gamma\rangle = \langle R^n f, g_\gamma\rangle - \langle R^n f, g_{\gamma_n}\rangle \ \langle g_{\gamma_n}, g_\gamma\rangle. \tag{43}$$

Since we previously stored $\langle R^n f, g_\gamma\rangle$ and $\langle R^n f, g_{\gamma_n}\rangle$, this update requires only the computation of $\langle g_{\gamma_n}, g_\gamma\rangle$. Dictionaries are generally built so that few of these inner products are non-zero, and non-zero inner products are computed with a small number of operations. Let us suppose that such inner product computations are done in $O(I)$ operations and that there are $O(Z)$ non-zero inner products for any g_{γ_n}. Computing $\{\langle R^{n+1} f, g_\gamma\rangle\}_{\gamma\in\mathbf{\Gamma}}$ thus requires $O(IZ)$ operations. Hence, the total numerical complexity of computing P matching pursuit iterations is $O(PIZ)$. An efficient implementation of a discrete Gabor dictionary [10] has $I = 1$ and $Z = N$, so that P iterations require $O(NP)$ operations. Since the dictionary has $O(N \log N)$ vectors, the total memory needed for the algorithm is $O(N \log N)$.

For an orthogonal matching pursuit algorithm, once the vector g_{γ_n} is selected, we must compute the orthogonal vector u_n with the Gram-Schmidt Equation (28). We suppose that for $p < n$, we have already computed the expansion coefficient of each u_p in $(g_{\gamma_k})_{0\le k\le p}$,

$$u_p = \sum_{k=0}^{p} b_{k,p} g_{\gamma_k} \ . \tag{44}$$

From

$$u_n = g_{\gamma_n} - \sum_{p=0}^{n-1} \frac{\langle g_{\gamma_n}, u_p \rangle}{||u_p||^2} u_p \,, \tag{45}$$

we can compute the expansion

$$u_n = \sum_{k=0}^{n} b_{k,n} g_{\gamma_k} \,. \tag{46}$$

If the inner product of any two elements in $\mathcal{D}$ is calculated in $O(I)$ operations, the $(b_{k,n})_{0 \le k \le n}$ are obtained in $O(nI + n^2)$ operations. We then compute the inner product of the new residue $R^{n+1}f$ with any $g_\gamma \in \mathcal{D}$, from the orthogonal updating Equation (31)

$$\langle R^{n+1} f, g_\gamma \rangle = \langle R^n f, g_\gamma \rangle - \langle R^n f, g_{\gamma_n} \rangle \, \langle u_n, g_\gamma \rangle. \tag{47}$$

Since

$$\langle u_n, g_\gamma \rangle = \sum_{k=0}^{n} b_{k,n} \langle g_{\gamma_k}, g_\gamma \rangle, \tag{48}$$

computing $(\langle R^{n+1} f, g_\gamma \rangle)_{\gamma \in \Gamma}$ requires $O(nIZ)$ operations. The total number of operations to compute P orthogonal matching pursuit iterations is therefore $O(P^3 + P^2 IZ)$ operations. For a dictionary of discrete Gabor signals with signals of size N, since $I = 1$, $Z = N$, and $P \le N$, the number of operations is $O(NP^2)$. It also requires $O(N \log N + P^2)$ memory to store the inner products $(\langle R^n f, g_\gamma \rangle)_{\gamma \in \Gamma}$ and the expansion coefficients $(b_{k,p})_{0 \le k,p \le n}$.

For P iterations, the non-orthogonal matching pursuit algorithm is P times faster than the orthogonal one. When P is large, which is the case in many signal processing applications, the orthogonal pursuit algorithm is much slower than the non-orthogonal one, and requires much more memory. When P remains small, the orthogonal pursuit is more advantageous because it converges faster.

§6 Matching pursuit with time-frequency dictionaries

For dictionaries of time-frequency atoms, a matching pursuit yields an adaptive time-frequency transform. It decomposes any function $f(t) \in L^2(\mathbb{R})$ into the sum of complex time-frequency atoms that best match its residues. This section describes the properties of this particular matching pursuit decomposition. We derive a new type of time-frequency energy distribution by summing the Wigner distributions of the time-frequency atoms.

Since a time-frequency atom dictionary is complete, Theorem 2 proves that a matching pursuit decomposes any function $f \in L^2(\mathbb{R})$ into

$$f = \sum_{n=0}^{+\infty} \langle R^n f, g_{\gamma_n} \rangle g_{\gamma_n}, \tag{49}$$

where $\gamma_n = (s_n, u_n, \xi_n)$ and

$$g_{\gamma_n}(t) = \frac{1}{\sqrt{s_n}} g\left(\frac{t-u_n}{s_n}\right) e^{i\xi_n t}. \tag{50}$$

These atoms are chosen to best match the residues of f.

We derive a new time-frequency energy distribution from the decomposition of any $f(t)$ within a time-frequency dictionary, by adding the Wigner distribution of each selected atom. Recall that the cross Wigner distribution of two functions $f(t)$ and $h(t)$ is defined by

$$W[f,h](t,\omega) = \frac{1}{2\pi} \int_{-\infty}^{+\infty} f(t+\frac{\tau}{2}) \bar{h}(t-\frac{\tau}{2}) e^{-i\omega\tau} d\tau. \tag{51}$$

The Wigner distribution of $f(t)$ is $Wf(t,\omega) = W[f,f](t,\omega)$. Since the Wigner distribution is quadratic, we derive from the atomic decomposition in Equation (49) of $f(t)$ that

$$\begin{aligned} Wf(t,\omega) = & \sum_{n=0}^{+\infty} |\langle R^n f, g_{\gamma_n}\rangle|^2 W g_{\gamma_n}(t,\omega) \\ & + \sum_{n=0}^{+\infty} \sum_{m=0, m\neq n}^{+\infty} \langle R^n f, g_{\gamma_n}\rangle \overline{\langle R^m f, g_{\gamma_m}\rangle} W[g_{\gamma_n}, g_{\gamma_m}](t,\omega). \end{aligned} \tag{52}$$

The double sum corresponds to the cross terms of the Wigner distribution. It contains the terms that one usually tries to remove in order to obtain a clear picture of the energy distribution of $f(t)$ in the time-frequency plane. We therefore keep only the first sum and define

$$Ef(t,\omega) = \sum_{n=0}^{+\infty} |\langle R^n f, g_{\gamma_n}\rangle|^2 W g_{\gamma_n}(t,\omega). \tag{53}$$

A similar decomposition algorithm over time-frequency atoms was derived independently by Qian and Chen [12] in order to define this energy distribution in the time-frequency plane. From the well known dilation and translation properties of the Wigner distribution and the Equation (50) of a time-frequency atom, we derive that for $\gamma = (s, \xi, u)$

$$W g_\gamma(t,\omega) = Wg\left(\frac{t-u}{s}, s(\omega-\xi)\right), \tag{54}$$

and hence

$$Ef(t,\omega) = \sum_{n=0}^{+\infty} |\langle R^n f, g_{\gamma_n}\rangle|^2 Wg\left(\frac{t-u_n}{s_n}, s_n(\omega-\xi_n)\right). \tag{55}$$

The Wigner distribution also satisfies

$$\int_{-\infty}^{+\infty}\int_{-\infty}^{+\infty} Wg(t,\omega)dt\ d\omega = \|g\|^2 = 1, \tag{56}$$

so the energy conservation Equation (18) implies

$$\int_{-\infty}^{+\infty}\int_{-\infty}^{+\infty} Ef(t,\omega)dt\ d\omega = \|f\|^2. \tag{57}$$

We can thus interpret $Ef(t,\omega)$ as an energy density of f in the time-frequency plane (t,ω). Unlike the Wigner and the Cohen class distributions, it does not include cross terms.

If $g(t)$ is the Gaussian window

$$g(t) = 2^{1/4}e^{-\pi t^2}, \tag{58}$$

then

$$Wg(t,\omega) = 2e^{-2\pi(t^2+(\frac{\omega}{2\pi})^2)}, \tag{59}$$

so $Ef(t,\omega)$ remains positive. The time-frequency atoms $g_\gamma(t)$ are then called Gabor functions. The time-frequency energy distribution $Ef(t,\omega)$ is a sum of Gaussian blobs whose locations and variances along the time and frequency axes depend upon the parameters (s_n, u_n, ξ_n).

Figure 1(a) is a signal f of 512 samples that is built by adding chirps, truncated sinusoidal waves and waveforms of different time-frequency localizations. No Gabor functions have been used to construct this signal. Figure 1(b) shows the time-frequency energy distribution $Ef(t,\omega)$. Since $Ef(t,\omega) = Ef(t,-\omega)$, we only display its values for $\omega \geq 0$. Each Gabor time-frequency atom selected by the matching pursuit is an elongated Gaussian blob in the time-frequency plane. We clearly see appearing two chirps that cross each other, with a localized time-frequency waveform on the top of their crossing point. We can also detect closely spaced Diracs and truncated sinusoidal waves having close frequencies. Several isolated localized time-frequency components also appear in this energy distribution.

Figure 2(a) is the graph of a speech recording corresponding to the word "greasy", sampled at 16 kHz. From the time-frequency energy displayed in Figure 2(b), we can see the low-frequency component of the "g" and the quick burst transition to the "ea". The "ea" has many harmonics that are lined up but we can also see localized high-frequency impulses that correspond to the pitch. The "s" component has a time-frequency energy spread over a high-frequency interval. Most of the signal energy is characterized by a few time-frequency atoms. For $n = 250$ atoms, $\frac{\|R^n f\|}{\|f\|} = .169$, although the signal has 5782 samples, and the sound recovered from these atoms is of excellent quality.

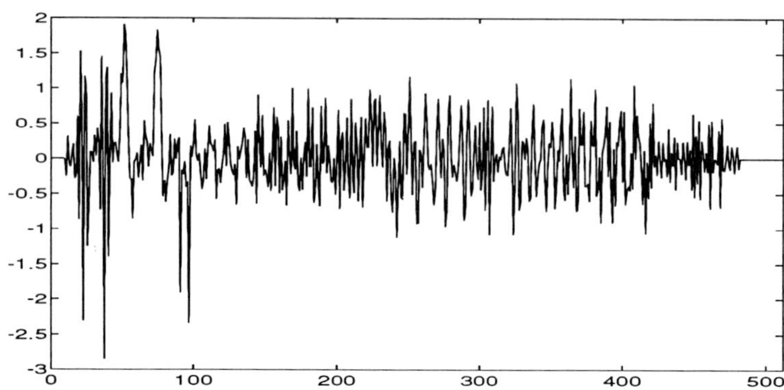

Figure 1(a). Signal of 512 samples built by adding chirps, truncated sinusoidal waves and waveforms of different time-frequency localizations.

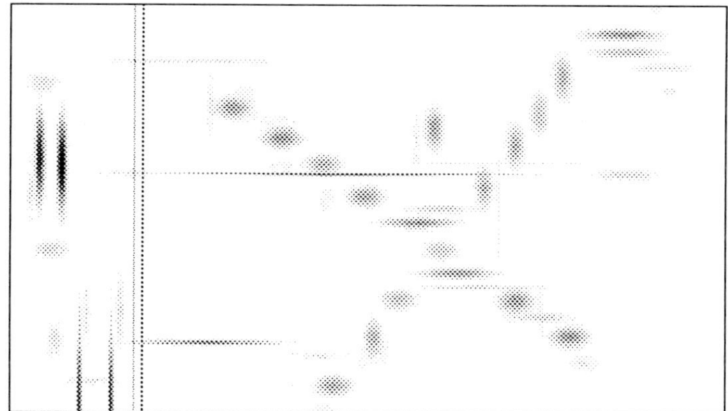

Figure 1(b). Time-frequency energy distribution $Ef(t,\omega)$ of the signal shown in (a). The vertical axis is frequency. The highest frequencies are on the top. The darkness of this time-frequency image increases with the value $Ef(t,\omega)$.

Let us now compare the decay rate of the residues for non-orthogonal matching pursuits versus orthogonal matching pursuits. The top curve in Figure 3(a) gives the decay of $\log_{10} \frac{\|R^n f\|}{\|f\|}$ as a function of the number of iterations n, for a non-orthogonal matching pursuit. For $n \geq 130$, the decay rate is almost constant. This confirms the exponential decay proved by Theorem 2. The

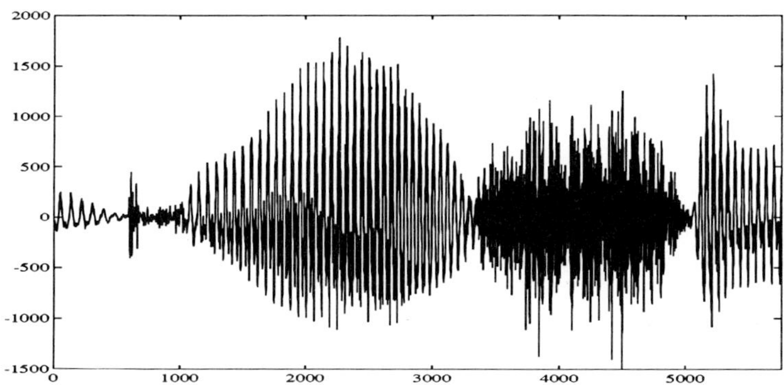

Figure 2(a). Speech recording of the "greasy," sampled at 16 kHz.

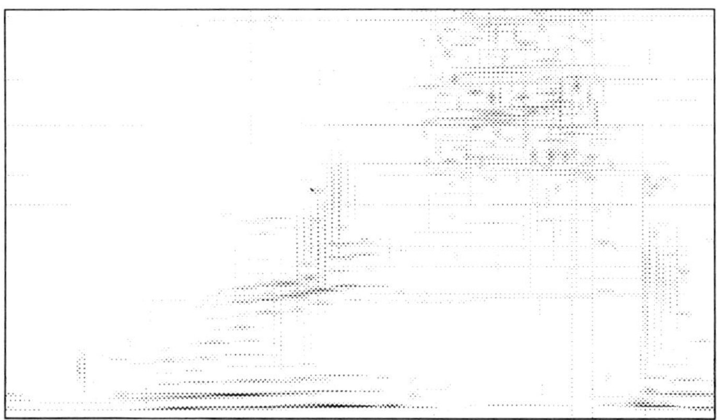

Figure 2(b). Time-frequency energy distribution of the speech recording shown in (a). We see the low-frequency component of the"g," the quick burst transition to the "ea" and the harmonics of the "ea." The "s" has energy spread over high frequencies.

bottom curve in Figure 3(a) gives the decay of $\log_{10} \frac{\|R^n f\|}{\|f\|}$ as a function of the number of iterations n, for an orthogonal matching pursuit. For the first 180 iterations we cannot distinguish the two curves, which means that the decay of the residues for the two algorithms is nearly identical. This indicates that the atoms selected by the non-orthogonal pursuit are nearly orthogonal. After this point, we see that the residues of the orthogonal pursuit decay much faster and

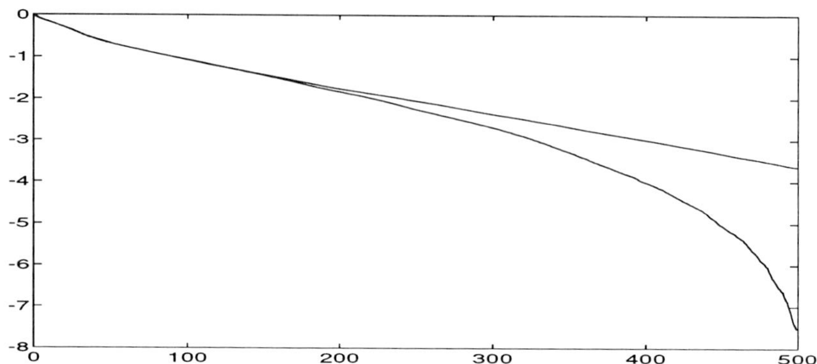

Figure 3(a). The top and bottom curves give respectively the decay of $\log_{10} \frac{\|R^n f\|}{\|f\|}$ for a non-orthogonal and an orthogonal matching pursuit, applied to the signal in Figure 1(a).

for $n = N = 512$, the number of samples of the signal, the residue is zero. The next section analyzes more precisely this divergence of the convergence rates of the orthogonal and non-orthogonal pursuits. The point of separation of the rates corresponds to a stage at which the residues have no more "coherent" structures. At this stage, the residue $R^n f$ has statistical properties that are very close to a the realization of a stationary white noise. For many information processing applications, we stop the decomposition when these coherent structures have disappeared, so for such applications the orthogonal pursuit does not offer much advantage.

§7 Chaos in matching pursuit and noise removal

A non-orthogonal matching pursuit can require an infinite number of iterations to converge, even in finite dimensions. We know already that the norms of these residues converge to zero; we now examine in more detail the asymptotic behavior of the residues. Numerical experiments suggest that matching pursuits are chaotic maps, and we can prove this for a particular dictionary. Experiments also show that these matching pursuits possess invariant measures, and we use this property to develop a noise removal algorithm.

To study the asymptotic properties of the residues, we first renormalize them to prevent their convergence to zero. We define the renormalized residue

$$\tilde{R}^n f = \frac{R^n f}{\|R^n f\|}. \tag{60}$$

Renormalized matching pursuit is the map defined by

$$M(\tilde{R}^n f) = \tilde{R}^{n+1} f. \tag{61}$$

Since

$$\|R^{n+1}f\|^2 = \|R^n f\|^2 - |\langle R^n f, g_{\gamma_n}\rangle|^2, \tag{62}$$

we obtain that

$$M(\tilde{R}^n f) = \frac{\tilde{R}^n f - \langle \tilde{R}^n f, g_{\gamma_n}\rangle g_{\gamma_n}}{\sqrt{1 - |\langle \tilde{R}^n f, g_{\gamma_n}\rangle|^2}} . \tag{63}$$

At each iteration the renormalized matching pursuit map removes the largest dictionary component of the residue and renormalizes it. This action is much like that of a left-shift operator acting on a base-N decimal number: the shift operator removes the most significant (leftmost) digit of the expansion and then "renormalizes" the expansion by multiplying by N. Let Σ_N be the set of all base N decimals. The left-shift map $L_N : \Sigma_N \to \Sigma_N$ is formally defined by

$$L_N(0.s_1 s_2 s_3 \ldots) = 0.s_2 s_3 s_4 \ldots \tag{64}$$

where $0.s_1 s_2 \ldots$ is the base-N decimal $\sum_{k=1}^{\infty} \frac{s_k}{N^k}$. The left shift map is known to be a chaotic map, which suggests that renormalized matching pursuits share this property.

Additional evidence that renormalized matching pursuits are chaotic can be found in the fact that the map has "sensitive dependence" on the initial signal f when f is close to a dictionary element. Consider two signals f_1 and f_2 defined by

$$f_1 = (1 - \epsilon)g + \epsilon h_1 \tag{65}$$

and

$$f_2 = (1 - \epsilon)g + \epsilon h_2 \tag{66}$$

where g is the closest dictionary element to f_1 and f_2, $\|h_1\| = \|h_2\| = 1$, and $\langle h_1, g\rangle = \langle h_2, g\rangle = 0$. Then $\|f_1 - f_2\| = \epsilon\|h_1 - h_2\|$ can be made arbitrarily small, while $\|\tilde{R}f_1 - \tilde{R}f_2\| = \|h_1 - h_2\|$ is of order 1. The map thus separates points near dictionary elements.

We examined the renormalized matching pursuit map for a dictionary in 3 dimensions [4]. Through symmetry operations this map can be reduced to a 1-dimensional map, which we prove is topologically equivalent to a left-shift map. Thus, renormalized matching pursuit in this case is a chaotic map whose properties are completely understood.

Numerical experiments with several types of dictionaries, including Gabor dictionaries, provide evidence that the renormalized matching pursuit map possesses an invariant measure and that it is mixing. The mixing property means that if we start with a collection of test signals, then after sufficient iterations of renormalized matching pursuit, the density of the residues in the signal space will be close to the invariant density function. To express this result in another way, consider a random process which yields signals with a probability measure on the signal space given by our invariant measure. Then the mixing property implies that after sufficient iterations of the map, the residues will look like realizations of this process.

The renormalized matching pursuit map continually sets to zero the largest of the dictionary components of the current residue, and redistributes the energy from this removed component over the remainder of the residue. We expect this system to be near equilibrium when all dictionary components of the residue have roughly the same magnitude–when this is so, the operations of setting the largest component to zero and of redistributing the energy have little effect. The residues converge to realizations of a random process for which the distribution of the dictionary components is flat. This invariant process can be interpreted as a generic noise with respect to our dictionary, which we call dictionary noise.

For a Gabor dictionary we observe that after several iterations the residues have statistical properties that are close to realizations of a white stationary process. To better understand this phenomenon, we first note that the Gabor dictionary is invariant under translation and frequency modulation, which means that for any $g_\gamma(t) \in \mathcal{D}$, and $(u, \xi) \in \mathbb{R}^2$, there exists a $\phi \in \mathbb{R}$ such that $e^{i\phi}e^{i\xi t}g_\gamma(t-u) \in \mathcal{D}$. One can prove [4] that for a translation and modulation invariant dictionary, if there exists an invariant measure, then this measure is invariant with respect to operators that translate signals in time or frequency. This invariant measure therefore corresponds to a white stationary process. A detailed analysis of the invariant measure was performed performed for a dictionary composed of a discrete Dirac basis plus a discrete Fourier basis. This dictionary is clearly invariant by translations and frequency modulations. We constructed a stochastic differential equation model of the evolution of the renormalized residues and solved the corresponding Fokker-Planck equation to compute the properties of the invariant measure. We obtained excellent agreement with numerical data [4].

An important measure of the flatness of the signal f with respect to a dictionary is the correlation ratio

$$\lambda(f) = \sup_{\gamma \in \Gamma} \frac{|\langle f, g_\gamma \rangle|}{\|f\|} \ . \tag{67}$$

The correlation ratio is the fraction of the signal energy which will be removed by one iteration of a matching pursuit when the optimality ratio $\alpha = 1$. Signals which possess structures closely resembling dictionary elements will have large values of $\lambda(f)$. As the matching pursuit proceeds, these structures are removed, and $\lambda(R^k f)$ decreases. When signal residues are as flat as realization of a dictionary noise, the convergence rate of the pursuit is near a minimum. It is often not worth continuing the matching pursuit iterations since the signal includes no structures that strongly correlate to dictionary elements. Let λ_∞ be the expected value of the correlation ratio for the dictionary noise. This constant is computed through numerical experiments over many iterations of the map. We call the coherent structures of a signal f the first m vectors $(g_{\gamma_n})_{0 \le n < m}$ that correlate with the residue better than the average correlation ratio λ_∞. For $0 \le n < m$ we have

$$\lambda(R^n f) > \lambda_\infty \tag{68}$$

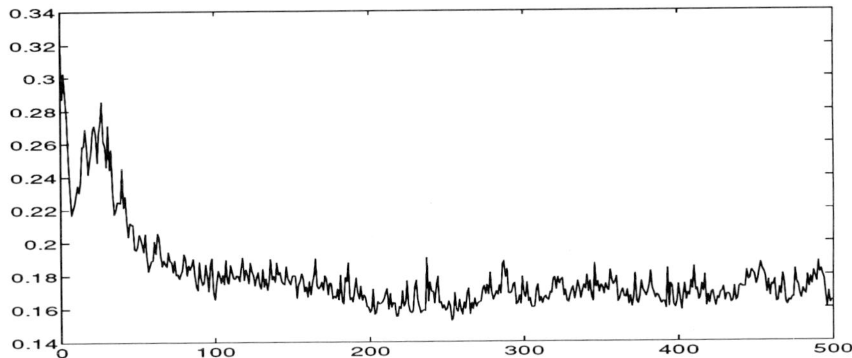

Figure 3(b). $\lambda(R^n f)$ as a function of the number of iterations n, for the signal in Figure 1(a).

and for m we have

$$\lambda(R^m f) \leq \lambda_\infty. \tag{69}$$

The decay rate of the energy of the residues depends only upon the correlation ratio. If we set the optimality factor $\alpha = 1$, then Equation (11) implies that

$$\frac{\|R^{n+1} f\|}{\|R^n f\|} = (1 - \lambda^2(R^n f))^{\frac{1}{2}}. \tag{70}$$

Figure 3(b) displays the correlation ratio of the residues $\lambda^2(R^n f)$ for the signal in Figure 1(a). After 180 iterations, $\lambda(R^n f)$ fluctuates around the mean λ_∞. It is at that same point that the decay rate of the orthogonal pursuit residues, shown in Figure 3(a), becomes faster than the decay rate of the non-orthogonal pursuit. This indicates that the orthogonal pursuit converges significantly faster than the non-orthogonal pursuit only when the residues are already close to the realization of a dictionary noise. For information processing applications, it is often not useful to continue the decomposition beyond this point, so for such applications the orthogonal pursuit does not offer much advantage over the non-orthogonal pursuit.

The ability to find the coherent structures of a signal can be used to remove noise from signals. In this approach, the noise is entirely defined by the choice of the dictionary and corresponds to the invariant measure of the matching pursuit map. For a Gabor dictionary, the dictionary noise is white and stationary, but this is not necessarily true for other dictionaries. The dictionary must be chosen so that its elements correlate as closely as possible the signal inner structures, but so that they avoid high correlations with realizations of the noise to be removed. During the matching pursuit decomposition, we test the correlation ratio in Equation (68) and stop when condition in Equation

(69) is true, *i.e.*, when we have selected all coherent structures. Figure 4(a) shows a signal obtained by adding a Gaussian white noise to the speech recording given in Figure 2(a), with a signal to noise ratio of 1.5 dB. Figure 4(b) is the time-frequency energy distribution of this noisy signal. The white noise generates time-frequency atoms spread across the whole time-frequency plane, but we can still distinguish the time-frequency structures of the original signal because their energy is better concentrated in this plane. This signal contains $m = 76$ coherent structures which are displayed in the time-frequency energy distribution of Figure 5(a). Figure 5(b) is the signal reconstructed from these coherent time-frequency atoms. The SNR of the reconstructed signal is 6.8 dB. The white noise has been removed and the recovered signal has a good auditory quality because the main time-frequency structures of the original speech signal have been retained.

§8 Conclusion

Matching pursuits provide extremely flexible signal representations since the choice of dictionaries is not limited. For information processing or compact signal coding, it is important to have strategies to adapt the dictionary to the class of signal that is decomposed. Time-frequency dictionaries include vectors that are spread between the Fourier and Dirac bases. They are regularly distributed on the unit sphere of the signal space and are thus well adapted to decomposing signals for which we have little prior information. When enough prior information is available, one can adapt the dictionary to the probability distribution of the signal class within the signal space **H**. Learning a dictionary is equivalent to finding the important inner structures of the signals that are decomposed. Classical algorithms for the optimization of code books, such as LBG [7], do not converge to satisfying solutions in such high dimensional spaces. We are currently studying the problem of adapting the dictionary to specific signal properties.

Acknowledgments. Geoffrey Davis was supported by an ONR/ASEE graduate fellowship. This work was supported by the AFOSR grant F49620-93-1-0102, ONR grant N00014-91-J-1967 and the Alfred Sloan Foundation.

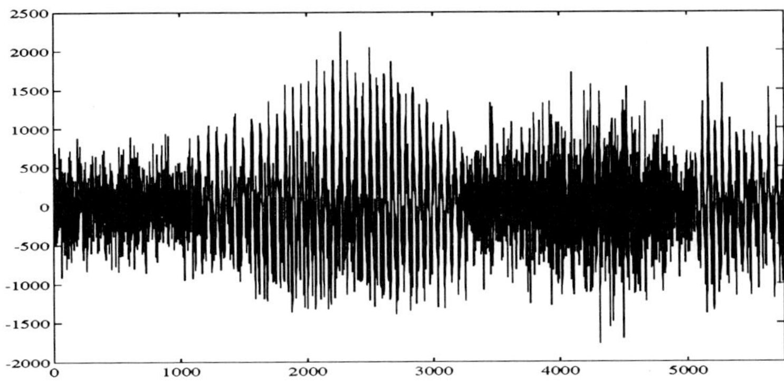

Figure 4(a). Signal obtained by adding a Gaussian white noise to the speech recording shown in Figure 2(a). The signal to noise ratio is 1.5db.

Figure 4(b). Time-frequency energy distribution of the noisy speech signal. The energy distribution of the white noise is spread across the whole time-frequency plane.

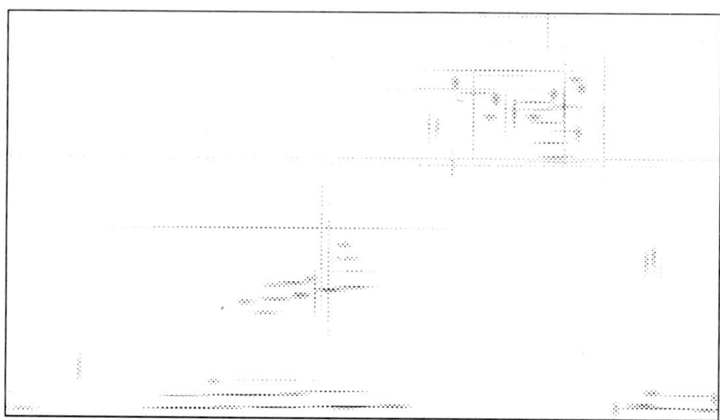

Figure 5(a). Time-frequency energy distribution of the $m = 76$ coherent structures of the noisy speech signal shown in Figure 3(a).

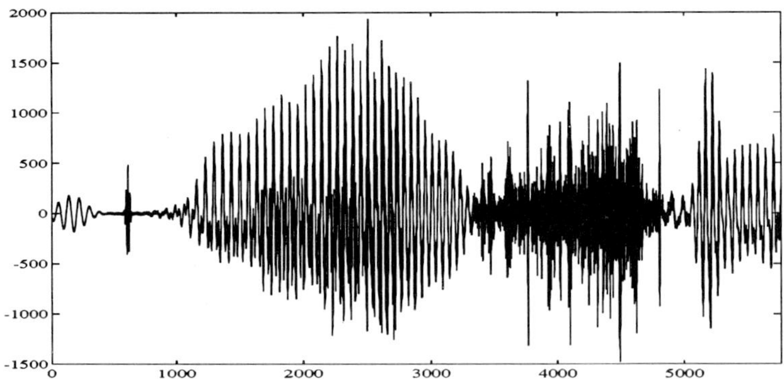

Figure 5(b). Time-frequency energy distribution of the $m = 76$ coherent structures of the noisy speech signal shown in Figure 3(b): signal reconstructed from the 76 coherent structures shown in (a). The white noise has been removed.

References

1. Chen, S., S. A. Billings, and W. Luo, Orthogonal least squares methods and their application to non-linear system identification, *International J. of Control* **50** (5) (1989), 1873–1896.
2. Cohen, L., Time-frequency distributions: a review, *Proceedings of the IEEE* **77** (7) (1989), 941–979.
3. Daubechies, I., *Ten Lectures on Wavelets*, CBMS-NSF Series in Appl. Math., SIAM, 1991.
4. Davis, G., S. Mallat, and M. Avellaneda, *Chaos in Adaptive Approximations*, Technical Report, Computer Science, New York University, April 1994.
5. Friedman, J. H. and W. Stuetzle, Projection pursuit regression, *J. of Amer. Stat. Assoc.* **76** (1981), 817–823.
6. Garey, M. R. and D. S. Johnson, *Computers and Intractability: A Guide to the Theory of NP-Completeness*, W. H. Freeman and Co., New York, 1979.
7. Gray, R., Vector quantization, *IEEE ASSP Magazine*, April 1984.
8. Huber, P. J., Projection pursuit, *The Annals of Statistics* **13** (2) (1985), 435–475.
9. Jones, L. K., On a conjecture of Huber concerning the convergence of projection pursuit regression, *The Annals of Statistics* **15** (2) (1987), 880–882.
10. Mallat, S. and Z. Zhang, Matching pursuit with time-frequency dictionaries, *IEEE Trans. on Signal Processing*, December 1993.
11. Pati, Y. C., R. Rezaiifar, and P. S. Krishnaprasad, Orthogonal matching pursuit: recursive function approximation with applications to wavelet decomposition, *Proc. of the* 27th *Annual Asilomar Conf. on Signals, Systems, and Computers*, November 1993.
12. Qian, S. and D. Chen, Signal representation via adaptive normalized Gaussian functions, *IEEE Trans. on Signal Processing* **36** (1) (1994).

Geoffrey Davis, Stéphane Mallat, and Zhifeng Zhang
Courant Institute of Mathematical Sciences
Computer Science Department
New York University
251 Mercer Street
New York, NY 10012, USA

Getting Around the Balian-Low Theorem Using Generalized Malvar Wavelets

Bruce W. Suter and Mark E. Oxley

Abstract. The Balian-Low Theorem represents a significant constraint for windowed complex exponential bases. A new complex-valued generalized Malvar wavelet will be defined, and some of its properties will be stated. This new theory will permit us to obtain something close to an exponential basis.

§1 Introduction

Time-frequency analysis is an active topic in signal processing, which has recently attracted many researchers' attentions. An important issue for time-frequency analysis is the time-frequency localization property. An excellent tutorial paper on this property is Daubechies' paper [4] where the tradeoffs between time and frequency localizations are studied with both the wavelet transform and the windowed Fourier transform. Recall the definition of the windowed Fourier transform. Let $g(x)$ (x will denote the time variable) be a window function and $s(x)$ be a signal, its windowed Fourier transform is given by

$$[T_g s]_{m,n} = \int_{-\infty}^{\infty} e^{-i2\pi mx} g(x-n)s(x)dx,$$

where $m, n \in \mathbf{Z}$. Note that $g(x-n)$ is the window translated by n time units and m denotes the frequency. If $g(x)$ is localized in time then the windowed Fourier transform of s yields the frequency content of s (represented by m) in the vicinity of the time $x = n$. Thus, one seeks a window function g such that the windowed Fourier transform of any signal s yields the localization of the frequency content localized in time, which is called the time-frequency localization problem. A significant constraint of the time-frequency localization

Wavelets: Theory, Algorithms, and Applications
Charles K. Chui, Laura Montefusco, and Luigia Puccio (eds.), pp. 295–310.

ISBN 0-12-174575-9

property is indicated in the following theorem which was discovered independently by Balian [1] and Low [2]. We preface it with some notations. Let $g(x)$ denote a window function and $G(\omega)$ denote its Fourier transform. For each $m, n \in \mathbf{Z}$, let $g_{mn} = \{\exp(2\pi imx)g(x-n)\}$.

Theorem 1. (Balian-Low Theorem) *If $\{g_{m,n}\}$ constitutes a frame for $L^2(\mathbf{R})$, then either $xg(x) \notin L^2(\mathbf{R})$ or $\omega G(\omega) \notin L^2(\mathbf{R})$.*

This theorem tells us that if one chooses a window function g such that g is well localized in time, then the Fourier transform G of the window g can not be well localized in frequency. The above argument was made rigorous and extended by Coifman and Semmes to include the case of non-tight frames (see Daubechies [4,5]). An independent and more elegant proof of the Balian-Low Theorem was given by Battle in [3]. For more general lattice expansion cases, see Daubechies and Janssen [5]. To overcome the negative aspects of the above results, one modifies several characteristics on the function g_{mn}. Specifically, this is accomplished via either (i) the theory of frames by using irregular lattice or (ii) the modification of the complex exponentials in g_{mn}. The collection $\{g_{m,n}\}$ is not a frame in general. For more detail, see Benedetto [39]. In this paper, we will show that one needs an orthonormal basis to get around the Balian-Low constraint. Consequently, we will focus on the second approach (ii). Existing methods in modifying the complex exponentials involve the replacement of them by sinusoids. In this approach there have been two kinds of constructions with different motivations. One is the so-called Wilson bases motivated from quantum mechanics [7]. The construction by Wilson [7] was simplified by Daubechies, Jaffard, and Journé in [6]. The other kind is the so-called Lapped Orthogonal Transform (LOT) or Malvar wavelets motivated by the desire to eliminate the blocking effects in image coding [11-14,16-18]. Recently, from the perspective of system theory, Vaidyanathan and Chen [25,26] have investigated lapped transforms and filter bank theory. The Lapped Orthogonal Transform (LOT) introduced by Cassereau [17] was further developed by Malvar [11-14] so that in a modulated lapped transform the aliasing errors of subband coding are cancelled and a perfect reconstruction is allowed. The framework for the generalization of this work was provided by Coifman and Meyer in [15] through a functional specification of windows, which permits windows with variable lengths overlapping. It has been illustrated in [8,23] that the smooth wavelet bases constructed by Meyer and Lemarié is a special case of Malvar wavelets in frequency domain. Auscher [9], Auscher, Weiss, and Wickerhauser in [8], Coifman and Wickerhauser in [35], Wesfried and Wickerhauser in [36], and Wickerhauser in [10] also discussed local sinusoid transforms.

Up to this stage, the complex exponentials in g_{mn} have been replaced only by sinusoids. Most recently, Suter and Oxley [19-21] developed the so-called Generalized Lapped Orthogonal Transforms (GLOT) or generalized Malvar wavelets which are constructed from a family of real orthogonal bases with different real local bases for different time intervals. Possible local bases include trigonometric, orthogonal polynomials, Daubechies wavelets on the interval, etc. Recent results indicate the ability to perform asymmetrical overlaps using real-valued Malvar wavelets [37]. Moreover, Xia and Suter [24] have recently generated nonseparable multidimensional generalized Malvar wavelets. We note that the above two constructions only allow the replacement of the complex exponential $e^{i2\pi mx}$ by real functions. In another words, the lapped transforms are real-valued. However, Young and Kingsburg in [22] studied motion estimation in frequency domain by utilizing Complex Lapped Transform (CLT) which has advantages over real-valued LOT. Also, they mentioned that their CLT is not orthogonal.

In this paper, we will review the generalized Malvar wavelets obtained in [19-21] and we will introduce complex-valued Malvar wavelets. With a slight modification of the complex exponentials $e^{i2\pi mx}$ in g_{mn} we can still have the orthogonality property. Thus, we are allowed to construct an orthogonal Complex Lapped Transform.

This paper is organized as follows. In Section 2, we review Wilson bases and generalized Malvar wavelets. In Section 3, we introduce complex-valued Malvar wavelets. In Section 4, we discuss the implementations and possible applications. Finally, in Section 5 we have conclusions and remarks.

§2 Wilson bases and generalized Malvar wavelets

In this section, we first review Wilson bases obtained by Daubechies, Jaffard, and Journé [6] and Auscher [9]. We then review generalized Malvar wavelets developed in [19-21]. For the relationship between Wilson bases and Malvar wavelets, see Meyer [23]. For more about wavelets, see [23,28,29].

2.1 Wilson bases

We start with a result of Daubechies, Jaffard, and Journé in [6]. For a window function $g(x)$, integers $l \in \mathbf{Z}^+ = \{0,1,2,...\}$ and $n \in \mathbf{Z}$, we define a family of functions or wavelets as

$$\psi_{l,n}(x) = \begin{cases} g(x-n), & \text{if } l=0, \\ \sqrt{2}\cos(2\pi lx)g(x-\frac{n}{2}), & \text{if } l>0 \text{ and } l+n \text{ even}, \\ \sqrt{2}\sin(2\pi lx)g(x-\frac{n}{2}), & \text{if } l>0 \text{ and } l+n \text{ old}. \end{cases} \tag{2.1}$$

Then,

Theorem 2. (Daubechies, Jaffard, and Journé [6]). *Let $G(\omega)$ be the Fourier transform of $g(x)$. Suppose that G is a real-valued function that satisfies*

$$|G(\omega)| \leq C(1+|\omega|)^{-1-\epsilon},$$

for some $\epsilon > 0$. Assume that

$$\sum_l G(\omega - l)G(\omega - l - 2j) = \delta(j), \quad j \in \mathbf{Z}, \omega \in \mathbf{R},$$

where δ is the Kronecker delta. Then, $\{\psi_{l,n}\}$ forms an orthonormal basis for $L^2(\mathbf{R})$, called a Wilson basis.

In [9], Auscher shows that a compactly supported $g(x)$ in Theorem 2 can be obtained. For more detail, see [6,8,9].

2.2 Generalized Malvar wavelets, transforms, and examples

In this section we review some results obtained in [19-21] for real bases.

2.2.1 Generalized Malvar wavelets

We begin by partitioning the real line $\mathbf{R}$. Let $\mathbf{Z}$ denote the integers and let $\mathbf{N}$ denote the natural numbers. Let $\{a_j | j \in \mathbf{Z}\}$ denote a strictly increasing sequence such that $a_j \to -\infty$ as $j \to -\infty$ and $a_j \to +\infty$ as $j \to +\infty$. The interval $(a_j, a_{j+1}]$ denotes the jth subinterval and $\mathbf{R} = \bigcup_{j\in\mathbf{Z}}(a_j, a_{j+1}]$ so that the collection $\{(a_j, a_{j+1}] | j \in \mathbf{Z}\}$ forms a partition of $\mathbf{R}$. For each $j \in \mathbf{Z}$ let $\{f_{j,k} | k \in \mathbf{N}\}$ denote a real weighted orthonormal basis defined on the interval $I_j = [a_j, a_{j+1}]$ where orthogonality is measured with respect to the weight function p_j, that is,

$$\int_{a_j}^{a_{j+1}} f_{j,k}(x) f_{j,l}(x) p_j(x) dx = \delta_{k,l}.$$

At each point a_j we center an interval, namely $(a_j - \epsilon_j, a_j + \epsilon_j)$ with $\epsilon_j > 0$ such that $\epsilon_{j+1} + \epsilon_j \leq a_{j+1} - a_j$. Observe the redundancy with the overlapping intervals $\tilde{I}_j = (a_j - \epsilon_j, a_{j+1} + \epsilon_{j+1})$ and $\mathbf{R} = \bigcup_{j\in\mathbf{Z}} \tilde{I}_j$. We define the extension functions $\tilde{f}_{j,k}$ by constructing the even extension about a_j of $f_{j,k}$ on $(a_j - \epsilon_j, a_j)$ and the odd extension about a_{j+1} of $f_{j,k}$ on $(a_{j+1}, a_{j+1} + \epsilon_{j+1})$. Specifically, $\tilde{f}_{j,k}$ can be defined as

$$\tilde{f}_{j,k}(x) = \begin{cases} 0, & -\infty < x \leq a_j - \epsilon_j \\ f_{j,k}(2a_j - x), & a_j - \epsilon_j < x < a_j \\ f_{j,k}(x), & a_j \leq x \leq a_{j+1} \\ -f_{j,k}(2a_{j+1} - x), & a_{j+1} < x < a_{j+1} + \epsilon_{j+1} \\ 0, & a_{j+1} + \epsilon_{j+1} \leq x < \infty. \end{cases} \tag{2.2}$$

Let $\tilde{p}_j$ denote the even extensions of p_j about both endpoints; specifically,

$$\tilde{p}_j(x) = \begin{cases} 0, & -\infty < x \leq a_j - \epsilon_j \\ p_j(2a_j - x), & a_j - \epsilon_j < x < a_j \\ p_j(x), & a_j \leq x \leq a_{j+1} \\ p_j(2a_{j+1} - x), & a_{j+1} < x < a_{j+1} + \epsilon_{j+1} \\ 0, & a_{j+1} + \epsilon_{j+1} \leq x < \infty. \end{cases} \tag{2.3}$$

Let w_j denote a window function which is positive on the interval $(a_j - \epsilon_j, a_{j+1} + \epsilon_{j+1})$ having a peak amplitude of A_j. We will call this open interval the positive support of w_j and denote it $supp_+(w_j)$. It is important to note that the intersection of the positive support of two nonadjacent windows is the null set since $\epsilon_{j+1}+\epsilon_j \leq a_{j+1}-a_j$. That is, $supp_+(w_{j-1}) = (a_{j-1}-\epsilon_{j-1}, a_j+\epsilon_j)$ an open interval and $supp_+(w_{j+1}) = (a_{j+1}-\epsilon_{j+1}, a_{j+2}+\epsilon_{j+2})$, an open interval. So $supp_+(w_{j-1}) \bigcap supp_+(w_{j+1}) = \emptyset$ since $a_j+\epsilon_j \leq a_{j+1}-\epsilon_{j+1}$. The amplitude-normalized window $\hat{w}_j(x)$ is given by $\hat{w}_j(x) = w_j(x)/A_j$. We choose amplitude-normalized windows $\hat{w}_j(x)$ with the following properties:

$$\begin{array}{llcll}
(a) & \hat{w}_j(x) & = & 1 & \text{for} \quad x \in (a_j+\epsilon_j, a_{j+1}-\epsilon_{j+1}) \\
(b) & \hat{w}_j(x) & = & 0 & \text{for} \quad x \notin (a_j-\epsilon_j, a_{j+1}+\epsilon_{j+1}) \\
(c) & \hat{w}_j(a_j-\sigma) & = & \hat{w}_{j-1}(a_j+\sigma) & \text{for} \quad \sigma \in [-\epsilon_j, \epsilon_j] \\
(d) & \hat{w}_j^2(x)+\hat{w}_{j-1}^2(x) & = & 1 & \text{for} \quad x \in [a_j-\epsilon_j, a_j+\epsilon_j].
\end{array}$$

We now define the functions $u_{j,k}$ for each $j \in \mathbf{Z}$ and $k \in \mathbf{N}$ by

$$u_{j,k}(x) = \hat{w}_j(x)\tilde{f}_{j,k}(x)\sqrt{\tilde{p}_j(x)}.$$

The following theorem states a very useful property of the set of functions $\{u_{j,k}|j \in \mathbf{Z}, k \in \mathbf{N}\}$.

Theorem 3. (Suter and Oxley [21]). $\{u_{j,k}|j,k \in \mathbf{Z}\}$ *is an orthonormal basis for* $L^2(\mathbf{R})$.

A proof of this theorem is given in Section 3 as a special case of the complex-valued Malvar wavelets.

2.2.2 Generalized Malvar wavelet transform

We define the Generalized Malvar Wavelet Transform (GMWT) to be a transform, $\mathbf{T} : L^2(\mathbf{R}) \to \ell^2(\mathbf{Z}\times\mathbf{N})$, which maps functions in $L^2(\mathbf{R})$ into doubly indexed sequences of real numbers, specifically, $\ell^2(\mathbf{Z}\times\mathbf{N})$ by

$$[\mathbf{T}s]_{j,k} = \int_{a_j-\epsilon_j}^{a_{j+1}+\epsilon_{j+1}} s(x)u_{j,k}(x)dx.$$

T is also called Generalized Lapped Orthonormal Transform (GLOT). The next theorem states some properties concerning $\mathbf{T}$.

Theorem 4. (Suter and Oxley [21]). *The GMWT* $\mathbf{T}$ *satisfies the following properties:*
(i) $\mathbf{T}$ *is defined on all of* $L^2(\mathbf{R})$;
(ii) $\mathbf{T}$ *is a linear transformation;*
(iii) $\mathbf{T}$ *is a continuous transformation and*

$$\| \mathbf{T} \| = \sup\{\| \mathbf{T}s \|_{\ell^2} : \| s \|_{L^2} \leq 1\} = 1;$$

(iv) $\mathbf{T}$ *is a unitary transformation, that is,*

$$\langle \mathbf{T}r, \mathbf{T}s\rangle_{\ell^2} = \langle r, s\rangle_{L^2}$$

for every $r, s \in L^2(\mathbf{R})$; *and*

(v) $\mathbf{T}$ *is invertible and given* $\alpha \in \ell^2(\mathbf{Z}\times\mathbf{N})$ *then*

$$[\mathbf{T}^{-1}\alpha](x) = \sum_{j\in\mathbf{Z}}\sum_{k\in\mathbf{N}} \alpha_{j,k}u_{j,k}(x) \qquad \text{a.e. } x \in \mathbf{R}.$$

Proof: To prove property (i), observe that $[\mathbf{T}s]_{j,k} = \langle s, u_{j,k}\rangle$ so that by Bessel's inequality

$$\| \mathbf{T}s \|_{\ell^2}{}^2 = \sum_{j\in\mathbf{Z}}\sum_{k\in\mathbf{N}} |[\mathbf{T}s]_{j,k}|^2 = \sum_{j\in\mathbf{Z}}\sum_{k\in\mathbf{N}} |\langle s, u_{j,k}\rangle_{L^2}|^2 \le \| s \|_{L^2}{}^2 < \infty \,.$$

Thus $\mathbf{T}$ is defined for all $s \in L^2(\mathbf{R})$. Property (ii) is clear by integration properties. Property (iv) follows from Parseval's Equality and taking $r = s$ yields property (iii). Property (v) is the direct application of the (Generalized) Fourier Series Theorem. ■

§3 Complex-valued Malvar wavelets

In this section, we first state a generalized theorem on complex-valued Malvar wavelets with variable length windows. We then go to a special case which is close to the windowed exponentials. In the following, local bases $f_{j,k}$ on $[a_j, a_{j+1}]$ are complex-valued and the weight function p_j and window function $\hat{w}_j$ are non-negative. The orthonormality for the local bases $f_{j,k}$ are in the following sense

$$\langle f_{j,k}, f_{j,l}\rangle_{L^2(I_j)} = \int_{a_j}^{a_{j+1}} f_{j,k}(x)\overline{f_{j,l}(x)}p_j(x)dx = \delta_{k,l}, \tag{3.1}$$

where the over bar denotes the complex conjugate. The functions $\tilde{f}_{j,k}$ and $\tilde{p}_j$ denote the extensions of $f_{j,k}$ and p_j similar to Equations (2.2) and (2.3), respectively. Let $\hat{w}_j$ be the window functions satisfying (a)-(d) in Subsection 2.2.1. Let

$$u_{j,k}(x) = \hat{w}_j(x)\tilde{f}_{j,k}(x)\sqrt{\tilde{p}_j(x)}.$$

Then, we have the following result.

Theorem 5. $\{u_{j,k} | j, k \in \mathbf{Z}\}$ *forms an orthonormal basis for* $L^2(\mathbf{R})$.

Proof: : Observe that $u_{j,k}$ has compact support and by construction satisfies $\|u_{j,k}\|_{L^2(\mathbf{R})} < \infty$ for each $j \in \mathbf{Z}$ and $k \in \mathbf{N}$. Therefore, $\{u_{j,k} | j \in \mathbf{Z}, k \in \mathbf{N}\} \subset L^2(\mathbf{R})$.

Next we demonstrate that $\{u_{j,k} | j \in \mathbf{Z}, k \in \mathbf{N}\}$ is an orthonormal set. By the nonoverlapping property of the positive support intervals it is clear that

$\langle u_{j,k}, u_{i,l}\rangle = 0$ if $|i-j| \geq 2$ for all $k, l \in \mathbf{N}$. Therefore, we need only consider two cases: (I) $i = j$ and (II) $|j-i| = 1$.

Case I. We will show $\langle u_{j,k}, u_{j,l}\rangle = \delta_{k,l}$. To see this, observe that

$$\langle u_{j,k}, u_{j,l}\rangle = \int_{a_j-\epsilon_j}^{a_{j+1}+\epsilon_{j+1}} \hat{w}_j^2(x)\tilde{f}_{j,k}(x)\overline{\tilde{f}_{j,l}(x)}\tilde{p}_j(x)dx \quad .$$

From the definitions of the even and odd extensions one sees that

$$\begin{aligned}
&\langle u_{j,k}, u_{j,l}\rangle \\
&= \int_{a_j-\epsilon_j}^{a_j} \hat{w}_j^2(x)f_{j,k}(2a_j-x)\overline{f_{j,l}(2a_j-x)}p_j(2a_j-x)dx \\
&+ \int_{a_j}^{a_j+\epsilon_j} [\hat{w}_j^2(x)-1]f_{j,k}(x)\overline{f_{j,l}(x)}p_j(x)dx \;+\; \int_{a_j}^{a_{j+1}} f_{j,k}(x)\overline{f_{j,l}(x)}p_j(x)dx \\
&+ \int_{a_{j+1}-\epsilon_{j+1}}^{a_{j+1}} [\hat{w}_j^2(x)-1]f_{j,k}(x)\overline{f_{j,l}(x)}p_j(x)dx \\
&+ \int_{a_{j+1}}^{a_{j+1}+\epsilon_{j+1}} \hat{w}_j^2(x)f_{j,k}(2a_{j+1}-x)\overline{f_{j,l}(2a_{j+1}-x)}p_j(2a_{j+1}-x)dx \\
&= A+B+C+D+E.
\end{aligned}$$

Note that $C = \delta_{k,l}$ by choice. For A, let $x = a_j - \sigma$ and for B, let $x = a_j + \sigma$. We have

$$A+B = \int_0^{\epsilon_j} [\hat{w}_j^2(a_j-\sigma)+\hat{w}_j^2(a_j+\sigma)-1]f_{j,k}(a_j+\sigma)\overline{f_{j,l}(a_j+\sigma)}p_j(a_j+\sigma)d\sigma.$$

By property (c) $\hat{w}_j(a_j-\sigma) = \hat{w}_{j-1}(a_j+\sigma)$ and so property (iv) implies

$$\hat{w}_j^2(a_j-\sigma)+\hat{w}_j^2(a_j+\sigma)-1 = 0 \quad \text{for} \quad -\epsilon_j \leq \sigma \leq \epsilon_j,$$

hence $A+B = 0$. Similarly, in integral D, let $x = a_{j+1} - \sigma$ and in E, let $x = a_{j+1} + \sigma$. We have

$$\begin{aligned}
D+E = \int_0^{\epsilon_{j+1}} &[\hat{w}_j^2(a_{j+1}-\sigma)+\hat{w}_j^2(a_{j+1}+\sigma)-1] \\
&\times f_{j,k}(a_{j+1}-\sigma)\overline{f_{j,l}(a_{j+1}-\sigma)}p_j(a_{j+1}-\sigma)d\sigma.
\end{aligned}$$

Properties (c) and (d) imply $D+E = 0$. Consequently,

$$\langle u_{j,k}, u_{j,l}\rangle = \delta_{k,l}.$$

Case II. We will show $\langle u_{i,k}, u_{j,l}\rangle = 0$ for $i = j-1$ for every $j \in \mathbf{Z}, k, l \in \mathbf{N}$, and then the case $i = j+1$ will follow easily.

By construction, $u_{j-1,k}$ and $u_{j,l}$ are possibly nonzero only on $(a_j-\epsilon_j, a_j+\epsilon_j)$, hence,

$$\langle u_{j-1,k}, u_{j,l}\rangle = \int_{a_j-\epsilon_j}^{a_j+\epsilon_j} \hat{w}_{j-1}(x)\hat{w}_j(x)\tilde{f}_{j-1,k}(x)\overline{\tilde{f}_{j,l}(x)}\sqrt{\tilde{p}_{j-1}(x)\tilde{p}_j(x)}dx.$$

From the definitions of the extensions

$$\begin{aligned}\langle u_{j-1,k}, u_{j,l}\rangle \\ &= \int_{a_j-\epsilon_j}^{a_j} \hat{w}_{j-1}(x)\hat{w}_j(x)f_{j-1,k}(x)\overline{f_{j,l}(2a_j-x)}\sqrt{p_{j-1}(x)p_j(2a_j-x)}dx \\ &- \int_{a_j}^{a_j+\epsilon_j} \hat{w}_{j-1}(x)\hat{w}_j(x)f_{j-1,k}(2a_j-x)\overline{f_{j,l}(x)}\sqrt{p_{j-1}(2a_j-x)p_j(x)}dx.\end{aligned}$$

In the first integral , let $x = a_j - \sigma$ and in the second integral let $x = a_j + \sigma$, then

$$\begin{aligned}\langle u_{j-1,k}, u_{j,l}\rangle = \int_0^{\epsilon_j} &[\hat{w}_{j-1}(a_j-\sigma)\hat{w}_j(a_j-\sigma) - \hat{w}_{j-1}(a_j+\sigma)\hat{w}_j(a_j+\sigma)] \\ &\times [f_{j-1,k}(a_j-\sigma)\overline{f_{j,l}(a_j+\sigma)}\sqrt{p_{j-1}(a_j-\sigma)p_j(a_j+\sigma)}]d\sigma.\end{aligned}$$

Property (c) implies

$$\hat{w}_{j-1}(a_j-\sigma)\hat{w}_j(a_j-\sigma) - \hat{w}_{j-1}(a_j+\sigma)\hat{w}_j(a_j+\sigma) = 0,$$

therefore, $\langle u_{j-1,k}, u_{j,l}\rangle = 0$. Cases I and II demonstrate that $\{u_{j,k}|j \in \mathbf{Z}, k \in \mathbf{N}\}$ is an orthonormal set.

Lastly, we prove that $\{u_{j,k}|j \in \mathbf{Z}, k \in \mathbf{N}\}$ is a basis for $L^2(\mathbf{R})$. To do this, we will show that given $s \in L^2(\mathbf{R})$ there exists a set of scalars $\{\alpha_{j,k} \mid j \in \mathbf{Z}, k \in \mathbf{N}\}$ such that $s = \sum_{j\in\mathbf{Z}}\sum_{k\in\mathbf{N}} \alpha_{j,k}u_{j,k}$ in the $L^2(\mathbf{R})$ sense. Let $s \in L^2(\mathbf{R})$ and define $s_j(x) = s(x)\hat{w}_j(x)$ for each $j \in \mathbf{Z}$. Since s_j has positive support on $(a_j-\epsilon_j, a_{j+1}+\epsilon_{j+1})$ we "fold" s_j on $(a_j-\epsilon_j, a_j)$ and $(a_{j+1}, a_{j+1}+\epsilon_{j+1})$ into the interval $[a_j, a_{j+1}]$ by defining

$$h_j(x) = \begin{cases} 0, & \text{for } -\infty < x < a_j \\ s_j(x) + s_j(2a_j - x), & \text{for } a_j \le x \le a_j + \epsilon_j \\ s_j(x), & \text{for } a_j + \epsilon_j < x < a_{j+1} - \epsilon_{j+1} \\ s_j(x) - s_j(2a_{j+1} - x), & \text{for } a_{j+1} - \epsilon_{j+1} \le x \le a_{j+1} \\ 0, & \text{for } a_{j+1} < x < \infty. \end{cases}$$

Now h_j is supported on $[a_j, a_{j+1}]$. Since, by assumption, $\{f_{j,k}|k \in \mathbf{N}\}$ is a complex-valued orthonormal basis with respect to the weight p_j on $[a_j, a_{j+1}]$ then $\{f_{j,k}\sqrt{p_j}|k \in \mathbf{N}\}$ is a complex-valued orthonormal basis with respect to the weight 1 on $[a_j, a_{j+1}]$. Consequently, there exist complex numbers $\{\alpha_{j,k} | k \in \mathbf{N}\}$ such that

$$h_j(x) = \sum_{k=1}^{\infty} \alpha_{j,k}\, f_{j,k}(x) \sqrt{p_j(x)} \tag{3.2}$$

where convergence and equality is in the $L^2(I_j)$ sense, and $\alpha_{j,k}$ is given by the inner product rule

$$\begin{aligned}
\alpha_{j,k} = \langle h_j, f_{j,k}\sqrt{p_j}\rangle_{L^2(I_j)} &= \int_{a_j}^{a_{j+1}} h_j(x)\overline{f_{j,k}(x)}\sqrt{p_j(x)}\, dx \\
&= \int_{a_j-\epsilon_j}^{a_{j+1}+\epsilon_{j+1}} s_j(x)\overline{\tilde{f}_{j,k}(x)}\sqrt{\tilde{p}_j(x)}\, dx \\
&= \int_{a_j-\epsilon_j}^{a_{j+1}+\epsilon_{j+1}} s(x)\overline{u_{j,k}(x)}\, dx \\
&= \langle s, u_{j,k}\rangle \quad .
\end{aligned}$$

Applying the extension rules in Equation (2.2) at both endpoints to Equation (3.2) yields

$$\tilde{h}_j(x) = \sum_{k=1}^{\infty} \alpha_{j,k}\tilde{f}_{jk}(x)\sqrt{\tilde{p}_j(x)}$$

where $\tilde{h}_j$ has an odd extension about $x = a_j$ and an even extension about $x = a_{j+1}$. That is, the use of the extension rules in Equation (2.2) applied to h_j defined in Equation (3.2) yields

$$\tilde{h}_j(x) = \begin{cases} 0, & -\infty < x < a_j - \epsilon_j \\ s_j(x) + s_j(2a_j - x), & a_j - \epsilon_j \le x \le a_j + \epsilon_j \\ s_j(x), & a_j + \epsilon_j < x < a_{j+1} - \epsilon_{j+1} \\ s_j(x) - s_j(2a_{j+1} - x), & a_{j+1} - \epsilon_{j+1} \le x \le a_{j+1} + \epsilon_{j+1} \\ 0, & a_{j+1} + \epsilon_{j+1} < x < \infty. \end{cases}$$

Multiplying $\tilde{h}_j(x)$ by $\hat{w}_j(x)$ and summing over j produces

$$\begin{aligned}
\sum_{j\in\mathbf{Z}} \hat{w}_j(x)\tilde{h}_j(x) &= \sum_{j\in\mathbf{Z}}\sum_{k\in\mathbf{N}} \alpha_{j,k}\hat{w}_j(x)\tilde{f}_{j,k}(x)\sqrt{\tilde{p}_j(x)} \\
&= \sum_{j\in\mathbf{Z}}\sum_{k\in\mathbf{N}} \alpha_{j,k}u_{jk}(x).
\end{aligned} \tag{3.3}$$

To complete the proof, we will show that the right-hand side of (3.3) is equal to $s(x)$. If x is a point where $s(x)$ is defined and $x \in [a_J + \epsilon_J, a_{J+1} - \epsilon_{J+1}]$ for some fixed $J \in \mathbf{Z}$ then $\sum_{j\in\mathbf{Z}} \hat{w}_j(x)\tilde{h}_j(x) = \hat{w}_J(x)h_J(x) = 1 \cdot s(x) = s(x)$. If

$x \in (a_J - \epsilon_J, a_J + \epsilon_J)$ then

$$\begin{aligned}
&\sum_{j\in\mathbf{Z}} \hat{w}_j(x)\tilde{h}_j(x) \\
&= \hat{w}_{J-1}(x)\tilde{h}_{J-1}(x) + \hat{w}_J(x)\tilde{h}_J(x) \\
&= \hat{w}_{J-1}(x)[s_{J-1}(x) - s_{J-1}(2a_J - x)] + \hat{w}_J(x)[s_J(x) + s_J(2a_J - x)] \\
&= \hat{w}^2_{J-1}(x)s(x) - \hat{w}_{J-1}(x)\hat{w}_{J-1}(2a_J - x)s(2a_J - x) \\
&+ \hat{w}^2_J(x)s(x) + \hat{w}_J(x)\hat{w}_J(2a_J - x)s(2a_J - x) \\
&= s(x) - [\hat{w}_{J-1}(x)\hat{w}_{J-1}(2a_J - x) - \hat{w}_J(x)\hat{w}_J(2a_J - x)]s(2a_J - x).
\end{aligned}$$

Let $x = a_J + \sigma$ where $-\epsilon_J \le \sigma \le \epsilon_J$ then the term within the brackets becomes

$$\begin{aligned}
&\hat{w}_{J-1}(a_J + \sigma)\hat{w}_{J-1}(a_J - \sigma) - \hat{w}_J(a_J + \sigma)\hat{w}_J(a_J - \sigma) \\
&= \hat{w}_J(a_J - \sigma)\,[\hat{w}_{J-1}(a_J - \sigma) - \hat{w}_J(a_J + \sigma)] = 0,
\end{aligned}$$

where the last equality follows from the window property (c). Thus,

$$\sum_{j\in\mathbf{Z}} \hat{w}_j(x)\tilde{h}_j(x) = s(x).$$

From Equation (3.3), this implies $\sum_{j\in\mathbf{Z}} \hat{w}_j(x)\tilde{h}_j(x) = s(x)$ almost everywhere for $x \in \mathbf{R}$. This proves that $\{u_{j,k} | j \in \mathbf{Z}, k \in \mathbf{N}\}$ is a basis for $L^2(\mathbf{R})$ and concludes our proof that this set is an orthonormal basis for $L^2(\mathbf{R})$. ■

Similar to the real generalized Malvar wavelet transform in Subsection 2.2.2, we have the following Complex Generalized Malvar Wavelet Transform (CGMWT) which is a transform $\mathbf{T} : L^2(\mathbf{R}) \to \ell^2(\mathbf{Z} \times \mathbf{N})$, mapping complex-valued functions in $L^2(\mathbf{R})$ into doubly indexed sequences of complex numbers, specifically, $\ell^2(\mathbf{Z} \times \mathbf{N})$ by

$$[\mathbf{T}s]_{j,k} = \int_{a_j - \epsilon_j}^{a_{j+1}+\epsilon_{j+1}} s(x)\overline{u_{j,k}(x)}dx.$$

Theorem 6. *The CGMWT* $\mathbf{T}$ *satisfies the following properties:*

(i) $\mathbf{T}$ *is defined on all of* $L^2(\mathbf{R})$;
(ii) $\mathbf{T}$ *is a linear transformation;*
(iii) $\mathbf{T}$ *is a continuous transformation and*

$$\| \mathbf{T} \| = \sup\{\| \mathbf{T}s \|_{\ell^2} : \| s \|_{L^2} \le 1\} = 1;$$

(iv) $\mathbf{T}$ *is a unitary transformation; and*
(v) $\mathbf{T}$ *is invertible and given* $\alpha \in \ell^2(\mathbf{Z} \times \mathbf{N})$ *then*

$$[\mathbf{T}^{-1}\alpha](x) = \sum_{j\in\mathbf{Z}} \sum_{k\in\mathbf{N}} \alpha_{j,k}u_{j,k}(x) \qquad \text{a.e. } x \in \mathbf{R}.$$

The proof of Theorem 6 is similar to the proof of Theorem 4, therefore we omit it here. We now apply the CGMWT theory to get around the Balian-Low constraint. To show this we choose a special case of a CGMWT, specifically, choose $a_j = j$ for all $j \in \mathbf{Z}$ and $\epsilon_j = \epsilon$ for all $j \in \mathbf{Z}$ such that $0 < \epsilon \leq 1/2$. This implies that we are choosing equal intervals all of length 1 and the overlaps are all the same amount. Take $f_{j,k}(x) = \frac{1}{\sqrt{2}} \exp(i2\pi k[x-j]) = \frac{1}{\sqrt{2}} \exp(i2\pi kx)$ for each $j \in \mathbf{Z}$ and $k \in \mathbf{Z}$. We know that the set $\{\frac{1}{\sqrt{2}} \exp(i2\pi kx) | k \in \mathbf{Z}\}$ is orthonormal on $[0,1]$ with respect to the weight function $p_j(x) = 1$. It follows that $\tilde{f}_{j,k}$ can be expressed as

$$\tilde{f}_{j,k}(x) = \begin{cases} 0, & -\infty < x \leq j-\epsilon \\ \frac{1}{\sqrt{2}} \exp(-i2\pi kx), & j-\epsilon < x < j \\ \frac{1}{\sqrt{2}} \exp(i2\pi kx), & j \leq x \leq j+1 \\ -\frac{1}{\sqrt{2}} \exp(-i2\pi kx), & j+1 < x < j+1+\epsilon \\ 0, & j+1+\epsilon \leq x < \infty. \end{cases} \tag{3.4}$$

We choose special window functions $\hat{w}_j$ to be translations of a single window function. Specifically, for each $j \in \mathbf{Z}$,

$$\hat{w}_j(x) = g(x-j),$$

where g has positive support on $(-\epsilon, 1+\epsilon)$, that is,

$$g(x) = \begin{cases} 0, & -\infty < x \leq -\epsilon \\ g_-(x), & -\epsilon < x < \epsilon \\ 1, & \epsilon \leq x \leq 1-\epsilon \\ g_+(x), & 1-\epsilon < x < 1+\epsilon \\ 0, & 1+\epsilon \leq x < \infty, \end{cases}$$

where g_- and g_+ satisfy the properties:

$$\begin{array}{lll} (a) & g_-(\sigma) = g_+(1-\sigma) & \text{for } \sigma \in [-\epsilon, \epsilon] \\ (b) & g_-^2(x) + g_+^2(x-1) = 1 & \text{for } x \in (-\epsilon, 1+\epsilon). \end{array}$$

Property (a) is a symmetric property and property (b) is an unit energy property. An example of a window satisfying these properties is given by

$$\hat{w}_n(x) = g(x-n) = \begin{cases} 0, & -\infty < x < n-\epsilon \\ \sin\left(\frac{\pi}{4\epsilon}\{x-[n-\epsilon]\}\right), & n-\epsilon \leq x \leq n+\epsilon \\ 1, & n+\epsilon < x < n+1-\epsilon \\ \cos\left(\frac{\pi}{4\epsilon}\{x-[n+1-\epsilon]\}\right), & n+1-\epsilon \leq x \leq n+1+\epsilon \\ 0, & n+1+\epsilon < x < \infty. \end{cases}$$

Combining these choices yields the collection $\{u_{j,k}(x) = \tilde{f}_{j,k}(x) g(x-j) | j \in \mathbf{Z}, k \in \mathbf{N}\}$ which, by Theorem 5, is an orthonormal basis for $L^2(\mathbf{R})$.

One can see that the window function g is localized in time (x denotes time variable), and, $xg(x) \in L^2(\mathbf{R})$. The Balian-Low Theorem implies that $G(\omega)$, the Fourier transform of g, is not localized in frequency and $\omega G(\omega) \notin L^2(\mathbf{R})$. While this is true, it does describe the constraint on the (discrete) windowed Fourier transform or short time Fourier transform

$$[Fs]_{j,k} = \int_{-\infty}^{\infty} e^{-i2\pi kx} g(x-j)s(x)dx = \int_{j-\epsilon}^{j+\epsilon} e^{i2\pi kx} g(x-j)s(x)ds.$$

But, the way around this constraint is to *not* use the short-time Fourier transform. Instead, we use the CGMWT

$$\begin{aligned}[Ts]_{j,k} &= \int_{-\infty}^{\infty} \tilde{f}_{j,k}(x)g(x-j)s(x)dx \\ &= \frac{1}{\sqrt{2}}\int_{j-\epsilon}^{j} e^{-i2\pi kx} g(x-j)s(x)dx + \frac{1}{\sqrt{2}}\int_{j}^{j+1} e^{i2\pi kx} g(x-j)s(x)dx \\ &- \frac{1}{\sqrt{2}}\int_{j+1}^{j+1+\epsilon} e^{-i2\pi kx} g(x-j)s(x)dx,\end{aligned}$$

which is very "close" to being a short-time Fourier transform. With the CGLOT, given a signal s, we can determine its time local behavior and its "frequency" local behavior simultaneously. The k variable denotes frequency in the short time Fourier transform, but it is not exactly frequency in the CGLOT. It is close though.

§4 Discrete time implementation of CGMWT

In this section we consider the implementation of CGMWT with a discrete time signal sampled from a finite energy continuous-time signal.

Signal Decomposition:

Let $s(x)$ be an arbitrary signal in $L^2(\mathbf{R})$. We use the above basis $u_{j,k}(x)$. According to the orthonormal expansion,

$$\alpha_{n,m} = \langle s, u_{n,m}\rangle = \int_{n-\epsilon}^{n+1+\epsilon} s(x)g(x-n)\tilde{f}_{n,m}(x)dx = \int_{n}^{n+1} h_n(x)\, e^{j2\pi mx}dx$$

where $s(x)$ is the input signal and $g(x)$ and $\tilde{f}(x)$ are as defined above and the folded signal h_n is defined as:
$h_n(x) =$

$$\begin{cases} s(x)g(x-n) + s(2n-x)g(n-x), & n \le x \le n+\epsilon \\ s(x), & n+\epsilon < x < (n+1)-\epsilon \\ s(x)g(x-n) - s(2(n+1)-x)g(n+2-x), & (n+1)-\epsilon \le x \le (n+1). \end{cases}$$

Then we have the following decomposition for discrete-time signals.

(i) Define a new array – Assume signal $s(x)$ is sampled at a fixed rate $\delta > 0$. Let $H_{n,l} = h_n(n + \delta l)$. Where, $H_{n,0} = h_n(n)$ is the folded data at the start of the interval, and $H_{n,N} = h_n(n+1)$ is the folded data at the end of the interval. The trapezoidal rule can then be used to approximate the integration:

$$\alpha_{n,m} = \frac{1}{2}H_{n,0} + \sum_{l=1}^{N-1} H_{n,l}e^{i2\pi ml/N} + \frac{1}{2}H_{n,N}$$

From the h_n equations, it is clear that $H_{n,N}$ will equal zero. Now let:

$$\beta_{n,l} = \begin{cases} H_{n,0}, & l = 0 \\ 2H_{n,l}, & 1 \le l \le N-1. \end{cases}$$

(ii) Perform an Inverse FFT – Perform an N point inverse Fourier transform of β. The spectral coefficients, α, are calculated with a Fourier transform of the windowed and scaled input signal:

$$\alpha_{n,m} = \frac{1}{2N}\sum_{l=0}^{N-1} \beta_{n,l}e^{i2\pi ml/N} = \frac{1}{2}\mathcal{F}^{-1}[\beta].$$

Signal Reconstruction:

(i) Perform a Forward FFT – The signal may now be reconstructed, given the coefficients $\{\alpha_{n,m}\}$.

$$\beta_{n,l} = 2 \cdot \mathcal{F}[\alpha_{n,m}];$$

(ii) Scale the Values – Scaled the β values back to $H_{n,L}$:

$$H_{n,l} = \begin{cases} \beta_{n,0}, & l = 0 \\ (\frac{1}{2})\beta_{n,l}, & 1 \le l \le N-1; \end{cases}$$

(iii) Unfold – Finally, the reconstructed signal, $S_{n,l}$, is calculated by dividing back out the original window, $g(x)$:

$$S_{n,l} = \begin{cases} \frac{H_{n,0}}{2g(0)}, & x = n \\ g(x-n)H_{n,l} - g(n-x)H_{n-1,N-l}, & n < x < n+\epsilon \\ H_{n,l}, & n+\epsilon \le x \le (n+1)-\epsilon \\ g(x-n)H_{n,l} + g((n+2)-x)H_{n+1,N-l}, & (n+1)-\epsilon < x < n+1. \end{cases}$$

§5 Conclusions

In this paper, we generalized Malvar wavelets from sinusoids to general local bases, such as, orthogonal polynomials. This generalization allows us to construct different local bases in different time intervals according to the nature of signals. We also generalized the real-valued Malvar wavelets to complex-valued Malvar wavelets. With this generalization, we are able to construct orthogonal bases from windowed complex exponentials besides real sinusoids, which overcomes the negative constraint of the Balian-Low Theorem.

Acknowledgments. This work was supported by the Air Force Office of Scientific Research under Grants AFOSR-616-92-0019 and AFOSR-616-93-0019.

References

1. Balian, R, Un principe d'incertitude fort en théorie du signal ou en mécanique quantique, *C. R. Acad. Sc. Paris* **292** série 2 (1987), 1357–1362.
2. Low, F., Complete sets of wave-packets, in *A Passion for Physics-Essays in Honor of Geoffrey Chew*, Word Scientific, Singapore, 1985, 17–22,
3. Battle,G., Heisenberg proof of Low's theorem, *Lett. Math. Phys.* **15** (1988), 175–179.
4. Daubechies, I., The wavelet transform, time-frequency localization, and signal analysis, *IEEE Trans. Inform. Theory* **36** (1990), 961–1005.
5. Daubechies, I. and A. J. E. M. Janssen, Two theorems on lattice expansions, *IEEE Trans. Inform. Theory* **39** (1993), 3–6.
6. Daubechies, I., S. Jaffard, and J.-L. Journé, A simple Wilson orthonormal basis with exponential decay, *SIAM J. Math. Anal.* **22** (1991), 554–572.
7. Wilson, K. G., Generalized Wannier functions, Cornell University, 1987, preprint.
8. Auscher, P., G. Weiss, and W. V. Wickerhauser, Local sine and cosine basis of Coifman and Meyer and the construction of smooth wavelets, in *Wavelets: A Tutorial in Theory and Applications*, C. K. Chui (ed.), Academic Press, Boston, 1992, 237–256.
9. Auscher, P., Remarks on the local Fourier bases, in *Wavelets: Mathematics and Applications*, J. J. Benedetto and M. W. Frazier (eds.), CRC Press, Ann Arbor, 1993, 204–218.
10. Wickerhauser, M. V., Smooth localized orthonormal bases, *C. R. Acad. Sci. Paris* **316** Serie I (1993), 423–427.
11. Malvar, H. S. and D. H. Staelin, The LOT: Transform coding without blocking effects, *IEEE Trans. ASSP* **37** (1989), 553–559.
12. Malvar, H. S., Lapped transforms for efficient transform/subband coding, *IEEE Trans. ASSP* **38** (1990), 969–978.
13. Malvar, H. S., Extended lapped transforms: Properties, applications and fast algorithms, *IEEE Trans. Signal Processing* **40** (11) (1992), 2703–2714.
14. Malvar, H. S., *Signal Processing with Lapped Transforms*, Artech House, Boston, MA, 1992.
15. Coifman, R. et Y. Meyer, Remarques sur l'analyse de Fourier à fenêtre, *C. R. Acad. Sci. Paris* **312** Serie I (1991), 259–261.
16. JPrincen, J. P. and A. B. Bradley, Analysis/synthesis filter bank design based on time domain aliasing cancellation, *IEEE Trans. ASSP* **34** (1986), 1153–1161.
17. Cassereau, P., A new class of optimal unitary transforms for image processing, M.S. thesis, Dept. Elec. Eng. Comput. Sci., Massachusetts Inst. Technol., Cambridge, May 1985.
18. Akanasu, A. N. and F. E. Wadas, On lapped orthogonal transforms, *IEEE Trans. Signal Processing* **40** (1992), 439–443.

19. Suter, B. W. and M. E. Oxley, On windows and orthonormal bases, in *Progress in Wavelet Analysis and Applications*, Y. Meyer and S. Roques (eds.), Editions Frontieres, Paris, France, 1993, 237–242.
20. Suter, B. W. and M. E. Oxley, On variable length windows and weighted orthonormal bases, in *Proceedings of the AFIT/AFOSR Workshop on The Role of Wavelets in Signal Processing Applications*, B. W. Suter, M. E. Oxley, and G. T. Warhola (eds.), (Defense Technical Information Center, DTIC Number ADA256127), Wright-Patterson AFB, Ohio, March 1992, 166–178.
21. Suter, B. W. and M. E. Oxley, On variable overlapped windows and weighted orthonormal bases, Technical Report, Air Force Institute of Technology, Wright-Patterson AFB, Ohio, April 1992, to appear in *IEEE Trans. Signal Processing.*
22. Young, R. W. and N. G. Kingsburg, Frequency-domain motion estimation using a complex lapped transform, *IEEE Trans. Image Processing* **2** (1) (1993), 2–17.
23. Meyer, Y., *Wavelets: Algorithms and Applications*, SIAM, Philadelphia 1993.
24. Xia, X.-G. and B. W. Suter, A family of multidimensional nonseparable Malvar wavelets, Technical Report, Air Force Institute of Technology, Wright-Patterson AFB, Ohio, 1993.
25. Vaidyanathan, P. P. and T. Chen, Role of anticausal inverses in multirate filter-banks–Part I: System-theoretic fundamentals, 1993, preprint.
26. Vaidyanathan, P. P. and T. Chen, Role of anticausal inverses in multirate filter-banks–Part II: The FIR case, factorizations, and biorthonormal lapped transform, 1993, preprint.
27. Vaidyanathan, P. P., *Multirate Systems and Filter Banks*, Prentice Hall, Englewood Cliffs, NJ, 1993.
28. Daubechies, I., *Ten Lectures on Wavelets*, SIAM, Philadelphia, 1992.
29. Chui, C. K., *An Introduction to Wavelets*, Academic Press, New York, 1992.
30. Daubechies, I., Wavelets on the interval, in *Progress in Wavelet Analysis and Applications*, Y. Meyer and S. Roques (eds.), Editions Frontieres, Paris, France, 1993, 95–107.
31. Cohen, A., I. Daubechies, B. Jawerth, and P. Vial, Multiresolution analysis, wavelets and fast algorithms on an interval, *C. R. Acad. Sci. Paris* **316** Série I (1993), 417–421.
32. Cohen, A., I. Daubechies, and P. Vial, Wavelets and fast algorithms on the interval, AT&T Bell Laboratories, preprint.
33. Meyer, Y., Ondelettes sur l'intervalle, *Rev. Math. Iberoamericana*, 1992.
34. Herley, C., J. Kovačević, K. Ramchandran, and M. Vetterli, Tilings of time frequency plane: Construction of arbitrary orthogonal bases and fast tiling algorithms, *IEEE Trans. Signal Processing*, (Special Issue on Wavelets and Signal Processing), December 1993.
35. Coifman, R. R. and M. V. Wickerhauser, Entropy-based algorithms for

best basis selection, *IEEE Trans. Inform. Theory*, (Special Issue on Wavelets), **38** (2) (1992), 713–718.

36. Wesfried, E. and M. V. Wickerhauser, Adapted local trigonometric transforms and speech processing, *IEEE Trans. Signal Processing*, (Special Issue on Wavelets and Signal Processing), December 1993.
37. Suter, B. W. and M. E. Oxley, On lapped transform with asymmetrical overlaps, in *Proc. ICASSP 94*, Adeladie, Australia, April, 1994, to appear.
38. Coifman, R. R., Private communication, 1992.
39. Benedetto, J. J., Irregular sampling and frames, in *Wavelets: A Tutorial in Theory and Applications*, C. K. Chui (ed.), 1992, 445–507,

Bruce W. Suter
Department of Electrical and Computer Engineering
Air Force Institute of Technology
2950 P Street
Wright-Patterson Air Force Base, OH 45433-7765
bsuter@afit.af.mil

Mark E. Oxley
Department of Mathematics and Statistics
Air Force Institute of Technology
2950 P Street
Wright-Patterson Air Force Base, OH 45433-7765
moxley@afit.af.mil

Time Scale Energetic Distribution

Guy Courbebaisse, Bernard Escudié, and Thierry Paul

Abstract. This paper introduces the "affine Wigner distribution" P_X and presents its relationship with the affine Wigner operator. In the derivation of this expression, we find it convenient to use a family of coherent states introduced by one of the authors. In addition, a computational scheme is given, and the analytic distribution

$$P_X^{\alpha=0}(t,\eta) = \int_{I\!R} \Xi^*(u)\Xi\left(-\eta\left(t+\frac{\eta t_0^2}{u+\eta t}\right)\right) du$$

is obtained, where $X = d\Xi/dt$. The so-called "Altes class" of signals seems to match well with this representation which uses the primitive of the studied signals. The main feature of this method is to use the time representation of the signal. Some other time-scale distributions such as Bertrand distribution, $D2$ Flandrin distribution, bilinear Grossmann distribution, Unterberger distribution, to name a few, use the frequency representation of the signal.

§1 Introduction

Wavelet techniques have evolved since their introduction by J. Morlet and A. Grossmann [10]. They are now well-known as time-scale or time-relative frequency representations. Later, a different time-scale representation [4] has been defined as affine Wigner representation [8]. Recently, a new class of quadratic time-frequency representations has been proposed (see [15,16]). Parallel to this development, physicists look for similar distributions such as the passive Unterberger distribution (see [5,6,7,20]). This paper is related to the results derived from those that deal with quantum mechanics, mainly the definition of "affine Wigner functions" and the coherent states [18]. For signal processing considerations, one has to deal with time representations.

Wavelets: Theory, Algorithms, and Applications
Charles K. Chui, Laura Montefusco, and Luigia Puccio (eds.), pp. 311–322.

ISBN 0-12-174575-9

§2 Analytic signal and Hilbert space

2.1 Analytic signal

The signals to be processed are generated by physical systems which provide real-valued data:

$$S(t) \in I\!R, \quad t \in I\!R. \tag{1}$$

The Fourier transform of such signals has the well-known properties:

$$S(t) \rightleftharpoons s(\nu), \quad s \in \mathbb{C}, \nu \in I\!R. \tag{2}$$

If $S(t) \in I\!R$, then

$$s(\nu) = s^{\#}(\nu) := s^*(-\nu). \tag{3}$$

If a signal is of finite energy, then one has the Parseval identity:

$$E_S = \langle S, S \rangle = \int_{I\!R} |S(t)|^2 \, dt = E_s = \langle s, s \rangle = \int_{I\!R} |s(\nu)|^2 \, d\nu < \infty. \tag{4}$$

The analytic signal, as defined by Ville [19], is related to S by:

$$X(t) \rightleftharpoons x(\nu) = 2U(\nu)s(\nu), \quad X(t), x(\nu) \in \mathbb{C}, \quad t \in I\!R, \quad \nu \in I\!R^+, \tag{5}$$

where U is the Heaviside step function, and

$$X(t) = 2\int_0^\infty s(\nu)e^{2i\pi\nu t} d\nu. \tag{6}$$

$X(t)$ can be analytically continued to the upper half-plane $\{\theta = t + i\tau \in \mathbb{C}, \tau > 0\}$, *i.e.*, all the singularities of $X(\theta)$ are located in the closed lower complex half plane (see Figure 1).

Classical examples of analytic signals are provided by functions of the form

$$X(\theta) = \frac{1}{1 - 2i\pi\frac{\theta}{T}} \rightleftharpoons x(\nu) = Te^{-\nu T}U(\nu),$$

where $T > 0$ and the pole is $\tau_1 = -i\frac{T}{2\pi}$, as well as products of such functions. As a complex-valued signal, $X(t)$ can be written as

$$X(t) = S(t) + iQ(t), \tag{7}$$

where $Q(t)$ is the quadrature signal

$$Q(t) \rightleftharpoons q(\nu) = -i\,\mathrm{sgn}(\nu)s(\nu) \tag{8}$$

which may be expressed as an Hilbert transform

$$Q(t) = \frac{1}{\pi}\ \mathrm{p.v.} \int_{I\!R} \frac{S(u)}{t-u} du. \tag{9}$$

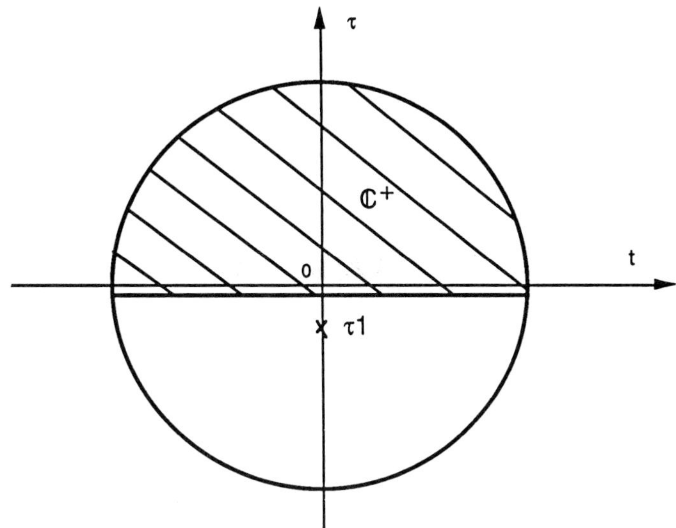

Figure 1. Pole location and the upper half complex plane $\mathbb{C}^+$.

2.2 Hilbert space of analytic functions

$F(\theta)$ and $X(\theta)$ are functions which are analytic in the domain $Im(\theta) > 0$, or $\theta \in \mathbb{C}^+$ [21].

$$F(\theta) = F(t + ia) = F\left(t + \frac{i}{\eta}\right) = \int_0^\infty \xi^{\alpha+1} x(\xi) e^{2i\pi\theta\xi} d\xi \tag{10}$$

which implies that

$$F(\theta) = (2i\pi)^{-(\alpha+1)} \int_0^\infty (2i\pi\xi)^{\alpha+1} x(\xi) e^{2i\pi\theta\xi} d\xi = (2i\pi)^{-(\alpha+1)} \frac{d^{\alpha+1}}{d\theta^{\alpha+1}} X(\theta), \tag{11}$$

where $a = 1/\eta$ is a scale or dilation factor which is dimensionless, and θ is a relative time variable defined as the ratio of two time or duration variables. We define the space $\mathcal{H}_\alpha$, as in ([2,11]),

$$\mathcal{H}_\alpha = \left\{ F : \mathbb{C}^+ \to \mathbb{C},\ F \text{ analytic },\ \int_0^\infty \int_{-\infty}^\infty |F(x+iy)|^2 y^\alpha dx dy < \infty \right\} \tag{12}$$

where $\mathcal{H}_\alpha$ is a Hilbert space. The following correspondence T_ℓ, with $\ell > 0$, is established in [18], between $L^2(\mathbb{R}^+, dq)$ and $\mathcal{H}_{2\ell-1}$:

$$\Psi \in L^2(\mathbb{R}^+, dq) \longrightarrow F(z) = [T_\ell \Psi](z)\,;\ z = x + iy, \tag{13}$$

where

$$F(z) = \int_0^\infty q^\ell e^{izq}\Psi(q)dq. \tag{14}$$

Here, T_ℓ is unitary between $L^2(I\!R^+, dq)$ and $\mathcal{H}_{2\ell-1}$. With each element of $SL(2, I\!R)$, we have

$$\begin{pmatrix} a & b \\ c & d \end{pmatrix} \in SL(2, I\!R) \tag{15}$$

is associated the operator

$$Y_\alpha\left(\begin{pmatrix} a & b \\ c & d \end{pmatrix}^{-1}\right) \tag{16}$$

acting in $\mathcal{H}_\alpha$ as

$$\left(Y_\alpha\left(\begin{pmatrix} a & b \\ c & d \end{pmatrix}^{-1}\right)F\right)(z) = (cz+d)^{-\alpha-2}F\left(\frac{az+b}{cz+d}\right). \tag{17}$$

This family of operators gives a unitary representation of $SL(2, I\!R)$. If we take into account the subgroup of upper triangular matrices of $SL(2, I\!R)$, isomorphic to the $ax+b$ group via the parametrization

$$\begin{pmatrix} \sqrt{a} & \frac{b}{\sqrt{a}} \\ 0 & \frac{1}{\sqrt{a}} \end{pmatrix}, \tag{18}$$

then the restriction of the representation Y_α gives the following representation of the $ax+b$ group

$$\left(V_\alpha(a,b)F\right)(z) = a^{-\frac{\alpha+2}{2}}F\left(\frac{z-b}{a}\right). \tag{19}$$

§3 Analytic signal and wavelet representation

3.1 A quadrature property

Let $X(t) = S(t) + iQ(t)$ be the analytic signal related to S. The quadrature signal related to $X(t)$ is defined by

$$-Q[X(t)] = \left(X(t') *_{t'} P\left(\frac{-1}{\pi t'}\right)\right) \rightleftharpoons \operatorname{sgn}(\nu)[1+\operatorname{sgn}(\nu)]s(\nu) = -z(\nu). \tag{20}$$

The result is that

$$\frac{i}{\pi}\int_{I\!R}\frac{X(t')}{t'-t}dt' = X(t) \tag{21}$$

which is nothing but the Cauchy formula.

3.2 Wavelet properties

The result presented above may be interpreted as a wavelet transform of $X(t)$, namely

$$X(\theta) = \frac{i}{\pi}\left(\int_{I\!R} \frac{X(u)}{u - t - \frac{i}{\eta}} du\right)_{t=\theta} . \tag{22}$$

$G(u)$ is the wavelet kernel defined as

$$\Theta(t, \eta) = \sqrt{\eta}\int_{I\!R} X(u) G^*(\eta(u - t)) du. \tag{23}$$

As defined by Equation (22) G is not an admissible wavelet. To achieve this property we define

$$G_\alpha = G^{\alpha+1}(u) = \left(\frac{1}{u - i}\right)^{\alpha+1} \Rightarrow g(0) = 0; \quad \int_{I\!R} G(u) du = g(0). \tag{24}$$

As a result we point out that

$$\frac{d^{\alpha+1} X(\theta)}{d\theta^{\alpha+1}} = \frac{i}{\pi}\eta \int_{I\!R} \frac{X(u)}{\eta(u - t) - i} du, \tag{25}$$

which is the wavelet

$$g(\nu) = \nu^\alpha T e^{-\nu T} \rightleftharpoons G(u) = \frac{1}{(2i\pi)^\alpha} \frac{d^\alpha}{du^\alpha}\left[\left(1 - 2i\pi\frac{u}{T}\right)^{-1}\right] \tag{26}$$

that was previously studied by one of the authors [18].

§4 The affine Wigner representation

4.1 Parity operator in the space $\mathcal{H}_\alpha$

The $SL(2, R)$ group acts on $\mathbb{C}^+$ by the action

$$z \longrightarrow \frac{az + b}{cz + d} . \tag{27}$$

As a special element of the maximal compact subgroup of $SL(2, I\!R)$, *i.e.*, the group of orthogonal matrices

$$SO(2) = \left\{\begin{pmatrix} \cos\theta & -\sin\theta \\ \sin\theta & \cos\theta \end{pmatrix}, \ \theta \in [0, 2\pi]\right\} \subset SL(2, I\!R) \tag{28}$$

is obtained and the element

$$\begin{pmatrix} 0 & -1 \\ 1 & 0 \end{pmatrix} \in SO(2) \tag{29}$$

is chosen. This corresponds to the transformation A defined by $z \to -1/z$. The transformation A leaves i invariant, and A is the symmetry with respect to i, being equivalent to the transformation $(t, \nu) \to (-t, -\nu)$ (see [3]).

As a parity operator in the $\mathcal{H}_\alpha$ space, the I operator is given by [18] as follows.

$$I(F(\theta)) = \theta^{-(\alpha+2)} F\left(\frac{-1}{\theta}\right), \quad \theta \in \mathbb{C}^+, \tag{30}$$

where I is an unitary involutive self-adjoint operator in $\mathcal{H}_\alpha$ [18].

4.2 Affine Wigner operator and affine Wigner representation

The construction of the affine Wigner operator is due to the action of the $ax+b$ group, given by $V_\alpha(a,b)$, which displaces the operator I [18]. This allows us to define the affine Wigner operator as

$$I_{[a,b]} = V_\alpha^{-1}(a,b) I V_\alpha(a,b) \tag{31}$$

since

$$V_\alpha^{-1}(a,b) = Y_\alpha \begin{pmatrix} \sqrt{a} & \frac{b}{\sqrt{a}} \\ 0 & \frac{1}{\sqrt{a}} \end{pmatrix}, \tag{32}$$

where Y_α is given by (17) and I is defined by

$$I = Y_\alpha \begin{pmatrix} 0 & -1 \\ 1 & 0 \end{pmatrix}. \tag{33}$$

The affine Wigner operator is defined in [18] by

$$I_{[a,b]} F(\theta) = (a\theta + b)^{-(\alpha+2)} F\left(\frac{-b\theta - \frac{1+b^2}{a}}{a\theta + b}\right). \tag{34}$$

Under this condition, there exists a unitary transformation which maps $\mathcal{H}_\alpha \rightarrow L^2(I\!R, dx)$, which is the Hilbert space for analytic signals. The time-scale or affine Wigner representation is defined by the mean value:

$$P_\alpha(b,a) = \langle F | I_{[a,b]} | F \rangle. \tag{35}$$

Taking $X(t)$ and Equation (34) into account, we have

$$\frac{\partial}{\partial\theta} X\left(\frac{-b\theta - \frac{1+b^2}{a}}{a\theta + b}\right) = \frac{1}{(a\theta+b)^2}\frac{dX}{d\theta} \quad \text{if } \alpha = 0. \tag{36}$$

Let us now define

$$I_{[a,b]} X_{\alpha+1}(\theta) = \Xi\left(\frac{-bt - \frac{1+b^2}{a}}{at + b}\right)_{t=Re[\theta]} \tag{37}$$

where

$$X(t) = \frac{d\Xi(t)}{dt},$$

and

$$P_X^{\alpha=0}(b,a) = \int_{I\!R} \Xi^*(t) \Xi\left(\frac{-bt - \frac{1+b^2}{a}}{at + b}\right) dt, \tag{38}$$

where $X(t)$ is the analytic signal related to signal $S(t)$ [7]. X is an analytic function of $\theta \in \mathbb{C}^+$. We consider only the real values of the relative time variable t, assuming that $X(t)$ is analytic on the real axis $t = Re(\theta)$.

The general affine Wigner distribution is defined by

$$P_\alpha(b,a) = \int_{C+} F^*(\theta) F\left(\frac{-b\theta - \frac{1+b^2}{a}}{a\theta + b}\right)(a\theta + b)^{-(\alpha+2)} d\theta \tag{39}$$

where F is an element of $\mathcal{H}_\alpha$.

4.3 The time-scale affine Wigner representation

The previous result may be interpreted as a time-scale or affine Wigner representation of finite energy signal as

$$P_\alpha(b, t_0, a) = \int_{I\!R} F^*(u) F\left(\frac{-bu - \frac{t_0^2+b^2}{a}}{au+b}\right)(au+b)^{-(\alpha+2)}du, \tag{40}$$

where t_0 is a time reference and t_0^2/a plays the same role as $1/a$. With $a = 1/\eta$ we obtain

$$P_\alpha(t, t_0, \eta) = \eta^{\alpha+2} \int_{I\!R} F^*(u) F\left(\eta \frac{-bu - \eta(t_0^2 + t^2)}{au+b}\right)(u+\eta t)^{-(\alpha+2)}du. \tag{41}$$

Under the constraint that $\alpha = 0$, the affine Wigner representation is expressed as follows

$$P_0(t, t_0, \eta) = \frac{\eta}{t_0^2} \int_{I\!R} F^*\left(\eta\left(\frac{t_0^2}{t'} - t\right)\right) F(-\eta(t' + t))dt',$$

where

$$t' = \frac{\eta t_0^2}{u + \eta t}. \tag{42}$$

The so-called "time warping" transformation is logarithmic, namely

$$w = \log\left(\frac{t'}{t_0}\right) \rightarrow t' = t_0 e^w, \tag{43}$$

where w is a dimensionless variable. The affine Wigner representation may be displayed as follows:

$$P_0(t, t_0, \eta) = \frac{\eta}{t_0} \int_{I\!R} F^*(\theta' + \eta t_0 e^{-w}) F(\theta' - \eta t_0 e^w) e^w dw \tag{44}$$

where $-\theta' = \eta t$ and P_0 is a bilinear representation, with a symmetric structure given by

$$\eta t_0 e^w \rightarrow -\eta t_0 e^{-w}, \tag{45}$$

and a "time warping" transformed variable as defined by Equation (43). This property may be interpreted as time transformation similar to those defined for frequency variables of the hyperbolic class [15].

§5 Time-scale affine representation of finite energy signals

5.1 A class of "scale invariant" signals

$P_X^0(t, t_0, \eta)$ is related to $\Xi(t)$ by: $d\Xi/dt = X(t)$. To compute P_X^0, one has to derive $\Xi(t)$ from $X(t)$. There exists a class of "scale invariant" signals defined by R. A. Altes [1] as

$$x(g^m \nu) = x(\nu)\left(\frac{\nu}{\nu_0}\right)^{-m} k \tag{46}$$

where ν_0 is a reference for frequency, k being a constant

$$x\left(\frac{\nu}{g^m}\right) = \left(\frac{2i\pi}{b_0}\right)^m \nu^m x(\nu) \tag{47}$$

and

$$x(\nu) = A \exp\left\{-\frac{\ln^2\left(\frac{\nu}{\nu_0}\right)}{2\ln(g)}\right\} \exp\left\{-\frac{2i\pi b \ln\left(\frac{\nu}{\nu_0}\right)}{\ln(g)}\right\} U(\nu), \tag{48}$$

where $U(\nu)$ is the Heaviside function. $x(\nu) \rightleftharpoons X(t)$ may be derived so as to get the main property:

$$X_{-(\alpha+1)}(t) \rightleftharpoons b_0^{-(\alpha+1)} x\left(g^{\alpha+1}\nu\right) = x(\nu)\left(\frac{\nu}{\nu_0}\right)^{\alpha+1} \frac{k}{b_0^{\alpha+1}}\,. \tag{49}$$

The representation $P_X^0(t, t_0, \eta)$ can be written as

$$P_X^0(t, t_0, \eta) = (bg)^{-(\alpha+1)} \int_{I\!R} X^*(t') X\left(-\eta\left(t + \frac{\eta t_0^2}{t' + \eta t}\right)\right) dt' \tag{50}$$

with the constraint $d\Xi/dt = X(t)$. Under asymptotic conditions such that

$$X(t) = A(t) e^{i\Phi(t)}, \quad \nu_i(t) = \frac{1}{2\pi}\frac{d\Phi}{dt}, \tag{51}$$

$$\frac{1}{A}\left|\frac{dA}{dt}\right| \ll \nu_i(t)\,, \quad \sqrt{\left|\frac{d\nu_i}{dt}\right|} \ll \nu_i\,, \tag{52}$$

$P_X^0(t, t_0, \eta)$ for this class of "scale invariant" signals may be computed easily with the stationary phase method.

5.2 Properties of the $P_X^\alpha(t, t_0, \eta)$ representation

The operator $I_{[a,b]}$ acts explicitly in $L^2(I\!R, dx)$

$$I_{[a,b]}x(\xi) = e^{2i\pi\frac{b}{a}\xi}\int_0^\infty J_\ell\left(2\sqrt{\frac{xy}{a}}\right)x(ay)e^{-2i\pi by}dy\,, \tag{53}$$

where $\ell = \alpha+1, \alpha > -1$ (see [3]). If we take into account

$$\int_0^\infty\int_{I\!R} P_X^\alpha\left(t, t_0, a=\frac{1}{\eta}\right)\frac{da}{a^2}dt \;=\; cE_X, \tag{54}$$

(see [4]), where the constant c can be chosen to be 1 and derive the zero order moment, then we obtain

$$\int_{I\!R}\frac{1}{\eta}J_\ell(2\xi\eta)x(\xi)x^*(\xi)d\xi,\;\; \eta = \frac{1}{a}. \tag{55}$$

This quantity depends on the Bessel function $J\ell(2\xi\eta)$ as well as on the spectral density $|x(\xi)|^2$.

5.3 $P_X^\alpha(t, t_0, \eta)$ and asymptotic signals

Under the asymptotic conditions, there exists an easy derivation of $\Xi(t)$ given by $X(t)$ in the form of Equation (51), namely

$$\frac{d\Xi}{dt} = \left(\dot{A}_\Xi + i\dot{\Phi}_\Xi A_\Xi\right)e^{i\Phi_\Xi}, \approx i\dot{\Phi}_\Xi A_\Xi e^{i\Phi_\Xi}, \tag{56}$$

where we may consider

$$F(t) = A_\Xi(t)e^{i\Phi_\Xi(t)}.$$

This implies that

$$\Xi(t) \approx 2\pi\nu_i(t)A(t)e^{i\left(\Phi(t)-\frac{\pi}{2}\right)} \tag{57}$$

and $P_X^0(t, t_0, \eta)$ is derived as

$$P_X^0(t, t_0, \eta) = 4\pi^2\int_{I\!R}\nu_i(u)\nu_i\left(-\eta\left(t+\frac{\eta t_0^2}{u+\eta t}\right)\right)A(u)A\left(-\eta\left(t+\frac{\eta t_0^2}{u+\eta t}\right)\right)$$

$$\times e^{-i\left(\Phi(u)-\Phi\left(-\eta\left(t+\frac{\eta t_0^2}{u+\eta t}\right)\right)\right)}du. \tag{58}$$

This result shows that $P_X^0(t, t_0, \eta)$ may be easily computed by the stationary phase method under asymptotic conditions.

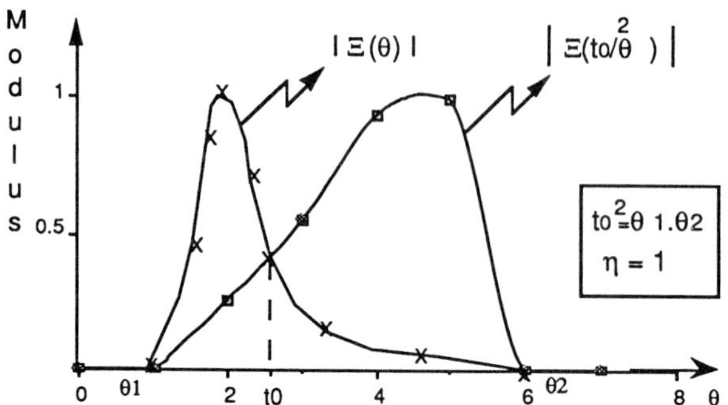

Figure 2. Effect of the affine Wigner operator on $\Xi(\theta)$.

§6 Computation of $P_X^{\alpha}(t, t_0, \eta)$

6.1 Time reference for finite duration signals

For asymptotic conditions the analytic signal may be represented, as in [17], as the exponential model

$$S(t) = A(t)\cos\Phi(t) = Re\,X(t), \tag{59}$$

where $A(t)$ may be considered as a "finite duration" signal, provided that the "duration bandwith" product: $BT \gg 1$ [1]; this signal may be represented as a band limited one, namely

$$\nu_m \le \nu_i(t) \le \nu_M\,;\ \nu_M - \nu_m = B. \tag{60}$$

Figure 2 displays the time axis and the duration T defined as

$$P_X^0(t, t_0, \eta) = \int_{I\!R} \Xi^*(u)\Xi\left(-\eta\left(t + \frac{\eta t_0^2}{u + \eta t}\right)\right) du, \tag{61}$$

where $t_0^2 = \theta_1\theta_2, \theta = u + \eta t$. It follows that

$$P_X^0(t, t_0, \eta) = \int_{I\!R} \Xi^*(\theta - \eta t)\Xi\left(-\eta\left(t + \frac{\eta t_0^2}{\theta}\right)\right) d\theta, \tag{62}$$

where variables θ_1 and θ_2 are defined as those for the time scale representation derived by A. Grossmann et al. [13]. These variables are such that the "duration" of $\Xi(\theta)$ and $\Xi(t_0^2/\theta)$ is the same [5], as shown in Figure 2.

6.2 A way to compute $P_X^0(t, t_0, \eta)$

A way to compute $P_z^0(t, t_0, \eta)$ is presented in the above where

$$P_X^0(t, t_0, \eta) = \int_{\theta_1 - \eta t'}^{-\theta_1 \eta^2 + \eta t'} \Xi^*(\theta - \eta t) \Xi\left(-\eta\left(t + \frac{\eta\theta_1\theta_2}{\theta}\right)\right) d\theta, \tag{63}$$

with $t = t' < 0$ such that

$$0 \leq t' \leq \frac{\theta_1}{2\eta} + \frac{\eta\theta_2}{2}. \tag{64}$$

This method requires computation of the discrete sum for sampled signals

$$P_X^0(t' = m\Delta t, t_0, \eta) = \sum_{k_0}^{k_m} \Xi^*(k\Delta t - \eta m\Delta t) \Xi\left(-\eta m\Delta t - \frac{\eta^2 k_1 k_2}{k}\right) \tag{65}$$

where $\theta = k\Delta t$, $\theta_1 = k_1\Delta t$, $\theta_2 = k_2\Delta t$, $t' = m\Delta t$, and Δt is the sampling period defined according to the Nyquist theorem.

§7 Conclusion

Time-scale representations of finite energy signals may be defined in different ways. Taking into account the properties of analytic signals, we are able to define a symmetric involutive operator acting on the $ax + b$ group, analogous to the symmetry operator (or parity) used on the Weyl-Heisenberg group. The time-scale representation defined as the mean value of this operator appears to be a possible tool for signal processing, mainly for asymptotic amplitudes and frequency modulated signals. In addition, the computation is optimized, since the time representation of the signal is used. This is a distinct difference from the other affine time-frequency distributions [9] where the Mellin transform [14] must be computed to implement the affine time-frequency distributions, taking into acount the hyperbolic sampling signals [14].

References

1. Altes, R. A., Method of wideband signal design for radar and sonar systems, Phys. Doct. Thesis, Dept. Elec. Eng., Univ. of Rochester, NY, 1970.
2. Bargmann, V., *Comm. Pure Appl. Math.* **14** (1961), 187.
3. Bargmann, V., *Comm. Pure Appl. Math.* **20** (1) (1961), 1.
4. Bertrand, J. and P. Bertrand, Affine time-frequency distribution, in *Time-Frequency Signal Analysis - Methods and Applications*, B. Boashash (ed.), Longman-Cheshire, Melbourne, Australia, 1991.
5. Courbebaisse, G., Transformée bilinéaire temps-échelle des signaux asymptotiques d'énergie finie, 14ème Coll. GRETSI sur le traitement du signal et des images, Juan les Pins(F), 1993, 41–44.

6. Escudié, B. and A. Grossmann, Une représentation bilinéaire en temps et en échelle des signaux d'énergie finie, 13ème Coll. GRETSI sur le traitement du signal et des images, Juan les Pins (F), 1991, 33–36.
7. Escudié, B., A. Grossmann, and R. Kronland-Martinet, Bilinear time scale representations of finite energy signals and properties related to signal processing, *ISSSE 92* Paris (CNAM), 1992, 485–487.
8. Flandrin, P. and O. Rioul, Affine smoothing of the Wigner-Ville distribution, *IEEE ICASSP*, Albukerque, NM, 1990, 2455–2458.
9. Flandrin, P., Sur une classe générale d'extensions affines de la distribution de Wigner-Ville, 13ème Coll. GRETSI sur le traitement du signal et des images, Juan les Pins (F), 1991, 17–20.
10. Grossmann, A. and J. Morlet, *SIAM J. Math. Anal.* **15** (NJ4) (1984), 723–736.
11. Lang, S., $SL_2(I\!R)$, Addison Wesley, Publ. Company, 1975.
12. Mamode, M. and B. Escudié, Estimation optimale de la date d'arrivée d'un signal et conditions de découplage vis à vis des paramètres perturbateurs: exemple de signaux sonar animaux à grand produit durée-bande, 8ème Coll. sur le Traitement du Signal, Nice (F), 1981, 91–97.
13. Martin, W. and B. Picinbono, Annales Télécomm. **t.38** (NJ5/6) (1983), 179–189.
14. Ovarlez, J. Ph., La transformation de Mellin: un outil pour l'analyse des signaux à large bande, Thèse, Univ. Paris VI, 1992.
15. Papandreou, A., F. Hlawatsch, and G. Faye Boudreaux-Bartels, Quadratic time-frequency representations: the new hyperbolic class and its intersection with the affine class, *Proc. 6th IEEE-SP Workshop on Statistical Signal & Array Processing*, Victoria, Canada, October 1992.
16. Papandreou, A., F. Hlawatsch, and G. Faye Boudreaux-Bartels, The hyperbolic class of quadratic time-frequency representations, Part 1, constant-Q warping, the hyperbolic paradigm, properties members, submitted to *IEEE Trans. Signal Proc.*, December 1993.
17. Paul, Th., Ondelette et mécanique quantique, Phys. Doct. Thesis, Univ. Aix-Marseille II, Luminy, December 1985.
18. Paul, Th., Affine coherent states and the radial Schrpdinger equation, Radial harmonic oscillator and hydrogen atom, CPT-84/p.1710, preprint.
19. Paul, Th., Functions analytic on the half-plane as quantum mechanical states, *J. Math. Phys.* **25** (NJ11) (1984), 3252–3263.
20. Unterberger, A., The calculus of pseudo differential operators of Fuchs type, *Comm. Part. Diff. Eq.* **9** (1984), 1179–1236
21. Ville, J., Câbles et transmissions NJ1, Tome 2, 1948, 61–74.

Guy Courbebaisse and Bernard Escudié
ICPI - CPE - LTS 25 Rue du Plat
69288 LYON Cedex 02 France

Thierry Paul
CEREMADE Univ. Paris IX Dauphine
75775 Paris Cedex 16 France

Part V

Wavelets and Fractals

Some Mathematical Results about the Multifractal Formalism for Functions

Stephane Jaffard

Abstract. Our purpose is to investigate the mathematical validity of the multifractal formalism for functions. We will give some general results showing that it always yields an upper bound for the spectrum of singularities (the function that associates to each positive α the Hausdorff dimension of the set where a function F is approximately Hölder of order α), and we will study some examples and counterexamples of its validity, for instance, the Riemann function $\varphi(x) = \sum n^{-2} \sin n^2 x$.

§1 Introduction

Our purpose in this paper is to explain the main mathematical results on the so-called multifractal formalism for functions. These results are either new or proved in [19] or [21].

The study of multifractal functions has proved important in several domains of physics. The examples include plots of random walks, development of interfaces in reaction-limited growth processes, and turbulent velocity signals at inertial range (see [3]). The relevant mathematical tool studied in this context is the spectrum of singularities (the function that associates to each positive α the Hausdorff dimension of the set where F is approximately Hölder of order α). The most important example where one would like to determine the spectrum of singularities of a function is probably the velocity of fully developed turbulence. The reason is that turbulent flows are not spatially homogeneous: the irregularity of the velocity seems to differ widely from point to point. This phenomenon, called "intermittency," suggests that the determination of the spectrum of singularities of the velocity of the fluid might be a nontrivial function, universal (*i.e.*, independant of boundary conditions in the limit of small viscosity), and thus would yield an important information on the nature of turbulence.

Wavelets: Theory, Algorithms, and Applications
Charles K. Chui, Laura Montefusco, and Luigia Puccio (eds.), pp. 325–361.

ISBN 0-12-174575-9

Obviously, it is almost impossible to compute numerically a spectrum from the mathematical definition since it involves the successive determination of several intricate limits, and a blind application of the formulas would yield enormous, totally unstable calculations. The only method is to find some "reasonable" assumptions under which the spectrum could be derived using only "averaged quantities" (which should be numerically stable) extracted from the signal. Such formulas for the spectrum can be guessed using similarities with statistical physics or a heuristic calculation that we will recall in the following. Frisch and Parisi proposed (see [11]), in one dimension, a formula using the L^p modulus of continuity of the velocity along one axis. Arneodo, Bacry, and Muzy (in [2,3,26]) proposed, in one dimension too, other formulas based on the wavelet transform of the signal, and they proved their validity when the function considered is the primitive of a multinomial measure, or a C^∞ perturbation of such a measure.

In this paper we will give some general results concerning the multifractal formalism for functions: we will show that, for any function, it yields an upper bound of its spectrum of singularities, but we will also show on some explicit counterexamples that, in general, it does not yield the exact spectrum. Then we will introduce a natural definition of "selfsimilar functions" for which the multifractal formalism holds. An example of these functions is obtained by considering the primitives of selfsimilar measures, but they include also widely oscillating, several dimensional functions, two requirements which are obviously neede, for instance, in a realistic modelization of turbulence. We will finally consider the Riemann function

$$\varphi(x) = \sum n^{-2} \sin n^2 x,$$

and show that it satisfies the multifractal formalism. This example is important for historical reasons (the determination of the pointwise regularity of φ has been a longstanding problem which has been completely settled only very recently, see [21]), but also because it indicates that the multifractal formalism must be valid under much weaker conditions than selfsimilarity.

Before describing the multifractal formalism, we need to recall some definitions and notations concerning the Hölder regularity of functions.

Suppose that α is a positive real number; a function $F : \mathbb{R}^m \to \mathbb{R}$ is $C^\alpha(x_0)$ if there exists a polynomial P of degree less than α such that

$$|F(x) - P(x - x_0)| \leq C|x - x_0)|^\alpha. \tag{1.1}$$

Note that the constant term of $P(x - x_0)$ is $F(x_0)$, and if F is $C^{[\alpha]}$ in a neighborhood of x_0, the polynomial P is exactly the Taylor expansion of F at x_0 of order α. Nonetheless, Equation (1.1) can hold for a large α even if F is not differentiable in a neighborhood of x_0 (consider the chirp $x^n \sin(x^{-n})$ in a neighborhood of 0 for a large n).

The function F belongs to $\Gamma^\alpha(x_0)$ if (see [19])

$$\begin{aligned} &F \notin C^\eta(x_0) \quad \forall\, \eta > \alpha, \\ &F \in C^\eta(x_0) \quad \forall\, \eta < \alpha. \end{aligned} \tag{1.2}$$

A function F is C^α (or $C^\alpha(\mathbb{R}^m)$) if Equation (1.1) holds for any x in $\mathbb{R}^m$, the constant C being uniform; F is Lipschitz if it is $C^1(\mathbb{R}^m)$ (using *this* definition, which means in this case that $\nabla f \in L^\infty$). We will also need the following slightly different definition (already introduced in [22]) which somehow asserts that the fact that F is singular at x_0 can be checked on a "large" set near x_0.

Definition 1. *Let α such that $-m < \alpha \leq 1$; a point x_0 is a strong α-singularity of F if F is $C^\alpha(x_0)$ and if there exist C, C' such that $\forall\ j, \exists A_j, B_j$*

$$\text{measure } (A_j) \geq C2^{-mj}, \text{ measure } (B_j) \geq C2^{-mj}$$
$$|x - x_0| \leq 2^{-j} \quad \forall\, x \in A_j \cup B_j,$$
$$\sup_{A_j} F \leq C'2^{-\alpha j} + \inf_{B_j} F.$$

Recall that the wavelet transform of a function F is defined by

$$C(a,b) = \frac{1}{a^m} \int F(t)\psi(\frac{t-b}{a})dt$$

where ψ is a radial function, with moments of order less than K vanishing and with derivatives of order less than K having fast decay (with a K "large enough" depending on the properties of F that we want to analyze). We will recall all the properties of the wavelet transform that we need; nonetheless, let us mention that all the properties that we will need are detailed in [22] or [25].

Definition 2. *The spectrum of singularities of a function F is, for each $\alpha \geq 0$ the number $d(\alpha)$ which is the Hausdorff dimension of the set of points x_0 where F is $\Gamma^\alpha(x_0)$.*

Note that $d(\alpha)$ is defined point by point for each value of α, and not as usual in the almost everywhere sense (the example of Riemann's function will clearly demonstrate the importance of the value of the spectrum at one particular point). This definition should be commented; it is not clear right now if the "good" definition of the dimension is the Hausdorff or the Packing dimension; it will, however, be clear in the following that the box (or capacity) dimension is not the good notion.

We are now in a position to describe the methods used by Frisch and Parisi on one side and Arneodo, Bacry, and Muzy on the other in order to determine the spectrum of singularities of functions.

The *Structure Function* method requires first the computation of

$$S_q(l) = \int_{\mathbb{R}^m} |F(x+l) - F(x)|^q dx.$$

Suppose that the scaling $S_q(l) \sim |l|^{\zeta(q)}$ holds. The spectrum of singularities is computed using the formula

$$d(\alpha) = \inf_q (q\alpha - \zeta(q) + m). \tag{1.3}$$

The *Wavelet Transform Integral* method requires one to compute

$$\tilde{Z}(a,q) = \int_{\mathbf{R}^m} |C(a,b)|^q db,$$

and, if $\tilde{Z}(a,q) \sim a^{\eta(q)}$,

$$d(\alpha) = \inf_q (q\alpha - \eta(q) + m). \tag{1.4}$$

The *Wavelet Transform Maxima* method requires first the computation of

$$Z(a,q) = \sum_{\ell \in \mathcal{L}(a)} \sup_{(b,a')\in\ell} |C(a',b)|^q,$$

where ℓ is a line of maxima of the wavelet transform considered on $[a,0]$, and $\sup_{(b,a')\in\ell}$ means that the supremum is taken for (b,a') on this line of maxima (so that $a' \leq a$). If $Z(a,q) \sim a^{\theta(q)}$, then

$$d(\alpha) = \inf_q (qh - \theta(q)). \tag{1.5}$$

Numerically, according to [3], the best method seems to be the last one, probably because the restriction of the computation to the maxima insures that small errors are less likely taken into account since, at the maxima, they are relatively less important. More generally, methods that involve the wavelet transform are numerically more stable, probably because they involve only averaged quantities and not the direct values of the function.

The multifractal formalism may be surprising at first sight because it relates pointwise behaviors to norms (*i.e.*, global) estimates. Before giving some mathematical explanations for it, it may be enlightening to give the heuristic classical argument from which it is derived. Though this argument cannot be transformed into a correct mathematical proof, it shows at least why these formulas can be expected to hold, and a careful study of its defects shows under which type of additional conditions it should be mathematically correct.

Let us calculate the contribution of singularities of order α to the integral

$$\int_{\mathbf{R}^m} |F(x+l) - F(x)|^q dx.$$

Near a singularity of order α, we have, in a small box of size l,

$$|F(x+l) - F(x)|^q \sim l^{\alpha q}.$$

If the dimension of these singularities is $d(\alpha)$, it means that there are about $l^{-d(\alpha)}$ such boxes, each of volume l^m, so that the total contribution to the integral is $l^{\alpha q+m-d(\alpha)}$. The real order of magnitude of the integral is given by the largest contribution, so that

$$\eta(q) = \inf_{\alpha}(\alpha q + m - d(\alpha)). \tag{1.6}$$

This formula is not the one we are looking for since we know $\eta(q)$ and we look for $d(\alpha)$. However, if Equation (1.6) holds and if d is concave (we will see that in general this assumption needs not be verified, however in many cases it is), $d(\alpha)$ is recovered by an inverse Legendre transform formula which yields Equation (1.4). Of course, if d is not concave, one expects Equation (1.4) to yield only the convex hull of the spectrum.

Obviously, Equation (1.6) is more likely to hold because the concavity problem does not appear for Equation (1.6) (a straightforward application of Young's formula shows that $\eta(q)$ is always concave).

The problem of the validity of Equation (1.6) has been investigated by I. Daubechies and J. Lagarias in [7], where it is shown to hold for some functions solutions of two-scales equations. It is, nonetheless, easy to construct counterexamples to this formula (see [19]). But, this heuristic argument gives a justification to the use of Equation (1.3). The fact that it also gives a justification to Equation (1.4) is because $S_q(l)$ and $\eta(q)$ are of the same order of magnitude for any function (a straightforward consequence of Sobolev-type imbeddings that we will recall in Section 3; we will study in details the wavelet maxima method in Section 6.

Let us come back to the heuristic argument in favor of the multifractal formalism we gave. A careful inspection shows that it supposes that the same type of scalings near a given point hold at all scales; if there were big smooth bumps at certain scales, which are not followed by smaller ones, these bumps would bring a contribution to the integral without generating singularities and the formula would fail. Actually this remark is the key to several points of this paper: counterexamples to the multifractal formalism are constructed following that idea, and it will hold in situations where we clearly impose such local scaling properties. (it will be explicitely stated in the definition of selfsimilar functions but it will be a little less obvious in the case of Riemann's function).

Since the scalings assumed above do not necessarily hold, we will use the following definitions for ζ, η, and θ. Let

$$\zeta(q) = \liminf \frac{\log S_q(l)}{\log l};$$

$$\eta(q) = \liminf \frac{\log \tilde{Z}(a,q)}{\log a};$$

$$\theta(q) = \liminf \frac{\log Z(a,q)}{\log a}.$$

The first valid result without any specific assumption on the function F is the following.

Theorem 1. *If $q > 1$ and $\zeta(q) < q$, then $\xi(q) = \eta(q)$ for any function F. In general, the functions $\zeta(q)$ or $\eta(q)$ are not related to $\theta(q)$.*

If F is a function of one real variable, the box dimension of the graph of F is $\eta(1)$.

The following upper bound holds for any function F such that $\eta(p) > m \ \forall\, p$;

$$d(\alpha) \leq \inf_p (m - \eta(p) + \alpha p). \tag{1.7}$$

and without any assumption on η, if $D(\alpha)$ is the packing dimension of the strong $\alpha-$singularities,

$$D(\alpha) \leq \inf_p (m - \eta(p) + \alpha p).$$

In general, Equation (1.7) cannot be an equality. More precisely, let $d(\alpha)$ be a Riemann integrable positive function on $\mathbb{R}^+$. There exist F_1 and F_2 which share the same function η, but the spectrum of F_1 is $d(\alpha)$ and F_2 is C^∞ except at the origin.

Let us comment these results. We will show on some counterexamples that a smooth function (with a large $\eta(p)$) may nonetheless be such that $\theta(p)$ can be arbitrarily small (the case $\theta(p) = -\infty \ \forall\, p > 0$ can even happen).

Remark that the assertion concerning the counterexamples is stronger than the mere failure of the Legendre transform formulas. It asserts that there is not enough information in the function η to determine the spectrum. It also shows that, mathematically, "any" function can be a spectrum. It is surprising to notice that, in the applications it does not seem to be the case. The spectra computed numerically have always the same shape: roughly speaking, the upper part of an ellipse, the shape we will find for selfsimilar functions. There can be several explanations to this analogy. Either these signals satisfy some "scaling-invariance" properties which makes them fit in the framework (perhaps generalized in some ways) of selfsimilar functions, or a pessimistic explanation could be that while these spectra are being (perhaps wrongly) calculated using a Legendre transform formula, it is the convex hull of the true spectrum that is actually calculated, instead of the spectrum itself. Hence this generic shape results.

We will show that the "Wavelet Transform Maxima" method, after the following slight modification, yields the correct spectrum for selfsimilar functions: basically, the lines of maxima can be too close to each other; in that case we have to restrict the sum and keep for each interval of width a only the line passing through this interval that yields the largest contribution.

Let us now define the selfsimilar functions by analogy with the selfsimilar sets. We first recall the following definition of [17].

A set Ω is strictly selfsimilar if it is a finite union of disjoint subsets $\Omega_1, \ldots, \Omega_d$ which can be deduced from Ω by similitudes. For instance, the triadic Cantor set or the Van Koch curve are selfsimilar while, in two dimensions, the Sierpinski gaskets are selfsimilar. These sets have been widely studied, and so were the measures supported on them ([4, 17,9,3] for instance). They play an important role in the modelization of several physical phenomena.

Let us give a first intuitive definition of a selfsimilar function. Suppose that F is continuous and compactly supported and let Ω be the bounded open subset of $\mathbf{R}^n$ such that $\bar{\Omega} = \text{supp}(F)$. F is said to be selfsimilar if there exists disjoint open subsets $\Omega_1, \ldots, \Omega_d$ of Ω such that the graph of F restricted to each Ω_i is a "contraction" of the graph of F, modulo a certain error, supposed to be smooth. Suppose first that "smooth" means Lipschitz. Then let us formalize this definition.

There should exist contractive similitudes (*i.e.*, products of an isometry by an homothety $x \to \mu_i x$ of ratio $\mu_i < 1$.)) $(S_i)_{i=1,\ldots,d}$ such that, if $S_i(\Omega) = \Omega_i$,

1. $\Omega_i \subset \Omega \quad \forall\, i$,
2. $\Omega_i \cap \Omega_j = \phi \quad$ if $i \neq j$,
3. There exists g_j Lipschitz on $\bar{\Omega}_j$ such that

$$F(x) = \lambda_j F(S_j^{-1}(x)) + g_j(x) \quad \forall\, x \in \Omega_j. \tag{1.8}$$

Equation (1.8) does not tell how F behaves outside the Ω_i; we make the assumption that it is smooth, *i.e.*, lipschitz, outside $\bigcup \Omega_i$.

Since $F(S_j^{-1}(x)) = 0$ if $x \notin \Omega_j$,

$$F(x) = \sum_{i=1}^{d} \lambda_j F(S_j^{-1}(x)) + g(x),$$

where $g = g_j$ on $\Omega_j, g = F$ outside $\bigcup \Omega_j$. g is obviously continuous since F is continuous; and since it is Lipschitz on $\bigcup \bar{\Omega}_j$ and outside $\bigcup \Omega_j$, it is uniformly Lipschitz.

Remark that this equation holds for any Lipschitz function F (it defines g) so that it is interesting only if F is not uniformly lipschitz, and in that case, we will be interested in determining the points where F is C^α for $\alpha < 1$.

We will slightly generalize this model by assuming that g is C^k and not necessarily compactly supported, but that the derivatives of g of order less than k have fast decay. The same remark shows that, in this case, we should suppose that F is not C^k, and we will be interested in determining where F is C^α for $\alpha < k$. We will thus use the following definition.

Definition 3. *A function $F : \mathbb{R}^m \to \mathbb{R}$ is selfsimilar (of order k) if the three following conditions hold:*

1. *There exists a bounded open set Ω and $S_1, \ldots, S_\alpha$ contractive similitudes such that*

$$S_i(\Omega) \subset \Omega$$

$$S_i(\Omega) \cap S_j(\Omega) \neq 0 \quad \text{if} \quad i \neq j. \tag{1.9}$$

The S_i are contractive similitudes (i.e., products of an isometry by an homothety $x \to \mu_i x$ of ratio $\mu_i < 1$).

2. *There exists a C^k function g such that g and its derivatives of order less than k have fast decay and F satisfies*

$$F(x) = \sum_{i=1}^{d} \lambda_i F(S_i^{-1}(x)) + g(x). \tag{1.10}$$

3. *The function F is not uniformly C^k in a certain closed subset of Ω.*

The first condition was first introduced by Hutchinson (in [17]) in order to study selfsimilar sets ; it is called the "open set condition." A stronger condition is sometimes required, namely,

$$S_i(\bar{\Omega}) \cap S_j(\bar{\Omega}) = \phi \quad \text{if } i \neq j$$

called the "separated open set condition." This assumption is, for instance, made by Arneodo, Bacry, and Muzy in [3] in order to prove that the multifractal formalism holds for primitives of measures; we will need this condition in a generalized case considered in Section 6.

For any $k \in \mathbb{R}^+$, the last condition is equivalent to the existence of sequences $a_n \to 0$, $b_n \in K$, $C_n \to \infty$ such that

$$| C(a_n, b_n) | \geq C_n a_n^k,$$

where $C(a, b)$ is the wavelet transform of F which will be defined in the next section. This equivalent condition is a straightforward consequence of the wavelet characterization of the spaces $C^s(\mathbb{R}^m)$ that we will recall below, and of the localization of the wavelets.

We will see that solutions of Equation (1.10) are sometimes not functions but distributions.

Let

$$\alpha_{min} = \inf_{i=1,\ldots,d} \left(\frac{\log \lambda_i}{\log \mu_i} \right), \quad \text{and } \alpha_{max} = \sup_{i=1,\ldots,d} \left(\frac{\log \lambda_i}{\log \mu_i} \right),$$

and let τ be the function defined by

$$\sum_{i=1}^{d} \lambda_i^a \mu_i^{-\tau(a)} = 1.$$

The following result shows which part of the spectrum of a selfsimilar function can be recovered by the multifractal formalism.

Theorem 2. *Suppose that F is selfsimilar. If $\alpha_{min} > 0$, the function $d(\alpha)$ vanishes outside the interval $[\alpha_{min}, \alpha_{max}]$ and is analytic and concave on this interval. Its maximal value d_{max} satisfies*

$$\sum \mu_i^{d_{max}} = 1.$$

Let α_0 be the value for which this maximum is attained. Suppose first that g is C^∞. If $\alpha \leq \alpha_0$, $d(\alpha)$ can be obtained by computing the Legendre transform either of $\eta(q) - m$ or of $\zeta(q) - m$.

Moreover, suppose also g is only C^k. Let p_0 be defined by $\eta(p_0) = kp_0$ and let $\alpha_1 < \alpha_0$ be the value of the inverse Legendre transform of $\eta(q) - m$ at p_0. Then if $\alpha \leq \alpha_1$, $d(\alpha)$ can be obtained by computing the Legendre transform either of $\eta(q) - m$ or of $\zeta(q) - m$.

Without any assumption on α_{min}, if $\sum \mid \lambda_j \mid \mu_j^m < 1$, the same results hold when we replace $d(\alpha)$ by $D(\alpha)$, if g and the λ_i are positive and if, furthermore, the separated open set condition holds.

If $d_{max} = m$, or equivalently if $\overline{\cup S_i(\Omega)} = \overline{\Omega}$, the wavelet maxima method yields the whole spectrum of selfsimilar functions, provided that we keep only the largest maximum of a box of size Ca for a constant C large enough; i.e., $d(\alpha)$ can be obtained by computing the Legendre transform of $\theta(q) - m$ for any $\alpha < k$.

The following corollary shows that the previous result can be extended to a larger class than the selfsimilar functions.

Corollary 1. *Let A be a pseudo-differential elliptic operator of order s whose symbol σ satisfies*

$$\mid \partial_x^\alpha \partial_\xi^\beta \sigma(x,\xi) \mid \leq C(1+ \mid \xi \mid)^{s-\beta} \quad \forall\, \alpha, \beta.$$

Suppose that h is a selfsimilar function (satisfying $h(x) = \sum \lambda_i h(S_i^{-1}(x)) + g(x)$) and, if s is negative, suppose furthermore that the moments of g of order at most $|s|$ vanish. Let $F = A(h)$. Then if F is $C^\varepsilon(\mathbf{R}^m)$ for $\varepsilon > 0$, its spectrum is a concave function whose increasing part is given as above by the formula

$$d(\alpha) = \inf_q \; (\alpha q - \eta(q) + m).$$

We will also consider the following example which does not fit into the framework of selfsimilar functions

$$\varphi(x) = \sum_1^\infty \frac{1}{n^2} \sin \pi n^2 x.$$

Let us recall the main historical steps of the determination of the Hölder regularity of φ at every point.

Hardy and Littlewood proved that φ is nowhere $C^{3/4}$ except perhaps at the rational points of the form $(2p+1)/(2q+1)$, $p, q \in \mathbf{Z}$ (see [15]). Their proof is interesting in many respects. For instance, it anticipates wavelet methods. They remark that the function

$$C(a,b) = \frac{a}{2}(\theta(b+ia) - 1)$$

where θ is the Jacobi function $\theta(z) = \sum_{n\in\mathbf{Z}} e^{i\pi n^2 z}$), is the convolution of φ with contractions (by a factor a) of

$$\psi(x) = \frac{1}{\pi(x-i)^2}.$$

Since ψ has a vanishing integral, an Abel type theorem (which we will state precisely in Equation (2.3) shows that, if φ is smooth at x_0, $C(a,b)$ must have a certain decay when $a \to 0$ and $b \to x_0$. This method thus yields upper bounds for the function

$$\alpha(x_0) = \sup\{\beta :\ \varphi \in C^\beta(x_0)\}.$$

Actually it yields the following more precise result which relates the pointwise behavior of φ at x_0 to the Diophantine approximation properties of x_0. Suppose that x_0 is not a rational number. Let p_n/q_n be the sequence of its approximations by continued fractions and we define

$$\tau(x_0) = \sup\{\tau : \mid x_0 - \frac{p_m}{q_m} \mid\leq \frac{1}{q_m^\tau}\}$$

for infinitely many m's such that p_m and q_m are not both odd. Then

$$\alpha(x_0) \leq \frac{1}{2} + \frac{1}{2\tau(x_0)},$$

a result which is actually stated by J. J. Duistermaat in [8] where a more direct proof is given.

Converse results giving information about the pointwise behavior of φ from estimates on its convolutions are more difficult to obtain since they are of tauberian type. We will state such a result in Equation (2.4) or (2.5) for instance, and we will show how to use it in the case of Riemann's function in Section 7.

Finally Gerver proved the differentiability at the rational points of the form $(2p+1)/(2q+1)$ [9] (where we now know that φ is exactly $C^{3/2}$, see [13]). The analysis of the behavior of φ near such a rational point x_0 has been considerably sharpened since. Y. Meyer exhibited a complete "chirp" asymptotic expansion which describes the oscillations of φ:

$$\varphi(x) = u(x) + \sum_{n\geq 0}(x-x_0)^{\frac{3}{2}+n} v_+^n(\frac{1}{x-x_0}) \ \text{ if } \ x \geq x_0,$$

$$\varphi(x) = u(x) + \sum_{n\geq 0}|x-x_0|^{\frac{3}{2}+n} v_-^n(\frac{1}{|x-x_0|}) \ \text{ if } \ x \leq x_0,$$

where u is C^∞, the $v_\pm^n$ are 2π periodic with a vanishing integral and are $C^{\frac{1}{2}+n}$ (see [13]). In fact, we will actually show that these points are the only ones where a chirp expansion exists.

The results of Hardy and Gerver left open the problem of the determination of the exact regularity of φ at irrational points (hence they also left open the determination of its spectrum of singularities). We will see that the multifractal formalism for functions is valid in that case. The interesting point here is that φ has a very different structure from the other cases where the multifractal formalism is known to hold. The main results concerning φ are stated in the following theorem.

Theorem 3. *Suppose that x is not a rational number and let p_n/q_n be the sequence of its approximations by continued fractions. Let*

$$\tau(x) = \sup\{\tau : \mid x - \frac{p_m}{q_m} \mid \leq \frac{1}{q_m^\tau}\}$$

for infinitely many m's such that p_m and q_m are not both odd. Then

$$\alpha(x) = \frac{1}{2} + \frac{1}{2\tau(x)}.$$

The spectrum of singularities of φ is given by

$$\begin{aligned} d(\alpha) &= 4\alpha - 2 \text{ if } \alpha \in [\frac{1}{2}, \frac{3}{4}] \\ &= 0 \quad \text{if } \alpha = 3/2 \\ &= -\infty \quad \text{otherwise;} \end{aligned}$$

and if $\alpha \leq 3/4$, $d(\alpha)$ satisfies Equation (1.4).

We will now explain how all these results can be obtained. We first explore the relations between the size of the wavelet transform and the local regularity of the function.

§2 Regularity, singularities, and two-microlocalization

The results of Theorems 1 and 2 relate the pointwise behavior of a function to estimates on its wavelet transform. Our purpose in this section is to recall existing results on this point, and to introduce new ones. We first recall basic properties of the wavelet transform.

Let ψ be a radial function in $C^{k+1}(\mathbb{R}^m)$, with moments of order less than $k+1$ vanishing and supported in $B(0,1)$. Recall that

$$C(a,b)(F) = \frac{1}{a^m} \int_{\mathbb{R}^m} F(t)\psi\left(\frac{t-b}{a}\right) dt.$$

Then

$$F(t) = C(\psi) \int_{a>0} \int C(a,b)(F)\dot{\psi}\left(\frac{t-b}{a}\right) db \frac{da}{a^2} \tag{2.1}$$

where $C(\psi)$ is a constant depending only on ψ.

The relations between the regularity of F and the size of the wavelet transform are given by the following results for $s > 0$ (see [18]).

1. Suppose $s \notin \mathbb{N}$. Then $F \in C^s(\mathbb{R}^m)$ if and only if

$$|C(a,b)(F)| \leq Ca^s. \tag{2.2}$$

2. If $F \in C^s(x_0)$, then

$$|C_{a,b}(F)| \leq Ca^s \left(1 + \frac{|b - x_0|}{a}\right)^s. \tag{2.3}$$

3. If Equation (2.3) holds and if $F \in C^\varepsilon(\mathbb{R}^m)$ for an $\varepsilon > 0$, there exists a polynomial P such that

$$|F(x) - P(x - x_0)| \leq C|x - x_0|^s \log\left(\frac{1}{|x - x_0|}\right) \tag{2.4}$$

where $| x - x_0 | \leq 1/2$.

4. If

$$|C_{a,b}(F)| \leq Ca^s \left(1 + \frac{|b - x_0|}{a}\right)^{s'} \quad \text{for } s' < s, \tag{2.5}$$

then F is $C^s(x_0)$.

Due partly to physical motivations (the study of the velocity of turbulent fluids, for instance,), we do not want to restrict the discussion to bounded functions; we want to be able to consider points where F has a singularity (*i.e.*, in a neighborhood of which it is unbounded).In this section, we will obtain results similar to Equation (2.3) or (2.4) for singularities;, but first, we will adopt a definition of singularities which, as we will see, is not clear in all cases.

The following definition is a straightforward generalization of Equation (1.1) to negative exponents.

Definition 3. *Suppose that $-m < s \leq 0$. F is $C^s(x_0)$ if*

$$|F(x)| \leq C|x - x_0|^s. \tag{2.6}$$

We have to make the assumption $-m < s \leq 0$ because, if $s \leq m$, F might not be locally integrable and thus it might not be a distribution. In that case, no computation on F (like defining wavelet coefficients) would make sense.

We will now link Equation (2.6) to conditions on the wavelet transform of F.

Proposition 1. *Let $-m < s \leq 0$; if $f \in C^s(x_0)$ then*

$$|C(a,b)(F)| \leq Ca^s(1 + \frac{|b - x_0|}{a})^s.$$

Conversely, suppose that $\exists\ s' < s$ *such that*

$$|C(a,b)(F)| \leq Ca^s(1 + \frac{|b - x_0|}{a})^{s'};$$

then $f \in C^s(x_0)$.

Proposition 1 yields a justification for Definition 4, and thus also for the definition of strong α-singularities when α is negative.

The problem of defining Hölder exponents for $s \leq -m$ is not straightforward. As mentioned before, we cannot consider only conditions like $|F(x)| \leq C|x - x_0|^s$ since F might not be a distribution. Some authors (see [3,23]) proposed the following definition :

$$F \in C^s(x_0) \Longleftrightarrow (-\Delta)^{-\frac{[s]}{2}} F \in C^{s-[s]}(x_0).$$

There are two problems with this definition. The first one is that it is not consistent with the definition for $s > 0$. Let us take an example. Consider

$$F(x) = x^{1/2}\cos(1/x),$$

the primitive of F is $O(x^{5/2})$ at the origin. Nonetheless, we would not consider F as a $C^{3/2}$ function at the origin. Furthermore, this definition is not consistent with the "natural" definition in Equation (2.6) when $-n < \alpha \leq 0$, for essentially the same reasons (we leave this verification to the reader).

We define, for negative values of s,

$$C^s(\mathbb{R}^m) = \{F : \mid C(a,b) \mid\leq Ca^s\}.$$

The condition $F \in C^s(x_0)$ for a negative s implies a kind of global regularity for F (the fact that $f \in C^s(\mathbb{R}^m)$). For $s \leq -m$, we will suppose that this regularity holds, which will insure that F is a distribution. We adopt the following definition.

Definition 4. *Suppose that* $s \leq -m$; *a function* F *belongs to* $C^s(x_0)$ *if* $F \in C^s(\mathbb{R}^m)$ *and in a neighborhood of* x_0, $\mid F(x) \mid\leq C \mid x - x_0 \mid^s$.

This definition coincides with Definition 3 when $-m < s \leq 0$, since in that case

$$\mid F(x) \mid\leq C \mid x - x_0 \mid^s \Longrightarrow f \in C^s(\mathbb{R}^m).$$

Lemma 1. *If* $s \leq -m$, *the following implications hold*

$$F \in C^s(x_0) \Rightarrow |C(a,b)(F)| \leq Ca^s(1 + \frac{|b - x_0|}{a})^s,$$

$$|C(a,b)(F)| \leq Ca^s(1 + \frac{|b - x_0|}{a})^{s'} \quad \textit{for an} \quad s' < s \Rightarrow F \in C^s(x_0).$$

This lemma shows that our definition of pointwise singularities is consistent with all the usual properties of the pointwise Hölder condition. It is interesting to check if some distributions which "should " belong to these generalized Hölder spaces satisfy these conditions. For instance, the distribution $p.p.(1/x)$ defined by

$$\langle p.p.(1/x) \mid \phi\rangle = \lim_{\epsilon\to 0} \int_{\mathbf{R}-[-\varepsilon,\varepsilon]} \frac{\phi(x)}{x} dx$$

is $C^{-1}(0)$ and $f.p.(1/x^2)$ defined by

$$\langle f.p.(1/x^2) \mid \phi\rangle = \lim_{\epsilon\to 0} \left(\int_{\mathbf{R}-[-\varepsilon,\varepsilon]} \frac{\phi(x)}{x^2} dx - \frac{2\phi(0)}{\varepsilon} \right)$$

is $C^{-2}(0)$.

The size of the wavelet coefficients and the existence of strong α-singularities are related as follows.

Proposition 2. *Suppose that F is $C^\alpha(x_0)$ and that there exists a sequence $a_n \to 0$ verifying $a_n/a_{n+1} \leq C$ and wavelet coefficients $C(a_n, b)$ in a cone pointing towards x_0 (i.e., $\mid b - x_0 \mid\leq Ca_n$) such that*

$$\mid C(a_n, b) \mid\geq Ca_n^\alpha.$$

Then x_0 is a strong α-singularity of F.

§3 Some functional norm estimates

We will first show the link between quantities like $S_p(l)$ or $\tilde{Z}(a,q)$ and Sobolev or Besov-type norms. Let us recall a few definitions and characterizations.

Supose that $s \in \mathbb{R}$, $p > 0$, and $q > 1$. A function F belongs to the homogeneous Besov space $\dot{B}_p^{s,q}$ if

$$\int_{a>0} \left[\int |C(a,b)|^p db\right]^{q/p} \frac{da}{a^{sq+1}} < +\infty.$$

Recall that $\eta(p)$ is the infimum of all numbers τ verifying, for a small enough,

$$\tilde{Z}(a,p)(= \int |C(a,b)|^p db) \leq Ca^\tau.$$

Thus, if $p > 1$, $\eta(p) = \sup\{\tau \ : \ F \in B_p^{\tau/p,\infty}\}$

We will now give a similar characterization for the function $\zeta(p)$. Recall that the spaces $H^{s,p}$, introduced by Nikolskii (see [1,27]), are defined (for $s \notin \mathbb{N}$) as follows.

Given s a non-integer, $s = m + \sigma$, where $s \geq 0$, m an integer, and $0 < \sigma < 1$. Then for any multi-index α such that $|\alpha| = m$,

$$\int \frac{|\partial^\alpha F(x+h) - \partial^\alpha F(x)|^p}{|h|^{\sigma p}} dx \leq C \tag{3.1}$$

where $p \geq 1$ and $F \in H^{s,p}$ if $F \in L^p$. Recall that $\zeta(p)$ is the largest real number such that

$$S_p(h) \left(= \int |F(x+h) - F(x)|^p dx \right) \leq C h^{\zeta(p)}$$

for h small enough. Thus, if $p \geq 1$, $\zeta(p) = \sup\{s : F \in H^{s/p,p}\}$.

Of course, we see here that the formula in the so-called "Structure Function Method" must be modified as follows in order to be coherent with Equation (3.1). If $\zeta(p) < 1$, the formula holds; if it is equal to 1, one should use the same formula but with the gradient of F; and so on until $\zeta(p)$ falls between two integers. Note that this procedure is obviously difficult to handle numerically if $\xi(p)$ is large.

We will also need the spaces $L^{p,s}$ to be defined for $s > 0$, by

$$f \in L^{p,s} \Leftrightarrow f \in L^p \quad \text{and} \quad (-\Delta)^{s/2} f \in L^p \ .$$

The classical Sobolev-type imbeddings between these spaces show that

$$\eta(p) = \sup\{s : \ F \in H^{s/p,p}\} = \sup\{s : \ F \in B_p^{s/p,p}\} = \sup\{s : \ F \in L^{p,s/p}\} \tag{3.2}$$

Suppose that we use an orthonormal basis of wavelets indexed by the dyadic cubes (if the wavelet is $2^{mj/2}\psi(2^j x - k)$, the corresponding cube is $\lambda = k2^{-j} + 2^{-j}[0,1]^m$). We denote these wavelets by ψ_λ and the corresponding wavelet coefficients by C_λ.

The wavelet characterization of $L^{p,s}$ is the following (see [25]),

$$f \in L^{p,s} \Leftrightarrow \int \left(\sum |C_\lambda|^2 \, 2^{(2s+n)j} \, \chi_\lambda(x) \right)^{p/2} dx \ < \ +\infty \tag{3.3}$$

where χ_λ is the characteristic function of the dyadic cube λ and 2^{-j} the length of a side of this cube.

We have identified the function η (or ζ) in terms of Sobolev-type norms when $p > 1$. However, it happens that the infimum in Equation (1.3) or (1.4) is attained for values of $p \leq 1$. Therefore, we have to introduce some functional spaces adapted to this case; they will be derived from the real Hardy $\mathcal{H}^p$ spaces.

Recall (see [25]) that a possible definition of the Hardy spaces is the set of distributions whose wavelet coefficients satisfy

$$\int \left(\sum_{j,k} |C_\lambda|^2 \, 2^{nj} \, \chi_\lambda(x) \right)^{p/2} dx \ < \ +\infty.$$

The wavelets we use are $C^{m(p^{-1}-1)}$ and have vanishing moments up to the order $m(p^{-1}-1)$. The Hardy spaces are the direct generalization of the spaces $L^{p,0}$ ($= L^p$) if one uses Equation (3.3) as a definition of these spaces.

We now define the "Sobolev-Hardy" space $L^{p,s}$ for $p \leq 1$ by

$$f \in L^{p,s} \Leftrightarrow f \in \mathcal{H}^p \quad \text{and} \quad (-\Delta)^{s/2} f \in \mathcal{H}^p ,$$

and

$$\eta(p) = \sup\{s \ : \ F \in B_p^{s/p,p}\} = \sup\{s \ : \ F \in L^{p,s/p}\}.$$

We asume the last two definitions of Equation (3.2) hold. The following proposition is a consequence of the wavelet characterization of these spaces.

Proposition 3. *Using the generalizations of the space $L^{p,s}$ defined above, the following characterizations hold:*

$$\eta(p) = \sup\{s \ : \ F \in B_p^{s/p,p}\} = \sup\{s \ : \ F \in L^{p,s/p}\}, \quad \forall\, p > 0,$$

$$\eta(p) = \zeta(p) = \sup\{s \ : \ F \in H^{s/p,p}\}, \quad \forall\, p > 1,$$

and $\eta(2) = \sup\{s \ : \ \int \mid \hat{F}(\xi) \mid^2 (1+ \mid \xi \mid^2)^{s/2} d\xi \leq C\}$.

The last result concerning the interpretation of $\eta(2)$ remains to be proved. Since $B_2^{s,2} = L^{2,s}$, and

$$B_p^{s+\varepsilon,q} \subset B_p^{s,q'} \subset B_p^{s-\varepsilon,q''}, \quad \forall\, q, q', \text{and } q'',$$

so that

$$\begin{aligned} \eta(2) &= \sup\{s : F \in L^{2,s/2}\} \\ &= \sup\{s \ : \ \int \mid \hat{F}(\xi) \mid^2 (1+ \mid \xi \mid^2)^{s/2} d\xi \leq C\}. \end{aligned}$$

Note that this result differs slightly from [2] where the interpretation given for $\eta(2)$ is

$$\mid \hat{F}(\xi) \mid^2 \sim \mid \xi \mid^{-\eta(2)-2} .$$

This interpretation is correct provided that such a scaling holds. We will see in Section 4 an interpretation of $\eta(1)$ of a very different nature.

§4 Upper bounds and counterexamples

A problem arises when we attempt to determine the mathematical definition of "dimension" to use. The physical literature is often unclear about this point. While sometimes it uses the name of Hausdorff dimension, in computing, it uses computing coverings by boxes of the same size. Naturally, a given set of points (a potential "set of singularities" of our function) can have very different dimensions, depending on the definition considered. We will see that the "good definition" depends on the kind of singularities we look for. For α-singularities, we get a bound on the Hausdorff dimension, and for strong α-singularities, on the packing dimension. An important difference between the two settings is that in the first one we necessarily have to suppose some minimal uniform regularity for F, which is not required in the second one. It will be clear that the "box dimension" (also called capacity or potential dimension) is never the good definition. This raises an important problem concerning the numerical determination of spectrums of singularities since the capacity dimension is the only one which can be computed numerically because it involves one covering at each scale and not the best one to be found among an infinity of coverings.

Let us now recall the definitions of the Hausdorff dimension and Hausdorff measure.

Let $A \subset \mathbb{R}^n$ and R_ε, the set of all coverings of A by sets of diameter at most ε. Let

$$M(\varepsilon, d) = \inf_{r \in R_\varepsilon} \sum_{A_i \in r} (\operatorname{diam} A_i)^d.$$

Then, by definition, the d-dimensional Hausdorff measure is

$$d - \text{Measure}(A) = \limsup_{\varepsilon \to 0} M(\varepsilon, d) .$$

The Hausdorff dimension of A is

$$D = \inf \{d \;/\; d - \text{Measure}(A) = 0\} = \sup \{d \;/\; d - \text{Measure}(A) = +\infty\} .$$

Note that if the coverings are done using only balls or only dyadic cubes, we obtain an equivalent quantity for the d-Measure, and, consequently, D is not changed.

The following result gives an upper bound for the dimensions of singularities and is proved using the wavelet characterization of Besov spaces.

Proposition 4. *Let $s - \frac{m}{p} > 0$ and $p > 0$. If $F \in B_p^{s,\infty}$,*

$$d(\alpha) \leq m - (s - \alpha)p.$$

Thus, if $\eta(p)$ satisfies $\eta(p) > m$, $\forall\, p$, we have

$$d(\alpha) \leq \inf_p (m - \eta(p) + \alpha p).$$

This proposition is reminiscent of [4] where Brown, Michon, and Peyrière proved similar results for measures (in dimension 1).

If $s \leq m/p$, a function in $L^{p,s}$ or $B_p^{s,p}$ can be infinite on a dense set, and thus be smooth at no point (see [29], p. 159), so that no such result can hold if we do not make the assumption $s - \frac{m}{p} > 0$.

A rather natural question that follows is whether or not similar bounds (or equalities) for $d(\alpha)$ hold if we replace the Hausdorff dimension by the packing dimension (see [4,22,28]) for its definition and basic properties). Let us motivate this problem.

Keeping in mind that the multifractal formalism was introduced for the study of turbulence, we consider the results that exist in this area. Caffarelli, Kohn and Nirenberg in [5] obtain a bound on the dimension of (possible) singularities in Navier-Stokes equations, which turn out to be a bound on the packing dimension. They also show that these singularities are actually "strong singularities" following the definition we gave (with $\alpha = 0$).

Let us be more precise. The main reason to obtain bounds for dimensions of strong singularities is that when global regularity conditions (which insure that F is continuous) do not hold, no result such as Proposition 4, can be obtained. Actually, even in the strict framework of selfsimilar functions, we will see that no such bounds exist. Since, in the applications, we clearly want to be able to consider unbounded functions, we need to obtain some positive results in that case.

Let us first recall the definition of the Packing dimension and Box dimension of a subset E of $\mathbb{R}^n$ (see [33]).

Let $J > 0$ and Λ_J be the set of dyadic cubes of size 2^{-J} that contain a point of E.

We define

$$m_d(E) = \lim_{J \to +\infty} \sum_{\lambda \in \Lambda_J} 2^{-dJ} = \Lambda_j^{\sharp} 2^{-dJ}.$$

Then

$$\mathrm{mes}_d(E) = \inf_{E \subset \cup E_n} \sum_n \mathrm{mes}_d(E_n).$$

The box dimension of E is the value of d for which $m_d(E)$ falls from $+\infty$ to 0. This dimension is also called *potential dimension* by physicists; it is the only one which is numerically easy to compute.

The packing dimension of E is the value of d for which $\mathrm{mes}_d(E)$ falls from $+\infty$ to 0.

Proposition 5. *Let $F \in W^{s,p}$ and α such that $-m < \alpha \leq 1$. The packing dimension of the strong α-singularities of F is bounded by $m-(s-\alpha)p$, so that, if $D(\alpha)$ is the packing dimension of strong α-singularities, and $-m < \alpha \leq 1$, then*

$$D(\alpha) \leq \inf_p (m - \eta(p) + \alpha p). \tag{4.1}$$

For the reasons mentioned above, such a result is more satisfactory than Proposition 4 since we do not have to make the assumption of a minimal Hölder regularity of F, an assumption which may prove wrong for some applications to turbulence. Actually, since the solutions of Navier-Stokes equations in dimension 3 are suspected to have "true" singularities, *i.e.*, to be unbounded, the numerical estimation of $D(\alpha)$ for $\alpha = 0$ using Equation (4.1) could be a way to check if a stronger result than the one obtained by Caffarelli, Kohn, and Nirenberg in [5] might hold.

Of course, a way to avoid this problem could be to consider primitives or perhaps iterated primitives of the velocity, but such quantities would have less physical meaning.

Let us now show that if F is a function of one real variable, the box dimension of the graph of F is exactly $2-\eta(1)$, provided $\eta(1)$ is between 0 and 1. It is a straightforward consequence of the following result stated in [10].

Proposition 6. *Suppose $0 < \gamma < 1$ and $F : [0,1] \to \mathbb{R}$ is continuous. Then the box dimension of the graph of F is exactly $2-\gamma$, if and only if*

$$F \in \bigcap_{\alpha<\gamma} B_1^{\alpha,\infty} \setminus \bigcup_{\beta>\gamma} B_1^{\beta,\infty}.$$

Thus the result holds because η can also be defined by

$$\eta(p) = \sup\{s : F \in B_p^{s/p,\infty}\}$$

(a straightforward consequence of the imbeddings between Besov spaces).

Let $s > 0$. Recall that $f \in \Gamma^s(x_0)$ if

$$f \in \bigcap_{\varepsilon>0} C^{s-\varepsilon}(x_0) \text{ but } f \notin \bigcup_{\varepsilon>0} C^{s+\varepsilon}(x_0).$$

For each $s > 0$, let $E^{(s)}$ be the set of real numbers x_0 such that f belongs to $\Gamma^s(x_0)$ and let $d(s)$ be the Hausdorff dimension of $E^{(s)}$. We define $\mathcal{C}$ as the class of functions that can be written as the supremum of a countable set of functions of the form $c1_{[a,b]}(x)$ (where we can have $a = b$). Thus, Riemann integrable functions belong to $\mathcal{C}$, but also, for instance, the indicatrix function of the rationals (but not the indicatrix function of the irrationals).

Proposition 7. *Let $d(s) : (0,+\infty) \to [0,m]$ be a function in $\mathcal{C}$. There exist two continuous functions F_1 and $F_2 : \mathbb{R}^m \to \mathbb{R}$ which share the same function η, such that $d(s)$ is the spectrum of singularities of F_1, and F_2 is C^∞ except at the origin (so that its spectrum vanishes everywhere).*

The function F_1 can be constructed explicitely by defining its coefficients on an orthonormal wavelet basis (see [19]).

The construction of F_2 is very easy. One remarks that at each level j, the number of non-vanishing wavelet coefficients of F_1 is a $o(2^j)$ (actually, it is much smaller). We now consider a function F_2 which has, at each level, j the same non-vanishing wavelet coefficients as F_1, but situated at different dyadic intervals. We group them in the smallest possible interval I_j centered at the origin. Thus, these two functions share the same $B_p^{s,\infty}$ norm, and moreover, the same function η. However, for $x \neq 0$, there is a finite number of non-vanishing wavelet coefficients in a certain interval centered at x, because the length of I_j tends to 0. Hence, F_2 is C^∞ at x.

In our opinion, the idea of the construction of F_2 reveals an important point. It shows that any criterium which is invariant after a permutation of the positions of the wavelet coefficients at the same scale cannot yield the spectrum of singularities, essentially because it cannot contain the geometrical information which is relevant in the definition of the dimension. As a result, formulas involving these quantities can be true only in some very specific cases where we have an *a-priori* knowledge of this geometry, which is clearly the case for selfsimilar functions, and, in a less obvious way, also the case for Riemann's function because of the very simple transformations of the derivative of Riemann's function (Jacobi's theta function) under the action of the linear fractionals which belong to the Theta modular group. We will specify these points in Section 7.

§5 Basic properties of selfsimilar functions

Before starting the study of selfsimilar functions, let us give a few examples.

1) *Primitives of multinomial measures in dimension 1*

Let μ be a probability measure supported by $[0,1]$ and suppose that for any interval I,

$$\mu(S_i(I)) = \lambda_i \mu(I), \quad \forall\, i = 1, \dots, d$$

with $\sum \lambda i = 1$, the S_i as above, and $\Omega = (0,1)$. Let

$$F(x) = \int_0^x d\mu - x \qquad \forall\, x \in [0,1].$$

Then F vanishes at 0 and 1 and is smooth outside the intervals $S_i([0,1])$. One immediately checks that F is continuous and

$$F(x) = \lambda_i F(S_i^{-1}(x)) + g_i(x) \quad \forall\, x \in S_i([0,1]),$$

with g_i linear. Therefore, F is selfsimilar.

For any probability measure μ on $\mathbb{R}$, the scaling index of μ at x_0 is the supremum of all $\alpha's$ such that

$$\mu([x_0 - \varepsilon, x_0 + \varepsilon]) \leq \varepsilon^\alpha \quad \forall\, \varepsilon > 0.$$

One easily checks that μ has a scaling index α at x_0 if and only if its primitive $F(x) = \mu([0,x])$ is C^α at x_0 (see [3,22]). This property allowed Arneodo, Bacry, and Muzy in [3] to determine the spectrum of singularities of the primitives of multinomial measures when the separated open set condition holds. This remark shows that when F is the primitive of a one dimensional measure, some results derived in this paper are a consequence of similar results concerning measures (for $\alpha \in [0,1]$) that were obtained by Brown, Michon and Peyrière in [4]. Actually, multifractal measures play an important role in the study of selfsimilar functions since the determination of the dimensions of the singularities are obtained by constructing such measures supported by these sets of singularities.

2) *Lacunary trigonometric series*

Let

$$F_\alpha(x) = \sum_{j=0}^{\infty} 2^{-\alpha j} \sin 2\pi 2^j x$$

for $x \in [0,1]$ and $0 < \alpha \leq 1$. Define

$$\begin{aligned} g(x) &= \sin 2\pi x \quad \text{ if } x \in [0,1] \\ &= 0 \quad \text{elsewhere.} \end{aligned}$$

Obviously, we have

$$F_\alpha(x) = 2^{-\alpha} F_\alpha(2x) + 2^{-\alpha} F_\alpha(2x-1) + g(x),$$

and the F_α are selfsimilar.

3) *Several dimension examples*

In dimension 1, the "geometry" contained in the transforms S_i is poor. In several dimensions, sometimes it is no longer the case. Let us consider two examples. First, let $\Omega = [-1,1]^2$, $i,j \in \{\frac{1}{2}, -\frac{1}{2}\}$, and $S_{i,j}(x) = \frac{1}{2}x + (i,j)$.

The $S_{i,j}$ map the square Ω on its four subsquares of half size. If the homothety had a ratio smaller than $\frac{1}{2}$, by iterating the $S_{i,j}$, we would obtain a two-dimensional Cantor set. There exist more "exotic" examples ; for instance, if

$$S_1^{-1}(X) = \begin{pmatrix} 1 & -1 \\ 1 & 1 \end{pmatrix} X, S_2^{-1} = \begin{pmatrix} 1 & -1 \\ 1 & 1 \end{pmatrix} X + (1,0),$$

the S_i map a "fractal dragon" on its two selfsimilar components (see [6,13]).

Let us also mention two examples of functions which, though not selfsimilar in the sense we gave, satisfy functional equations similares to Equation (1.10).

First, the scaling function of the function φ used in the construction of orthonormal wavelet bases satisfies (see [6,9])

$$\varphi(x) = \sum a_k \varphi(2x-k).$$

Here, Equation (1.9) does not hold except for some non-smooth functions φ like characteristic functions of sets, in which case there exist examples similar to the fractal dragon we mentioned above (see [6,13]). I. Daubechies and J. Lagarias recently proved in [7] that a converse formula to the multifractal formalism holds for these functions: the function η is the Legendre transform of $d(\alpha)$.

Our second example is the Brownian bridge on $[0,1]$ which satisfies

$$B(t) = \frac{1}{\sqrt{2}} B_1(2t) + \frac{1}{\sqrt{2}} B_2(2t-1) + \xi \Lambda(x), \tag{5.1}$$

where B_1 and B_2 are two Brownian bridges that have the same law as B; ξ is a gaussian; $\Lambda(x) = \sup(x, 1-x)$ on $[0,1]$; and the three terms of the right hand side of Equation (5.1) are independant. We are in a situation where Equation (1.10) holds "in law."

Let us determine in which sense Equation (1.10) has solutions and examine some basic regularity properties of these solutions. This regularity will depend upon the assumption we make on the λ_i.

Iterating Equation (1.10), we obtain, for any N,

$$\begin{aligned} F(x) = & \sum_{n=0}^{N-1} \sum_{(i_1,\dots,i_n)} \lambda_{i_1} \dots \lambda_{i_n} g\left(S_{i_n}^{-1} \dots S_{i_1}^{-1}(x)\right) \\ & + \sum_{(i_1,\dots,i_N)} \lambda_{i_1} \dots \lambda_{i_n} F\left(S_{i_N}^{-1} \dots S_{i_1}^{-1}(x)\right). \end{aligned}$$

Thus, a (formal) solution of Equation (1.10) is given by

$$F(x) = \sum_{n=0}^{N-1} \sum_{(i_1,\dots,i_n)} \lambda_{i_1} \dots \lambda_{i_n} g\left(S_{i_n}^{-1} \dots S_{i_1}^{-1}(x)\right). \tag{5.2}$$

Note that, here, F is written as a superposition of similar structures at different scales, which is reminiscent of some possible models of turbulence ([11,32]). This formula resembles a wavelet decomposition (except that g needs not have cancellation).

When the (formal) Equation (5.2) converges in a certain functional space, we will have a solution of Equaton (1.10) in this space. Actually, it is easy to check that Equation (5.2) converges almost everywhere if the separated open set condition holds. We will particularly be interested in three cases: first, when does Equation (1.10) have solutions that are actually functions (*i.e.*, locally in L^1); secondly, in which cases do the solutions have some global C^α smoothness (we will see the importance of this case because it is only in this setting that the fractal formalism works without any modification); and finally, in which distribution spaces does the series converge when we make no assumption on the λ_i. We will see that a good setting to study the last question is supplied by the real Hardy spaces.

Recall that

$$\alpha_{\min} = \inf_{j=1,\dots,d} \frac{\log \mid \lambda_j \mid}{\log \mu_j} \quad \text{and} \quad \alpha_{\max} = \sup_{j=1,\dots,d} \frac{\log \mid \lambda_j \mid}{\log \mu_j}$$

Theorem 4. *Suppose that* $\sum \mid \lambda_j \mid \mu_j^m < 1$. *In this case, Equation* (1.10) *has a unique distribution solution, which is an* L^1 *function and is given by the series in Equation* (5.2). *If, furthermore,* $0 < \alpha_{min} < k$, *this function is* $C^{\alpha_{min}}$.

Suppose that $\sum \mid \lambda_j \mid \mu_j^m \geq 1$. *Then Equation* (1.10) *may have several distribution solutions. Let* $p < 1$ *such that* $\sum \mid \lambda_j \mid^p \mu_j^m < 1$. *If* g *is* C^k *with* $k > m(p^{-1} - 1)$, *and if the moments of* g *of order less than* k *vanish, Equation* (5.2) *converges in the Hardy real space* $\mathcal{H}^p$, *so that Equation* (1.10) *has at least one solution in that space of distributions. Furthermore, these results are optimal.*

The existence of solutions is proved by showing that Equation (5.2) is a Cauchy sequence in the corresponding functional space. A similar argument yields uniqueness. Non-uniqueness is made clear by remarking that the difference of two solutions of Equation (1.10) yields an equation which is satisfied by some multinomial measures. For instance, (see [9]) consider the example of the canonical measure on the triadic Cantor set, which satisfies

$$\mu(x) = \frac{3}{2}\mu(3x) + \frac{3}{2}\mu(3x - 2).$$

Let us start with some preliminary results concerning the geometry of the mappings S_i.

If A is a subset of $\mathbb{R}^n$, let us define the mapping S by

$$S(A) = \bigcup_{i=1}^{d} S_i(A);$$

and let K be the set defined by

$$K = \bigcap_{n \in \mathbb{N}} S^n(\bar{\Omega}).$$

K is called the invariant compact of S ; its Hausdorff dimension is d_{max} (defined in Theorem 2). The singularities of F are situated on K, or more precisely, if x does not belong to K, F is C^k in a neighborhood of x.

The points of K can be represented as the limit points of the branches of a tree constructed in the "time-frequency half-space" as follows.

The root is conventionally the point $(0, 1) \in \mathbb{R}^m \times \mathbb{R}^+$. This root is linked to the d first nodes which are the $(S_j(0), \mu_j)$. This point $(S_j(0), \mu_j)$ is linked to the $(S_j S_k(0), \mu_j \mu_k), \ldots$

If $\mathbb{R}^m$ is identified to $\mathbb{R}^m \times \{0\}$, clearly, the branch indexed by a sequence $i \in \{1, \ldots, d\}^{\mathbb{N}}$ goes towards the point x_i (and it is the only one which does so if the mapping x_i is one-to-one).

We will see that this tree is related to the wavelet transform of F. More precisely, the order of magnitude of the wavelet transform of F near (x_i, μ_i) is λ_i.

For a given branch indexed by $i = (i_1, \ldots, i_n)$, let

$$\alpha(i) = \frac{Log(\lambda_{i_1} \ldots \lambda_{i_n})}{Log(\mu_{i_1} \ldots \mu_{i_n})}.$$

The following result yields the exact regularity of f at each point of K when $\alpha_{min} > 0$.

Proposition 8. *Suppose that $\alpha_{min} > 0$ and let $x \in K$. Then F is $\Gamma^{\alpha(x)}$ at x where*

$$\alpha(x) = \liminf_{j \to \infty} \inf_{i \in B_j(x)} \frac{Log(\lambda_i)}{Log(\mu_i)}.$$

Here, $B_j(x)$ is the set of branches ending at $(S_{i_1} \ldots S_{i_n}(0), \mu_{i_1} \ldots \mu_{i_n})$, and such that $\mu_{i_1} \ldots \mu_{i_n} \sim 2^{-j}$ and

$$|S_{i_1} \ldots S_{i_n}(0) - x| \leq 10 \text{ diam } (\Omega) \mu_{i_1} \ldots \mu_{i_n}$$

(This requirement means that the end-point of the branch is, in the time-frequency half-space, in a certain cone over x).

This result is obtained as follows. Taking the wavelet transform of Equation (1.10), one obtains the following renormalization formula for the wavelet transform of F (where $C(a, b)$ and $\omega(a, b)$ are the wavelet transforms of F and g, respectively)

$$C(a, t) = \sum_{j=1}^{d} \lambda_j \omega\left(\frac{a}{\mu_j}, S_j^{-1}(t)\right) + \sum_{j=1}^{d} \lambda_j \, C\left(\frac{a}{\mu_j}, S_j^{-1}(t)\right). \tag{5.3}$$

By recursion, one deduces that the order of magnitude of the wavelet transform of F near (x_i, μ_i) is λ_i.

This estimate shows that the wavelet transform of F is "large" near the tree defined above; and thus the ramifications of this tree of wavelet maxima reflects the "dynamics" of selfsimilarity, as guessed by Arneodo, Bacry, and Muzy in [3].

The regularity of F at each point is then a straightforward consequence of Equations (2.2) and (2.3).

It is remarkable that the results do not depend on the function g. If it were replaced by another function, the new F would have the same regularity at every point. It is only the global smoothness of g that is important. It defines a value beyond which one can no more calculate the regularity of F.

Let τ be the function defined implicitely by

$$\sum_{i=1}^{d} \lambda_i^a \mu_i^{-\tau(a)} = 1.$$

Let $\alpha < k$ and define $d(\alpha)$ as the Hausdorff dimension of the set of points x where F is $\Gamma^\alpha(x)$. Then $d(\alpha)$ vanishes outside $[\alpha_{min}, \alpha_{max}]$, and on this interval,

$$d(\alpha) = (\inf_a a\alpha - \tau(a)).$$

This last result is obtained by the classical method for determining Hausdorff dimensions (one constructs probability measures supported by the set of singularities, which have locally some specific scalings, see [9] for this method). In order to Prove Theorem 2, one checks that $\int |C(a,b)|^q db$ scales like $a^{m+\tau(q)}$.

Note that the fact that a function F is $C^\alpha(x_0)$ for $\alpha > 1$ does not imply anything about the regularity at x of partial derivatives of F. In particular, F need *not* be $C^{\alpha-1}(x_0)$. (The most classical example of this phenomenon are the chirps studied in [22]). Therefore the spectrum of singularities of $\partial_l F$ is *a priori* unrelated to the spectrum of F (and, thus, the spectrum of F is also unrelated to the spectrum of primitives of F). However, if F is selfsimilar, everything happens as if this "pathology" did not exist:

Suppose that $\alpha_{min} > 1$. Then $\tilde{F} = \partial_l F$ is selfsimilar and satisfies

$$\tilde{F}(x) = \sum \frac{\lambda_i}{\mu_i} F(S_i^{-1}(x)) + \partial_l g(x).$$

Also, the function $\tilde{\tau}$ associated to $\tilde{F}$ satisfies

$$\sum \left(\frac{\lambda_i}{\mu_i}\right)^a \mu_i^{-\tilde{\tau}(a)} = 1.$$

Consequently, $\tilde{\tau}(a) = \tau(a) + a$ and $\tilde{d}(\alpha) = d(\alpha - 1)$, and the spectrum is translated by 1, as expected.

Let us make a few comments on Theorem 2.

We clearly need here $q > 0$, because if $q < 0$, the parts where the wavelet transform is vanishing would bring an arbitrarily large contribution. We will come back to this problem in the following section.

Consider now the formula where the integral is replaced by a sum on the wavelet maxima of $C(a,b)$. If, for a fixed a, they are separated by a distance of the order a, the two quantities

$$\int |C(a,b)|^q db \quad \text{and} \quad a^m \sum_{\max} |C(a,b)|^q$$

are of the same order of magnitude. Thus, in that case, the verification of the fractal formalism when using the wavelet maxima reduces to the verification for the integral formula.

If the maxima are not spaced by at least Ca, one possibility is to keep the largest maximum on a box of side a in the formula. One then has the previous case.

Thus, with perhaps the modification we mentioned, the multifractal formalism holds when using the two wavelet methods. Hence, it is also valid for the structure function method, since we showed that $\zeta(q) = \eta(q)$ for $q > 1$. The restriction $q > 1$ shows, however, that it might not yield the whole left-hand side of the spectrum but rather a smaller part corresponding to the region where the infimum in the Legendre transform formula is attained for $q > 1$.

Up to now, we made the assumption $\alpha_{min} > 0$ (which means $|\lambda_i| < 1$), implying that $F \in C^{\alpha_{min}}$. We were only interested in exponents larger than $\alpha > 0$. However, we would like to consider negative exponents, which, as mentioned before, should be pertinent in some applications. We already discussed the definition of negative exponents in Section 2. The multifractal formalism holds using a slightly different definition for the spectrum of singularities.

Let us suppose that

$$\sum_{i=1}^{d} |\lambda_i||\mu_i|^m < 1, \tag{5.4}$$

so that F verifying Equation (1.10) belongs to L^1.

If $\sum_{i=1}^{d} |\lambda_i||\mu_i|^m \geq 1$, F can clearly be unbounded. In fact, the following result holds.

Lemma 2. *Suppose that one of the λ_j satisfies $\lambda_j > 1$ and that $g(x) \geq 0\ \forall x \in \Omega$ and g does not vanish identically. Then $\forall\ x \in K$, F is unbounded in any neighborhood of x, so that $d(\alpha) = 0$, $\forall \alpha > -n$.*

In this case, we see that the spectrum of singularities of F for $\alpha > -n$ is trivial: $d(\alpha) = 0\ \forall \alpha > -n$. This example suggests that if only Equation (5.4) holds, the spectrum defined by the Hausdorff dimension of α-singularities is not the right quantity to consider, but rather, the computation of the packing dimension of strong singularities.

Proposition 9. *If F is selfsimilar and $x \in K$, F has a strong singularity of order $\alpha(x)$ at x but no higher strong singularity. The packing dimension of the strong α-singularities of f is given by*

$$D(\alpha) = \inf(a\alpha - \eta(a) + m).$$

§6 The wavelet maxima method

Our purpose in this section is two-fold. First, we will show that the wavelet maxima method can yield a function $\theta(q)$ which is much smaller than $\eta(q) - 1$ so that in general the multifractal formalism cannot hold using this method. (We will actually see that the counterexamples we give extend to the very specific setting of selfsimilar functions, where the multifractal formalism works when using the wavelet transform integral method). Secondly, we will show that, after a slight modification, it yields the right spectrum, even for negative values of p, in strong contrast with other methods. For positive p's, the reason why the wavelet maxima method sometimes works while it does not work other times is easy to understand intuitively if we relate it to the wavelet transform integral method. $\int_{\mathbb{R}^m} |C(a,b)|^q db$ and $a \sum_{\ell \in \mathcal{L}(a)} \sup_{(b,a') \in \ell} |C(a',b)|^q$ have the same order of magnitude if the spacing between the maxima is about a, and then the second term is a Riemann sum of the first one. It follows that the counterexamples will construct will have maxima with a spacing much closer than a, and if we slightly modify the wavelet maxima method by imposing that we select only one maximum (or say C maxima) in an interval of length a, then the multifractal formalism will hold.

In order to give some insight about the pitfalls of the wavelet maxima method, we start by describing an example where the maxima accumulate in certain regions. As often, this example involves chirps.

Proposition 10. *Suppose that ψ is compactly supported on $[0,l]$, and that*

$$\exists\, \varepsilon > 0, m \in \mathbb{N} \,:\, \psi(x) = x^m \;\forall\, x \in [0,\varepsilon]$$

(it is the case if ψ is a spline wavelet).

Using ψ as an analyzing wavelet, there exists a function F compactly supported and arbitrarily smooth, and a sequence $a_n \to 0$ such that, for all n, the function $C(a_n,b)$ has an infinity of maxima. There exists a C^∞ function F such that $\theta(q)$ can be chosen arbitrarily small (actually one can choose $\theta(q) = -\infty\; \forall q$).

In that case, if ψ has n vanishing moments, $\eta(p) = np$. This counterexample can be easily extended to selfsimilar functions.

Proposition 11. *Suppose that F is selfsimilar. If in the wavelet maxima method, we keep only the largest maximum in a cube of size a, then the two quantities*

$$\int |C(a,b)|^q db \quad \text{and} \quad a^m \sum_{\max} |C(a,b)|^q$$

are of the same order of magnitude; and thus the multifractal formalism will hold in that case.

This result is quite straightforward since we estimated $\int |C(a,b)|^q db$ precisely by computing its order of magnitude near the wavelet maxima. We showed that its value is about λ_i^q near the tree T and that far from it, it is smaller. This shows that there exists at least one maximum near each point of the tree. The estimation of $a^m \sum_{\max} |C(a,b)|^q$ is similar to the estimation of the corresponding integral. Thus, the verification of the fractal formalism when using the wavelet maxima reduces to the verification for the integral formula, and the multifractal formalism holds when using the two wavelet methods.

Let us consider the case $q < 0$, but for the quantity $\sum_{\max} |C(a,b)|^q$. An important difference with the exact computations of [3] appears. Recall that the authors were interested in the case where F is the primitive of a multinomial measure supported by a Cantor set. The wavelet maxima are situatued on the "tree over the Cantor set" since F is piecewise constant outside this set, and for a small enough, the wavelet transform vanishes there. The same proof as above shows that Theorem 2 will hold for $q < 0$ (with the same restriction concerning the distance between the maxima).

In the general case we consider in this paper, F is a C^k function outside K for which we do not have special information (since g is C^k but arbitrary). Thus, there may be extremely small wavelet maxima "far away" (at the scale a) from K (*i.e.*, at a distance $\gg a$). There is no formula involving negative value of q that can reasonably work. Let us show an example of this phenomenon.

First remark that if $\Omega \neq K$, it is easy to construct examples where g and F will be locally constant so that $F(x+h) - F(x)$ will vanish on an open subset for h small enough, and similarly, $C(a,b)$ will identically vanish on an open subset, for a small enough. $\xi(q)$ and $\eta(q)$ take the value $+\infty$ for negative values of p so that, in all generality, computing $\xi(q)$ and $\eta(q)$ for negative values of p does not make sense. The same problem appears for formulas involving wavelet maxima. Of course, in the regions where $C(a,b)$ vanishes, there are no more maxima. However, it is easy to construct g with lines of very small maxima as follows.

Let ψ be a C^{k+2} function, with moments of order $k+1$ vanishing and supported by $[3/2, 2]$, and let

$$h(x) = \sum_{j \geq 0} 2^{-2kj} \psi(2^{2j} x)$$

be supported by $[0,2]$, ψ the analyzing wavelet, and $a = 2^{-j}$. Then $C(a,b)$ vanishes outside the interval $\mid b \mid \leq 2^{-j}/2$ (if $\mid b \mid \leq 2^{-j}/8$), but $C(a,0) = C2^{-kj}$. Hence, $C(a,b)$ has a line of maxima which goes through the interval $\mid b \mid \leq 2^{-j}/2$ and the supremum on this line is of the order of magnitude of 2^{-kj}.

Suppose that $F = g$ in a neighborhood of 0 (which we can always suppose if $\Omega \neq K$). Then $Z(a,q)$ is larger than Ca^{-kq}. We see that no bound of $Z(a,q)$ independant of k can be found. If the multifractal formalism held (since $d(\alpha)$ is independant of k), the order of magnitude of $Z(a,q)$ would be finite; therefore, a contradiction.

Let us give a last remark on Theorem 2. Suppose first that g is C^∞ if $\alpha \leq \alpha_0$, the infimum in the Legendre transform is attained for $q > 0$.This is straightforward since $\tau(0) = -d_{max}$ and τ is convex and increasing. We cannot calculate directly $d(\alpha)$ up to α_0 except if we use a C^∞ wavelet with all vanishing moments, but, following [2], it can be done by using a sequence of smoother and smoother wavelets and determining larger and larger parts of the spectrum. The case where g is C^k is still easier to check.

We now want to show that we can recover the left part of the spectrum corresponding to negative values of q by using a slight modification of the wavelet maxima method when

$$\overline{\cup S^j(\Omega)} = \overline{\Omega}$$

(this implies that there exists no region where F is smooth). The validity of this condition can actually be checked on the part of the spectrum computed for $q > 0$ since at the maximum ($q = 0$), $d(\alpha) = d_{max}$ which satisfies

$$\sum \mu_j^{d_{max}} = 1,$$

but the condition $\overline{\cup S^j(\Omega)} = \overline{\Omega}$ can be rewritten $\sum \mu_j^m = 1$ since $Vol(\Omega_j) = \mu_j^m Vol(\Omega)$. Hence, $\overline{\cup S^j(\Omega)} = \overline{\Omega}$ is equivalent to $d_{max} = m$ which is easy to check.

In that case, the tree T leaves no void in the region of the upper half plane above Ω (meaning that for any (a, b) in this region, we can find a point of the tree in a domain $[a/C, Ca] \times [b - Ca, b + Ca]$), but, after perhaps increasing the constant C, we can also find a point $(\mu_i, S^i(t))$ where the order of magnitude of the wavelet transform of F is λ_i. If we modify the wavelet maxima method by imposing that we first take the largest local maximum on the box $[a/C, Ca] \times [b - Ca, b + Ca]$ (which amounts to considering a kind of maximal function), we see that for a given frequency a,

$$\sum_{max} |C(a, b)|^q \sim \sum_{a \sim \mu_i} |\lambda_i|^q.$$

We have,

$$\limsup_{a \to 0} a^{-\tau(q)} \sum_{max} |C(a, b)|^q db \geq C > 0,$$

and

$$\limsup_{a \to 0} \frac{a^{-\tau(q)}}{|\log a|} \sum_{max} |C(a, b)|^q db \leq C' < +\infty.$$

Hence, the multifractal formalism in that case.

§7 Riemann's function

Recall that our main motivation is to understand the range of validity of the multifractal formalism. We showed that if the wavelet transform of F satisfies some kind of (linear) renormalization given by Equation (5.3), the multifractal formalism holds. The case of Riemann's function is interesting because the multifractal formalism will be a consequence of nonlinear renormalizations given by fractional linear transforms.

Using Cauchy's formula, we obtain that (using the wavelet $\psi(x) = (x - i)^{-2}$) the wavelet transform of Riemann's function is $2ia(\theta(b + ia) - 1)/2$. To determine Hölder exponents between $1/2$ and $3/4$, because of Equations (2.3) and (2.5), we can add a term ia and the study of the pointwise regularity of φ reduces to obtaining estimates similar to Equation (2.5) for the function

$$C(a, b) = a\theta(b + ia).$$

As in the case of selfsimilar functions, we have to find a way to estimate with precision the size of the wavelet transform everywhere. Here the approximate selfsimilarity of the wavelet transform is replaced by an explicit description of its transformations under the action of the Theta modular group which we now recall.

It is obtained by composing the two transforms

$$x \to x+2 \text{ and } x \to -1/x.$$

It is composed of the fractional linear transformations

$$\gamma(x) = \frac{rx+s}{qx-p}$$

where $rp + sq = -1$, r, s, p, and q are integers and the matrix

$$\begin{pmatrix} r & s \\ p & q \end{pmatrix} \quad \text{is of the form} \begin{pmatrix} \text{even odd} \\ \text{odd even} \end{pmatrix} \quad \text{or} \quad \begin{pmatrix} \text{odd even} \\ \text{even odd} \end{pmatrix}. \tag{7.1}$$

When γ belongs to the Theta modular group, θ is transformed following the formula (see [2,22])

$$\theta(z) = \theta(\gamma(z))e^{im\pi/4}q^{-1/2}(z - \frac{p}{q})^{-1/2} \tag{7.2}$$

where m is an integer, depending on r, s, p, and q.

Suppose that ρ is not a rational number. Let p_n/q_n be the sequence of its approximations by continued fractions. The idea of the proof of Theorem 3 is to use Equation (7.2), allowing us to deduce the behavior of $\theta(z)$ near p_n/q_n (hence near ρ) from its behavior near 0 or 1. Because of Equation (7.1), we will have to separate two cases depending whether or not p_n and q_n are both odd. So we first derive some straightforward estimates for θ near 0 and 1.

First remark that

$$|\theta(z) - 1| \leq \frac{1}{2} \text{ if } Imz \geq 1.$$

It follows that

$$|\theta(z) - 1| \leq 2\sum_{n\geq 1} e^{-\pi n^2 Imz} \leq \frac{2e^{-\pi Imz}}{1 - e^{-\pi Imz}} \leq \frac{1}{2}.$$

We also have

$$|\theta(z)| \leq C|Imz|^{-1/2} \text{ if } Im(z) \leq 1 \tag{7.3}$$

because $|\theta(z)| \leq \sum e^{-n^2 Im(z)}$. The sum for $n \leq Im(z)^{-1/2}$ is bounded trivially by $Im(z)^{-1/2}+1$ and the same bound holds for $n > Im(z)^{-1/2}$ (by comparison with an integral).

Let us now obtain the behavior of θ near the point 1. Recall that θ satisfies (see [2])

$$\theta(1+z) = \sqrt{\frac{i}{z}}\left(\theta(-\frac{1}{4z}) - \theta(-\frac{1}{z})\right),$$

so that

$$\theta(1+z) = 2\sqrt{\frac{i}{z}}\,(A(4z) - A(z)),$$

where $A(z) = \sum_1^\infty e^{-i\pi n^2/z}$. If $Im(\frac{-1}{z}) \geq 1$,

$$|A(z)| \leq 2\exp(-\pi Im(\frac{-1}{z})),$$

we have,

$$|\theta(1+z)| \leq C|z|^{-1/2}\exp(-\pi Im(\frac{-1}{z})). \tag{7.4}$$

Formula (7.2) together with Equations (7.3) and (7.4) allows us to estimate θ in the neighborhood of any rational number (hence also in the neighborhood of any irrational number if we approximate it as well as possible by rationals, which is precisely the job performed by continued fractions). (See [21]).

Proposition 12. *Let* $\frac{p_n}{q_n}$ *be the sequence of approximations of* ρ *by continued fractions. Let* τ_n *be defined by*

$$|\rho - \frac{p_n}{q_n}| = (\frac{1}{q_n})^{\tau_n}.$$

For each n, *if*

$$3|\rho - \frac{p_n}{q_n}| \leq |b - \rho + ia| \leq 3|\rho - \frac{p_{n-1}}{q_{n-1}}| \tag{7.5}$$

the following estimates hold.

If p_n *and* q_n *are not both odd but* p_{n-1} *and* q_{n-1} *are both odd,*

$$|C(a,b)| \leq Ca^{\frac{1}{2}+\frac{1}{2\tau_n}}(1+\frac{|b-\rho|}{a})^{\frac{1}{2\tau_n}}. \tag{7.6}$$

If p_n *and* q_n *are not both odd and* p_{n-1} *and* q_{n-1} *are not both odd,*

$$|C(a,b)| \leq Ca^{\frac{1}{2}+\frac{1}{2\tau_n}}(1+\frac{|b-\rho|}{a})^{\frac{1}{2\tau_n}}, \tag{7.7}$$

or

$$|C(a,b)| \leq Ca^{\frac{1}{2}+\frac{1}{2\tau_{n-1}}}(1+\frac{|b-\rho|}{a})^{\frac{1}{2\tau_{n-1}}}.$$

If p_n *and* q_n *are both odd,*

$$|C(a,b)| \leq Ca^{\frac{1}{2}+\frac{1}{2\tau_{n-1}}}(1+\frac{|b-\rho|}{a})^{\frac{1}{2\tau_{n-1}}}. \tag{7.8}$$

Furthermore, if p_n and q_n are not both odd, these estimates are optimal, meaning that there exists a point in the domain of Equation (7.5) *where Equations* (7.6) *or* (7.7) *are equalities.*

If $\eta(\rho) = \limsup \tau_n(\rho)$ where the lim sup bears only on the n's such that p_n and q_n are not both odd, the result implies that $\varphi \in \Gamma^{\frac{1}{2}+\frac{1}{2\eta(\rho)}}(\rho)$. Let us show how to derive the spectrum of singularities from this result.

Recall that if

$$E_\tau = \{\rho : \ |\rho - \frac{p_n}{q_n}| \leq \frac{C}{q_n^\tau} \ \text{ for infinitely many } n'\text{s}\},$$

the Hausdorff dimension of E_τ is $2/\tau$ and the $\mathcal{H}^{2/\tau}$ measure of E_τ is positive (a direct consequence of Propositions 10.4 and 8.5 of [5]).

Let

$$F_\tau = \{\rho : \ |\rho - \frac{p_n}{q_n}| \leq \frac{C}{q_n^\tau} \}.$$

for infinitely many n's such that p_n and q_n are not both odd, and let

$$G_\tau = \{\rho : \ |\rho - \frac{p_n}{q_n}| \leq \frac{C}{q_n^\tau} \}$$

for infinitely many n's such that p_n and q_n are both odd. Of course, because of the best approximation properties of continued fractions, we have

$$E_\tau = F_\tau \cup G_\tau.$$

We will need the following lemma proved in [24].

Lemma 3. *Let $\rho \in \mathbb{R}$. If p and q have no common factor and*

$$|q\alpha - p| < \frac{1}{2q},$$

then p/q is a continued fraction approximation of ρ.

This lemma implies that if p/q is a continued fraction which approximates ρ and such that p and q are odd and $|\rho - p/q| \leq q^{-\tau}$ with $\tau > 2$, $p/2q$ is a continued fraction which approximates $\rho/2$.

Let us prove that the $\mathcal{H}^{2/\tau}$ measure of F_τ is positive. If the $\mathcal{H}^{2/\tau}$ measure of G_τ vanishes, we have nothing to prove. Otherwise, the remark we just made shows that if $\tau > 2$ and $x \in G_\tau$, then $x/2 \in F_\tau$. In other words, if the $\mathcal{H}^{2/\tau}$ measure of G_τ is positive, the $\mathcal{H}^{2/\tau}$ measure of F_τ is also positive (if $\tau = 2$, $F_\tau = \mathbb{R}$).

Consider the set

$$F_\tau - \bigcup_{\tau' > \tau} E_{\tau'}.$$

The $\mathcal{H}^{2/\tau}$ measure of $\bigcup_{\tau'>\tau} E_{\tau'}$ vanishes since F_τ has a $\mathcal{H}^{2/\tau}$ measure positive and $F_\tau - \bigcup_{\tau'>\tau} E_{\tau'}$ has dimension $2/\tau$.

If $\rho \in F_\tau - \bigcup_{\tau'>\tau} E_{\tau'}$, since $\rho \in F_\tau$, Equation (2.3) implies that φ is not smoother than $\frac{1}{2} + \frac{1}{2\tau}$ at ρ, and since $\rho \notin \bigcup_{\tau'>\tau} E_{\tau'}$, φ is $C^{\frac{1}{2}+\frac{1}{2\tau}-\varepsilon}(\rho)$ $\forall\varepsilon > 0$. Therefore, $\varphi \in \Gamma^{\frac{1}{2}+\frac{1}{2\tau}}(\rho)$ and the dimension of $\{\rho : \ \varphi \in \Gamma^{\frac{1}{2}+\frac{1}{2\tau}}(\rho)\}$ is at least $2/\tau$.

Suppose that ρ is such that $\varphi \in \Gamma^{\frac{1}{2}+\frac{1}{2\tau}}(\rho)$. Then φ is $C^{\frac{1}{2}+\frac{1}{2\tau}-\varepsilon}(\rho)$ $\forall\varepsilon > 0$ and thus $\rho \in E_{\tau'}$ $\forall\tau' < \tau$. We have

$$\{\rho : \ \varphi \in \Gamma^{\frac{1}{2}+\frac{1}{2\tau}}(\rho)\} \subset \bigcup_{\tau'>\tau} E_{\tau'},$$

and the dimension of $\{\rho : \ \varphi \in \Gamma^{\frac{1}{2}+\frac{1}{2\tau}}(\rho)\}$ is bounded by $2/\tau$; hence, the second part of Theorem 3 is proved.

Let us now check that the multifractal formalism is true for Riemann's function. Let

$$S_n(x) = \sum_{m=1}^{n} e^{im^2\pi x}.$$

In [25], Z. Zalcwasser proves that

$$\begin{aligned}\int_0^1 |S_n(x)|^p dx &\sim n^{p/2} \text{ if } 0 < p < 4\\ &\sim n^2 \log(n+1) \text{ if } p = 4\\ &\sim n^{p-2} \text{ if } p > 4.\end{aligned}$$

Taking D-adic blocs (for a D large enough), we have

$$\begin{aligned}\| \sum_{D^j \le m^2 < D^{j+1}} e^{im^2\pi x} \|_{L^p} &\sim D^{j/4} \text{ if } 0 < p < 4\\ &\sim D^{j(p-2)/2p} \text{ if } p > 4.\end{aligned}$$

Let $\Phi = \sum \frac{1}{n^2} e^{in^2\pi x}$. We have $\Phi' \in B_p^{-\frac{1}{4},\infty}$ if $p < 4$ and $\Phi' \in B_p^{-\frac{1}{2}+\frac{1}{p},\infty}$ if $p > 4$. Then $\Phi \in B_p^{\frac{3}{4},\infty}$ if $p < 4$ and $\Phi \in B_p^{\frac{1}{2}+\frac{1}{p},\infty}$ if $p > 4$, and these estimates are optimal. Because of the continuity of the Hilbert transform on Besov spaces, the same result holds for φ. With $\eta(p)$ defined by

$$\eta(p) = \sup\{s : \ \varphi \in B_p^{s/p,\infty}\},$$

we have

$$\eta(p) = 3p/4 \text{ if } 0 < p \leq 4$$
$$= 1 + p/2 \text{ if } p \geq 4.$$

If $\alpha < 1/2$, $\inf_p(\alpha p - \eta(p) + 1) = -\infty$ and if $1/2 \leq \alpha \leq 3/4$, $\inf_p(\alpha p - \eta(p) + 1) = 4\alpha - 2$. We recover the increasing part of the spectrum and show the validity of the multifractal formalim in that case.

The fact that Equations (7.6) and (7.8) cannot be improved in a cone

$$Im(z - \rho) \geq CRe(z - \rho)$$

yields a slightly more precise information than the fact that φ is not smoother than $\frac{1}{2} + \frac{1}{2\eta(\rho)}$ at ρ because it shows that φ has no chirp expansion at an irrational point ρ (see [13]). The only points where φ has a chirp expansion are the rationals of the form odd/odd. It also shows that fractional integrals of φ of order s will be exactly $\Gamma^{s+\frac{1}{2}+\frac{1}{2\eta(\rho)}}(\rho)$. Actually, from Proposition 12 and the chirp characterization given in [20], one easily obtains the following corollary.

Corollary 2. *Let*

$$\varphi_s(x) = \sum_1^{\infty} \frac{1}{n^{2+2s}} \sin \pi n^2 x.$$

If $s \in (-1/2, +\infty)$ and if ρ is not a rational quotient of two odd numbers, $\varphi_s \in \Gamma^{s+\frac{1}{2}+\frac{1}{2\eta(\rho)}}(\rho)$ and the spectrum of singularities of φ_s is given by

$$d(\alpha) = 4(\alpha - s) - 2 \text{ if } \alpha \in [s + \frac{1}{2}, s + \frac{3}{4}]$$
$$= 0 \quad \text{if } \alpha = 2s + 3/2;$$
$$= -\infty \quad \text{otherwise.}$$

§8 Some concluding remarks

Let us now try to understand reasons for the possible failure of the multifractal formalism. Consider again the example of Riemann's function. A big difference appears with selfsimilar functions if we want to check the multifractal formalism for $\alpha \geq 3/4$; *i.e.*, for the decreasing part of the spectrum. Recall that $d(3/2) = 0$ and otherwise, $d(\alpha) = -\infty$. Since the multifractal formalism involves a Legendre transform formula, we do not expect it to yield the exact spectrum but rather its convex hull. However, it is not the case here because of the chirp behavior corresponding to the point 3/2.

Recall that if x_0 is the quotient of two odd numbers,

$$\varphi(x) = u(x) + \sum_{n \geq 0} (x - x_0)^{\frac{3}{2}+n} v_+^n\left(\frac{1}{x - x_0}\right) \text{ if } x \geq x_0,$$

$$\varphi(x) = u(x) + \sum_{n \geq 0} |x - x_0|^{\frac{3}{2}+n} v_-^n\left(\frac{1}{|x - x_0|}\right) \text{ if } x \leq x_0,$$

where u is C^∞, the $v_\pm^n$ are 2π periodic with a vanishing integral and are $C^{\frac{1}{2}+n}$. Such a behavior is characterized by the following conditions for the wavelet transform (using a wavelet with enough smoothess and cancellation),

$$|C(a, b)| \leq C a^s \left(1 + \frac{|b - x_0|}{a}\right)^{-s'}$$

for all (s, s') such that $s + s' \leq 1/2$ and $s' + 2s \leq 3/2$ (Theorem 4.2 of [22]).

In particular, it implies that $\forall D > 0$, in the cone

$$|b - x_0| \leq Da,$$

the wavelet transform has a fast decay

$$|C(a, b)| \leq C(D, n) a^n \ \ \forall\, n > 0.$$

Thus the wavelet maxima method would yield

$$\sum_{b \in \, max} |C(a, b)|^p \geq a^{np}$$

so that $\eta(p) = +\infty \ \forall p < 0$. The wavelet maxima method cannot yield the right part of the spectrum of Riemann's function, and more generally, the same remark holds for any function which has a chirp behavior near at least one point. Let us show another property of chirps incompatible with the multifractal formalism.

If $F \in B_p^{s,q}$, $(-\Delta)^{a/2} F \in B_p^{s-a,q}$ $\forall a \in \mathbf{R}$, so that, if the multifractal formalism holds, for F and for $(-\Delta)^{a/2} F$, it implies that the spectrum of $(-\Delta)^{a/2} F$ is simply the spectrum of F translated by a. This is in agreement with the (strictly speaking, wrong) assertion that the Hölder regularity of F at x_0 is shifted by 1 when one differentiates F. Of course, if the points with a chirp behavior have a positive dimension, the spectrum won't be shifted as expected (this is the case for the right hand side of Riemann's function as we already noticed); these arguments show that the multifractal formalism can hold only with the assumption that there are not too many chirp-like points.

References

1. Adams, R. A., *Sobolev Spaces*, Pure and Applied Mathematics **65**, Academic press, 1978.
2. Arneodo, A., E. Bacry, and J. F. Muzy, Wavelet analysis of fractal signals, direct determination of the singularity spectrum of fully developped turbulence data, 1991, preprint.

3. Bacry, E., A. Arneodo, and J. F. Muzy, Singularity spectrum of fractal signals from wavelet analysis: exact results, 1991, preprint.
4. Brown, G., G. Michon, and J. Peyrière, On the multifractal analysis of measures, 1991, preprint.
5. Caffarelli, L., R. Kohn, and L. Nirenberg, Partial regularity of suitable weak solutions of the Navier-Stokes equations, *Comm. on Pure Appl. Math.* **35** (1982), 771.
6. Cohen, A., Thèse de l'Université Paris Dauphine, 1990.
7. Daubechies, I. and J. Lagarias, On the thermodynamic formalism for multifractal functions, 1993, preprint
8. Duistermaat, J. J., Selfsimilarity of Riemann's non-differentiable function, 1991, preprint.
9. Falconer, K., *Fractal Geometry*, John Wiley and Sons, 1990.
10. Frazier, M., B. Jawerth, and G. Weiss, *Littlewood-Paley Theory and the Study of Function Spaces*, CBMS 79 (AMS), 1991.
11. Frisch, U. and G. Parisi, Fully developped turbulence and intermittency, *Proc. Int. Summer School Phys.*, Enrico Fermi, North Holland, 1985, 84–88.
12. Gerver, J., The differentiability of the Riemann function at certain rational multiples of π, *Amer. J. Math.* **92** (1970).
13. Grochenig, K. and W. R. Madych, Multiresolution analysis, Haar bases and selfsimilar tilings, 1990, preprint.
14. Hardy, G. H., Weierstrass's non-differentiable function, *Trans. Amer. Math. Soc.* **17** (1916), 301–325.
15. Hardy, G. H. and J. E. Littlewood, Some problems of Diophantine approximation, *Acta Mathematica* **37** (1914), 193–239.
16. Holschneider, M. and P. Tchamitchian, Pointwise analysis of Riemann's "non differentiable" function, *Inventiones Mathematicae* **105** (1991), 157–176.
17. Hutchinson, J., Fractals and selfsimilarity, *Indiana Univ. Math. J.* **30** (1981), 713–747.
18. Jaffard, S., Pointwise smoothness, two-microlocalization and wavelet coefficients, *Publicacions Matematiques* **35** (1991), 155–168.
19. Jaffard, S., Multifractal formalism for functions, part I: results valid for all functions; part II: selfsimilar functions, 1993, preprint.
20. Jaffard, S., Construction de fonctions multifractales ayant un spectre de singularités prescrit, *C.R.A.S. Série 1* **315** (1992), 19–24.
21. Jaffard, S., The spectrum of singularities of Riemann's function, 1994, preprint.
22. Jaffard, S. and Y. Meyer, Pointwise behavior of functions, 1993, preprint.
23. Mallat, S. and W. L. Hwang, Singularity detection and processing with wavelets, 1991, preprint.
24. Mandelbrot, B., Intermittent turbulence in selfsimilar cascades: Divergence of high moments and dimension of the carrier, *J. Fluid Mechanics* **62** (1974), 331.

25. Meyer, Y., *Ondelettes et Opérateurs*, Hermann, 1990.
26. Muzy, J. F., A. Arneodo, and E. Bacry, Direct determination of the singularity spectrum of fully developed turbulence data, 1991, preprint.
27. Nicolskii, S. M., Extension of functions of several variables preserving differential properties, *A.M.S. Translations (2)* **83** (1969), 159–188.
28. Tricot, C., Thèse, Université d'Orsay.
29. Valiron, G., *Théorie des Fonctions*, Masson, 1942.
30. Van der Porten, An Introduction to Continued Fractions in Diophantine Analysis, *London Math. Series, Lecture notes series*, **109**, Cambridge University Press, 99–138.
31. Zalcwasser, Z., Sur les polynomes associés aux fonctions modulaires θ, *Studia Mathematica* **7** (1938), 16–35.
32. Ziemin,V. D., Hierarchic models of turbulence, *Izvestiya Atmospheric and oceanic Physics* **17** (N12) (1981).

Stephane Jaffard
CERMA
ENPC, La Courtine
93167 Noisy le grand, FRANCE
and CMLA
ENS Cachan, 61 av. du Président Wilson,
94235 Cachan Cedex, FRANCE
sj@cerma.enpc.fr

Fractal Wavelet Dimensions and Time Evolution

Matthias Holschneider

Abstract. In this paper we want to give a new definition of fractal dimensions as small scale behavior of the q-energy of wavelet transforms. This is a generalization of previous multi-fractal approaches. With this particular definition we will show, that the 2- dimension (=correlation dimension) of the spectral measure determines the long time behavior of the time evolution generated by a bounded self-adjoint operator acting on some Hilbert space $\mathcal{H}$. It will be proved that for ϕ, $\psi \in \mathcal{H}$ we have

$$\liminf_{T\to\infty} \frac{\log \int_0^T d\omega \left|\left\langle \psi \mid e^{-iA\omega}\phi\right\rangle\right|^2}{\log T} = -\kappa^+(2)$$

and that

$$\limsup_{T\to\infty} \frac{\log \int_0^T d\omega \left|\left\langle \psi \mid e^{-iA\omega}\phi\right\rangle\right|^2}{\log T} = -\kappa^-(2),$$

where $\kappa^{\pm}(2)$ are the upper and lower correlation dimensions of the spectral measure associated with ψ and ϕ. A quantitative version of the RAGE theorem shall also be given.

§1 Introduction

Let μ be a finite (signed) measure. A well-known theorem of Wiener states that

$$\lim_{T\to\infty} \frac{1}{T} \int_0^T d\omega \; |\widehat{\mu}(\omega)|^2 = \sum_{x\in\mathbb{R}} |\mu\{x\}|^2,$$

where the Fourier transform is given by

$$\widehat{\mu}(\omega) = \int d\mu(t)\, e^{-i\omega t}.$$

Wavelets: Theory, Algorithms, and Applications
Charles K. Chui, Laura Montefusco, and Luigia Puccio (eds.), pp. 363–381.

ISBN 0-12-174575-9

Note that the sum is finite since μ is finite. Now let A be a a self-adjoint operator acting on some Hilbert-space $\mathcal{H}$. For any state $\phi \in \mathcal{H}$, we shall be interested in the long time behavior of the unitary evolution $\phi \to e^{-itA}\phi$. More precisely, we look at ($\psi \in \mathcal{H}$)

$$\frac{1}{T}\int_0^T dt\, \left|\langle \psi \mid e^{-itA}\phi\rangle\right|^2 \quad \text{as } T \to \infty\,. \tag{1.1}$$

By the usual functional calculus the above integral is equal to

$$\frac{1}{T}\int_0^T d\omega\, \left|\widehat{\mu}_{\psi,\phi}(\omega)\right|^2,$$

where $\mu_{\psi,\phi}$ is the spectral measure associated with ψ and ϕ. Therefore, in the case of an operator having only pure point-spectrum, the long time behavior of the averaged time-evolution in Equation (1.1) of a non-zero ϕ is given by Wiener's theorem:

$$\frac{1}{T}\int_0^T dt\, \left|\langle \psi \mid e^{-itA}\phi\rangle\right|^2 \sim 1 \quad (T \to \infty)$$

On the other hand, let $\phi \in \mathcal{H}_{\text{cont}}$ belong to the continuous spectrum of A (see [4] for the notation). Then the celebrated RAGE theorem (see *e.g.*, [4]) that states that, for any compact operator B, we have

$$\frac{1}{T}\int_0^T dt\, \left\|B\, e^{-itA}\phi\right\|^2 \to 0 \quad (\text{as } T \to \infty).$$

In particular, upon setting $B : \varphi \to \langle \psi \mid \varphi\rangle\, \psi$, we see that the mean evolution in Equation (1.1) for $\phi \in \mathcal{H}_{\text{cont}}$ is given by

$$\frac{1}{T}\int_0^T dt\, \left|\langle \psi \mid e^{-itA}\phi\rangle\right|^2 \to 0 \quad (T \to \infty). \tag{1.2}$$

In this paper we are concerned with a quantitative version of the last equation. It will turn out that the speed of decay towards 0 is determined by the fractal correlation dimension of the spectral measure.

The rest of the paper is organized as follows. We start by presenting our main analysis tool, which is the wavelet transform. Next, we introduce the wavelet-dimensions for any tempered distribution. The last part considers the relation between the correlation dimension and the averaged time evolution.

§2 Introduction to wavelet transforms

Let us consider the Schwarz space $S(\mathbb{R})$ that consists of those functions that together with all their derivatives decay at infinity faster than any polynomial. A topology is induced by the semi-norms

$$\|s\|_{S(\mathbb{R}),n,m} = \sup_{t\in\mathbb{R}} \left|t^n \partial_t^m s(t)\right|.$$

Let $S_+(\mathbb{R})$ be the subset of those functions in $S(\mathbb{R})$ whose Fourier transform is supported by the positive frequencies only. It is a closed sub-space of $S(\mathbb{R})$ and we equip it with the topology induced by $S(\mathbb{R})$. Since the Fourier transform of any function $s \in S_+(\mathbb{R})$ is smooth at $\omega = 0$ and identically vanishing for $\omega \leq 0$, all moments for s vanish

$$\widehat{s}(\omega) = O(\omega^n)\ (\omega \to 0) \quad \Longleftrightarrow \quad \int_{-\infty}^{+\infty} dt\ t^n\ s(t) = 0 \quad \text{for all } n \in \mathbb{N}_0. \tag{2.1}$$

The wavelet transform [3,5,16] of $s \in S_+(\mathbb{R})$ with respect to the wavelet $g \in S_+(\mathbb{R})$ is given by

$$\mathcal{W}_g s(b,a) = \int_{-\infty}^{+\infty} dt\ \frac{1}{a}\ \overline{g}\left(\frac{t-b}{a}\right) s(t) = \int_{-\infty}^{+\infty} dt\ \overline{g}_{b,a}(t) s(t), \quad a > 0, b \in \mathbb{R}.$$

It can easily be shown (*e.g.*, [10,16]) that for $s,\ g \in S_+(\mathbb{R})$ the wavelet transform of $s \in S_+(\mathbb{R})$ with respect to $g \in S_+(\mathbb{R})$ is a highly localized function over the half-plane $\mathbb{H} = \mathbb{R} \times \mathbb{R}^+$. More precisely let $S(\mathbb{H})$ be the space of all functions $\mathcal{T}$ over the half-plane that are localized such that

$$\|\mathcal{T}(b,a)\|_{S(\mathbb{H}),n} = \sup_{(b,a)\in\mathbb{H}} |\mathcal{T}(b,a)|\ (a + 1/a)^n\ (1 + |b|)^n < \infty \quad \text{for all } n > 0.$$

These are actually semi-norm and $S(\mathbb{H})$ is a Fréchet space with the induced topology. The wavelet transform is then a continuous map from $S_+(\mathbb{R})$ to $S(\mathbb{H})$.

Now, let us consider the wavelet synthesis $\mathcal{M}_h : S(\mathbb{H}) \to S_+(\mathbb{R})$ with respect to $h \in S_+(\mathbb{R})$. It is point-wise defined for $\mathcal{T} \in S(\mathbb{H})$ by

$$\mathcal{M}_h \mathcal{T}(t) = \int_0^{\infty} \frac{da}{a} \int_{-\infty}^{+\infty} db\ \mathcal{T}(b,a)\ \frac{1}{a}\ h\left(\frac{t-b}{a}\right).$$

The wavelet synthesis is again a continuous map; it is essentially the inverse of the wavelet transform:

$$c_{g,h}^{-1}\, \mathcal{M}_h\, \mathcal{W}_g = \mathbf{1}, \quad s(t) = c_{g,h}^{-1} \int_0^{\infty} \frac{da}{a} \int_{-\infty}^{+\infty} db\ \mathcal{W}_g s(b,a)\ \frac{1}{a}\ h\left(\frac{t-b}{a}\right) \tag{2.2}$$

with any function $h \in S_+(\mathbb{R})$ satisfying

$$c_{g,h} = \int_0^{\infty} \frac{d\omega}{\omega}\ \widehat{h}(\omega)\ \overline{\widehat{g}}(\omega) \quad 0 < |c_{g,h}| < \infty. \tag{2.3}$$

The function h is called a reconstruction wavelet for g. In particular, every wavelet in $S_+(\mathbb{R})$ is its own reconstruction wavelet up to some constant trivial factor.

Upon reconstructing with g and analyzing with some other function h, we obtain a simple relation between the wavelet transform with respect to h and the one with respect to g.

$$\mathcal{W}_h s(b,a) = \int_0^{\infty} \frac{da'}{a'} \int_{-\infty}^{+\infty} db' \frac{1}{a'} \Pi_{g,h}\left(\frac{b-b'}{a'}, \frac{a}{a'}\right) \mathcal{W}_g s(b',a'), \tag{2.4}$$

with $\Pi_{g,h}(b,a) = c_{g,h}^{-1} \mathcal{W}_h g(b,a)$. For $g = h$ this is the so-called reproducing kernel equation.

2.1 The wavelet transform of distributions

Let η be a distribution in $S'(\mathbb{R})$. By the usual duality approach we can define the wavelet transform with respect to $g \in S_+$ via its action on $\mathcal{T} \in S(\mathbb{H})$,

$$(\mathcal{W}_g \eta)(\mathcal{T}) = \frac{1}{c_{g,h}} \eta(\mathcal{M}_h \mathcal{T}).$$

It can be shown [10], that we may identify $\mathcal{W}_g \eta$ with a function over the half-plane given point-wise by

$$\mathcal{R}(b,a) = \eta(\overline{g}_{b,a}).$$

The identification is made through the (absolutely convergent) "scalar product"

$$(\mathcal{W}_g \eta)(\mathcal{T}) = \langle \mathcal{R} \mid \mathcal{T} \rangle_{\mathbb{H}} = \int_0^\infty \frac{da}{a} \int_{-\infty}^{+\infty} db \, \overline{\mathcal{R}}(b,a) \, \mathcal{T}(b,a).$$

The function $\mathcal{R}$ is a smooth function over the half-plane of at most polynomial growth in $b/(a+1/a)$ as $|b|$ or $a + 1/a \rightarrow \infty$. For simplicity we shall denote this function $\mathcal{R}$ over the half-plane again by $\mathcal{W}_g \eta$. The action of $\eta \in S'(\mathbb{R})$ on $s \in S_+(\mathbb{R})$ may now be written as absolutely convergent integral over the half-plane [10]

$$\eta(s) = c_{g,h}^{-1} \, \langle \mathcal{W}_g \eta \mid \mathcal{W}_h s \rangle_{\mathbb{H}}$$

and Equation (2.4) is still point-wise valid in this case. For further reference we note that in Fourier space we have (in the sense of distributions)

$$\mathcal{W}_g \eta(b,a) = \frac{1}{2\pi} \int_0^\infty d\omega \; \overline{\hat{g}}(a\omega) \; e^{ib\omega} \; \hat{\eta}(\omega).$$

This shows, in particular, that if $\hat{\eta}$ is a function and $\hat{\eta}(\omega) = O(\omega^n)$ as $\omega \rightarrow 0$, then

$$\mathcal{W}_g \eta(b,a) = O(1/a^{n+1}) \quad (a \rightarrow \infty) \tag{2.5}$$

uniformly in b.

2.2 Wavelet analysis of local regularity

The wavelet transform may be seen as a sort of mathematical microscope [1,8] whose position is given by b, whereas a is the length-scale at which the object is examined. In particular, it has been proved to be very useful in characterizing the local fractality (=local regularity) of functions [13] or even arbitrary distributions. For instance global Hölder regularity of degree α is characterized by a uniform decrease of the wavelet coefficients at small scale:

$$s(t) - s(u) = O(|t-u|^\alpha), \text{ with } 0 < \alpha < 1 \iff \mathcal{W}_g s(b,a) = O(a^\alpha). \tag{2.6}$$

If $\alpha > 1$ and $n < \alpha < n+1$ with $n \in \mathbb{N}$, then one has to replace s by $\partial_t^n s$ and the statement is still valid. However, even point-wise information is available. Suppose s satisfies at some point t_0 at

$$s(t_0 + t) - s(t_0) = P_n(t) + O(t^\alpha), \qquad (t \rightarrow 0)$$

with some polynomial P_n of degree n and some $\alpha \in (n, n+1]$. Then

$$\mathcal{W}_g s(t_0 + b, a) = O(a^\alpha + b^\alpha). \tag{2.7}$$

Vice versa, if one knows that globally the function is of Hölder regularity with some arbitrary small regularity exponent $\epsilon > 0$ and if the wavelet transform satisfies Equation (2.7), with $\alpha \notin \mathbb{N}_0$, then there is a polynomial P_n of degree n such that [11,13,14]:

$$s(t_0 + t) - s(t_0) = P_n(t) + O(\log t \; t^\alpha).$$

Thus, the small-scale behavior of the wavelet transform reflects well the local fractality (=local regularity).

2.3 A functional calculus

Although we shall not make direct use of the results listed below, we state them as instructive motivation for the rest of the paper. Consider a self-adjoint operator A acting in some Hilbert space and let $R_z = (A - z)^{-1}$ be the resolvent of A. By the standard functional calculus

$$a^{n-1} \; R_z^n(A) = \frac{1}{a} \int \frac{dE_\lambda}{\left(\frac{\lambda - b}{a} + i\right)^n}, \qquad (z = b + ia)$$

with dE_λ being the spectral family. Therefore, the powers of the resolvent can be seen as a wavelet transform of the spectral family with respect to the wavelet $h(t) = (t+i)^{-n}$ or by taking matrix elements as wavelet transform of the spectral measures $d\mu_{\phi,\psi}(\lambda) = \langle \phi \mid dE_\lambda \psi \rangle$.

$$\langle \psi \mid R^n_{b+ia}(A)\phi \rangle = a^{1-n} \mathcal{W}_h d\mu_{\phi,\psi}(b, a)$$

The half-plane of the wavelet transform is now the complex-half plane where the imaginary part corresponds to the analyzed scale and the real part to the position of the mathematical microscope. Note, however, that h is not in $S_+(\mathbb{R})$ but this fact actually poses no extra difficulties. Upon replacing t by A in the inversion Formula (2.2) one might hope to get an expression for $s(A)$ and this can actually be established in a mathematically correct way [12]. In particular, it has been proved that in a weak sense, we have for $n \geq 2$, $g \in S_+(\mathbb{R})$, and A a self-adjoint operator (upon setting $s(t) = e^{-iTt}$)

$$e^{-iTA} = \frac{2\pi}{ic} \int_0^\infty da a^{n-1} \int_{-\infty}^{+\infty} db \, \overline{g}(aT) \, e^{-iTb} \, R^n_{b+ia}(A),$$

with $c = \int_0^\infty d\omega \omega^{n-2} \, \overline{\widehat{g}}(\omega) e^{-\omega}$ and the integral over the half-plane being $\lim_{\rho \to \infty} \int_{1/\rho}^{\rho} \frac{da}{a} \int_{-\rho}^{\rho} db$. This shows explicitly how the behavior for large T is determined by the behavior of R_z at $\Im z \sim 1/T$. Indeed, suppose that the support of $\widehat{g}$ is contained in an interval around $\omega = 1$. Then the integral over the complex z half-plane actually runs only over a strip around $\Im z \sim 1/T$. Thus the long time behavior $t \to \infty$ of the time evolution is linked to the small scale behavior of some wavelet transform of the spectral measure, which, in turn, is closely connected with local fractal properties (= local regularity properties) of the measure itself. This relation can actually be quantified as we shall see in the following sections.

§3 The definition of the wavelet dimensions

For the remaining part of the paper, we are only interested in local properties; *i.e.*, we suppose that all analyzed distributions $\eta \in S'(\mathbb{R})$ are well behaved at infinity. More precisely we require that for all $s \in S(\mathbb{R})$ we have $s * \eta \in S(\mathbb{R})$, a condition to be assumed throughout. It is based on purely technical nature and may be relaxed considerably from case to case. In particular, note that the Fourier transform of such a distribution actually is a smooth function (but not localized). In addition, we assume that $\eta \neq 0$. For every scale a we look at the mean q-energy at scale a

$$G_g(a,q) = \|\mathcal{W}_g\eta(\cdot,a)\|_q^q = \int_{-\infty}^{+\infty} db\ |\mathcal{W}_g\eta(b,a)|^q \qquad \text{with } q \geq 1.$$

Note that for $q = 2$ this actually is an energy since (*e.g.*, [5])

$$\int_{-\infty}^{+\infty} dt\ |s(t)|^2 = \frac{1}{c_{g,g}} \int_0^{\infty} \frac{da}{a} \int_{-\infty}^{+\infty} db\, |\mathcal{W}_g s(b,a)|^2 . \tag{3.1}$$

At a small scale, a scaling behavior of the form $G(a,q) \sim a^{\kappa(q)}$ can be observed –at least for affine self-similar measures, *e.g.*, the triadic Cantor set with Bernoulli measure—giving rise [9] to the definition of the fractal dimensions $\kappa(q)$. However, we shall use a slightly modified definition. We set

$$\Gamma_g(a,q) = \int_a^1 \frac{d\alpha}{\alpha}\ G_g(\alpha,q).$$

For every $q \geq 1$ this is a monotone function of a. Therefore, the limit $a \to 0$ exists but may be infinite. This will always be the case if η is singular enough. In the opposite case when the limit is finite, we subtract the constant $\int_0^1 \frac{d\alpha}{\alpha} G(\alpha,q)$ and we set

$$\Gamma_g(a,q) = \int_0^a \frac{d\alpha}{\alpha}\ G_g(\alpha,q).$$

To summarize, we have

$$\Gamma_g(a,q) = \min\left\{\int_a^1 \frac{d\alpha}{\alpha}\ G_g(\alpha,q), \int_0^a \frac{d\alpha}{\alpha}\ G_g(\alpha,q)\right\}.$$

Note that $G_g(a,q) \sim a^{\kappa(q)}$ implies $\Gamma_g(a,q) \sim a^{\kappa(q)}$ unless $\kappa(q) = 0$. The generalized dimensions $\kappa^+(q)$ and $\kappa^-(q)$ are now defined as follows

$$\kappa^+(q) = \limsup_{a\to 0} \frac{\log \Gamma_g(a,q)}{\log a}, \quad \kappa^-(q) = \liminf_{a\to 0} \frac{\log \Gamma_g(a,q)}{\log a}. \tag{3.2}$$

In the case where $\eta = 0$, we set $\kappa^{\pm}(q) = 0$. We shall refer to these numbers as the upper and lower q-wavelet dimensions. Note that in this form the dimensions are defined for any distribution, in particular for measures and functions. The approach using integrated wavelet transform to define fractal dimensions was introduced in [9], where it was also shown that for affine self-similar measures these dimensions coincide with the generalized fractal dimensions defined in [7]. More precisely, we have $\kappa(q) = \tau(q) - (q-1)$ where $\tau(q)$ is defined by the box-counting procedure as described in [7] (see also Proposition 5.4). Later variations of this approach have been considered [17].

The fractal dimensions $\kappa^-(q)$ are actually well defined as is shown by the following theorem.

Theorem 3.1. *The dimensions $\kappa^{\pm}(q)$ do not depend on the analyzing wavelet $g \in S_+(\mathbb{R})$ for $q \geq 1$, provided $g \neq 0$ and $|\kappa^{\pm}(q)| < \infty$.*

Proof: We start by rephrasing expressions like Equation (3.2). We need only the first identity of the following lemma for this proof, while theremaining ones are for later reference.

Lemma 3.2. *Let $s(t) > 0$ be a positive function defined for $t > 0$. Then we have*

$$\liminf_{t\to 0} \frac{\log s(t)}{\log t} = \sup\{\gamma \in \mathbb{R} : s(t) = O(t^\gamma), (t \to 0)\},$$

$$\limsup_{t\to 0} \frac{\log s(t)}{\log t} = \inf\{\gamma \in \mathbb{R} : t^\gamma = O\big(s(t)\big), (t \to 0)\},$$

$$\liminf_{t\to \infty} \frac{\log s(t)}{\log t} = \sup\{\gamma \in \mathbb{R} : t^\gamma = O\big(s(t)\big), (t \to \infty)\},$$

$$\limsup_{t\to \infty} \frac{\log s(t)}{\log t} = \inf\{\gamma \in \mathbb{R} : s(t) = O(t^\gamma), (t \to \infty)\}.$$

(The proof is well-known; we will not repeat here.)

To prove the independency of $\kappa^-(q)$ on g, it is enough to show that $\Gamma_g(a,q) = O(a^\gamma)$ implies $\Gamma_h(a,q) = O(a^\gamma)$ for any pair of wavelets $g, h \in S_+(\mathbb{R})$. This shall be done as follows.

We start by modifying a little the definition of the partition function Γ_g. Here, we have to distinguish two cases according to whether or not

$$\lim_{a\to 0} \int_a^1 \frac{d\alpha}{\alpha} G_g(\alpha, q)$$

is infinite. Consider first the case that it is infinite. Note that we may suppose for all $m > 0$,

$$\widehat{\eta}(\omega) \leq O(\omega^m) \quad (\omega \to 0). \tag{3.3}$$

Indeed, the fractal dimensions of η and $\eta + s$ with $s \in S(\mathbb{R})$ are the same because of the fast decay of $\mathcal{W}_g s$ at small scale $a \to 0$ (see Equation (2.6) for α arbitrarily large). Therefore, if we chose some $\psi \in S(\mathbb{R})$ satisfying for all m at $\psi(\omega) = 1 + O(\omega^m)$ as $\omega \to 0$, we may replace η by $\eta - \psi * \eta$ without modifying the dimensions. This latter distribution satisfies Equation (3.3). By Equation (2.5), the wavelet transform decays fast as $a \to \infty$, and thus we may replace $\int_a^1$ by $\int_a^\infty$ in the definition of Γ_g. In other words, we may consider

$$\Gamma_g(a, q) = \int_a^\infty \frac{d\alpha}{\alpha} \, \|\mathcal{W}_g \eta(\cdot, \alpha)\|_q^q .$$

We are ready to study the estimates. From Equation (2.4), it follows that with

$$K_{a',a}(b) = \frac{1}{a'} \Pi_{g,h}\left(\frac{b}{a'}, \frac{a}{a'}\right);$$

the passage from $\mathcal{W}_g \eta$ to $\mathcal{W}_h \eta$ reads

$$\mathcal{W}_h \eta(\cdot, a) = \int_0^\infty \frac{da'}{a'} K_{a',a} \, * \, \mathcal{W}_g \eta(\cdot, a').$$

However, we have to make sure that $K_{a',a}$ is well defined. The only possible obstruction to this is the constant $c_{g,h}$ as defined in Equation (2.3) that may vanish. (Note that g never goes to ∞ and $h \in S_+(\mathbb{R})$.) However it cannot vanish for all the dilated and translated versions $g_{\beta,\alpha} = \alpha^{-1} g([\cdot - \beta]/\alpha)$ of g since this would imply that the wavelet transform of h with respect to g vanishes, which is impossible for $h \neq 0$. Replacing g by one of its dilated and translated versions $g_{\beta,\alpha}$ is equivalent to replacing $\mathcal{W}_g(b, a)$ by

$$\mathcal{W}_{g_{\beta,\alpha}} \eta(b, a) = \frac{1}{\alpha} \mathcal{W}_g \eta\left(\frac{b - \beta}{\alpha}, \frac{a}{\alpha}\right),$$

and as a result, the dimensions computed with $g_{\beta,\alpha}$ instead of g are the same. We may suppose that $c_{g,h} \neq 0$. Using Minkowkis and Hölder's inequalities we write

$$\begin{aligned} \|\mathcal{W}_h \eta(\cdot, a)\|_q^q &\leq \left\{ \int_0^\infty \frac{da'}{a'} \, \|K_{a',a} \, * \, \mathcal{W}_g \eta(\cdot, a')\|_q \right\}^q \\ &\leq \left\{ \int_0^\infty \frac{da'}{a'} \, \|K_{a',a}\|_1 \, \|\mathcal{W}_g \eta(\cdot, a')\|_q \right\}^q . \end{aligned}$$

Now we have

$$\|K_{a',a}\|_1 = \int_{-\infty}^{+\infty} db \, \frac{1}{a'} \left| \Pi_{g,h}\left(\frac{b}{a'}, \frac{a}{a'}\right) \right| = H(a/a'),$$

with

$$H(a) = \int_{-\infty}^{+\infty} db \; |\Pi_{g,h}(b,a)| \, .$$

This is a non-negative function that is rapidly decaying as $a + 1/a$ gets large. Using Jensen's inequality, we have

$$\|\mathcal{W}_h \eta(\cdot, a)\|_q^q \leq \left\{ \int_0^\infty \frac{da'}{a'} H(a/a') \right\}^{q-1} \int_0^\infty \frac{da'}{a'} \; H(a/a') \; \|\mathcal{W}_g \eta(\cdot, a')\|_q^q \; .$$

By the high localization of H, the first integral is a finite constant and

$$\begin{aligned}
\Gamma_h(\epsilon, q) &\leq O(1) \int_\epsilon^\infty \frac{da}{a} \int_0^\infty \frac{da'}{a'} \; H(a/a') \; \|\mathcal{W}_g \eta(\cdot, a')\|_q^q \\
&= O(1) \int_0^\infty \frac{da'}{a'} \; H(1/a') \int_\epsilon^\infty \frac{da}{a} \; \|\mathcal{W}_g \eta(\cdot, aa')\|_q^q \\
&= O(1) \int_0^\infty \frac{da'}{a'} \; H(1/a') \int_{\epsilon a'}^\infty \frac{da}{a} \; \|\mathcal{W}_g \eta(\cdot, a)\|_q^q \, .
\end{aligned}$$

We have

$$\Gamma_h(\epsilon, q) \leq O(1) \int_0^\infty \frac{da}{a} H(\epsilon/a) \, \Gamma_g(a, q).$$

The same type of relation holds when g and h are exchanged. The theorem will be an immediate result of the next lemma that deals with the mean values.

Lemma 3.3. *Let $s(t)$, $t > 0$, be a non-negative, monotone function of at most polynomial growth near 0,*

$$s(t) = O(t^{-m}), \quad (t \to 0), \quad \textit{for some } m > 0.$$

Near ∞, we assume that it is bounded

$$s(t) = O(1), \quad t \geq 1.$$

Let further $H(t)$ be a non-negative function that is arbitrarily well polynomially localized, namely

$$H(t) = O((t + 1/t)^{-n}), \quad \textit{for all } n > 0.$$

Then for the mean values

$$r(t) = \int_0^\infty \frac{da}{a} \, H(t/a) \, s(a),$$

we have the following

$$\begin{aligned}
\liminf_{t \to 0} \frac{\log r(t)}{\log t} &= \liminf_{t \to 0} \frac{\log s(t)}{\log t} \\
\limsup_{t \to 0} \frac{\log r(t)}{\log t} &= \limsup_{t \to 0} \frac{\log s(t)}{\log t} .
\end{aligned}$$

Proof: Suppose $s(t) \leq O(t^\gamma)$ as $t \to 0$. Then

$$r(t) \leq O(t^\gamma) \int_0^\infty \frac{da}{a} H(1/a)\, a^\gamma.$$

The integral on the right converges due to the high localization of H and thus $r(t) = O(t^\gamma)$ too as was to be shown for the first estimation.

We next come to the second estimation. Suppose that the right hand side of the second inequality is ∞. In this case clearly there is nothing to prove. We may suppose now that there is a $\lambda > 0$ and a constant $c > 0$ such that

$$s(t) \geq c\, t^\lambda, \quad (0 < t < 1). \tag{3.4}$$

We choose an ϵ, $0 < \epsilon < 1$, and keep it fixed. For $0 < t < 1$ we split the integral defining $r(t)$ into three parts.

$$r(t) = \left\{ \int_0^{t^{1+\epsilon}} + \int_{t^{1+\epsilon}}^{t^{1-\epsilon}} + \int_{t^{1-\epsilon}}^\infty \right\} \frac{da}{a} H(t/a)\, s(a) = X_1 + X_2 + X_3.$$

In the last term we may estimate $s(t) = O(1)$ and we have

$$X_3 \leq O(1) \int_{1/t^\epsilon}^\infty \frac{da}{a} H(1/a).$$

Since $H(t)$ is arbitrarily well polynomially localized, it follows that $X_3 = O(t^n)$ for all $n > 0$. In X_1 we may estimate $s(t) \leq t^{-m}$ and we have

$$X_1 \leq O(1)\, t^{-m} \int_0^{t^\epsilon} \frac{da}{a} H(1/a)\, a^{-m}.$$

Since H is arbitrary well polynomially localized, the integral is rapidly decaying and, hence, $X_1 = O(t^n)$ for all $n > 0$. What remains is the main contribution which is the middle term X_2. Since $s(t)$ is monotonically increasing as $t \to 0$, we may estimate $s(a) \leq (\geq) s(t^{1+\epsilon})$ or $s(a) \leq (\geq) s(t^{1-\epsilon})$ for $t^{1+\epsilon} \leq a \leq 1$, depending on whether s is non-increasing or non-decreasing. We obtain

$$X_2 \leq (\geq) c'\, s(t^{1\pm\epsilon}) \int_0^\infty \frac{da}{a} H(1/a), \quad \text{respectively.}$$

The last integral is again convergent and, hence, $X_2 \leq (\geq) c'' s(t^{1\pm\epsilon})$. Because of Equation (3.4) it follows that this estimate also holds for the sum of the three contributions, and therefore, for all $\epsilon > 0$, we have

$$r(t) \leq (\geq) c'''\, s(t^{1\pm\epsilon}), \quad \text{respectively.}$$

Since ϵ was arbitrary the lemma is proved. ■

By exchanging the roles of g and h, the theorem is proved. The proof remains the same if $\lim_{a\to 0}\Gamma_g(a,q)<\infty$. ■

Remark. The proof requires the high localization of the kernel $\Pi_{g,h}$, which, in turn, reflects that the wavelets must be very regular with all moments vanishing. For a fixed q and a given $\epsilon>0$, however, only some regularity is needed and also some moments have to vanish in order to compute the dimensions $\kappa^-(q)$ without error and the dimensions $\kappa^+(q)$ with error at most ϵ.

§4 Time evolution and the dimension $\kappa(2)$

In this section we want to establish a relation between the behavior to the Fourier transform of η and the fractal wavelet dimension $\kappa(2)$. Some evidence for such a relation has been given in [15]. This question is of relevance whenever one considers the long time behavior $T\to\infty$ of e^{-iTA} where A is a self adjoint operator. Indeed, in this case, we have

$$\langle \psi \mid e^{-iTA}\phi\rangle = \widehat{\mu}_{\psi,\phi}(T)$$

where $\mu_{\psi,\phi}$ is the spectral measure associated to ψ and ϕ. We are mainly interested in the time averages of the form

$$\int_0^T d\omega\, \left|\langle \psi \mid e^{-i\omega A}\phi\rangle\right|^2 = \int_0^T d\omega\, |\widehat{\mu}_{\psi,\phi}(\omega)|^2 .$$

The next theorem shows how this time average is related to the dimension $\kappa^{\pm}(2)$. We will prove the following theorem not only for measures but also for essentially any tempered distribution.

Theorem 4.1. *Let $\eta\in S'(\mathbb{R})$, $\eta\neq 0$, satsify at $s*\eta\in S(\mathbb{R})$ for all $s\in S(\mathbb{R})$. Suppose that $\eta\notin L^2(\mathbb{R})$. Then $\kappa^-(2)\le 0$, and it follows that*

$$-\kappa^+(2) = \liminf_{T\to\infty}\frac{\log\int_0^T d\omega\,|\widehat{\eta}(\omega)|^2}{\log T} \le \limsup_{T\to\infty}\frac{\log\int_0^T d\omega\,|\widehat{\eta}(\omega)|^2}{\log T} = -\kappa^-(2). \tag{4.1}$$

If $\eta\in L^2(\mathbb{R})$, then $\kappa^-(2)\ge 0$. The long-time behavior is trivial in the sense that Equation (4.1) goes to a finite constant. The speed of convergence is given by

$$-\kappa^+(2) = \liminf_{T\to\infty}\frac{\log\int_T^\infty d\omega\,|\widehat{\eta}(\omega)|^2}{\log T} \le \limsup_{T\to\infty}\frac{\log\int_T^\infty d\omega\,|\widehat{\eta}(\omega)|^2}{\log T} = -\kappa^-(2).$$

This shows that if the the spectral measure of a bounded self-adjoint operator A has a certain 2-wavelet-dimension, then the long time behavior of

e^{-iTA} is governed by this number. In particular, for all $\gamma > -\kappa^+(2)$, there is a constant $c > 0$ such that for T large enough,

$$\int_0^T d\omega \, |\langle \psi \mid e^{-i\omega A}\phi\rangle|^2 \geq cT^\gamma.$$

This shows that the decay in Equation (1.2) may be very slow if the spectral measure has an upper 2-wavelet-dimension that is close to -1, which would be the 2-wavelet-dimension of the pure point spectrum. (For a similar relation see [18] and [6]. For a different approach see Equation [2].)

Proof: Let $g \in S_+(\mathbb{R})$ and consider $\mathcal{W}_g\eta$. As we may, we suppose that $\widehat{g}$ is compactly supported. Again we may suppose in addition that $\widehat{\eta}(\omega) = O(\omega^m)$ for all m and that the wavelet transform decays fast at large scale. Suppose $\eta \notin L^2(\mathbb{R})$. This implies (see Equation (3.1)) that

$$\int_0^\infty \frac{da}{a} \int_{-\infty}^{+\infty} db \, |\mathcal{W}_g\eta(b,a)|^2 = \infty$$

and hence,

$$\Gamma_g(a,2) = \int_a^\infty \frac{d\alpha}{\alpha} \int_{-\infty}^{+\infty} db \, |\mathcal{W}_g\eta(b,\alpha)|^2 .$$

We now consider $\kappa^-(2) \leq 0$ and thus $\int_\alpha^1 \frac{da}{a} \int_{-\infty}^{+\infty} db \, |\mathcal{W}_g\eta(b,a)|^2 \to \infty$, as $\alpha \to 0$. A direct application of Parsevals equation shows that we have

$$\int_{-\infty}^{+\infty} db \; |\mathcal{W}_g\eta(b,a)|^2 = \int_0^\infty d\omega \; |\widehat{g}(a\omega)|^2 \; |\widehat{\eta}(\omega)|^2 .$$

We then let

$$H(\omega) = \int_\omega^\infty \frac{da}{a} \, |\widehat{g}(a)|^2 = \int_1^\infty \frac{da}{a} \; |\widehat{g}(a\omega)|^2 .$$

By a simple exchange of integration, we have

$$\begin{aligned}\int_{1/T}^\infty \frac{da}{a} \int_{-\infty}^{+\infty} db \; |\mathcal{W}_g\eta(b,a)|^2 &= \int_{1/T}^\infty \frac{da}{a} \int_0^\infty d\omega \; |\widehat{g}(a\omega)|^2 \; |\widehat{\eta}(\omega)|^2 \\ &= \int_0^\infty d\omega \; H(\omega/T) \; |\widehat{\eta}(\omega)|^2 .\end{aligned}$$

And finally, we have

$$\int_0^\infty d\omega \; H(\omega/T) \; |\widehat{\eta}(\omega)|^2 = \int_{1/T}^\infty \int_{-\infty}^{+\infty} db \; |\mathcal{W}_g\eta(b,a)|^2 = \Gamma_g(T^{-1},2).$$

Since H is non-negative and of compact support (since $\widehat{g}$ is), we can find numbers $\lambda > 0$ and $\Lambda > 0$ such that

$$\lambda\chi_{[0,\lambda]}(\omega) \leq H(\omega) \leq \Lambda_n\chi_{[0,\Lambda]}(\omega),$$

where χ_I is the characteristic function of I. We have

$$\lambda \int_0^{\lambda T} d\omega \ |\widehat{\eta}(\omega)|^2 \leq \Gamma_g(T^{-1}, 2) \leq \Lambda \int_0^{\lambda T} d\omega \ |\widehat{\eta}(\omega)|^2 ,$$

and the theorem follows. The proof for the case $\eta \in L^2(\mathbb{R})$ is similar. We only have to use

$$\Gamma_g(a, 2) = \int_0^a \frac{da}{a} \int_{-\infty}^{+\infty} db \, |\mathcal{W}_g \eta(b, a)|^2$$

and to adapt the limits of integration accordingly. ■

As an easy corollary of this theorem, we have the following quantitative version of the RAGE theorem. Let A be a bounded self-adjoint operator acting on some Hilbert space $\mathcal{H}$ and let dE_λ be its spectral family. Let further B be a Hilbert-Schmidt operator with singular decomposition

$$B : s \mapsto \sum_{n \in \mathbb{N}} \gamma_n \langle \phi_n \mid s \rangle \psi_n$$

with respect to the orthonormal sets $\{\phi_n\}$ and $\{\psi_n\}$. Since

$$\left\| B e^{-i\omega A} \varphi \right\|^2 = \sum_{n \in \mathbb{N}} |\gamma_n|^2 \left| \langle \phi_n \mid e^{-i\omega A} \varphi \rangle \right|^2 ,$$

we have

$$|\gamma_n|^2 \left| \langle \phi_n \mid e^{-i\omega A} \varphi \rangle \right|^2 \leq \left\| B e^{-i\omega A} \varphi \right\|^2 .$$

One can use Theorem 4.1 to obtain quantitative estimates for the speed of decay of

$$\frac{1}{T} \int_0^T d\omega \left\| B e^{-i\omega A} \varphi \right\|^2$$

in terms of the 2-wavelet-dimensions of the spectral measure $d\mu_{\phi_n, \varphi}(\lambda) = \langle \phi_n \mid dE_\lambda \varphi \rangle$.

§5 Appendix: some estimates and explicit formulas

We note some easy estimates of the dimensions $\kappa^-(q)$. We start by considering the effect of deriving a distribution.

Proposition 5.1. *Let $\eta \in S'(\mathbb{R})$ be a distribution such that for all $s \in S(\mathbb{R})$ we have $s * \eta \in S(\mathbb{R})$. Then if we replace η by $\partial\eta$, we have to replace $\kappa^\pm(q)$ by $\kappa(q)^\pm - q$.*

Proof: This follows from the identity

$$\mathcal{W}_{\partial g} \eta(b, a) = -a \mathcal{W}_g \partial\eta(b, a). \tag{5.1}$$

■

If some global regularity of s is known we have a lower bound for the dimensions of s.

Proposition 5.2. *If s is a function of the Hölder regularity with exponent α*

$$|s(t) - s(u)| = O(|t - u|),$$

then

$$\kappa^+(q) \geq \kappa^-(q) \geq q\alpha.$$

Proof: This follows from Equation (2.7) and the rephrasing of the definition of the dimensions given in Lemma 3.2. ■

Another estimate of this type is well adapted to the situation where the support of the distribution does not contain any interval but rather has some Cantor-set-like structure. The fractality of the support together with the local regularity exponent may be used to obtain an estimate for $\kappa^-(q)$. As an example, let us look at δ and $\partial\delta$. Both are supported by the origin with the former being more regular. This is clearly mirrored in the dimensions where the former gives $\kappa^{\pm}(q) = -q$ whereas the latter is $\kappa^{\pm}(q) = -2q$. First we need an suitable definition of dimension for a set $\Lambda \subset \mathbb{R}$. For all $\epsilon > 0$ consider the set of points at distance from Λ smaller or equal to ϵ,

$$M(\Lambda, \epsilon) = \{t \in \mathbb{R} : |t - u| \leq \epsilon \text{ for some } u \in \Lambda\}.$$

Suppose that Λ is compact. Then M is of finite Lebesque measures, denoted by $|M|$. The cover dimension $D(\Lambda)$ is now defined through the small scale behavior of $M(\Lambda, \epsilon)$.

$$D(\Lambda) = \liminf_{\epsilon \to 0} \frac{\log |M(\Lambda, \epsilon)|}{\log \epsilon},$$

Proposition 5.3. *Let $\eta \subset S'(\mathbb{R})$ have compact support $\Lambda \subset \mathbb{R}$. Suppose in addition that uniformly in b, we have*

$$\mathcal{W}_g \eta(b, a) = O(a^\alpha), \qquad (a \to 0).$$

Further suppose that η is supported by some compact set Λ with lower cover dimension $D(\Lambda)$. Then we have

$$\kappa^+(q) \geq \kappa^-(q) \geq D(\Lambda) + \alpha q.$$

Proof: Following the remark after the proof of Theorem 3.1, for every q, we may compute the dimensions with a wavelet that is regular enough and has sufficient moments vanishing but that need not be in $S_+(\mathbb{R})$. In particular, we may suppose that the wavelet we use has compact support, that is, say, contained in $[-1/2, +1/2]$. For a given scale a, the support of $\mathcal{W}_g\eta(\cdot, a)$ is then contained in $M(\Lambda, a)$. For all $\epsilon > 0$ we, therefore, may estimate

$$\int_{-\infty}^{+\infty} db\ |\mathcal{W}_g\eta(b,a)|^q = O(1) \int_{M(\Lambda,a)} db\ a^{-\alpha q} = O(a^{D(\Lambda)-\alpha q-\epsilon}).$$

The Proposition is proved because of Lemma 3.2. ∎

There is at least one class of distributions with support on a Cantor set, where all the dimensions can be computed explicitely. They are the affine self-similar measures that we shall present now. Consider $N > 1$ finite, closed, pairwise disjoint intervals I_n. The disjointness ensures that every interval is separated by a gap from its neighbors. Denote by I the smallest interval that contains all the I_n. Let $T_n : \mathbb{R} \to \mathbb{R}$ be affine maps whose restriction to I is onto I_n. If we write $T_n : t \to \lambda_n t + \beta_n$, we assume $\lambda_n > 0$. Now let N positive numbers p_n with $\sum p_n = 1$ be given. They are usually called the *a priory* probabilities since the distribution we shall construct is a probability measure. It can be shown that there is exactly one probability measure that satisfies the following self-similarity equation for any Borel set K

$$\mu(K) = \sum_{n=1}^{N} p_n \mu(T_n^{-1}K).$$

It is called a affine self-similar measure with length-scales λ_n and probabilities p_n.

Proposition 5.4. *The q-wavelet dimensions of an affine self-similar measure with length-scales λ_n and probabilities p_n are solutions of the following implicit equation $\kappa(q) = \kappa^{\pm}(q)$*

$$\sum_{n=1}^{N} p_n^q\, \lambda_n^{1-q-\kappa(q)} \equiv 1.$$

Proof: Let $T_n : t \to \lambda_n t + \beta_n$. We then have, by direct computation,

$$\mathcal{W}_g\mu(b,a) = \sum_{n=1}^{N} p_n \mathcal{W}_g\mu(T_n^{-1}b, a/\lambda_n)/\lambda_n.$$

From the remark after the proof of Theorem 3.1, for every q we may assume that the wavelet we use for the computation of the dimensions is compactly supported. Since the support of μ is contained in the union of the non-intersecting I_n, for small enough a, the support of the wavelet will, with at most one of the

I_n, have a non-empty intersection. Therefore, given a small enough, for each b, the sum consists of at most one term and we may write

$$\begin{aligned}G(a,q) &= \int_{-\infty}^{+\infty} db\ |\mathcal{W}_g\mu(b,a)|^q \\ &= \sum_{n=1}^{N} \int_{-\infty}^{+\infty} db\ |p_n \mathcal{W}_g\mu(T_n^{-1}b, a/\lambda_n)/\lambda_n|^q \\ &= \sum_{n=1}^{N} p_n\, \lambda_n^{1-q} G(a/\lambda_n, q).\end{aligned}$$

Instead of considering μ, we shall look at some mth primitive of μ and use Proposition 5.1 afterwards. Because of Equation (5.1), we obtain an equation of the same type for the new q-energy that we shall again denote by $G(a,q)$,

$$G(a,q) = \sum_{n=1}^{N} p_n^q\, \lambda_n^{mq+1-q}\, G(a/\lambda_n, q).$$

For m sufficiently large, we have

$$\Gamma(a,q) = \int_0^a \frac{d\alpha}{\alpha}\, G(\alpha, q),$$

and obtain a scaling equation for Γ ,

$$\Gamma(a,q) = \sum_{n=1}^{N} p_n^q\, \lambda_n^{mq+1-q}\, \Gamma(a/\lambda_n, q).$$

The proposition follows now from the next lemma.

Lemma 5.5. *Let $s(t)$ be a positive function defined for $t \in (0,1]$ that is bounded away from 0 and infinity*

$$\forall\, t \in (0,1] \quad 0 < \inf_{u\in[t,1]} s(u) \le \sup_{u\in[t,1]} s(u) < \infty,$$

and that satisfies pointwise at

$$s(t) = \sum_{n=1}^{N} \alpha_n\ s(t/\beta_n) \quad t \in (0, \min\{\beta_n\}],$$

with some constants $\alpha_n > 0$, $\beta_n \in (0,1)$. Then

$$\limsup_{t\to 0} \frac{\log s(t)}{\log t} = \liminf_{t\to 0} \frac{\log s(t)}{\log t} = \delta$$

is the unique solution of

$$\sum_{n=1}^{N} \alpha_n \, \beta_n^{-\delta} = 1.$$

Proof: We first want to show that

$$\sum_{n=1}^{N} \alpha_n < 1 \Rightarrow \lim_{t \to 0} s(t) = 0,$$
$$\sum_{n=1}^{N} \alpha_n > 1 \Rightarrow \lim_{t \to 0} s(t) = \infty.$$

The lemma follows then by considering $t^{-\gamma} s(t)$ that satisfies the same type of equation as s but where α_n is replaced by $\sum_{n=1}^{N} \alpha_n \beta_n^{-\gamma}$. Let us consider

$$s^+(t) = \min\{ \sup_{u \in [t,1]} s(u), \sup_{u \in [0,t]} s(u)\},$$
$$s^-(t) = \max\{ \inf_{u \in [t,1]} s(u), \inf_{u \in [0,t]} s(u)\}.$$

They are the smallest (largest) monotone functions that majorize (minorize) s for $t \in (0, 1]$. For $t \in (0, 1]$, we have

$$0 < s^-(t) \le s^+(t) < \infty,$$

since suppose $s^+(t) = \infty$, say, then $\sup_{u \in [t,1]} s(u) = \infty$ which contradicts our hypothesis on s. We may estimate for $0 < t \le \min\{\beta_n\}$

$$s(t) = \sum_{n=1}^{N} \alpha_n \; s(t/\beta_n) \le \sum_{n=1}^{N} \alpha_n \; s^+(t/\beta_n),$$
$$s(t) = \sum_{n=1}^{N} \alpha_n \; s(t/\beta_n) \ge \sum_{n=1}^{N} \alpha_n \; s^-(t/\beta_n).$$

The righthand sides are again monotone functions that majorize (minorize) s and thus since $s^{\pm}$ is optimal, we have for $0 < t < \min\{\beta_n\}$

$$s^+(t) \le \sum_{n=1}^{N} \alpha_n \; s^+(t/\beta_n),$$
$$s^-(t) \ge \sum_{n=1}^{N} \alpha_n \; s^-(t/\beta_n).$$

By monotony we may write

$$s^+(t) \leq \left(\sum_{n=1}^{N} \alpha_n\right) \; s^+(t/\beta^+),$$
$$s^-(t) \geq \left(\sum_{n=1}^{N} \alpha_n\right) \; s^+(t/\beta^-)$$

where $\beta^\pm$ are given by either $\min\{\beta_n\}$ or $\max\{\beta_n\}$ respectively depending on wether $s^\pm$ is non-decreasing or non-growing. Since $\beta^\pm \in (0,1)$ we are done. ■

This shows the proposition. ■

Acknowledgments. Special thanks to A. Grossmann, R. Seiler, J. M. Combes, and M. Demuth for some stimulating discussions. Many thanks also to the Max-Planck Arbeitsgruppe "partielle Differentialgleichungen" at the University of Potsdam where part of the work has been carried out.

References

1. Arnéodo A., G. Grasseau, M. Holschneider, On the wavelet transform of multifractals, *Phys. Rev. Letters* **61** (1988), 2281-2284.
2. Avron J. & B. Simon (1981), Transient and recurrent spectrum, *J. Funct. Anal.* **43** (1981), 1.
3. Combes, J. M., A. Grossmann, and Ph. Tchamitchian, (eds.), *Wavelets*, Springer-Verlag Berlin, 1989.
4. Cycon H. L., R. G. Froese, W. Kirsch, and B. Simon, *Schrödinger Operators*, Springer Verlag, 1987.
5. Grossmann A., J. Morlet, and T. Paul, Transforms associated to square integrable group representations II: examples, *Ann. Inst. Henri Poincar1827, Physique thèorique* **45** (1986), 293–309.
6. Guarneri I., On an estimate concerning quantum diffusion in the presence of a fractal spectrum, *Europhys. Lett.*, to appear.
7. Hentschel H. G. E. and I. Procaccia, The infinite number of generalized dimensions of fractals and strange attractors, *Physica* **8D** (1983), 435–444.
8. Holschneider M., On the wavelet transformation of fractal objects, *J. Stat. Phys.* **50** (1988), 953–993.
9. Holschneider M., L'analyse d'objets fractals et leur transforme en ondelettes, Thèse de doctorat, Université de Provence (Aix-Marseille I), 1988.
10. Holschneider M., Localization properitès of wavelet transforms, *J. Math. Phys.* **34** (7) (1993), 3227–3244.
11. Holschneider M., More on the analysis of local regularity through wavelet transforms, *J. Stat. Phys.*, submitted.

12. Holschneider M., Functional calculus using wavelet transforms, *J. Math. Phys.*, 1994, to appear.
13. Holschneider M. and Ph. Tchamitchian, Pointwise regularity of Riemanns "nowhere differentiable" function, *Inventiones Mathematicae* **105** (1991), 157–175.
14. Jaffard S., Estimations Hölderiennes ponctuelles des fonctions au moyen de leurs coefficients d'ondelettes, *CRAS Sèr. I* **308** (1989), 7.
15. Ketzmerick R., G. Petschel, and T. Geisel, Slow decay for temporal correlations in quantum systems with cantor spectra, Univ. Frankfurt, preprint.
16. Meyer Y., *Ondelettes et Opèrateurs*, 3 volumes, Hermann, Paris, 1990.
17. Muzy J. F., E. Bacry E., and A. Arnéodo, Wavelets and multifractal formalism for singular signals, *Phys. Rev. Lett* **67** (1992), 3515–3518.
18. Strichartz R., Fourier asymptotics of fractal measures, *J. Funct. Anal.* **89** (1990), 154.

Matthias Holschneider
CPT CNRS Luminy
Case 907
F-13288 Marseille
France
hols@marcptnx1.univ-mrs.fr

Part VI

Numerical Methods and Algorithms

Multiscale Methods for Pseudo-Differential Equations on Smooth Closed Manifolds

Wolfgang Dahmen, Siegfried Prössdorf, and Reinhold Schneider

Abstract. This paper is concerned with the numerical solution of pseudo-differential equations on smooth manifolds. Some relevant concepts and basic properties of a suitable multiscale framework are presented which facilitate an efficient and stable matrix compression for the fast solution of linear systems arising from Galerkin discretizations.

§1 Introduction

This work should be viewed as a continuation of previous investigations of multiscale and wavelet methods for the numerical solution of integral- and pseudo-differential equations documented in [20,21,22,23]. Motivated by the startling paper [3] the objective in [20,21] was to provide a common analysis for generalized Petrov-Galerkin schemes in connection with a large class of pseudo-differential operators. This covers collocation as well as classical Galerkin schemes. The class of operators contains essentially all classical pseudo-differential operators, in particular, also those arising in connection with boundary integral methods. The ability of treating such a general framework, however, relies to some extent on confining the discussion to periodic boundary conditions so that Fourier transform techniques can be fully exploited. In particular, the stability of the schemes can then be characterized in a very convenient way which is, of course, important for analyzing the wavelet compression techniques for the fast solution of the resulting systems of equations. Nevertheless, since most of the arguments are actually local in nature the question arises, whether the results remain valid in more realistic situations. This motivated implementing a multiscale method for the double layer potential equation on closed polyhedral domains in $\mathbb{R}^3$ [23]. In spite of the technical complications due to the lack of periodicity and although the domain is in this case no longer smooth the numerical experiments confirmed the theoretical predictions from the model problem case.

Wavelets: Theory, Algorithms, and Applications
Charles K. Chui, Laura Montefusco, and Luigia Puccio (eds.), pp. 385–424.

ISBN 0-12-174575-9

Since a significant part of the constructions in [23] is yet to be justified by rigorous arguments we feel that a reasonable step in this direction is to consider next an intermediate situation. While we give up on periodic boundary conditions we will still assume that the underlying domains for the operator equations are compact connected C^∞-Riemannian manifolds without boundary so that the concept of pseudo-differential operators is still an appropriate framework to deal with. Of course, this covers the important case when the domain is a smooth boundary of some bounded domain in $\mathbb{R}^{n+1}$. In many applications it would then actually suffice to assume that the boundary is C^k for some $k > 1$. We will also confine our attention to Galerkin methods, *i.e.*, test and trial functions will be regular L_2 functions. We will take particular advantage of the concept of biorthogonality which actually offers more flexibility in adapting compression and convergence rates than orthogonal multiscale bases. We will employ a type of estimate proposed in [37] which in this case leads to an improvement over the general results obtained in [21]. Nevertheless, it is by no means clear how to actually construct practicable biorthogonal multiscale bases on general smooth manifolds. This task is expected to depend on each application at hand and may actually turn out to be the main obstacle. All we can do so far is to propose a general concept of multiresolution decomposition of spaces based on classical tools in differential geometry like charts and atlas in order to be then able to bring out in a more concrete way what the essential ingredients and properties of such constructions should be in the present context. In doing so we also take the opportunity to review some principal ideas and concepts on which the application of multiscale techniques to integral- and pseudo-differential equations are presently based upon. The paper is organized as follows.

In Section 2 we describe the class of problems to be treated and list those properties of the operators which are relevant for the subsequent analysis. Section 3 is devoted to working out and discussing the essential features of multiscale decompositions on smooth manifolds which are desirable in the present context. At the present stage much of this has still the character of a wish list. As mentioned above, the feasability of those requirements can probably not be established in the present generality but will depend highly on the concrete case at hand. Therefore the present discussion should mainly help to bring the essential properties into focus which one would have to care about in any application of the type considered here. Specifically, it will be seen that for our present purposes it is important to ensure that the multiscale basis functions realize moment conditions of sufficiently high order. Therefore we discuss, in particular, principal strategies for raising the order of moment conditions once some multiscale decomposition is given. In Section 4 we propose a Galerkin scheme based on these multiscale decompositions. Finally, in Sections 5 and 6 we derive the estimates our compression techniques are based on and discuss the corresponding overall efficiency of the scheme.

§2 Pseudo-differential equations on smooth manifolds

It is well-known that exterior boundary value problems for elliptic partial differential equations can be reduced to an integral equation over the boundary of the domain. Moreover, when the boundary is smooth the corresponding integral operators turn out to be certain classical pseudo-differential operators.

This motivates the following general setting to be considered throughout the remainder of this paper. Let Γ denote an orientable connected compact Riemannian C^∞-manifold without boundary endowed with a smooth Riemannian metric g. The volume element is given in local coordinates by $ds_x = \sqrt{g(x,x)}dx_1 \wedge \cdots \wedge dx_n$ so that one can define on Γ the inner product

$$\langle u, v\rangle = \int_\Gamma u(x)\overline{v(x)}ds_x \ . \tag{2.1}$$

Since Γ is a compact manifold it admits a partition

$$\Gamma = \bigcup_{m=0}^{m_\Gamma} \overline{\Gamma_m}$$

into disjoint open sets Γ_m and there exists a finite covering $\Gamma = \cup_{m=0}^{m_\Gamma}\tilde{\Gamma}_m$, $\Gamma_m \subset\subset \tilde{\Gamma}_m$, along with local charts $\kappa_m \in C_0^\infty(\tilde{\Gamma}_m, \mathbb{R}^n)$ which map Γ_m onto some *reference domain* $\Sigma \subset RR^n$. Here Σ could be the standard n-simplex or the $n-$cube $[0,1]^n$. We call the above partition a *triangulation* of Γ. Obviously the collection $\{(\tilde{\Gamma}_m; \kappa_m) : m = 0, \ldots, m_\Gamma\}$ forms an atlas of Γ. Furthermore, one can construct a partition of unity on Γ consisting of functions $\chi_m \in C_0^\infty(\tilde{\Gamma}_m, \mathbb{R})$ where

$$\chi_m = 1 \ \text{ on } \ \Gamma_m, \quad \sum_{m=0}^{m_\Gamma} \chi_m = 1.$$

The present paper is concerned with discussing general concepts for the numerical solution of pseudo-differential equations

$$Au = f \ , \tag{2.2}$$

where A belongs to a class of opeartors to be defined next. Locally, pseudo-differential operators are described in terms of pseudo-differential operators on $\mathbb{R}^n$. We recall the following definition from [31,32]. A pseudo-differential operator belongs to the class $\Psi^r(\mathbb{R}^n) = \Psi^r_{1,0}(\mathbb{R}^n)$ if it is a linear operator of the form

$$Au(x) = \int_{\mathbb{R}^n} \Big(\int_{\mathbb{R}^n} e^{2\pi i\langle\xi, x-y\rangle}\sigma(x,\xi)u(y)dy\Big)d\xi \ , \quad u \in C_0^\infty(\mathbb{R}^n) \ , \tag{2.3}$$

where the symbol $\sigma(x,\xi)$ belongs to the symbol class $S^r_{1,0}(\mathbb{R}^n\times\mathbb{R}^n)$ containing all $\sigma(x,\xi)\in C^\infty(\mathbb{R}^n\times\mathbb{R}^n)$ such that for each multi-indices $\alpha,\beta\in\mathbb{N}_0^n$ there exists some constant $c_{\alpha,\beta}$ such that

$$|D_x^\beta D_\xi^\alpha\sigma(x,\xi)|\le c_{\alpha,\beta}(1+|\xi|)^{r-|\alpha|}\ ,\ x,\xi\in\mathbb{R}^n. \tag{2.4}$$

An operator A is a pseudo-differential operator in $\Psi^r_{1,0}(\Gamma)$ if for any partition of unity $\{\chi_m\in C_0^\infty(\tilde\Gamma_m): m=0,\dots,m_\Gamma\}$, relative to an atlas $\{\tilde\Gamma_m,\kappa_m\}$, all transported operators

$$A_{m,m'}:=\kappa_m^*\circ\chi_m A\chi_{m'}\circ(\kappa_{m'}^*)^{-1}\ , \tag{2.5}$$

where $\kappa_m^*u(x):=u(\kappa_m^{-1}x)$, are pseudo-differential operators in $\Psi^r_{1,0}(\mathbb{R}^n)$.

The operator A admits a decomposition

$$A=A_\infty+\sum_{(m,m')\in\Xi}\chi_m A\chi_{m'}, \tag{2.6}$$

where $\Xi:=\{(m,m'):\operatorname{supp}\chi_m\cap\operatorname{supp}\chi_{m'}\neq\emptyset\}$. Furthermore, A_∞ has a $C^\infty-$kernel $K_{A_\infty}\in C^\infty(\Gamma\times\Gamma)$.

We will make crucial use of the following well-known estimates for the Schwartz kernel of a pseudo-differential operator in $\Psi^r_{1,0}(\mathbb{R}^n)$.

Lemma 2.1. *The Schwartz kernel K_A of $A\in\Psi^r_{1,0}(\mathbb{R}^n)$ satisfies for $x\neq y$, $x,y\in\mathbb{R}^n$, the estimate*

$$|\partial_x^\alpha\partial_y^\beta K_A(x,y)|\le c_{\alpha,\beta}\operatorname{dist}(x,y)^{-(n+r+|\alpha|+|\beta|)}\ ,\ n+r+|\alpha|+|\beta|>0\ . \tag{2.7}$$

Defining as usual by

$$\hat v(y):=\int_{\mathbb{R}^n} v(x)e^{-i\langle x,y\rangle}\,dx$$

the Fourier transform of v, $H^s(\mathbb{R}^n)$ denotes the ordinary Sobolev space of functions $v\in L_2(\mathbb{R}^n)$ for which

$$\|v\|_{H^s}:=\|(1-|\cdot|^2)^{s/2}\hat v\|_{L_2(\mathbb{R}^n)}$$

is finite. We may now introduce the Sobolev space $H^s(\Gamma)$ as the completion of $C^\infty(\Gamma)$ with respect to the norm

$$\|u\|_s^2:=\sum_{m=0}^{m_\Gamma}\|\kappa_m^*(\chi_m u)\|_{H^s}^2\ , \tag{2.8}$$

where, of course, $\kappa_m^*(\chi_m u):=\kappa_m^*\chi_m\cdot\kappa_m^* u$. It is well known that $H^s(\Gamma)$ does not depend on the particular atlas and that through Equation (2.8) any other atlas gives rise to an equivalent norm.

It is important to know that $A \in \Psi^r_{1,0}(\Gamma)$ establishes a bounded linear map

$$A : H^t(\Gamma) \to H^{t-r}(\Gamma). \tag{2.9}$$

To formulate our final requirements on the operator A we exploit the fact that the symbols in $S^r_{1,0}$ can be split as a sum of the principal part σ_0 representing an operator of order r and a second part σ_1 which represents an operator of order $r' < r$. We will always assume in the following that

- A is *strongly elliptic*, *i.e.*, the Garding inequality

$$\mathrm{Re}\sigma_0(x,\xi) \geq c|\xi|^r, \quad \xi \in \mathbb{R}^n, \tag{2.10}$$

holds uniformly for all the transported operators $A_{m,m'}$, $(m,m') \in \Xi$.

- A is injective

$$\mathrm{Ker}A = \{0\}\ . \tag{2.11}$$

We will use the well-known fact that the conditions in Relations (2.10) and (2.11) imply the bounded invertibility of the operators $A, A^* : H^s(\Gamma) \to H^{s-r}(\Gamma)$ for $s \in \mathbb{R}$.

§3 Multiscale decompositions

The objective of this section is to develop an appropriate multilevel framework for discretizing operator equations of the type described in the previous section.

3.1 Outline and goal

Galerkin schemes for the solution of an equation

$$Au = f$$

of the form described in the previous section can be conveniently formulated as *projection methods*. More precisely, given a nested sequence $\mathcal{S}$ of sufficiently regular trial spaces S_j in $L_2(\Gamma)$ spanned by *single scale bases* $\Phi_j = \{\varphi^j_k : k \in \Delta_j\}$, we assume that

$$Q_j v := \sum_{k\in\Delta_j} \langle v, \tilde{\varphi}^j_k\rangle \varphi^j_k \tag{3.1}$$

are uniformly bounded projectors from $L_2(\Gamma)$ onto S_j, *i.e.*, $\tilde{\Phi}_j$ is a *dual* collection

$$\langle \varphi^j_k, \tilde{\varphi}^j_{k'}\rangle = \delta_{k,k'}, \quad k, k' \in \Delta_j. \tag{3.2}$$

Clearly,

$$Q^*_j v := \sum_{k\in\Delta_j} \langle v, \varphi^j_k\rangle \tilde{\varphi}^j_k$$

is the adjoint of Q_j and the Galerkin method relative to S_j consists in finding $u^j \in S_j$ such that

$$Q^*_j A u_j = Q^*_j f. \tag{3.3}$$

Although the basis functions in Φ_j may have small support the stiffness matrices relative to Φ_j will have no zero entries which excludes the employment of typical efficient sparse linear solvers. The objective is therefore to look for other bases of S_j such that the corresponding stiffness matrices are possibly close to sparse matrices. A hint how to construct such bases comes from the telescoping expansion

$$Q_j^* A Q_j = \sum_{l,l'=0}^{j} (Q_l^* - Q_{l-1}^*) A (Q_{l'} - Q_{l'-1}), \quad (Q_{-1} = 0) \tag{3.4}$$

since the components on the right hand side of Equation (3.4) involve *approximation errors* for the operators Q_j and Q_j^*. The corresponding matrix representations become available once one knows the representation

$$(Q_{j+1} - Q_j)v = \sum_{k \in \nabla_j} \langle v, \tilde{\psi}_k^j \rangle \psi_k^j, \tag{3.5}$$

which incidentally requires the operators $Q_{j+1} - Q_j$ to be also projectors. Thus one seeks for a basis Ψ^j of the complement $W_j := (Q_{j+1} - Q_j)S_{j+1}$ of S_j in S_{j+1}. Making any practical use of such ideas will hinge on the transformation $\mathbf{T}_j$ taking the coefficients of any $v_j \in S_j$ relative to the *multiscale basis*

$$\Psi_j := \bigcup_{l=-1}^{j-1} \Psi^l, \tag{3.6}$$

where $\Psi^{-1} := \Phi_0$, into those relative to the single scale basis Φ_j. Therefore there are two essential requirements one has to deal with:

- **Efficiency:** Carrying out $\mathbf{T}_j$ should require only $\mathcal{O}(\dim S_j)$ operations.
- **Stability:** The transformations $\mathbf{T}_j$ should be *uniformly stable, i.e.*, they should satisfy

$$\|\mathbf{T}_j\| \|\mathbf{T}_j^{-1}\| = \mathcal{O}(1), \quad j \to \infty. \tag{3.7}$$

These issues as well as aspects concerning the identification of representations of Equation (3.5) will be discussed next in some generality to isolate the essential features which matter in any concrete realization.

3.2 Some basic concepts

We begin by recalling a few relevant general concepts discussed in [15].

Data fitting, data compression or the numerical solution of an operator equation are problems which can ultimately be cast into the task of approximating some element f in a function space $\mathcal{F}$ from some (at least locally) finite dimensional subspace $S_j \subset \mathcal{F}$. To be a little more specific, we will always assume here that $\mathcal{F}$ is a reflexive Banach space such that L_p-spaces for $1 < p < \infty$ and, in particular L_2-spaces are typical candidates. However, no matter how large the dimension of S_j might be, *i.e.*, how fine the level of discretization might be, it would be difficult to recapture the intrinsic features of the infinite dimensional function space $\mathcal{F}$ hosting the solution f. A long experience in approximation theory and numerical analysis indicates that this can be accomplished with much more success if one does not only work with a single scale of resolution represented by S_j but exploits information gained from the interaction between all preceding scales. The asymptotics of such interactions often turn out to express the essential features of $\mathcal{F}$ adequately. Typical examples are multigrid methods and the concept of wavelets. Thus one is lead to consider a whole sequence $\mathcal{S}$ of nested (finite dimensional) subspaces

$$S_0 \subset S_1 \subset \ldots \subset S_j \subset \ldots \mathcal{F} \tag{3.8}$$

of $\mathcal{F}$ such that

$$\overline{\bigcup_{j \in \mathbb{N}_0} S_j} = \mathcal{F}. \tag{3.9}$$

Instead of looking only for *approximations* to f from each S_j the key is to gain such approximations from a *representation* of f making use of all of $\mathcal{S}$. As suggested by the initial comments a convenient way to get such representations is to consider a sequence $\mathcal{Q}$ of uniformly bounded linear projectors Q_j from $\mathcal{F}$ onto S_j. Since

$$\|Q_n f - f\|_{\mathcal{F}} \leq (1 + \|Q_n\|_{\mathcal{F}}) \inf_{f_n \in S_n} \|f - f_n\|_{\mathcal{F}}, \tag{3.10}$$

Equation (3.9) readily implies that the telescoping expansion

$$f = \sum_{j \in \mathbb{N}_0} (Q_j - Q_{j-1}) f \tag{3.11}$$

(with $Q_{-1} = 0$) converges strongly in $\mathcal{F}$. Clearly $(Q_j - Q_{j-1})f$ may be viewed as the detail added to a current approximation when progressing up the levels of discretization and one seeks to characterize f in terms of these details.

There are two issues that come to mind in this context. Firstly, as mentioned already above, one needs a handy representation of the detail $(Q_j - Q_{j-1})f$. Therefore

$$W_j := (Q_j - Q_{j-1})\mathcal{F} \tag{3.12}$$

should be a direct summand of S_j in S_{j+1} so that one can look for a basis of W_j to represent the detail. Thus $Q_j - Q_{j-1}$ should also be a linear projector which is easily seen to be equivalent to the condition

$$Q_l Q_j = Q_l \quad \text{for all} \quad l \leq j. \tag{3.13}$$

For instance orthogonal projectors or Lagrange interpolation projectors satisfy Equation (3.13).

Secondly, the successive details should be sufficiently independent to recapture and separate all essential information on each scale. This could be expressed by requiring that in addition to Equation (3.11),

$$\|f\|_{\mathcal{F}} \sim \left(\sum_{j \in \mathbb{N}_0} \|(Q_j - Q_{j-1})f\|_{\mathcal{F}}^q \right)^{1/q} \tag{3.14}$$

holds for some $1 \le q \le \infty$ (with the usual interpretation for $q = \infty$). It will also turn out that Equation (3.14) is intimately related to the above mentioned stability of the transformations $\mathbf{T}_j$. When $\mathcal{F}$ is a Hilbert space $q = 2$ would be a natural choice associated with orthogonal projectors Q_j. In this case, Equation (3.14) is obviously true. However, orthogonality between levels which is often hard to realize is not quite necessary. To describe some circumstances under which Equation (3.14) holds (actually for arbitrary q) let us assume that there exists some one parameter family of nonnegative subadditive functionals $\omega_{\mathcal{F}}(\cdot, t), t \ge 0$, with the following properties:

$$\omega_{\mathcal{F}}(f,t) \le c\, \|f\|_{\mathcal{F}}, \quad \text{for all} \quad f \in \mathcal{F}. \tag{3.15}$$

$$\lim_{t \to 0+} \omega_{\mathcal{F}}(f,t) = 0 \quad \text{for all} \quad f \in \mathcal{F}. \tag{3.16}$$

$$\omega_{\mathcal{F}}(f_1 + f_2, t) \le \omega_{\mathcal{F}}(f_1, t) + \omega_{\mathcal{F}}(f_2, t), \quad f_1, f_2 \in \mathcal{F}. \tag{3.17}$$

Moreover, for each $\lambda > 0$ there exists some $\lambda' > 0$ such that for all $t > 0$, $f \in \mathcal{F}$

$$\omega_{\mathcal{F}}(f, \lambda t) \le \lambda'\, \omega_{\mathcal{F}}(f,t). \tag{3.18}$$

Thus $\omega_{\mathcal{F}}$ has the properties of a *modulus of continuity.*

One can quantify now the approximation properties of a given dense sequence $\mathcal{S}$ of closed subspaces of $\mathcal{F}$ by requiring that a *Whitney-type estimate*

$$\inf_{f_j \in S_j} \|f - f_j\|_{\mathcal{F}} \le c\, \omega_{\mathcal{F}}(f, \rho^{-j}) \tag{3.19}$$

holds for some fixed $\rho > 1$. Here ρ is essentially the factor by which the meshsize decreases from one level to the next one.

In addition we will have to impose a mild regularity assumption on $\mathcal{S}$ in terms of the following *inverse* or *Bernstein estimate*

$$\omega_{\mathcal{F}}(f_j, t) \le c \left(\min\{1, t\rho^j\}\right)^{\gamma} \|f_j\|_{\mathcal{F}} \tag{3.20}$$

which is to hold for some $\gamma > 0$ and all $f_j \in S_j$, uniformly in $j \in \mathbb{N}_0$.

These notions are illustrated by the following example.

Example 3.1 Let D be some domain in $\mathbb{R}^n$ and

$$\mathcal{F} = L_p(D), \quad \mathcal{F}' = L_{p'}(D) \quad \text{for} \quad 1 < p < \infty, \qquad \frac{1}{p} + \frac{1}{p'} = 1. \tag{3.21}$$

Let

$$(\Delta_h^l f)(x) := \sum_{j=0}^{l} \binom{l}{j} (-1)^{l-j} f(x + jh)$$

be the lth order difference operator in the direction h. Setting

$$D_{h,l} := \{x \in D : x + jh \in D, \; j = 0, \ldots l\},$$

the lth order L_p-modulus of continuity of f is defined as

$$\omega_l(f, t, D)_p := \sup_{|h| \le t} \|\Delta_h^l f\|_{L_p(D_{h,l})},$$

where $|\cdot|$ is a norm on $\mathbb{R}^n$. In this case

$$\omega_{L_p(D)}(f, t) := \omega_l(f, t, D)_p \tag{3.22}$$

is known to satisfy Relations (3.15), (3.16), and (3.17) and hence is an admissible choice. For instance, Inequality (3.19) is valid for Equation (3.22) with $l \ge 1$ when S_j is spanned by compactly supported functions which sum to one on D.

Remark 3.1. In general, Inequality (3.19) is typically valid for some $l \in \mathbb{N}$ when S_j contains all polynomials of degree less than l and is spanned by compactly supported functions and the diameter of the supports is uniformly proportional to ρ^{-j} for some $\rho > 1$. For various versions of different moduli see, *e.g.*, [16]. Likewise such an inverse estimate for Equation (3.22) is known to hold for essentially all finite element spaces, spline spaces, or multiresolution settings of interest for some $l \in \mathbb{N}$ and some $\gamma > 0$. Usually it is established first for the basis functions and then extended to all elements in S_j by stability.

Now let $\mathcal{F}^*$ denote the dual of $\mathcal{F}$ and let Q_j^* be the adjoint of Q_j. Since the Q_j^* are also projectors their ranges $\tilde{S}_j$ are closed subspaces of $\mathcal{F}'$. It is easy to see that Equation (3.13) on $\mathcal{Q}$ is equivalent to the fact that the spaces $\tilde{S}_j$ are also nested. The following result was proved in [15].

Theorem 3.1. *For $\mathcal{F}$, S as above let the uniformly bounded sequence of linear projectors $\mathcal{Q}$ satisfy Equation* (3.13). *If both S and $\tilde{S}$ satisfy a direct and inverse estimate of Relations* (3.19) *and* (3.20) *for some $\gamma, \gamma^* > 0$, respectively, then the norm equivalence of Relation* (3.14) *holds for all $1 \le q \le \infty$.*

The results in [15] also indicate that the norm equivalence Relation (3.14) extends in a natural way to other spaces enclosing $\mathcal{F}$ by duality. To describe this we will fix in the following for simplicity $q=2$, $\rho=2$ and let $B^t(\mathcal{F})$ denote the space of all functions in $\mathcal{F}$ for which

$$\|v\|_{B^t(\mathcal{F})} := \|v\|_{\mathcal{F}} + \left(\sum_{j\in\mathbb{N}_0} 2^{jt2}\omega_{\mathcal{F}}(f,2^{-j})^2\right)^{1/2} \tag{3.23}$$

is finite. Thus $B^t(\mathcal{F})$ is essentially a Besov space. It is easy to show that under the assumptions of Theorem 3.1 one has for every $t < \min\{\gamma, l\}$

$$S_j \subset B^t(\mathcal{F}), \quad j \in \mathbb{N}_0. \tag{3.24}$$

Moreover, the norm equivalence

$$\|v\|_{B^t(\mathcal{F})} \sim \left(\sum_{j\in\mathbb{N}_0} 2^{jt2}\|(Q_j - Q_{j-1})f\|_{\mathcal{F}}^2\right)^{1/2} \tag{3.25}$$

holds for $0 < t < \gamma$ [15]. Defining now the *shift-operators* (see also [33])

$$\Upsilon^s u := \sum_{j\in\mathbb{N}_0} 2^{js}(Q_j - Q_{j-1})u, \tag{3.26}$$

and observing that by Equation (3.13),

$$\sum_{j\in\mathbb{N}_0} 2^{js}(Q_j - Q_{j-1})\Upsilon^t u = \sum_{j\in\mathbb{N}_0} 2^{j(s+t)}(Q_j - Q_{j-1})u,$$

one can show that

$$\|\Upsilon^s u\|_{B^t(\mathcal{F})} \sim \|u\|_{B^{t+s}(\mathcal{F})}, \quad -\gamma^* < t+s < \gamma, \tag{3.27}$$

provided that

$$\left(B^t(\mathcal{F}^*)\right)^* = B^{-t}(\mathcal{F}). \tag{3.28}$$

In fact, one first employs Relation (3.25) and Theorem 3.1 to confirm Relation (3.27) for $0 \le t+s < \gamma$. Then one uses Equation (3.28) in combination with a duality argument. One immediately infers then from Relation (3.27) the following fact.

Theorem 3.2. *Suppose that in addition to the hypotheses of Theorem 3.1 Equation (3.28) holds. Then*

$$\|u\|_{B^t(\mathcal{F})} \sim \left(\sum_{j\in\mathbb{N}_0} 2^{jt2}\|(Q_j - Q_{j-1})u\|_{\mathcal{F}}^2\right)^{1/2}, \quad -\gamma^* < t < \gamma. \tag{3.29}$$

Finally, we recall from Remarks 3.12 and 3.14 in [15] the following useful facts.

Theorem 3.3. *Suppose the assumptions of Theorem 3.2 hold.*
(i) *One has*

$$\|u_j\|_{B^t(\mathcal{F})} \leq c\, 2^{j(t-s)} \|u_j\|_{B^s(\mathcal{F})}, \quad u_j \in S_j,\ -\gamma^* < s \leq t < \gamma,$$
$$\|u_j\|_{B^t(\mathcal{F}^*)} \leq c\, 2^{j(t-s)} \|u_j\|_{B^s(\mathcal{F}^*)}, \quad u_j \in \tilde{S}_j,\ -\gamma < s \leq t < \gamma^*,$$

where the constant c is independent of j.

(ii)
$$\|Q_j\|_{B^t(\mathcal{F})} = \mathcal{O}(1), \quad j \to \infty, \quad \gamma^* < t < \gamma,$$
$$\|Q_j^*\|_{B^t(\mathcal{F}^*)} = \mathcal{O}(1), \quad j \to \infty, \quad \gamma < t < \gamma^*.$$

(iii) *One has*
$$\|Q_j u - u\|_{(B^t(\mathcal{F}^*))^*} \leq c\, 2^{-jt} \|u\|_{\mathcal{F}}, \quad u \in \mathcal{F},$$

if and only if

$$\|Q_j^* u - u\|_{\mathcal{F}^*} \leq c\, 2^{-jt} \|u\|_{B^t(\mathcal{F}^*)}, \quad u \in B^t(\mathcal{F}^*).$$

3.3 Multiscale transformations

In the present context one would have to realize such a multiscale concept for the case $\mathcal{F} = L_2(\Gamma)$ which we will now mainly focus on. Without loss of generality we will assume that $\rho = 2$, *i.e.*, the meshsize is halved when moving up one level.

One typically defines the spaces S_j as linear span of some *fine scale* basis $\Phi_j = \{\varphi_k^j : k \in \Delta_j\}$. A least requirement is that these bases are uniformly stable, *i.e.*,

$$\|\mathbf{c}\|_{l_2(\Delta_j)} \sim \| \sum_{k\in\Delta_j} c_k \varphi_k^j \|_{L_2(\Gamma)}, \tag{3.30}$$

holds uniformly in $j \in \mathbb{N}_0$. In particular, this implies the normalization

$$\|\varphi_k^j\|_{l_2(\Delta_j)} \sim 1. \tag{3.31}$$

Furthermore, according to Remark 3.1 the bases should satisfy

$$\operatorname{diam}\ \operatorname{supp} \varphi_k^j \leq c\, 2^{-j}, \quad j \in \mathbb{N}. \tag{3.32}$$

Nestedness and uniform stability imply that the φ_k^j satisfy a *refinement* relation

$$\varphi_k^j = \sum_{q\in\Delta_{j+1}} m_{q,k}^j \varphi_q^{j+1}, \quad k \in \Delta_j, \tag{3.33}$$

where the matrices

$$\mathbf{M}_{j,0} := (m^j_{q,k})_{q\in\Delta_{j+1},k\in\Delta_j}$$

define uniformly bounded mappings from $l_2(\Delta_j)$ into $l_2(\Delta_{j+1})$.

According to our initial remarks one should next look for appropriate projectors Q_j of the form

$$Q_j f = \sum_{k\in\Delta_j} \langle f, \tilde{\varphi}^j_k\rangle \varphi^j_k, \tag{3.34}$$

where the collection $\tilde{\Phi}_j$ is *biorthogonal* or *dual* to Φ_j, *i.e.*,

$$\langle \varphi^j_k, \tilde{\varphi}^j_{k'}\rangle = \delta_{k,k'}, \quad k,k' \in \Delta_j. \tag{3.35}$$

As mentioned above, Equation (3.13) implies that $\tilde{\Phi}_j$ is also refinable, *i.e.*, there exists a matrix $\mathbf{G}_{j,0} = (g^j_{q,k})_{q\in\Delta_{j+1},k\in\Delta_j}$ such that

$$\tilde{\varphi}^j_k = \sum_{q\in\Delta_{j+1}} g^j_{q,k}\tilde{\varphi}^{j+1}_q, \quad k\in\Delta_j, \tag{3.36}$$

where [15]

$$\overline{g}^j_{q,k} = \langle \varphi^{j+1}_q, \tilde{\varphi}^j_k\rangle. \tag{3.37}$$

It is easy to check that this means

$$\mathbf{M}_{j,0}\mathbf{G}^*_{j,0} = I, \tag{3.38}$$

where $\mathbf{G}_{j,0} = (g^j_{q,k})_{q\in\Delta_{j+1},k\in\Delta_j}$.

The objective is now to characterize the complements W_j of Equation (3.12). One way to go about this task is to look first for *some initial* complement of S_j in S_{j+1}. To this end, we may assume without loss of generality that

$$\Delta_{j+1} = \Delta_j \cup \nabla_j, \quad \Delta_j \cap \nabla_j = \emptyset. \tag{3.39}$$

One then has to find an additional set of functions $\Psi^j = \{\psi^j_k : k\in\nabla_j\}$ of the form

$$\psi^j_k = \sum_{q\in\Delta_{j+1}} m^j_{q,k}\varphi^{j+1}_q, \quad k\in\nabla_j, \tag{3.40}$$

and the matrices $\mathbf{M}_{j,1} = (m^j_{q,k})_{q\in\Delta_{j+1},k\in\nabla_j}$ have to be chosen so that there exists $\mathbf{G}_j = (\mathbf{G}_{j,0}, \mathbf{G}_{j,1}) = (g^j_{q,k})_{q,k\in\Delta_{j+1}}$ such that

$$\varphi^{j+1}_q = \sum_{k\in\Delta_j} g^j_{q,k}\varphi^j_k + \sum_{k\in\nabla_j} g^j_{q,k}\psi^j_k, \quad q\in\Delta_{j+1}. \tag{3.41}$$

Since whenever

$$\sum_{k\in\Delta_j} c^j_k\varphi^j_k + \sum_{k\in\nabla_j} d^j_k\psi^j_k = \sum_{k\in\Delta_{j+1}} c^{j+1}_k\varphi^{j+1}_k$$

one has [5]

$$\frac{1}{\|\mathbf{G}_j\|}\|(\mathbf{c}^j,\mathbf{d}^j)\|_{l_2(\Delta_j\cup\nabla_j)} \le \|\mathbf{c}^{j+1}\|_{l_2(\Delta_{j+1})} \le \|\mathbf{M}_j\|\ \|(\mathbf{c}^j,\mathbf{d}^j)\|_{l_2(\Delta_j\cup\nabla_j)}, \tag{3.42}$$

one can summarize the above circumstances as follows [5].

Proposition 3.1. *Suppose the Φ_j are uniformly stable and satisfy Equation (3.33). Then the following statements are equivalent.*

(i) *The collections*

$$\Phi_j \cup \Psi^j$$

where the ψ_k^j are given by Equation (3.40) and satisfy Equation (3.41), are uniformly stable bases of S_{j+1}.

(ii) *The matrices $\mathbf{M}_j, \mathbf{G}_j$ are uniformly bounded automorphisms of $l_2(\Delta_{j+1})$ and satisfy*

$$\mathbf{M}_j\mathbf{G}_j^* = \mathbf{G}_j^*\mathbf{M}_j = I, \tag{3.43}$$

or equivalently

$$\mathbf{M}_{j,0}\mathbf{G}_{j,0}^* + \mathbf{M}_{j,1}\mathbf{G}_{j,1}^* = I, \quad \mathbf{G}_{j,i}^*\mathbf{M}_{j,i'} = \delta_{i,i'}, i, i' \in \{0,1\}.$$

Any matrix $\mathbf{M}_{j,1}$ for which (i) or (ii) holds is called a *stable completion* of $\mathbf{M}_{j,0}$. Given a stable completion on each level, any function

$$f_j = \sum_{k\in\Delta_j} c_k^j \varphi_k^j \in S_j$$

in single-scale representation can now be expanded in terms of the multiscale basis

$$\Psi_j := \cup_{l=-1}^{j-1}\Psi^l,$$

where $\Psi^{-1} := \Phi_0$, as

$$f_j = \sum_{l=-1}^{j-1}\sum_{k\in\nabla_l} d_k^l \psi_k^l.$$

The transformation

$$\mathbf{c}^j = \mathbf{T}_j\mathbf{d}^{(j)}, \tag{3.44}$$

which takes the multiscale coefficients

$$\mathbf{d}^{(j)} := (\mathbf{c}^0, \mathbf{d}^0, \ldots, \mathbf{d}^{j-1})$$

into the single scale coefficients $\mathbf{c}^j$, can be realized through a cascade scheme

$$\begin{array}{ccccccccc}
 & \mathbf{M}_{0,0} & & \mathbf{M}_{1,0} & & \mathbf{M}_{2,0} & & \mathbf{M}_{j-1,0} & \\
\mathbf{c}^0 & \rightarrow & \mathbf{c}^1 & \rightarrow & \mathbf{c}^2 & \rightarrow & \cdots & \rightarrow & \mathbf{c}^j \\
 & \mathbf{M}_{0,1} & & \mathbf{M}_{1,1} & & \mathbf{M}_{2,1} & & \mathbf{M}_{j-1,1} & \\
 & \nearrow & & \nearrow & & \nearrow & \cdots & \nearrow & \\
\mathbf{d}^0 & & \mathbf{d}^1 & & \mathbf{d}^2 & & & &
\end{array} \tag{3.45}$$

whose stationary variants are familiar from the wavelet concept (see, *e.g.*, [8,24,28,34,35]). The corresponding inverse transformation $\mathbf{T}_j^{-1}$ is given by a similar scheme

$$\begin{array}{ccccccccc} & \mathbf{G}^*_{j-1,0} & & \mathbf{G}^*_{j-2,0} & & \mathbf{G}^*_{j-3,0} & & & \\ \mathbf{c}^j & \longrightarrow & \mathbf{c}^{j-1} & \longrightarrow & \mathbf{c}^{j-2} & \longrightarrow & \cdots & \mathbf{c}^0 \\ & \mathbf{G}^*_{j-1,1} & & \mathbf{G}^*_{j-2,1} & & \mathbf{G}^*_{j-3,1} & & \\ & \searrow & & \searrow & & \searrow & \cdots & \\ & & \mathbf{d}^{j-1} & & \mathbf{d}^{j-2} & & \cdots & \mathbf{d}^0 . \end{array} \tag{3.46}$$

While in signal analysis applications typically both transformations $\mathbf{T}_j$ and $\mathbf{T}_j^{-1}$ are needed only the $\mathbf{T}_j$ have to be executed in the present context. Thus the success of such techniques relies essentially on the efficiency and stability of the transformations $\mathbf{T}_j$ (see Subsection 3.1 above).

Remark 3.2. Carrying out $\mathbf{T}_j$ should require only $\mathcal{O}(\dim S_j)$ operations. When $\#\Delta_{j+1}/\#\Delta_j \geq \rho$ for some $\rho > 1$, the above scheme shows that this is the case provided that the matrices $\mathbf{M}_j$ are uniformly sparse, by which we mean that all columns and rows contain only a uniformly bounded number of non-zero entries. In view of Equations (3.32) and (3.40), the functions ψ_k^j will then also satisfy Equation (3.32).

Remark 3.3. The uniform stability of the transformations $\mathbf{T}_j$ in the sense of (3.7) above is known to be equivalent to the fact that

$$\Psi := \cup_{j=-1}^{\infty} \Psi^j$$

is a *Riesz-basis* of $L_2(\Gamma)$ and that there exists a *biorthogonal* Riesz-basis $\tilde{\Psi}$ (see, *e.g.*, [8,11,15]). This means that every $v \in L_2(\Gamma)$ has a unique expansion

$$v = \sum_{j=-1}^{\infty} \sum_{k \in \nabla_j} d_k^j(v) \psi_k^j, \tag{3.47}$$

where $d_k^j(v) = \langle v, \tilde{\psi}_k^j \rangle$, and that

$$\|\mathbf{d}(v)\|_{l_2(\mathcal{J})} \sim \|v\|_{L_2(\Gamma)} \tag{3.48}$$

where

$$\mathcal{J} = \cup_{j=-1}^{\infty} \nabla_j .$$

This explains the relevance of the norm equivalence of Equation (3.14). If one can manage to identify a basis Ψ^j of complements of Equation (3.12) Theorem 3.1 in combination with Proposition 3.1 provides conditions that ensure Equation (3.7). In fact, Proposition 3.1 concerns stability on a *single level* while Theorem 3.1 addresses stability *across all levels*. In order to make this more precise we wish to mention a potential tool that might help to fulfil the conditions arising in this context.

A mechanism that could be used to identify a basis Ψ^j of a space W_j of Equation (3.12) for given Q_j is offered by the following result from [5]. Let $\mathcal{L}(U,V)$ denote the space of bounded linear mappings from U into V.

Proposition 3.2. *Let $\breve{\mathbf{M}}_{j,1}$ be some stable completion of $\mathbf{M}_{j,0}$ with corresponding left inverses $\breve{\mathbf{G}}^*_{j,0}, \breve{\mathbf{G}}^*_{j,1}$. Then for any matrices*

$$\mathbf{L} \in \mathcal{L}(l_2(\nabla_j), l_2(\Delta_j)), \quad \mathbf{K} \in \mathcal{L}(l_2(\nabla_j), l_2(\nabla_j)), \tag{3.49}$$

such that

$$\mathbf{K}^{-1} \in \mathcal{L}(l_2(\nabla_j), l_2(\nabla_j)), \tag{3.50}$$

the matrix

$$\mathbf{M}_{j,1} := \mathbf{M}_{j,0}\mathbf{L} + \breve{\mathbf{M}}_{j,1}\mathbf{K} \tag{3.51}$$

is also a stable completion of $\mathbf{M}_{j,0}$ and the corresponding inverse is given by

$$\mathbf{G}_{j,0} := \breve{\mathbf{G}}_{j,0} - \breve{\mathbf{G}}_{j,1}(\mathbf{K}^*)^{-1}\mathbf{L}^*, \quad \mathbf{G}_{j,1} := \breve{\mathbf{G}}_{j,1}(\mathbf{K}^*)^{-1}. \tag{3.52}$$

Conversely, for any two stable completions $\mathbf{M}_{j,1}, \breve{\mathbf{M}}_{j,1}$ of $\mathbf{M}_{j,0}$ there exist matrices $\mathbf{L}, \mathbf{K}$ satisfying Equations (3.49) and (3.50) such that these completions are related by Equation (3.51).

An important application of this result can be stated as follows [5].

Proposition 3.3. *Let $\mathcal{Q}$ be defined by Equation (3.34) where the Φ_j and $\tilde{\Phi}_j$ are dual. Suppose $\breve{\mathbf{M}}_{j,1}$ is any stable completion of $\mathbf{M}_{j,0}$ with corresponding left inverses $\breve{\mathbf{G}}^*_{j,0}, \breve{\mathbf{G}}^*_{j,1}$. Then for any $\mathbf{K}$, satisfying Equations (3.49) and (3.50),*

$$\mathbf{M}_{j,1} := (I - \mathbf{M}_{j,0}\mathbf{G}^*_{j,0})\breve{\mathbf{M}}_{j,1}\mathbf{K}^* \tag{3.53}$$

is also a stable completion of $\mathbf{M}_{j,0}$ with the following properties:

(i) *One has*

$$\mathbf{M}_{j,0}\mathbf{G}^*_{j,0} + \mathbf{M}_{j,1}\mathbf{G}^*_{j,1} = I, \quad \mathbf{G}^*_{j,i}\mathbf{M}_{j,i'} = \delta_{i,i'}, i, i' \in \{0,1\}$$

where $\mathbf{G}_{j,0}$ is given by Equations (3.36) and (3.37) and

$$\mathbf{G}_{j,1} = \breve{\mathbf{G}}_{j,1}\mathbf{K}^{-1}.$$

(ii) *The functions*

$$\psi^j_k = \sum_{q\in\Delta_{j+1}} m^j_{q,k}\varphi^{j+1}_q, \quad \tilde{\psi}^j_k = \sum_{q\in\Delta_{j+1}} g^j_{q,k}\tilde{\varphi}^{j+1}_q, \quad k \in \nabla_j, \tag{3.54}$$

satisfy

$$\langle \psi^j_k, \tilde{\psi}^{j'}_{k'}\rangle = \delta_{j,j'}\delta_{k,k'}, \quad j, j' \geq -1, \ k \in \nabla_j, k' \in \nabla_{j'}, \tag{3.55}$$

as well as

$$\langle \varphi_k^j, \tilde{\psi}_{k'}^j \rangle = \langle \tilde{\varphi}_k^j, \psi_{k'}^j \rangle = 0, \quad k \in \Delta_j, \; k' \in \nabla_j. \tag{3.56}$$

Hence

$$(Q_{j+1} - Q_j)v = \sum_{k \in \nabla_j} \langle v, \tilde{\psi}_k^j \rangle \psi_k^j. \tag{3.57}$$

We are now ready to state the following theorem.

Theorem 3.4. *Suppose that the collections $\Phi_j, \tilde{\Phi}_j, j \in \mathbb{N}_0$, are dual to each other, satisfy both Equations (3.31) and (3.32), and are both refinable, i.e., Equations (3.33) and (3.36) hold. Moreover suppose that the matrices $\check{\mathbf{M}}_{j,1}$ form some uniformly stable completion of $\mathbf{M}_{j,0}$. Let $\mathbf{K}_j, \mathbf{K}_j^{-1}$ be any uniformly bounded mappings from $l_2(\nabla_j)$ onto $l_2(\nabla_j)$. Denoting by $\mathcal{S}$, $\tilde{\mathcal{S}}$ the sequence of spaces generated by $\Phi_j, \tilde{\Phi}_j$, respectively, assume finally that $\mathcal{S}$ and $\tilde{\mathcal{S}}$ satisfy a Whitney-type estimate of Equation (3.19) for some $l, l^* \in \mathbb{N}$ as well as an inverse estimate of Equation (3.20) for some $\gamma, \gamma^* > 0$ relative to the L_2-modulus of smoothness defined in Example 3.1. Then the collection $\Psi = \{\psi_k^j : k \in \nabla_j, j = -1, 0, \ldots\}$ where the ψ_k^j are defined by Equation (3.54) with respect to the stable completion*

$$\mathbf{M}_{j,1} := (I - \mathbf{M}_{j,0}\mathbf{G}_{j,0}^*)\check{\mathbf{M}}_{j,1}\mathbf{K}_j^*$$

form a Riesz-basis of $L_2(\Gamma)$. The dual Riesz-basis $\tilde{\Psi}$ is given by

$$\tilde{\psi}_k^j = \sum_{q \in \Delta_{j+1}} g_{q,k}^j \tilde{\varphi}_q^{j+1}, \quad k \in \nabla_j, \tag{3.58}$$

where $\mathbf{G}_j$ is defined in Proposition 3.3 (i).

Proof: One easily concludes from Equations (3.31) and (3.32) for Φ_j and $\tilde{\Phi}_j$ that the projectors Q_j, defined by Equation (3.34), are uniformly bounded and that the $\Phi_j, \tilde{\Phi}_j$ are uniformly stable. Thus, the $\mathbf{G}_{j,0}$ of Equation (3.36) are uniformly bounded mappings. Since the $\check{\mathbf{M}}_{j,1}$ are assumed to be uniformly stable completions Proposition 3.3 implies that the $\mathbf{M}_{j,1}$ are uniformly stable completions and the $\mathbf{G}_{j,1} = \check{\mathbf{G}}_{j,1}\mathbf{K}_j^{-1}$ are also uniformly bounded. By Proposition 3.1 the complement bases Ψ^j formed by the ψ_k^j are uniformly stable. By Equations (3.11), (3.55), and (3.57), every $v \in L_2(\Gamma)$ has a unique expansion

$$v = \sum_{(k,j) \in \mathcal{J}} \langle v, \tilde{\psi}_k^j \rangle \psi_k^j$$

(see Equation (3.47)). Finally, the refinability of $\tilde{\Phi}_j$ implies the commutativity condition of Equation (3.13) on the projectors Q_j. Thus all hypotheses in Theorem 3.1 are satisfied, so that the uniform stability of the Ψ^j for each scale combined with Theorem 3.1 yields Equation (3.48) which completes the proof. ∎

Remark 3.4. One can choose $\mathbf{K}_j = I$ in Theorem 3.4 so that the ψ_k^j and $\tilde{\psi}_k^j$ all have compact support provided the initial completions $\breve{\mathbf{M}}_{j,1}, \breve{\mathbf{G}}_{j,1}$ are uniformly sparse. The fact that the $\mathbf{G}_{j,0}$ are sparse follows from Equations (3.32) and (3.35) for the $\tilde{\Phi}_j$. Typical examples of sparse initial completions are obtained in connection with *hierarchical bases* for piecewise linear finite elements or more generally for *nodal finite element bases* [19,41], see also [5] for more examples.

Remark 3.5. In practical applications the Φ_j are usually chosen so as to satisfy Equations (3.31) and (3.32) as well as the refinement Equations (3.33). It is usually a much harder problem to find appropriate dual collections $\tilde{\Phi}_j$ which are also refinable and satisfy analogous conditions. This problem is rather well understood when dealing with classical shift-invariant spaces defined on all of $\mathbb{R}$ or $\mathbb{R}^n$ [11,12]. A strategy for the general case is to deal first with the algebraic problem of finding sparse $\mathbf{G}_{j,0}$ satisfying Equation (3.38), *e.g.*, making again use of Proposition 3.2. One can then try to construct $\tilde{\Phi}_j$ through a limiting process of *subdivision* type [15]. The convergence however will depend again on the choice of $\mathbf{G}_{j,0}$. It seems that this dependence is only understood in the clasical shift-invariant stationary case $\mathbf{G}_{j,0} = \mathbf{G}_0$ [11]. One should however note that the $\tilde{\Phi}_j$ do not have to be known explicitly in our context.

3.4 A local construction

We will discuss next possible ways of constructing suitable multiresolution sequences $\mathcal{S}$ on Γ. Of course, Equations (2.5) and (2.6) suggest transporting appropriate multiresolution sequences defined on $\mathbb{R}^n$ by means of a fixed atlas.

Since in many applications of the present type the trial functions need not be very regular, in fact, often not even continuous, the following simple strategy comes to mind first. Suppose the reference domain $\Sigma \subset \mathbb{R}^n$ is self-similar. It is known how to construct (discontinuous) orthonormal *multi-wavelets* on Σ (see, *e.g.*, [1,36,38]) which clearly form a Riesz-basis for $L_2(\Sigma)$. Denoting the corresponding nested multiresolution spaces by $V_j(\Sigma)$, we may then define

$$S_j := \{v : \Gamma \to \mathbb{C} : v\,|_{\Gamma_m} = f \circ \kappa_m,\ f \in V_j(\Sigma)\}, \tag{3.59}$$

so that

$$S_j \simeq \prod_{m=0}^{m_\Gamma} (\kappa_m^*)^{-1} V_j(\Sigma). \tag{3.60}$$

It is clear that $\mathcal{S}$ satisfies Equations (3.8) and (3.9). As for the stability of the corresponding transported multiscale bases, it is convenient to introduce the following modified inner product on Γ. For $u, v \in S_j$ let $\tilde{u}_m, \tilde{v}_m$ be those functions in $V_j(\Sigma)$ such that $u\,|_{\Gamma_m} = \tilde{u}_m(\kappa_m \cdot)$, $v\,|_{\Gamma_m} = \tilde{v}_m(\kappa_m \cdot)$. The inner product

$$\langle u, v\rangle_\Gamma := \sum_{m=0}^{m_\Gamma} \langle u, v\rangle_{\Gamma_m}, \tag{3.61}$$

where

$$\langle u, v\rangle_{\Gamma_m} := \int\limits_{\Sigma} \tilde{u}_m(x)\overline{\tilde{v}_m(x)}dx, \tag{3.62}$$

induces a norm which is equivalent to $\|\cdot\|_{L_2(\Gamma)}$ and, by construction, carries any orthogonality or biorthogonality relation in $V_j(\Sigma)$ over to S_j.

The same construction works, of course, for any multiresolution sequence $\mathcal{V}(\Sigma)$ and corresponding multiscale decompositions relative to any suitable bounded reference domain Σ. Specifically, for $\Sigma = [0,1]^n$ one can take tensor products of orthogonal or biorthogonal wavelets relative to the interval $[0,1]$ (see, *e.g.*, [7,13,25]). This gives wavelets of arbitrarily high regularity on Σ. Therefore, the elements of S_j have the same degree of smoothness in the interior of each tile Γ_m.

If the elements of S_j are required to have some *global* regularity on Γ a somewhat different approach has to be taken. In this case let $\mathcal{V}$ denote some multiresolution sequence on a sufficiently large neighborhood of Σ or even on all of $\mathbb{R}^n$. By $V_j(\Sigma)$ we denote now the space spanned by those functions in V_j whose support intersects Σ. Let $\Theta_j = \{\theta^j_k : k \in \triangleleft_j\}$ be the set of those basis functions. Accordingly, let $\Lambda_j = \{\lambda^j_k : k \in \triangleright_j\}$ be corresponding bases of orthogonal or biorthogonal wavelets. We will always assume that Θ_j, Λ_j as well as corresponding biorthogonal sets $\tilde{\Theta}_j, \tilde{\Lambda}_j$ consist only of compactly supported functions (see Equation (3.32)). Various examples can be found in the literature. Recall that

$$\kappa_m(\Gamma_m) = \Sigma \subset\subset \kappa_m(\tilde{\Gamma}_m) := \Sigma_m. \tag{3.63}$$

Taking if necessary j sufficiently large we can always ensure that the elements of Θ_j, Λ_j or $\tilde{\Theta}_j, \tilde{\Lambda}_j$ are supported in some domain $\tilde{\Sigma}$ satisfying

$$\Sigma \subset \tilde{\Sigma} \subset \Sigma_m \cap \Sigma_{m'}, \quad m, m' \in \{0, 1, \ldots, m_\Gamma\}. \tag{3.64}$$

As for the construction of Φ_j it is clear that whenever $\operatorname{supp}\theta^j_{k'} \subset \Sigma$ the functions φ^j_k, defined by

$$\varphi^j_k \circ \kappa_m^{-1} := \theta^j_{k'}, \quad k := (k', m),$$

belong to Φ_j. In addition Φ_j must contain functions which are not fully supported in a single tile Γ_m. To be more specific, suppose that $\theta^j_{k'}$ is not fully supported in Σ and that $\operatorname{supp}\theta^j_{k'} \cap \partial\Sigma$ is mapped by κ_m^{-1} into $\Gamma_m \cap \Gamma_{m'}$. Then there has to exist some θ^j_l such that the function φ^j_k of the form

$$\varphi^j \mid_{\Gamma_m} = \theta^j_{k'}(\kappa_m \cdot) \mid_{\Gamma_m}, \quad \varphi^j \mid_{\Gamma_{m'}} = \theta^j_l(\kappa_{m'} \cdot) \mid_{\Gamma_{m'}} \tag{3.65}$$

has the desired degree of regularity on $\Gamma_m \cup \Gamma_{m'}$. The resulting space S_j will then be isomorphic to a subset of $\prod_{m=0}^{m_\Gamma} (\kappa^*_m)^{-1} V_j(\Sigma)$.

Let us illustrate this concept by a simple example which has been used essentially in [23] in connection with the numerical solution of the double layer potential equation on polyhedra. Suppose now that Γ is the smooth boundary of some bounded domain in $\mathbb{R}^3$. Let $\mathcal{T}$ be the canonical triangulation of $\mathbb{R}^2$ which is induced by the integer lattice $\mathbb{Z}^2$ and the integer translates of lines with slope one. Thus, the triangle Σ with vertices $(0,0),(1,0),(1,1)$ belongs to $\mathcal{T}$. Let θ be the Courant hat function, *i.e.*, the unique continuous function which is piecewise linear with respect to $\mathcal{T}$ and satisfies the interpolation conditions

$$\theta(k) = \delta_{0,k}, \qquad k \in \mathbb{Z}^2. \tag{3.66}$$

For some fixed $q \in \mathbb{N}_0$ let

$$V_j(\Sigma) := \operatorname{span}\{\theta_k^j := 2^{j+q}\theta(2^{j+q}\cdot -k) : k \in \triangleleft_j\}, \tag{3.67}$$

where

$$\triangleleft_j := 2^{-j-q}\mathbb{Z}^2 \cap \Sigma.$$

The complementing bases

$$\Lambda_j := \{\lambda_k^j := 2^{j+q}\theta(2^{j+q+1}\cdot -k) : k \in \triangleright_j\}, \tag{3.68}$$

where

$$\triangleright_j := (2^{-j-q-1}\mathbb{Z}^2 \cap \Sigma) \setminus \triangleleft_j,$$

gives rise to the hierarchical basis (see [41]). The Λ_j are easily seen to form a stable decomposition of $V_{j+1}(\Sigma)$ for each j. Although it does *not* give rise to stable multiscale transformations in the sense of Equation (3.7) it provides a very useful initial decomposition in the sense of Proposition 3.2. Obviously, taking q sufficiently large, Equation (3.64) is satisfied. Moreover, since both the functions θ_k^j, λ_k^j are defined through the interpolation conditions of Equation (3.66) the transports of those θ_k^j, λ_k^j whose support overlaps the boundary of Σ can always be extended by interpolation on tile boundaries to continuous functions on Γ whenever the mappings κ_m satisfy

$$\kappa_m(\xi) = \kappa_{m'}(\xi), \quad \xi \in \overline{\Gamma}_m \cap \overline{\Gamma}_{m'}.$$

Thus the κ_m have to define a *continuous parametric surface* many concrete examples of which can be found in the *Computer Aided Geometric Design* literature. The corresponding decompositions of the spaces S_{j+1} induced by Λ_j will be stable on each level but, as mentioned before, will not give rise yet to a Riesz-basis. However, we will see below how to generate from such initial decompositions other ones with more favorable properties with regard to the present type of applications.

3.5 Moment conditions

The Whitney-type estimate of Equation (3.19) for $\omega_{\mathcal{F}}(\cdot, 2^{-j}) = \omega_l(\cdot, 2^{-j}, D)_2$, $D \subseteq \mathbf{R}^n$ is usually established for $\mathcal{S}$ (or $\tilde{\mathcal{S}}$) if all polynomials of degree $l-1$ (or l^*-1) can be represented as linear combinations of the φ_k^j (or $\tilde{\varphi}_k^j$). We say that then Φ_j is *exact* of degree $l-1$. Thus when $\tilde{\Lambda}_j$ is exact of degree d^*, say, we know, since by Equation (3.56) $\langle \lambda_k^j, \tilde{\theta}_{k'}^j \rangle = 0$ for $k' \in \triangleleft_j, k \in \triangleright_j$, that the *moment conditions*

$$\int_{\tilde{\Sigma}} x^\alpha \overline{\lambda_k^j(x)} dx = 0, \quad k \in \triangleright_j, \ |\alpha| \leq d^*, \tag{3.69}$$

hold. This will be used later as the basis for matrix compression. Again Proposition 3.2 offers a mechanism for *raising* the order of vanishing moments for the primary spaces $V_j(\Sigma)$. To this end, let

$$\mu_k^{j,\alpha} := \int_{\mathbf{R}^n} x^\alpha \overline{\theta_k^j(x)} dx, \quad k \in \triangleleft_j,$$

where the θ_k^j are now supposed to satisfy the refinement Equation (3.33) relative to $\mathbf{M}_{j,0}$. Suppose further that $\breve{\mathbf{M}}_{j,1}$ is some stable completion of $\mathbf{M}_{j,0}$ which could, for instance, be obtained through hierarchical bases of Equation (3.68). Defining now another stable completion $\mathbf{M}_{j,1}$ of $\mathbf{M}_{j,0}$ by Equation (3.51) but with $\mathbf{K} = I$, we obtain, in view of Equation (3.40), as new complementing basis functions

$$\lambda_k^j = \sum_{q \in \triangleleft_j} L_{q,k} \theta_q^j + \breve{\lambda}_k^j, \quad k \in \triangleright_j, \tag{3.70}$$

i.e., they are linear combinations of the old complementing basis functions $\breve{\lambda}_k^j$ and the single scale basis functions θ_k^j. Thus, requiring that Equation (3.69) holds is equivalent to saying that

$$\mathbf{L}^* \mu^{j,\alpha} + \breve{\mathbf{M}}_{j,1}^* \mu^{j+1,\alpha} = 0, \quad |\alpha| \leq d^*, \tag{3.71}$$

where $\mu^{j,\alpha} = (\mu_k^{j,\alpha})_{k \in \triangleleft_j}$. The objective is to find sparse $\mathbf{L} = \mathbf{L}_j$ satisfying Equation (3.71) for a given d^*. In the stationary case, *i.e.*, when the θ_k^j have the form $\theta(2^j \cdot - k)$ the $\mu_k^{j,\alpha}$ turn out to be values of polynomials at grid points so that uniformly sparse matrices $\mathbf{L}_j$ can indeed be found to ensure that Equation (3.64) still holds for the new complementing bases.

Remark 3.6. The numerical experiments in [23] indicate that in practice the realization of sufficiently high vanishing moments has a more significant effect than the stability of the $\mathbf{T}_j$ in a strict sense, in particular, when working with a moderate number of levels.

Suppose now that we have constructed some sufficiently regular multiresolution $\mathcal{S}$ on Γ and for each j some initial stable completion $\mathbf{M}_{j,1}$ of the refinement matrices $\mathbf{M}_{j,0}$ along with the corresponding complementing bases $\breve{\Psi}^j$. By Equation (3.64), we know that whenever $\operatorname{supp} \breve{\psi}^j_k \cap \Gamma_m \neq 0$ one has $\operatorname{supp} \breve{\psi}^j_k \subset \tilde{\Gamma}_m$. Therefore we can partition the sets ∇_j into subsets $\nabla_{j,m}$, $m = 0, \ldots m_\Gamma$, such that $\operatorname{supp} \breve{\psi}^j_k \subset \tilde{\Gamma}_m$ whenever $k \in \nabla_{j,m}$. For each $m \in \{0, \ldots m_\Gamma\}$ one has to determine then a sparse matrix $\mathbf{L}_{j,m}$ such that the functions

$$\psi^j_k := \breve{\psi}^j_k + \sum_{k' \in \Delta_j} (\mathbf{L}_{j,m})_{k',k} \varphi^j_{k'}, \quad k \in \nabla_{j,m},$$

are still supported in $\tilde{\Gamma}_m$ and satisfy

$$\int_{\mathbf{R}^n} x^\alpha \overline{(\kappa^*_m \psi^j_k)(x)} dx = 0, \quad |\alpha| \leq d^*, \ k \in \nabla_{j,m}. \tag{3.72}$$

3.6 Summary and assumptions

So far we have discussed some principal features of multiscale decompositions on Γ highlighting possible strategies for realizing them as well as difficulties one typically encounters in this context. We conclude this section with summarizing those requirements on a multiresolution $\mathcal{S}$ the subsequent analysis will be based upon. Let $\mathcal{S}$ satisfy Equations (3.8) and (3.9) relative to $\mathcal{F} = L_2(\Gamma)$. By $\Phi_j, \tilde{\Phi}_j$ we will always denote uniformly stable bases which consist of compactly supported functions of Equation (3.32) and which are dual to each other (see Equation (3.35)).

Projectors: Viewing $\Phi_j, \tilde{\Phi}_j$ as column vectors and denoting by $\langle v, \tilde{\Phi}_j \rangle$, $\langle \Phi_j, v \rangle$ the row, column vectors with entries $\langle v, \tilde{\varphi}^j_k \rangle$, $\langle \varphi^j_k, v \rangle$, $k \in \Delta_j$, respectively, the operators

$$Q_j := \langle \,\cdot\, , \tilde{\Phi}_j \rangle \Phi_j, \quad Q^*_j := \langle \,\cdot\, , \Phi_j \rangle \tilde{\Phi}_j \tag{3.73}$$

define, under the above assumptions, uniformly bounded projectors from $L_2(\Gamma)$ onto $S_j, \tilde{S}_j$, respectively, which are dual to each other.

Multiscale bases, stability, norm equivalences: We will employ the notation

$$\mathcal{J}_{-1} := \{(0,k) : k \in \Delta_0\}, \quad \mathcal{J}_l := \{(l,k) : k \in \nabla_l\}, \ l \in \mathbb{N}_0,$$

as well as

$$\mathcal{J}^j := \bigcup_{l=-1}^{j} \mathcal{J}_l, \quad \mathcal{J} := \mathcal{J}^\infty.$$

The sets

$$\Psi = \{\psi^j_k : (j,k) \in \mathcal{J}\}, \quad \tilde{\Psi} = \{\tilde{\psi}^j_k : (j,k) \in \mathcal{J}\}$$

will denote corresponding biorthogonal Riesz-bases such that

$$Q_{j+1} - Q_j = \langle \cdot , \tilde{\Psi} \rangle \Psi^j, \quad Q^*_{j+1} - Q^*_j = \langle \cdot , \Psi^j \rangle \tilde{\Psi}^j, \tag{3.74}$$

where as above $\Psi^j = \{\psi^j_k : k \in \nabla_j\}$ and analogously for $\tilde{\Psi}^j$.

Note that under the present assumptions the Besov spaces $B^t(\mathcal{F})$ for $\mathcal{F} = L_2(\Gamma)$ coincide with the Sobolev spaces $H^t(\Gamma)$ so that, in particular, Equation (3.28) holds. In this context the roles of the moduli $\omega_{\mathcal{F}}, \omega_{\mathcal{F}^*}$ are played by the L_2 moduli of smoothness $\omega_{d+1}(\cdot, t, \Gamma)_2, \omega_{d^*+1}(\cdot, t, \Gamma)_2$, respectively. Assuming that Equations (3.19) and (3.20) hold with some positive constants γ, γ^* relative to $S_j, \tilde{S}_j$, Theorem 3.2 implies now that

$$\|v\|_t^2 \sim \sum_{(j,k)\in\mathcal{J}} 2^{2jt} |\langle v, \tilde{\psi}^j_k \rangle|^2, \tag{3.75}$$

$$\|v\|_t^2 \sim \sum_{(j,k)\in\mathcal{J}} 2^{2jt} |\langle v, \psi^j_k \rangle|^2, \tag{3.76}$$

hold for

$$\begin{aligned} -\min\{\gamma^*, d^*+1\} &< t < \min\{\gamma, d+1\}, \\ -\min\{\gamma, d+1\} &< t < \min\{\gamma^*, d^*+1\}, \end{aligned} \tag{3.77}$$

respectively, where here and in the following we write briefly

$$\|\cdot\|_t := \|\cdot\|_{H^t(\Gamma)}, \quad \|\cdot\|_0 := \|\cdot\|_{L_2(\Gamma)}.$$

In particular, the transformations $\mathbf{T}_j$ from Equation (3.44) are uniformly stable.

Approximation properties, inverse inequalities: Under the above circumstances Theorem 3.3 yields the following approximation properties:

$$\|v - Q_j v\|_\tau \leq c\, 2^{j(\tau-t)} \|v\|_t, \quad v \in H^t(\Gamma), \tag{3.78}$$

for $-d^* - 1 \leq \tau < \min\{\gamma, d+1\}$, $\tau \leq t$, $-\min\{\gamma^*, d^*+1\} < t \leq d+1$, as well as

$$\|v - Q^*_j v\|_\tau \leq c\, 2^{j(\tau-t)} \|v\|_t, \quad v \in H^t(\Gamma), \tag{3.79}$$

for $-d - 1 \leq \tau < \min\{\gamma^*, d^*+1\}$, $\tau \leq t$, $-\min\{\gamma, d+1\} < t \leq d^*+1$.

Likewise, Theorem 3.3 suggests to require the inverse estimates

$$\|v_j\|_t \leq c\, 2^{j(t-\tau)} \|v_j\|_\tau, \quad v_j \in S_j, \tag{3.80}$$

for $\tau \leq t < \min\{\gamma, d+1\}$, as well as

$$\|v_j\|_t \leq c\, 2^{j(t-\tau)} \|v_j\|_\tau, \quad v_j \in \tilde{S}_j, \tag{3.81}$$

for $\tau \le t < \min\{\gamma^*, d^*+1\}$. Note that in both cases the parameter τ is not bounded from below. We will sketch the argument only for the second case of Equation (3.81). In fact, one readily derives from Theorem 3.3 that Equation (3.81) holds for $-\min\{\gamma, d+1\} < \tau \le t < \min\{\gamma^*, d^*+1\}$. We will employ then a bootstrapping argument to extend the range of τ. To this end, let $\langle\cdot,\cdot\rangle_s$ denote the scalar product on $H^s(\Gamma)$ such that $\|u\|_s^2 = \langle u, u\rangle_s$. Suppose that we have already proved the validity of

$$\|v_j\|_t \le c\, 2^{j(t-s)}\|v_j\|_s, \quad v_j \in \tilde{S}_j,$$

for some $s \le t < \min\{\gamma^*, d^*+1\}$. For $\tau < s$, $s-\tau < \min\{\gamma^*, d^*+1\} - s$, we then conclude that

$$\|v_j\|_\tau = \sup_{u\in H^{2s-\tau}(\Gamma)} \frac{|\langle v_j, u\rangle_s|}{\|u\|_{2s-\tau}} \ge \sup_{u_j\in\tilde{S}_j} \frac{|\langle v_j, u_j\rangle_s|}{\|u_j\|_{2s-\tau}}\,.$$

Invoking the previous estimate yields

$$\|v_j\|_\tau \ge c \sup_{u_j\in\tilde{S}_j} \frac{|\langle v_j, u_j\rangle_s|}{\|u_j\|_s 2^{j(s-\tau)}} \ge c\, 2^{j(\tau-s)}\|v_j\|_s\,,$$

which proves the claim for any $\tau < s$, $s-\tau < \min\{\gamma^*, d^*+1\} - s$. Repeating this argument yields Equation (3.81) and similarly Equation (3.80).

Moment conditions: Finally we will assume that the ψ_k^j have vanishing moments in the sense that

$$\int_{\mathbf{R}^n} x^\alpha \overline{(\kappa_m^* \psi_k^j)(x)}dx = 0, \quad |\alpha| \le d^*, k\in\nabla_{j,m},\ m = 0,\ldots,m_\Gamma. \tag{3.82}$$

§4 Galerkin scheme

This section is concerned with describing and analysing multiscale-Galerkin schemes for the numerical solution of the pseudo-differential operator Equation (2.2) where the operator A of type (1,0)) (see [31,32,39]) belongs to $\Psi_{1,0}^r(\Gamma)$ and is elliptic.

We adhere to the notation of the previous section and in addition to the hypotheses collected in Subsection 3.6 we will always assume that

$$\gamma, \gamma^* > \max\{0, \frac{r}{2}\}. \tag{4.1}$$

Popular methods for solving (2.2) numerically are Petrov-Galerkin methods. For the particular case that Γ is an n-dimensional torus, these methods have been investigated in [20,21,22] in the context of multi-resolution analysis.

Here we confine ourselves to the case of Galerkin schemes, *i.e.*, we seek for a solution $u_j \in S_j$ of the variational problem

$$\langle Au_j, v_j \rangle = \langle f, v_j \rangle \ , \quad \text{for all} \quad v_j \in S_j \ . \tag{4.2}$$

As mentioned earlier (see Equation (3.3)), it is convenient to reformulate the Galerkin method as a projection method

$$Q_j^* A u_j = Q_j^* f \ , \tag{4.3}$$

where as before Q_j^* is the L^2 adjoint of Q_j (see Equation (3.73)).

According to Hildebrandt and Wienholtz [30], the Galerkin scheme is $(\frac{r}{2}, -\frac{r}{2})$-stable, if A satisfies Equations (2.10) and (2.11). In this case the following well known properties are valid. There exist a constant $c > 0$ and $N_0 \in \mathbb{N}$ such that for all $j > N_0$

$$\|Q_j^* A u_j\|_{-\frac{r}{2}} \geq c \|u_j\|_{\frac{r}{2}} \ , \ u_j \in S_j \ , \tag{4.4}$$

$$\|Q_j^* A^* u_j\|_{-\frac{r}{2}} \geq c \|u_j\|_{\frac{r}{2}} \ , \ u_j \in S_j \ . \tag{4.5}$$

It is well known that the solutions of Equation (4.3) converge with optimal order in $H^{\frac{r}{2}}(\Gamma)$, (see, *e.g.*, [9]). Moreover, an Aubin Nitsche type duality argument yields $(s, s-r)$ stability defined below.

Lemma 4.1. *Suppose that A satisfies Equations* (2.10) *and* (2.11). *Then for any $r - d - 1 \leq s \leq \frac{r}{2}$ there exist constants $c_s > 0$ such that the Galerkin scheme is $(s, s-r)$-stable, i.e.,*

$$\|Q_j^* A u_j\|_{s-r} \geq c_s \|u_j\|_s \ , \ u_j \in S_j \ . \tag{4.6}$$

The difference $u - u_j$ between the solution u of $Au = f$ and the approximate solution u_j of Equation (4.2) *satisfies the error estimate*

$$\|u - u_j\|_t \leq c\, 2^{j(t-\tau)} \|u\|_\tau \ , \tag{4.7}$$

for $-d - 1 + r \leq t < \min\{\gamma, d+1\}$, $t \leq \tau$, $\frac{r}{2} \leq \tau \leq d+1$.

Proof: Let us denote by $u \in H^s(\Gamma)$ the solution of $Au = f$, $f \in H^{s-r}(\Gamma)$ and by u_j the solution of the Galerkin scheme in Equation (4.2). We consider only the extreme case $s = r - d - 1$. The well known Aubin-Nitsche trick (see, *e.g.*, [9]) yields optimal order convergence in $H^{r-d-1}(\Gamma)$. More precisely, we have

$$\|u - u_j\|_{r-d-1} \leq c\, 2^{j(r-d-1-t)} \|u\|_t \ , \tag{4.8}$$

if $\frac{r}{2} \leq t \leq d+1$. Consequently, setting $t = \frac{r}{2}$ we derive

$$\|u - u_j\|_{r-d-1} \leq c\, 2^{-j(d+1-\frac{r}{2})} \|u\|_{\frac{r}{2}},$$

and, therefore,

$$\|(A^{-1} - A_j^{-1})f\|_{r-d-1} \leq c\, 2^{-j(d+1-\frac{r}{2})} \|f\|_{\frac{-r}{2}} \ , \tag{4.9}$$

whenever $f \in H^{\frac{-r}{2}}(\Gamma)$, where

$$A_j := Q_j^* A Q_j .$$

Choosing $f = f_j = A_j v_j \in Q_j^* H^{\frac{-r}{2}}(\Gamma) = \tilde{S}_j$, we apply the inverse estimate of Equation (3.81) to the right hand side of Equation (4.9) to conclude that

$$\|(A^{-1} - A_j^{-1})f_j\|_{r-d-1} \leq c\, 2^{-j(d+1-\frac{r}{2})} \|f_j\|_{\frac{-r}{2}} \leq c\ \|f_j\|_{-d-1} \ . \tag{4.10}$$

Since $A^{-1} : H^{-q}(\Gamma) \to H^{r-q}(\Gamma)$ is bounded we have already established the estimate

$$\|A_j^{-1} f_j\|_{r-d-1} \leq c\ \|f_j\|_{-d-1} \ , \quad f_j \in \tilde{S}_j \ . \tag{4.11}$$

Inserting $f_j = A_j v_j = Q_j^* A v_j$, $v_j \in S_j$, we have confirmed $(r - d - 1, -d - 1)$-stability

$$\|A_j v_j\|_{-d-1} \geq c\ \|v_j\|_{r-d-1} \ , \ v_j \in S_j \ . \tag{4.12}$$

The $(s, s - r)$-stability can be proved by the same arguments, provided that $r - d - 1 \leq s \leq \frac{r}{2}$. ■

Representing the operator A_j by a matrix requires fixing a basis of S_j. In view of Equations (3.4) and (3.5), the stiffness matrix relative to the multiscale basis Ψ_j has the form

$$\mathbf{A}_j = (\langle A\psi_k^l, \psi_{k'}^{l'} \rangle)_{(l,k),(l',k') \in \mathcal{J}^j} . \tag{4.13}$$

$\mathbf{A}_j$ will be referred to as *wavelet representation* of the operator $A_j := Q_j^* A Q_j$ [20,21]. The importance of this representation with regard to the numerical solution of Equation (4.2) relies on two issues, namely *preconditioning* and *compression.* To facilitate a convenient analysis of this problem let us introduce the operators $F_j : \tilde{S}_j \to l_2(\mathcal{J}^j)$, defined by

$$(F_j u)_{(l,k)} := \langle u, \psi_k^l \rangle, \quad (l, k) \in \mathcal{J}^j \ . \tag{4.14}$$

The adjoint operator F_j^* of F_j is given by

$$F_j^*(\mathbf{d}) = \sum_{(l,k) \in \mathcal{J}^j} d_k^l \psi_k^l, \quad \mathbf{d} \in l_2(\mathcal{J}^j). \tag{4.15}$$

Furthermore we will employ the operators defined by

$$F_j^{-1}(\mathbf{d}) = \sum_{(l,k) \in \mathcal{J}^j} d_k^l \tilde{\psi}_k^l, \quad ((F_j^*)^{-1} f)_{(l,k)} = \langle f, \tilde{\psi}_{(l,k)} \rangle \ , \quad (l, k) \in \mathcal{J}^j. \tag{4.16}$$

These operators relate the basis free description of the projection method to the corresponding stiffness matrices. It is important to note that the stability of the multiscale bases of Equation (3.75), Equation (3.76) can be rephrased as follows:

$$\|F_j u\|_{l_2(\mathcal{J}^j)} \sim \|u\|_0, \quad u \in \tilde{S}_j, \qquad \|(F_j^*)^{-1}u\|_{l_2(\mathcal{J}^j)} \sim \|u\|_0, \quad u \in S_j. \tag{4.17}$$

Moreover, one easily confirms that

$$\mathbf{A}_j = F_j A F_j^* = F_j Q_j^* A Q_j F_j^* \ , \quad Q_j^* A Q_j = F_j^{-1} \mathbf{A}_j (F_j^*)^{-1} \ . \tag{4.18}$$

In analogy to Equation (3.26) we introduce now the one paramenter group of operators Υ_j^s, $s \in \mathbf{R}$, defined by

$$\Upsilon_j^s u = \sum_{l=-1}^{j} 2^{ls}(Q_{l+1} - Q_l)u \tag{4.19}$$

and

$$\Upsilon^s u = \sum_{l=-1}^{\infty} 2^{ls}(Q_{l+1} - Q_l)u \ . \tag{4.20}$$

Note next that for a given matrix $\mathbf{A}_j = \big(a_{(l,k),(l',k')}\big)_{(l,k),(l',k') \in \mathcal{J}^j}$ we have

$$(\Upsilon_j^t)^* F_j^{-1} \mathbf{A}_j (F_j^*)^{-1} \Upsilon_j^s = F_j^{-1} \mathbf{D}_j^t \mathbf{A}_j \mathbf{D}_j^s (F_j^*)^{-1} \ , \tag{4.21}$$

where $(\mathbf{D}_j^s)_{(l,k),(l',k')} = 2^{ls}\delta_{(l,k),(l',k')}$.

An important advantage of a stable multiscale basis is the following *preconditioning* of the matrix $\mathbf{A}_j$.

Theorem 4.1. *Under the above assumptions the matrices*

$$\mathbf{B}_j := \mathbf{D}_j^{-r/2} \mathbf{A}_j \mathbf{D}_j^{-r/2} \tag{4.22}$$

have uniformly bounded spectral condition numbers (see [15,17,21,22]*).*

Proof: Our assumptions on A imply that

$$\|v_j\|_{\frac{r}{2}} \sim \|Q_j^* A v_j\|_{-\frac{r}{2}}, \quad v_j \in S_j,$$

so that, by Equation (3.27), we obtain for $w_j := \Upsilon_j^{\frac{r}{2}} v_j$

$$\|w_j\|_0 \sim \|(\Upsilon_j^{-\frac{r}{2}})^* Q_j^* A Q_j \Upsilon_j^{-\frac{r}{2}} w_j\|_0.$$

Now we infer from Equation (4.17)

$$\|(F_j^*)^{-1} w_j\|_{l_2(\mathcal{J}^j)} \sim \|F_j (\Upsilon_j^{-\frac{r}{2}})^* Q_j^* A Q_j \Upsilon_j^{-\frac{r}{2}} F_j^* (F_j^*)^{-1} w_j\|_{l_2(\mathcal{J}^j)}.$$

The assertion follows now from Equations (4.18) and (4.21). ■

§5 Optimal convergence order

5.1 Basic estimates

The stiffness matrices $\mathbf{A}_j$ defined in the previous section will generally not be sparse. To facilitate a rapid solution of Equation (4.2) our goal is to approximate $\mathbf{A}_j$ by suitable sparse matrices. This approximation will hinge upon estimates for the entries $\langle A\psi_k^l, \psi_{k'}^{l'}\rangle$ for $(l,k),(l',k') \in \mathcal{J}^j$. In fact, using approriate versions of Schur's lemma these results will be used to estimate the norms of the compressed matrices and their inverses. This, in turn, will allow us to establish error bounds for a fixed required accuracy as well as asymptotic error bounds suggested by the convergence rates (see, *e.g.*, [21,22]).

Let us denote by $\Omega_{(l,k)}$ the set

$$\Omega_{(l,k)} := \{x \in \Gamma : \psi_k^l(x) \neq 0\} , \tag{5.1}$$

for $(l,k) \in \mathcal{J}^j$.

Lemma 5.1. *Let $n + 2(d^* + 1) + r > 0$ and*

$$\mathrm{dist}(\Omega_{(l,k)}, \Omega_{(l',k')}) > 0 . \tag{5.2}$$

Then there exists a constant c depending only on r, n, d^* such that the coefficients of the matrices $\mathbf{A}_j$, $j \in \mathbb{N}$, defined by Equation (4.13) satisfy

$$|a_{(l',k'),(l,k)}| = |\langle A\psi_k^l, \psi_{k'}^{l'}\rangle| \leq c\, \frac{2^{l(-\frac{n}{2}-d^*-1)}2^{l'(-\frac{n}{2}-d^*-1)}}{[\mathrm{dist}(\Omega_{(l,k)}, \Omega_{(l',k')})]^{n+r+2(d^*+1)}} , \tag{5.3}$$

uniformly in $(l,k) \in \mathcal{J}_l$, $(l',k') \in \mathcal{J}_{l'}$.

Proof: In view of the decomposition Equation (2.6) we investigate the transported operators $\mathcal{A} = \kappa_{m'}^* \circ \chi_{m'} A \chi_m \circ (\kappa_m^*)^{-1}$ on local charts for which $m, m' \in \Xi$ (see Equation (2.6)). More precisely, we estimate the quantities

$$\langle \chi_{m'} A(\chi_m \psi_k^l), \psi_{k'}^{l'}\rangle = \int_{\mathbf{R}^n} \mathcal{A}(\kappa_m^* \psi_k^l)(x)\overline{(\kappa_{m'}^* \psi_{k'}^{l'})(x)} J(x)dx , \tag{5.4}$$

where $J \in \mathbb{C}^\infty(\mathbb{R}^n)$. Since

$$J(x)\mathcal{A}$$

is a pseudo-differential operator in $\Psi_{1,0}^r(\mathbb{R}^n)$ with Schwartz kernel $K_\mathcal{A}$ and since $\mathrm{dist}(\Omega_{(l,k)}, \Omega_{(l',k')}) > 0$ we apply Taylor's theorem to estimate the integral

$$\begin{aligned} &\int_{\mathbf{R}^n} \mathcal{A}(\kappa_m^* \psi_k^l)(x)\overline{(\kappa_{m'}^* \psi_{k'}^{l'})(x)} J(x)dx \\ &= \int_{\mathbf{R}^n} K_\mathcal{A}(x,y)(\kappa_m^* \psi_k^l)(x)\overline{(\kappa_{m'}^* \psi_{k'}^{l'})(y)} dxdy . \end{aligned} \tag{5.5}$$

Due to Equation (3.82), the moments corresponding to the Taylor polynomials

$$\sum_{\alpha,\beta} c_{\alpha,\beta}(x_0, y_0)(x - x_0)^\alpha (y - y_0)^\beta$$

of $K_{\mathcal{A}}$ with respect to a point $(x_0, y_0) \in \kappa_m \Omega_{(l,k)} \times \kappa_{m'} \Omega_{(l',k')}$ vanish if $|\alpha| \leq d^*$ or $|\beta| \leq d^*$. Thus all Taylor polynomials of degree less than $2d^*+2$ are canceled. An application of Lemma 2.1 yields

$$\begin{aligned}
&|\int_{\mathbf{R}^{2n}} K_{\mathcal{A}}(x,y)(\kappa_m^* \psi_k^l(y))(\overline{\kappa_{m'}^* \psi_{k'}^{l'}(x)})dxdy| \\
&\leq c \sum_{|\alpha|+|\beta|=2d^*+2} \sup_{x\in\kappa_m\Omega_{(l,k)}, y\in\kappa_{m'}\Omega_{(l',k')}} |\partial_x^\alpha \partial_y^\beta K_{\mathcal{A}}(x,y)| \\
&\int_{\mathbf{R}^{2n}} |(x-x_0)^\alpha (y-y_0)^\beta (\kappa_m^* \psi_k^l(x))\overline{\kappa_{m'}^* \psi_{k'}^{l'}(y)}|dxdy \ , \\
&\leq c 2^{(l+l')(-\frac{n}{2}-d^*-1)}(\mathrm{dist}(\Omega_{(l,k)}, \Omega_{(l',k')}))^{-n-r-2(d^*+1)} \ . \qquad (5.6)
\end{aligned}$$

In the last step we have used that the manifold Γ is a compact Riemannian manifold as well as Lemma 2.1.

Since the kernel of A_∞ is smooth, similar arguments confirm that

$$|\langle A_\infty \psi_k^l, \psi_{k'}^{l'} \rangle| \leq c\, 2^{-(l+l')(d^*+1+\frac{n}{2})} \qquad (5.7)$$

is valid for all $(l,k),(l',k') \in \mathcal{J}^j$. ■

The above type of estimate has been used by von Petersdorf and Schwab [37] to adapt the compression to the optimal order of convergence of the Galerkin scheme for first kind boundary integral operators. Their method improves in this case upon the corresponding results in [21].

In Section 6, however, we will employ another type of estimates containing the difference of two levels yielding optimal order of complexity for a compression with fixed accuracy.

As in [21] we propose a level depending truncation, *i.e.*, we define the compressed matrix $\mathbf{A}_j^\epsilon$ by

$$(a_{(l,k),(l',k')})_j^\epsilon := \begin{cases} a_{(l,k),(l',k')} & \text{if } \mathrm{dist}(\Omega_{(l,k)}, \Omega_{(l',k')}) \leq \mathcal{B}_{l,l'}, \\ 0 & \text{otherwise.} \end{cases} \qquad (5.8)$$

The following Lemmata establish a few technical preliminaries which will be used later in connection with a Schur Lemma argument, where in particular the truncation bandwidth $\mathcal{B}_{l,l'}$ will be tuned properly.

Lemma 5.2. *Let* $r_{(l,k),(l',k')} = ((a_{(l,k),(l',k')})_j - (a_{(l,k),(l',k')})_j^\epsilon)$ *then, for*

$$\mathcal{B}_{l,l'} \geq \max\{2^{-l}, 2^{-l'}\} \ , \qquad (5.9)$$

the estimate

$$\sum_{k\in\nabla_l} |r_{(l,k),(l',k')}| \leq c2^{-(l+l')(d^*+1+\frac{n}{2})}\mathcal{B}_{l,l'}^{-2(d^*+1)-r}2^{ln} \tag{5.10}$$

holds.

Proof: We wish to majorize the sum

$$\sum_{k\in\nabla_l} |r_{(l,k),(l',k')}| = \sum_{\{k\in\nabla_l:\operatorname{dist}(\Omega_{(l,k)},\Omega_{(l',k')})>\mathcal{B}_{l,l'}\}} |a_{(l,k),(l',k')}| \tag{5.11}$$

by an integral. On account of Equations (5.3) and (5.9) one obtains

$$\begin{aligned}
&\sum_{k\in\nabla_l} |r_{(l,k),(l',k')}| \\
&\leq c\, 2^{-(l+l')(d^*+1+\frac{n}{2})}\times \\
&\qquad \sum_{\{k\in\nabla_l:\operatorname{dist}(\Omega_{(l,k)},\Omega_{(l',k')})>\mathcal{B}_{l,l'}\}} [\operatorname{dist}(\Omega_{(l,k)},\Omega_{(l',k')})]^{-n-2(d^*+1)-r} \\
&\leq c\, 2^{-(l+l')(d^*+1+\frac{n}{2})}2^{ln} \int\limits_{|x|>\mathcal{B}_{l,l'}} |x|^{-n-2(d^*+1)}dx \\
&\leq c\, 2^{-(l+l')(d^*+1+\frac{n}{2})}2^{ln}\mathcal{B}_{l,l'}^{-2(d^*+1)-r}\ . \quad \blacksquare
\end{aligned} \tag{5.12}$$

An appropriate choice of $\mathcal{B}_{l,l'}$ provides the following estimate.

Lemma 5.3. *Let $s, \tilde{s} \in [0, d^* + 1 + \frac{r}{2}]$, we choose $a > 1$ and suppose that*

$$\mathcal{B}_{l,l'} \geq a j^{\frac{2}{2(d^*+1)+r}} 2^{\frac{j(s+\tilde{s}-r)-l(d^*+1+\tilde{s})-l'(d^*+1+s)}{2(d^*+1)+r}}\ . \tag{5.13}$$

Then the estimate

$$\sum_{l=0}^{j}\sum_{k\in\nabla_l} 2^{\frac{-ln}{2}}2^{-l'\tilde{s}}2^{-ls}|r_{(l,k),(l',k')}| \leq c\, 2^{\frac{-l'n}{2}}a^{-r-2(d^*+1)}j^{-1}2^{-j(s+\tilde{s}-r)} \tag{5.14}$$

is valid.

Proof: The asserted estimate is a consequence of Lemma 5.2 and Equation (5.13)

$$\begin{aligned}
&\sum_{l=0}^{j}\sum_{k\in\nabla_l} 2^{\frac{-ln}{2}}2^{-l'\tilde{s}}2^{-ls}|r_{(l,k),(l',k')}| \\
&\leq c\sum_{l=0}^{j} 2^{\frac{-l'n}{2}}2^{-l(s+d^*+1)}2^{-l'(\tilde{s}+d^*+1)}\mathcal{B}_{l,l'}^{-r-2(d^*+1)} \\
&\leq c\, 2^{\frac{-l'n}{2}}a^{-r-2(d^*+1)}j^{-1}2^{-j(s+\tilde{s}-r)}\ . \quad \blacksquare
\end{aligned} \tag{5.15}$$

For the level depending thresholding, we assume $s, \tilde{s} > \frac{r}{2}$ and define the compressed matrix $\mathbf{A}_j^\epsilon$ by Equation (5.8) with

$$\mathcal{B}_{l,l'} \geq \max\{a\,2^{-l}, a\,2^{-l'}, aj^{\frac{2}{2(d^*+1)+r}} 2^{\frac{j(s+\tilde{s}-r)-l'(d^*+1+\tilde{s})-l(d^*+1+s)}{2(d^*+1)+r}}\}. \tag{5.16}$$

Let $\|\mathbf{A}\|$ denote the spectral norm of the matrix $\mathbf{A}$.

Theorem 5.1. *Let $s, \tilde{s} > \frac{r}{2}$ and $r_{(l,k),(l',k')}$ be defined by Lemma 5.2 and Equation (5.16). Let $\mathbf{R}(s, \tilde{s})$ be the matrix with entries $2^{-l'\tilde{s}}2^{-ls}|r_{(l,k),(l',k')}|$. Then the spectral norm of the matrix $\mathbf{R}(s, \tilde{s})$ as a mapping on $l_2(\mathcal{J}^j)$ is bounded by*

$$\|\mathbf{R}(s, \tilde{s})\| \leq c\, a^{-r-2(d^*+1)} 2^{-j(s+\tilde{s}-r)} j^{-1}. \tag{5.17}$$

Proof: This is an application of Schur's lemma (see, *e.g.*, [21]). Taking Lemma 5.3 into account, we conclude that

$$\begin{aligned}\|\mathbf{R}(s, \tilde{s})\| &\leq \max_{\{0\leq l'<j, k'\in\nabla_{l'}\}} \sum_{l=0}^{j} \sum_{k\in\nabla_l} 2^{\frac{(l'-l)n}{2}} 2^{-l'\tilde{s}} 2^{-ls} |r_{(l,k),(l',k')}| \\ &+ \max_{\{0\leq l<j, k\in\nabla_l\}} \sum_{l'=0}^{j} \sum_{k'\in\nabla_{l'}} 2^{\frac{(l-l')n}{2}} 2^{-l'\tilde{s}} 2^{-ls} |r_{(l,k),(l',k')}| \quad (5.18)\\ &\leq c\, a^{-r-2(d^*+1)} 2^{-j(s+\tilde{s}-r)} j^{-1}. \quad \blacksquare\end{aligned}$$

Recalling Equation (4.18), we have the operator A_j^ϵ corresponding to the compressed matrix $\mathbf{A}_j^\epsilon$ given by

$$A_j^\epsilon = F_j^{-1} \mathbf{A}_j^\epsilon (F_j^*)^{-1}. \tag{5.19}$$

We may now formulate the following consistency result.

Theorem 5.2. *If $-(d+1)+r \leq s < \min\{\gamma, d+1\}$, $-\min\{\gamma, d+1\} < t \leq d+1$ and $d \leq d^* + r$ then*

$$\|(A_j - A_j^\epsilon)u\|_{s-r} \leq c\, a^{-r-2(d^*+1)} 2^{j(s-t)} \|u\|_t. \tag{5.20}$$

Proof: We show only the extreme case $t = d+1$ which yields the optimal convergence rate $2(d+1) - r$. Using $\mathbf{R}(s, \tilde{s})$ given by Theorem 5.1, where $s = \tilde{s} = d+1$, we obtain

$$\|(A_j - A_j^\epsilon)u\|_{-d-1} \leq \inf_{v\in H^{d+1}(\Gamma)} \frac{\langle Q_j^*(A_j - A_j^\epsilon)Q_j)u, v\rangle}{\|v\|_{d+1}}. \tag{5.21}$$

Thus we further conclude that

$$\begin{aligned}\|(A_j - A_j^\epsilon)u\|_{-d-1} &\le c\frac{\langle(\Upsilon_j^{-(d+1)})^* Q_j^*(A_j - A_j^\epsilon)Q_j\Upsilon_j^{-(d+1)}\Upsilon_j^{(d+1)}u, \Upsilon_j^{d+1}v\rangle}{\|v\|_{d+1}} \\ &\le c\,\|F_j(\Upsilon_j^{-(d+1)})^* Q_j^*(A_j - A_j^\epsilon)Q_j\Upsilon_j^{-(d+1)}F_j^*\| \\ &\times \|\Upsilon_j^{d+1}u\|_0 \frac{\|\Upsilon_j^{d+1}v\|_0}{\|v\|_{d+1}} \\ &\le c\, j\, \|\mathbf{R}(s,\tilde{s})\|\|u\|_{d+1}, \qquad (5.22)\end{aligned}$$

where we have applied the approximation property in Equation (3.78) to derive

$$\|\Upsilon_j^{d+1}u\|_0 \le c\sqrt{j}\|u\|_{d+1}$$

for u in the last step. The assertion follows from an application of Theorem 5.1 to the right-hand side of Equation (5.22). The estimates for the remaining s, t follow by similar arguments. ■

The present result can be slightly improved when $s < d+1$ or $\tilde{s} < d+1$ by employing the modulus of continuity as in [21].

For $s, \tilde{s} > \frac{r}{2}$, choosing the parameter a sufficiently large, Theorem 5.2 combined with a perturbation argument yields stability in the energy norm (see, *e.g.*, [21]). Similarly, taking Lemma 4.1 into account, one can prove stability for the compressed operator

$$\|A_j^\epsilon u_j\|_{s-r} \ge c_s\|u_j\|_s \quad , \quad u_j \in S_j \ , \qquad (5.23)$$

for $r - d - 1 \le s \le r/2$.

Theorem 5.3. *For all $r - d - 1 \le s \le r/2$ and $\epsilon > 0$, there exist $a \ge 1$, $N_0 \in \mathbb{N}$, depending on s, ϵ, and a compressed matrix $\mathbf{A}_j^\epsilon$ as defined above such that for $j \ge N_0$*

$$\|(A_j - A_j^\epsilon)u\|_{s-r} \le \epsilon\|u\|_s \qquad (5.24)$$

and

$$\|A_j^\epsilon u_j\|_{s-r} \ge c_s\|u_j\|_s \quad , \quad u_j \in S_j \ . \qquad (5.25)$$

We are now in a position to establish asymptotic convergence rates for the solution of the compressed scheme. The proof follows by analogous arguments as in [21].

Theorem 5.4. *Let $A \in \Psi_{0,1}^r(\Gamma)$ satisfy Equations (2.10) and (2.11), $r-d-1 \le s < \min\{\gamma, d+1\}$, $s \le t$, $t \le d+1$, and $f \in H^{t-r}(\Gamma)$. Then there exist $\epsilon > 0$ and a matrix $\mathbf{A}_j^\epsilon$ such that the compressed scheme $A_j^\epsilon u_j^\epsilon = Q_j^* f$ has a unique solution u_j^ϵ which differs from the exact solution of the equation $Au^* = f$ in the $H^s(\Gamma)$ norm by*

$$\|u^* - u_j^\epsilon\|_s \le c\, 2^{j(s-t)}\|u^*\|_t \ , \qquad (5.26)$$

where the constant c does not depend on j.

Proof: By the $(s, s-r)$-stability we obtain

$$\begin{aligned}\|u^* - u_j^\epsilon\|_s &\leq \inf_{u_j \in S_j} [\|u^* - u_j\|_s + c\|A_j^\epsilon(u_j - u_j^\epsilon)\|_{s-r}] \\ &\leq \inf_{u_j \in S_j} [\|u^* - u_j\|_s + \\ &\quad + c(\|f - Q_j^* f\|_{s-r} + \|A_j u_j - Au^*\|_{s-r} + \|(A_j^\epsilon - A_j)u_j\|_{s-r})] \ .\end{aligned} \tag{5.27}$$

The first terms in the sum can be estimated by use of a direct estimate Equation (3.79) and inverse properties of Equation (3.80). The last term has been dealt with in Theorem 5.2. ■

5.2 Efficiency

In this section we treat only the extreme case $s = \tilde{s} = d+1$. To preserve some flexibility in choosing a suitable truncation we make use of the following observation

$$\begin{aligned}& j^{\frac{2}{2(d^*+1)+r}} 2^{\frac{j(2(d+1)-r)-(l+l')(d^*+1+d+1)}{2(d^*+1)+r}} \\ &\leq j^{\frac{2}{2(d^*+1)+r}} 2^{-j} 2^{\frac{j(2(d+1)+2(d^*+1))-(l+l')(d^*+1+d+1)}{2(d^*+1)+r}} \\ &\leq j^{\frac{2}{2(d^*+1)+r}} 2^{-j} 2^{\frac{(j-l)(d^*+1+d+1)}{2(d^*+1)+r}} 2^{\frac{(j-l')(d^*+1+d+1)}{2(d^*+1)+r}} \\ &\leq j^{\frac{2}{2(d^*+1)+r}} 2^{-j} 2^{(j-l)M} 2^{(j-l')M'} \ ,\end{aligned} \tag{5.28}$$

provided that

$$M, M' \geq \frac{(d^*+1+d+1)}{2(d^*+1)+r} \ . \tag{5.29}$$

Thus, in view of Equation (5.16), we may work in the following with the last simpler bound in Equation (5.28). The next result indicates the importance of being able to choose $d^* > d$ so that even when $r < 0$ the constants M, M' can be arranged to be less than one. This flexibility is offered by biorthogonal multiscale bases while orthogonal decompositions limit such options and seem to be less favorable for efficient compression.

Theorem 5.5. *Let $M, M' < 1$ satisfy Equation (5.29) and choose*

$$\mathcal{B}_{l,l'} \sim \max\{j^{\frac{2}{2(d^*+1)+r}} 2^{-j} 2^{(j-l)M} 2^{(j-l')M'}, 2^{-l}, 2^{-l'}\} \ .$$

Then the matrix $\mathbf{A}_j^\epsilon$ has

$$\mathcal{O}(j^{\max\{\frac{2n}{2(d^*+1)+r}, 2\}} 2^{jn}) = \mathcal{O}(N_j (\log N_j)^{\max\{\frac{2}{2(d^*+1)+r}, 2\}})$$

non-zero coefficients, where $N_j = \dim S_j$. In case $M, M' = 1$, the compressed matrix $\mathbf{A}_j^\epsilon$ has $\mathcal{O}(j^{\frac{2n}{2(d^*+1)+r}+2} 2^{jn}) = \mathcal{O}(N_j (\log N_j)^{\frac{2}{2(d^*+1)+r}+2})$ non-zero coefficients.

Proof: Let $l \geq l'$, we count the number $\#\mathbf{A}^{\epsilon}_{l,l'}$ of non-zero elements of the block matrices $\mathbf{A}^{\epsilon}_{l,l'}$ when

$$\mathcal{B}_{l,l'} \sim j^{\frac{2}{2(d^*+1)+r}} 2^{-j} 2^{(j-l)M} 2^{(j-l')M'} \tag{5.30}$$

Since in each row of $\mathbf{A}^{\epsilon}_{l,l'}$ we have at most $\sim [\mathcal{B}_{l,l'} 2^{l'}]^n$ non-zero entries we conclude that

$$\begin{aligned} \#\mathbf{A}^{\epsilon}_{l,l'} &\leq c 2^{ln} [\mathcal{B}_{l,l'} 2^{l'}]^n \\ &\leq c[2^{l+l'} j^{\frac{2}{2(d^*+1)+r}} 2^{-j} 2^{(j-l)M} 2^{(j-l')M'}]^n \\ &\leq c[j^{\frac{2}{2(d^*+1)+r}} 2^{j} 2^{(j-l)(M-1)} 2^{(j-l')(M'-1)}]^n \,. \end{aligned} \tag{5.31}$$

Summing over all $-1 \leq l, l' \leq$ yields in case $M, M' < 1$

$$\begin{aligned} \#\mathbf{A}^{\epsilon}_{j} &\leq \sum_{l,l'=-1}^{j} \#\mathbf{A}^{\epsilon}_{l,l'} \\ &\leq c j^{\frac{2n}{2(d^*+1)+r}} 2^{jn} \sum_{l,l'=-1}^{j} [2^{(j-l)(M-1)} 2^{(j-l')(M'-1)}]^n \\ &\leq c j^{\frac{2n}{2(d^*+1)+r}} 2^{jn} \,. \end{aligned} \tag{5.32}$$

Counting elements $\langle A\psi^l_k, \psi^{l'}_{k'} \rangle$ for wich $\mathcal{B}_{l,l'} \sim \max\{2^{-l}, 2^{-l'}\}$ gives

$$\mathcal{O}(2^{ln}[\mathcal{B}_{l,l'} 2^{l'}]^n) = \mathcal{O}(2^{ln})$$

entries, for fixed l, l'. This proves the desired result. The result for $M, M' = 1$ is an immediate consequence of Equation (5.31). ■

§6 Optimal compression

In order to avoid the logarithmic terms in the complexity estimates, we need an estimate for all entries of the matrix $\mathbf{A}_j$ exhibiting a decay in $|l - l'|$. To obtain this kind of estimate we need some preparations. First of all we change our point of view and do not strive any more for asymtotically optimal convergence rates. Instead we address the following different problem: Given a tolerance $\epsilon > 0$, we wish to find a compressed matrix such that the error between the solution u_j of the Galerkin scheme and the solution u^{ϵ}_j of the compressed Galerkin scheme stays below ϵ, for instance, in the energy norm.

Our next estimates require first extracting constants which, due to the nature of the wavelet basis, simply means to replace all those entries by zero which contain a scalar product with functions on the coarsest grid.

Here we will assume in addition that the dual collections $\tilde{\Phi}_j$ consists of functions which are Hölder continuous of order $m+\rho > 0$, $m \in \mathbb{N}_0), \rho \in (0,1)$. More precisely, we require that

$$|D^\alpha \varphi_k^l(x) - D^\alpha \varphi_k^l(x')| \leq c 2^{l(\frac{n}{2}+\rho)}[\mathrm{dist}\{x,x'\}]^{\rho+m} \ , \ |\alpha| \leq m \ . \tag{6.1}$$

Note that in the stationary case this property holds automatically when $\tilde{\varphi}$ is a continuously refinable function. In the same fashion as in Lemma 5.1 one confirms the following result.

Lemma 6.1. *Let $n + d^* + 1 + r > 0$, then there exists a constant c depending only on r, n, d such that the coefficients $\langle A\psi_{k'}^l, \varphi_k^l \rangle$ satisfy*

$$|\langle A\psi_{k'}^l, \varphi_k^l\rangle| \leq c \, \frac{2^{lr}}{[1 + 2^l \mathrm{dist}(\Omega_{(l,k')}, \mathrm{supp}\varphi_k^l)]^{n+r+d^*+1}} \ , \quad k \in \Delta_l, k' \in \nabla_l \ , \tag{6.2}$$

uniformly in $l \in \mathbb{N}$.

Proof: In analogy to the proof of Lemma 5.1 one shows in case

$$\mathrm{dist}(\Omega_{(l,k')}, \mathrm{supp}\varphi_k^l) > c\, 2^{-l} \ , \quad k \in \Delta_l, k' \in \nabla_l \ ,$$

that

$$\begin{aligned} |\langle A\psi_{k'}^l, \varphi_k^l\rangle| &\leq c \, \frac{2^{-l\frac{n}{2}} 2^{-l(\frac{n}{2}+d^*+1)}}{[\mathrm{dist}(\Omega_{(l,k')}, \mathrm{supp}\varphi_k^l)]^{n+r+d^*+1}} \\ &\leq c \, \frac{2^{lr}}{[1 + 2^l \mathrm{dist}(\Omega_{(l,k')}, \mathrm{supp}\varphi_k^l)]^{n+r+d^*+1}} \ . \end{aligned} \tag{6.3}$$

If $\mathrm{dist}(\Omega_{(l,k')}, \mathrm{supp}\varphi_k^l) \leq c\ 2^{-l}$ we apply the continuity of $A : H^s(\Gamma) \to H^{s-r}(\Gamma)$ to conclude that

$$|\langle A\psi_{k'}^l, \varphi_k^l\rangle| \leq c\, 2^{lr} \ , \quad k \in \Delta_l \ , \ k' \in \nabla_l \ . \tag{6.4}$$

■

The next lemma sets the ground for neglecting levels which are far apart from each other.

Lemma 6.2. *Let $n + d^* + 1 + r > 0$ and*

$$\mathrm{dist}(\Omega_{(l,k)}, \Omega_{(l',k')}) \leq c \max\{2^{-l}, 2^{-l'}\} \ . \tag{6.5}$$

Then there exists a constant c, such that the coefficients $2^{-\frac{lr}{2}}\langle A\psi_k^l, \psi_{k'}^{l'}\rangle 2^{-\frac{l'r}{2}}$ satisfy

$$|2^{-\frac{lr}{2}}\langle A\psi_k^l, \psi_{k'}^{l'}\rangle 2^{-\frac{l'r}{2}}| \leq c\, 2^{-|l-l'|(\frac{n}{2}+\rho)} \ , \ l, l' \geq 0 \ , \tag{6.6}$$

where ρ is the Hölder exponent from Equation (6.1).

Proof: Let us assume that $l' \leq l$ and $r \leq 0$. The assumption that the basis functions φ_k^l have some Hölder regularity (see Equation (6.1)) is crucial in the following argument. Setting $A^\sharp := (I - Q_0^*)A(I - Q_0)$, and noting that

$$Q_l^*(A^\sharp \psi_{k'}^l)(x) = \sum_{k \in \Delta_l} \langle A^\sharp \psi_{k'}^l, \varphi_k^l \rangle \tilde{\varphi}_k^l(x) \, ,$$

Lemma 6.1 yields

$$|Q_l^* A^\sharp \psi_k^l(x)| \leq c\, 2^{l(\frac{n}{2}+r)}[1 + 2^l \mathrm{dist}\{x, x_k^l\}]^{-n-r-d^*-1} \, , \tag{6.7}$$

for any fixed $x_k^l \in \mathrm{supp}\, \varphi_k^l$. We infer from the Hölder regularity Equation (6.1) that

$$|\psi_{k'}^{l'}(x) - \psi_{k'}^{l'}(x')| \leq c\, 2^{l'(\frac{n}{2}+\rho)}[\mathrm{dist}\{x, x'\}]^\rho$$

holds. Keeping this in mind and using biorthogonality, we obtain, for $l, l' \geq 0$, $l' \leq l$,

$$\begin{aligned}
&|2^{-\frac{lr}{2}} \langle A\psi_k^l, \psi_{k'}^{l'} \rangle 2^{-\frac{l'r}{2}}| = |2^{-\frac{lr}{2}} \langle A^\sharp \psi_k^l, \psi_{k'}^{l'} \rangle 2^{-\frac{l'r}{2}}| = |2^{-\frac{lr}{2}} \langle Q_{l+1}^* A^\sharp \psi_k^l, \psi_{k'}^{l'} \rangle 2^{-\frac{l'r}{2}}| \\
&\leq c\, 2^{\frac{-r(l+l')}{2}} \Big| \int_\Gamma (\psi_{k'}^{l'}(x) - \psi_{k'}^{l'}(x'))(Q_{l+1}^* A^\sharp \psi_k^l)(x) ds_x \Big| \\
&\leq c\, 2^{\frac{-r(l+l')}{2}} \int_\Gamma 2^{l'(\frac{n}{2}+\rho)} 2^{l(\frac{n}{2}+r)} [\mathrm{dist}\{x, x'\}]^\rho (1 + 2^{l+1}\mathrm{dist}\{x, x_h^{l+1}\})^{-n-r-d^*-1} ds_x \\
&\leq c\, 2^{-|l-l'|(\frac{n-r}{2}+\rho)} \, .
\end{aligned} \tag{6.8}$$

Here we have exploited the fact that $(I - Q_l^*)$ annihilates constant functions.

In the case $r > 0$ we require Hölder continuity of order $m + \rho > \frac{r}{2}$. The proof follows by a modifiction of the above arguments. More details can be found in [21]. ■

Combining Lemma 6.2 with Lemma 5.1 provides the following result.

Lemma 6.3. *Suppose that $n + r + d^* + 1 > 0$. Then there exists a constant c such that, with $l, l' \geq 0$,*

$$|2^{-l'\frac{r}{2}} 2^{-l\frac{r}{2}} \langle A\psi_k^l, \psi_{k'}^{l'} \rangle| \leq c \, \frac{2^{-|l-l'|(\frac{n}{2}+\rho)}}{(1 + 2^{\min\{l,l'\}}\mathrm{dist}\,(\Omega_{(l,k)}, \Omega_{(l',k')}))^{n+d^*+1+r}} . \tag{6.9}$$

To define the compression strategy, we set

$$\mathcal{B}_{l,l'} = \begin{cases} c\max\{\epsilon_1^{-1}2^{-l}, \epsilon_1^{-1}2^{-l'}\} & \text{if } \ |l - l'| < \epsilon_2 \, , \ l, l' \geq 0 \, , \\ \infty & \text{if } \ |l - l'| \geq \epsilon_2 \, , \ l, l' \geq 0, \\ 0 & \text{if } \ l = -1 \text{ or } l' = -1. \end{cases} \tag{6.10}$$

Deleting those matrix entries $\langle A\psi_{(l,k)}, \psi_{(l',k')} \rangle$ for which $\mathcal{B}_{l,l'} \geq \epsilon_1$ results in the compressed matrix $\mathbf{A}_j^\epsilon$ and considering again the operator $A_j^\epsilon = F_j^{-1}\mathbf{A}_j^\epsilon (F_j^*)^{-1}$, it is clear that the compressed matrix $\mathbf{A}_j^\epsilon$ has at most $\mathcal{O}(2^{jn}) = \mathcal{O}(N_j)$ non-zero coefficients.

In analogy with the preceding section and [21] we can prove an approximation and stability result for the operator A_j^ϵ. As a consequence, we derive an error estimate for the solution u_j^ϵ of the compressed scheme $A_j^\epsilon u_j^\epsilon = Q_j^* f$. The proof follows analogously to the reasoning in [21].

Theorem 6.1. *Let $A \in \Psi_{1,0}^r(\Gamma)$ satisfy Equations (2.10) and (2.11) and let $f \in H^{-\frac{r}{2}}(\Gamma)$. For all $\epsilon > 0$, there exist $a \geq 1$, $N_0 \in \mathbb{N}$, and a compressed matrix $\mathbf{A}_j^\epsilon$, defined by Equation (6.10), such that for $u \in H^{-\frac{r}{2}}(\Gamma)$*

$$\|(A_j - A_j^\epsilon)u\|_{-\frac{r}{2}} \leq c\,\epsilon \|u\|_{\frac{r}{2}} \tag{6.11}$$

and

$$\|A_j^\epsilon u_j\|_{-r/2} \geq c\|u_j\|_{r/2} \quad , \quad u_j \in S_j, \tag{6.12}$$

holds for $j \geq N_0$. Moreover, the compressed scheme $A_j^\epsilon u_j^\epsilon = Q_j^* f$ has a unique solution u_j^ϵ which differs from the exact solution of the equation $Au^* = f$ in the $H^s(\Gamma)$ norm by

$$\|u^* - u_j^\epsilon\|_{-\frac{r}{2}} \leq \epsilon \|u^*\|_{\frac{r}{2}} \ , \tag{6.13}$$

where the constant c does not depend on j.

Proof: We will briefly sketch the proof. Let $\mathbf{A}_j^{\epsilon_1}$ to be defined by setting all coefficients in $\mathbf{A}_j$ equal to zero for which

$$\operatorname{dist}(\Omega_{(l,k)}, \Omega_{(l',k')}) > \epsilon_1^{-1} \max\{2^{-l}, 2^{-l'}\}.$$

As a first step, we estimate the spectral norm $\|2^{-(l+l')\frac{r}{2}}(\mathbf{A}_j - \mathbf{A}_j^{\epsilon_1})\|$. To this end, let us define

$$r^{\epsilon_1}_{(l,k),(l',k')} = 2^{(l+l')\frac{-r}{2}}\left((a_{(l,k),(l',k')})_j - (a_{(l,k),(l',k')})_j^{\epsilon_1}\right).$$

Then, for $\mathcal{B}^{\epsilon_1}_{l,l'} := \epsilon^{-1} c \max\{2^{-l}, 2^{-l'}\}$, we apply Lemma 5.1 and majorize the sum

$$\sum_{k \in \nabla_l} |r^{\epsilon_1}_{(l,k),(l',k')}| = \sum_{\{k \in \nabla_l : \operatorname{dist}(\Omega_{(l,k)}, \Omega_{(l',k')}) > \mathcal{B}^{\epsilon_1}_{l,l'}\}} |a_{(l,k),(l',k')}| 2^{(l+l')\frac{-r}{2}} \tag{6.14}$$

by an integral as in the proof of Lemma 5.2

$$\begin{aligned}
&\sum_{k \in \nabla_l} |r^{\epsilon_1}_{(l,k),(l',k')}| \\
&\leq c\, 2^{-(l+l')(d^*+1+\frac{n+r}{2})} \times \\
&\qquad \sum_{\{k \in \nabla_l : \operatorname{dist}(\Omega_{(l,k)}, \Omega_{(l',k')}) > \mathcal{B}^{\epsilon_1}_{l,l'}\}} (\operatorname{dist}(\Omega_{(l,k)}, \Omega_{(l',k')}))^{-n-2(d^*+1)-r} \\
&\leq c\, 2^{-(l+l')(d^*+1+\frac{n+r}{2})} 2^{ln} (\mathcal{B}^{\epsilon_1}_{l,l'})^{-2(d^*+1)-r} \\
&\leq c\, \epsilon_1^{2(d^*+1)+r} 2^{(2(d^*+1)+r)\min\{l,l'\}} 2^{ln} 2^{-(l+l')(d^*+1+\frac{n+r}{2})} \\
&\leq c\, \epsilon_1^{2(d^*+1)+r} 2^{ln} 2^{-n\frac{l+l'}{2}} 2^{-|l-l'|(d^*+1+\frac{r}{2})} \ .
\end{aligned} \tag{6.15}$$

Now we continue as in Lemma 5.3,

$$\sum_{l=0}^{j}\sum_{k\in\nabla_l} 2^{\frac{-ln}{2}}|r^{\epsilon_1}_{(l,k),(l',k')}| \leq c\sum_{l=0}^{j} 2^{\frac{-l'n}{2}}2^{-|l-l'|(\frac{r}{2}+d^*+1)}\epsilon_1^{2(d^*+1)+r}$$
$$\leq c\,2^{\frac{-l'n}{2}}\epsilon_1^{r+2(d^*+1)} . \tag{6.16}$$

In a similar way as in Theorem 5.1 we now infer from Equation (6.16) that

$$\|2^{-(l+l')\frac{r}{2}}(\mathbf{A}_j - \mathbf{A}_j^{\epsilon_1})\| \leq \max_{\{0\leq l'<j,k'\in\nabla_{l'}\}}\sum_{l=0}^{j}\sum_{k\in\nabla_l} 2^{\frac{(l'-l)n}{2}}|r^{\epsilon_1}_{(l,k),(l',k')}|$$
$$+ \max_{\{0\leq l<j,k\in\nabla_l\}}\sum_{l'=0}^{j}\sum_{k'\in\nabla_{l'}} 2^{\frac{(l-l')n}{2}}|r^{\epsilon_1}_{(l,k),(l',k')}|$$
$$\leq c\,\epsilon_1^{2(d^*+1)+r}. \tag{6.17}$$

As we have already observed, the matrix $\mathbf{A}_j^{\epsilon_1}$ has still $\mathcal{O}(j^2 2^{jn})$ non-zero coefficients. Next we discard all coefficients in $\mathbf{A}_j^{\epsilon_1}$ arising from levels far apart from each other, more precisely those for which $|l-l'| \geq \epsilon_2$, $l, l' \geq 0$, to obtain a second matrix $\mathbf{A}_j^{\epsilon_2}$. It is easy to see that this matrix has only $\mathcal{O}(2^{jn})$ non-zero entries. Note that we do not affect any coefficients containing a scalar product with a function without vanishing moments. Investigating the perturbation

$$\sum_{k\in\nabla_l}|r^{\epsilon_2}_{(l,k),(l',k')}| = \sum_{\{k\in\nabla_l:\mathrm{dist}(\Omega_{(l,k)},\Omega_{(l',k')})\leq\mathcal{B}^{\epsilon_1}_{l,l'}\}} |a_{(l,k),(l',k')}|2^{(l+l')\frac{-r}{2}}$$
$$\leq \sum_{\{k\in\nabla_l:\mathrm{dist}(\Omega_{(l,k)},\Omega_{(l',k')})>\mathcal{B}^{\epsilon_1}_{l,l'}\}} 2^{-|l-l'|(d^*+1+\frac{n+r}{2})} , \tag{6.18}$$

we estimate the number of terms in the sum of Equation (6.18) by

$$c\,(\mathcal{B}^{\epsilon_1}_{l,l'})^n 2^{ln} \leq c\,2^{ln}2^{-n\min\{l,l'\}}.$$

Therefore, we have

$$\sum_{k\in\nabla_l}|r^{\epsilon_2}_{(l,k),(l',k')}| \leq c\,2^{-|l-l'|(d^*+1+\frac{n+r}{2})}2^{ln}2^{-n\min\{l,l'\}} . \tag{6.19}$$

Applying now Schur's lemma yields

$$\|\mathbf{A}_j^{\epsilon_2} - \mathbf{A}_j^{\epsilon_1}\| \leq \max_{\{0\leq l'<j,k'\in\nabla_{l'}\}}\sum_{\{l=0,|l-l'|\geq\epsilon_2\}}^{j}\sum_{k\in\nabla_l} 2^{\frac{(l'-l)n}{2}}|r^{\epsilon_1}_{(l,k),(l',k')}|$$
$$+ \max_{\{0\leq l<j,k\in\nabla_l\}}\sum_{\{l'=0,|l-l'|\geq\epsilon_2\}}^{j}\sum_{k'\in\nabla_{l'}} 2^{\frac{(l-l')n}{2}}|r^{\epsilon_1}_{(l,k),(l',k')}|$$

$$\leq c \Big(\sum_{l \leq l' - \epsilon_2} \epsilon_1^{2(d^*+1)+r} 2^{(l-l')(\rho)} + \sum_{l > l' + \epsilon_2} \epsilon_1^{2(d^*+1)+r} 2^{(l'-l)(\rho)} \Big)$$

$$\leq c\, 2^{-\epsilon_2 \rho} \epsilon_1^{2(d^*+1)+r} \leq c\, \epsilon\,. \tag{6.20}$$

In this way we have obtained a compressed matrix $\mathbf{A}_j^\epsilon$ with $\mathcal{O}(N_j) = \mathcal{O}(2^{jn})$ non-zero coefficients. In view of Equation (6.20), the corresponding operator $A_j^\epsilon = F_j^{-1} \mathbf{A}_j^\epsilon (F_j^*)^{-1}$ satisfies the following estimate

$$\begin{aligned} \|(A_j - A_j^\epsilon)u_j\|_{\frac{-r}{2}} &\leq c\, \|(\Upsilon_j^*)^{\frac{-r}{2}} (A_j - A_j^\epsilon) \Upsilon_j^{\frac{-r}{2}} \Upsilon_j^{\frac{r}{2}} u_j\|_0 \\ &\leq c\, \|2^{-(l+l')\frac{r}{2}} (\mathbf{A}_j^{\epsilon_2} - \mathbf{A}_j)\| \|\Upsilon_j^{\frac{r}{2}} u_j\|_0 \leq c\, \epsilon \|u_j\|_{\frac{r}{2}}\,. \end{aligned}$$

The parameters $\mathcal{B}_{l,l'}^{\epsilon_1}, E_2$ can be choosen such that ϵ becomes appropriately small. A perturbation argument (see [21]), ensures the stability of the perturbated operator A_j^ϵ and the reasoning in the proof of Theorem 5.4 yields the desired result in Equation (6.13). ■

Acknowledgments. The second author has been supported by a grant of Deutsche Forschungsgemeinschaft under grant number Pr 336/2-1.

References

1. Alpert, B., G. Beylkin, R. Coifman, and V. Rokhlin, Waveletlike bases for the fast solution of second-kind integral equations *SIAM J. Sci. Statist. Comp.* **14** (1993), 159–184.
2. Andersson, L., N. Hall, B. Jawerth, and G. Peters: Wavelets on closed subsets of the real line, in *Recent Advances in Wavelet Analysis* L.L. Schumaker, G. Webb (eds.), Academic Press, Boston,1993, 1–63.
3. Beylkin, G., R. Coifman, and V. Rokhlin, Fast wavelet transforms and numerical algorithms I, *Comm. Pure and Appl. Math.* **44** (1991), 141–183.
4. de Boor, C., R. A. DeVore, and A. Ron, On the construction of multivariate (pre–) wavelets, *Constructive Approximation* **9** (1993), 123–166.
5. Carnicer, J. M., W. Dahmen, J. M. Peña, Local decomposition of refinable spaces, in preparation.
6. Cavaretta, A. S., W. Dahmen, and C. A. Micchelli, Stationary Subdivision, *Memoirs of the American Math. Soc.* **93** (453) (1991).
7. Chui, C. K. and E. Quak, Wavelets on a bounded interval, in *The Numerical Methods of Approximation Theory*, D. Braess and L. Schumaker (eds.), Birkhäuser, Basel, 1992.
8. Chui, C. K., *An Introduction to Wavelets*, Academic Press, 1992.
9. Ciarlet, P. G., *The Finite Element Method for Elliptic Problems*, North Holland, 1978.

10. Cohen, A., Rapport Scientifique, Université Paris IX Dauphine, June 1992.
11. Cohen, A. and I. Daubechies, Non–separable bidimensional wavelet bases, AT&T Bell Laboratories, New Jersey, 1991, preprint.
12. Cohen, A., I. Daubechies, and J.-C. Feauveau, Biorthogonal bases of compactly supported wavelets, *Comm. Pure and Appl. Math.* **45** (1992), 485–560.
13. Cohen, A., I. Daubechies, and P. Vial, Wavelets on the interval and fast wavelet transforms, *Appl. and Comp. Harmonic Anal.* **1** (1993), 54–81.
14. Coifman, R., Adapted multiresolution analysis, computation, signal processing and operator theory, *Proc. of the Int. Congress of Mathematicians*, Kyoto, 1990, 879–887.
15. Dahmen, W., Some remarks on multiscale transformations, stability and biorthogonality, in *Curves and Surfaces*, P. J. Laurent, A. LeMéhauté, L. L. Schumaker (eds.), 1994, to appear.
16. Dahmen, W., R. A. DeVore, and K. Scherer, Multidimensional spline approximation, *SIAM J. Numer. Anal.* **17** (1980), 380–402.
17. Dahmen, W. and A. Kunoth, Multilevel preconditioning, *Numer. Math.* **63** (1992), 315–344.
18. Dahmen, W. and C. A. Micchelli, Biorthogonal wavelet expansions, in preparation
19. Dahmen, W., P. Oswald, and X. Q. Shi, C^1 conforming hierarchical bases, *J. Comp. and Appl. Math.* **9** (1993), 263–281.
20. Dahmen, W., S. Prössdorf, and R. Schneider, Wavelet approximation methods for pseudodifferential equations I: Stability and convergence, *Math. Z.*, to appear.
21. Dahmen, W., S. Prössdorf, and R. Schneider, Wavelet approximation methods for pseudodifferential equations II: Matrix compression and fast solution, *Advances in Comp. Math.*, **1** (1993), 259–335.
22. Dahmen, W., S. Prössdorf, and R. Schneider, Multiscale methods for pseudo-differential equations, in *Recent Advances in Wavelet Analysis*, L.L. Schumaker, G. Webb (eds.), Academic Press, Boston, 1993, 191–235.
23. Dahmen, W., B. Kleemann, S. Prössdorf, and R. Schneider, A multiscale method for the double layer potential equation on a polyhedron, in *Advances of Computational Mathematics*, H. P. Dikshit, C. A. Micchelli (eds.), World Scientific, New Jersey, 1994.
24. Daubechies, I., Orthonormal bases of compactly supported wavelets, *Comm. Pure and Appl. Math.* **41** (1988), 909–996.
25. Jaffard, S., Wavelet methods for fast resolution of elliptic problems, *SIAM J. Numer. Anal.* **29** (1992), 965–987.
26. Jia, R. Q. and C. A. Micchelli, Using the refinement equation for the construction of pre-wavelets II: Powers of two, in *Curves and Surfaces*, P. J. Laurent, A. LeMéhauté, L. L. Schumaker (eds.), Academic Press, New York, (1991), 209–246.
27. Harten, A. and I. Yad-Shalom, Fast multiresolution algorithms for matrix-vector multiplication, ICASE Report No. 92–55, October 1992.

28. Harten, A., Multiresolution representation of data, CAM Report 93–13, UCLA, 1993.
29. Harten, A., Multiresolution algorithms for the numerical solution of Hyperbolic conservation laws, Courant Math. and Comp. Lab. Report, 1993.
30. Hildebrandt, S. and E. Wienholtz, Constructive proofs of representation theorems in seperable Hilbert spaces *Comm. and Pure Appl. Math.* **17** (1964), 369–373.
31. Hörmander, L., *Analysis of Linear Partial Differential Operators,* Grundlehren Series, Springer Verlag, 1985.
32. Kumano-go, H., *Pseudodifferential Operators,* MIT Press, Boston, 1981.
33. Kunoth, A., Multilevel preconditioning, Ph. D. Thesis, FU Berlin, 1994.
34. Meyer, Y. *Ondelettes et Opérateurs 2 : Opérateur de Caldéron-Zygmund,* Hermann, Paris, 1990.
35. Mallat, S. G., Multiresolution approximation and wavelet orthonormal basis of $L_2(\mathbb{R})$, *Trans. Amer. Math. Soc.* **315** (1989), 69–87.
36. Micchelli, C. A. and Y. Xu, Using the refinement equation for the construction of wavelets on invariant sets, April 1994, preprint.
37. von Petersdorf, T. and C. Schwab, Wavelet Approximation for first kind boundary integral equations on polygons, Techn. Note BN-1157, Institute for Physical Science and Technology, University of Maryland at College Park, February 1994.
38. von Petersdorf, T., R. Schneider, and C. Schwab, Multi-wavelet approximation for the double layer potential equation, in preparation.
39. M. A. Shubin, *Pseudodifferential Operators and Spectral Theory*, Springer Verlag, 1985.
40. Triebel, H., *Interpolation Theory, Function Spaces and Differential Operators*, North-Holland, Amsterdam, 1978.
41. Yserentant, H., On the multilevel splitting of finite element spaces. *Numer. Math.* **49** (1986), 379–412.

Wolfgang Dahmen
Institut fü r Geometrie und Praktische Mathematik
RWTH Aachen
52056 Aachen, Germany
dahmen@igpm.rwth-aachen.de

Siegfried Prössdorf
Institut für Angewandte Analysis und Stochastik
10117 Berlin, Germany
proessdorf@iaas-berlin.d400.de

Reinhold Schneider
Fachbereich Mathematik TH Darmstadt
64289 Darmstadt, Germany
schneider@mathematik.th-darmstadt.de

Wavelet Methods for the Numerical Solution of Boundary Value Problems on the Interval

Silvia Bertoluzza, Giovanni Naldi, and Jean Christophe Ravel

Abstract. We will describe some results on the numerical solution of differential equations on the interval, in particular with regard to the treatment of boundary conditions. We will concentrate on the elliptic case where we will test some methods specially suited to treat Dirichlet's boundary conditions. Among such methods we will describe a Galerkin method based on the wavelets on the interval of [9] and a wavelet collocation method. For both methods we give some error estimates, along with some numerical results.

§1 Introduction

In recent years, the application of methods based on wavelets or on multiresolution analisys to the numerical solution of PDEs, have been studied in several papers both from the theoretical and the computational point of view. Wavelet methods have been used in order to get a class of self adaptive schemes which have been successfully applied to the numerical solution of equations of nonlinear type ([16,17]).

There are several reasons that seem to make of such methods a good instrument for the numerical solution of PDEs.

First of all they allow to construct local high order schemes. This, in particular, implies that few degrees of freedom (d.o.f.) are necessary when (and where) the solution is smooth and one can concentrate an high percentage of the degrees of freedom near singularities. Moreover Gibbs phenomena will be concentrated near singularities and will not spread all over the domain.

The key for self adaptivity are the good localization properties that wavelets display both in space and frequency. In an evolution problem, thanks to such a property one has the possibility of predicting easily, from the solution at a given time-step, a restricted range of d.o.f. that will be enough to represent the solution at the next time-step up to a certain precision. In this way one

Wavelets: Theory, Algorithms, and Applications
Charles K. Chui, Laura Montefusco, and Luigia Puccio (eds.), pp. 425–448.

ISBN 0-12-174575-9

will have to solve a much smaller linear system at each time-step. Moreover it has been shown in several papers ([7,15]) that there exists a preconditioning technique based on a diagonal preconditioner such that the condition number of the preconditioned matrix is uniformly bounded.

Another good reason to use wavelet based methods for PDEs is that we have an whole set of fast algorithms available. Among such algorithms, which are mainly based on the fast wavelet transform, we recall for instance the algorithms for fast matrix-vector and fast matrix-matrix multiplication ([7]). The fast wavelet transform itself may be used in the framework of a multigrid type method, to build computationally efficient restriction and prolongation operators ([10]).

On the other hand, some difficulties arise when applying such a method to problems in bounded domains, in particular, the problem of an efficient treatment of boundary conditions of Dirichlet type, even in the homogeneous case. In fact, wavelets are originally a basis on the whole real line (or for periodic functions). If one wants to use them in the solution of a boundary value problem on the interval by simply taking the restriction of each basis function to the interval, some instability problems appear so one needs to pay extra care.

This is not the only difficulty to overcome in the application of wavelet based methods to PDEs. Other problems come for instance from the fact that until now the non-linear terms have been computed in the physical space and projected back to the wavelet coefficient space via some quadrature formula, a technique that slows the whole procedure, reducing drastically its efficiency. Non-rectangular domains in higher dimensions are also difficult to treat, since wavelets in higher dimensions are mainly constructed as tensor product of one-dimensional wavelets. However, in this paper, we will concentrate on the problem of treating Dirichlet boundary conditions. Several techniques are available and already experimented in the finite elements framework to impose boundary conditions either exactly ([20]) or approximately. For instance the Dirichlet problem may be reduced to a Neumann problem where the boundary conditions are imposed as Lagrange multipliers ([1]) or by means of penalty techniques ([2]).

We will study in this paper two methods based on two different approaches, in both of which the approximate solution satisfies the boundary condition exactly. The first technique, which is described in Section 2, is based on the multiresolution analysis on the interval constructed in [9]. This is a clever way to take the restriction to the interval of a compactly supported MRA on the line. In practice one keeps the basis functions whose support is internal to the interval and constructs in addition some edge functions which allow to have the same order of accuracy that we have on the line, and to impose in an easy way a boundary condition of Dirichlet type without introducing instabilities. After constructing such approximation spaces, one simply uses a classical Galerkin method in order to compute the approximate solution.

In Section 3, we present a second approach, based not on the Daubechies orthonormal scaling functions, but on their autocorrelation function. Such function generates a new multiresolution analysis in which an interpolation operator is easily defined (and computed). In such an approximation space we look for an approximate solution by means of a collocation technique ([5]), (that is we look for the function in the approximation space that verifies the differential equation exactly at some "collocation points"). Theoretical results on collocation schemes are not so easily obtained as in the Galerkin case. We also include, in this section, a stability and convergence result for one particularly simple formulation of the collocation scheme. Unfortunatly in such a case we have a minimal order of convergence. It seems reasonable to think that an higher order may be attained by modifying the scheme in a suitable way, as shown in Subsection 3.2, but we did not yet prove any theoretical results on such modified collocation schemes.

The two methods described seem, at first, completely unrelated. However this is not true. In fact, it is possible to prove (see Remark 2) that the two stiffness matrices are equal excepted for the terms involving edge functions (or points). In effect, if the two methods (Galerkin method with a Daubechies orthonormal basis and collocation with its autocorrelation function) were applied to a problem on the whole real line, one would get exactly the same discrete scheme (with two different continuous interpretation) ([5]). In a certain sense we can then say that the two methods are two different ways to impose boundary conditions within the same discrete scheme.

We remark that the good features of wavelet methods described previously (that is possiblity of applying adaptive techniques, preconditioning, etc.) apply also to the methods described here, though we did not consider these items in this paper, where we concentrated just on the problem of boundary conditions.

1.1 Notations and model problem

Let us first introduce some notation: with $\|\cdot\|_s$ and $|\cdot|$ we will indicate respectively the norm and seminorm of $H^s(\mathbb{R})$ while $\|\cdot\|_{s,\Omega}$ and $|\cdot|_{s,\Omega}$ will indicate respectively the $H^s(\Omega)$ norm and seminorm. Somewhere we will write $\|\cdot\|_{s,\mathbb{R}}$ to underline that we are taking the norm all over $\mathbb{R}$. $\langle\,\cdot,\cdot\,\rangle$ will stand for the scalar product in $L^2(\mathbb{R})$.

For the sake of simplicity we will concentrate on the following model problem.

$$\begin{cases} -u'' = f & \text{in } (0,1) \\ u(0) = a, u(1) = b\ . \end{cases} \tag{1.1}$$

In the following, ϕ will be any of the compactly supported scaling functions introduced in [12] and verifying the orthonormality properties $\langle\phi(x),\phi(x-n)\rangle = \delta_{0n}$. L will be the length of the support of ϕ, R, its regularity (*i.e.*, $\phi \in H^R$), while $M+1$ will be the number of zero moments of the associated wavelet function ψ (that is $\int x^k\psi(x)\,dx = 0$, $k = 0,..,M$).

ϕ will be supposed to verify the following dilation equation

$$\phi(x) = \sum_{k=0}^{L} h_k \sqrt{2}\phi(2x - k). \tag{1.2}$$

Lastly, C will stand for several constant, independent of j.

§2 Galerkin approach

2.1 Multiresolution analysis on the interval

The first approach that we want to describe for the solution of our problem is a Galerkin method based on the multiresolution analysis on the interval constructed in [9]. We first begin by describing such a construction. We start with any multiresolution analysis on $L^2(\mathbb{R})$ in [12,13] with M+1 vanishing moments and supp $\phi =$ supp $\psi = [0, L]$. We want to construct an increasing sequence V_j of closed subspaces of $L^2((0,1))$ retaining all the good properties of the multiresolution analysis of $L^2(\mathbb{R})$. Among such properties we recall accuracy, *i.e.*, we want the subspaces V_j to contain the restriction to $(0,1)$ of polynomials up to order M.

A first solution is the one proposed by Y. Meyer ([19]); it consists in defining $V_j(0,1)$ as

$$V_j(0,1) = V_j|_{(0,1)}\,.$$

An orthonormal basis is constructed by orthonormalizing the family

$$\{\phi_{jk}|_{(0,1)}, \quad k = -L+1, 2^j - 1\}$$

which is proven to be linearly independent. Such a space $V_j(0,1)$ works well when applied to a problem with Neumann's type boundary conditions but when applied to the solution of a Dirichlet BVP it appears to generate some instability problems.

The construction of [9] which we will now briefly describe, seems better suited for our kind of problems. Let us first introduce some notation: we write

$$M_{nk} := \int_{\mathbb{R}} x^n \phi(x - k) \tag{2.1}$$

$$\widetilde{p_j^n} := 2^{j/2}(2^j x)^n \tag{2.2}$$

$$\bar{p}_j^n := 2^{j/2}(2^j(x-1))^n. \tag{2.3}$$

In such a construction, one begins by including in V_j all the interior scaling functions that are ϕ_{jk}, $k = 0, 2^J - L$.

In order to have the same accuracy as the multiresolution on the whole line, we have to add in a certain sense the polynomials up to order M to our multiresolution analysis. This is done by defining some edge functions which will enable us to recover polynomials and, hence accuracy. More precisely we define on the lefthand side

$$\begin{aligned}\widetilde{\phi}_j^n &= \widetilde{p}_j^n - \sum_{k=0}^{+\infty} \langle\, \widetilde{p}_j^n, \phi_{jk}\,\rangle \phi_{jk} \\ &= \widetilde{p}_j^n - \sum_{k=0}^{+\infty} M_{nk}\phi_{jk} = \sum_{k=-\infty}^{-1} M_{nk}\phi_{jk},\end{aligned} \tag{2.4}$$

and on the righthand side

$$\begin{aligned}\bar{\phi}_j^n &= \bar{p}_j^n - \sum_{-\infty}^{2^j-L} \langle\, \bar{p}_j^n, \phi_{jk}\,\rangle \phi_{jk}, \\ &= \bar{p}_j^n - \sum_{k=-\infty}^{-L} M_{nk}\phi_{j,2^j+k} = \sum_{k=1-L}^{+\infty} M_{nk}\phi_{j,2^j+k}\end{aligned} \tag{2.5}$$

where the last equalities in Equations (2.4) and (2.5) are obtained by a simple change of variable.

We can now construct our space $V_j(0,1)$ as

$$V_j(0,1) = span < \phi_{jk},\ k = 0, 2^j - L,\ \widetilde{\phi}_j^n,\ n = 0, M,\ \bar{\phi}_j^n,\ n = 0, M >$$

with the restriction that j has to be large enough that the construction on the right and the construction on the left do not interfere with each other, namely, $2^{-j}L < 1/2$.

Then the spaces $V_j(0,1)$ form a multiresolution analysis of $L^2(0,1)$ in the following sense:

$$V_j(0,1) \subset V_{j+1}(0,1) \subset L^2(0,1)$$
$$\overline{\cup V_j(0,1)} = L^2(0,1).$$

The set $\{\phi_{jk},\ k = 0, 2^j - L,\ \widetilde{\phi}_j^n,\ n = 0, M,\ \phi_j^n,\ n = 0, M\}$ forms a Riesz's basis for $V_j(0,1)$. In particular, one can prove the following theorem [9].

Theorem 2.1. *There exist constants a_{nl}, b_{nk}, and c_{nl}, d_{nk} independent of j such that*

$$\widetilde{\phi}_j^n = \sum_{l=0}^{M} a_{nl}\widetilde{\phi}_{j+1}^l + \sum_{k=0}^{2L-3} b_{nk}\phi_{j+1,k} \tag{2.6}$$

and

$$\bar{\phi}_j^n = \sum_{l=0}^{M} c_{nl}\bar{\phi}_{j+1}^l + \sum_{k=0}^{2L-3} d_{nk}\phi_{j+1,k}. \tag{2.7}$$

Practically the functions $\widetilde{\phi}_j^k$, ϕ_{jk}, and $\bar{\phi}_j^k$ play the role of scaling functions of our multiresolution analysis. In the interior of the interval, we have the same

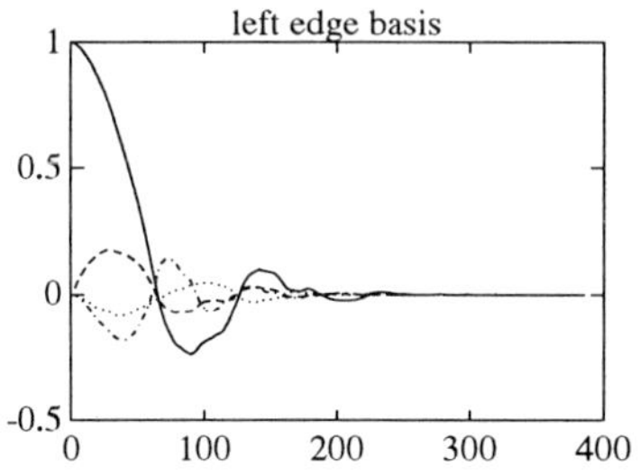

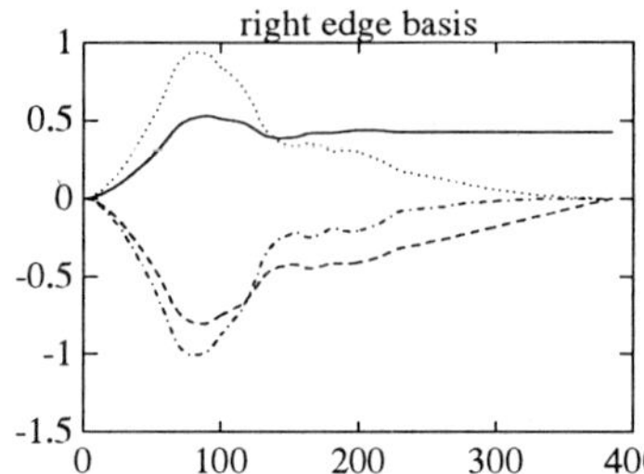

Figure 1. In this fugure we show the basis functions $\widetilde{\phi}^n$ and $\bar{\phi}^n$ obtained by the construction described in Subsection 2.1 for the Daubechies scaling functions used in the numerical tests (see Section 4).

basis functions as we have on the real line. Close to the boundary we have some modified functions (namely the $\widetilde{\phi}$'s on the left and the $\bar{\phi}$'s on the right) that allow us to recover polynomials. At this stage one could orthonormalize the $\widetilde{\phi}$'s and the $\bar{\phi}$'s (which are already orthogonal to the ϕ's). However, we are not particularly interested to an orthonormal basis but rather to a basis well suited to impose boundary values. The $\widetilde{\phi}$'s and $\bar{\phi}$'s are then a good choice since one obtains trivially

$$\widetilde{\phi}^0_j(0) = 1, \qquad \widetilde{\phi}^n_j(0) = 0, \quad n \neq 1 \tag{2.8}$$

$$\bar{\phi}^0_j(1) = 1, \qquad \bar{\phi}^n_j(1) = 0, \quad n \neq 1. \tag{2.9}$$

In [9], the construction is extended to wavelet functions and then to having a complete multiresolution analysis on the interval. Many of the results that we have on wavelet on the line carry over to such a construction of wavelet on the interval. In particular, regarding accuracy the following estimate can be easily proven.

Proposition 2.2. *Let $P_j : L^2(0,1) \to V_j(0,1)$ be the L^2 projection. Then $\forall\ t$ $t \leq s$, $0 \leq t \leq R$, $0 \leq s \leq M+1$, $f \in H^s(0,1)$ implies*

$$||f - P_j f||_{t,(0,1)} \leq 2^{-j(s-t)} |f|_{s,(0,1)}.$$

2.2 The Galerkin method

Let us now come back to the numerical solution of our model problem:

$$\begin{cases} -u'' = f & \text{in } (0,1) \\ u(0) = a, & u(1) = b. \end{cases} \tag{2.10}$$

In a Galerkin type approach, we first introduce the spaces $V_j^{bc} = \{u \in V_j(0,1)|\ u(0) = a, u(1) = b\}$ of functions in V_j satisfying the boundary conditions, and $V_j^0 = \{u \in V_j(0,1)|u(0) = u(1) = 0\}$. It is easy to verify that a function in V_j^{bc} may be expressed as

$$u_j = a\widetilde{\phi}_j^0 + b\bar{\phi}_j^0 + \sum_{n=1}^{M} \alpha_{jn}\widetilde{\phi}_j^n + \sum_{n=1}^{M} \beta_{jn}\bar{\phi}_j^n + \sum_{k=0}^{2^j-L} \gamma_{jk}\phi_{jk} \tag{2.11}$$

while the functions of V_j^0 have the form

$$v_j = \sum_{n=1}^{M} \alpha_{jn}\widetilde{\phi}_j^n + \sum_{n=1}^{M} \beta_{jn}\bar{\phi}_j^n + \sum_{k=0}^{2^j-L} \gamma_{jk}\phi_{jk}. \tag{2.12}$$

We look for an approximate solution $u_j \in V_j^{bc}$ verifying

$$\int_0^1 u_j' v_j' = \int_0^1 f v_j, \quad \forall\ v_j \in V_j^0. \tag{2.13}$$

The problem then reduces to a linear system of the form

$$M\underline{u} = \underline{b} \tag{2.14}$$

where the unknown $\underline{u}$ and the righthandside $\underline{b}$ are, respectively,

$$\underline{u} = \begin{pmatrix} \alpha_{j1} \\ \cdot \\ \cdot \\ \alpha_{jM} \\ \gamma_{j0} \\ \cdot \\ \cdot \\ \cdot \\ \gamma_{2^j-L} \\ \beta_{j1} \\ \cdot \\ \cdot \\ \beta_{jM} \end{pmatrix} \quad \text{and} \quad \underline{b} = \begin{pmatrix} \int_0^1 f\widetilde{\phi}_j^1 - a\int_0^1 (\widetilde{\phi}_j^0)'(\widetilde{\phi}_j^1)' \\ \cdot \\ \cdot \\ \int_0^1 f\widetilde{\phi}_j^M - a\int_0^1 (\widetilde{\phi}_j^0)'(\widetilde{\phi}_j^M)' \\ \int_0^1 f\phi_{j0} \\ \cdot \\ \cdot \\ \cdot \\ \int_0^1 f\phi_{j,2^j-L} \\ \int_0^1 f\bar{\phi}_j^1 - b\int_0^1 (\bar{\phi}_j^0)'(\bar{\phi}_j^1)' \\ \cdot \\ \cdot \\ \int_0^1 f\bar{\phi}_j^m - b\int_0^1 (\bar{\phi}_j^0)'(\bar{\phi}_j^M)' \end{pmatrix} \tag{2.15}$$

The approximate solution is then obtained by plugging in Equation (2.11) the coefficients obtained by solving the linear system of Equation (2.14).

The approximate solution verifies the standard estimates for Galerkin methods. In particular, we have $\forall f \in H^s(0,1)$, $s \leq M+1$,

$$||u - u_j||_1 \leq C||u - P_j u||_1 \leq C2^{-j(s-1)}|u|_s\ . \tag{2.16}$$

2.3 Computation of the stiffness matrix

Let us now come to the problem of computing the stiffness matrix M in an efficient way. As is usual with wavelets, one tries to compute some fundamental quantities which will be used as bricks for the computations in each problem.

In particular, we will concentrate here on an efficient computation of the stiffness matrix relative to the test problem in Eqaution (2.10). The technique we are going to show applies, however, to a much larger class of differential equations [11].

The stiffness matrix in Equation (2.14) has the following structure

$$M = \begin{pmatrix} B & C^T & 0 & 0 & 0 \\ C & & & & 0 \\ 0 & & D & & 0 \\ 0 & & & & E \\ 0 & 0 & 0 & E^T & G \end{pmatrix}$$

where

$$B = (b_{nm}); \qquad b_{nm} = \int_0^1 \frac{d}{dx}\tilde{\phi}_j^m \frac{d}{dx}\tilde{\phi}_j^n$$

$$C = (c_{nm}); \qquad c_{nm} = \int_0^1 \frac{d}{dx}\tilde{\phi}_j^m \frac{d}{dx}\phi_{jn}$$

$$D = (d_{nm}); \quad d_{nm} = \int_0^1 \frac{d}{dx}\phi_{jm} \frac{d}{dx}\phi_{jn}$$

$$G = (g_{nm}); \qquad g_{nm} = \int_0^1 \frac{d}{dx}\bar{\phi}_j^m \frac{d}{dx}\bar{\phi}_j^n$$

$$E = (e_{nm}); \qquad e_{nm} = \int_0^1 \frac{d}{dx}\bar{\phi}_j^m \frac{d}{dx}\phi_{jn}.$$

The elements of D are easily computed by remarking that, since supp $\phi_{jn} \subset [0,1]$ we have

$$\int_0^1 \phi'_{jn}\phi'_{jm} = \int_{\mathbf{R}} \phi'_{jn}\phi'_{jm} = 2^{2j} \int_{\mathbf{R}} \phi'(x)\phi'(x-(n-m))\,dx.$$

The last integral can be computed following [6].

We will briefly describe here the computation of the elements of B which (together with G) is the most complicated case, since it involves the computation of a definite integral. We have

$$\begin{aligned}
b_{nm} &= \int_0^1 \frac{d}{dx}\tilde{\phi}_j^m \frac{d}{dx}\tilde{\phi}_j^n \\
&= \int_0^{L2^{-j}} \frac{d}{dx}\tilde{\phi}_j^m \frac{d}{dx}\tilde{\phi}_j^n \\
&= \int_0^{L2^{-j}} \left(\sum_{l=-\infty}^{-1} M_{ml}\phi'_{jl}\right)\left(\sum_{k=-\infty}^{-1} M_{nk}\phi'_{jk}\right) \\
&= \int_0^{L2^{-j}} \left(\sum_{l=-L+1}^{-1} M_{ml}\phi'_{jl}\right)\left(\sum_{k=-L+1}^{-1} M_{nk}\phi'_{jk}\right) \\
&= \sum_{l=-L+1}^{-1}\sum_{k=-L+1}^{-1} M_{ml}M_{nk}\int_0^{L2^{-j}} \phi'_{jl}\phi'_{jk} \\
&= \sum_{l=-L+1}^{-1}\sum_{k=-L+1}^{-1} M_{ml}M_{nk}2^{2j}\int_0^{L2^{-j}} 2^j\phi'(2^jx-l)\phi'(2^jx-k) \\
&= \sum_{l=-L+1}^{-1}\sum_{k=-L+1}^{-1} 2^{2j}M_{ml}M_{nk}\int_0^{L} \phi'(x-l)\phi'(x-k).
\end{aligned}$$

Hence we see that the dependence in j appears only through the factor 2^{2j}. This is easily verified also for the matrices C, E, and G. Then these matrices are computed and stored. As j changes the matrix M is computed by assembling the rescaled submatrices B, C, D, E, and G.

To compute the submatrices one need to compute integrals of the type

$$\int_k^{k+1} \phi'(x-n)\phi'(x-m), \quad k,n,m \in \mathbf{Z}.$$

Such a problem reduces by a simple change of variable to the problem of computing

$$\int_0^1 \phi'(x-n)\phi'(x-m), \quad n,m \in \mathbf{Z}.$$

The computation of such integrals is discussed in [11]. Let us briefly describe their result. First of all we remark that, denoting $\chi = \chi_{[0,1]}$ we have

$$\int_0^1 \phi'(x-n)\phi'(x-m) = \int_{\mathbf{R}} \chi(x)\phi'(x-n)\phi'(x-m).$$

We now apply the two refinement equations

$$\chi(x) = \chi(2x) + \chi(2x-1)$$

and

$$\phi'(x) = \sum_{n=0}^{N} h_n 2\sqrt{2}\phi'(2x-n).$$

With a suitable change of variable we get

$$\int_{\mathbf{R}} \chi(x)\phi'(x-n)\phi'(x-m) =$$
$$8\sum_{i=0}^{1}\sum_{k=0}^{N}\sum_{l=0}^{M} h_k h_l \int_{\mathbf{R}} \chi(y)\phi'(y-(2n+k-i))\phi'(y-(2m+k-i))\,dy.$$

Such a relation may be seen as an eigenvalue-eigenvector relation and the eigenverctor corresponding to the eigenvalue 1/8 which is unique (see [11]) gives us the integrals we are looking for.

§3 Collocation approach

3.1 The rough method

First of all let us briefly describe the multiresolution setting in which we will state the collocation method. Once again we start by any of the compactly supported MRA of [12,13]. This time we will construct a new MRA on the line by taking the autocorrelation function of the scaling function ϕ. We then introduce the function

$$\theta(x) = \phi(\cdot) * \phi(-\cdot)(x) = \int_{-\infty}^{+\infty} \phi(y)\phi(y-x)\,dy. \tag{3.1}$$

Such a function has been studied in several works ([4,7,14]). Its main feature, which makes of it a good candidate for the design of a collocation method, is what we could call *interpolation property* at integers, that is

$$\theta(0) = 1, \qquad \theta(n) = 0, \quad n \neq 0 \tag{3.2}$$

which descends trivially from the orthonormality of the system $\{\phi(x-n)\}$. Defining $\widetilde{V}_j$ as

$$\widetilde{V}_j = span < \theta_{jk}, \quad k \in \mathbf{Z} >$$

where

$$\theta_{jk}(x) = 2^{j/2}\theta(2^j x - k), \tag{3.3}$$

we have a new multiresolution analysis of $\mathbb{R}$. In this multiresolution analysis it is easy to define an interpolation operator. More precisely $\forall f \in H^s$, $s > 1/2$, we define

$$I_j f = \sum_k f(x_{jk})\theta(2^j x - k), \qquad x_{jk} = 2^{-j}k. \tag{3.4}$$

In such a framework we have also a natural correspondence between the functions θ_{jk} and the dyadic points $x_{jk} = 2^{-j}k$. Each function θ_{jk} corresponds to the point x_{jk} where it assumes the value $2^{j/2}$ and with respect to which it is symmetric.

It is possible to prove that such an interpolation operator gives us an approximation of f with the same order of accuracy that we would have had using the L^2 projection. More precisely we have ([4])

$$||f - I_j f||_s \le C2^{-j(r-s)}|f|_r, \quad \forall\, r < 2M + 1 \text{ and } \quad \forall\, s < 2R. \tag{3.5}$$

Let us now describe in which way we intend to use these function in the solution of our model problem. We chose a uniform dyadic grid of $(0,1)$, *i.e.*, we consider the points x_{jk}, $k = 0, 2^j$. The first idea is to take the functions corresponding to such points, that is $\theta_{jk}, k = 0, 2^j$, and look for a solution of the form

$$u = \sum_{k=0}^{2^j} u_k \theta_{jk}, \tag{3.6}$$

verifying the boundary conditions (that is $u_0 = 2^{-j/2}a$ and $u_N = 2^{-j/2}b$) and veryfying the equation exactly at the grid points.

In order to be able to give some error estimate for the method we are going to describe why we need an estimate on the interpolation error $||f - \bar{I}_j f||_{1,(0,1)}$ where $\bar{I}_j f = \sum_{k=0}^{N} f(x_{jk})\theta(2^j x - k)$. If we take the constant function $f = c$, it is easy to verify that we have $||f - \bar{I}_j f||_{1,(0,1)} = const$, independent of j. That is, the space introduced does not seem sufficient for the solution of our problem. We will then modify the functions θ_{j0} and θ_{jN} in order to have the constants interpolated exactly. If we introduce the functions

$$\widetilde{\theta}_{j0} = \sum_{k\le 0} \theta_{jk}, \qquad \text{and} \qquad \widetilde{\theta}_{jN} = \sum_{k \ge N} \theta_{jk}, \tag{3.7}$$

it is easy to verify the following properties:

$$\widetilde{\theta}_{j0}(0) = 2^{j/2}, \qquad \widetilde{\theta}_{j0}(x_{jk}) = 0, \quad k > 0$$

$$\widetilde{\theta}_{jN}(1) = 2^{j/2}, \qquad \widetilde{\theta}_{jN}(x_{jk}) = 0, \quad k < N.$$

Moreover defining

$$\widetilde{I}_j f = 2^{-j/2}\left(f(0)\widetilde{\theta}_{j0} + \sum_{k=1}^{N-1} f(x_{jk})\theta_{jk} + f(1)\widetilde{\theta}_{jN} \right). \tag{3.8}$$

We have

$$||c - \widetilde{I}_j c||_{1,(0,1)} = 0.$$

We can then prove the following interpolation estimate.

Lemma 3.1 $f \in H^{3/2+\epsilon}(0,1)$ *implies*

$$||f - \widetilde{I}_j f||_{1,(0,1)} \leq C(f) 2^{-j/2} \tag{3.9}$$

Proof:
Step 1.

We construct an extension of f that is $\mathcal{E} : H^{3/2+\epsilon}(0,1) \longrightarrow H^{3/2+\epsilon}(\mathbb{R})$ continuous verifying

$$\mathcal{E}f|_{(0,1)} = f.$$

We have

$$\|f - \widetilde{I}_j\|_{1,(0,1)} \leq \|f - I_j\mathcal{E}f\|_{1,(0,1)} + \|I_j\mathcal{E}f - \widetilde{I}_j f\|_{1,(0,1)}. \tag{3.10}$$

We remark that

$$||f - I_j\mathcal{E}f||_{1,(0,1)} = ||\mathcal{E}f - I_j\mathcal{E}f||_{1,(0,1)} \leq$$
$$C2^{-j/2}||\mathcal{E}f||_{3/2,\mathbf{R}} \leq C2^{-j/2}||f||_{3/2,(0,1)} \; . \tag{3.11}$$

Step 2.

We need now to bound $||I_j\mathcal{E}f - \widetilde{I}_j f||_{1,(0,1)}$. We have

$$||I_j\mathcal{E}f - \widetilde{I}_j f||_{1,(0,1)} =$$
$$|| \sum_{k=-L}^{0} 2^{-j/2}(\mathcal{E}f(x_{jk}) - f(0))\theta_{jk} + \sum_{k=N}^{N+L} 2^{-j/2}(\mathcal{E}f(x_{jk}) - f(1))\theta_{jk}||_{1,(0,1)},$$

where the sums are restricted to $k = 0, -L$ and $k = N, N + L$ because we take the norm over $(0, 1)$. Let us bound the first sum:

$$|| \sum_{k=-L}^{0} 2^{-j/2}(\mathcal{E}f(x_{jk}) - f(0))\theta_{jk}||_{1,(0,1)}$$
$$\leq 2^j || \sum_{k=-L}^{0} 2^{-j/2}(\mathcal{E}f(x_{jk}) - f(0))\theta_{jk}||_{0,(0,1)}$$
$$\leq 2^{j/2} B \sqrt{\sum_{k=-L}^{0} |\mathcal{E}f(x_{jk}) - f(0)|^2} \; .$$

We take advantage of the embedding of $H^{3/2+\epsilon}$ in $W^{1,\infty}$ and write

$$|\mathcal{E}f(x_{jk}) - f(0)| = \left| \int_0^{x_{jk}} (\mathcal{E}f_1)'(x)\, dx \right|$$
$$\leq |x_{jk}| ||\mathcal{E}f||_{1,\infty,\mathbf{R}} \leq |k| 2^{-j} ||f||_{3/2+\epsilon,(0,1)} \; .$$

This implies

$$|| \sum_{k=-L}^{0} 2^{-j/2}(\mathcal{E}f(x_{jk} - f(0))\theta_{jk}||_{1,(0,1)} \leq C2^{-j/2}||f||_{3/2+\epsilon,(0,1)} \ .$$

Bounding the second sum in the same way, we obtaiin

$$\|I_j\mathcal{E}f - \widetilde{I}_jf\|_{1,(0,1)} \leq C2^{-j/2}\|f\|_{3/2+\epsilon,(0,1)} \ . \tag{3.12}$$

Substituting Equations (3.11) and (3.12) into Equation (3.10) gives us the thesis. ■

Now we can introduce the following scheme: find u_j of the form

$$u_j = 2^{-j/2}\left(a\widetilde{\theta}_{j0} + \sum_{n=1}^{N-1} \alpha_{jn}\theta_{jn} + b\widetilde{\theta}_{jN}\right) \tag{3.13}$$

verifying

$$-u_j''(x_{jk}) = f(x_{jk}), \qquad k = 1, ..., N-1. \tag{3.14}$$

Remark that u_j verifies by definition the boundary conditions.

The stiffness matrix will the have the form

$$M = \begin{pmatrix} 1 & 0 & \cdots & 0 \\ -\widetilde{\theta}_{j0}''(x_1) & -\theta_{j1}''(x_1) & \cdots & -\widetilde{\theta}_{jN}''(x_1) \\ \cdot & \cdot & & \cdot \quad \cdot \\ \cdot & \cdot & & \cdot \\ -\widetilde{\theta}_{j0}''(x_{N-1}) & -\theta_{j1}''(x_{N-1} & \cdots & -\widetilde{\theta}_{jN}''(x_{N-1}) \\ 0 & 0 & \cdots & 1 \end{pmatrix}.$$

Remark. The first important remark which will play an essential role in the theorem that we are going to prove is the following. Due to the definition of the function θ the entries of the matrix verify

$$-\theta_{jn}''(x_{jk}) = \int_{\mathbf{R}} \phi_{jk}'\phi_{jn}' = \int_{-L2^{-j}}^{1+L2^{-j}} \phi_{jk}'\phi_{jn}' \ . \tag{3.15}$$

We now want to prove the following theorem.

Theorem 3.2. *The schema expressed in Equation* (3.13–3.14) *is stable and verifies the following error estimate.* $f \in H^{1/2+\epsilon}(0,1)$ *implies*

$$||u_j - u||_{1,(0,1)} \leq C(f)2^{-j/2} \ . \tag{3.16}$$

Proof:

Step 1.

We will look at the scheme as an approximate Galerkin scheme where the approximation comes from two different sources: approximation of the domain by means of a larger domain and numerical integration. More precisely we will aproximate

$$\int_0^1 f\theta_{jk} \approx \int_{-L2^{-j}}^{1+l2^{-j}} f\theta_{jk} \approx 2^{-j/2} f(x_{jk}).$$

Let us then introduce some notation:

$$\Omega_j =]-L2^{-j}, 1+L2^{-j}[$$

$$\mathcal{V}_j^0 = \{\sum_{k=1}^{N-1} \alpha_{jk}\theta_{jk}\} \subset H_0^1(\Omega_j)$$

$$\mathcal{V}_j^{bc} = a\widetilde{\theta}_{j0} + b\widetilde{\theta}_{jN} + \mathcal{V}_j^0$$

$$\widetilde{a}_j(u,v) = \int_{\Omega_j} u'v'$$

$$a_j : \mathcal{V}_j^{bc} \times \mathcal{V}_j^0 \to \mathbb{R}; \qquad a_j(v_j, \theta_{jk}) = -2^{-j/2} v''(x_{jk})$$

$$f_j : \mathcal{V}_j^0 \to \mathbb{R}; \qquad 2^{-j/2} f_j(\theta_{jk}) = f(x_{jk}).$$

Using this notation our discrete scheme expressed in Equations (3.13–3.14) may be rewritten in the following form:

$$\begin{cases} \text{find } u_j \in \mathcal{V}_j^{bc} \text{ such that } \forall\, v_j \in \mathcal{V}_j \text{ we have} \\ a_j(u_j, v_j) = f_j(v_j). \end{cases} \tag{3.17}$$

To give an estimate on the error we will use the following lemma (see [8], p. 256).

Lemma 3.3. *If $\exists\, \alpha$ such that $\forall\, v_j \in \mathcal{V}_j$ we have (uniform coercivity)*

$$\alpha ||v_j||_{1,\Omega_j}^2 \le a_j(v_j, v_j). \tag{3.18}$$

Then

$$\begin{aligned} ||\mathcal{E}u - u_j||_{1,\Omega_j}^2 \le \Bigg(& ||\mathcal{E}u - \widetilde{I}_j u||_{1,\Omega_j} \\ & + \sup_{w_j \in \mathcal{V}_j} \frac{|\widetilde{a}_j(\widetilde{I}_j u, w_j) - a_j(\widetilde{I}_j u, w_j)|}{||w_j||_{1,\Omega_j}} \\ & + \sup_{w_j \in \mathcal{V}_j} \frac{|f_j(\widetilde{I}_j u) - \widetilde{a}_j(\mathcal{E}u, w_j)|}{||w_j||_{1,\Omega_j}} \Bigg). \end{aligned} \tag{3.19}$$

Step 2.

In order to apply Lemma 3.3 to our case we need to verify that the bilinear form a_j verifies the hypothesis in Equation (3.18). In order to do that we will need the following lemma (see [5]).

Lemma 3.4. *Let $\{c_k\} \in l^2$, $s \leq r$. Then there exists a constant C independent of j such that*

$$||\sum_k c_k\theta_{jk}||_s \leq C||\sum_k c_k\phi_{jk}||_s.$$

Now we have:

$$a_j(v_j, v_j) = a_j(\sum_k c_k\theta_{jk}, \sum_k c_k\theta_{jk}) = \widetilde{a}_j(\sum_k c_k\phi_{jk}, \sum_k c_k\phi_{jk})$$
$$\geq \alpha \left\| \sum_k c_k\phi_{jk} \right\|^2_{1,\mathbf{R}} \geq L\alpha \left\| \sum_k c_k\theta_{jk} \right\|^2_{1,\mathbf{R}} \geq L\alpha||v_j||^2_{1,\Omega_j}.$$

So we can apply Lemma 3.1.

Step 3.

We now need to give an estimate on the three terms on the righthandside of Equation (3.19). The first one is easily bounded using the same techniques of Lemma 3.1. We have

$$||\mathcal{E}u - \widetilde{I}_j u||_{1,\Omega_j} \leq C2^{-j/2}||u||_{3/2+\epsilon,(0,1)} \quad . \tag{3.20}$$

In order to estimate the other two terms we need the following lemma (see [3]).

Lemma 3.5. *Let $r \leq 2M$ and $v \in H^r(0,1)$. Let also $\mathcal{E} : H^r(0,1) \to H^r(\mathbf{R})$ be a continuous extension operator. Then we have*

$$\sum_{k=0}^{N} \left| \int_{\Omega_j} \mathcal{E}v(x)\theta_{jk}(x)\, dx - 2^{-j/2}v(x_{jk}) \right|^2 \leq C2^{-2jr}|v|_{r,(0,1)} \quad . \tag{3.21}$$

Now we come to the second term. We have, letting $w_j = \sum_{k=1}^{N-1} c_k\theta_{jk}$,

$$|\widetilde{a}_j(\widetilde{I}_j u, w_j) - a_j(\widetilde{I}_j u, w_j)|$$
$$\leq \left(\sum_{k=0}^{N-1} c_k^2\right)^{1/2} \left(\sum_k |\widetilde{a}_j(\widetilde{I}_j u, \theta_{jk} - a_j(\widetilde{I}_j u, \theta_{jk})|^2\right)^{1/2}$$
$$\leq ||w_j||_{1,\Omega_j} \left(|\sum_{k=0}^{N-1} \int_{\Omega_j} (-\widetilde{I}_j u)''\theta_{jk} + 2^{-j/2}\widetilde{I}_j u(x_{jk})|^2 \right)^{1/2} .$$

Applying Lemma 3.5 with $v = -(\widetilde{I}_j u)''$ and $r = 1/2 + \epsilon$, we have

$$\frac{|\widetilde{a}_j(\widetilde{I}_j u, w_j) - a_j(\widetilde{I}_j u, w_j)|}{||w_j||_{1,\Omega_j}} \leq C2^{-j/2}||\widetilde{I}_j u||_{5/2+\epsilon,(0,1)} \leq ||u||_{5/2+\epsilon,(0,1)}.$$

As for the third term, we have

$$|f_j(w_j) - \tilde{a}_j(\mathcal{E}u, w_j)| \le ||w_j||_{1,\Omega_j}$$
$$\le ||w_j||_{1,\Omega_j} \left(\sum_{k=0}^{N-1} | - 2^{j/2} u''(x_{jk}) + \int_{\Omega_j} (\mathcal{E}u)'' \theta_{jk}|^2 \right)^{1/2}$$

and this is also bounded using Lemma 3.5. We have

$$\frac{|f_j(w_j) - \tilde{a}_j(\mathcal{E}u, w_j)|}{||w_j||_{1,\Omega_j}} \le C 2^{-j/2} ||u||_{5/2+\epsilon,(0,1)} \ .$$

And with this the proof is complete. ■

Remark. The hypothesis $f \in H^{1/2+\epsilon}(0,1)$ is the minimal regularity hypothesis we need if we want the scheme in Equations (3.13–3.14) to be well defined. The error estimate of Theorem 3.2 is optimal in the sense that a higher order approximation of the solution cannot be achieved within the space in which we are working.

3.2 Higher order methods

As we saw in the previous section, the rough collocation method is stable and convergent but has an extremely poor order of convergence. Since such a method is the restriction to the interval of an high order method on the line, one would like to preserve the high order of convergence in our case. We describe here two possible modifications of the method. For such modified methods did not yet give any theoretical estimate, but the numerical tests show that the methods perform well.

Method 1.

We look for a function of the form

$$u_j = \sum_{k=-L+1}^{2^j+L-1} \alpha_k \theta_{jk} \ . \tag{3.22}$$

(That is, we consider in our development all the k's such that supp $\theta_{jk} \cap [0,1] \neq \emptyset$). The unknowns are $2^j + 2L - 1$, so we need $2^j + 2L - 1$ collocation points. We have the natural $2^j + 1$ dyadic points $x_k = k2^{-j}$, $k = 0, ..., 2^j$ and we add some points near the boundaries. More precisely we add the points $(2k+1)2^{-(j+1)}$ with $k = 0, ..., L-2$ near the left boudary and $1 - (2k+1)2^{-(j+1)}$ with $k = 0, .., L-2$ near the right boundary.

Method 2.

The second method is based on the use of some modified interpolating functions which are constructed in order to achieve an interpolating operator on the interval of the same accuracy as the one on the line, following more or less the same principle of the construction of multiresolution on the interval of the first section. That is, we add the polynomials to the space in which we are solving our discrete problem. Such functions were constructed in [14]. Let us now briefly review how such functions are constructed.

We consider a function f defined on $\mathbb{R}$ and we are interested in finding (at least) approximatively $I_j|_{[0,1]}$. In order to do that we would need the values of f in all those dyadic points x_k such that supp $\theta_{jk} \cap [0,1] \neq \emptyset$, *i.e.*, we need to know $f(x_k)$ for $k = -L+1, ... 2^j+L-1$. Let us now suppose that we know f only in $[0,1]$ so that its values at x_k, $k = -L+1, ..., -1$ and $k = 2^j+1, ..., 2^j+L-1$ are not not known. We may then use some values which are extrapolated from the values in those dyadic points which are internal to the interval $[0,1]$. More precisely we define P_1 and P_2 to be the polynomials of degree L that interpolate f respectively at $x_0, x_1, ..., x_L$ and at $x_{2^j-L}, x_{2^j-L+1}, ..., x_{2^j}$.

$$\begin{aligned} P_1(x_k) &= f(x_k), \qquad & k = 0, ..., L \\ P_2(x_k) &= f(x_k), \qquad & k = 2^j - L, ..., 2^j. \end{aligned}$$

We have

$$P_1(x_n) = \sum_k a_{nk} f(x_k), \qquad n = -L, -1, \qquad k = 0, L \tag{3.23}$$

$$P_2(x_n) = \sum_k b_{nk} f(x_k), \quad n = 2^j+1, 2^j+L, \quad k = 2^j - L, 2^j \tag{3.24}$$

with $a_{nk} = l_k^1(x_n)$, $b_{nk} = l_k^2(x_n)$, where l_k^1 and l_k^2 are the Lagrange polynomials relative to the two L-uples of interpolation points $\{x_0, ..., x_L\}$ and $\{x_{2^j-L}, ..., x_{2^j}\}$.

$$l_k^1 = \prod_{\substack{i=0 \\ i\neq k}}^{L} \frac{x - x_i}{x_k - x_i}, \qquad l_k^2 = \prod_{\substack{i=2^j-L \\ i\neq k}}^{2^j} \frac{x - x_i}{x_k - x_i}.$$

We can now define

$$\mathcal{I}_j f = 2^{-j/2} \left(\sum_{n=-L}^{-1} P_1(x_n)\theta_{jn} + \sum_{k=0}^{2^j} f(x_k)\theta_{jk} + \sum_{n=2^j+1}^{2^j+L} P_2(x_n)\theta_{jn} \right). \tag{3.25}$$

Combining Equations (3.23), (3.24), and (3.25) we have

$$\begin{aligned} \mathcal{I}_j f = &\sum_{k=0}^{L} 2^{-j/2} f(x_k) \left(\theta_{jk} + \sum_{n=-L}^{-1} a_{nk}\theta_{jn} \right) + \sum_{k=L+1}^{2^j-L-1} 2^{-j/2} f(x_k)\theta_{jk} + \\ &\sum_{k=2^j-L}^{2^j} 2^{-j/2} f(x_k) \left(\theta_{jk} + \sum_{n=2^j+1}^{2^j+L} b_{nk}\theta_{jn} \right). \end{aligned}$$

Thus we may define some modified interpolating functions by means of

$$\theta^l_{jk} = \theta_{jk} + \sum_{n=-L}^{-1} a_{nk}\theta_{jn}, \qquad \theta^r_{jk} = \theta_{jk} + \sum_{n=2^j+1}^{2^j+L} b_{nk}\theta_{jn},$$

so that we have

$$\mathcal{I}_j f = 2^{-j/2}\left(\sum_{k=0}^{L} f(x_k)\theta^l_{jk} + \sum_{k=L+1}^{2^j-L-1} f(x_k)\theta_{jk} + \sum_{k=2^j-L}^{2^j} f(x_k)\theta^r_{jk}\right). \quad (3.26)$$

Remark that the functions θ^l_{jk} and θ^r_{jk} still verify the interpolation property $\theta^l_{jk}(x_n) = \delta_{nk}$ and $\theta^r_{jk}(x_n) = \delta_{nk}$.

Moreover, the operator $\mathcal{I}_j$ is well defined for functions in $H^s(0,1)$ $s > 1/2$ and it verifies the following error estimate: if $f \in H^{L+1}(0,1)$ then

$$||f - \mathcal{I}_j f||_{s,(0,1)} \le 2^{-j(L+1-s)}|f|_{L+1,(0,1)}.$$

The approximate solution u_j may now be developed using the 2^j+1 basis functions of the development in Equation (3.26). We may then collocate the equation in the 2^j+1 points $x_0, ..., x_{2^j}$. We will then have the following linear system

$$u_0 = 2^{-j/2}a$$

$$-\sum_{k=0}^{L} u_k \left(\theta^l_{jk}\right)''(x_n) - \sum_{k=L+1}^{2^j-L-1} u_k \left(\theta_{jk}\right)''(x_n) - \sum_{k=2^j-L}^{2^j} u_k \left(\theta^r_{jk}\right)''(x_n) = f(x_n), \qquad n = 1, ..., 2^j-1$$

$$u_{2^j} = 2^{-j/2}b.$$

§4 Numerical results

We want now to test the methods we described in the previous section. In particular, we are interested in comparing the orders of convergence and the behavior in problems in which the solution presents a boundary layer. With such a goal we tested the two methods we described on two problems.

Problem 1.

$$\begin{cases} -u'' = f & \text{in } (0,1) \\ u(0) = 0, & u(1) = 0 \end{cases}$$

with f chosen such that the true solution of the problem is $u = x(1-x)\sin^2(6x)$.

Problem 2.

$$\begin{cases} -.01u'' + u' = 1 & \text{in } (0,1) \\ u(0) = 0, \quad u(1) = 0. \end{cases}$$

For the numerical test we chose ϕ in the Galerkin method to be the Daubechies scaling function with $M = 3$ introduced in [12] and θ of the collocation method to be its autocorrelation function. In this way, as we already remarked, the stiffness matrices differ only for the lines and columns involving the boundaries. The edge-functions $\widetilde{\phi}_n$ and $\bar{\phi}_n$ are shown in Figure 1.

As for the collocation method, we tested both the rough method and the high order Method 1. In Tables 1 and 2, we report for each method the values of the L^2 and L^∞ norms of the errors for different values of j in the case of Problem 1.

A comparison of the results of the three methods is shown in Figure 2. Figure 3 shows clearly the different orders of the three methods, where the errors as functions of the number of degrees of freedom are drawn in log-log scale. We remark that though no theoretical estimate is available, the high order Method 1 behaves very well with respect to the other two methods.

Table 1. L^2-errors for the three methods.

J	Collocation low order	Collocation high order	Galerkin
4	.0515	$1.0678\ 10^{-3}$	$9.87\ 10^{-2}$
5	$2.6578\ 10^{-2}$	$1.5529\ 10^{-5}$	$1.5137\ 10^{-2}$
6	$1.3366\ 10^{-2}$	$5.7322\ 10^{-8}$	$1.7371\ 10^{-4}$
7	$6.6915\ 10^{-3}$	$5.5889\ 10^{-9}$	$2.7282\ 10^{-6}$

Table 2. L^∞-errors for the three methods.

J	Collocation low order	Collocation high order	Galerkin
4	.0819	$1.2472\ 10^{-2}$	$1.624\ 10^{-1}$
5	$4.1839\ 10^{-2}$	$1.2550\ 10^{-4}$	$1.5137\ 10^{-2}$
6	$2.0993\ 10^{-2}$	$6.3678\ 10^{-8}$	$3.0143\ 10^{-4}$
7	$1.0503\ 10^{-2}$	$7.2994\ 10^{-9}$	$7.0575\ 10^{-6}$

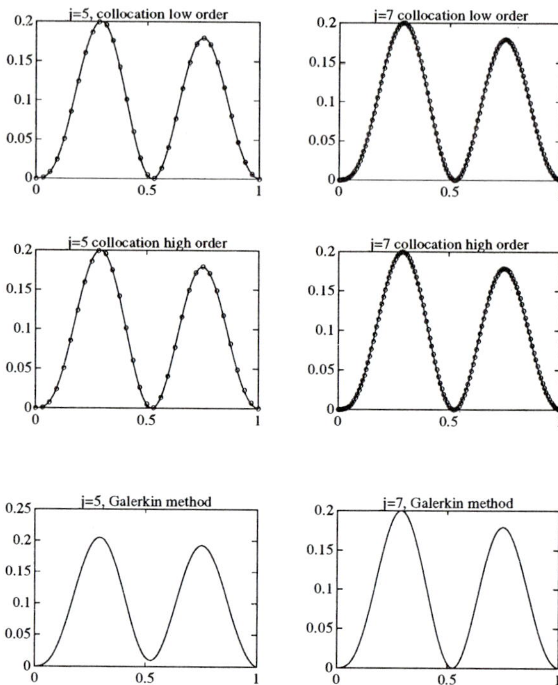

Figure 2. In this figure we show the results of the three methods on test Problem 1.

Since the basis constructed in Section 2 is not symmetric (see Figure 1), the Galerkin method behaves differently at the two boundaries. In Figure 4, we plotted the pointwise error for the Galerkin method applied to Problem 1. We can observe how the error is concentrated near the right boundary. Figure 5 shows the solutions of test Problem 2 obtained by means of the three methods. Also for this problem the collocation methods perform well. In particular with regard to Gibb's phenomena, they behave better than the Galerkin scheme. Also for this problem we have to consider the effects of the asymmetry of the Galerkin basis. In order to do that we tested such a method on the solution of the following problem.

Problem 3.

$$\begin{cases} -.01u'' - u' = 1 \quad \text{in} \quad (0,1) \\ u(0) = 0, \quad u(1) = 0\,. \end{cases}$$

The results of such a test are shown in Figure 6. It is especially interesting to compare the size of Gibb's phenomena for the two different test problems. In particular we remark that the solution of the problem with a boundary layer close to 0 behaves clearly better than the solution of the problem with

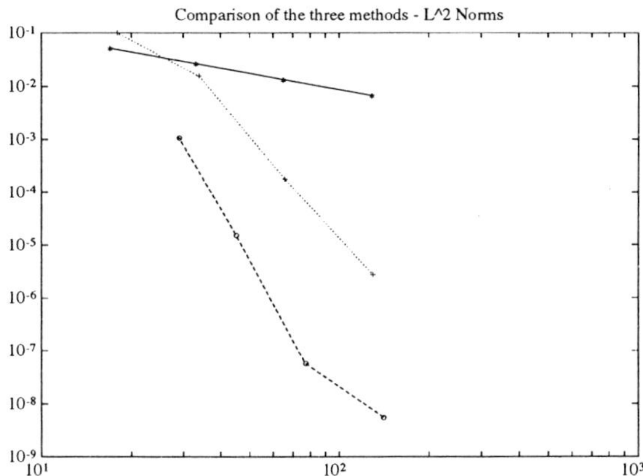

Figure 3. In this figure we compare the three L^2-errors for the three methods applied to test problem 1. The error are plotted in log-log scale. $*$: Low order collocation method. $+$: Galerkin method. o : High order collocation method.

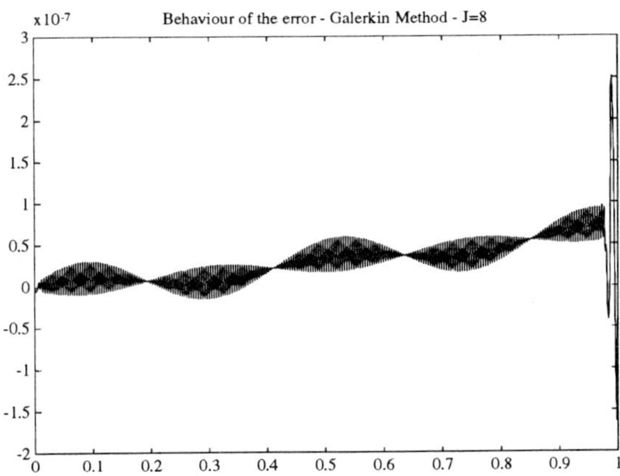

Figure 4. In this figure we plotted the pointwise error for the Galerkin method applied to test Problem 1. Remark the asymmetric behavior of the error, which concentrates mainly at the right boundary.

a boundary layer close to 1. It seems sensible to take this observation into consideration in the choice of a method.

Such effects of asymmetry were enhanced by our choice of the scaling function. In fact the function we choose is heavily asymmetric. It seems sensible that such effects would be reduced by choosing a more symmetric basis (see [13]). However,such asymmetry problems do not appear in the collocation schemes since they are symmetric by construction.

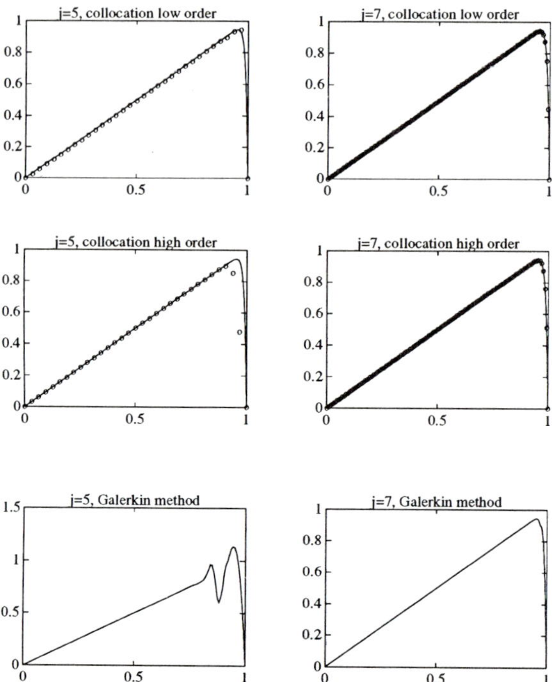

Figure 5. In this figure we plotted the results of the three method applied to test Problem 2. We remark that the Galerkin method presents much stronger Gibb's phenomena than the other two methods.

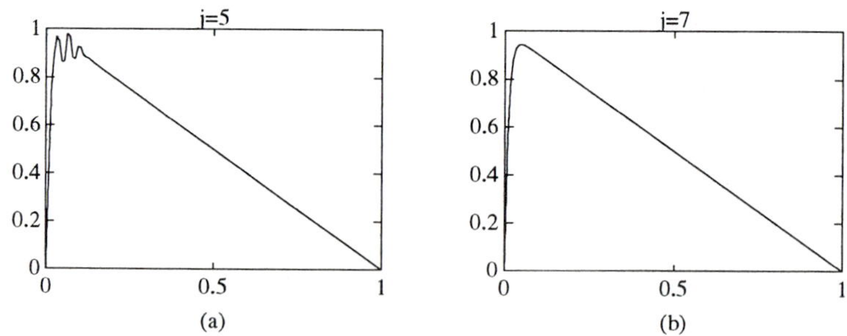

Figure 6. In this figure we plotted the solution of the Galerkin method applied to test Problem 3. Remark that the Gibb's phenomena are much weaker than those appearing fo the same method applied to test Problem 2.

Acknowledgments. J. C. Ravel is supported by EEC contract n. ERBCHBICT920130, project Human Capital and Moblility.

References

1. Babuska, I., The finite element method with Lagrangian multiplier, *Numer. Math.* **20** (1973).
2. Babuska, I., The finite element method with penalty *Math. Comp.* **26** (1973).
3. Bertoluzza, S., Some error estimates for wavelet expansion, *Math. Models and Meth. in Appl. Sc.* **2** (4) (1992).
4. Bertoluzza, S.and Naldi, G., Some remarks on wavelet interpolation, in *Matematica Aplicada e Computacional*, to appear.
5. Bertoluzza, S. and G. Naldi, Apartial differential wquations,*I.A.N.* (887) (1993), preprint.
6. Beylkin, G., R. Coifman, and V. Rokhlin, Fast wavelet transforms and numerical algorithms I *Comm. Pure and Appl. Math.* **XLIV** (2) (1991).
7. Beylkin, G. and N. Saito, Multiresolution representation using the autocorrelation functions of compactly supported wavelets, in *Progress in wavelet Analysis and Applications, Proc. of the Int. Conf. on Wavelets and Applications*, Toulouse, 1992.
8. Ciarlet, P. G., *The Finite Elements Method for Elliptic Problems* North Holland, Amsterdam, 1978.
9. Cohen, A., I. Daubechies, and P. Vial, P., Wavelets on the interval and fast wavelet transforms, *Appl. and Comp. Harmonic Anal.* **1** (1) (1993), 54–81.
10. Dahlke, S. and A. Kunoth, A., A Biorthogonal wavelet approach for solving boundary value problems, Institut fur Geometrie un Praktische Mathematik - RWTH Aachen, preprint.
11. Dahmen, W. and C. A. Micchelli, Using the refinement equation for evaluating integrals of wavelets, *SIAM J. Numer. Anal.* **30** (2) (1993).
12. Daubechies, I., Orthonormal bases of compactly supported wavelets, *Comm. Pure and Appl. Math* **41** (1988).
13. Daubechies, I., *Ten Lectures on Wavelets*, Jones & Bartlett, 1992.
14. Donoho, D., Interpolating wavelet transform, Department of Statistics, Stanford University, 1992, preprint.
15. Jaffard, S., Wavelet methods for fast resolution of elliptic problems, *SIAM J. Numer. Anal.* **29** (4) (1992).
16. Liandrat, J. and Ph. Tchamitchian, Resolution of the 1D regularized Burgers equation using a spatial wavelet approximation, ICASE Report (90-83) (1990).
17. Maday, Y., V. Perrier, and J. C. Ravel, Adaptivité dynamique sur bases d'ondelettes pour l'approximation d'équations aux dérivées partielles, *C. R. Acad Sci. Paris* **t.312, Série I** (1991).
18. Meyer, Y., *Ondelettes et Opérateurs*, in two volumes, Hermann, Paris, 1990.

19. Meyer, Y., *Ondelettes sur l'Intervalle*, Rev. Mat. Iberoam., submitted.
20. Strang, G. and G. J. Fix, *An Analysis of the Finite Element Method*, Prentice-Hall, Englewood Cliffs, 1973.

Silvia Bertoluzza
Istituto di Analisi Numerica
Consiglio Nazionale delle Ricerche
27100 Pavia
Italy
aivlis@dragon.ian.pv.cnr.it

Giovanni Naldi
Dipartimento di Matematica
Università di Pavia
27100 Pavia
Italy
naldi@dragon.ian.pv.cnr.it

Jean Christophe Ravel
Laboratoire d'Analyse Numérique
Université Pierre et Marie Curie
75252 Paris Cédex
France

On the Nodal Values of the Franklin Analyzing Wavelet

Dimitri Karayannakis

Abstract. We derive upper and lower bound estimates for the value of the Franklin analyzing wavelet at its "exceptional" node $\lambda = 3/2$; we thus confirm a conjecture of E. Berkson about the sign of this value. All calculations, involving double hypergeometric series and elliptic integrals, can be carried out on an ordinary 9 D scientific pocket-calculator.

§1 Introduction

The Franklin wavelets, introduced in 1986 independently by P. G. Lemarié and G. Battle, form a new orthonormal basis of the space $L^2(0,1)$; their structure, though rather simple, is substantially rich since they are generated by a single function (the Franklin analyzing wavelet) with estimates that are relatively easy to handle. It is worth noticing that similar constructions were already used in pure mathematics as well as in mathematical applications, before the systematic development of wavelets during the last few years. To name a few of these applications, we mention constructive quantum field theory and signal analysis; for an extensive presentation of the theory of wavelets see, *e.g.*, [5].

Definition. *The Franklin analyzing wavelet (F.a.w.) is the function* $g : \mathbb{R} \to \mathbb{R}$ *given by*

$$g(\lambda) = \int_{\mathbf{R}} \exp[(2\lambda - 1)\pi i x\ ]\ \omega(x)\ \frac{\sin^2\left(\frac{\pi x}{2}\right)}{\left(\frac{\pi x}{2}\right)^2} dx,$$

where

$$\omega(x) = \frac{\sin^2\left(\frac{\pi x}{2}\right)}{1 - \frac{2}{3}\sin^2\left(\frac{\pi x}{2}\right)} \left[\frac{\sin^4\left(\frac{\pi x}{2}\right)}{1 - \frac{2}{3}\sin^2\left(\frac{\pi x}{2}\right)} + \frac{\cos^4\left(\frac{\pi x}{2}\right)}{1 - \frac{2}{3}\cos^2\left(\frac{\pi x}{2}\right)}\right]^{-1/2}.$$

Wavelets: Theory, Algorithms, and Applications
Charles K. Chui, Laura Montefusco, and Luigia Puccio (eds.), pp. 449–458.

ISBN 0-12-174575-9

The function g is piecewise linear with nodes at the points $\lambda = m/2$, m integer. Since g is evidently symmetric with respect to the axis $\lambda = 1/2$, we can restrict our attention at the nodes with $m \geq 1$. In [2, Theorem (6.1)] E. Berkson proved the following property concerning the nodal values of the F.a.w.

Theorem. *If n is a positive integer and $< t >$ denotes the largest integer not exceeding t, then the sign of $g(n/2)$ is given by the formula*

$$\operatorname{sgn} g(n/2) = (-1)^{<n/2>}, \qquad \textit{for} \qquad n \neq 3.$$

Berkson also conjectured [2, p. 393] that the node $\lambda = 3/2$ constitutes an exception to his theorem since computerized integration, via Simpson's rule, suggested $g(3/2)$ to be small and positive.

In the present work we give a rigorous proof that confirms Berkson's conjecture; in fact, we entrap $g(3/2)$ in a relatively narrow positive open interval of length less than 0.12. Our computations avoid equating approximants since we only have to utilize upper and lower bounds (which we indicate by the upper indices $+$ and $-$ respectively); hence, the resulting inequalities are rigorously accurate.

Following [2, Equation (3.5)] we shall use as a starting point the equation

$$g\left(\frac{n+1}{2}\right) = \frac{\sqrt{3}}{\pi}\int_0^\pi \cos n\theta \left(\frac{1-\cos\theta}{2+\cos\theta}\right)\sqrt{\frac{4-\cos^2\theta}{1+2\cos^2\theta}}\,d\theta, \quad n \text{ integer } \geq 0. \tag{1}$$

In our case $n = 2$; therefore we are interested in obtaining lower and upper bounds for the number

$$g(3/2) = \frac{\sqrt{3}}{\pi}I, \quad I = \int_0^\pi \cos 2\theta \left(\frac{1-\cos\theta}{2+\cos\theta}\right)\sqrt{\frac{4-\cos^2\theta}{1+2\cos^2\theta}}\,d\theta. \tag{2}$$

§2 An alternative form for g(3/2)

For reasons that will become apparent later on, we prefer to rewrite the integral I in Equation (2) under the form

$$I = \int_0^\pi \cos 2\theta \left(\frac{1-\cos\theta}{\sqrt{1+2\cos^2\theta}}\right)\sqrt{\frac{2-\cos\theta}{2+\cos\theta}}\,d\theta. \tag{3}$$

If we partition $[0,\pi]$ into the intervals $[0,\pi/2]$ and $[\pi/2,\pi]$, use the two double angle-formulae for $\cos 2\theta$ and then employ inverse sine and cosine substitutions, we can easily check that Equation (3) is equivalent to

$$I = \left[\int_0^1 \frac{2x^2-1}{(1+2x^2)^{1/2}}\left(\frac{1-x}{1+x}\right)^{1/2}\left(\frac{2-x}{2+x}\right)^{1/2} dx\right] + \left[\int_0^1 \frac{2x^2-1}{(1+2x^2)^{1/2}}\left(\frac{1+x}{1-x}\right)^{1/2}\left(\frac{2+x}{2-x}\right)^{1/2} dx\right], \tag{4}$$

(where the latter improper integral automatically converges, being equal to a properly existing integral).

Simple manipulations and the substitution $x = \sqrt{t}$, $0 \le t \le 1$, in Equation (4) result in the following form of I

$$I = I_1 + \frac{3}{2} I_2 - \sqrt{2} I_3, \tag{5}$$

where

$$I_1 = \int_0^1 t^{\frac{5}{2}-1}(1-t)^{\frac{1}{2}-1}[1-(-2)t]^{-\frac{1}{2}}\left(1-\frac{1}{4}t\right)^{-\frac{1}{2}} dt \tag{6a}$$

$$I_2 = \int_0^1 t^{\frac{3}{2}-1}(1-t)^{\frac{1}{2}-1}[1-(-2)t]^{-\frac{1}{2}}\left(1-\frac{1}{4}t\right)^{-\frac{1}{2}} dt \tag{6b}$$

$$I_3 = \int_0^1 \left[(4-t)(1-t)(t-0)\left(t-\left(-\frac{1}{2}\right)\right)\right]^{-\frac{1}{2}} dt. \tag{6c}$$

For Equations (6a) and (5b) we use Formula 231(5) in [4], while for (6c) we utilize (255.00) in [3].

After the necessary calculations, we see that Equations (6) lead to

$$I_1 = \frac{3}{8}\pi F_1(5/2, 1/2, 1/2, 3; -2, 1/4) \tag{7a}$$

$$I_2 = \frac{1}{2}\pi F_1(3/2, 1/2, 1/2, 2; -2, 1/4) \tag{7b}$$

$$I_3 = \frac{2}{\sqrt{6}} F(90°/60°). \tag{7c}$$

In Equation (7), by $F_1(\alpha, \beta, \gamma, \delta; x, y)$ we have denoted the double hypergeometric series of the first kind in the two real variables x, y, namely

$$F_1(\alpha, \beta, \gamma, \delta; x, y) = \sum_{m=0}^{\infty}\sum_{n=0}^{\infty} \frac{(\alpha)_{m+n}(\beta)_m(\gamma)_n}{(\delta)_{m+n} m! n!} x^m y^n,$$

where for $t > 0$ and k an integer≥ 0, the expression $(t)_k$ indicates the number $\Gamma(t+k)/\Gamma(t)$ (*i.e.*, we used Pochhammer's symbol). In addition, by $F(90°/60°)$ we have denoted the complete elliptic integral of the first kind with a 60° modular angle.

Remark 1. Although the natural region of convergence for a general F_1 is the rectangle $(-1,1)\times(-1,1)$, the hypergeometric series in Equations (7) converge since I_1 and I_2 exist as finite numbers. As we shall see below, the series in Equations (7) can be put into a more convenient form such that the values of x, y are positive and belong to the region of convergence; the new series, therefore, will be easier to handle when estimates are required.

Remark 2. From the tabulated values of the complete elliptic integrals of the first kind (*e.g.*, see [1]) we have that $F^+(90°/60°) = 2.156515648$ and $F^-(90°/60°) = 2.156515647$; we shall indicate these numbers and the number $F(90°/60°)$ simply by F^+, F^-, and F, respectively.

Employing now, in succession, the functional relations 239(1,3) in [4], from Equations (7a) and (7b), we obtain, respectively, the equations

$$F_1(5/2,1/2,1/2,3;-2,1/4) = \frac{1}{\sqrt{3}}F_1(1/2,1/2,2,3;3/4,1/4) \tag{8a}$$

$$F_1(5/2,1/2,1/2,2;-2,1/4) = \frac{1}{\sqrt{3}}F_1(1/2,1/2,1,2;3/4,1/4). \tag{8b}$$

Combining Equations (5), (6), (7), and (8), and after substituting the specific values of the parameters into the corresponding series, and taking into account that $\Gamma(k) = (k-1)!$ for any integer $k \geq 1$, we obtain the following alternative formula for the nodal value $g(3/2)$

$$g(3/2) = \frac{3}{4}\mathrm{S} - \frac{2}{\pi}F \tag{9}$$

where

$$\mathrm{S} = \sum_{m=0}^{\infty} \frac{(1/2)_m}{m!}\left(\frac{3}{4}\right)^m \sigma_m \tag{10a}$$

with

$$\sigma_m = \sum_{n=0}^{\infty} a_{mn}\left(\frac{1}{4}\right)^n, \qquad a_{mn} = \frac{2n+m+3}{(n+m+2)!}\left(\frac{1}{2}\right)_{m+n}. \tag{10b}$$

It is evident that Equation (9) leads to the following (strict) inequalities concerning upper and lower bounds of $g(3/2)$

$$\frac{3}{4}\mathrm{S}^- - \frac{2}{\pi^-}F^+ < g(3/2) < \frac{3}{4}\mathrm{S}^+ - \frac{2}{\pi^+}F^-, \tag{11}$$

where $\pi^- = 3.141592653$ and $\pi^+ = 3.141592654$.

Remark 3. Based on Equation (11), in order to rigorously prove Berkson's conjecture, it will suffice to verify that $\mathrm{S} > \eta = 1.830507340$ $\left(\text{since } \eta > \frac{8}{3}\frac{F^+}{\pi^-}\right)$.

§3 A lower bound for $\sum$

In this section we derive an estimate for S^- and thus for $g(3/2)^-$. For the calculations here, as well as for those in Section 5 where we estimate S^+, the main ingredients are the functional property $\Gamma(1+x) = x\Gamma(x)$, $x > 0$, as well as the classical Taylor's expansion $\sum_{1}^{\infty} \frac{t^k}{k} = -\ln(1-t)$, for $t = \frac{1}{4}$ and $t = \frac{3}{4}$.

Our strategy will be to secure, by explicitly writing out few isolated terms of S, the crucial bound of η (Remark 3) and substitute the remaining double sum by an expression involving sums which can be estimated or even explicitly calculated. To this matter we shall need the following lemma.

Lemma 1. *Let k, m, n denote integers. Then*

(i) *For* $k \geq 3$, $\frac{15}{16}(k-1)! \leq \left(\frac{1}{2}\right)_k \leq \frac{5}{16}k!$

(ii) *For* $k \geq 4$, $\frac{35}{32}(k-1)! \leq \left(\frac{1}{2}\right)_k \leq \frac{35}{128}k!$

(iii) *For* $k \geq 5$, $\frac{315}{256}(k-1)! \leq \left(\frac{1}{2}\right)_k \leq \frac{63}{256}k!$

(iv) *For* $n \geq 1, m \geq 3$, $\frac{2n+m+3}{(n+m)(n+m+1)(n+m+2)} > \frac{1}{n^2m^2}$.

Proof: (i)-(iii) are proved by induction and (iv) after direct calculations.

It is now a direct consequence of Equation (10) combined with Lemma 1 that

$$\mathrm{S} = \sigma_0 + \frac{3}{8}\sigma_1 + \frac{27}{128}\sigma_2 + S, \tag{12}$$

where

$$S = \sum_{m=3}^{\infty} \frac{(1/2)_m}{m!} \left(\frac{3}{4}\right)^m \sigma_m, \tag{12$'$}$$

and

$$\sigma_0 > \left(\frac{19907}{12288}\right) + \left(\frac{35}{32}\right) s_0 \tag{13a}$$

$$\sigma_1 > \left(\frac{9571}{24576}\right) + \left(\frac{315}{256}\right) s_1 \tag{13b}$$

$$\sigma_2 > \left(\frac{289}{1536}\right) + \left(\frac{315}{296}\right) s_2, \tag{13c}$$

with

$$\begin{aligned} s_0 &= \sum_{n=4}^{\infty} \frac{2n+3}{n(n+1)(n+2)} \left(\frac{1}{4}\right)^n, \\ s_1 &= \sum_{n=4}^{\infty} \frac{2}{(n+1)(n+3)} \left(\frac{1}{4}\right)^n, \\ s_2 &= \sum_{n=3}^{\infty} \frac{2n+5}{(n+2)(n+3)(n+4)} \left(\frac{1}{4}\right)^n. \end{aligned} \tag{13d}$$

(Here the fractions in Equations (13a)-(13b) are the sums of the first four terms and in Equation (12c) of the first three, in the corresponding expressions for σ_m, $0 \leq m \leq 2$, given in Equation 9(b).)

The decomposition $\frac{2n+3}{n(n+1)(n+2)} = \frac{3}{2} \cdot \frac{1}{n} - \frac{1}{n+1} - \frac{1}{2} \cdot \frac{1}{n+2}$, along with the formula $\sum_1^\infty \frac{(1/4)^n}{n} = \ln(4/3)$, leads to the equation

$$s_0 = \left(\frac{3867}{1280}\right) - \frac{21}{2}\ln\left(\frac{4}{3}\right). \tag{14}$$

Combining Inequalities (13) and Equation (14) and neglecting the positive numbers s_1 and s_2, we automatically have

$$\sigma_0 + \frac{3}{8}\sigma_1 + \frac{27}{128}\sigma_2 > \theta \tag{15}$$

where

$$\theta = \left(\frac{251171}{49152}\right) - \frac{735}{64}\ln\left(\frac{4}{3}\right) > 1.80623826. \tag{15'}$$

It is now easy to check that we can write the double sum in (12′) under the form

$$S = S_1 + S_2 + S_3, \tag{16}$$

where

$$S_1 = \sum_{m=3}^{\infty}\left[\left(\frac{1}{2}\right)_m\right]^2 \frac{(m+3)}{m!(m+2)!}\left(\frac{3}{4}\right)^m, \tag{16'a}$$

$$S_2 = \sum_{m=3}^{\infty}\frac{1}{4}\frac{(1/2)_m(1/2)_{m+1}}{m!(m+3)!}(m+5)\left(\frac{3}{4}\right)^m, \tag{16'b}$$

$$S_3 = \sum_{m=3}^{\infty}\frac{(1/2)_m}{m!}\left(\frac{3}{4}\right)^m \sum_{n=2}^{\infty} a_{mn}\left(\frac{1}{4}\right)^n \tag{16'c}$$

(with a_{mn} as in Equation (9b)).

Neglecting the positive number S_3 and employing Lemma 1 (i)-(ii), we have

$$S > S_1 + S_2 > \tilde{S}_1 + \tilde{S}_2, \tag{17}$$

where

$$\begin{aligned} \tilde{S}_1 &= \left(\frac{15}{16}\right)^2 \sum_{m=3}^{\infty}\frac{(m+3)}{m^2(m+1)(m+2)}\left(\frac{3}{4}\right)^m, \\ \tilde{S}_2 &= \frac{1}{4}\cdot\frac{15}{16}\cdot\frac{35}{32}\sum_{m=3}^{\infty}\frac{(m+5)}{m(m+1)(m+2)(m+3)}\left(\frac{3}{4}\right)^m. \end{aligned} \tag{17'}$$

Employing only the sum of the two first terms for each series in Equation (17′), we have

$$S > \frac{67277}{2752512} > 0.024442036. \tag{18}$$

Combining Equations (12), (15), (15′), and (18) we conclude that

$$\mathrm{S} > 1.830680296 > \eta, \tag{18'}$$

and thus Berkson's conjecture, that $g(3/2) > 0$, is now rigorously established.

§4 Refinements of the estimates for S^- and $g(3/2)^-$

Recalling now Inequalities (13b) and (13c), we can calculate explicitly the values of s_1 and s_2 as we did with s_0 in Equation (14); in addition we will improve Equation (17) by refining the estimate for $\tilde{S}_1$, calculating $\tilde{S}_2$ and, finally, by estimating S_3 using Lemma 1 (iv).

We have the following decompositions

$$\begin{aligned} S_1 &= \sum_{n=4}^{\infty} \frac{(1/4)^n}{n+1} - \sum_{n=4}^{\infty} \frac{(1/4)^n}{n+3}, \\ S_2 &= \frac{1}{2} \sum_{n=3}^{\infty} \frac{(1/4)^n}{n+2} + \sum_{n=3}^{\infty} \frac{(1/4)^n}{n+3} - \frac{3}{2} \sum_{n=3}^{\infty} \frac{(1/4)^n}{n+4}. \end{aligned} \tag{19}$$

Using translation of indices along with the formula $\sum_{n=1}^{\infty} \frac{(1/4)^n}{n} = \ln(\frac{4}{3})$ we have

$$S_1 = \left(\frac{66283}{3840}\right) - 60 \ln\left(\frac{4}{3}\right), \quad S_2 = \left(\frac{11489}{128}\right) - 312 \ln\left(\frac{4}{3}\right). \tag{19'}$$

Combining Inequalities (13) and Equation (19′) we can refine (slightly) the estimate in Equation (15′) as follows

$$\sigma_0 + \frac{3}{8}\sigma_1 + \frac{27}{128}\sigma_2 > \theta + \frac{3}{8} \cdot \frac{315}{256} S_1 + \frac{27}{128} \cdot \frac{315}{296} S_2 > 1.806592526. \tag{20}$$

Again, by decomposing, we can write equations (17′), respectively, under the form

$$\begin{aligned} \tilde{S}_1 = \frac{225}{256} \Bigg[& \frac{1}{2} \sum_{m=3}^{\infty} \frac{(3/4)^m}{m} - \sum_{m=3}^{\infty} \frac{(3/4)^m}{m+1} \\ & + \frac{1}{2} \sum_{m=3}^{\infty} \frac{(3/4)^m}{m+2} + 3 \sum_{m=3}^{\infty} \frac{(3/4)^m}{m^2(m+1)(m+2)} \Bigg] \end{aligned} \tag{21}$$

and

$$\begin{aligned} \tilde{S}_2 = \frac{225}{2048} \Bigg[& \frac{5}{6} \sum_{m=3}^{\infty} \frac{(3/4)^m}{m} - 2 \sum_{m=3}^{\infty} \frac{(3/4)^m}{m+1} \\ & + \frac{3}{2} \sum_{m=3}^{\infty} \frac{(3/4)^m}{m+2} - \frac{1}{3} \sum_{m=3}^{\infty} \frac{(3/4)^m}{m+3} \Bigg]. \end{aligned} \tag{21'}$$

Based on the formula $\sum_{m=1}^{\infty} \frac{(3/4)^m}{m} = \ln 4$, we can explicitly calculate all the sums involved in Equations (21) and (21′) except of $\sum_{m=3}^{\infty} \frac{(3/4)^m}{m^2}$ whose presence prevents us from calculating explicitly $\tilde{S}_1$. Nevertheless, since in the crude estimate of Equation (18), we had taken indirectly into account only the first two terms of $\tilde{S}_1$, we will improve $\tilde{S}_1$ by considering in Equation (21) the exact sums of the first three series and summing up the first three terms of the fourth.

We then find that

$$\tilde{S}_1 > \frac{225}{256}\left[\frac{1}{9}\ln 2 - \left(\frac{238339}{4300800}\right)\right] > 0.018983479, \tag{22}$$

and

$$\tilde{S}_2 = \frac{525}{2048}\left[\frac{71}{81}\ln 4 - \left(\frac{15617}{17280}\right)\right] > 0.079822899. \tag{22'}$$

Finally, using successively in (16′c) parts (i), (iii), and (iv) of Lemma 1, we obtain the inequality $S_3 > \tilde{S}_3$ where

$$\tilde{S}_3 = \frac{4725}{4096}\left[\sum_{m=3}^{\infty}\frac{(3/4)^m}{m^3}\right]\cdot\left[\sum_{n=2}^{\infty}\frac{(1/4)^n}{n^2}\right]. \tag{23}$$

Employing only the sums of the first three terms in each series of Equation (23) we obtain

$$S_3 > 0.000419155. \tag{23'}$$

Combining Equations (20), (22), and (23′), we find that

$$\mathrm{S} > 1.905818093. \tag{24}$$

We conclude, based on Equation (11), that

$$g(3/2) > 0.056483064. \tag{25}$$

§5 An upper bound for S and g(3/2)

We shall need the following self-evident lemma.

Lemma 2. *For* $m > 0$, $n \geq 0$, $\dfrac{2n+m+3}{(n+m+1)(n+m+2)} < \dfrac{1}{m}$.

We will combine later on Lemma 1 and Lemma 2 to obtain in Equation (31) an upper bound for S_3 (defined in Equation (16′c)). We start by obtaining upper estimates for σ_0, σ_1, and σ_2 as well as for S_1 and S_2 using Lemma 1 and working similarly as Sections 3 and 4.

We have the following inequalities:

$$\sigma_0 < \left(\frac{19907}{12288}\right) + \frac{35}{128}\sum_{n=4}^{\infty}\left(\frac{1}{4}\right)^n\frac{2n+3}{(n+1)(n+2)} < 1.620533067 \tag{26}$$

$$\sigma_1 < \left(\frac{9571}{24576}\right) + \frac{63}{128}\sum_{n=4}^{\infty}\frac{(1/4)^n}{n+3} < 0.146173883 \tag{27}$$

$$\sigma_2 < \left(\frac{289}{1536}\right) + \frac{63}{256}\sum_{n=3}^{\infty}\left(\frac{1}{4}\right)^n \frac{2n+5}{(n+3)(n+4)} < 0.117826907 \tag{28}$$

$$S_1 < \frac{25}{256}\sum_{m=3}^{\infty}\left(\frac{3}{4}\right)^m \frac{m+3}{(m+1)(m+2)} < 0.03743197 \tag{29}$$

$$S_2 < \frac{175}{2048}\sum_{m=3}^{\infty}\left(\frac{3}{4}\right)^m \frac{m+5}{(m+2)(m+3)} < 0.026078105 \tag{30}$$

$$S_3 < \left(\frac{5}{16}\right)\cdot\left(\frac{635}{256}\right)\cdot\left(\sum_{m=3}^{\infty}\frac{\left(\frac{1}{4}\right)^m}{m}\right)\cdot\left(\sum_{n=2}^{\infty}\left(\frac{1}{4}\right)^n\right) < 0.106363710. \tag{31}$$

Since S$< \sigma_0^+ + \frac{3}{8}\sigma_1^+ + \frac{27}{128}\sigma_2^+ + S_1^+ + S_2^+ + S_3^+$, by combining Equations (26)-(31), we obtain the estimate

$$\text{S} < 2.054407642. \tag{32}$$

Based on Equation (11), we see now that

$$g(3/2) < 0.167925233. \tag{33}$$

Remark 4. It is evident that the method itself provides the strategy for its own improvement. Keeping the exact calculations and the basic inequalities intact, one could improve the estimates for S^- and S^+ by allowing a much larger number of terms to be involved there where only three or four such terms were taken into account. For example, if we allow a program concerning the sums $\tilde{S}_1$ (in Equation (21)) and S_3 (Equations (23) and (31)), to run on a super-computer, we can obtain $g(3/2)$ in an interval of length less than 0.01 in a relatively short time.

References

1. M. Abramowitz and I. A. Stegun, (eds.), *Handbook of Mathematical Functions (with Formulas, Graphs, and Mathematical Tables)*, Dover Publ., Inc., New York, 1972.
2. E. Berkson, *"On the structure of the graph of the Franklin analyzing wavelet,"* Proc. of the Special Year in Mod. Anal. at the Uo I (Urbana), Vol. I, Cambridge University Press, 1989, p. 366.
3. P. F. Byrd and M. D. Friedman, *Handbook of Elliptic Integrals for Engineers and Physicists*, Springer-Verlag, Berlin, 1954.

4. A. Erdélyi et al., *Higher Transcendental Functions*, Vol. I, McGraw Hill, New York, 1953.
5. Y. Meyer, "*Wavelets and Operators,*" Proc. of the Special Year in Mod. Anal. at University of Illinois (Urbana), Vol. I, Cambridge University Press, 1989, p. 256.

Dimitri Karayannakis
Department of Physics,
University of Crete
Iraklion, Crete, Greece

Parallel Numerical Algorithms with Orthonormal Wavelet Packet Bases

Laura Bacchelli Montefusco

Abstract. The representation in wavelet packet bases of certain classes of operators as, for example, Calderón-Zygmund or pseudo-differential operators, may be seen as a method of conversion, for a given accuracy, to sparse form. Moreover, this form presents an equal-sized band block structure. Taking advantage of this structure, we present here coarse-grained parallel algorithms for some basic problems of large-scale numerical linear algebra. It is shown that, starting from an efficient parallel wavelet packet matrix decomposition, these algorithms can reach good efficiency on distributed memory multiprocessors as their arithmetic work-load is very well balanced and they require only few nearest neighbour communications. Timing and performance analyses from a first experimentation on a hypercube iPSC/860 are presented.

§1 Introduction

Wavelet analysis, born as an alternative to the classical windowed Fourier analysis, in the last few years has found a wide range of fields of applications. This paper concerns the application of wavelet and wavelet packet bases to a particular field of the numerical analysis which is known as parallel numerical linear algebra.

In fact, for large-scale problems, the discretization of certain linear operators as, for example, integral operators, leads to large, dense matrices which are difficult to handle in numerical analysis. On the contrary, if the operators are expressed by means of a suitable wavelet basis, the corresponding matrices present (for a given accuracy) a band-diagonal form and can be applied to arbitrary vectors with order $O(N)$ operations. The sparsity of these representations is a consequence of the localization of wavelets in both space and wave number domains. This is briefly the basic idea of the fast algorithms given in [2] where the operators taken into consideration are Calderón-Zygmund and

Wavelets: Theory, Algorithms, and Applications
Charles K. Chui, Laura Montefusco, and Luigia Puccio (eds.), pp. 459–493.

ISBN 0-12-174575-9

pseudo-differential operators and the orthonormal wavelet basis functions used are those constructed by I. Daubechies in [6].

The authors of [2] present two different schemes for the fast numerical application of the considered operators to arbitrary vectors, which correspond to two different decompositions of the operator matrices. The first (standard form) is a straightforward realization of the operator matrices in the wavelet basis and can be obtained with order $O(NlogN)$ operations; the second (non-standard form) is not directly interpretable by means of a basis change but follows from a deep analysis of David-Journe's T(1) theorem (see [13]) and is shown to be an order $O(N)$ procedure. Both of these schemes convert (with finite precision and whenever regularity permits) dense matrices into highly sparse matrices, which serve, therefore, as good starting points for further approximate calculations. However, there is, in these schemes, a drawback which can be serious if one takes into consideration the large dimensions of the problems for which they have been proposed, and that is, they are not suited for parallel computing. In fact the band-diagonal blocks of the decomposed matrices are of decreasing size, thus preventing the realization of efficient parallel algorithms due to the consequent unbalanced work-load.

With the aim of eliminating this undesired unbalance and for constructing a decomposition scheme which corresponds to a straightforward realization of the operator matrix in a suitable basis, in the present paper, we have considered the orthonormal wavelet packet bases given in [5]. In these bases, the transformed matrices present an equal-sized block structure, each representing the contribution from different "frequency bands." If the matrices are the discretization of Calderón-Zygmund or pseudo-differential operators, the different blocks show a diagonal dominant band structure: the larger the number of vanishing moments of the mother wavelet the stronger this diagonal dominance. These are the theoretical results which follow from [13]. From the numerical point of view, the decomposition seems to work in a more general setting.

As one would logically expect, the better balance of the block structure of the wavelet packet decomposition is obtained at a slightly higher computational cost. In fact, realization of the operator matrix in these bases, that is the matrix decomposition algorithm is an order $O(kN^2)$ procedure, where the constant k depends on the number of decomposition steps, as well as on the number of vanishing moments of the mother wavelet. The cost can be reduced to $O(kN)$ if, working with wavelets having a sufficiently large number of vanishing moments, one knows *a priori* the position of the singularities (as for Calderón-Zygmund operators) and takes into account the finite precision of calculations.

Once the operator matrix has been transformed by means of wavelet packet decomposition and the elements below a fixed accuracy ϵ have been discarded, we have the compressed form of the operator that corresponds to a sparse form of the matrix; this is an equal-sized band diagonal form which represents a good starting point to construct efficient parallel algorithms.

In this paper we have considered three fundamental large-scale computation problems: matrix-vector, matrix-matrix multiplication, and the solution of large linear systems by means of the preconditioned conjugate gradient method. For these problems we have shown that, starting from an efficient parallel wavelet packet matrix decomposition algorithm, we can easily construct coarse grained parallel algorithms where the arithmetic work-load is very well balanced and which reach good efficiency on distributed memory multiprocessors as they require only very few nearest neighbor communications.

The paper is organized as follows: in Section 2 we briefly review the basic properties of compactly supported orthogonal wavelet and wavelet-packet bases and we give some notation needed to present the corresponding decomposition and reconstruction algorithms in compact form. These algorithms are used in Section 3 to express linear operators in wavelet and wavelet-packet bases and, for special classes of operators, to compress them by setting to zero the entries of the operator matrices below a fixed accuracy. In Section 4 we propose a parallel algorithm for the wavelet packet matrix decomposition and present a performance evaluation obtained by running its implementation on a hypercube iPSC/860. Finally in Sections 5 and 6 we present some coarse-grained parallel numerical algorithms that, starting from the compressed form of matrices obtained by means of the previous parallel decomposition, seem to be very effective for the problems being considered. Numerical and timing results from a first experimentation on the above hypercube are presented.

§2 Orthogonal wavelet and wavelet-packet bases

In this section we briefly review the main properties of compactly supported orthogonal wavelets and wavelet packets and define our notation. For a more detailed exposition and proofs the reader is referred to [6,7,12].

2.1 Compactly supported orthogonal wavelets

Let us consider $L^2(\mathbb{R})$, the space of complex-valued square integrable function on R. A multiresolution approximation of $L^2(\mathbb{R})$ is given by a nested sequence of closed subspaces V_j

$$\ldots \subset V_{-2} \subset V_{-1} \subset V_0 \subset V_1 \subset V_2 \subset \ldots$$

such that

$$\overline{\bigcup_{j\in Z} V_j} = L^2(\mathbb{R}), \quad \bigcap_{j\in \mathbf{Z}} V_j = \{0\},$$

$$f(x) \in V_j \iff f(2x) \in V_{j+1} \quad \forall j.$$

For all j, there exists a $\phi(x)$, called *scaling function* such that

$$\{\phi_{j,k}(x) = 2^{j/2}\phi(2^j x - k) : k \in Z\}$$

is an orthonormal basis of V_j.

The orthogonal projection of a function $f \in L^2(\mathbb{R})$ into V_j is given by

$$f_j = \mathbf{P}_j f = \sum_k \langle f, \phi_{jk} \rangle \phi_{j,k} = \sum_k c_k^j \phi_{jk} \tag{2.1}$$

and can be interpreted as an approximation to f at resolution 2^{-j}. By defining W_j as the orthogonal complement of V_j in V_{j+1}

$$V_{j+1} = V_j \oplus W_j, \tag{2.2}$$

the space $L^2(\mathbb{R})$ is represented as a direct sum

$$L^2(\mathbb{R}) = \bigoplus_{j \in Z} W_j. \tag{2.3}$$

Moreover, from the previous assumption on V_j it follows that there exists a function $\psi(x)$, called *mother wavelet*, such that

$$\{\psi_{j,k}(x) = 2^{j/2}\psi(2^j x - k) : k \in Z\}$$

is an orthonormal basis of W_j. The spaces W_j are called wavelet subspaces of $L^2(\mathbb{R})$ relative to the scaling function $\phi(x)$ and the orthogonal projection of a function $f \in L^2(\mathbb{R})$ into W_j, given by

$$g_j = \mathbf{Q}_j f = \sum_k < f, \psi_{jk} > \psi_{j,k} = \sum_k d_k^j \psi_{jk}, \tag{2.4}$$

can be interpreted as the details lost going from the approximation at resolution $2^{-(j+1)}$ to the poorer approximation at resolution 2^{-j}.

From Equation (2.3) it follows that $\{\psi_{jk}, k \in Z, j \in Z\}$ is an orthonormal basis of $L^2(\mathbb{R})$, and every $f \in L^2(\mathbb{R})$ has a unique wavelet decomposition

$$f = \sum_j \sum_k < f, \psi_{jk} > \psi_{jk} = \sum_j \sum_k d_k^j \psi_{jk}. \tag{2.5}$$

The functions $\phi(x)$ and $\psi(x)$ satisfy the well-known "two scale equations," that, in the compactly supported case, can be expressed as

$$\phi(x) = \sqrt{2} \sum_{k=0}^{L-1} h_k \phi(2x - k) \tag{2.6}$$

$$\psi(x) = \sqrt{2} \sum_{k=0}^{L-1} w_k \phi(2x - k).$$

with

$$w_k = (-1)^k h_{L-k-1}, \qquad k = 0, \ldots, L-1.$$

For the wavelets in [6] L is related to the number of vanishing moments of $\psi(x)$; that is, if

$$\int_{-\infty}^{+\infty} x^m \psi(x) dx = 0 \qquad m = 0, \ldots, M-1, \tag{2.7}$$

we have $L = 2M$.

From Equation (2.2) it follows that every function $f_{j+1} \in V_{j+1}$, whose expression is

$$f_{j+1} = \sum_k c_k^{j+1} \phi_{j+1,k}\,, \tag{2.8}$$

can be written as the sum of its components in V_j and W_j

$$f_{j+1} = f_j + g_j = \sum_k c_k^j \phi_{jk} + \sum_k d_k^j \psi_{jk} \tag{2.9}$$

where the coefficients in Equations (2.8) and (2.9) are given in [7].

Decomposition

$$\begin{cases} c_k^j = \sum_l h_{l-2k} c_l^{j+1} \\ d_k^j = \sum_l w_{l-2k} c_l^{j+1}. \end{cases} \tag{2.10}$$

Reconstruction

$$c_k^{j+1} = \sum_l h_{k-2l} c_l^j + \sum_l w_{k-2l} d_l^j. \tag{2.11}$$

On varying index j, Formulas (2.10) and (2.11) represent the numerical algorithms for wavelet decomposition and reconstruction. This wavelet decomposition-reconstruction algorithm can be expressed in compact form if we introduce a matricial notation.

In fact, let

$$\Phi_0 = \begin{bmatrix} \cdots \\ \cdots \\ \cdots \\ \phi_{0,-1}(x) \\ \phi_{0,0}(x) \\ \phi_{0,1}(x) \\ \cdots \\ \cdots \\ \cdots \end{bmatrix} \quad \text{and} \quad \Psi_0 = \begin{bmatrix} \cdots \\ \cdots \\ \cdots \\ \psi_{0,-1}(x) \\ \psi_{0,0}(x) \\ \psi_{0,1}(x) \\ \cdots \\ \cdots \\ \cdots \end{bmatrix} \tag{2.12}$$

be bi-infinite vectors and let **H** and **W** be bi-infinite Hurwitz matrices associated with the compactly supported bi-infinite real vectors **h** and **w**, that is

$$\mathbf{H} = (H_{i,j})_{i,j \in Z} = (h_{i-2j})_{i,j \in Z},$$

$$\mathbf{W} = (W_{i,j})_{i,j\in Z} = (w_{i-2j})_{i,j\in Z}. \tag{2.13}$$

Then the two-scale relations can be written as

$$\begin{cases} \Phi_0 = \sqrt{2}\mathbf{H}^T\Phi_1 \\ \Psi_0 = \sqrt{2}\mathbf{W}^T\Phi_1 \end{cases} \tag{2.14}$$

and the decomposition-reconstruction algorithm is as follows (assuming we do only s decomposition steps starting from an approximation $f_n \in V_n$).

Decomposition

$$\begin{cases} \mathbf{c}^j = \mathbf{H}^T\mathbf{c}^{j+1} \\ \qquad\qquad\qquad j = n-1, \ldots, n-s \\ \mathbf{d}^j = \mathbf{W}^T\mathbf{c}^{j+1}. \end{cases} \tag{2.15}$$

Reconstruction

$$\mathbf{c}^{j+1} = \mathbf{H}\mathbf{c}^j + \mathbf{W}\mathbf{d}^j \qquad j = n-s, \ldots, n-1. \tag{2.16}$$

The compact form can then be obtained by observing that, due to the orthogonality of the basis functions and of the subspace decomposition, matrices $\mathbf{H}$ and $\mathbf{W}$ satisfy the following relations:

$$\begin{cases} \mathbf{H}^T\mathbf{H} = \mathbf{I} \qquad \mathbf{H}^T\mathbf{W} = \mathbf{0} \\ \mathbf{W}^T\mathbf{W} = \mathbf{I} \qquad \mathbf{W}^T\mathbf{H} = \mathbf{0} \end{cases} \tag{2.17}$$

$$\mathbf{H}\mathbf{H}^T + \mathbf{W}\mathbf{W}^T = \mathbf{I}. \tag{2.18}$$

If we now consider the bi-infinite two-block matrices

$$\mathbf{T} = \left[\begin{array}{c|c} \mathbf{H} & \mathbf{W} \end{array}\right] \qquad \text{and} \qquad \mathbf{T}^T = \left[\begin{array}{c} \mathbf{H}^T \\ \hline \mathbf{W}^T \end{array}\right],$$

one step of the wavelet decomposition-reconstruction algorithm is given by

$$\begin{bmatrix} \mathbf{c}^j \\ \mathbf{d}^j \end{bmatrix} = \mathbf{T}^T \begin{bmatrix} \mathbf{c}^{j+1} \end{bmatrix}, \tag{2.19}$$

and

$$\begin{bmatrix} \mathbf{c}^{j+1} \end{bmatrix} = \mathbf{T} \begin{bmatrix} \mathbf{c}^j \\ \mathbf{d}^j \end{bmatrix} \tag{2.20}$$

Moreover, from Equations (2.17) and (2.18) it follows that

$$\left.\begin{array}{l}\mathbf{T}\mathbf{T}^{\mathrm{T}} = \mathbf{I} \\ \mathbf{T}^{\mathrm{T}}\mathbf{T} = \mathbf{I}\end{array}\right\} \Longrightarrow \mathbf{T}^{\mathrm{T}} = \mathbf{T}^{-1},$$

that is, matrix $\mathbf{T}$ is orthogonal (unitary in the complex case).

In general, if we define

$$\mathbf{T}^T_{k(k=3)} = \left[\begin{array}{c|c|c} \begin{array}{c} \mathbf{H}^T \\ \hline \mathbf{W}^T \end{array} & 0 & \\ \hline \mathbf{0} & \mathbf{I} & \mathbf{0} \\ \hline \multicolumn{2}{c|}{\mathbf{0}} & \mathbf{I} \end{array}\right], \tag{2.21}$$

then the s decomposition steps can be seen as obtained through s pre-multiplications of the starting vector $\mathbf{c}^n$ by the orthogonal matrices $\mathbf{T}^T_k$; that is,

$$\begin{bmatrix} \mathbf{c}^{n-s} \\ \hline \mathbf{d}^{n-s} \\ \hline \vdots \\ \hline \mathbf{d}^{n-2} \\ \hline \mathbf{d}^{n-1} \end{bmatrix} = \mathbf{T}^T_s \ldots \mathbf{T}^T_2\, \mathbf{T}^T_1 \begin{bmatrix} \\ \mathbf{c}^n \\ \\ \end{bmatrix}, \tag{2.22}$$

and consequently the reconstruction is

$$\begin{bmatrix} \\ \\ \mathbf{c}^n \\ \\ \\ \end{bmatrix} = \mathbf{T}_1\,\mathbf{T}_2\ldots\mathbf{T}_s \begin{bmatrix} \mathbf{c}^{n-s} \\ --- \\ \mathbf{d}^{n-s} \\ --- \\ \vdots \\ --- \\ \mathbf{d}^{n-2} \\ --- \\ \mathbf{d}^{n-1} \end{bmatrix}. \tag{2.23}$$

Remark. In pratice we assume that vector $\mathbf{c}^n$ is of finite length $N = 2^n$ (that means $dim(V_n) = 2^n$). In this case, as in the applications which follow, matrices $\mathbf{H}$ and $\mathbf{W}$ are finite and must be modified according to [4].

2.2 Compactly supported orthogonal wavelet packets

Starting from the multiresolution approximation of $L^2(\mathbb{R})$ generated by the scaling function $\phi(x)$ given in the previous paragraph and from the corresponding wavelet $\psi(x)$, we now define a new orthonormal basis of $L^2(\mathbb{R})$ called wavelet-packet basis [3] [5].

We set

$$\begin{cases} \psi_0(x) = \phi(x) \\ \psi_1(x) = \psi(x), \end{cases} \tag{2.24}$$

and define the wavelet-packets relative to $\psi_0(x) = \phi(x)$ as

$$\begin{cases} \psi_{2l}(x) = \sqrt{2}\displaystyle\sum_{k=0}^{2M-1} h_k \psi_l(2x-k) \\ \qquad\qquad l = 0, 1, \ldots \\ \psi_{2l+1}(x) = \sqrt{2}\displaystyle\sum_{k=0}^{2M-1} w_k \psi_l(2x-k). \end{cases} \tag{2.25}$$

The family $\psi_n(x)$, $n \in Z_+$, so obtained preserves the orthogonality property of the scaling function $\psi_0(x) = \phi(x)$ [3]. If we now consider the family of subspaces

$$U_j^n = \text{clos}_{L^2(\mathbb{R})}\{\psi_{jk}^n(x) = 2^{j/2}\psi_n(2^j x - k) : k \in Z\}, \quad j \in Z, \quad n \in Z_+, \tag{2.26}$$

we see immediately that

$$\begin{cases} U_j^0 = V_j \\ U_j^1 = W_j \end{cases}$$

and from $V_{j+1} = V_j \oplus W_j$, it follows that $U_{j+1}^0 = U_j^0 \oplus U_j^1$; moreover, we have

$$U_{j+1}^n = U_j^{2n} \oplus U_j^{2n+1}. \tag{2.27}$$

By using Equations (2.26) and (2.27) we see that the wavelet packets spaces U_j^n represent a further decomposition of the wavelet spaces W_j according to

$$\begin{aligned} W_j = U_j^1 &= U_{j-1}^2 \oplus U_{j-1}^3 \\ &= U_{j-2}^4 \oplus U_{j-2}^5 \oplus U_{j-2}^6 \oplus U_{j-2}^7; \\ &\ldots\ldots\ldots\ldots \end{aligned} \tag{2.28}$$

$$W_j = U_{j-k}^{2k} \oplus U_{j-k}^{2k+1} \oplus \ldots \oplus U_{j-k}^{2^{k+1}-1},$$

that is, the jth frequency band corresponding to W_j is further partitioned into 2^k "sub-bands".

The analog of Formulas (2.15) and (2.16) which gives the wavelet packet (WP) decomposition-reconstruction algorithm, starting from $f_n \in V_n = U_n^0$ and performing s decomposition-reconstruction steps are:

decomposition

$$\begin{cases} \mathbf{u}_j^{2l} = \mathbf{H}^T \mathbf{u}_{j+1}^{2l}, & k = 1, \ldots, s, \quad j = n-k, \\ \mathbf{u}_j^{2l+1} = \mathbf{W}^T \mathbf{u}_{j+1}^{2l}, & l = 0, 1, \ldots, 2^{k-1} - 1, \end{cases} \tag{2.29}$$

reconstruction

$$\mathbf{u}_{j+1}^l = \mathbf{H}\mathbf{u}_j^{2l} + \mathbf{W}\mathbf{u}_j^{2l+1}, \qquad \begin{array}{l} k = s, \ldots, 1, \quad j = n-k, \\ l = 0, 1, \ldots, 2^{k-1} - 1. \end{array} \tag{2.30}$$

The graphical representation of this algorithm is given in Figure 1.

As previously done for the wavelet decomposition-reconstruction algorithm, Equations (2.29) and (2.30) can also be expressed in compact form by suitably defining certain decomposition matrices. Let $\mathbf{Q}_k^T$ be the block diagonal matrix with 2^{k-1} blocks given by $\mathbf{T}^T$. For example, for $k = 3$, we

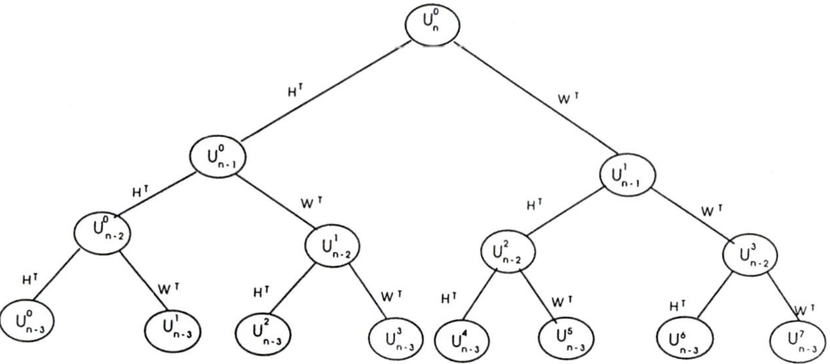

Figure 1. Wavelet packet decomposition tree.

have

$$\mathbf{Q}^T_{k(k=3)} = \left[\begin{array}{c|c} \begin{array}{c|c} \begin{matrix}\mathbf{H}^T\\ -\\ \mathbf{W}^T\end{matrix} & 0 \\ \hline \mathbf{0} & \begin{matrix}\mathbf{H}^T\\ -\\ \mathbf{W}^T\end{matrix}\end{array} & \mathbf{0} \\ \hline \mathbf{0} & \begin{array}{c|c} \begin{matrix}\mathbf{H}^T\\ -\\ \mathbf{W}^T\end{matrix} & 0 \\ \hline \mathbf{0} & \begin{matrix}\mathbf{H}^T\\ -\\ \mathbf{W}^T\end{matrix}\end{array}\end{array}\right] \tag{2.31}$$

From Equations (2.17) and (2.18), it follows that

$$\begin{cases} \mathbf{Q}_k^T\mathbf{Q}_k = \mathbf{I}, & k = 1, 2, \ldots \\ \mathbf{Q}_k\mathbf{Q}_k^T = \mathbf{I}. \end{cases} \tag{2.32}$$

That is, each $\mathbf{Q}_k$ is an orthogonal matrix.

The s steps WP decomposition algorithm is, therefore, in compact form:

$$\begin{bmatrix} \mathbf{u}_{n-s}^{0} \\ --- \\ \mathbf{u}_{n-s}^{1} \\ --- \\ \vdots \\ \vdots \\ \vdots \\ \vdots \\ --- \\ \mathbf{u}_{n-s}^{2^s-1} \end{bmatrix} = \mathbf{Q}_s^T \ldots \mathbf{Q}_2^T\, \mathbf{Q}_1^T \begin{bmatrix} \\ \mathbf{u}_n^0 \\ \\ \end{bmatrix}, \tag{2.33}$$

while the reconstruction algorithm is

$$\begin{bmatrix} \\ \mathbf{u}_0^n \\ \\ \end{bmatrix} = \mathbf{Q}_1\, \mathbf{Q}_2 \ldots \mathbf{Q}_s \begin{bmatrix} \mathbf{u}_{n-s}^{0} \\ --- \\ \mathbf{u}_{n-s}^{1} \\ --- \\ \vdots \\ \vdots \\ \vdots \\ \vdots \\ --- \\ \mathbf{u}_{n-s}^{2^s-1} \end{bmatrix} \tag{2.34}$$

Another more compact form of the s-step WP decomposition-reconstruction algorithm can be obtained by considering the matrix:

$$\mathbf{Q}^{sT} = \mathbf{Q}_s^T \mathbf{Q}_{s-1}^T \ldots \mathbf{Q}_1^T; \tag{2.35}$$

then $\mathbf{Q}^{sT}$ is a rectangular block matrix

$$\mathbf{Q}^{sT} = \begin{bmatrix} \mathbf{F}_0^T \\ \hdashline \mathbf{F}_1^T \\ \hdashline \vdots \\ \vdots \\ \vdots \\ \vdots \\ \hdashline \mathbf{F}_{2^s-1}^T \end{bmatrix}, \tag{2.36}$$

$\mathbf{F}_i$ being rectangular Hurwitz-type matrices generated (excluding the edges where they must be modified according to [4]) by shifting the filters $\mathbf{f}_i, i = 0, ..., 2^s - 1$, by 2^s places. These filters characterize the spaces U^i_{n-s} and are obtained by suitably convolving the basic filters $\mathbf{h}$ and $\mathbf{w}$.

More precisely, if we consider the dyadic expansion of the index i of $\mathbf{f}_i$ $0 \le i \le 2^s - 1$,

$$i = \sum_{j=1}^{s} \epsilon^i_j 2^{j-1}, \qquad \epsilon^i_j \in \{0, 1\}, \tag{2.37}$$

and we set

$$\begin{matrix} \mathbf{Z}_0 = \mathbf{H} \\ \mathbf{Z}_1 = \mathbf{W} \end{matrix} \quad \text{or equivalently} \quad \begin{matrix} \mathbf{z}_0 = \mathbf{h} \\ \mathbf{z}_1 = \mathbf{w}, \end{matrix}$$

then

$$\mathbf{F}_i^T = \mathbf{Z}^T_{\epsilon^i_1} \mathbf{Z}^T_{\epsilon^i_2} \ldots \mathbf{Z}^T_{\epsilon^i_s}, \tag{2.38}$$

or

$$\mathbf{f}_i = \mathbf{z}_{\epsilon^i_1} * \mathbf{z}_{\epsilon^i_2} * \ldots * \mathbf{z}_{\epsilon^i_s}. \tag{2.39}$$

For filters $\mathbf{h}$ and $\mathbf{w}$ of l_1 elements we obtain, after s convolution steps, filters $\mathbf{f}_i$, $i = 0, ..., 2^s - 1$, all of equal length given by

$$l_s = (2^s - 1)(l_1 - 1) + 1. \tag{2.40}$$

By means of these filters we can express the function $\psi^i_j(x)$ directly in terms of the scaling function $\psi_0(x) = \phi(x)$; that is, we have, since $j = n - s$

$$\psi^i_{n-s}(x) = 2^{s/2} \sum_{k=0}^{l_s - 1} f_{i_k} \phi(2^s x - k), \qquad i = 0, \ldots, 2^s - 1. \tag{2.41}$$

Moreover, the filters satisfy the orthogonality conditions:

$$\mathbf{F}_i^T\mathbf{F}_i = \mathbf{I}, \qquad \mathbf{F}_i^T\mathbf{F}_j = \mathbf{0}, \qquad j \neq i, \quad i = 0, \ldots, 2^s - 1, \tag{2.42}$$

and

$$\sum_{i=0}^{2^s-1} \mathbf{F}_i\mathbf{F}_i^T = \mathbf{I}. \tag{2.43}$$

This ensures that matrix $\mathbf{Q}^s$ is an orthogonal matrix. We can thus express the s-step WP decomposition-reconstruction algorithm as:

decomposition

$$\begin{bmatrix} \mathbf{u}_{n-s}^0 \\ \hline \mathbf{u}_{n-s}^1 \\ \hline \vdots \\ \vdots \\ \vdots \\ \vdots \\ \hline \mathbf{u}_{n-s}^{2^s-1} \end{bmatrix} = \begin{bmatrix} \mathbf{F}_0^T \\ \hline \mathbf{F}_1^T \\ \hline \vdots \\ \vdots \\ \vdots \\ \vdots \\ \hline \mathbf{F}_{2^s-1}^T \end{bmatrix} \begin{bmatrix} \\ \\ \mathbf{u}_n^0 \\ \\ \\ \end{bmatrix}, \tag{2.44}$$

reconstruction

$$\begin{bmatrix} \\ \\ \mathbf{u}_n^0 \\ \\ \\ \end{bmatrix} = \left[\begin{array}{c|c|c|c} \mathbf{F}_0 & \mathbf{F}_1 & \cdots & \mathbf{F}_{2^s-1} \end{array}\right] \begin{bmatrix} \mathbf{u}_{n-s}^0 \\ \hline \mathbf{u}_{n-s}^1 \\ \hline \vdots \\ \vdots \\ \vdots \\ \vdots \\ \hline \mathbf{u}_{n-s}^{2^s-1} \end{bmatrix}. \tag{2.45}$$

This form of the algorithm, even if not optimal from the computational point of view, is compact, easily generalizable and well suited for parallel implementation.

Remark. An important point regards the number of vanishing moments of the functions ψ_j^i or, equivalently, the number of vanishing moments of the filters $\mathbf{f}_i$, $i = 0, ...2^s - 1$, after s decomposition steps.

In fact, it is well-known that, if the mother wavelet $\psi(x)$ satisfies Equation (2.7), then the coefficients w_k, $k = 0, ..., 2M - 1$ satisfy

$$\sum_{k=0}^{2M-1} k^m w_k = 0 \qquad m = 0, 1, \ldots, M-1. \tag{2.46}$$

By remembering that the number M of vanishing moments of the filter corresponds to a zero of multiplicity M in $\omega = 0$ for its Fourier transform, it follows immediately that convolving two filters having respectively M_1 and M_2 vanishing moments, we obtain a resulting filter which has $M = M_1 + M_2$ vanishing moments. We can, therefore, conclude that the number of vanishing moments of the filters $\mathbf{f}_i, i = 0, ..., 2^s - 1$, varies from 0 to $s \cdot M$, according to definition in Equation (2.39) of the filters.

Remark. The decomposition Formula (2.44) together with relation in Equations (2.41) - (2.43) lets us easily see the analogy between wavelet packets and multiplicity R, k-regular, compactly supported orthonormal wavelet bases [8]. These bases are in fact characterized by a scaling vector $\mathbf{h}_0$ and $R - 1$ wavelet vectors $\mathbf{h}_l$, $l = 1, 2, ..., R - 1$, all of finite length N_R.

The scaling function and the wavelets associated with the scaling and wavelet vectors are given respectively by

$$\psi_0(x) = \sqrt{R} \sum_{k=0}^{N_R - 1} h_{0k} \psi_0(Rx - k), \tag{2.47}$$

$$\psi_l(x) = \sqrt{R} \sum_{k=0}^{N_R - 1} h_{lk} \psi_0(Rx - k), \tag{2.48}$$

and the scaling and wavelet vectors satisfy the following equations for l=1,2,..., R-1,

$$\sum_k h_{l,k} h_{m,k-Rm} = \delta(l - m)\delta(m), \tag{2.49}$$

$$\sum_k h_{l,k} = \sqrt{R}\delta(l). \tag{2.50}$$

These definitions compared with Equations (2.41)-(2.43) show that wavelet packets represent a particular choice of R-multiplicity wavelets where $R = 2^s$ and the scaling and wavelet vectors $\mathbf{f}_0$ and $\mathbf{f}_i, i = 1, ..., 2^s - 1$ are obtained by suitably convolving the basic filters $\mathbf{h}$ and $\mathbf{w}$.

§3 Linear operators in wavelet and wavelet-packet bases

Let $\mathbf{A}$ be the linear map

$$\mathbf{A} : V_n \longrightarrow V_n.$$

We consider

$$\mathbf{A}\mathbf{x} = \mathbf{y}, \quad \mathbf{x}, \mathbf{y} \in V_n.$$

As previously seen, vectors in V_n can be decomposed into their components in the subspaces $W_{n-1}, W_{n-2}, ..., W_{n-s}, V_{n-s}$ (W decomposition) or U^0_{n-s}, $U^1_{n-s}, \ldots, U^{2^s-1}_{n-s}$ (WP decomposition). Each decomposition corresponds to a suitable change of basis. This change must also be made for the linear map; that is, A must be decomposed into components (*i.e.*, submatrices) representing the action of A between these subspaces; that is obtained by expressing the original matrix in the same basis as the vectors.

3.1 Wavelet matrix decomposition

As previously seen, s steps of wavelet decomposition of a vector can be obtained by applying s times Matrix (2.21); that means that, if we define

$$\mathbf{T}^{sT} = \mathbf{T}^T_{n-s} ... \mathbf{T}^T_1, \tag{3.1}$$

the relation

$$\mathbf{A}\mathbf{x} = \mathbf{y}$$

can be expressed in the wavelet basis as

$$\mathbf{T}^{sT}\,\mathbf{A}\,\mathbf{T}^s\,\mathbf{T}^{sT}\,\mathbf{x} = \mathbf{T}^{sT}\,\mathbf{y}, \tag{3.2}$$

where the matrix

$$\overline{\mathbf{A}}^s = \mathbf{T}^{sT}\,\mathbf{A}\,\mathbf{T}^s, \tag{3.3}$$

has the form

$$\overline{\mathbf{A}}^s = \left[\begin{array}{c|c|c} \Omega^2 & \Gamma^2 & \Gamma^2_1 \\ \hline \Delta^2 & \Theta^2 & \Gamma^2_2 \\ \hline \Delta^2_1 & \Delta^2_2 & \Theta^1 \end{array}\right] \tag{3.4}$$

where

$$\Omega^0 = \mathbf{A}, \quad \Omega^i = \mathbf{H}^T \Omega^{i-1} \mathbf{H}, \quad \Theta^i = \mathbf{W}^T \Omega^{i-1} \mathbf{W}, \quad i = 1, \ldots, s;$$

$$\Gamma_1^1 = \Omega^1, \quad \Gamma_1^i = \mathbf{H}^T \Gamma_i^{i-1}; \quad \Gamma^i = \mathbf{H}^T \Omega^{i-1} \mathbf{W},$$

$$\Gamma_2^i = \mathbf{W}^T \Gamma_1^{i-1}; \quad \Delta^i = \mathbf{W}^T \Omega^{i-1} \mathbf{H},$$

$$\Delta_1^1 = \Omega^1, \quad \Delta_1^i = \Delta_1^{i-1} \mathbf{H}; \quad \Delta_2^i = \Delta_1^{i-1} \mathbf{W}.$$

This is the standard form of Beylkin-Coifman-Rokhlin [2].

As pointed out by the authors of [2], when matrix $\mathbf{A}$ is the discretization of Calderón-Zygmund or pseudo differential operators, *i.e.*, operators that are smooth away from the main diagonal, the elements of matrix realization in the wavelet basis decay as does the $(M+1)^{th}$ power of the inverse distance from the diagonals of the single building blocks (see estimate (4.26) of [2]).

Therefore, for wavelets with sufficiently large number of vanishing moments, if we discard all elements that are smaller than a predetermined precision ϵ, we can decompose the transformed matrix to $O(NlogN)$ elements which are localized in bands around the diagonals of the building blocks.

The computational complexity of this decomposition depends on the *a priori* information available about the operator of which matrix $\mathbf{A}$ represents the discretized form. If no information is available, the computational complexity of the wavelet decomposition is $O(N^2)$. However, if the structure of the singularities is known, the decomposition procedure requires order $(NlogN)$ operations if only coefficients exceeding the threshold ϵ are evaluated.

3.2 Wavelet packet matrix decomposition

The realization of operator matrices in wavelet basis presents a block structure whose dimensions decrease according to the halving of the dimensions of the wavelet spaces W_j. This can be a serious drawback if one considers using this algorithm for reducing the complexity of applying large dense matrices to arbitrary vectors. In fact, large scale computations often require the use of parallel computing, while the work-load which is inherently unbalanced in this decomposition prevents the realization of efficient parallel algorithms.

The wavelet packet basis, which corresponds to the decomposition of the original space $V_n = U_n^0$ (which is assumed to be of finite dimension $N = 2^n$) as

$$V_n = \bigoplus_{i=0}^{2^s-1} U_{n-s}^i, \quad s = 0, 1, \ldots, n, \tag{3.5}$$

where $U_{n-s}^i, i = 0, ..., 2^s - 1$ are orthogonal spaces of the same dimension $N_s = 2^{n-s}$, seems to be more suited for the realization of efficient parallel algorithms.

In fact, the expression of

$$\mathbf{A}\mathbf{x} = \mathbf{y}$$

in the wavelet packet basis is

$$\mathbf{Q}^{s^T} \mathbf{A} \mathbf{Q}^s \mathbf{Q}^{s^T} x = \mathbf{Q}^{s^T} \mathbf{y}, \tag{3.6}$$

where the matrix

$$\mathbf{A}^s = \mathbf{Q}^{sT}\mathbf{A}\mathbf{Q}^s \tag{3.7}$$

has the form

$$\mathbf{A}^{s(s=3)} = \left[\begin{array}{c|c|c|c} \mathbf{F}_0^T\mathbf{A}\mathbf{F}_0 & \mathbf{F}_0^T\mathbf{A}\mathbf{F}_1 & \mathbf{F}_0^T\mathbf{A}\mathbf{F}_2 & \mathbf{F}_0^T\mathbf{A}\mathbf{F}_3 \\ \hline \mathbf{F}_1^T\mathbf{A}\mathbf{F}_0 & \mathbf{F}_1^T\mathbf{A}\mathbf{F}_1 & \mathbf{F}_1^T\mathbf{A}\mathbf{F}_2 & \mathbf{F}_1^T\mathbf{A}\mathbf{F}_3 \\ \hline \mathbf{F}_2^T\mathbf{A}\mathbf{F}_0 & \mathbf{F}_2^T\mathbf{A}\mathbf{F}_1 & \mathbf{F}_2^T\mathbf{A}\mathbf{F}_2 & \mathbf{F}_2^T\mathbf{A}\mathbf{F}_3 \\ \hline \mathbf{F}_3^T\mathbf{A}\mathbf{F}_0 & \mathbf{F}_3^T\mathbf{A}\mathbf{F}_1 & \mathbf{F}_3^T\mathbf{A}\mathbf{F}_2 & \mathbf{F}_3^T\mathbf{A}\mathbf{F}_3 \end{array}\right] \tag{3.8}$$

Matrix $\mathbf{A}^s$ presents an equal sized block structure and, for each block $\mathbf{A}_{kl}^s$, an estimate similar to Equation (4.31) of [2] can be given (of course if the underlying operator satisfies the decay estimates in Equations (4.5), (4.6), and (4.30) of [2]); that is

$$\left|a_{k,l,i,j}^s\right| \leq \frac{C_{M_{kl}}}{1+|i-j|^{M_{kl}+1}} \tag{3.9}$$

where M_{kl} is the number of vanishing moments of the corresponding wavelet packet basis functions and $C_{M_{kl}}$ is a constant depending on the operator and on M_{kl}. This behavior of the elements of the matrix expressed in wavelet packet basis allows us to easily convert it into compressed form. In fact, by discarding all elements below a fixed tolerance we obtain the compressed matrix which presents a band diagonal block structure (with the exception, of course, for block $\mathbf{A}_{00}^s$). Examples of truncated wavelet packet (TWP) matrix decomposition are given in Figures 2 and 3.

In Figure 2, we show the compression of the matrix

$$\mathbf{A}_{1,ij} = \begin{cases} \dfrac{1}{i-j}, & i \neq j, \quad i,j = 1,\ldots,N, \\ 0, & i = j, \end{cases} \tag{3.10}$$

obtained by using the wavelet packets generated by a function $\psi_1(x)$ with four vanishing moments and by performing 3 decomposition steps. All entries in the resulting matrix below 10^{-6} are put to zero and the non-zero elements are shown in black in the figure. In Figure 3, we give the compressed form of the matrix

$$\mathbf{A}_{2,ij} = \frac{1}{0.25+(i-j)^2}, \qquad i,j = 1,\ldots,N. \tag{3.11}$$

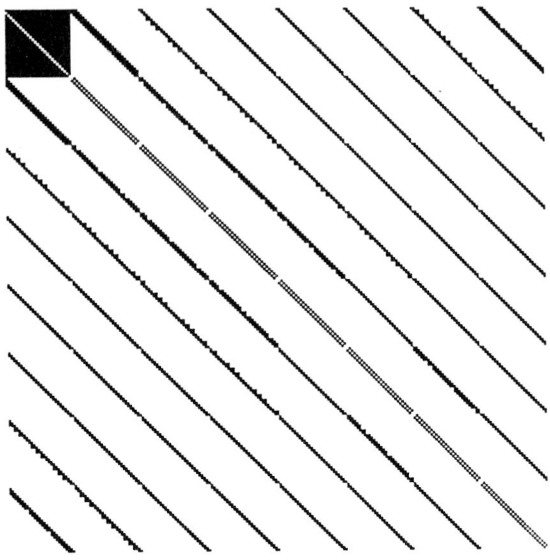

Figure 2. TWP decomposition of matrix $\mathbf{A}_1$ for N=1024, M=4, s=3. Entries above the threshold of 10^{-6} are shown in black.

In this example, the mother wavelet used has two vanishing moments, the decomposition steps are three, and only the entries above the threshold of 10^{-5} are displayed.

The approximation error of the presented compressed form can be evaluated as follows.

For each block $\mathbf{A}^s_{kl}$, we indicate with $\mathbf{A}^s_{B,kl}$, its compressed form obtained, after fixing a band width $B \geq 2M_{kl}$, by putting to zero the entries on place (i,j) with $|i-j| > B$. It is then easily shown, using the results of [13] and [2], that

$$||\mathbf{A}^s_{kl} - \mathbf{A}^s_{B,kl}|| \leq \frac{C_{M_{kl}}}{M_{kl} B^{M_{kl}}}. \tag{3.12}$$

The global estimate is then obtained by repeatedly applying the following.

Lemma. *Let* $\mathbf{A}_i$, $i = 1, \ldots, 4$ *be real* $m * m$ *matrices such that* $||\mathbf{A}_i|| \leq \epsilon_i$, $i = 1, \ldots 4$, *for positive constants* ϵ_i, $i = 1, \ldots 4$. *Then*

$$\left\| \begin{matrix} \mathbf{A}_1 & \mathbf{A}_2 \\ \mathbf{A}_3 & \mathbf{A}_4 \end{matrix} \right\| \leq 2\epsilon \tag{3.13}$$

where $\epsilon = \max_i \epsilon_i$.

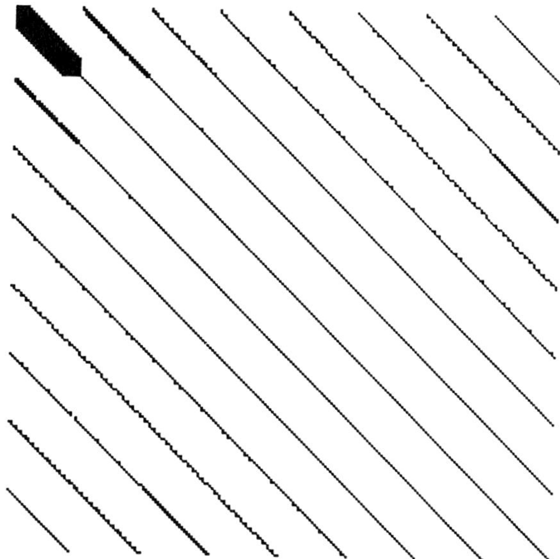

Figure 3. TWP decomposition of matrix $\mathbf{A}_2$ for N=1024, M=2, s=3. Entries above the threshold of 10^{-5} are shown in black.

After s decomposition steps from Equations (3.12) and (3.13), we, therefore, obtain

$$\|\mathbf{A}^s - \mathbf{A}_B^s\| \leq \frac{2^s C_M}{M B^M}. \tag{3.14}$$

Hence, by suitably choosing the bandwith B, we can evaluate the compressed form with the desired approximation error.

Remark. In the estimate in Equation (3.14), we have used (for the sake of simplicity) the number M of vanishing moments of the mother wavelet, even if we have said that the number of vanishing moments of the wavelet packets $\psi_{n-s}^i, i = 1, ..., 2^s - 1$ ranges from M to $s \cdot M$ (with the exclusion obviously of ψ_{n-s}^0). That means that, in practice, the error of the compressed form is less than the theoretical estimate in Equation(3.14) and it depends not so strongly on the number of decomposition steps (which is usually a low number, that is $2^s << N$). Moreover, in order to obtain the same accuracy for each block, the band width B has to be adapted to the number of vanishing moments of the basis functions, with consequent reduction in the calculations.

The wavelet packet decomposition procedure of Equation (3.7) requires order $2r_s N^2$ operations if all elements are calculated and no *a priori* information is available regarding the location of the coefficients corresponding to singularities in the operator. The operation count can be reduced if the above information is available. In fact, if only a band of width B is evaluated for each block, the number of operations required is $O(C_{sB}N)$ with $C_{sB} = 2^{s+1}B$.

From the above observations it follows that the wavelet packet decomposition is efficient to convert matrices to sparse form only if we consider a low number of decomposition steps.

§4 Parallel algorithms for matrix realization in wavelet packet basis

In the following, we will consider large-scale linear problems whose matrices are obtained as the discretization of integral operators of Calderón-Zygmund type or of pseudo differential operators (so-called "sparsifiable matrices") and try to solve them by means of efficient parallel algorithms, which exploit modern multiprocessor architectures in the best possible way. In particular, we will consider distributed-memory, message passing multiprocessors, with special emphasis on the hypercube. In fact, hypercube architecture is very flexible since many other interconnection topologies (rings, grids, trees, etc.) can be embedded in it. This makes the hypercube an ideal test bed for experimentation with parallel algorithms.

Algorithms for distributed-memory, message-passing multiprocessors must, in order to reach good efficiency, take the following requirements into account:

a. The arithmetic work-load must be well balanced among the processors; in fact, the parallel arithmetic time, t_{p_a}, is given by

$$t_{p_a} = \max_i t_{ia} \tag{4.1}$$

where t_{ia} is the time spent by the ith processor to execute the arithmetic calculations.

b. Particular care must be paid to the degree of concurrency of the algorithm. In other words, good synchronization between arithmetic computation and communication avoids a processor being idle while waiting for the data necessary for its computation.

c. The communication requirements must be kept as low as possible. In fact, in a distributed-memory system, communication is accomplished by passing messages among processors; hence, the need for global communication becomes a critical factor affecting algorithm efficiency.

Keeping these three fundamental requirements in mind, we have considered the sparse block forms $\overline{A}^s$ and A^s of the considered matrices realized in wavelet and wavelet packet basis. It is evident that $\overline{A}^s$ does not satisfy the previous points a and b; so we have discarded it as not suited for parallel algorithms.

On the contrary, the wavelet packet decomposition A^s, even if obtained at a higher cost, seems to represent a good starting point to construct efficient coarse-grained parallel algorithms. Nevertheless, before discussing any of these algorithms, we begin by considering an efficient parallel WP-decomposition algorithm as this is a necessary starting point for future developments.

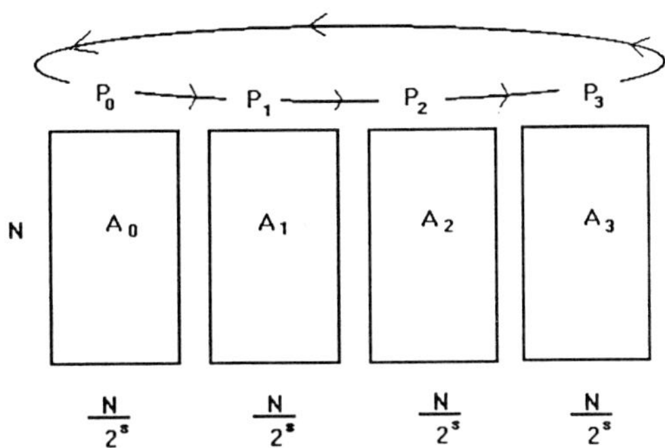

Figure 4. Scheme of the "column oriented" mapping of matrix **A** onto a ring of 2^s processors.

4.1 A parallel WP-decomposition algorithm

Let

$$\mathbf{A}^s = \mathbf{Q}^{sT}\mathbf{A}\mathbf{Q}^s \tag{4.2}$$

where **A** is an $N * N$ sparsifiable matrix, $\mathbf{Q}^{sT}$ the orthogonal wavelet packet decomposition matrix (2.36) characterized by the filters $\mathbf{f}_i$, $i = 0, ..., 2^s - 1$, and s the number of decomposition steps. We assume to work on a distributed memory multiprocessor with $p = 2^s$ processors.

The parallel decomposition algorithm we consider is different depending on the need to evaluate only the truncated form $\mathbf{A}^s_B$ or the complete decomposition $\mathbf{A}^s$. In the first case, it is probably possible to map directly onto each processor the band of the matrix needed for evaluation of the entries of part of $\mathbf{A}^s_B$ due to it, with consequent elimination of any communication among the processors. On the contrary, in the second case, the choice of mapping matrix onto the processors is of basic importance and the topology for which the communication requirements can best be met. For this reason and with the aim of minimizing communication, maximizing concurrency and having a balanced arithmetic work-load across the processors, we have chosen to map a contiguous block of N/p columns (or rows) onto each processor; the related topology which allows efficient communication (neighbor to neighbor), is a ring embedded in the hypercube. This can be efficiently realized through the binary reflected Grey code, which is a suitable ordering of the processors such that consecutive processors in the ring are physically connected in the hypercube (see Figure 4).

In both cases the algorithm consists of two steps: in the first, matrix A is reduced to horizontal blocks of dimension $(N/2^s) * N$ by means of premultiplication by matrix $\mathbf{Q}^{sT}$. This premultiplication can be done concurrently by the 2^s processors; each processor evaluates $\mathbf{F}_i{}^T\mathbf{A}$ with need of communication only if the complete decomposition has to be done. In the second step each processor executes the postmultiplication by $\mathbf{F}_i$, $i = 0, ..., 2^s - 1$, producing the desired block structure of $\mathbf{A}^s$ (or $\mathbf{A}^s_B$). During this step, in neither case is there any need of communication as each processor already possesses the filters $\mathbf{f}_i$, $i = 0, ..., 2^s - 1$, and the necessary horizontal block of the matrix.

The complete parallel algorithm can be described as follows.

First step: for $k = 1, \ldots, p-1$, each node P_i

a - evaluates the premultiplication of the currently residing block of N/p columns of $\mathbf{A}$ by the filter $\mathbf{F}_i^T$;
b - sends the block to its right-hand neighbor;
c - receives a new block from its left-hand neighbor.

Second step: for $k = 0, \ldots, p-1$, each node P_i

d - evaluates the post-multiplication $(\mathbf{F}_i^T\mathbf{A})\mathbf{F}_k$.

This algorithm is called "column oriented" since the mapping of the matrix has been done by blocks of columns. At the end of the algorithm, a block row of $\mathbf{A}^s$ is resident in each processor. Similarly, a "row oriented" algorithm can be realized starting from a mapping of blocks of rows and executing as the first step the postmultiplications by filters $\mathbf{F}_k$, $k = 0, \ldots, 2^s - 1$ and, as the second step, the premultiplications. In this second case, at the end of the algorithm, in each processor a block column of $\mathbf{A}^s$ is resident.

4.2 Performance analysis

The theoretical estimate of the performance of the presented algorithms is very simple. By indicating with S_1 and E_1, the speedup and efficiency of the parallel algorithm for the complete wavelet packet decomposition, and with S_2 and E_2 the same quantities for the truncated wavelet packet decomposition, we have

$$S_i = \frac{T_{is}}{T_{ip}}, \quad E_i = \frac{S_i}{p}, \quad i = 1, 2 \tag{4.3}$$

where T_{is} and T_{ip}, $i = 1, 2$, are, respectively, the scalar and parallel times for the WP and TWP decomposition. The scalar time is given by

$$T_{1s} = 2l_s N^2 \tau_a, \quad T_{2s} = C_{sB} N \tau_a\,, \tag{4.4}$$

where τ_a is the time to do a floating point multiplication, l_s is the length of the filters $\mathbf{f}_k$, $i = 0, \ldots, 2^s - 1$, and C_{sB} is a constant dependent on the number of decomposition steps s and the band width B. For the TWP decomposition, parallel time is given only by arithmetic time

$$T_{2p} = 2l_s \frac{N^2}{p} \tau_a \tag{4.5}$$

producing a "perfect" speedup and efficiency, while in the expression of parallel time for the WP decomposition we have two terms more, representing communication overheads due to $(p-1)$ neighbor to neighbor communications of rectangular blocks having $(N/P) * N$ elements.

Thus, we have

$$T_{1p} = 2l_s \frac{N^2}{p}\tau_a + (p-1)\frac{N^2}{p}\tau_{com} + (p-1)\tau_{start}, \tag{4.6}$$

with

τ_{comm} =time to communicate a single real word between physically connected processors;

τ_{start} = startup time for a message.

The need for communication together with the consequent synchronization overhead, slightly reduces the efficiency of the algorithm as the number of processors, and, therefore, of decomposition steps, increases. However, it still remains on good levels as shown in Table 1 which presents actual timing results for this algorithm obtained by running its implemetation on a hypercube iPSC/860.

Table 1. Timing results (in sec$*10^{-3}$)and performance evaluation of the parallel WP decomposition algorithm with $p = 2^s$ processors and $M = 2$.

N	*steps*	T_{1s}	T_{1p}	*Speedup*	*Efficiency*
256	2	2397	680	3.525	0.88
"	3	4623	762	6.06	0.76
512	2	9974	2556	3.90	0.97
"	3	19202	2788	6.88	0.86
1024	2	39896	10387	3.84	0.96
"	3	79797	10958	7.33	0.91

It is worth noting that, for good efficiency of the algorithm it is of great importance for the length of filter $\mathbf{f}_i$, $i = 0, ..., 2^s - 1$, to be always the same; this produces, in fact, a uniform arithmetic work-load across the processors.

Remark. When the position of the singularities of the operator is known *a priori* and the truncated form $\mathbf{A}_B^s$ of the matrix in wavelet packet basis is evaluated, the parallel TWP algorithm presented, which makes use only of a band of the matrix of a suitable width directly mapped onto each processor, does not consider that block $\mathbf{A}_{00}^s$ is not strictly banded but presents the same structure as the original matrix. The direct evaluation of $\mathbf{A}_{00}^s = \mathbf{F}_0^T \mathbf{A} \mathbf{F}_0$ would introduce a strong unbalance in the work-load of the processors with consequent strong reduction of efficiency. For this reason, we have decided to take advantage of the approximate quadrature formulas given in [2,16] and, while the processor zero evaluates only the diagonal part of $\mathbf{A}^s_{00}$ togheter with the other blocks of the first row of $\mathbf{A}^s_B$, the host provides the approximation of the missing entries of $\mathbf{A}^s_{00}$.

Remark. The speedup and efficiency presented in Table 1 are both relative (that is referred to the same algorithm running on a single processor), not obtained by considering the best scalar algorithm. In fact Algorithm (2.44) is not optimal from the computational point of view but is easily generalizable to the case of higher multiplicity wavelets of which wavelet packets represent a particular choice.

§5 Parallel algorithms for TWP compressed matrices

In this section we will consider two fundamental problems of parallel numerical linear algebra: matrix-vector and matrix-matrix multiplication. The matrices taken into consideration are so-called sparsifiable matrices, that is, they are the discretization of certain linear operators such as Calderón-Zygmund or pseudo-differential operators. For both these problems we propose coarse-grained parallel algorithms based on their TWP compressed form, which can reach, for a given accuracy, good efficiency on distributed memory multiprocessors.

5.1 Parallel evaluation of matrix-vector multiplication

The algorithm presented starts from the consideration that the application of a large sparsifiable matrix $\mathbf{A}$ to an arbitrary vector can be obtained (with finite precision) with an order $O(n)$ procedure if we express both in wavelet-packet basis. That is, instead of evaluating

$$\mathbf{y} = \mathbf{A}\mathbf{x}, \tag{5.1}$$

we first consider

$$\mathbf{y}^s = \mathbf{A}^s_B \mathbf{x}^s, \tag{5.2}$$

with $\mathbf{A}^s_B = trunc_B(\mathbf{Q}^{s^T}\mathbf{A}\mathbf{Q}^s)$ and $\mathbf{x}^s = \mathbf{Q}^{s^T}\mathbf{x}$ and then we evaluate

$$\mathbf{y} = \mathbf{Q}^s \mathbf{y}^s. \tag{5.3}$$

As previously seen, the TWP compressed form $\mathbf{A}^s_B$ is easily obtained by means of an efficient parallel algorithm. $\mathbf{A}^s_B$ is either directly calculated by means of a parallel TWP-decomposition algorithm, or obtained after a complete parallel WP decomposition by discarding all elements less than a predetermined tolerance. In both cases, working with $p = 2^s$ processors and after s decomposition steps, at the end of the column-oriented parallel algorithm, each block row of the compressed matrix is already resident in each processor. We then only need to add to the algorithm presented in the previous section a parallel multiplication algorithm which proceeds as follows.

Each processor $P_i, i = 0, \ldots, p-1$

1)- evaluates $\mathbf{F}_i^T\mathbf{x}$ and broadcasts the result $\mathbf{x}_i^s$ to all other processors;

2)- multiplies the currently residing block row of $\mathbf{A}_B^s$ into the vector $\mathbf{x}^s$, thus obtaining $\mathbf{y}_i^s$;

3)- evaluates $\mathbf{F}_i\mathbf{y}_i^s$ and sends it to the host which accumulates all contributions to obtain $\mathbf{y} = \sum_{i=0}^{p-1} \mathbf{F}_i\mathbf{y}_i^s$.

The performance analysis is straightforward. In fact, the only communication involved is the broadcast of vectors $\mathbf{x}_i^s$ in the first step, and the arithmetic work-load is equally distributed among the processors with the only exception of processor zero which has a little more work due to the fact that block $\mathbf{A}_{00}^s$ is not so strictly banded as the others.

Of course, this algorithm is efficient if the application of $\mathbf{A}$ to $\mathbf{x}$ has to be done a great number of times, as, for example, in the case of the iterative solution of large linear systems. In this case, the reduced complexity of the product $\mathbf{A}_B^s\mathbf{x}^s$ will compensate the effort in evaluating the compressed form $\mathbf{A}_B^s$.

5.2 Parallel multiplication of matrices in wavelet packet bases

Another important consequence of representing sparsifiable matrices in wavelet packet bases is that they can be efficiently and rapidly multiplied on parallel machines. In fact, if we consider the product of two s-step TWP decomposed matrices $\mathbf{X}$ and $\mathbf{Y}$, we obtain

$$\mathbf{Z}_B^s = \mathbf{X}_B^s \cdot \mathbf{Y}_B^s \tag{5.4}$$

which is the representation in the same wavelet packet basis as $\mathbf{X}$ and $\mathbf{Y}$ of $\mathbf{Z} = \mathbf{X} \cdot \mathbf{Y}$. Thus, if $\mathbf{Z}$ is itself a sparsifiable matrix, $\mathbf{Z}_B^s$ presents the same banded block structure as $\mathbf{X}_B^s$ and $\mathbf{Y}_B^s$, and each block is obtained as a sum of products of blocks of $\mathbf{X}_B^s$ and $\mathbf{Y}_B^s$ as follows.

$$\mathbf{Z}_{B,kl}^s = \sum_{j=0}^{2^s-1} \mathbf{X}_{B,kj}^s \cdot \mathbf{Y}_{B,jl}^s, \tag{5.5}$$

where elements in the product below the tolerance ϵ are set to zero in order to control the width of the bands of the product matrix. It is important to note that the number of operations required to evaluate $\mathbf{Z}_B^s$ is approximately $(2^sB)^2N$ instead of the N^3 necessary for the classical matrix product. (We have said approximately why we must take into account that blocks $\mathbf{X}_{B,00}^s$ and $\mathbf{Y}_{B,00}^s$ are, in general, dense and this would surely be true for block $Z_{B,00}^s$.)

It is now easy to see how $\mathbf{Z}_B^s$ can be efficiently evaluated on a distributed memory parallel machine.

We consider 2^s processors connected according to a ring topology and present a coarse-grained parallel multiplication algorithm which proceeds as follows.

Step 1

a) parallel column-oriented TWP decomposition of $\mathbf{X}$;
b) parallel row-oriented TWP decomposition of $\mathbf{Y}$.

After this step a block row of $\mathbf{X}_B^s$ and a block column of $\mathbf{Y}_B^s$ is present in each processor.

Step 2: for $k = 1, \ldots, p$, each processor P_i

a) executes the multiplication of the block row and block column currently residing in the processor;
b) sends the block column of $\mathbf{Y}_B^s$ to its righthand neighbor;
c) receives from its lefthand neighbor another block column of $\mathbf{Y}_B^s$.

After this step a block row of the product matrix is resident in each processor.

At the end of these two steps the product of the two matrices has been effected and the result is represented in wavelet packet basis, that is, it is already in sparse form. Further calculations can be done with fast algorithms, taking advantage of this compressed form.

A theoretical performance analysis of the parallel algorithm is presented without taking into account the first step of TWP decomposition, which has already been considered in the previous section. As communications are involved, the computation time for the parallel algorithm is the sum of two terms: the arithmetic time and the communication time (the latter including also some synchronization overhead). The first term, assuming that it is approximately the same for each processor, is given by

$$T_a = 2^s \cdot B^2 \cdot N \cdot \tau_a \tag{5.6}$$

while the second, due essentially to the "rolling" of the block columns of $\mathbf{Y}_B^s$ around the processor ring, is

$$T_{com} = (p-1) \cdot B \cdot N \cdot \tau_{com} + (p-1) \cdot \tau_{start}. \tag{5.7}$$

Thus, we have

$$T_p = 2^s \cdot B^2 \cdot N \cdot \tau_a + (p-1) \cdot B \cdot N \cdot \tau_{com} + (p-1) \cdot \tau_{start}, \tag{5.8}$$

where the last two terms represent communication overheads and will cause some loss in the efficiency of the algorithm. In practice, some further loss of efficiency is also due to the fact that, as block $\mathbf{Z}_{B,00}^s$ is generally expected to be full, in the first step of the multiplication algorithm, processor zero has to evaluate the full product $\mathbf{X}_{B,00}^s \cdot \mathbf{Y}_{B,00}^s$ instead of the band product.

This slight arithmetic work-load unbalance can be partially reduced by suitably pipelining calculations and communications. A typical example for which the presented parallel matrix multiplication algorithm could be very efficient is evaluation of the symmetric positive definite matrix

$$\mathbf{G} = \mathbf{A}^T\mathbf{A} \tag{5.9}$$

of the normal equation. In fact, with only the TWP decomposition of matrix $\mathbf{A}$ we easily obtain $\mathbf{G}_B^s \cong \mathbf{A}_B^s{}^T\mathbf{A}_B^s$ which is in compressed form, equally distributed by block row among the processors and, consequently, very suitable for solution with parallel iterative algorithms.

§6 Parallel preconditioned conjugate gradient method

It is well-known that the conjugate gradient method is one of the most efficient linear algebraic solvers for large-scale problems with symmetric positive definite matrices. Its convergence rate can be estimated in terms of the square root of the condition number of the coefficient matrix $\mathbf{A}$.

$$\|\mathbf{x}^k - \mathbf{x}\|_2 \le 2\sqrt{K(\mathbf{A})} \cdot \left(\frac{\sqrt{K(\mathbf{A})-1}}{\sqrt{K(\mathbf{A})+1}}\right) \cdot \|\mathbf{x}^0 - \mathbf{x}\|_2\,, \tag{6.1}$$

where

$$K(\mathbf{A}) = \|\mathbf{A}\|_2 \cdot \|\mathbf{A}^{-1}\|_2 = \frac{\lambda_{max}}{\lambda_{min}}. \tag{6.2}$$

If $\mathbf{A}$ is not well-conditioned, a preconditioning matrix $\mathbf{M}$ with the following properties is used.

1) $\mathbf{M}$ is a good approximation of $\mathbf{A}$ in the sense that the preconditioned matrix $\mathbf{M}^{-1}\mathbf{A}$ has a small condition number and its spectrum is clustered around 1;
2) linear systems with coefficient matrix $\mathbf{M}$ are relatively easy to solve.

In the parallel implementation of the PCG method the major bottleneck is often the parallelization of the preconditioner. Preconditioners that use purely local information are easily parallelizable but limited in their ability to improve the convergence rate. On the other hand, global coupling is more effective but not highly parallelizable. To overcome this difficulty, for a certain class of problems, multilevel preconditioners have been proposed in [1,17,18]. They are based on suitable coupling of the general multigrid idea with the use of nodal basis functions which leads to a certain orthogonal decomposition of the approximation space; preconditioning is therefore realized through a specific change of bases. A further generalization, which seems to present a good degree of parallelism, is that of multilevel filtering preconditioners [11]. In this case, preconditioning is achieved through a spectral decomposition obtained by means of suitably chosen band pass filters.

Due to the striking analogies of the above-mentioned concepts to those of multiresolution analysis and wavelet packet expansions, in this section, we present a parallel preconditioner based on the wavelet packet basis realization of **A**.

The main emphasis of this presentation is on its parallel efficiency which follows from its inherent high level of parallelism and from the efficient decomposition algorithm given in the previous sections. For a detailed analysis of the convergence behavior of preconditioners based on the wavelet packet decomposition of a certain type of matrix, the interested reader is referred to [14].

6.1 Optimal preconditioner from wavelet packet matrix realization

Let **A** be any $N * N$ matrix and $\mathbf{A}^s$ its expansion in the wavelet packet basis given by

$$\mathbf{A}^s = \mathbf{Q}^{sT}\mathbf{A}\mathbf{Q}^s. \tag{6.3}$$

Due to the orthogonality of matrix $\mathbf{Q}^s$, it is obvious that $\mathbf{A}^s$ has the same eigenvalues as **A**, since Equation (6.3) is a similarity transformation.

We now consider the block diagonal matrix

$$diag\mathbf{A}^{s(s=3)} = \left[\begin{array}{c|c|c|c} \mathbf{F}_0^T\mathbf{A}\mathbf{F}_0 & & & \\ \hline & \mathbf{F}_1^T\mathbf{A}\mathbf{F}_1 & & \\ \hline & & \mathbf{F}_2^T\mathbf{A}\mathbf{F}_2 & \\ \hline & & & \mathbf{F}_3^T\mathbf{A}\mathbf{F}_3 \end{array}\right] \tag{6.4}$$

and define the preconditioner $\mathbf{M}^s$ as

$$\mathbf{M}^s = \mathbf{Q}^s diag(\mathbf{A}^s)\mathbf{Q}^{sT}. \tag{6.5}$$

The following theorem states some results regarding the spectral properties of this preconditioner.

Theorem 1. *Let* **A** *be any* $N * N$ *matrix and* $\mathbf{A}^s$ *its espression in wavelet packet basis. If matrix* $\mathbf{M}^s$ *is given by*

$$\mathbf{M}^s = \mathbf{Q}^s diag(\mathbf{A}^s)\mathbf{Q}^{sT},$$

we have

$$\sigma_{max}(\mathbf{M}^s) \leq \sigma_{max}(\mathbf{A}), \tag{6.6}$$

and

$$tr(\mathbf{M}^s) = tr(\mathbf{A}). \tag{6.7}$$

Furthermore, if $\mathbf{A}$ *is Hermitian, we have*

$$\lambda_{min}(\mathbf{A}) \leq \lambda_{min}(\mathbf{M}^s) \leq \lambda_{max}(\mathbf{M}^s) \leq \lambda_{max}(\mathbf{A}). \tag{6.8}$$

In particular, if $\mathbf{A}$ *is positive definite, then* $\mathbf{M}^s$ *is also positive definite.*

The previous results show that $\mathbf{M}^s$ is a good candidate for preconditioning the conjugate gradient method as it preserves the positive definiteness of matrix $\mathbf{A}$, and its spectrum is bounded by spectrum $\mathbf{A}$. If we now denote with $\mathcal{D}_N^s$ the set of all $N * N$ block diagonal matrices where each block is a matrix of order $\frac{N}{2^s}$ and we set

$$\mathcal{M}^s = \left\{ \mathbf{Q}^s \Lambda^s \mathbf{Q}^{sT}, \Lambda^s \in \mathcal{D}_N^s \right\},$$

we can give a second theorem which characterizes some optimal approximation properties of the preconditioner $\mathbf{M}^s$.

Theorem 2. *For any arbitrary* $N * N$ *matrix* $\mathbf{A}$ *the preconditioner* $\mathbf{M}^s$*, uniquely determined by* $\mathbf{A}$ *as*

$$\mathbf{M}^s = \mathbf{Q}^s diag(\mathbf{Q}^{sT} \mathbf{A} \mathbf{Q}^{sT}),$$

is optimal, in the sense that it is the minimizer of

$$\|\mathbf{A} - \mathbf{C}^s\|_F \quad over \quad all \quad \mathbf{C}^s \in \mathcal{M}^s. \tag{6.9}$$

Furthermore the operator m_Q^s*, which maps every* $N * N$ *matrix* $\mathbf{A}$ *to the minimizer of* $\|\mathbf{A} - \mathbf{C}^s\|_F$ *over all* $\mathbf{C}^s \in \mathcal{M}^s$*, is a linear projection operator and has operator norms*

$$\|m_Q^s\|_2 = \|m_Q^s\|_F = 1. \tag{6.10}$$

The previous results have been stated for any arbitrary or symmetric positive definite matrix. In order to give an estimate of the behavior of $K(\mathbf{M}^{s^{-1}}\mathbf{A})$ or to analyze the convergence rate of the preconditioned conjugate gradient method, it is necessary to make some further hypotheses on matrix $\mathbf{A}$. In [14], it is shown that, for the special class of "sparsifiable" matrices, it is possible to prove that the condition number of the preconditioned matrix, $\mathbf{M}^{s^{-1}}\mathbf{A}$, is bounded by a constant which depends on the number M of vanishing moments of the mother wavelet $\psi(x)$, the number s of decomposition steps, but not on the matrix order.

The following tables will give some numerical results obtained for the classical matrix deriving from the discretization of the differential operator $Lv = -v''$, *i.e.*,

$$\mathbf{A}_{3,ij} = \begin{cases} 2 & \text{if } i = j, \\ -1 & \text{if } |i-j| = 1, \\ 0 & \text{otherwise,} \end{cases}$$

and for matrix $\mathbf{A}_2$ in the previous section.

Tables 2 and 3 show the behavior of the condition number of $\mathbf{M}^{s-1}\mathbf{A}$ when the matrix order N increases and for different values of M and s. The consequent number of the iterations of the WP-preconditioned conjugate gradient method is given in Tables 4 and 5 corresponding to the variation of the same quantities: N, M, and s.

Table 2. Condition number of the preconditioned systems $\mathbf{M}^{s-1}\mathbf{A}_2$ for different WP decomposition steps and different values of M.

N	*No pre–cond.*	$s=3$ $M=2$	$s=3$ $M=3$	$s=3$ $M=4$	$s=4$ $M=2$	$s=4$ $M=3$	$s=4$ $M=4$
128	132	2.5	1.8	1.3	2.9	1.9	1.4
256	262	2.5	1.8	1.3	3.0	1.9	1.4
512	522	2.6	1.8	1.3	3.0	1.9	1.4
1024	885	2.6	1.8	1.3	3.0	1.9	1.4

Table 3. Condition number of the preconditioned systems $\mathbf{M}^{s-1}\mathbf{A}_3$ for different WP decomposition steps and different values of M.

N	*No pre–cond.*	$s=3$ $M=2$	$s=3$ $M=3$	$s=3$ $M=4$	$s=4$ $M=2$	$s=4$ $M=3$	$s=4$ $M=4$
128	5587	7.7	3.3	1.8	11.5	3.8	1.9
256	22095	7.8	3.3	1.9	11.9	3.9	2.0
512	87829	7.8	3.3	1.9	12.1	4.0	2.0
1024	248386	7.8	3.3	1.9	12.1	4.0	2.0

Table 4. Number of iterations of the WP-preconditioned conjugate gradient method for different decomposition steps and different values of M (matrix $\mathbf{A}_2$).

N	*No precond.*	$M=2$ $s=3$	$M=2$ $s=4$	$M=4$ $s=3$	$M=4$ $s=4$
512	50	6	7	6	6
1024	61	6	7	6	6
2048	99	6	7	6	6

Table 5. Number of iterations of the WP-preconditioned conjugate gradient method for different decomposition steps and different values of M (matrix $\mathbf{A}_3$).

N	*No precond.*	$M=2$ $s=3$	$M=2$ $s=4$	$M=4$ $s=3$	$M=4$ $s=4$
512	469	11	13	11	13
1024	996	11	13	11	13
2048	1908	11	13	11	13

Finally, Table 6 shows that the great reduction of the number of iterations of the WP-preconditioned conjugate gradient method is not obtained at a higher computational cost. In fact, even if, in this first experimentation, we did not take into account the decay estimates given in the previous section and have evaluated the full blocks of diag($\mathbf{A}^s$), the computational time needed to obtain a solution according to the stopping criterion, $\frac{\|\mathbf{r}_k\|_2}{\|\mathbf{r}_0\|_2} < 10^{-6}$, is greatly reduced. A detailed analysis of the number of operations required to compute the optimal preconditioner $\mathbf{M}^s$ for different classes of matrix is given in [14].

Table 6. Scalar time in seconds for the solution $\mathbf{A}_2\mathbf{x} = \mathbf{b}$ with WP-preconditioned conjugate gradient method for different decomposition steps ($M = 2$).

N	*No precond.*	*PCG* $s=2$	*PCG* $s=3$	*PCG* $s=4$
512	8	1.62	1.75	2.73
1024	53	7.96	7.25	11.02
2048	285	43.07	31.68	46.34

6.2 The parallel algorithm

From the brief description given above, the inherent parallelism of preconditioner $\mathbf{M}^s$ is still evident. In fact, it is well-known that in the preconditioned conjugate gradient method the reduction of the number of iterations is obtained at the cost of solving the linear system

$$\mathbf{M}\overline{r}_k = r_k, \tag{6.10}$$

at each iteration, where $\mathbf{M}$ is the preconditioner, $\mathbf{r}_k$ the residual at the k-th iteration, and $\overline{r}_k$ the corrected residual.

If we now consider the preconditioner

$$\mathbf{M}^s = \mathbf{Q}^s diag(\mathbf{A}^s)\mathbf{Q}^{sT},$$

we can decide, instead of solving

$$\mathbf{M}^s\overline{\mathbf{r}}_k = \mathbf{r}_k, \tag{6.11}$$

to solve

$$diag(\mathbf{A}^s)\overline{\mathbf{r}}_k^s = \mathbf{r}_k^s \tag{6.12}$$

where

$$\overline{\mathbf{r}}_k^s = \mathbf{Q}^{sT}\overline{\mathbf{r}}_k, \qquad \mathbf{r}_k^s = \mathbf{Q}^{sT}\mathbf{r}_k, \tag{6.13}$$

and obtain the corrected residual as

$$\mathbf{r}_k = \mathbf{Q}^s\overline{\mathbf{r}}_k^s. \tag{6.14}$$

Solution of Equation (6.12) is easily and efficiently parallelizable, especially if we obtain the expression of $\mathbf{A}$ in wavelet packet basis by means of one of the parallel algorithms given in the previous sections. In any case, the diagonal blocks $\mathbf{A}_{ii}^s$, $i = 0, \ldots, 2^s - 1$, are already resident in the corresponding processors, with a consequent great reduction in the communication overhead. Moreover, in the parallel implementation of the conjugate gradient method, at each iteration, each processor already possesses the residual $\mathbf{r}_k$ so, without further need of communications, it can evaluate

$$\mathbf{r}_{k_i}^s = \mathbf{F}_i^T\mathbf{r}_k, \tag{6.15}$$

and solve the linear system

$$\mathbf{A}_{ii}^s\overline{\mathbf{r}}_{k_i}^s = \mathbf{r}_{k_i}^s. \tag{6.16}$$

The communication requirement of preconditioning is, therefore, limited to a *global sum* of all the contributions $\overline{\mathbf{r}}_{k_i} = \mathbf{F}_i\overline{\mathbf{r}}_{k_i}^s$ in order to obtain the corrected residual

$$\overline{\mathbf{r}}_k = \sum_{i=0}^{2^s-1} \overline{\mathbf{r}}_{k_i}. \tag{6.17}$$

Thus, the communication overhead of this preconditioner represents only a small part of the total communication of the parallel WP-preconditioned conjugate gradient method (see [15] for a detailed description of the parallel conjugate gradient method).

The arithmetic cost of this preconditioner for each processor is:

a) evaluation of $\mathbf{r}_{k_i}^s = \mathbf{F}_i^T\mathbf{r}_k$. This implies a computational cost given by:

$$cc \approx l_s * \frac{N}{p}\,; \tag{6.18}$$

b) solution of the linear system $\mathbf{A}^s_{ii}\overline{\mathbf{r}}^s_{k_i} = \mathbf{r}^s_{k_i}$, where $\mathbf{A}^s_{ii}$ is a symmetric, positive definite banded matrix. The corresponding computational cost is:

$$cc \approx \mathbf{B}^2 * \frac{N}{p}\,; \tag{6.19}$$

c) evaluation of $\overline{\mathbf{r}}_{k_i} = \mathbf{F}_i \overline{\mathbf{r}}^s_{k_i}$. The arithmetic cost is:

$$cc \approx l_s * \frac{N}{p}\,. \tag{6.20}$$

Thus, the total arithmetic overhead is

$$cc \approx O(k\frac{N}{p}), \tag{6.21}$$

with k constant depending on the mother wavelet used, on the number s of decomposition steps, and on the band width B.

This low arithmetic cost, together with the reduced communication requirement, makes this WP preconditioner very promising for a parallel implementation.

Of course, in the global evaluation of the efficiency of this preconditioner, it is necessary to take into account also the cost of the wavelet-packet decomposition. This cost is, however, partially absorbed by the decreased complexity in applying the decomposed matrix to the new search direction in the iterative algorithm.

Timing results, excluding the decomposition phase, are shown in Table 7. They are obtained by running the implemetation of the parallel WP-preconditioned conjugate gradient method on a hypercube iPSC/860 with embedded ring topology. Even in this first unoptimized experimentation, the performance evaluation shows that this parallel WP-preconditioner can reach good efficiency on distributed memory multiprocessors.

Table 7. Timing results (in sec$*10^{-3}$) and performance evaluation of the parallel WP preconditioned conjugate gradient method for the solution of $\mathbf{Ax} = \mathbf{b}$ for different numbers of processors (starting from the WP decomposed matrix with $M = 2$). Scalar times are obtained by running the same algorithm on a single processor.

N	*steps*	T_{1s}	T_{1p}	*Speedup*	*Efficiency*
256	2	1037	286	3.62	0.91
"	3	759	140	5.42	0.68
512	2	5035	1368	3.68	0.92
"	3	3301	469	7.04	0.88
1024	2	30708	8255	3.72	0.93
"	3	14356	1994	7.2	0.90

§7 Conclusion

We have presented some parallel numerical algorithms based on the wavelet packet decomposition of a matrix. We have then shown that, if matrix **A** is "sparsifiable", *i.e.*, it is the discretization of a certain type of integral or pseudo-differential operator, starting from an efficient parallel decomposition algorithm, the parallel algorithms presented can reach good efficiency on distributed memory multiprocessors. Since, as we pointed out previously, the wavelet packet basis can be seen as a particular choice of higher multiplicity wavelet basis, in the near future we intend to generalize these algorithms by making use of new filters $\mathbf{f}_i$ which, as shown in [9,10], can be algebraically constructed with higher numbers of vanishing moments and shorter support. This generalization can also be extended to the bi-orthogonal case.

Besides the promising experimental results, we think that the interest of this paper lies in the idea of spectral decomposition. In fact, just as it is well-known that the domain decomposition techniques are generally used for the construction of efficient parallel algorithms, so we think that this new idea of the spectral decomposition can, at least in some cases, be a more powerful tool for the construction and realization of parallel methods.

Acknowledgments. This research was supported by the Ministero della Ricerca Scientifica, 60% and 40% projects.

References

1. Bramble, J. H., J. E. Pasciak, and J. Xu, Parallel multilevel preconditioners, *Math. Comp.* **55** (1990), 1–22.
2. Beylkin, G., R. Coifman, and V. Rokhlin , Fast wavelet transform and numerical algorithms I, *Comm. Pure Appl. Math.* **44** (1991), 141–183.
3. Chui, C. K., *An Introduction to Wavelets*, Academic Press, Boston, 1992.
4. Cohen, A., I. Daubechies, and P. Vial, Wavelets on the interval and fast wavelet transform, *Appl. Comp. Harmonic Anal.* **1** (1993), 54–81.
5. Coifman, R. and Y. Meyer, Orthonormal wave packet bases, Yale University, 1990, preprint.
6. Daubechies, I., Orthonormal bases of compactly supported wavelets, *Comm. Pure Appl. Math.* **41** (1988), 909–996.
7. Daubechies, I., *Ten lectures on Wavelets*, CBMS-NSF Series in Applied Math. no. 61, SIAM Publ., 1992.
8. Gopinath, R. A. and C. S. Burrus, Wavelet transforms and filter banks, in *Wavelets: A Tutorial in Theory and Applications*, C. K. Chui (ed.), 1992, 603–654.
9. Kautsky, J., An algebraic construction of discrete wavelet transforms, *Aplikace Matematiky*, 1993, to appear.
10. Kautsky, J. and R. Turcajova, Pollen product factorization and construction of higher multiplicity wavelets, 1993, submitted to *Linear Algebra and its Applications*.

11. Jay Kuo, C. C., T. F. Chan, and C. Tong, Multilevel filtering elliptic preconditioner, *SIAM J. Matrix Anal. Appl.* **11** (3) (1990), 403–429.
12. Mallat, S., Multiresolution approximation and wavelet orthonormal bases of L^2, *Trans. Amer. Math. Soc.* **315** (1989), 69–88.
13. Meyer, Y., *Ondelettes et Opérateurs*, in two volumes, Hermann, Paris, 1990.
14. Montefusco, L. B., Optimal block preconditioning by means of wavelet packet decomposition, in progress.
15. Ortega, J. M., *Introduction to Parallel and Vector Solution of Linear Systems*, Plenum Press, New York, 1988.
16. Swelens, W. and R. Rossen, Calculation of the wavelet decomposition using quadrature formulae, *Wavelets: An Elemetary Treatment of Theory and Applications*, Tom H. Koornwinder (ed.), World Scientific, New Jersey, 1992, 139–160.
17. Tong, C. H., T. F. Chan, and C. C. Jay Kuo, A domain decomposition preconditioner based on a change to a multilevel nodal basis, *SIAM J. Sci. Stat. Comput.* **12** (66) (1991), 1486–1495.
18. Yserentant, H., Two preconditioners based on the multilevel splitting of finite element spaces, *Numer. Math.* **58** (1990), 163,184.

Laura Bacchelli Montefusco
Department of Mathematics
University of Bologna
40127 Bologna - Italy
montelau@dm.unibo.it

Representation of the Atomic Hartree-Fock Equations in a Wavelet Basis by Means of the BCR Algorithm

Patrick Fischer and Mireille Defranceschi

Abstract. The operator corresponding to the Hartree-Fock (HF) equation for cations with only one electron is decomposed onto an orthonormal wavelet basis. The Beylkin, Coifman, Rokhlin (BCR) algorithm is used to represent matrices in a sparse form called the Non-Standard (NS) form. For the sake of simplicity, only one-dimensional schemes of the operator are treated in this paper. The validity of the representation is tested by applying the operator to various functions, also expressed in a NS form, and the results are compared with the exact solution, the Slater function; the energy value is chosen as a comparative criterion.

§1 Introduction

Most often in quantum chemistry, molecular wave functions are determined in the so-called Hartree-Fock (HF) framework, which is an approximate expression for the Schrödinger equation where the interactions between electrons are treated in an average way [8]. The purpose of this paper is to show the potential of the Beylkin, Coifman, Rokhlin (BCR) algorithm [2] to represent the HF operator and to calculate some energetical quantities.

In the version restricted to double occupancy of spatial orbitals, the position space HF equation for a monoelectronic cation writes in atomic unit (a.u.) [10],

$$F\varphi(\mathbf{X}) = -\frac{\Delta}{2}\varphi(\mathbf{X}) - \frac{Z}{|\mathbf{X}|}\varphi(X) = \varepsilon\varphi(\mathbf{X}), \tag{1}$$

where the Fock operator F is the sum of a kinetic term $-\Delta(\mathbf{X})/2$ and a Coulombic potential term $V(\mathbf{X}) = -Z/|\mathbf{X}|$, with Z the nuclear charge. The wave function $\varphi(\mathbf{X})$ being radial, the Laplacian term can be expressed in spherical coordinates. Stating

$$\phi(|\mathbf{X}|) = |\mathbf{X}|\varphi(|\mathbf{X}|), \tag{2}$$

Wavelets: Theory, Algorithms, and Applications
Charles K. Chui, Laura Montefusco, and Luigia Puccio (eds.), pp. 495–506.

ISBN 0-12-174575-9

Equation (1) can be shortly written as:

$$-\frac{1}{2}\frac{\partial^2\phi(|\mathbf{X}|)}{\partial|\mathbf{X}|^2}-\frac{Z}{|\mathbf{X}|}\phi(|\mathbf{X}|)=\varepsilon\phi(|\mathbf{X}|). \tag{3}$$

To simplify calculations, Equation (3) is considered in a one-dimensional framework by stating $x=|\mathbf{X}|$ and $f(x)=\phi(|\mathbf{X}|)$. The operator under consideration becomes

$$Hf(x)=-\frac{1}{2}\frac{d^2f(x)}{dx^2}-\frac{Z}{|x|}f(x), \quad x\in\mathbb{R}, \tag{4}$$

with the function $f(x)$ being extended over the whole space $\mathbb{R}$ by anti-symmetry.

Orthonormal wavelets are used as basis functions to represent this operator [9]. They are characterized by a set of coefficients and they lead to a sparse matrix representation which allows fast calculations. In the present case, the effective wavelet transform is performed by cutting the operator into two parts which shall be treated separately. The kinetic term is analyzed in an iterative process while the potential term matrix is given by a quadrature formula; both will be described in the sequel.

Different approximations, called Slater Type Orbitals (STO), of the exact solution, the Slater function, are used to test the method. The error due to the numerical computations is determined by performing energetical quantities for each function. Energies being chosen as a comparative criterion, their theoretical values,

$$\varepsilon^{kin}=\frac{\langle Lf,f\rangle}{\langle f,f\rangle}=\text{kinetic energy, with } Lf(x)=-\frac{\Delta f}{2}(x), \tag{5a}$$

$$\varepsilon^{pot}=\frac{\langle Pf,f\rangle}{\langle f,f\rangle}=\text{potential energy, with } Pf(x)=-\frac{Z}{|x|}f(x), \tag{5b}$$

and

$$\varepsilon=\frac{\langle Hf,f\rangle}{\langle f,f\rangle}=\text{total energy, with } H=L+P, \tag{5c}$$

are computed in order to be compared with those determined by the algorithm.

§2 Description of the BCR algorithm

2.1 Wavelet transform of a function

Only compactly supported wavelets with vanishing moments constructed by Daubechies are used in this work. The advantage of these wavelets is that they lead to banded matrices with only few bands of nonzero values around the main diagonal. Their effective numerical construction can be found in [3]. We denote the scaling function by ϕ and the wavelet mother by Ψ. The wavelet basis is then given by:

$$\Psi_k^j(x)=2^{-j/2}\Psi(2^{-j}x-k), \qquad k,j\in\mathbf{Z}. \tag{6}$$

The multiresolution analysis [7] associated with such basis will be denoted by $\{V_j\}_{j\in\mathbf{Z}}$. For a fixed j, the basis corresponding to the subspace V_j is given by:

$$\phi_k^j(x) = 2^{-j/2}\phi(2^{-j}x - k), \qquad k \in \mathbf{Z}. \tag{7}$$

The scaling function $\phi(x)$ can also be written as:

$$\phi(x) = \sqrt{2}\sum_{k=0}^{L-1} h_k\, \phi(2x-k), \qquad h_k = \langle \phi, \phi_{-1,k}\rangle, \tag{8}$$

where the number L of coefficients is related to the number M of vanishing moments of $\Psi(x)$ and also related to other properties that can be imposed to $\phi(x)$. Consequently, the function Ψ can be also written as:

$$\Psi(x) = \sqrt{2}\sum_{k=0}^{L-1} g_k\, \phi(2x-k) \tag{9}$$

with $g_k = (-1)^k h_{L-k-1}, k = 0, \ldots, L-1$.

Functions verifying Equation (8) have their support included in $[0, \ldots, L-1]$. If there exists a coarsest scale ($j = n$) and a finest scale ($j = 0$), then Equations (6) and (7) can be rewritten as

$$\phi_{j,k}(x) = \sum_{\ell=0}^{L-1} h_\ell\, \phi_{j-1,2k+\ell}(x), \qquad j = 1, \ldots, n, \tag{10}$$

and

$$\Psi_{j,k}(x) = \sum_{\ell=0}^{L-1} g_\ell\, \phi_{j-1,2k+\ell}(x), \qquad j = 1, \ldots, n. \tag{11}$$

To perform the wavelet transform of a function $f(x)$, two sets of coefficients, d_k^j and s_k^j defined as

$$d_k^j = \int dx\; f(x)\Psi_{j,k}(x), \tag{12}$$

and

$$s_k^j = \int dx\; f(x)\phi_{j,k}(x), \tag{13}$$

have to be computed. Using Formulae (10) and (11) produces:

$$d_k^j = \sum_{\ell=0}^{L-1} g_\ell \int dx\; f(x)\phi_{j-1,2k+\ell}(x), \tag{14}$$

and

$$s_k^j = \sum_{\ell=0}^{L-1} h_\ell \int dx \; f(x)\phi_{j-1,2k+\ell}(x). \tag{15}$$

Starting with an initial set of coefficients s_k^0, $k = 0, \ldots, N-1$, d_k^j and s_k^j are determined by the following recursive formulae:

$$d_k^j = \sum_{\ell=0}^{L-1} g_\ell \; s_{2k+\ell}^{j-1} \,, \tag{16}$$

and

$$s_k^j = \sum_{\ell=0}^{L-1} h_\ell \; s_{2k+\ell}^{j-1} \,. \tag{17}$$

In Equations (16) and (17), d_k^j and s_k^j are considered as periodic sequences of period 2^{n-j}. The set d_k^j is composed of the coefficients corresponding to the decomposition of $f(x)$ on the basis $\Psi_{j,k}$, and s_k^j may be viewed as the set of averages between two scales.

The principle of applying Equations (16) and (17) usually is symbolized by the pyramid scheme,

$$\begin{array}{ccccccc} & \{s_k^0\} & & & & & \\ \swarrow & & \searrow & & & & \\ \{d_k^1\} & & \{s_k^1\} & & & & \\ & \swarrow & & \searrow & & & \\ & \{d_k^2\} & & \{s_k^2\} & & & \\ & & \swarrow & & \searrow & & \\ & & \{d_k^3\} & & \{s_k^3\} & & \\ & & & \swarrow & & \searrow & \end{array} \tag{18}$$

2.2 Non-Standard form of integral operators

Let us consider operators that can be written in an integral form,

$$T(f)(x) = \int dy \; K(x,y)f(y), \tag{19}$$

where $K(x,y)$ is the kernel associated to the operator T. In order to carry out computations, the representation of an operator consists in writing the corresponding kernel as a matrix. The Non-Standard (NS) form related to a wavelet basis leads to sparse matrices, and as a result, speeds up calculations. It is obtained by developing the kernel on the following two dimensional basis:

$$\{\Psi_{j,k}(x)\Psi_{j,k'}(y), \;\; \Psi_{j,k}(x)\phi_{j,k'}(y), \;\; \phi_{j,k}(x)\Psi_{j,k'}(y)\}_{j,k,k'\in \mathbf{Z}'}, \tag{20}$$

while the basis

$$\{\Psi_{j,j',k,k'}(x,y) = \Psi_{j,k}(x)\Psi_{j',k'}(y)\}_{j,j',k,k' \in \mathbf{Z}'}, \tag{21}$$

leads to the standard form (where matrices are full). Hence, the three following sets of coefficients

$$\alpha^j_{k,k'} = \int\int dxdy\ K(x,y)\Psi_{j,k}(x)\Psi_{j,k'}(y), \tag{22}$$

$$\beta^j_{k,k'} = \int\int dxdy\ K(x,y)\Psi_{j,k}(x)\phi_{j,k'}(y), \tag{23}$$

$$\gamma^j_{k,k'}(x) = \int\int dxdy\ K(x,y)\phi_{j,k}(x)\Psi_{j,k'}(y), \tag{24}$$

have to be computed. By applying Formulae (10) and (11), Equations (22), (23), and (24) may be rewritten as:

$$\alpha^j_{k,k'} = \sum_{\ell,\ell'=0}^{L-1} g_\ell\ g_{\ell'}\ r^{j-1}_{2k+\ell,2k'+\ell'}, \tag{25}$$

$$\beta^j_{k,k'} = \sum_{\ell,\ell'=0}^{L-1} g_\ell\ h_{\ell'}\ r^{j-1}_{2k+\ell,2k'+\ell'}, \tag{26}$$

$$\gamma^j_{k,k'} = \sum_{\ell,\ell'=0}^{L-1} h_\ell\ g_{\ell'}\ r^{j-1}_{2k+\ell,2k'+\ell'}, \tag{27}$$

where $r^j_{k,k'}$ is a fourth set of coefficients defined as

$$r^j_{k,k'} = \int\int dxdy\ K(x,y)\phi_{j,k}(x)\phi_{j,k'}(y), \tag{28}$$

which verifies the recursive rule:

$$r^j_{k,k'} = \sum_{\ell,\ell'=0}^{L-1} h_\ell\ h_{\ell'}\ r^{j-1}_{2k+\ell,2k'+\ell'}, \tag{29}$$

where $k, k' = 0, \ldots, 2^{n-j} - 1$, $j = 1, \ldots, n$.

If one denotes P_j to be the projection operator from $L^2(\mathbb{R})$ on the subspace V_j whose $\{\phi_{j,k}\}_{k\in\mathbf{Z}}$ is the basis, and Q_j the projection on W_j, then $\{\alpha^j_{k,k'}\}_{k,k'\in\mathbf{Z}}$, $\{\beta^j_{k,k'}\}_{k,k'\in\mathbf{Z}}$, $\{\gamma^j_{k,k'}\}_{k,k'\in\mathbf{Z}}$, $\{r^j_{k,k'}\}_{k,k'\in\mathbf{Z}}$ represent, respectively, the operators $A_j = Q_jTQ_j$, $B_j = Q_jTP_j$, $G_j = P_jTQ_j$, and $T_j = P_jTP_j$. The discretization T_0 of the operator T on the finest scale may be written as

$$T_0 = \sum_{j=1}^{n} A_j + B_j + G_j + T_n. \tag{30}$$

The matrix representation of an operator applied to a vector may be depicted by:

$$\begin{pmatrix} A_1 \vdots B_1 & & \\ \cdots\cdots & & \\ G_1 \vdots & & \\ & A_2 \vdots B_2 & \\ & \cdots\cdots & \\ & G_2 \vdots & \\ & & A_3 \vdots B_3 \\ & & \cdots\cdots \\ & & G_3 \vdots F_3 \end{pmatrix} * \begin{pmatrix} d^1 \\ \cdots \\ s^1 \\ \cdots \\ d^2 \\ \cdots \\ s^2 \\ \cdots \\ d^3 \\ \cdots \\ s^3 \end{pmatrix} = \begin{pmatrix} \widehat{d^1} \\ \cdots \\ \widehat{s^1} \\ \cdots \\ \widehat{d^2} \\ \cdots \\ \widehat{s^2} \\ \cdots \\ \widehat{d^3} \\ \cdots \\ \widehat{s^3} \end{pmatrix}. \quad (31)$$

In order to represent the HF operator in the NS form, illustrated in Equation (31), two different techniques will be used to compute the matrix coefficients for the Laplacian and the potential terms.

§3 Non-Standard form of the Hartree-Fock operator

The two parts of the HF operator defined in the introduction are analyzed separately with two different techniques. The NS form of the Laplacian term is worked out in an iterative process and the potential term by a quadrature formula.

3.1 Representation of the Laplacian term

The analysis of the operator,

$$Lf = -\frac{\Delta f}{2}, \quad (32)$$

is, in fact, a particular case of the representation of the derivative operator d^p/dx^p. The method to find the NS form of this operator is entirely described by Beylkin [1]. Only the most important points of this process are reminded. Starting with Equations (25), (26), and (27), coefficients $\{\alpha^j_{k,k'}\}_{k,k'\in\mathbf{Z}}$, $\{\beta^j_{k,k'}\}_{k,k'\in\mathbf{Z}}$, $\{\gamma^j_{k,k'}\}_{k,k'\in\mathbf{Z}}$, and $\{r^j_{k,k'}\}_{k,k'\in\mathbf{Z}}$ are performed with the kernel $K(x,y)$ defined by

$$\frac{d^p f(x)}{dx^p} = \int dy\ K(x,y)f(y). \quad (33)$$

By applying Equation (33) to Equations (22), (23), (24), and (28), one obtains

$$\alpha^j_{k,k'} = \int dx\ \Psi(2^{-j}x - k)2^{-pj}\ \Psi^{(p)}(2^{-j}x - k') = 2^{-pj}\alpha_{k-k'}, \quad (34)$$

$$\beta^j_{k,k'} = \int dx\ \Psi(2^{-j}x - k)2^{-pj}\ \phi^{(p)}(2^{-j}x - k') = 2^{-pj}\beta_{k-k'}, \quad (35)$$

$$\gamma^j_{k,k'} = \int dx \ \phi(2^{-j}x - k)2^{-pj} \ \Psi^{(p)}(2^{-j}x - k') = 2^{-pj}\gamma_{k-k'}, \tag{36}$$

$$r^j_{k,k'} = \int dx \ \phi(2^{-j}x - k)2^{-pj} \ \phi^{(p)}(2^{-j}x - k') = 2^{-pj}r_{k-k'}, \tag{37}$$

where

$$\alpha_\ell = \int dx \ \Psi(x - l) \ \Psi^{(p)}(x), \tag{38}$$

$$\beta_\ell = \int dx \ \Psi(x - l) \ \phi^{(p)}(x), \tag{39}$$

$$\gamma_\ell = \int dx \ \phi(x - l) \ \Psi^{(p)}(x), \tag{40}$$

$$r_\ell = \int dx \ \phi(x - l) \ \phi^{(p)}(x). \tag{41}$$

Coefficients $\{\alpha_\ell\}_{\ell\in\mathbf{Z}}$, $\{\beta_\ell\}_{\ell\in\mathbf{Z}}$, $\{\gamma_\ell\}_{\ell\in\mathbf{Z}}$ can be written as expressions involving the set $\{r_\ell\}_{\ell\in\mathbf{Z}}$ which verifies itself an iterative rule:

$$\alpha_\ell = 2^p \sum_{m,m'=0}^{L-1} g_m \ g_{m'} \ r_{2\ell+m-m'}, \tag{42}$$

$$\beta_\ell = 2^p \sum_{m,m'=0}^{L-1} g_m \ h_{m'} \ r_{2\ell+m-m'}, \tag{43}$$

$$\gamma_\ell = 2^p \sum_{m,m'=0}^{L-1} h_m \ g_{m'} \ r_{2\ell+m-m'}, \tag{44}$$

and

$$r_\ell = 2^p \sum_{m,m'=0}^{L-1} h_m \ h_{m'} \ r_{2\ell+m-m'}. \tag{45}$$

Using Proposition V.2 of the Beylkin's paper [1], one deduces that,

- $r_\ell = 2^p \left[r_{2\ell} + \frac{1}{2} \sum_{m=1}^{L/2} a_{2m-1}(r_{2\ell-2m+1} + r_{2\ell+2m-1}) \right], \quad \ell \in \mathbf{Z},$ (46)
- $\sum_{\ell\in\mathbf{Z}} \ell^p r_\ell = (-1)^p p!\ ,$ (47)
- $r_\ell \neq 0, \text{ for } -L+2 \leq \ell \leq L-2\ ,$ (48)
- $r_\ell = (-1)^p r_{-\ell}\ ,$ (49)

where

$$a_{2m-1} = 2 \sum_{i=0}^{L-2m} h_i \ h_{i+2m-1}. \tag{50}$$

By using Equation (47) to find an initial guessed set of coefficients, Equation (46) is iterated to compute $(r_\ell)_{\ell\in\mathbf{Z}}$. The convergence of the process is obtained in less than ten iterations, with an accuracy of 10^{-8}.

Restrictive conditions for the choice of the wavelets basis will be stated for the analysis of the potential term in the next subsection. Let us note that it can be easily proved that Equations (46) and (47) do not have any solution for the Laplacian operator if the number M of vanishing moments is equal to two.

3.2 Representation of the potential term

A quadrature formula proposed by Beylkin et al. [2] to compute Equation (24) without using the recursive scheme is presented in this subsection. The principle is first described for an one-dimensional function $f(x)$, and then the process is applied to the two-dimensional kernel $K(x,y)$ corresponding to the potential term.

If the scaling function $\phi(x)$ verifies the following conditions,

$$\int dx\ \phi(x+\tau)x^m = 0, \qquad m = 1,\ldots,M-1, \tau \in \mathbb{N}, \tag{51}$$

$$\int dx\ \phi(x) = 1, \tag{52}$$

and if the function $f(x)$ to analyze is smooth enough, then it is possible to compute s_k^j from an expansion of $f(x)$ into a Taylor series around $2^j(k+\tau)$. The result is

$$s_k^j = 2^{j/2} f(2^j(k+\tau)) + O(2^{j(M+1/2)}). \tag{53}$$

Coefficients $\{h_k\}_{k=0,\ldots,L-1}$ corresponding to the scaling function $\phi(x)$ which verifies Conditions (51) and (52) can be found in [2]. Application of this quadrature formula to a singular kernel is possible if its singularities are exactly located on the diagonal and if it is smooth out of the diagonal:

$$|K(x,y)| \leq \frac{1}{|x-y|}, \tag{54}$$

$$\left|\frac{\partial^m K(x,y)}{\partial x^m}\right| + \left|\frac{\partial^m K(x,y)}{\partial y^m}\right| \leq \frac{C_m}{|x-y|^{m+1}}, \quad m \geq 1. \tag{55}$$

By applying the quadrature formula to each variable x and y of a kernel $K(x,y)$ verifying Inequalities (54) and (55), one obtains

$$r_{k,k'}^j = 2^j K(2^j(k+\tau), 2^j(k'+\tau)) + O\left(\frac{1}{|k-k'|^{M+1}}\right). \tag{56}$$

Calculations of the NS form of the potential term is very simple if the corresponding kernel is known. The expected function $K(x,y)$ has to verify

$$Pf(x) = -\frac{Z}{|x|} f(x) = \int dy\ K(x,y) f(y). \tag{57}$$

The solution is a kernel involving the Dirac function

$$K(x,y) = -\frac{Z}{|x|}\,\delta(x-y). \tag{58}$$

Although $K(x,y)$ is a particular kernel defined only where Equations (54) and (55) are not, we can use the Formula (56) to compute $r^j_{k,k'}$. In fact, only $r^0_{k,k'}$ will be determined with the quadrature and then the recursive scheme in Equation (29) will be used to compute $r^j_{k,k'}$.

§4 Numerical applications

4.1 Application of an operator to arbitrary functions

To test the appropriateness of the method for the present problem, the NS form of the HF operator corresponding to the hydrogen atom ($Z = 1$) is applied to various functions that symbolize wave functions of chemical interest.

In such a representation, it suffices to multiply the matrix related to the operator by the vector related to the function. The result is consequently expressed as sets of coefficients $\{\mathbf{d}^j_k\}_{j,k\in\mathbf{Z}}$ and $\{\mathbf{s}^j_k\}_{j,k\in\mathbf{Z}}$, which allow the determination of the approximated operator T_0 applied to a function $f(x)$:

$$T_0(f)(x) = \sum_{j=1}^{n}\sum_{k=0}^{2^{n-j}}(\mathbf{d}^j_k\Psi_{j,k}(x) + \mathbf{s}^j_k\phi_{j,k}(x)). \tag{59}$$

As $\{\Psi_{j,k}\}_{j,k\in\mathbf{Z}}$ is an orthonormal basis of $L^2(\mathbb{R})$, the scalar product of a function $f1(x)$ and a function $f2(x)$ respectively represented by$\{d1^j_k\}_{j,k\in\mathbf{Z}}$ and $\{d2^j_k\}_{j,k\in\mathbf{Z}}$ is given by

$$\langle f1, f2\rangle = \sum_{j,k\in\mathbf{Z}} d1^j_k \cdot d2^j_k. \tag{60}$$

Consequently, the determination of the norm of a function can be made from its coefficients $\{d^j_k\}_{j,k\in\mathbf{Z}}$:

$$\|f\|_2 = \left(\sum_{j,k\in\mathbf{Z}} (d^j_k)^2\right)^{1/2}, \tag{61}$$

where $\|..\|_2$ denotes the norm in $L^2(\mathbb{R})$.

Therefore, if the result of the application of the operator H to a function f is denoted $\widehat{f}$, then the energy is given by:

$$\varepsilon_d = \sum_{j,k\in\mathbf{Z}} \widehat{d^j_k} \cdot d^j_k. \tag{62}$$

(Here, the function $f(x)$ is supposed to be a normalized function.)

In the numerical tests given in the next subsection, values of the energy computed with the method just described, ε_d, are compared with the theoretical values, ε, given in Equation (5) to assess the error due to the discretization for the various trial wave functions.

4.2 Numerical results

The sequence of calculations to be performed can be summarized as follows:

1. Computation of the matrices A_j, B_j and G_j, Equations (22), (23), and (24), by using the two techniques:
   - iterative process for the Laplacian term,
   - quadrature formula for the potential term.

2. Computation of the vectors d_j and s_j, Equations (12) and (13), related to various wave functions, and numerical evaluation of their norms (in order to verify the normalization factor).

3. Application of the matrices A_j, B_j and G_j to the vectors d_j and s_j to obtain the vectors $\mathbf{d}_j$ and $\mathbf{s}_j$ (Equation (31).)

4. Reconstruction, Equation (59), of the transformed function from the vectors $\mathbf{d}_j$ and $\mathbf{s}_j$.

5. Computation of the energies from the NS forms of the initial and the transformed functions.

COMMENTS:

a. Application of the quadrature formula requires a large set of coefficients $\{h_k\}$ because of Conditions (51) and (52) which have been imposed to the scaling function $\phi(x)$. In the present case, the number L of coefficients is equal to $3M$ instead of $2M$ (M being the number of vanishing moments) [2].

b. Calculations have been made for two sets of coefficients corresponding to $M = 4$ and $M = 6$. As results do not present significative differences only values for $M = 4$ are reported in this paper. The program does not give any solution for $M = 2$ since the iterative process used for the Laplacian operator does not converge.

Results obtained by a Fortran program based on the algorithm described above are reported in the table below. Four wave functions have been used to test the program: a Slater function which is the exact solution of the HF equation, and three Gaussian approximations constructed with, respectively, one (STO-1G), two (STO-2G), and three (STO-3G) Gaussian functions.

For each test, the program has been run for several values of N, and the results given in the table correspond to $N = 256$.

Three expectation values $\varepsilon_d^{kin}, \varepsilon_d^{pot}$, and ε_d, respectively related to the values of the kinetic, potential, and total energies are determined for each trial wave function. They are compared with their values ε^{kin}, ε^{pot}, and ε obtained analytically from Equations (5a), (5b), and (5c).

Table. Energy values (in a.u.) for various bases.

	STO − 1G	**STO − 2G**	**STO − 3G**	**SLATER**
ε^{kin}	0.4244132	0.4857612	0.4967535	0.5000000
ε_d^{kin}	0.4244099	0.4857904	0.4968063	0.5002572
Error	$7.78\ 10^{-6}$	$6.01\ 10^{-5}$	$1.06\ 10^{-4}$	$5.14\ 10^{-4}$
ε^{pot}	−0.8488264	−0.9715744	−0.9937322	−1.0000000
ε_d^{pot}	−0.8486610	−0.9711148	−0.9929722	−0.9985268
Error	$1.95\ 10^{-4}$	$4.73\ 10^{-4}$	$7.65\ 10^{-4}$	$1.47\ 10^{-3}$
ε	−0.4244132	−0.4858132	−0.4969787	−0.5000000
ε_d	−0.4242511	−0.4853243	−0.4961660	−0.4982696
Error	$3.82\ 10^{-4}$	$1.01\ 10^{-3}$	$1.65\ 10^{-3}$	$3.46\ 10^{-3}$

4.3 Comments

The first comment concerns total energies computed with the algorithm ε_d which are relatively close to their analytical values ε. The fact that the error due to the discretization is not the same one for each wave function is directly related to the asymptotic behavior of the functions used. Compact supported wavelets lead to a better analysis of Gaussian functions which have a smaller support (*i.e.*, STO-1G is better described then STO-2G, STO-3G, or the Slater function).

A second comment is that the relative good results obtained for the potential term put our mind at ease about the use of Equations (56). However, depending on the choice of various implicit parameters (choice of the wavelets, choice of the discretization step, ...) kinetic and potential terms can be independently accurately described. The results given in the table above correspond to the best compromise in minimizing the total energy ε.

§5 Conclusion

The aim of this work was to show the potential of orthonormal wavelets in a quantum chemistry problem. We mainly focused on one (BCR algorithm) of the possible representation of HF operator in a wavelet basis. Conclusive results obtained for the trivial case of the hydrogen atom encourage the development of a treatment of HF equations through the BCR algorithm in order to get a better analysis of wave functions.

Similarly to what had been done previously with the Fourier [5] and the continuous wavelet [6] transforms, an iterative process has been defined to solve these equations. The convergence of the method has been obtained in less than ten iterations and with high accuracy. The numerical nature of the method should allow for the expansion of the calculations to bigger systems than the one studied in this paper. Works along these lines are in progress.

In the light of these results, one can conclude that the representation of the HF operator in a wavelet basis thanks to the BCR algorithm is good enough to be used in further calculations. But improvements should be made to the method to make results competitive with position or even momentum ones.

Acknowledgments. The authors acknowledge their appreciation to Prof. P. L. Lions and Dr. A. Cohen for their useful comments and for reading the manuscript.

References

1. Beylkin, G., *Wavelets, Multiresolution Analysis and Fast Numerical Algorithms*, a draft of INRIA lectures.
2. Beylkin, G, R. Coifman, and V. Rokhlin, Fast wavelet transforms and numerical algorithms I, *Comm. Pure and Appl. Math.* **44** (1991), 141–183.
3. Daubechies, I., *Ten Lectures on Wavelets*, CBMS 61, SIAM, Philadelphia, 1992.
4. Fischer, P., Thesis of the Paris-Dauphine University, 1994.
5. Fischer, P., M. Defranceschi, and J. Delhalle, Molecular Hartree-Fock equations for iteration-variation calculations in momentum space, *Numer. Math.* **63** (1992), 67–82.
6. Fischer, P. and Defranceschi, Iterative process for Hartree-Fock equations thanks to a wavelet transform, in *Appl. Comp. Harm. Anal.*, 1994, to appear.
7. Mallat, S., Multiresolution approximation and wavelet orthonormal bases of $L^2(\mathbb{R})$, *Trans. Am. Math. Soc.* **315** (1989), 69–87.
8. McWeeny, R., *Methods of Molecular Quantum Mechanics*, 2nd Edition, Academic Press, New York, 1989.
9. Meyer, Y., *Ondelettes et Opérateurs I*, Hermann, Paris, 1990.
10. Szabo, A. and N. S. Ostlund, *Modern Quantum Chemistry*, McGraw Hill, New York, 1989.

Patrick Fischer
CEA/DSM/DRECAM/SRSIM, CE - Saclay,
F-91191 Gif-sur-Yvette Cedex, France
CEREMADE, Université de Paris - Dauphine,
Place du Maréchal de Lattre de Tassigny, F-75775 Paris Cedex, France

Mireille Defranceschi
CEA/DSM/DRECAM/SRSIM, CE - Saclay,
F-91191 Gif-sur-Yvette Cedex, France

Part VII

Applications

Efficiency Comparison of Wavelet Packet and Adapted Local Cosine Bases for Compression of a Two-dimensional Turbulent Flow

Mladen Victor Wickerhauser, Marie Farge, Eric Goirand, Eva Wesfreid, and Echeyde Cubillo

Abstract. We compare the efficiency of two rank-reduction methods for representing the essential features of a two-dimensional turbulent vorticity field. The two methods are both projections onto the largest components, in one case onto the wavelet packet best basis, in the other case onto the best basis of adapted local cosines. We compare the two methods in three ways: for efficiency of capturing enstrophy or square-vorticity, for faithfulness to the power spectrum, and for precision in resolving coherent structures. These properties are needed for subsequent segmentation into isolated coherent structures, or for measurement of statistical quantities related to coherent structures. We find that in all three respects the wavelet packet representation is superior to the local cosine representation.

§1 Background

1.1 Models of turbulence

The classical approach to turbulent flows considers either ensemble averages or equivalently, assuming ergodicity, time or space averages. In particular, the Kraichnan–Batchelor theory for two-dimensional turbulence [2,10] postulates homogeneous mixing within the flow and supposes that the whole vorticity field is involved in the turbulent cascade process.

In contrast to this approach, we think that a two-dimensional turbulent flow is generically inhomogeneous and can be better described as a superposition of coherent rotational vortices embedded in a random quasi-irrotational flow. We have observed, in numerical simulations of two-dimensional Navier–Stokes equations with random initial conditions, that isolated vortices result

Wavelets: Theory, Algorithms, and Applications
Charles K. Chui, Laura Montefusco, and Luigia Puccio (eds.), pp. 509–531.

ISBN 0-12-174575-9

from the condensation of enstrophy into localized, well-separated structures. These structures are stable as long as they do not interact with one another, but during close encounters they experience strong deformations, which then excite some internal degrees of freedom. This gives rise to a local *cascade* or transfer of enstrophy toward small scales and to its concomitant dissipation. Consequently, only a limited *active portion* of the vorticity field, correlated to the coherent vortices, is responsible for the turbulent cascade. The remainder, or *background portion* of the field, is passively advected and plays a negligible dynamical role.

Our *atomic* view may be compared with the vortex methods of Winckelmans and Leonard [22], Marchioro and Pulvirenti [15], and Saffman [18]. We generalize the simplest model used to approximate two-dimensional flows, that of superposed *point vortices*, by considering the flow to be a superposition of *atoms* that we choose from among a library of smooth localized functions such as wavelet packets [4] or localized cosine functions [3,14]. The additional parameters available to these atoms enable us to take into account the internal degrees of freedom of each vortex, which can be considered as a *molecule*.

1.2 Turbulent flow computation

The usual methodology to compute turbulent flow evolutions is to separate them into active components responsible for the nonlinear and chaotic dynamics, and passive components, advected by the active ones which they merely perturb. The active components must be explicitly computed, while the passive components can be modeled by some subgrid scale parametrization. The question we want to address is: which components in a two-dimensional turbulent flow are active and which are passive?

1.2.1 Wavenumber segmentation

The commonly used computational techniques, such as Galerkin methods, large eddy simulation, and nonlinear Galerkin methods, assume that there exists a gap in the energy spectrum of turbulent flows. Such a gap allows us unequivocally to distinguish low-wavenumber Fourier modes, which we consider to be active, from high-wavenumber Fourier modes, which we consider to be passive. In fact, this assumption is misleading, because such a spectral gap is never observed, neither for two-dimensional turbulent flows, nor for three-dimensional ones. We have to look for other functional bases where some kind of gap both exists and represents a physically meaningful distinction. Then the segmentation between active and passive modes will be more than a leap of faith.

In the case of three-dimensional turbulence, according to Kolmogorov's theory [9,17], energy cascades directly from low-wavenumbers to high-wavenumbers, where it is dissipated. In this case, the previous wavenumber segmentation is still relevant as long as there is no local inverse energy cascade from high-wavenumbers to low-wavenumbers. But in fact, such an inverse cascade seems to arise whenever the flow presents well-localized active structures, such as horseshoe vortices in boundary or mixing layers.

In the case of two-dimensional turbulence, wavenumber segmentation is always inappropriate. Kraichnan's theory [11,12] predicts, and numerical experiments have confirmed, that two-dimensional turbulent flows on average exhibit inverse energy cascades. Furthermore, any segmentation must also take into account the presence of coherent structures, since they are generic in two-dimensional turbulence and their mutual interactions are very probably responsible for the inverse energy cascade. Because such features are well localized in physical space, we propose to segment the field into components with definite physical space location as well as recognizable spectral content.

1.2.2 Phase space segmentation

The method we will follow to define a better segmentation is to search for a decomposition which puts the strongest concentration of enstrophy into the fewest *phase space atoms*. We choose the atoms to be smooth functions, well localized in both space and wavenumber coordinates. The additional dimensions available in phase space better allow us to discriminate individual coherent structures. We propose to characterize a coherent structure as the superposition of phase space atoms which share the same position. The degrees of freedom within each coherent structure are just the amplitudes of its component atoms.

We isolate structures by *coarse graining* in phase space. This is most easily accomplished if our representation yields very few large components, rather than large numbers of smaller components. We, therefore, compare techniques on the basis of their gain of concentration. This is similar to the goal of adapted waveform denoising [5], or else transform coding image compression [19,21], where effectiveness is measured by transform coding gain. In this paper we will compare transform coding into wavelet packets and local cosine phase space atoms. The wavelet packet representation has proved superior to the Fourier representation in capturing features which control the dynamics of two-dimensional flow evolution [7].

§2 Methods

2.1 Two-dimensional incompressible viscous flows

We begin with a high resolution direct numerical solution of the Navier–Stokes equations, which describe the dynamics of a two-dimensional incompressible viscous flow. In the periodic plane $S = (0, 2\pi) \times (0, 2\pi) \subset \mathbf{R}^2$ and in the absence of external forcing, these take the following form:

$$\begin{aligned} \tfrac{\partial \mathbf{u}}{\partial t} + (\mathbf{u} \cdot \nabla)\mathbf{u} + \nabla P - \nu \nabla^2 \mathbf{u} &= 0, \quad \text{in } S \times \mathbf{R}^+, \\ \nabla \cdot \mathbf{u} &= 0, \quad \text{in } S \times \mathbf{R}^+, \\ \mathbf{u}(\mathbf{x}, 0) &= \mathbf{u}_0(\mathbf{x}), \quad \text{in } S. \end{aligned}$$

Here $\mathbf{u}$ is the velocity field, P is the pressure field, and ν is the kinematic viscosity. We also impose periodic boundary conditions. We can rewrite these equations in terms of *vorticity* ω and *streamfunction* ψ, defined by

$$\mathbf{u} = \begin{pmatrix} u_1 \\ u_2 \end{pmatrix} = \begin{pmatrix} -\frac{\partial \psi}{\partial x_2} \\ \frac{\partial \psi}{\partial x_1} \end{pmatrix}; \qquad \omega = \frac{\partial u_2}{\partial x_1} - \frac{\partial u_1}{\partial x_2}.$$

The Navier–Stokes equations then become

$$\begin{aligned} \tfrac{\partial \omega}{\partial t} + J(\psi, \omega) - \nu \nabla^2 \omega = 0, &\quad (\mathbf{x}, t) \in S \times \mathbf{R}^+; \\ \omega = \nabla^2 \psi, &\quad (\mathbf{x}, t) \in S \times \mathbf{R}^+; \\ \omega(\mathbf{x}, 0) = \omega_0(\mathbf{x}), &\quad \mathbf{x} \in S. \end{aligned}$$

Again we have periodic boundary conditions. The Jacobian operator in terms of these new variables is:

$$J(\psi, \omega) = \frac{\partial \psi}{\partial x_1} \frac{\partial \omega}{\partial x_2} - \frac{\partial \psi}{\partial x_2} \frac{\partial \omega}{\partial x_1}.$$

We can expand ω and ψ in their Fourier series over the periodic domain S:

$$\begin{aligned} \omega(\mathbf{x}, t) &= \sum_{\mathbf{k}} \hat{\omega}(\mathbf{k}, t) e^{i\mathbf{k}\cdot\mathbf{x}}, &\qquad \hat{\omega}(\mathbf{k}, t) &= \frac{1}{2\pi} \int_{\mathbf{x}\in S} \omega(\mathbf{x}, t) e^{-i\mathbf{k}\cdot\mathbf{x}}\, d\mathbf{x}; \\ \psi(\mathbf{x}, t) &= \sum_{\mathbf{k}} \hat{\psi}(\mathbf{k}, t) e^{i\mathbf{k}\cdot\mathbf{x}}, &\qquad \hat{\psi}(\mathbf{k}, t) &= \frac{1}{2\pi} \int_{\mathbf{x}\in S} \psi(\mathbf{x}, t) e^{-i\mathbf{k}\cdot\mathbf{x}}\, d\mathbf{x}. \end{aligned}$$

We obtain a turbulent vorticity field by starting with a random initial vorticity field $\omega_0(\mathbf{x})$, then integrating for many time steps in the presence of time-periodic external forcing. Forcing is subsequently turned off and the same integration is continued until the vorticity field reaches a statistically steady state.

To integrate the Navier–Stokes equations, we use a pseudospectral Galerkin method which, at each time step, performs all differentiation in $\hat{\omega}, \hat{\psi}$ coordinates and all multiplication in ω, ψ coordinates. Both ω and ψ are represented as finite Fourier series, or superpositions of the Fourier modes at wavenumbers $0 \leq |\mathbf{k}| < k_r$, where k_r is the cutoff wavenumber which gives some fixed resolution. The time integration is done using an Adams–Bashforth scheme. For the subgrid scale model we use a hyperdissipation operator $-(-\nabla^2)^4$, which replaces the Laplacian operator in the Navier–Stokes equations.

The periodic plane S is sampled on 512^2 grid points in our simulation. The program ran for 38,000 time steps $\Delta t = 10^{-3}$ in units of T^1, which corresponds to 635 initial eddy-turnover times, starting from a random distribution of vorticity with energy $E = 0.5$ in ML^2T^{-2} units and an initial enstrophy $Z = 279$ in T^{-2} units. The vorticity field we analyze is one time slice, presumably a typical snapshot of a fully-developed turbulent flow, taken from the equilibrium end of this preparation.

2.2 Wavelet packet best basis

Wavelet packets are generalizations of the compactly-supported wavelets introduced by Daubechies, Mallat, and Meyer [6,13,16]. They constitute an overabundant set of basis functions with remarkable orthogonality properties, namely, that very many subsets form orthonormal bases. The one-dimensional functions were first described in [4]. Each basis element ψ is characterized by three attributes: scale s, wavenumber k, and position p, so we may label them ψ_{skp}. By the Heisenberg uncertainty principle, it is not possible to localize a function to arbitrary precision in both p and k; we must have $\delta p \cdot \delta k \geq 1$ in normalized units, where δp is the uncertainty in position and δk is the uncertainty in wavenumber. In our construction, we have $\delta p \approx 2^s$ and $\delta k \approx 2^{-s}$ in the same normalization, so that the product of the uncertainties is roughly as small as possible. Such functions, which cannot be significantly better localized in phase space, are evidently phase space atoms.

Fourier analysis with such waveforms or atoms consists of calculating the *wavelet packet transform* $w_{skp}(f) = \langle \psi_{skp}, f \rangle$. Certain subsets of the indices (s, k, p) give orthonormal bases B, and for these subsets we have the inversion formula:

$$f = \sum_{(s,k,p)\in B} \langle \psi_{skp}, f \rangle \psi_{skp}.$$

Wavelet packets are rarely constructed explicitly. More usually, we simply apply the fast discrete algorithm described in [4] to the sampled values of f, and thereby produce the coefficients $w_{skp}(f)$. The underlying functions ψ can, however, be developed as follows. We introduce two (short) finite sequences $\{h_n\}$ and $\{g_n\}$, called *conjugate quadrature filters*, which satisfy the relations:

$$\sum_n h_{2n} = \sum_n h_{2n+1} = \frac{1}{\sqrt{2}}, \qquad g_n = -(-1)^n h_{11-n}, \qquad \text{for all } n; \tag{1}$$

$$\sum_n h_n h_{n+2m} = \sum_n g_n g_{n+2m} = \begin{cases} 1, & \text{if } m = 0, \\ 0, & \text{otherwise;} \end{cases} \tag{2}$$

$$\sum_n h_n g_{n+2m} = 0, \qquad \text{for all } m \in \mathbf{Z}. \tag{3}$$

Next, we define a family of functions recursively for integers $k \geq 0$ by:

$$W_{2k}(x) = \sqrt{2} \sum_n h_n W_k(2x - n); \quad W_{2k+1}(x) = \sqrt{2} \sum_n g_n W_k(2x - n). \tag{4}$$

Note that W_0 satisfies a fixed-point equation. Conditions 1 through 3 ensure that a unique solution to this fixed-point problem exists, and that $\{W_k : k \in \mathbf{Z}\}$ forms an orthonormal basis for $L^2(\mathbf{R})$. The quadrature filter pair h, g can be chosen (see [6]) so that the solution has any prescribed degree of smoothness.

Equations 1 through 4 all have periodic analogs as well, which we use in the case of periodic boundary conditions. For the experiments in this article, we used periodic algorithm with the so-called "C 12" coefficients, using h_n and g_n as given in Table 1.

Table 1. C 12 coefficients for orthogonal wavelet packets.

n	Low-pass filter coefficient h_n	High-pass filter coefficient g_n
< 0	0	0
0	$1.6387336463179785 \times 10^{-2}$	$-7.2054944536811512 \times 10^{-4}$
1	$-4.1464936781966485 \times 10^{-2}$	$1.8232088709100992 \times 10^{-3}$
2	$-6.7372554722299874 \times 10^{-2}$	$5.6114348193659885 \times 10^{-3}$
3	$3.8611006682309290 \times 10^{-1}$	$-2.3680171946876750 \times 10^{-2}$
4	$8.1272363544960613 \times 10^{-1}$	$-5.9434418646471240 \times 10^{-2}$
5	$4.1700518442377760 \times 10^{-1}$	$7.6488599078264594 \times 10^{-2}$
6	$-7.6488599078264594 \times 10^{-2}$	$4.1700518442377760 \times 10^{-1}$
7	$-5.9434418646471240 \times 10^{-2}$	$-8.1272363544960613 \times 10^{-1}$
8	$2.3680171946876750 \times 10^{-2}$	$3.8611006682309290 \times 10^{-1}$
9	$5.6114348193659885 \times 10^{-3}$	$6.7372554722299874 \times 10^{-2}$
10	$-1.8232088709100992 \times 10^{-3}$	$-4.1464936781966485 \times 10^{-2}$
11	$-7.2054944536811512 \times 10^{-4}$	$-1.6387336463179785 \times 10^{-2}$
> 11	0	0

One-dimensional wavelet packets are defined from these W_k by the formula:

$$\psi_{skp}(x) = 2^{-s/2} W_k(2^{-s}x - p).$$

As described in [4], we obtain an orthonormal basis subset $\mathcal{I}$ by taking those functions $\{\psi_{skp} : (s,k,p) \in \mathcal{I}\}$ for which the half-open *dyadic intervals* $\{[\frac{k}{2^s}, \frac{k+1}{2^s}) : (s,k,p) \in \mathcal{I}\}$ form a disjoint cover of the unit interval.

Our library of basis functions in two dimensions consists of all possible tensor products of the ψ functions with both factors sharing the same scale s. The definitions and formulas for this two-dimensional case may be found in [21], elsewhere in this volume. Certain basis subsets can be described by disjoint tilings of the unit square, as follows. Let I be a half-open *dyadic square* $[\frac{k_x}{2^s}, \frac{k_x+1}{2^s}) \times [\frac{k_y}{2^s}, \frac{k_y+1}{2^s})$ and put $\psi_{I,(p_x,p_y)}(x,y) = 2^{-s}sW_{k_x}(2^{-s}x - p_x)W_{k_y}(2^{-s}y - p_y)$. Then every basis in our library, for functions on the $2^S \times 2^S$ grid, corresponds to a set of the form:

$$\{\psi_{I,(p_x,p_y)} : I \in \mathcal{I}, p_x \in \mathbf{Z}, p_y \in \mathbf{Z}, 0 \le p_x < 2^{S-s}, 0 \le p_y < 2^{S-s}\},$$

where $\mathcal{I}$ is a disjoint cover of the unit square by such dyadic squares I, for $0 \le s \le S$ and $0 \le k_x, k_y < 2^{s-1}$. Computation of inner products with all such functions is performed recursively, as is the search for the best-basis; the implementation of both algorithms is described in [20]. The entire procedure has complexity $O(N \log N)$ where N is the rank of the problem, and $N = 2^{2S}$ for the original grid-point formulation.

The *best basis* of wavelet packets for a fixed vorticity field is the one maximizing the coding gain or *entropy*, as described in [21]. The field is then approximated by ω^ϵ, a superposition of just the largest components. Call the best basis $\mathcal{I}_*$. We project onto the top few coefficients as follows:

$$\omega^\epsilon = \sum_{|c_I|>\epsilon} c_I \psi_I.$$

Here $I \in \mathcal{I}_*$, $c_I = \langle \omega, \psi_I \rangle$ and ϵ is some predetermined threshold. We also sum over all integer translates (p_x, p_y), even though that notation is suppressed for compactness.

In our four compression experiments, ϵ was chosen to ensure that ω^ϵ contained, 99.9, 99, 90, and 50 percent of the total *enstrophy* (or square vorticity) of the field. The remainder fields $\omega^r = \omega - \omega^\epsilon$ thus contain 0.1, 1, 10, and 50 percent of the enstrophy, respectively.

2.3 Local cosine best basis

The local cosine transform has much in common with the windowed Fourier transform [8], which is a well known decomposition into phase space atoms. Like the one-dimensional wavelet packet library, the library of one-dimensional local cosines contains atoms that can be characterized by scale s, wavenumber k, and position p. Since local cosines are real-valued, orthonormal analogs of windowed Fourier modes, we may use the nominal window width for s, the nominal left endpoint of the window for p, and the frequency index for k. Local cosines have the formula

$$\psi_{skp}(x) = \sqrt{\frac{2}{2^s}}\, b\left(\frac{x-p}{2^s}\right) \cos\left[\pi(k+\frac{1}{2})\frac{x-p}{2^s}\right].$$

Here $b = b(x)$ is a smooth function supported in the compact interval $[-\epsilon, 1+\epsilon]$ for $0 \le \epsilon < \frac{1}{2}$. This interval contains $[0,1]$ which we will say is its *nominal support*. The only special features are the use of odd half-integer frequencies in the cosine function, and the symmetry properties of the window function b:

$$b(-x)^2 + b(x)^2 = 1, \quad b(1-x)^2 + b(1+x)^2 = 1, \qquad -\epsilon < x < \epsilon.$$

These special properties insure orthogonality and make certain fast algorithms possible. Auscher et al. [1] explores the one-dimensional case in much greater detail.

Two-dimensional local cosines consist of tensor products $\psi_{skp}(x)\psi_{s'k'p'}(y)$, and their nominal supports are the cartesian product rectangles of the nominal supports of the x and y factors. For simplicity we will be using the same scale in both directions: $s = s'$ throughout this paper. Subsets of such functions can be indexed by dyadic squares, with the squares correspond to the nominal supports of the functions.

A two-dimensional *adapted local cosine basis*, as described in [21] elsewhere in this volume, corresponds to a disjoint cover by dyadic squares. The *best basis* for a given vorticity field is the one which maximizes coding gain. Descriptions and basic implementations of the algorithm to extract this basis may be found in [20].

2.4 Radial power spectrum

Given a sampled vorticity field $\omega(\mathbf{x}, t_0)$ at some instant t_0, we can compute its Fourier power spectrum $|\hat{\omega}(\mathbf{k}, t_0)|^2$ using the discrete Fourier transform or FFT. This will be a nonnegative function of two variables, but we are only interested in its mean behavior as $|\mathbf{k}| \to \infty$, regardless of direction. Thus, we will average over all rotations to get a *radial power spectrum density*:

$$\rho(k) = \frac{1}{2\pi k} \oint_{|\mathbf{k}|=k} |\hat{\omega}(\mathbf{k})|^2, \qquad k \geq 0.$$

We have suppressed the unneeded variable t_0. The *radial power spectrum* that we actually plot is an approximation of this density, found by summing the discrete samples of $|\hat{\omega}|^2$ lying in the annulus between two wavenumber radii k and $k+1$, then dividing by the multiplicity M_k or number of samples in each annulus:

$$\rho_{disc}(k) = \frac{1}{2\pi k\, M_k} \sum_{k \leq |\mathbf{k}| < k+1} |\hat{\omega}(\mathbf{k})|^2, \qquad k = 0, 1, \ldots.$$

The resolution of the discrete radial power spectrum is the same as that of the discrete Fourier transform.

From the slope of the log–log plot of ρ_{disc}, we can estimate the scaling law for enstrophy as a function of wavenumber. This can be used to test the faithfulness of segmentation methods, as described in [7].

§3 Results

We focused our attention on one vorticity field (at the top of Figures 1 to 4), representing what we believe is a generic time slice of a homogenous, isotropic, fully developed turbulent flow. Our experiment segmented it into high-enstrophy and low-enstrophy components in both the wavelet packet and local cosine best bases, and we reconstructed the different portions for visual inspection (Figures 1 to 4). We then computed the radial power spectra of the reconstructed pieces (Figures 5 to 7), and estimated the spectral slopes in the inertial and noise-like regions (Tables 2 and 3). and Finally, we found the rate of accumulation of enstrophy by components in the two methods (Figure 8 and Table 4).

3.1 Strong fields reconstructed after compression

Analysis and synthesis, either with local cosines or wavelet packets, is a perfect reconstruction algorithm, but we envision discarding most of the components before synthesis so as to compress the representation. The disappearance of some components introduces distortion and artifacts that become progrssively more visible at higher compression rates.

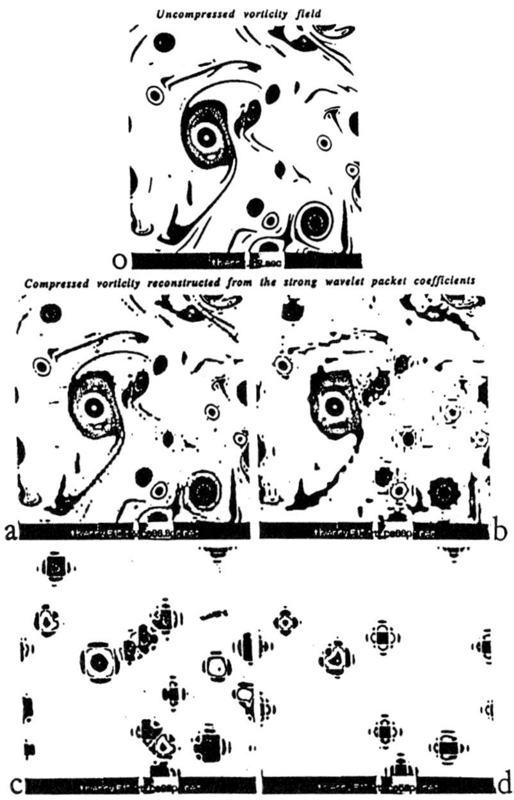

Figure 1. Reconstruction from the strongest wavelet packet coefficients: (0) uncompressed vorticity field corresponding to 262144 grid point coefficients; (a) compressed vorticity field reconstructed from the 3000 (11.4 per 1000) strongest wavelet packet coefficients, which retain 99.9% of the enstrophy; (b) compressed vorticity field reconstructed from the 739 (2.8 per 1000) strongest wavelet packet coefficients, which retain 99% of the enstrophy; (c) compressed vorticity field reconstructed from the 73 (0.28 per 1000) strongest wavelet packet coefficients, which retain 90% of the enstrophy; (d) compressed vorticity field reconstructed from the 14 (0.05 per 1000) strongest wavelet packet coefficients, which retain 50% of the enstrophy. (See Figure 1 in color in the color section of this volume.)

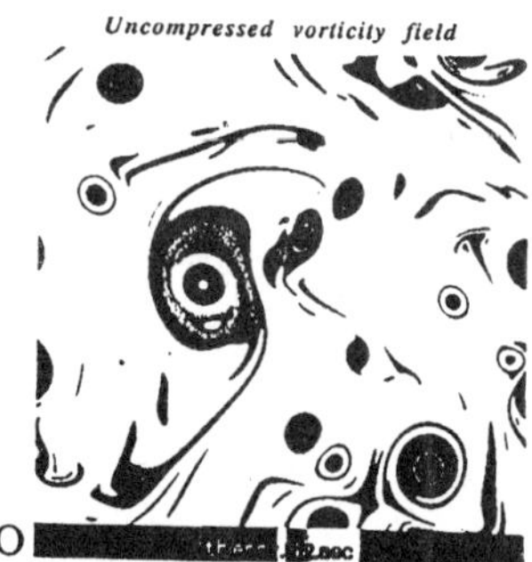

Figure 2. Reconstruction from the weakest wavelet packet coefficients: (0) uncompressed vorticity field corresponding to 262144 grid point coefficients; (a) compressed vorticity field reconstructed from the 259144 (98.8 per 100) weakest wavelet packet coefficients, which retain 0.1% of the enstrophy; (b) compressed vorticity field reconstructed from the 261405 (99.7 per 100) weakest wavelet packet coefficients, which retain 1% of the enstrophy; (c) compressed vorticity field reconstructed from the 262074 (99.97 per 100) weakest wavelet packet coefficients, which retain 10% of the enstrophy; (d) compressed vorticity field reconstructed from the 262130 (99.99 per 100) weakest wavelet packet coefficients, which retain 50% of the enstrophy. (See Figure 2 in color in the color section of this volume.)

Figure 3. Reconstruction from the strongest local cosine coefficients: (0) uncompressed vorticity field corresponding to 262144 grid point coefficients; (a) compressed vorticity field reconstructed from the 5114 (19.5 per 1000) strongest local cosine coefficients, which retain 99.9% of the enstrophy; (b) compressed vorticity field reconstructed from the 1479 (5.6 per 1000) strongest local cosine coefficients, which retain 99% of the enstrophy; (c) compressed vorticity field reconstructed from the 425 (1.6 per 1000) strongest local cosine coefficients, which retain 90% of the enstrophy; (d) compressed vorticity field reconstructed from the 70 (0.27 per 1000) strongest local cosine coefficients, which retain 50% of the enstrophy. (See Figure 3 in color in the color section of this volume.)

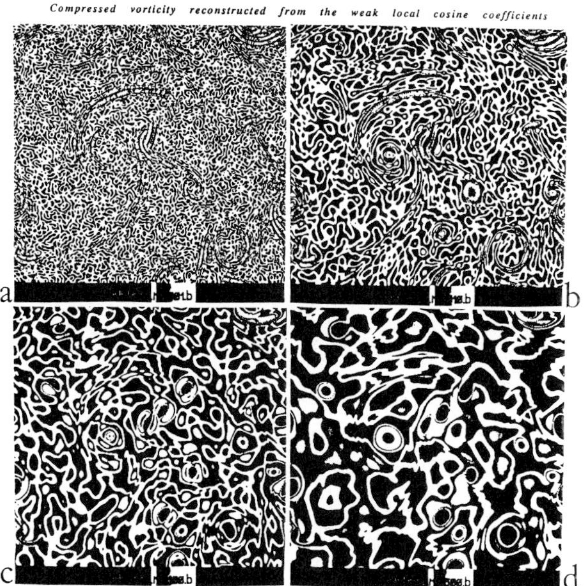

Figure 4. Reconstruction from the weakest local cosine coefficients: (0) uncompressed vorticity field corresponding to 262144 grid point coefficients; (a) compressed vorticity field reconstructed from the 257030 (98 per 100) weakest local cosine coefficients, which retain 0.1% of the enstrophy; (b) compressed vorticity field reconstructed from the 260665 (99.4 per 100) weakest local cosine coefficients, which retain 1% of the enstrophy; (c) compressed vorticity field reconstructed from the 261719 (99.8 per 100) weakest local cosine coefficients, which retain 10% of the enstrophy; (d) compressed vorticity field reconstructed from the 262074 (99.97 per 100) weakest local cosine coefficients, which retain 50% of the enstrophy. (See Figure 4 in color in the color section of this volume.)

In the local cosine case, the basis functions are registered on a fixed grid which may be inconveniently placed with respect to the principal features. Thus, it may take a superposition of several components to match the phase of a peak. If some of the components are small and disappear in the compression, there may be oscillations in the reconstructed compressed field near features of interest. These are visible in Figures 3a to 3d. Some of this oscillation also appears in the wavelet packet reconstructions in Figures 1a to 1d.

A *quasi-singular structure*, namely a sampled function near a discontinuity or cusp, is described as the superposition of all Fourier modes. The local cosine representation, like any Fourier compression, is unable to preserve nonsmooth functions. Thus the compressed image is undesirably smoother than the original, especially at high compression rates (Figures 3c and 3d). On the other hand, wavelet packets include wavelets which concentrate the energy of singularities into a few components. Those components vital for representing nonsmooth portions of the field survive the wavelet packet compression, so that even at high compression rates (Figures 1c and 1d) the quasi-singular features remain sharp and well-separated.

The discarded coefficients in both types of compressions will typically contain most of the high frequencies. These are registered at the unsmooth regions of the field, where they must superpose to match the rapid variation. These nonsmooth regions include vortex cores and the strongly sheared regions around vortices, or in between two interacting vortices. Since components are registered around these well localized phenomena, the field reconstructed from the discarded local cosine coefficients can be highly structured, as is evident in Figures 4a to 4d. At low compression rates such as those of Figures 4a and 4b, the discarded coefficients constitute a low-enstrophy, homogeneous, noise-like remainder field. Wavelet packets differ from local cosines in this respect chiefly because they are more efficient; they allow more small components to be discarded as noise, as seen in Figures 2a to 2d.

The local cosine components are not isotropic nor even very close to symmetric, since they are tensor products of nonsymmetric one-dimensional functions. We note, however, that we can use symmetric functions if we are willing to use cosines on the odd intervals and sines on the even intervals [3]. Similarly, compactly-supported one-dimensional orthogonal wavelet packets cannot be symmetric or antisymmetric, so they too must give rise to nonisotropic tensor product two-dimensional synthesis functions. However, if we use symmetric biorthogonal filters in the wavelet packet algorithm [20], or still other modifications, we can have compactly-supported synthesis functions which exhibit considerably more symmetry, though they will still have the tensor product character.

3.2 Weak fields reconstructed from the remainders

In both the wavelet packet and local cosine cases, for small compression ratio, the fields reconstructed from discarded weak components are homogeneous, but this homogeneity is lost as the compression ratio increases. At high compression ratios, there are so many components and so much energy in the weak field that it begins to exhibit the same inhomogeneity and anisotropy as the original field.

The requirement that the discarded components look like homogeneous noise imposes an upper limit on the compression ratio. We must not have organized features in the weak component field. The wavelet packet compression permits a higher upper bound than the local cosine compression, since its remainder or weak field remains disorganized up to higher compression ratios.

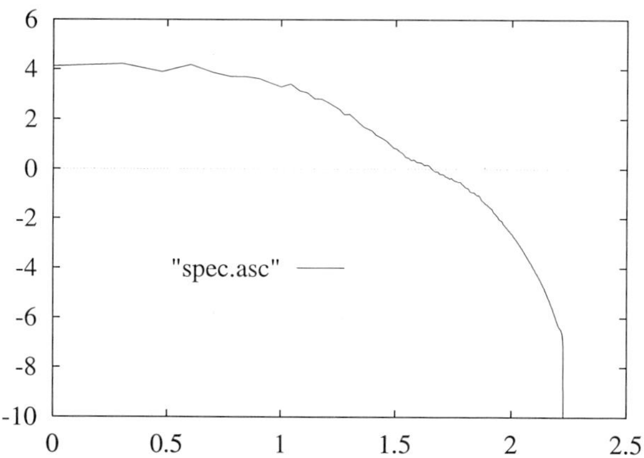

Figure 5. Power spectrum of the original vorticity field.

3.3 Spectra

Figure 5 shows the radial power spectrum of the original vorticity field. Figures 6a–d and 7a–d show the radial power spectra of vorticity fields compressed to varying degrees by both methods. In each plot, three spectra are superposed for comparison: that of the total field, that of the strong components, and that of the weak or discarded components. All plots use the same vertical scale $[6, -10]$, which is the common logarithm of the radial power spectral density, and the same horizontal scale $[0, 2.5]$, which is the common logarithm of wavenumber magnitude $|\mathbf{k}|$ and roughly corresponds to the wavenumber range $[0, 316]$. We use uniform scaling in order to allow a fair comparison for both slope and intensity.

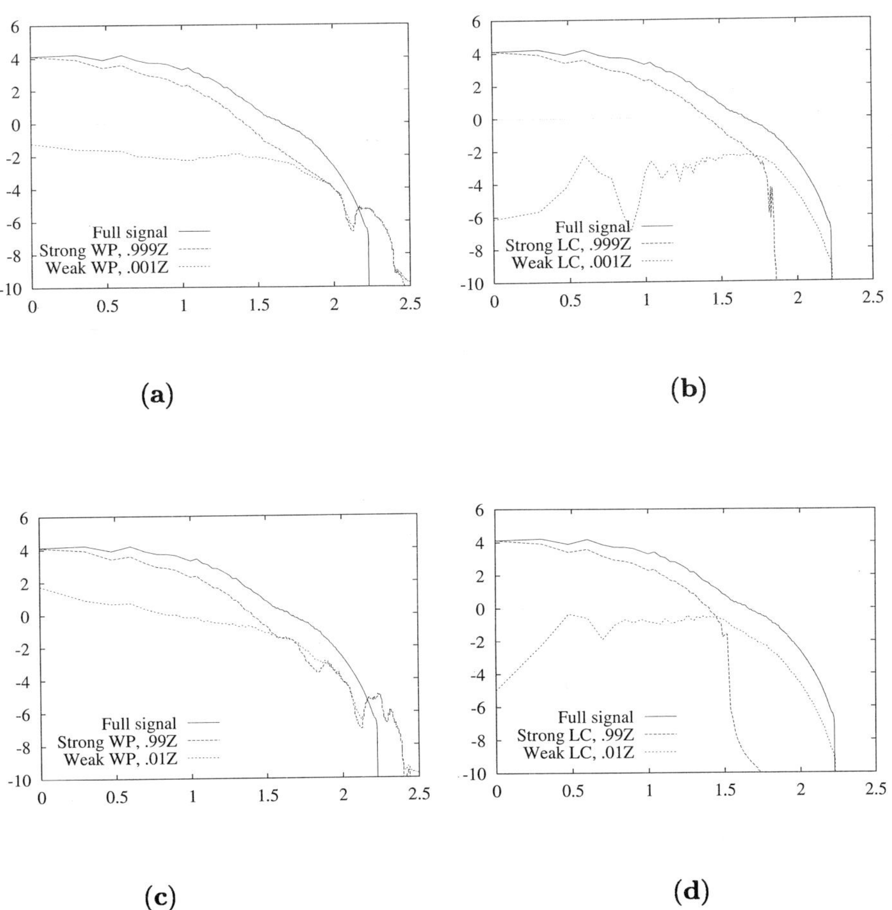

Figure 6. Superposed power spectra of the vorticity fields reconstructed from a segmentation into the "many" strongest and "few" weakest wavelet packet and local cosine components: (a) 3129 strongest wavelet packets, $.999Z$, plus 259,015 weakest wavelet packets, $.001Z$; (b) 5114 strongest local cosines, $.999Z$, plus 257,030 weakest local cosines, $.001Z$; (c) 803 strongest wavelet packets, $.99Z$, plus 261,341 weakest wavelet packets, $.01Z$; (d) 1479 strongest local cosines, $.99Z$, plus 260,665 weakest local cosines, $.01Z$.

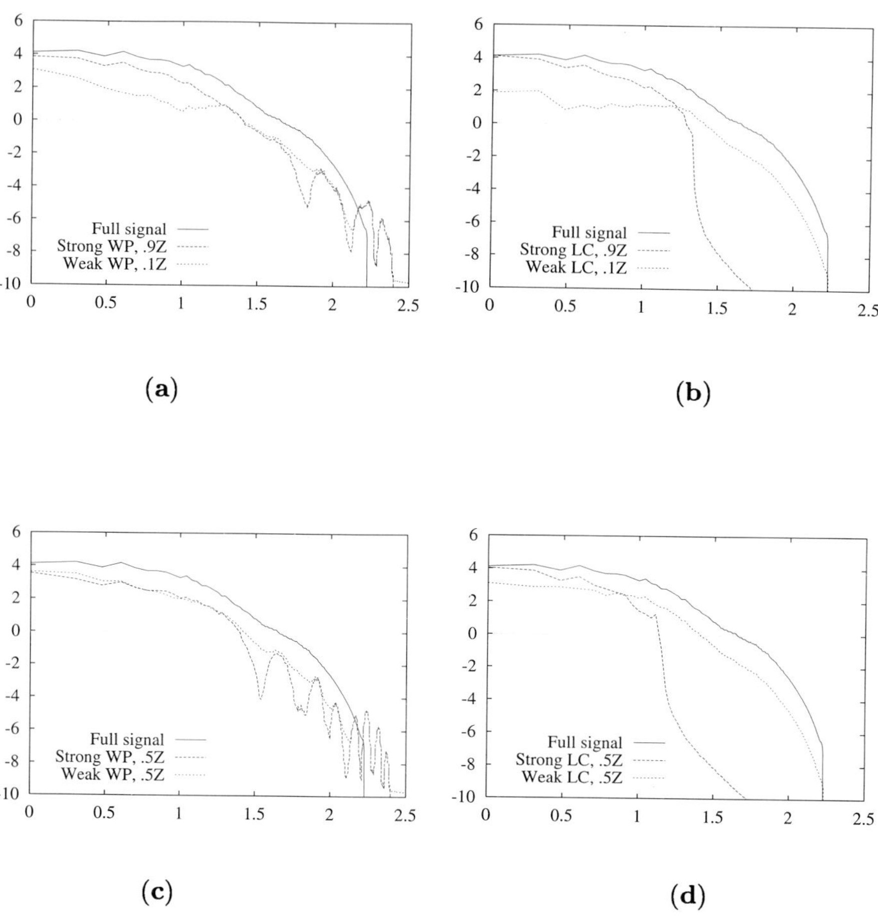

Figure 7. Superposed power spectra of the vorticity fields reconstructed from a segmentation into the "few" strongest and "many" weakest wavelet packet and local cosine components: (a) 79 strongest wavelet packets, $.9Z$, plus 262,065 weakest wavelet packets, $.1Z$; (b) 425 strongest local cosines, $.9Z$, plus 261,719 weakest local cosines, $.1Z$. (c) 16 strongest wavelet packets, $.5Z$, plus 262,065 weakest wavelet packets, $.5Z$; (d) 70 strongest local cosines, $.5Z$, plus 262,074 weakest local cosines, $.5Z$.

The radial power spectra computed from the strong wavelet packets oscillate rather wildly in the large wavenumbers, possibly because of the complicated spectra of individual wavelet packets. There may also be a kind of Gibbs phenomenon, in the wavenumber domain, as the long smooth spectral tails of the wavelet packets try to match the sharp spectral cutoff at $|\mathbf{k}| = 170$ in the original field. The local cosines match this cutoff much better; also, the spectrum of the total field does not oscillate near the cutoff.

Table 2. Spectral slopes in the presumed inertial range of various reconstructions from strong components.

Enstrophy fraction	Wavelet packet inertial range	WP slope	Local cosine inertial range	LC slope
$.999Z$	$[10, 50]$	-6.73	$[10, 50]$	-6.84
$.99Z$	$[10, 50]$	-6.60	$[10, 30]$	-6.99
$.9Z$	$[10, 50]$	-6.26	$[10, 18]$	-6.37
$.5Z$	$[10, 28]$	-6.58	$[8, 12]$	-7.72

Table 3. Spectral slopes of various reconstructions from weak components.

Enstrophy fraction	Wavelet packet noise-like range	WP slope	Local cosine noise-like range	LC slope
$.001Z$	$[1, 50]$	-0.46	$[1, 50]$	$+2.20$
$.01Z$	$[1, 50]$	-1.92	$[1, 30]$	$+1.68$
$.1Z$	$[1, 50]$	-2.73	$[1, 18]$	-0.53
$.5Z$	$[1, 28]$	-2.70	$[1, 12]$	-0.91

We take the inertial region to be the wavenumbers $10 \leq |\mathbf{k}| \leq 50$, or $1.0 \leq |\log_1 0\mathbf{k}| \leq 1.7$. In the inertial region the curve is approximated well by a line of slope -5.63. The radial power spectra of various reconstructions also have identifiable inertial ranges, and we list their slopes in those regions in Table 2. Notice that the local cosine compression's inertial range shrinks much faster than that of the wavelet packet compression, as the absence of discarded high-wavenumber modes eats into the spectrum from the right.

We have listed the slopes of the fields reconstructed from the weakest coefficients in Table 3, looking at the wavenumbers from $|\mathbf{k}| = 1$ up to the highest wavenumber taken to belong to the inertial range. Ideally these slopes would be 0, the radial power spectrum of white noise, but as the compression ratio increases and the discarded coefficients hold more of the enstrophy, they begin to assume more of the spectral characteristics of the original flow.

We can calculate the spectral slopes of the weak wavelet packet reconstructions in Figures 6a,c and 7a,c using rather high wavenumbers, but the weak local cosine spectra are sharply disturbed at the rather low wavenumbers where the strong-field spectrum has its cutoff, as seen in Figures 6b,d and 7b,d. The region where we can estimate the weak-field spectral slope is the chief difference between the two methods. Once that region is chosen, the slope decreases as compression ratio increases at nearly the same rate in both methods.

3.4 Enstrophy concentration

Table 4 shows that wavelet packets concentrate enstrophy better than local cosines, and that the stronger the compression, the better the wavelet packets behave in comparison with local cosines. Wavelet packets are about twice as efficient for retaining 99 and 99.9 percent of the enstrophy, and about five times as efficient for retaining 50 and 90 percent of the enstrophy.

Table 4. Fraction of coefficients needed to retain a given portion of the enstrophy in the vorticity field.

Enstrophy (percent)	Wavelet packets (per 1000)	Local cosines (per 1000)
99.9	11.94	19.51
99	3.06	5.64
90	0.30	1.62
50	0.06	0.27

Figure 8 shows the concentration of enstrophy into coefficients in the wavelet packet and local cosine transforms, respectively. The plots represent the enstrophy accumulated by retaining 1, 2, ..., 1000 of the coefficients after sorting them into decreasing order by amplitude. Certainly with the same number of coefficients retained, wavelet packets keep more enstrophy and therefore behave better in terms of enstrophy contraction.

3.5 Coherent structure identification

One of our main goals is to locate the centers and influence regions of coherent structures. This is done quite effectively by wavelet packets where the more we compress, the more we isolate strong features. The process of discarding weak wavelet packets kills off the small variations of the vorticity field and leaves only the strongest peaks; Figure 3c shows the culmination of this process, a few surviving vortex cores floating on a perfectly zero background. In the local cosine or any other Fourier-like techniques, the more we compress the smoother the signal will get and the more that quasi-singular features will be smeared one into another. This problem is evident in Figure 3d, but is foreshadowed

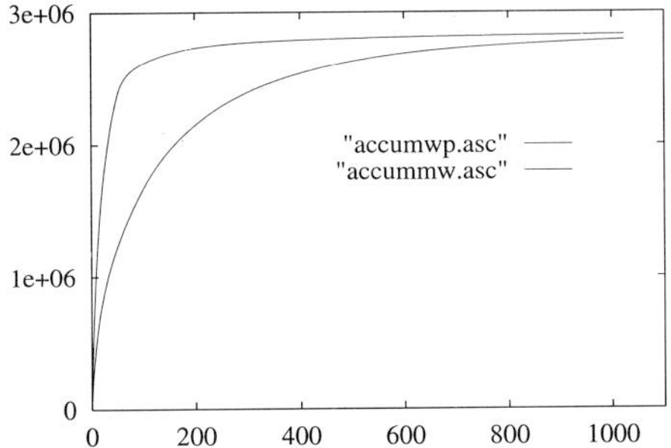

Figure 8. Accumulated enstrophy in partial sums of a decreasing-amplitude rearrangement of wavelet packet and local cosine representations.

even in a moderate compression such as Figure 3b. Another problem with local cosines is the parasitic oscillations we get at high compressions, which interfere with algorithms that localize peaks by finding maxima. In extreme cases, we may even encounter false peaks.

Another of our aims is to identify without ambiguity the dynamic portion of the field and to distinguish it from any passive background. We expect that this background flow should be a kind of noise: spatially homogeneous, isotropic, with a broad spectrum. Ideally, the rejected portion of the field should have amplitudes distributed like independent Gaussian random variables, so it is indistinguishable from measurement error. From the radial spectrum plots in Figures 6a,b, we see that both local cosine and wavelet packet compressions leave remainders that behave much like weak, uncorrelated noise, if the compression ratio stays below 100 or so. As the compresion ratio increases, the noise-like portion of the local cosine spectrum occupies fewer and fewer wavenumbers, until there is virtually no difference betwen the strong and weak spectra. In Figure 7d, the strong and weak spectra match almost perfectly above $|\mathbf{k}| = 12$. By contrast, the noise-like portion of the weak wavelet packet spectrum covers almost the same wavenumbers at low or high compressions, but it gets progressively less noise-like, in the sense of spectral slope, from Figure 6a to Figure 7c.

Both goals advance the efficiency of further computations involving the fields. When we obtain a satisfactory separation, we can follow the evolution of the two portions separately. We explicitly compute the time evolution of the coherent structures, which costs little because there are few of them. We replace the background by a synthetic noise having the same statistics, *i.e.*, the same spectral slope and similar higher order spectral moments such as skewness and flatness. This model may be simple and cheap to compute. If the background plays only a small role in the evolution, we will obtain reasonably accurate approximations to the "true" evolution at much lower cost.

3.6 Improvements in compression techniques

We may try to do the compression recursively until the discarded coefficients will represent only a noise. Starting with a high compression, we keep the top few strong coefficients. Then we highly recompress the reconstructed remainder field, keeping only the top few strong coefficients, and so on. This is the technique used in *adapted waveform de-noising* [5]. That method introduces a different basis for each stage in the compression so that at the end we do not have a single orthogonal basis representation of the field. However, the resulting "best component" decomposition of the coherent portion of the field can still be evolved by a low complexity computation, while the remainder can be even more accurately modeled because it has more of the properties of homogeneous noise for a given enstrophy than the remainder produced by a single best basis expansion.

Using the iterative compression technique, we can also try to recover some isotropy by rotating the initial field several times and performing a strong compression at each stage, then superposing all the strong coefficient reconstructed fields. Such methods, however, complicate the problem of computing and comparing the total number of retained coefficients. Further research is needed to circumvent the anisotropy of these separable compressions when the fields to be analyzed are a priori isotropic.

3.7 Conservation of conservation laws

Enstrophy $Z = \|\omega(\cdot, t)\|^2$ is conserved in the Navier–Stokes evolution. We are assured that $Z_{\mathbf{strong}} + Z_{\mathbf{weak}} = Z_{\mathbf{total}}$ for any segmentation $\omega_{\mathbf{total}} = \omega_{\mathbf{strong}} + \omega_{\mathbf{weak}}$ using orthogonal projections, and the conservation law ensures that the total enstrophy in the weak part will never grow more significant.

It is important to note, though, that energy $E = \|\mathbf{u}(\cdot, t)\|^2$ decreases as $t \to \infty$ because of the dissipation term in the Navier–Stokes equation. Having split a vorticity field into pieces, we are not assured that the energy in the two pieces will decay at equal rates. Furthermore, since we are using vorticity and stream function coordinates rather than velocity coordinates, the energy portions E_{strong} and E_{weak} will not add up to E_{total} in a general orthogonal segmentation. Thus it is possible for a significant part of the initial field's energy to wind up in the remainder portion after segmentation, and then for the retained portion's energy to decay faster so that with time the remainder portion's energy becomes even more significant, relatively speaking.

We do have conservation of energy across vorticity segmentations in the Fourier representation, however, where there is a relation $Z = |\mathbf{k}|^2 E$ for the Fourier mode at wavenumber $\mathbf{k}$. Since this holds at all times t, the energy in the remainder field will never grow more significant relative to the retained field. But no such relation holds for wavelet packets or other localized functions. This absence may be a major drawback of our atomic approach, and must be considered seriously if we want to use wavelets or wavelet packets to solve or simulate Navier–Stokes evolutions. Addressing such concerns, however, would go beyond the scope of our present paper.

§4 Perspective

Turbulence, either two-dimensional or three-dimensional, seems to be the random superposition of a set of metastable vortices, whose interactions give rise to its characteristic unpredictible behaviour. The goal for modeling or computing the evolution of turbulent flows is to take a coarse-graining point of view, namely to keep the essential information and discard details as noise.

The bases we have proposed to use in this paper, wavelet packets and local cosines, consist of phase space atoms which can be independently labeled by position, scale, and wavenumber. The Navier-Stokes evolution tends to aggregate these atoms into molecules corresponding to coherent structures. We can calculate the interaction matrix corresponding to the energy and enstrophy cascades in these atomic coordinates, where the number of significant components is much lower.

We gain by drastically reducing the number of degrees of freedom necessary to compute the turbulent flow evolution. Wavelet packets seem to be more efficient than local cosines at accomplishing this reduction. We think that this approach may be extended to the case of three-dimensional turbulent flows, where *vorticity tubes* will play the role of coherent structures. The three-dimensional case poses even greater computing challenges and might benefit even more from parameter reduction than our two-dimensional case.

Acknowledgments. The work of the last four authors was supported in part by the European Economic Community program *Human Capital and Mobility*, contract ERB-CHRX-CT92-001, and the NATO program *Collaborative Research*, contract CRG-930456. The first author was supported in part by AFOSR contract F49620-92-J-0106 and NSF grant DMS-9302828.

The vorticity field we use as example in this paper was computed by Thierry Philipovitch using the two-dimensional Navier-Stokes code written by Claude Basdevant. The visualizations were done in collaboration with Jean-François Colonna. These collaborations and grants are gratefully acknowledged.

References

1. Auscher, P., G. Weiss, and M. V. Wickerhauser, Local sine and cosine bases of Coifman and Meyer and the construction of smooth wavelets, in *Wavelets–A Tutorial in Theory and Applications*, C. K. Chui (ed.), Academic Press, Boston, 1992, 237–256.
2. Batchelor, G. K., Computation of the energy spectrum in homogeneous two-dimensional turbulence, *Physics of Fluids* **12** (Supplement II) (1969), 233–239.
3. Coifman, R. R. and Y. Meyer, Remarques sur l'analyse de Fourier à fenêtre, *Comptes Rendus de l'Académie des Sciences de Paris* **312** (1991), 259–261.
4. Coifman, R. R., Y. Meyer, S. R. Quake, and M. V. Wickerhauser, Signal processing and compression with wavelet packets, in *Progress in Wavelet*

Analysis and Applications, Y. Meyer and S. Roques (eds.), Proceedings of the International Conference "Wavelets and Applications," Toulouse, France, 8–13 June 1992, Editions Frontieres, Gif-sur-Yvette, France, 1993, 77–93.

5. Coifman, R. R. and M. V. Wickerhauser, Wavelets and adapted waveform analysis, in *Wavelets: Mathematics and Applications*, J. J. Benedeto and M. Frazier (eds.), Studies in Advanced Mathematics, CRC Press, Boca Raton, Florida, 1992, 399–423.
6. Daubechies, I., Orthonormal bases of compactly supported wavelets, *Comm. Pure and Appl. Math.* **61** (1988), 909–996.
7. Farge, M., E. Goirand, Y. Meyer, F. Pascal, and M. V. Wickerhauser, Improved predictability of two-dimensional turbulent flows using wavelet packet compression, *Fluid Dynamics Research* **10** (1992), 229–250.
8. Gabor, D., Theory of communication, *J. of the Institute of Electrical Engineers* **93** (III) (1946), 429–457.
9. Kolmogorov, A. N., On degeneration of isotropic turbulence in an incompressible viscous liquid, *Doklady Akademii Nauk SSSR* **31** (1941), 538–540. In Russian.
10. Kraichnan, R. H., Inertial ranges in two-dimensional turbulence, *Physics of Fluids* **10** (1967), 1417–1423.
11. Kraichnan, R. H., Statistical dynamics of two-dimensional flow, *J. of Fluid Mechanics* **67** part 1 (1975), 155–175.
12. Kraichnan, R. H. and D. Montgomery, Two-dimensional turbulence, *Report on Progress in Physics* **43** (1980), 547.
13. Mallat. S. G., A theory for multiresolution signal decomposition: The wavelet decomposition, *IEEE Trans. Pattern Anal. and Machine Intell.* **11** (1989), 674–693.
14. Malvar, H., Lapped transforms for efficient transform/subband coding, *IEEE Trans. ASSP* **38** (1990), 969–978.
15. Marchioro, C. and M. Pulvirenti, *Vortex Methods in 2D Fluid Dynamics*, Number 203 in Lecture Notes in Physics, Springer-Verlag, Berlin, 1984.
16. Meyer, Y., Orthonormal wavelets, in *Wavelets: Time-Frequency Methods and Phase Space*, J. M. Combes, A. Grossmann, and Ph. Tchamitchian (eds.), Springer-Verlag, Berlin, second edition, 1989, 21–37.
17. Monin. A. S. and A. M. Yaglom, *Statistical Fluid Mechanics; Mechanics of Turbulence*, J. L. Lumley (ed.), MIT Press, Cambridge, Massachusetts, 1971.
18. Saffman, P., Vortex interactions and coherent structures in turbulence, in *Transition and Turbulence*, R. E. Meyer (ed.), Academic Press, 1981.
19. Wickerhauser, M. V., High-resolution still picture compression, *Digital Signal Processing: a Review Journal* **2** (4) (1992), 204–226.
20. Wickerhauser, M. V., *Adapted Wavelet Analysis from Theory to Software*, AK Peters, Ltd., Wellesley, Massachusetts, 9 May 1994. 486+xii pages, with diskette, to appear.

21. Wickerhauser, M. V., Comparison of picture compression methods: Wavelet, wavelet packet, and local cosine transform coding, in this volume.
22. Winckelmans. G. S. and A. Leonard, Contributions to vortex particle methods for the computation of three-dimensional incompressible unsteady flows, *J. of Comp. Physics* **109** (1993), 247–273.

Mladen Victor Wickerhauser
Mathematics Department
Washington Unviersity
St. Louis, Missouri, USA
victor@math.wustl.edu

Marie Farge
Laboratoire de Metéorologie Dynamique du CNRS
Ecole Normale Supèrieure
24 rue Lhomond
75231 Paris Cedex 5, France

Eric Goirand
Mathematics Department
Washington Unviersity
St. Louis, Missouri, USA

Eva Wesfreid
CEREMADE
Université Paris–Dauphine
place du Maréchal de Lattre de Tassigny
5775 Paris, France

Echeyde Cubillo
Center for Theoretical Studies of Physical Systems
Clark Atlanta University
Atlanta, Georgia 30314, USA

Wavelet Spectra of Buoyant Atmospheric Turbulence

Meinhard E. Mayer, Lonnie Hudgins, and Carl A. Friehe

Abstract. We present a wavelet cospectral analysis of the boundary-layer flow in the atmosphere above ground. The data were collected during a period of increasing ground temperature, with production of turbulence by both buoyancy and wind shear. The cospectra show a bimodal structure similar to that previously found in shear-generated turbulence over the ocean. The method graphically depicts the strong correlation between momentum transport and heat flux–two of the off-diagonal components in the generalized Reynolds stress tensor. In addition, the cross-scalograms for the over-land data exhibit a pattern of regularly decreasing smaller-scale coherent structures trailing behind and in clear association with the large-scale ones.

§1 Introduction

Wavelet analysis has provided a means to examine the "structure" of turbulent flow (see, *e.g.*, [1–10]). In [6, 7] we applied wavelet analysis to the bivariate case and used the wavelet cross-transform and its "cross-scalogram" pictorial representation to study the off-diagonal terms in the generalized Reynolds stress tensor. This was applied to a set of steady-state atmospheric turbulence data to examine the Reynolds flux terms of horizontal momentum transfer and vertical heat flux. In turbulence, these are proportional to the averaged covariances of the instantaneous vertical velocity component with the horizontal velocity component, and instantaneous temperature, respectively. In the previous study, the wavelet cross-scalograms of the signal pairs showed intermittency and predominance of certain scales in the flow. The scalogram integrated over time reduces to the wavelet cospectrum. The wavelet cospectrum was superior to the Fourier cospectrum in resolving the large scales in

Wavelets: Theory, Algorithms, and Applications
Charles K. Chui, Laura Montefusco, and Luigia Puccio (eds.), pp. 533–541.

ISBN 0-12-174575-9

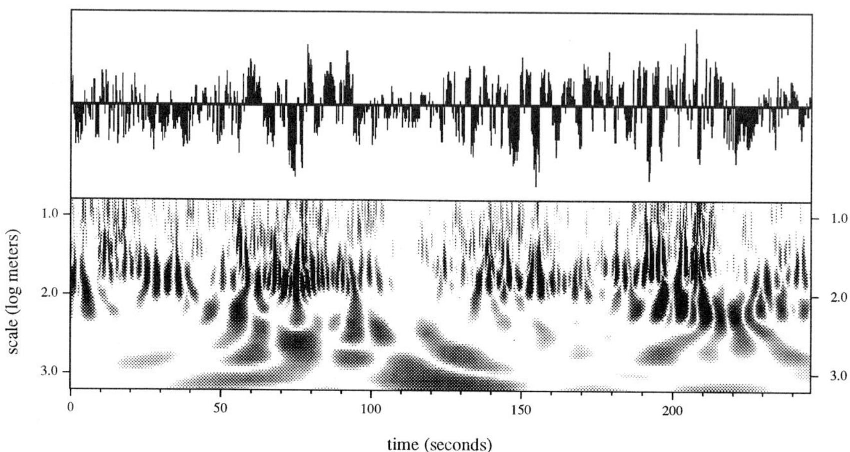

Figure 1. Time series (top) and cross-scalogram (bottom) of the momentum flux at 8 *m* in the boundary layer between ocean and atmosphere, from [6, 7]. Gray-scale does not distinguish > 0 from < 0.

the flow, since Fourier methods would have required a priori knowledge of the scales of interest.

The previous data set was obtained in homogeneous atmospheric high Reynolds number turbulence, at a height of 8 m above the ocean, in a strong and very steady wind. The turbulent fluctuations were predominantly generated by shear in the atmospheric surface layer; the production by buoyancy forces due to heat flux from the ocean to the air was small. A grey-scale version of the momentum-flux cross-scalogram from the previous analysis is shown in Figure 1.

In the present work, we apply the cross wavelet analysis technique to a set of atmospheric surface layer data where there is significant buoyant production of turbulence in addition to wind shear production. The data were obtained over land during mid-day.

§2 Summary of definitions

In distinction from most of the literature on wavelets, we use the *scale number* $s = \frac{1}{a}$, in place of the usual scaling of length, a, so that for Ψ an admissible wavelet, the scale transformation is $\Psi_s(t) = \sqrt{|s|}\ \Psi(st)$.

Definition. *The wavelet transform $\mathcal{W}_f(s,\tau)$ of $f \in L^2$ at time τ and scale number s is then defined by:*

$$\mathcal{W}_f(s,\tau) = \int f(t)\Psi_s(t-\tau)\,dt.$$

In terms of the variables s,τ the inverse wavelet transform is (for almost all t):

$$f(t) = \frac{1}{c_\Psi}\int\int ds\,d\tau\,\mathcal{W}_f(s,\tau)\overline{\Psi_s}(\tau - t).$$

Parseval's relation for wavelets,

$$\|f\|^2 = \frac{1}{c_\Psi}\int ds \int d\tau\,|\mathcal{W}_f(s,\tau)|^2,$$

or more generally

$$\langle f,g\rangle = \frac{1}{c_\Psi}\int ds \int d\tau\,\overline{\mathcal{W}_f}(s,\tau)\mathcal{W}_g(s,\tau),$$

brings us directly to definitions for wavelet spectral and cross-spectral densities.

2.1 Spectral and cross-spectral densities

Our definitions are related to earlier works of Gamage, Mahrt, Howell [5,8,9], and Meneveau [10].

Definition. *We define the wavelet power spectrum of f as:*

$$\mathcal{P}_f^w(s) = \int d\tau\,|\mathcal{W}_f(s,\tau)|^2.$$

For $f,g \in L^2$ the wavelet cross-spectrum is defined by:

$$\mathcal{C}_{fg}^w(s) = \int d\tau\,\overline{\mathcal{W}_f}(s,\tau)\mathcal{W}_g(s,\tau).$$

For real f, g, and Ψ, the cross-spectrum is identical to the cospectrum and we use the terms interchangeably. The density plot of the integrand in the τ, s-plane is called the *cross-scalogram*.

The relation of the wavelet cross-spectrum of f,g to the traditional Fourier cross-spectrum $\mathcal{C}_{fg}$ is:

$$\mathcal{C}_{fg}^w(s) = \tfrac{1}{2\pi}\int d\omega\,\mathcal{P}_{\Psi_s}(\omega)\mathcal{C}_{fg}(\omega);$$

it is the Fourier cross spectrum averaged by the power spectrum of the wavelet. Note that this is not simply Fourier smoothing, which convolves the Fourier spectrum with a smoothing kernel. The functional dependence on s is contained in the right-hand side via the scaling parameter of the *wavelet* Ψ_s.

§3 Data acquisition and sampling

The data analyzed here were obtained in the atmospheric boundary layer over flat land at a height of 4 m [8]. The three components of the velocity vector u, v, w were obtained with a 0.2 m path-length sonic anemometer. Temperature T was measured with a 75 μm diameter thermocouple bead. The sampling rate was 20 Hz.

The conditions included gradually rising air temperature due to heating of the land and the velocity field was reasonably steady with a mean wind speed of about 4 m/s. The velocity components u, v, w, and the temperature T were measured. The portion of the data set analyzed here was for a period of 54 minutes.

The 2^{16} samples were partitioned into 4×2^{14} subsamples of 14 minutes each, of which we discuss only one. After subtracting the means four sets of wavelet co-spectra were computed for the heat flux and momentum transport.

The wavelet used was a Daubechies D8 wavelet (versus a cubic spline wavelet used in the previous data set).

§4 Results

Figure 2 shows the time series for the three components of the velocity u, v, w and the temperature T for the whole sample of 54 minutes. The time series for the velocity components shows the typical random nature of turbulence, with perhaps a "switching" of v, the horizontal cross-wind component. The temperature time series shows the steady rise in temperature, with a definite asymmetry in the fluctuations, due to the heating of the air at the surface.

The cospectra for the momentum and heat fluxes (the time-integrals of the cross-scalograms [6,7]) are shown in Figure 3 for one of the 14-minute subsamples. These cospectra show the bimodal features also found in the previous over-ocean data set: the smaller scales are characterized by vigorous turbulent mixing which gives rise to a positive (upward) heat flux, and negative (downward) momentum flux. At larger scales there is a crossover of zero heat flux and fluxes of opposite signs of nearly the same magnitudes. These are believed to be manifestations of large-scale structures in the flow.

The cross-scalograms for momentum and heat fluxes for the same subsample of 14 minutes duration are shown in Figure 4. Unfortunately, the gray-scale representation does not allow the reader to distinguish positive from negative values, but shows clearly correlated structures in the two scalograms.

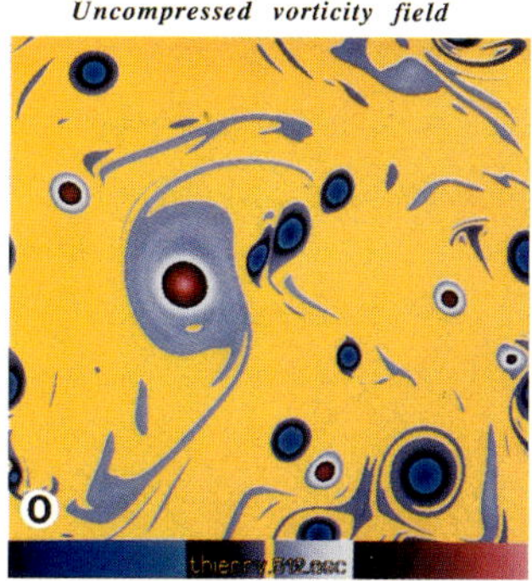

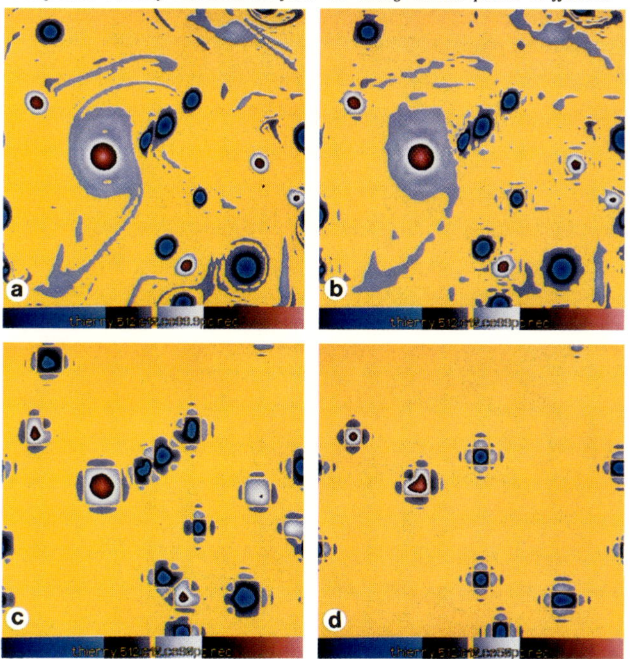

Figure 1. Reconstruction from the strongest wavelet packet coefficients: (0.) uncompressed vorticity field corresponding to 262144 grid point coefficients; (a.) compressed vorticity field reconstructed from the 3000 (11.4 per 1000) strongest wavelet packet coefficients, which retain 99.9% of the enstrophy; (b.) compressed vorticity field reconstructed from the 739 (2.8 per 1000) strongest wavelet packet coefficients, which retain 99% of the enstrophy; (c.) compressed vorticity field reconstructed from the 73 (0.28 per 1000) strongest wavelet packet coefficients, which retain 90% of the enstrophy; (d.) compressed vorticity field reconstructed from the 14 (0.05 per 1000) strongest wavelet packet coefficients, which retain 50% of the enstrophy.

Compressed vorticity reconstructed from the weak wavelet packet coefficients

Figure 2. Reconstruction from the weakest wavelet packet coefficients: For uncompressed vorticity field corresponding to 262144 grid point coefficients see Fig. 1 (0.). (a.) compressed vorticity field reconstructed from the 259144 (98.8 per 100) weakest wavelet packet coefficients, which retain 0.1% of the enstrophy; (b.) compressed vorticity field reconstructed from the 261405 (99.7 per 100) weakest wavelet packet coefficients, which retain 1% of the enstrophy; (c.) compressed vorticity field reconstructed from the 262074 (99.97 per 100) weakest wavelet packet coefficients, which retain 10% of the enstrophy; (d.) compressed vorticity field reconstructed from the 262130 (99.99 per 100) weakest wavelet packet coefficients, which retain 50% of the enstrophy.

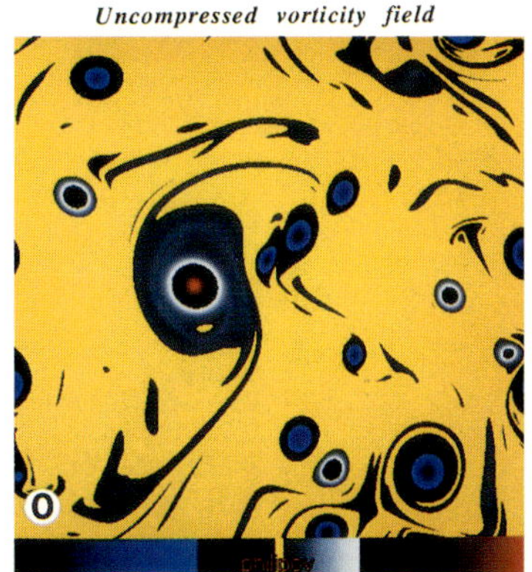

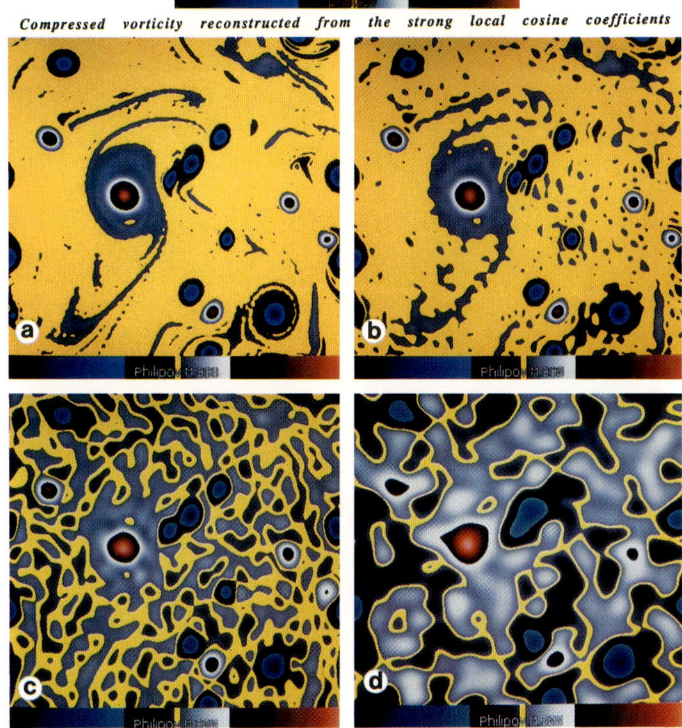

Figure 3. Reconstruction from the strongest local cosine coefficients: (0.) uncompressed vorticity field corresponding to 262144 grid point coefficients; (a.) compressed vorticity field reconstructed from the 5114 (19.5 per 100) strongest local cosine coefficients, which retain 99.9% of the enstrophy. (b.) compressed vorticity field reconstructed from the 1479 (5.6 per 1000) strongest local cosine coefficients, which retain 99% of the enstrophy; (c.) compressed vorticity field reconstructed from the 425 (1.6 per 1000) strongest local cosine coefficients, which retain 90% of the enstrophy; (d.) compressed vorticity field reconstructed from the 70 (0.27 per 1000) strongest local cosine coefficients, which retain 50% of the enstrophy.

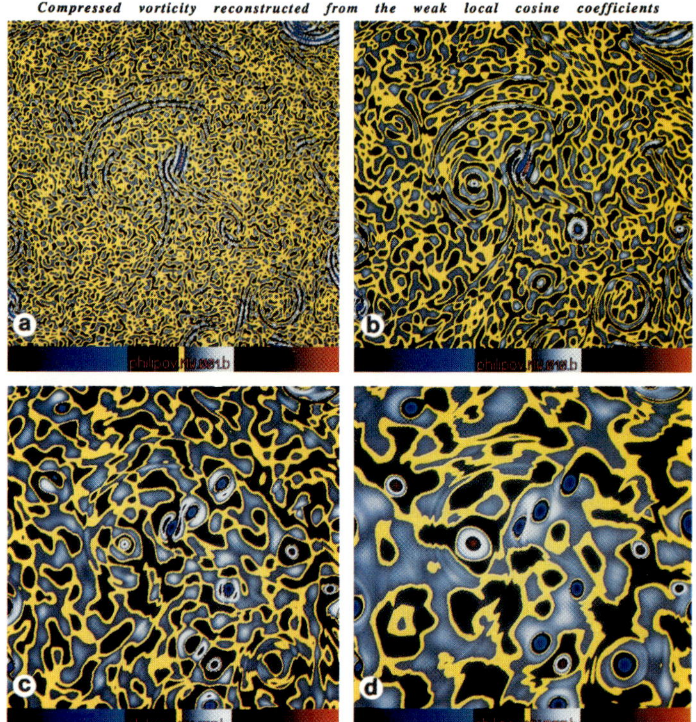

Figure 4. Reconstruction from the weakest local cosine coefficients: For uncompressed vorticity field corresponding to 262144 grid point coefficients see Fig. 1 (0.). (a.) compressed vorticity field reconstructed from the 257030 (98 per 100) weakest local cosine coefficients, which retain 0.1% of the enstrophy; (b.) compressed vorticity field reconstructed from the 260665 (99.4 per 100) weakest local cosine coefficients, which retain 1% of the enstrophy; (c.) compressed vorticity field reconstructed from the 261719 (99.8 per 100) weakest local cosine coefficients, which retain 10% of the enstrophy; (d.) compressed vorticity field reconstructed from the 262074 (99.97 per 100) weakest local cosine coefficients, which retain 50% of the enstrophy.

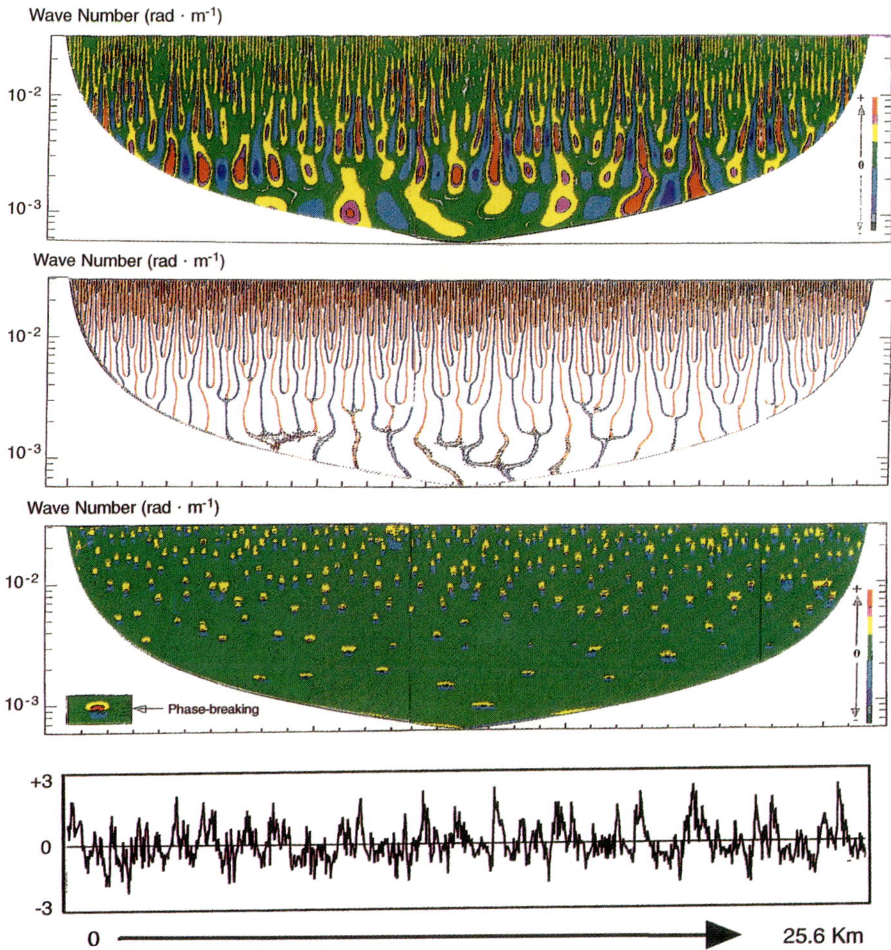

(Figure 1.2) This figure illustrates the wavelet analysis of a homogeneous series (vertical velocity of the air in $m{\cdot}s^{-1}$). The data are collected by an aircraft on a constant level run of 25.6 km in the Atmospheric Boundary Layer. The real part (top) of the wavelet transform (Morlet wavelet), the bifurcations of the turbulent structures (middle), and the regions of rapid changing of phases called phase-breakings (bottom), are represented in a space-wave-number frame. Color coding: from red to black for maximum to minimum.

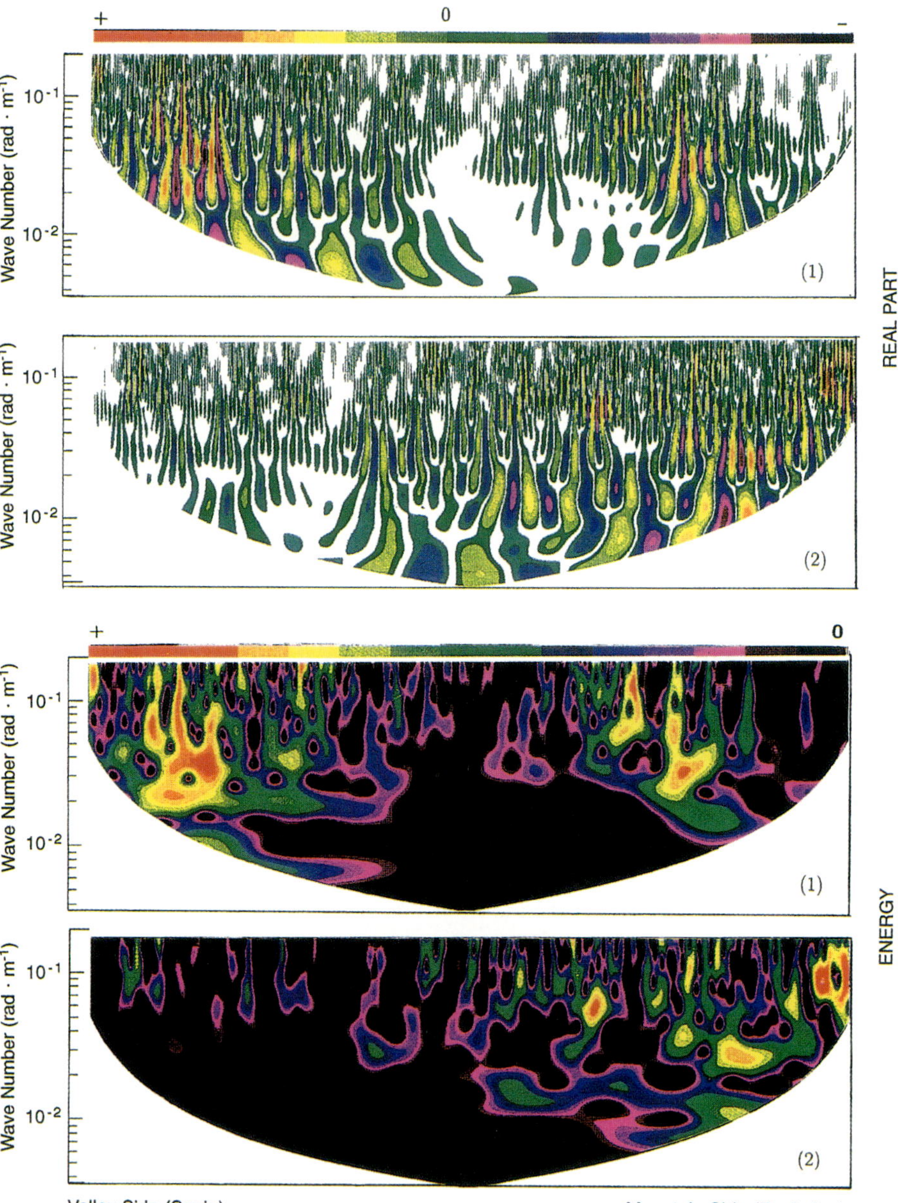

(Figure 3.2) This figure shows the wavelet analysis of two inhomogeneous turbulent series (vertical velocity of the air). The data are collected by an aircraft close to the Pyrénées chain. The two diagrams on the top represent the real part of the wavelet transform and those at the bottom represent the energy. These diagrams allow one to analyze, in a space-wave-number representation, the turbulent energy and to show the transfer from production to dissipation.

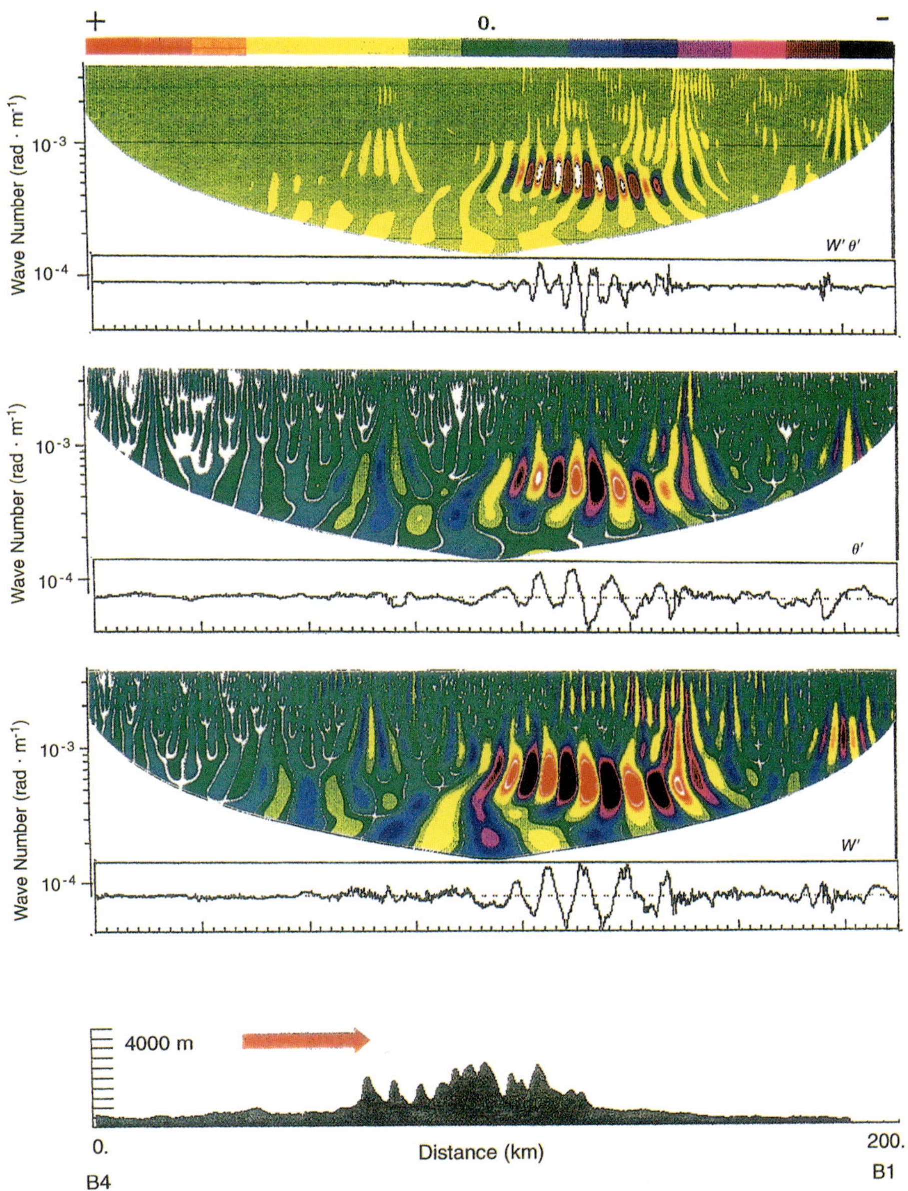

(Figure 3.3) The figure depicts the wavelet transform of the vertical velocity and the potential temperature of the air collected by an aircraft on a constant level run of 200 km length perpendicular to the Pyrénées chain. A lee-wave system degenerating to turbulence occurs downwind. The wavelet cross-spectrum (top) shows the evolution of the phases between the two parameters. It corresponds to a quadrature in the lee-wave part and to a positive turbulent flux at the end of the lee waves.

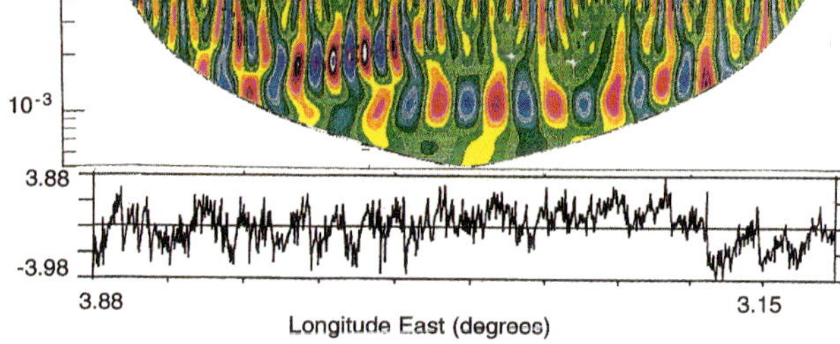

(Figure 4.2) The wavelet analysis of dynamical structures above the Mediterranean Sea in a deflecting wind around the Pyrénées chain is shown in this figure. The aircraft constant level run is perpendicular to the wind and this picture represents the real part of the wavelet transform applied to the transversal wind component. It emphasizes the dynamical structures along the wind direction.

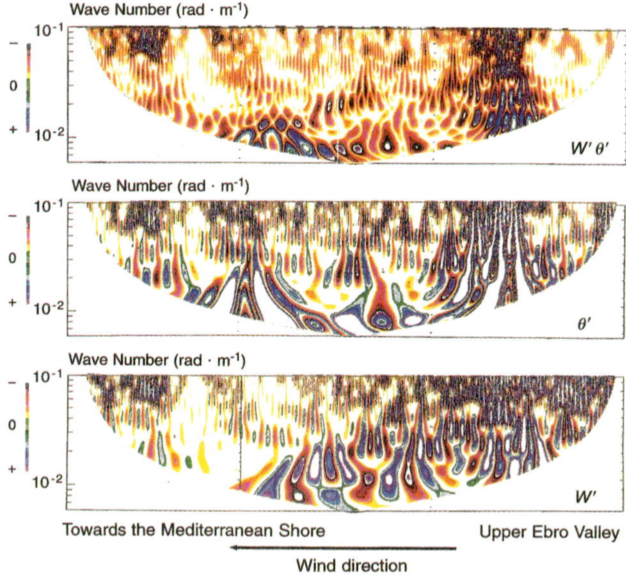

(Figure 4.3) The figure depicts the wavelet transform of the vertical velocity (bottom) and the potential temperature (middle) of the air along an aircraft constant level run of 100 km in the Ebro valley (Spain). The event studied is the Cierzo wind. These diagrams show dynamical and thermodynamical inhomogeneous turbulence The wavelet cross-spectrum (top) shows an important transfer in the upper part of the valley in the whole wave-number band.

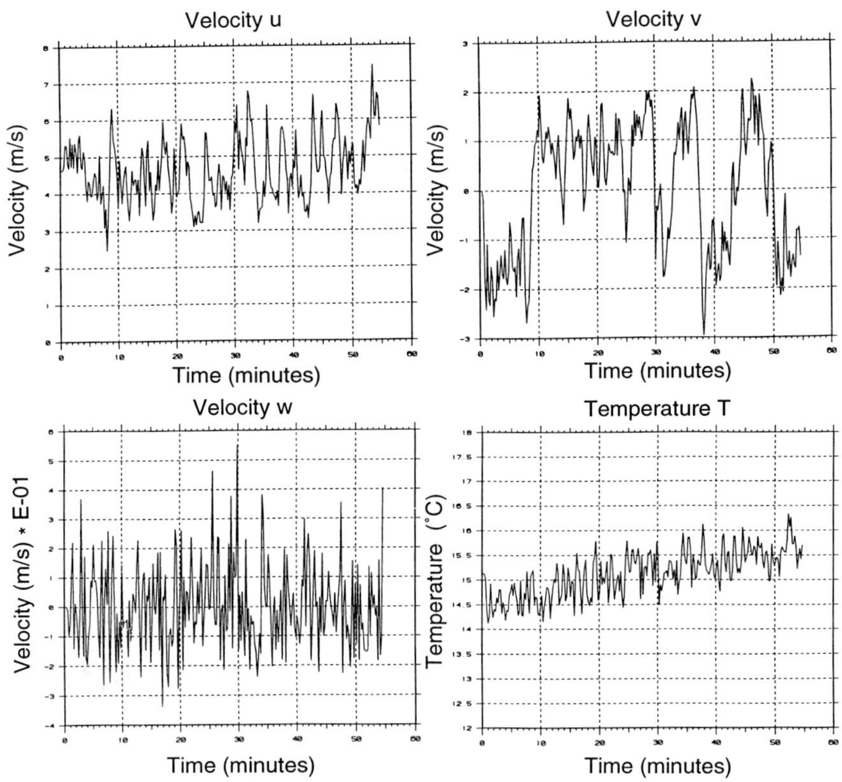

Figure 2. Velocity components and temperature versus time.

The superimposed $u - w$ and $w - T$ cross-scalograms show large coherent structures, correlation between the structures in the momentum flux and heat flow, and the linear "trailing" of smaller-scale structures behind the large-scale structures. The large scales at the bottom correspond to $\sim$ 1200 m. whereas the finest scales correspond to $\sim$10 m. Notice that in the trail of smaller-scale structures (behind the largest ones), the scales are decreasing by a factor of 2 in approximately 63 s. The horizontal axes of the scalograms are calibrated in time, but should be interpreted as distance in the fluid. This is accomplished via Taylor's hypothesis of "frozen turbulence," *i.e.*, $\mid x \mid = Ut$, where x is the distance in the fluid, and U is the mean convection velocity. Thus the 63 s correspond to a distance of approximately 278 m.

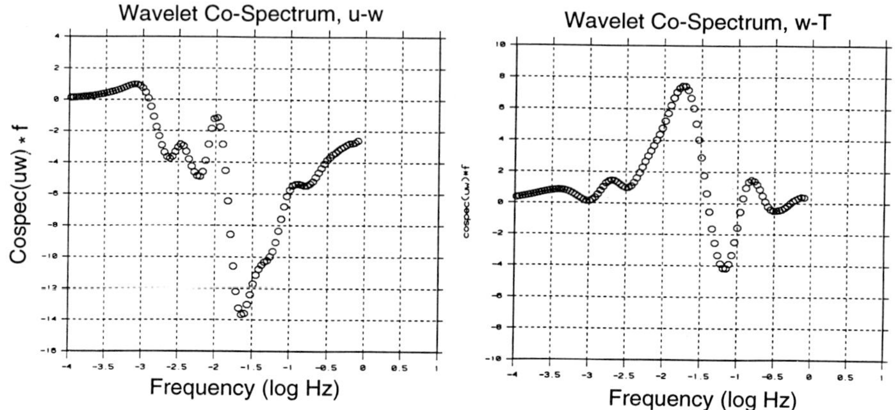

Figure 3. Cospectra of $u - w$ (momentum flux) and $w - T$ (heat flux).

Once again, a high degree of intermittency is observed in the coherent structures, similar to the ones seen previously in the over-water data.

The wavelet cross-scalograms also show a correlation between coherent structures in the heat flux and in the momentum transport. The correlation is quite good, and may allow for either quantity to be used as a "detector" of coherent structures. If one has both, then one should be able to use the additional information to improve the confidence in the localization of coherent structures.

We note that a similar pattern is not obvious in the scalogram of the over-ocean data (Figure 1). The major difference between the two data sets is the larger production of turbulence by buoyancy (vertical heat flux) in the over-land flow. Buoyancy could be altering the basic pattern in the turbulent fields in the near-surface flow.

The results presented here are *experimental* evidence of a clear association of smaller coherent structures with larger ones. Furthermore, the smaller ones follow *behind* the larger ones. We have what amounts to a snapshot in space, within the approximation of Taylor's hypothesis, of these objects convecting along together, in descending order of size. Our previous work with over-water data did not show this kind of association.

§5 Conclusions

This second application of the bivariate cross-scalogram technique to turbulence driven both by buoyancy and shear showed both similarities and differences compared to the shear-driven case. Both sets showed intermittency,

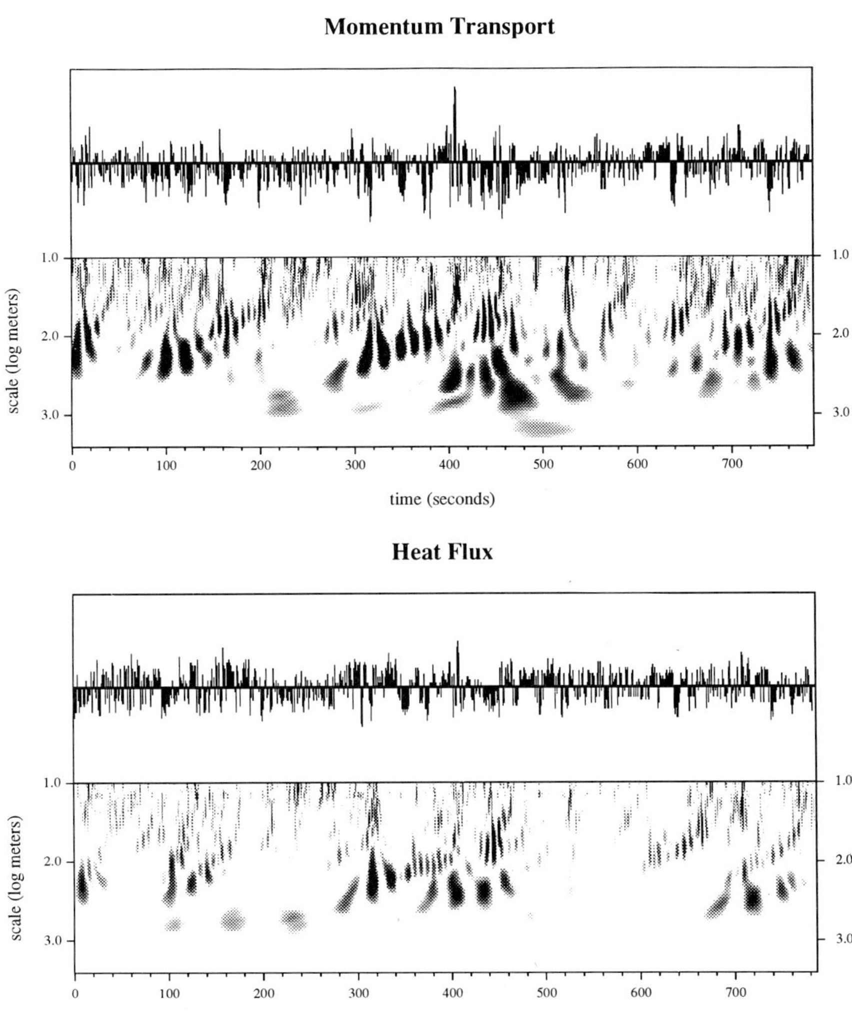

Figure 4. Time series of fluxes and cross-scalograms for the over-land data. Gray-scale does not distinguish > 0 from < 0.

large-scale coherent structures, and bimodal cospectra of Reynolds fluxes. A new feature of the present analysis is the linearly trailing smaller-scale structures following a large-scale event. Since the wavelet scale is logarithmic, the breakdown feature is exponential in space and is seen in both the momentum and heat flux.

The major questions now are the estimation of the amount of flux "carried" by these identifiable features and the exposition of the physical mechanisms that generate and control these patterns. Bivariate wavelet analysis appears to have been a good method for identification.

Future work based on the cross-scalogram technique may be helpful in resolving these questions.

Acknowledgments. C.A.F. would like to acknowledge the support of ONR grant N00014-J-0190, NSF grants ATM-8715619, 9024436, and the U.C. INCOR program.

References

1. Everson, R., L. Sirovich, and K. R. Sreenivasan, Wavelet Analysis of the Turbulent Jet, *Phys. Lett. A* **145** (1990), 314–322.
2. Farge, M., Wavelet transforms and their applications to turbulence. *Ann. Rev. Fluid Mech.* **24** (1992), 395–457.
3. Farge, M. and T. Philipovitch, Coherent structure analysis and extraction using wavelets, in *Progress in Wavelet Analysis and Applications*, pp. 477–490; Y. Meyer and S. Roques (eds.), Frontières, Gif, 1993.
4. Farge, M. and G. Rabreau, Transformée en ondelettes pour détecter et analyser les structures cohérentes dans les écoulements turbulents bidimensionnels, *C. R. Acad. Sci. Paris, Mécanique des fluides*, Série II **307** (1988), 1479–1486.
5. Gamage, N. and L. Mahrt, Scale estimation of coherent structures with variance statistics of the wavelet transform, Technical Report, Dept. of Atmospheric Sciences, OSU, February 1991.
6. Hudgins, L., C. A. Friehe, and M. E. Mayer, Wavelet and Fourier analysis of atmospheric turbulence, in *Progress in Wavelet Analysis and Applications*, pp. 491–498; Y. Meyer and S. Roques (eds.), Frontières, Gif, 1993.
7. Hudgins, L., C. A. Friehe, and M. E. Mayer, Wavelet transforms and atmospheric turbulence, *Phys. Rev. Lett.* **71** (20) (1993), 3279–3282.
8. Mahrt, L., Eddy asymmetry in the sheared heated boundary layer. *J. Atmospheric Sci.* **48** (1991), 472.
9. Mahrt, L. and J. Howell, Estimation of eddy characteristics from time series using localized transforms, in *Conference on Turbulent Shear Flows*, Technical University, Munich, Sepeptember 1991, 9–11.
10. Meneveau, C. Analysis of turbulence in the orthonormal wavelet representation. *J. Fluid Mechanics* **232** (1991), 469–520.
11. Oncley, S. P. Flux parameterization techniques in the atmospheric surface layer, Ph. D. Thesis, Department of Mechanical Engineering, University of California Irvine, 1989; 202 pp.

Meinhard E. Mayer
Department of Physics
University of California
Irvine, CA 92717-4575
mmayer@uci.edu

Lonnie Hudgins
Northrop Corporation
Electronics Systems Division
Hawthorne, California 90251-5032
lonnie@lynx.ps.uci.edu

Carl A. Friehe
Dept. of Mechanical Engineering
University of California
Irvine, CA 92717-3975
cfriehe@uci.edu

Experimental Study of Inhomogeneous Turbulence in the Lower Troposphere by Wavelet Analysis

Aimé Druilhet, Jean-Luc Attié, Leonardo de Abreu Sá, Pierre Durand, and Bruno Bénech

Abstract. In this paper we use wavelet transform to analyze aircraft data in the case of inhomogeneous turbulence. The data were gathered during a meteorological experiment (PYREX) to study the airflow around a mountain ridge: the Pyrénées. We present the spatial distribution and the various lengthscales of the energy fluxes computed from turbulence data (thermodynamics) for four specific flows : 1) Mechanical turbulence in the lee side of the mountain; 2) Mountain lee waves and small scale turbulence; 3) Coherent structures within Tramontana wind over the Mediterranean sea; and 4) Inhomogeneous and strongly turbulent valley wind (Cierzo wind). The last two local winds are generated by the Pyrénées mountain range.

§1 Introduction

In geophysical flows, turbulence plays a fundamental role. It is essentially turbulence that controls the transfer of energy, particularly momentum. Therefore, turbulence is localized in areas where strong shear occurs (such as jet streams, or flows above reliefs), or in areas with high energy gradients (such as cyclones).

Close to the ground, a transition boundary layer develops which then governs the main energy transfers between the ground and the atmosphere. This area which is strongly turbulent can be investigated "in situ", and, in fact, numerous instruments and techniques of measurement similar to the laboratory systems used in fluid mechanics have been developed.

For a long time, only homogeneous turbulence was investigated. The resulting knowledge was a parameterization of the turbulent flow characteristics. This parameterization is used in numerical models of the atmospheric flow such as weather forecasting, or in modeling the diffusion-transport of effluents.

Wavelets: Theory, Algorithms, and Applications
Charles K. Chui, Laura Montefusco, and Luigia Puccio (eds.), pp. 543–559.

ISBN 0-12-174575-9

Later, sensors and measurement techniques for atmospheric turbulence were adapted to be embarked onboard instrumented aircraft. With such mobile-type platforms, measurements could be performed in areas where experimental data, such as the average properties of heterogeneous surfaces or the flow distortion due to the relief, were absent. In these cases, the thermodynamic parameters are more complex because of the accumulation of various mechanisms which greatly disturb the steady state. These studies require the development of specific processing tools enabling, first of all, the precise description of the data samples as space-wavenumber structures. Secondly, these tools need to provide a means of isolating the various mechanisms simultaneously at work from the data gathered in inhomogeneous conditions.

The aim of this paper is to present some characteristics of turbulent signals analyzed with wavelet techniques. We will analyze mainly non-stationary signals collected with instrumented aircraft during the PYREX experiment [2,3], the goal of which was to study the atmospheric flow around a mountain range: the Pyrénées. For comparison, we will use as a reference the turbulence signals gathered in homogeneous conditions and analyzed in a more classical approach using a Fourier transform. We will then present the contribution of the wavelet transform to the study of homogeneous signals.

§2 Airborne measurements

Aircraft are particularly adapted to investigation of the atmosphere. Due to their mobility, they allow in situ study of the specific area where interesting atmospheric events occur. In situ measurements can be performed for various areas of atmospheric physics: turbulence, radiation, microphysics, and chemistry. The mean thermodynamic parameters are temperature, water vapor content, and the three wind components, computed from the measurement of a large set of raw parameters.

In France, the aircraft currently used for tropospheric turbulence studies works at about 100 m.s^{-1} airspeed. The measurement system (sensors, analog filtering, and digitization) provides signals with little noise up to a resolution of about 4 m.

In homogeneous atmospheric turbulence (in the atmospheric boundary layer for instance), the scales under study range from around ten meters up to one kilometer. A correct description of these scales requires a straight and constant level run of about 20 km. In disturbed flows, these runs could be as long as 200 km, like during the flights performed in the PYREX experiment.

Hence, according to the sampling rate and to the length of the record, a turbulent sample is defined by a time series of from 5000 to 50000 points. This resolution allows the description of a wide wavelength band (from 10 m to 20 km). The principal mathematical tool used is the Fourier transform, which allows spectra and cross-spectra computation, as well as numerical filtering and spectral coupling of the signals. The algorithm used is the Singleton' Fast Fourier Transform [6]. Inhomogeneous signals are studied, segment-by-segment, with the Short-Time Fast Fourier Transform and particularly by wavelet transform [1,4]. The wavelet used is essentially for determining the spectral structure of the signals, represented as a time-frequency or a space-wavenumber. The analyzing function is a complex Morlet-type wavelet [5].

§3 The "reference" case: homogeneous turbulence

3.1 The classical approach

Homogeneous turbulence can be found in the atmospheric boundary layer when mechanical production prevails, *i.e.*, with strong wind and weak sensible heat flux. The turbulence characteristics are solely dependent on the altitude above ground, *i.e.*, the vertical coordinate, which is the direction of the energy flux. These characteristics are independent of the position and also the orientation of the sample in the horizontal plane.

An example of the fluctuations having temperature (θ') and turbulent vertical velocity (w') is presented in Figure 1.1. The turbulence can be quantified by the various statistical moments. Among them, the second order moment $\overline{w'\theta'}$, is the vertical turbulent kinetic fluxes of sensible heat. One of the goals, which is partially achieved in atmospheric research, is the development of non-dimensional profiles for these quantities as a function of characteristic scales which are dependent on those parameters relevant to the turbulence, namely, heat and momentum.

The homogeneity of the samples gathered during airborne measurements is often degraded because of the natural complexity of the areas flown over. Homogeneity can be verified by the shape of the time-integral function used in the calculation of moment: for instance, a 2nd order moment is computed as follows:

$$\overline{X_i'X_j'} = \lim_{T\to\infty} \frac{1}{T} \int_0^T X_i'(t)X_j'(t)dt. \tag{1}$$

Let us define the time-integral function f:

$$f(\tau) = \int_0^\tau X_i'(t)X_j'(t)dt. \tag{2}$$

Convergence of the calculation of moment and the homogeneity of the sample can be verified by the linearity of f. In Figure 1.1, function f is plotted for the turbulent heat flux.

These turbulent samples can be characterized by a classification of the size of the various events that make up the samples. A more widely used method is the Fourier Transform, which gives us the turbulent energy distribution, as a function of the wavenumber (spectral structure). The turbulent series is described by the sum of the harmonic waves, each characterized by its amplitude and its phase. Although these two terms do not have a physical significance, they offer an accurate mathematical description of the sample.

Successive zooms of a turbulence sample give an impression of scale invariance. This characteristic is related to a high wavenumber subrange in which

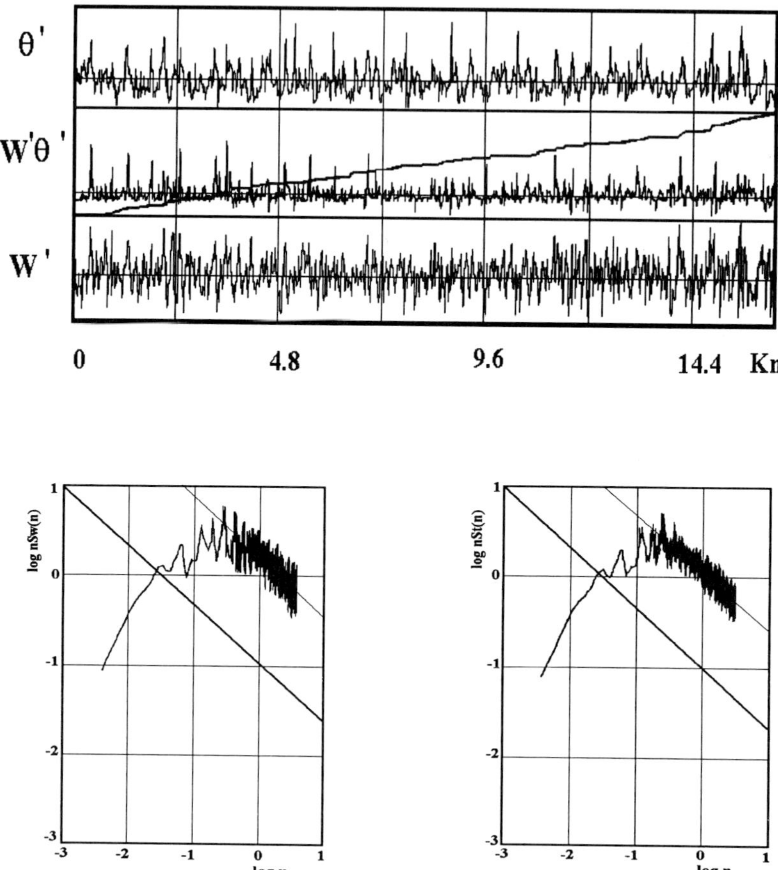

Figure 1.1. Analysis by Fourier transform of one case of homogeneous turbulence data: Temperature spectrum (left) and vertical velocity spectrum (right).

the energy decreases following a power law as a function of the wavenumber. This feature has given rise to an idea of hierarchy in turbulent vortices (inertial cascade) which perhaps does not correspond to a physical reality. The scale of invariance domain overlays the whole inertial subrange extending well beyond the highest wavenumber reached by airborne measurements. At aircraft altitudes, it ranges from some ten meters down to a few millimeters. In this subrange, turbulent kinetic energy is transferred, without production, towards the viscous dissipation range.

In parallel to the spectral structure of a single turbulent sample, through cross-spectral analysis we can characterize the common frequency band for two parameters. The cospectrum of w' and x' defines the frequencies (or wavelengths) which contribute to the vertical transfer of x.

3.2 Use of wavelet transform

The wavelet transform is useful even for homogeneous samples, although they are not intended for such application. It can give a more accurate description of the turbulence homogeneity. In particular, it allows for a more physical study of phase variability at various frequencies.

Using the wavelet transform in the study of homogeneous samples enables one to give a picture of the turbulence hierarchy in a space-wavenumber representation. An example of the turbulence structure is shown in Figure 1.2 where one can see: 1) the amplitude diagram; 2) the location of the extrema of the amplitude; and 3) the location of phase-breakings. An analysis of the phase demonstrates the fact that bifurcations occur simultaneously with phase-breakings. Better knowledge of turbulence homogeneity could be derived from a statistical study of the distribution of location and energy levels for these events in the space-wavenumber representation.

§4 Inhomogeneous samples

Homogeneous turbulences are probably uncommon events in the lower troposphere. Indeed, the smallest mechanical instability is often sufficient to generate coherent structures within the turbulent flow. These structures could range from helical cells to convective cells, according to the respective importance of the two sources of instability: mechanical or thermal. Laboratory studies tend to show that the bifurcations between the various types of structure are governed by threshold mechanisms. Later on, we will present in this paper an example of such structures within a flow over the sea.

More generally speaking, the relief ensures significant modification of the turbulent structure of the flow by the accumulation of mechanisms of widely varied scales. The resulting structure is highly inhomogeneous. Firstly, the characterization of samples and then the separating and quantifying of the various mechanisms, would seem to be the domains that are particularly suited to treatment of the wavelet transform.

To illustrate this, we have chosen some examples of turbulent samples measured on board instrumented aircraft in the proximity of the Pyrénées relief during the PYREX experiment. Figure 2 is a diagram of the experiment showing the main experimentation equipment. The regions of local wind are also represented. The following mechanisms will be analyzed later in this paper:

(1) inhomogeneous turbulence, close to the relief on the leeward side of the mountain; in this area, rotors are generally to be found;

(2) leewaves and associated turbulence above the relief;

(3) mechanical structure of the "Tramontana" wind; and

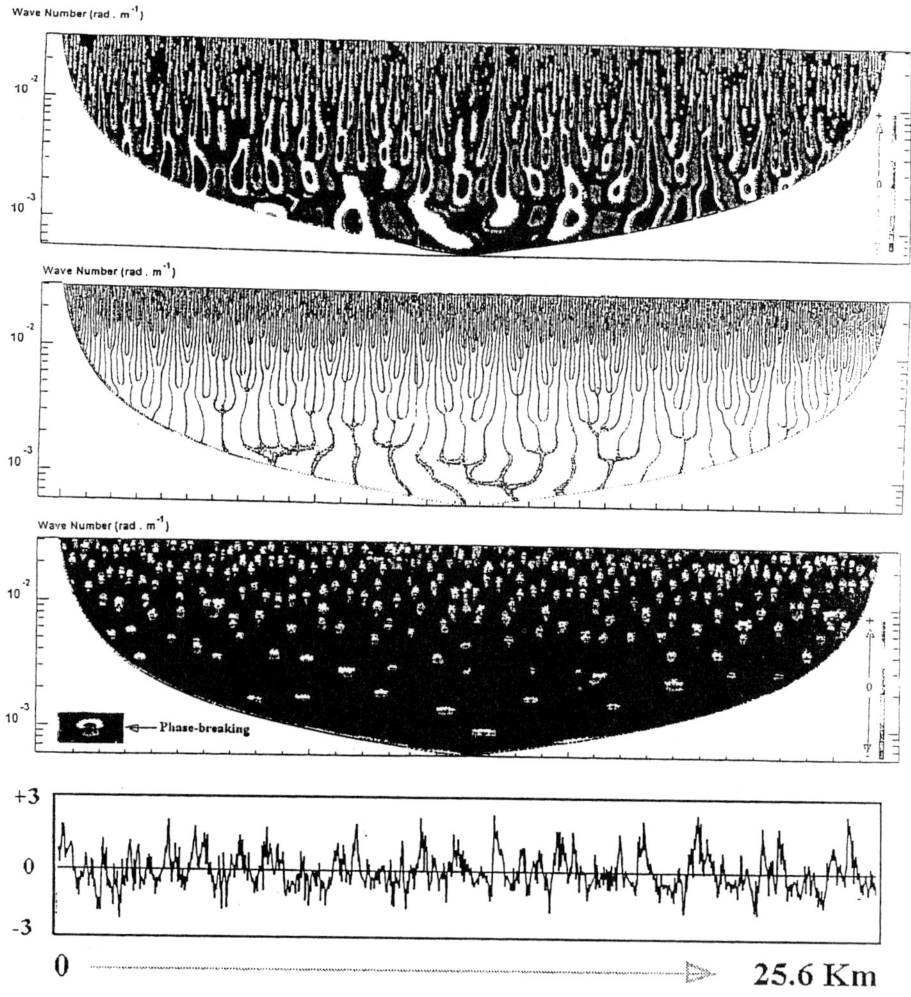

Figure 1.2. Real part of the wavelet transform (top), bifurcations corresponding to the extrema of the amplitude (middle), and phase-breakings (bottom) of a homogeneous turbulent sample (vertical velocity of the air). (See Figure 1.2 in color in the color section of this volume.)

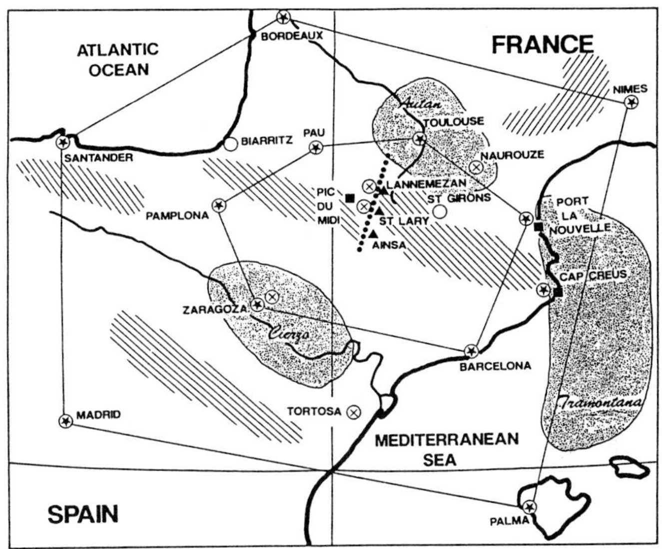

Figure 2. Synoptic view of the ground infrastructure during PYREX. The climatological range of the winds is indicated by half-tone. ⊛ Soundings; ▲ profilers; ⊗ sodar; • microbarographs; ■ constant level balloons (from Bougeault et al., 1993)[3].

(4) inhomogeneous turbulence and fluxes in a valley wind (the "Cierzo").

The first two topics relate to the turbulence in the close vicinity of the relief, resulting from the friction of the flow on the mountain and from the momentum flux modification. The last two relate to local winds generated by the Pyrénées mountain range. The formation and evolution of the winds depend on the flow around the relief.

§5 Lee waves and turbulence

5.1 Turbulence close to the mountain

Flight legs very close to the mountain were performed by the Merlin IV aircraft of Météo-France on November 16, 1990. The synoptic meteorological situation was characterized by a northerly wind. Strongly turbulent zones associated with the roughness are spread out on the downwind side of the mountain. Figure 3.1 shows the turbulent zones observed on the vertical velocity of the air close to the relief. Figure 3.2 presents the wavelet transform (real part and energy) performed on the flight legs 1 and 2. The wavelet transform provides a

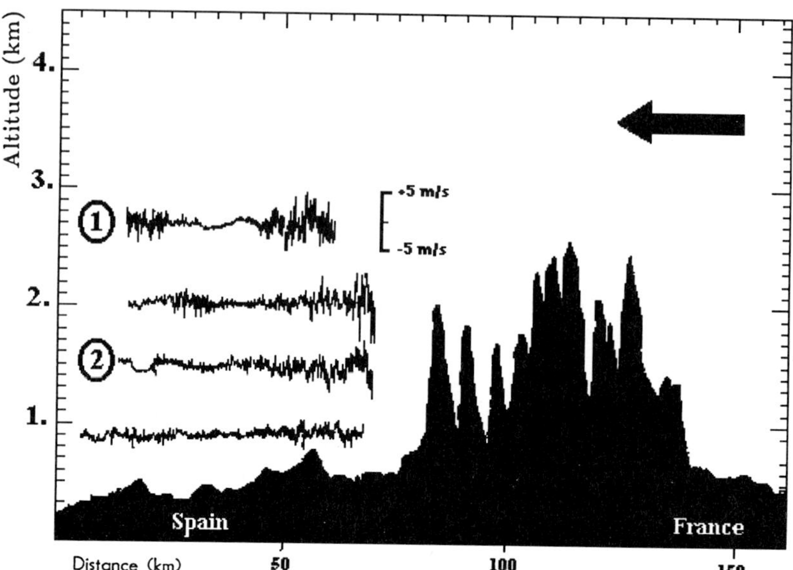

Figure 3.1. Aircraft measurements of vertical velocity of the air close to the Pyrénées on 16 November 1990 during the PYREX experiment.

spectral description of the turbulent zones and their characteristic scales. The energy diagram gives the evidence of the link between the zones of turbulent energy production and the zones of dissipation. These diagrams show the existence of areas, of around ten kilometers in horizontal length, in which the turbulence is very strong, with fluctuations in vertical velocity of more than 5 $m.s^{-1}$. The characteristic lengthscale of this turbulence, calculated from the wavenumber of maximum energy, is about 400 m. These turbulent areas are called "rotors" and they are related to the wave motion that develops in the upper stable layer which have very smooth borders (laminar flow).

Wavelet analysis provides important information on the evolution of the spectral characteristics of the turbulence produced in the rotor area. The evolution towards large wavenumbers enables localization of the areas where the energy will be dissipated. One can also deduce from these diagrams that dissipation is not homogeneously distributed in the turbulence areas. Therefore, it would seem that there does not exist any local equilibrium between the production and dissipation of turbulent kinetic energy.

5.2 Lee waves and turbulence

When clear of clouds, the middle troposphere is generally stable. Turbulence cannot develop nor can it stay. On the other hand, this condition of stability

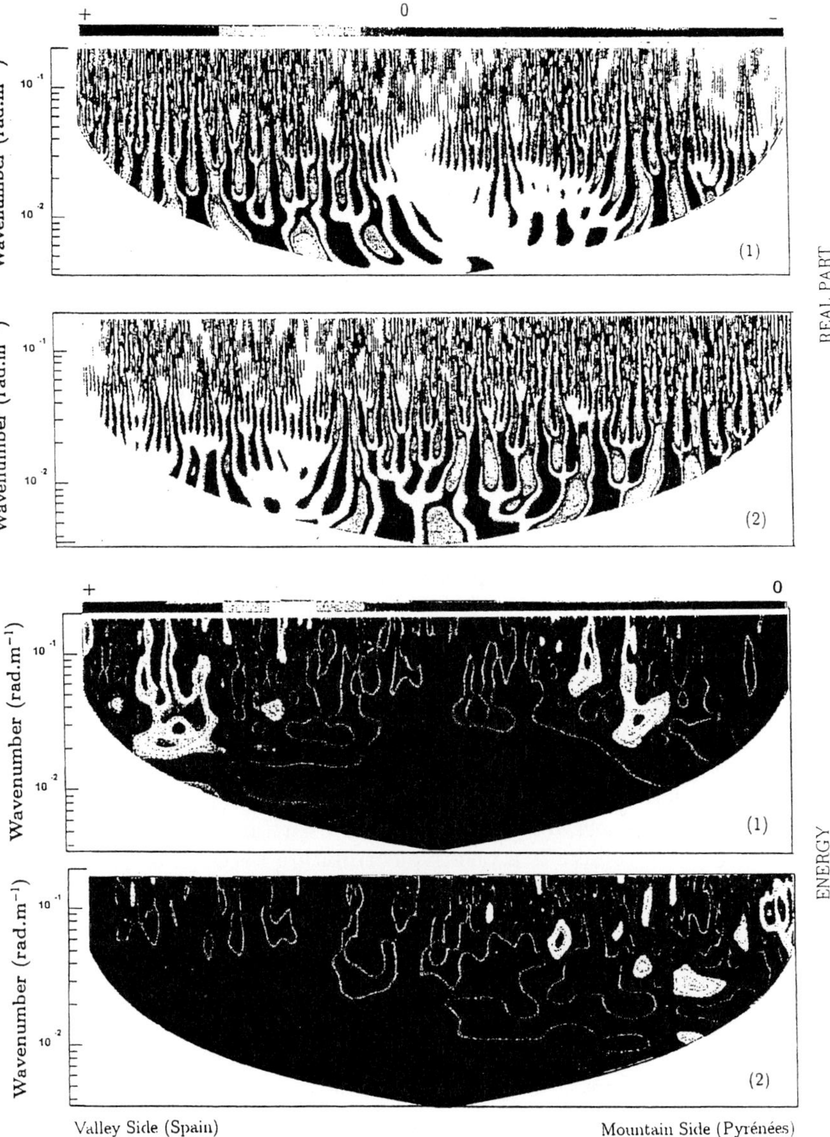

Figure 3.2. Wavelet transform performed on legs (1) and (2) shown on Figure 3.1: real part (top); energy (bottom). Color coding: from red to black for maximum to minimum. (See Figure 3.2 in color in the color section of this volume.)

is favorable to the development of wave motion triggered by the crossing of the mountain by the flow (lee waves). The wave motion often propagates vertically up to the tropopause. On the lee side of the mountain, it can extend over more than 100 km. During their propagation, the lee waves can get amplified to a point where they become unstable and degenerate into turbulence. They stabilize the turbulence zone which develops close to the mountain.

Figure 3.3 presents a sample of data measured by the Fokker 27 aircraft. It corresponds to a straight and constant level run, 200 km long performed at 4000 m altitude. On the downwind side of the mountain, an area of very clear lee waves is observable in the air temperature and vertical velocity signals. Their horizontal extension is limited to 5 wavelengths. The wavelet transform has been applied to the fluctuations in the vertical velocity w' and in the temperature θ'.

The amplitude diagrams give a remarkable illustration of the geographical evolution of the spectral characteristics of these signals, especially in regards to the lee waves. In particular, one can see, in Figure 3.3, the change in wavelength (L) as one goes further from the mountain, from 8 to 14 km, which corresponds to dL/dx=0.1 km/km. The information revealed is fundamental for the analysis of the conditions of wave persistence. Another extremely important piece of information brought out by this analysis is the localization of turbulence and the spectral characterization of turbulent zones associated with the development and the vanishing of lee waves. The wavelet transform demonstrates quite clearly the continuity in scale between lee waves and high frequency turbulence where the waves degenerate and the turbulent kinetic energy dissipates.

The real part of the wavelet cross-spectrum clearly shows up the sensible heat flux structure. At the lee-wave frequencies, w' and θ' are in quadrature, which is shown by the symmetrical oscillations of half a wavelength on the cross-spectrum diagram. On the contrary, the higher frequencies at the downwind end of the lee waves are not symmetrical, with most of the values being positive. What implies here is a positive correlation between w' and θ' and so a positive (upward) sensible heat flux.

§6 Local winds

6.1 Coherent structures within Tramontana wind

When a northerly wind bears down on the Pyrénées range, a violent regional wind blows up from the Roussillon over the septentrional basin of the Mediterranean, covering the Gulf of Lion, from the coasts of France and Spain, on towards the Balearic islands; this is the "Tramontana" wind. When moving over the sea away from the shore, the flow remains turbulent, but becomes more homogeneous. This airflow offers exceptional conditions of strong turbulence maintained by the wind stress and the buoyancy associated with surface momentum and heat flux.

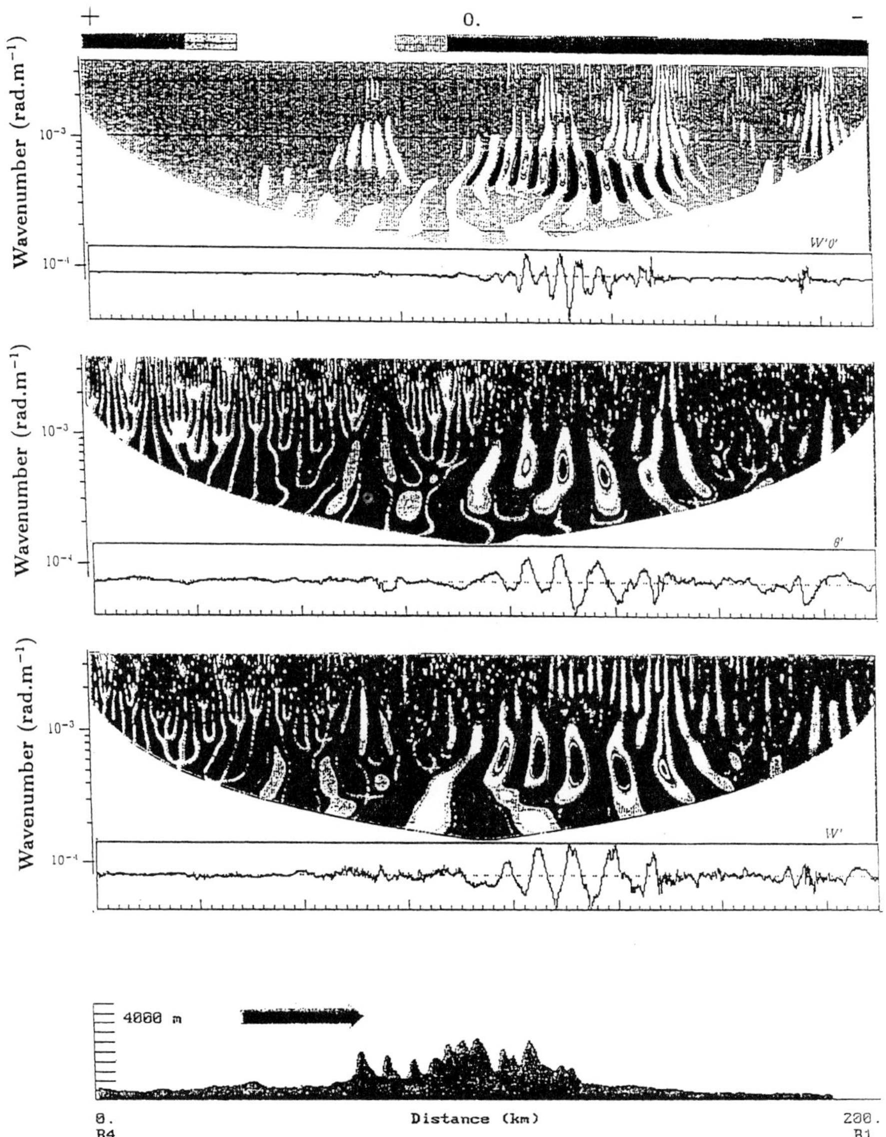

Figure 3.3. Real part of the wavelet transform of vertical velocity (bottom) and potential temperature (middle). On the upper part of the figure is represented the corresponding wavelet cross-spectrum: case of lee waves. Color coding: from red to black for maximum to minimum. (See Figure 3.3 in color in the color section of this volume.)

The Tramontana wind flow was studied during the PYREX experiment on November 4, 1990. Data were collected by the two instrumented aircraft, Merlin IV and Fokker 27, on legs located in an East-West vertical plane, from the open sea to the coast of Spain, which is perpendicular to the Tramontana wind.

During the study, boundary conditions over the sea were the following: sensible heat flux varied from 60 $W.m^{-2}$ to 30 $W.m^{-2}$ from open sea to coast and evaporation was as great as 280 $W.m^{-2}$. The wind strength decreased from the open sea to the coast, giving a simultaneous variation in the friction velocity from 0.25 to 0.1 $m.s^{-1}$. These surface parameters produced exceptional conditions of stability characterized by dominant mechanical production of turbulent kinetic energy and by an evaporation contribution to buoyancy of the same order of magnitude as that of the sensible heat flux.

The conditions are favorable for the development of coherent structures in the atmospheric boundary layer. The structure of the marine boundary layer is illustrated in Figure 4.1 by turbulent functions and by the measurements from the LEANDRE backscattered lidar embarked on board the Fokker 27 aircraft. The turbulent series was recorded at 940, 470, and 37 m altitudes on the East-West runs perpendicular to the mean flow.

The cross-section perpendicular to the airflow shows up coherent structures that develop along the mean wind. The lidar signal, proportional to aerosol concentration, exhibits the thickness of the boundary layer. We can clearly see, at the top of the boundary layer, a dome shape characterizing the top of dynamic structures whose altitude decreases when approaching the shore. Fluctuations in water vapor recorded close to the top of the boundary layer at 940 m, confirm this shape due to the entrainment of dry air from the upper stable layer. In the middle and at the bottom of the boundary layer, the diagnosis parameters are, respectively, the vertical and transversal wind velocities.

The wavelet transform (see Figure 4.2) is computed on the transversal velocity measured at low altitude. It demonstrates a series of oscillations of about 5 km in wavelength. Towards the end of this series, a second organization occurs with a shorter wavelength which increases with the distance from the shore. At lower lengthscales (greater wavenumbers) the turbulence structure appears more homogeneous and normally cascades towards the viscous dissipation range (local dissipation).

6.2 Turbulence and fluxes in a valley wind

For northerly flow conditions over the Pyrénées range, a strong wind develops in the upper Ebro valley and extends up to the Mediteranean shore. This wind, called "Cierzo," has been investigated in the principal valley axis with a straight and constant level flight run of 100 km long, *i.e.*, along the whole valley. The wavelet transform (see Figure 4.3) has been applied to the vertical velocity w' and to the air temperature θ'. Strong turbulence is shown in the upper (upwind) part of the valley. It is produced at middle and low frequencies

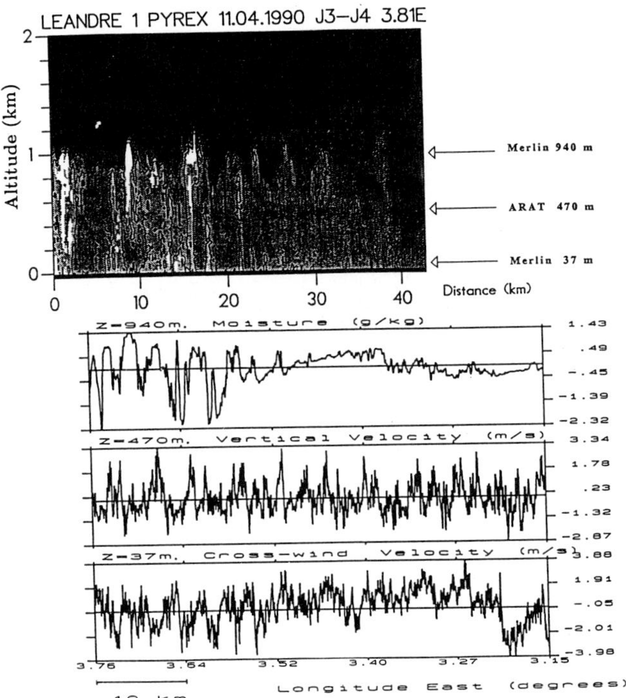

Figure 4.1. Lidar and in situ measurements along a flight track crossing the wind from east to west on 4 November. On the upper part is represented the lidar range corrected signal, obtained from the Fokker flying at 3200 m altitude. On the lower part are shown three detrended turbulent series obtained by the Fokker and the Merlin (from Bougeault et al., 1993)[3].

and is transferred up to the inertial subrange (local dissipation). In the lower (downwind) part of the valley, a turbulence area occurs in which the production range is clearly shifted towards higher frequencies. The temperature is more homogeneous with a maximum of energy located around the upper quarter of the valley.

The multiplication of the two diagrams yields the location of the zones of heat transfer. When energy is present at low frequencies of wavelength greater than 1 km, there is a quadrature between the two signals, as can be seen on symmetrical oscillations of the cross-spectrum diagram. Significant and organized heat transfer appears in the upper quarter of the valley. In the lower part of the valley, heat is transferred at higher frequencies. This analysis

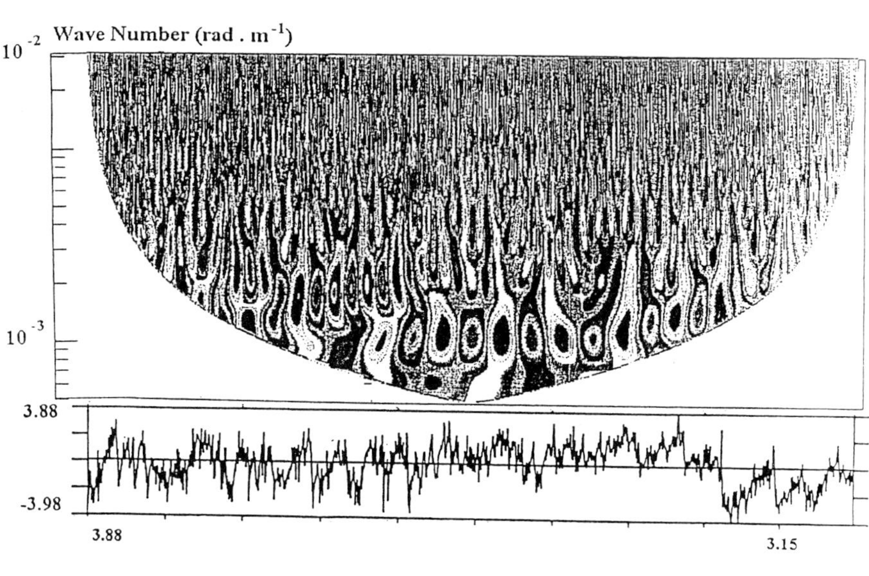

Figure 4.2. Real part of the wavelet transform applied to the transversal wind component: case of Tramontana wind. Color coding: from red to black for maximum to minimum. (See Figure 4.2 in color in the color section of this volume.)

enables one to localize areas of turbulence, to quantify the energy transfers, and to determine the characteristic lengthscales that contribute to the heat transfers.

§7 Conclusion

The classical determination of the characteristics of a homogeneous turbulence sample is achieved by the computation of the various turbulence moments and the use of the Fourier transform for spectral analyses. Similarly, the cross-characterization of two variables is made using the covariant moments and cross-spectral analyses.

Experimental studies of tropospheric turbulence are often performed with instrumented research aircraft. For many practical reasons, like homogeneity

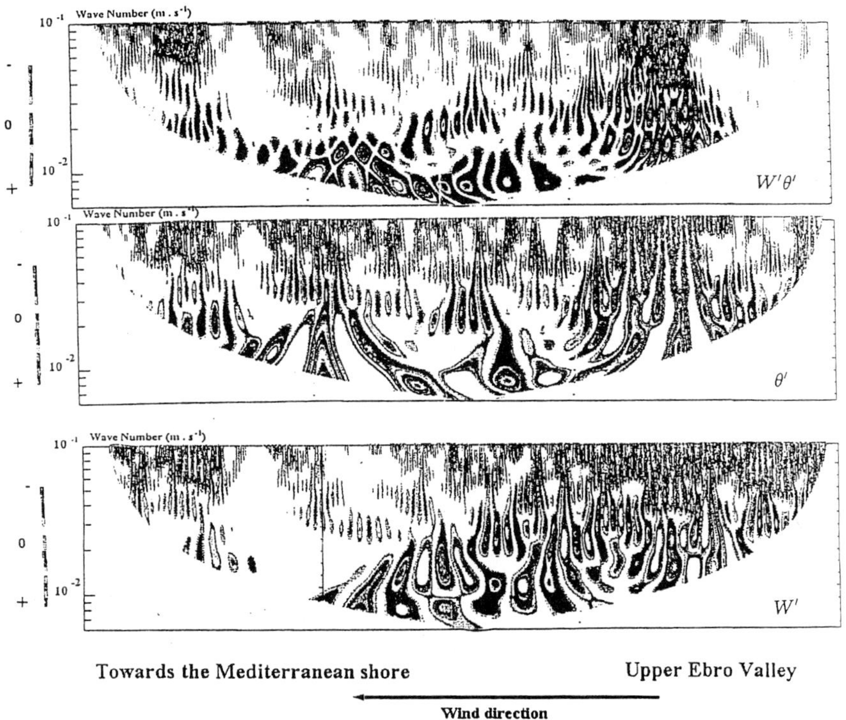

Figure 4.3. Real part of the wavelet transform of the vertical velocity (bottom), the temperature of the air (middle) and corresponding wavelet cross-spectrum (top): case of valley wind (the Cierzo). Color coding: from red to black for maximum to minimum. (See Figure 4.3 in color in the color section of this volume.)

defects of the flight area or disturbed meteorological conditions, specific techniques for the analysis of unsteady turbulent samples have to be developed. In this domain, the wavelet transform is a very powerful tool. With five examples, four of them being turbulence samples measured close the Pyrénées range, we have shown the capability of the wavelet transform method. In the case of homogeneous turbulence, a study of phase-breakings can lead to a quantitative analysis of the homogeneity of the sample.

In the examples of inhomogeneous turbulence, mountain turbulence, lee waves, internal circulations within tramontana wind, and turbulence and fluxes in a valley wind were found in the atmospheric airflow around the Pyrénées chain.

In addition, the wavelet transform also appears to be a powerful tool in several other turbulence related areas including the determination of the spatial distribution of the various events, the characterization of frequencies contributing to the transfer, the isolation of the physical mechanisms and quantification of their specific contribution, the identification of secondary structures within the turbulent flow, and, finally, the representation of the path followed by the turbulent kinetic energy towards viscous dissipation.

Acknowledgments. The PYREX experiment was made possible by the participation of a large number of institutes from France, Spain, and Germany. It was funded by Météo-France, the Instituto National de Meteorologica, the Institut National des Sciences de l'Univers (ARAT, PAMOS, and PAMOY programs), the Centre National d'Etudes Spatiales, Electricité de France, Région Midi Pyrénées, the Deutsche Forschungsanstalt für Luft and Raumfahrt.We thank too the division technique of the INSU, the Centre d'aviation Météorologique of Météo-France and the "Couselho Nacional de Desenvolvimento Cientifico e Tecnologico" (CNPq) of Brazil.

References

1. Arnéodo A., F. Argoul, E. Bacry, J. Elezgaray, E. Freysz, G. Grasseau, J. F. Muzy and B. Pouligny, Wavelet transform of fractals, *in Wavelets and applications*, Y. Meyer (ed.), Masson, 1992, 286–352.
2. Bougeault, P., A. J. Clar, B. Bénech, B. Carissimo, J. Pelon, and E. Richard, Momentum budget over the Pyrénées: the PYREX experiment *Bull. Amer. Meteor. Soc.* **71** (1990), 806–818.
3. Bougeault, P., A. J. Clar, J. L. Attié, I. Beau, B. Bénech, R. Benoit, P. Bessemoulin, J.L. Caccia, B. Carissimo, J.L. Champeaux, M. Crochet, A. Druilhet, P. Durand, A. Elkhalfi, A. Genoves, M. Georgelin, K.P. Hoinka, V. Klaus, E. Koffi, V. Kotroni, C. Mazaudier, J. Pelon, M. Petitdidier, Y. Pointin, D. Puech, E. Richard, T. Satomura, J. Stein and D.S. Tannhauser, The atmospheric momentum budget over a major mountain range: first results of the PYREX field program, *Annales Geophysicae* **11** (1993), 395–418.
4. Grossmann, A. and J. Morlet, Decomposition of Hardy functions into square integrable wavelets of constant shape, *SIAM J. Math. Anal.* **15** (1984), 723–736.
5. Farge, M., Wavelet transforms and their applications to turbulence, *Ann. Rev. Fluid Mech.* **24** (1992), 395–457.
6. Singleton, R.C., An algorithm for computing the mixed radix fast Fourier transform. *IEEE Trans. Audio and Electro Acoust.* **17** (1969), 93–103.

A. Druilhet (drua@aero.ups-tlse.fr)
J.L. Attié (attjl@aero.ups-tlse.fr)
P. Durand (durp@aero.ups-tlse.fr)
B. Bénech (drua@aero.ups-tlse.fr)

Laboratoire d'Aérologie
Université Paul Sabatier
118, route de Narbonne
F-31077 Toulouse
France

L. de Abreu Sá
Instituto Nacional de Pesquisas Espaciais
Saõ Jose dos Campos
Brazil
inpedct@fpsp.fapesp.br

Applications of Wavelet Transform for Seismic Activity Monitoring

Gabriella Olmo and Letizia Lo Presti

Abstract. This paper presents some applications of the wavelet transform in the field of seismic signal analysis and monitoring. These applications are devised as being part of a system whose ultimate objective is the automation of monitoring the seismic activity. In particular, the effectiveness of the transform for the arrival times estimation of the various seismic phases is shown, and its de-noising capability is exploited for making seismic traces smoother and more readable.

§1 The seismic monitoring automation

The issue of automating the seismic signal analysis is crucial. In fact, up to now the monitoring of the seismic activity requires constant human surveillance; data coming from the various seismometers are collected at control centers, where skilled experts continuously monitor them, 24 hours a day, in order to raise alarm procedures in case they detect the presence of a relevant event. This monitoring technique implies several consequences: high cost, due to the need for skilled experts; noticeable time delays before a seismic event is identified and actions are undertaken; high false alarm probability; and subjective interpretation of data which are frequenty blurred by large amounts of noise [11].

An automatic system could overcome most of these problems, providing an efficient monitoring of the seismic activity, ensuring an objective and stable data interpretation, noticeably reducing both the cost of the analysis and the time delays inherent in human monitoring. Moreover, the automatic system would be *portable*.

Obviously, the task of realizing a machine which is able to reproduce all the decisional processes of a human expert is not easy. Recently, Artificial Intelligence (AI) systems have been proposed as a novel approach to the automation

Wavelets: Theory, Algorithms, and Applications
Charles K. Chui, Laura Montefusco, and Luigia Puccio (eds.), pp. 561–572.

ISBN 0-12-174575-9

problem [11], and can provide new perspectives in this field. Knowledge-based systems can be very useful to support seismological data interpretation. Nevertheless, it must be kept in mind that the signals under study are both *non stationary* and *very noisy*; therefore, no information can be drawn from them without the support of adequately powerful *signal processing tools.*

All seismic data consist of recordings of seismic waves (*i.e.*, sound waves in the earth) as a function of time from the source to detectors scattered all over the surface [12]. Each record is called a *trace.* A trace can be virtually noise, or it can carry information about a seismic event; in this latter case, it is more properly called a *seismic signal.*

Seismic signals are collections of typically short duration energy bursts, named *phases*, each of which exhibits a peculiar pattern in terms of dominant frequency, amplitude, and polarization, and represent a different possible ray path between the source and the receiver.

An *event* is a collection of seismic signals which can be recognized as originated from the same physical source; the source itself is sometimes also called an event.

The ultimate objective of the seismic signal analysis is to estimate some significant parameters of the event such as hypocenter, epicenter, strength, and so on. Information about these parameters can be obtained by correlating several traces, coming from different seismometers, in order to identify and describe their possible common source. However, some important measures have to be performed also on the single trace; perhaps the most important is the estimation of the arrival times of the various phases, which allow experts to gain very significant information. In the following, we will focus on the analysis of each single trace; we have named this analysis step the *preprocessing phase.*

§2 The preprocessing phase

In the preprocessing phase, a number of very important tasks have to be performed over the trace. First of all, significant traces must be discriminated from merely noisy ones which must be rejected. This task can be called the *signal detection.*

Secondly, a possible enhancement of the signal-to-noise ratio (SNR) would be of great help; in fact, seismic signals are frequently blurred by large amounts of noise, with SNR values as low as 0 dB. A *de-noising* operation on the signal could make the traces more readable and the parameter estimation easier.

Finally, the estimation of significant parameters of the signal must be performed. As already mentioned, perhaps the most important parameter to be estimated is the arrival time of the different phases which build up the signal. This parameter is very important for subsequent estimation of the event parameters.

Up to now, the signal detection has been performed by means of traditional spectral estimation techniques and threshold detection. In other words, the spectral characteristics of the trace are evaluated by means of periodogram techniques. Moreover, the overall power in the significant frequency band is evaluated, and compared with some threshold value, settled in terms of the *average* noise level in that band. The discrimination between noise and signals is made on the basis of the obtained spectral characteristics and of the result of this threshold comparison. This very simple procedure has proven to be quite effective for the signal detection, given that the noise level is reasonable [12]. However, once the presence of the signal is claimed, the Fourier analysis is of little help for the parameter estimation. In fact, it provides a description of the *overall* spectral characteristics of the signal without maintaining any time reference. As a consequence, it is impossible to relate the frequency contributions to the time intervals when they actually occur. A *two-dimensional* signal processing tool must be employed in order to provide the spectral characteristics as a function of time.

A technique which is often applied in geophysics consists in partitioning the time axis into subintervals, and then computing in each interval a time–invariant operator, such as a Fourier transform, under the hypothesis that the signal is approximately stationary within the interval. By interpolating the operator between adjacent intervals, the net effect is to yield a time-varying linear operator over the entire time axis [12].

This technique corresponds to applying a Short Time (or Window) Fourier Transform (STFT) on the signal. The resulting spectral estimation refers to the time interval at hand, so that joint time-frequency information can be obtained. The partitioning method forms the basis of time-varying deconvolution of seismic traces. The results have established the method as worthwhile. The main shortcoming of this technique, however, is that the intervals are selected on the basis of individual judgment, and there is no built-in optimization procedure to guarantee that the best selection of intervals has indeed been made [12]. Therefore, this method cannot be employed in an automatic monitoring system. Moreover, a jointly good *time* and *frequency* resolution is highly recommendable for the seismic data analysis. A good time resolution is necessary for a very precise estimation of the arrival times; on the other side, a good frequency resolution would make possible a precise estimation of the frequency characteristics of all the different phases.

As it is well known, the *uncertainty principle* [6] settles a lower bound to the time-frequency resolution product which can be obtained using, for example, a STFT. Once a window function has been chosen, the time and frequency resolutions are fixed over all the time-frequency plane. This means that the STFT, being a *fixed resolution* analysis method, does not represent a flexible tool for non-stationary signals, and is suitable only for those signals whose energy is well localized both in time and in frequency.

As already mentioned, another reason why the arrival time estimation is critical is the large noise amount, which can seriously compromise the data interpretation, and which also makes questionable the signal discrimination on the basis of threshold decision on the received power. Therefore, a signal processing technique for some sort of *noise removal* is to be preferred.

From all these considerations stems the idea of applying the Wavelet

Transform (WT) [2,3,9,10] to seismic signals in order to achieve all the aforementioned goals. In fact, the WT represents a powerful analysis tool for non-stationary signals. An increasing number of applications of this technique can be devised, which efficiently exploit its peculiar characteristics, in particular, its multiple resolution capability (which allow one to achieve both good time and also good frequency resolution working at different scales) and its *de-noising capability.* These will be explained in the following sections.

§3 Applications of WT to geophysical signals: state of the art

Applications of the WT to geophysical signals have been devised since the very beginning of its story. For example, in [5] the application of multiresolution techniques is proposed especially in the field of oil and gas exploration. In fact, the high frequency components of the probe signals are very important for the detection and description of oil and gas layers, as they carry information about thinner layers. Consequently, in the data acquisition and analog-to-digital conversion process, the high frequencies must be carefully represented and preserved, a goal that is not easy to achieve as the attenuation characteristics of ground are such that higher frequencies are more attenuated. A good solution for this problem is the implementation of a two-dimensional sampling, *i.e.*, a separate sampling and analog-to-digital conversion in each significant subband. As this task should be performed without too much degradation of the time resolution, the idea to apply multiresolution techniques for the recording and sampling of such data stems rather naturally.

Another classical application of the WT is in the field of *data compression.* Many papers have been published on this subject and applications in seismology are sometimes presented as examples. In fact, due to the large amount of data which are continuously collected from seismometers, the problem of data compression is crucial in this field.

In this paper, we propose something which is different, *i.e.*, the use of the WT for the parameter evaluation and the de-noising of seismic traces. Of course, the application to the WT for a particular task does not prevent one to employ this technique also for the achievement of other objectives.

§4 The WT de-noising capability

Besides the advantages due to its multiresolution properties, the WT shows another important characteristic which can be exploited in the seismic signal domain, *i.e.*, the capability of discriminating a typical "noise" process from a typical "signal" one.

The de-noising capability of the WT is based on the different evolution across scales of its modulus maxima, depending on the signal regularity degree, defined in terms of the Lipschitz exponent. The theory is described in great detail in [7,8] and its major consequence is that the WT modulus maxima of a signal which can be described in terms of a negative Lipschitz exponent tend to vanish as the scale increases, whereas this does not happen if the signal is characterized by a positive exponent.

Based on the consideration that *typical* signals are Lipschitz positive (*i.e.*, they do not exhibit singularities which are worse than discontinuities), one can discriminate *true signals* (based on this definition) from those processes which can be described in terms of negative Lipschitz exponents (the most important is white noise). In fact, the WT modulus maxima of these latter processes tend to vanish as the scale increases [7,8]. As a result, when a signal-plus-noise is processed by means of the WT, those modulus maxima whose behavior is dominated by the noise can be discriminated from those which are dominated by the signal (and, therefore, will not vanish as the scale increases). Of course, not all the signal details are preserved by this de-noising technique, as many WT coefficients which are dominated by the noise actually carry information on the signal as well. However, if a signal detail gets lost, it is because it has been overcome by the noise, and consequently it could not be retrieved anyway. The reconstruction of the signal from its WT maxima is not, in general, *exact* [1,7,8] since the WT maxima do not represent a complete basis for the signal representation. However, it has been demostrated [1,7,8] that the possible ambiguity is of very limited extent.

In fact, the WT modulus maxima are related to the *signal sharp variation points*; therefore, they carry the most relevant information about the signal. Being based on the WT modulus maxima, the de-noising procedure turns out to be very efficient, from the computational point of view, while preserving most of the information. Several real seismic traces have been decomposed and then reconstructed only from their WT maxima coefficients with no appreciable difference observed.

In Figures 1–2, a seismic trace of 512 samples, as recorded at the Italian Institute of Geophysics, and its de-noised version are represented.

As it is shown in the figure, the de-noising is very effective and the signal pattern more easily recognizable, even though spurious oscillations cannot be eliminated. However, the de-noised signal is more compact and easier for further processing. It should be stressed that the de-noising procedure is by no way a *filtering* operation, being based on the different regularity characteristics of signal and noise; therefore, it preserves all the high frequency components of the signal.

§5 WT effectiveness for arrival times estimation: synthetic data

In order to evaluate the effectiveness of the WT for the arrival times estimation in ideal conditions, *i.e.*, in the absence of noise and of any other possible impairment, we have modeled seismic signals as sums of damped sinusoids with different starting point and dominant frequencies. As we are interested in achieving the best possible time resolution, only the first octave of the WT decomposition has been retained.

We have implemented the discrete WT following the classical scheme of subband coding [10,13]. We employ the orthonormal wavelet bases proposed by I. Daubechies in [4].

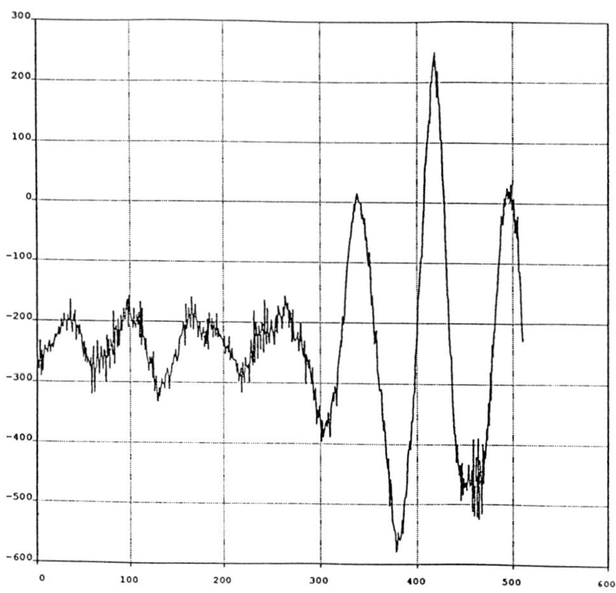

Figure 1. Original seismic trace.

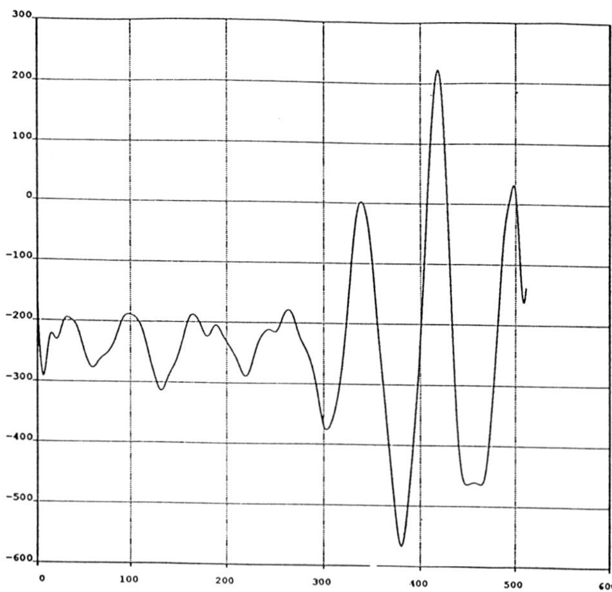

Figure 2. De-noised seismic trace.

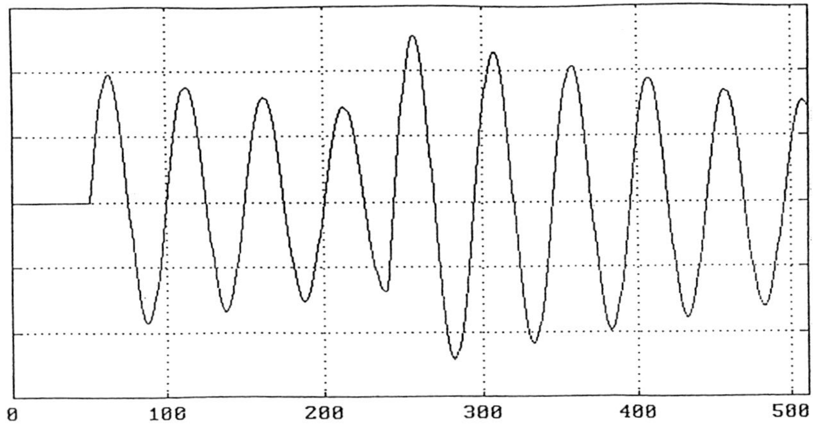

Figure 3. Sum of two damped sinusoids.

In Figure 3, a probe signal is represented. It is the sum of two damped sinusoids. In Figures 4–5 the first octave decomposition of this signal is shown when the D4 and D20 filters are employed, respectively [4].

From a theoretical point of view, shorter wavelets should perform better than longer ones in achieving the best time resolution. However, for our purposes, the longer D20 wavelet has proven to be more effective than the shorter D4. This can be explained easily considering that the longer the impulse response of the filters in the subband scheme decomposition, the better the frequency characteristics approximate ideal half-band filters (see Figure 6).

The high-pass filtered subsequence obtained at the first step of the subband decomposition, *i.e.*, the first octave of the WT, retains the information about the high frequency components of the original sequence which carry information about the abrupt variations in the sequence itself. Therefore, the more well-shaped is the high-pass filter, the more negligeable are the low-frequency components of the original spectrum which affect the high-pass subsequence, and the more separated are the high frequencies (*i.e.*, transient behavior) from the low frequencies (*i.e.*, stationary behavior) which could be retrieved in the low-pass filtered subsequence at the first step of the decomposition.

The results shown in Figure 5 are rather encouraging since the highest peaked waveform obtained can be very useful in an automatic phase identification system.

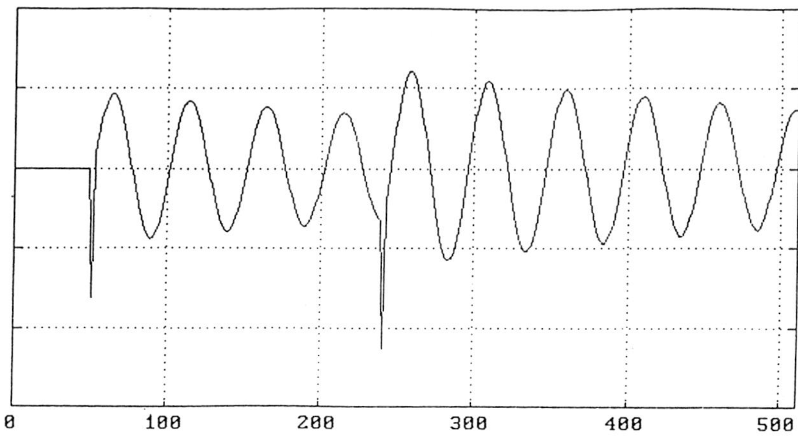

Figure 4. WT decomposition of the signal in Figure 3 (1st octave). D4 wavelet.

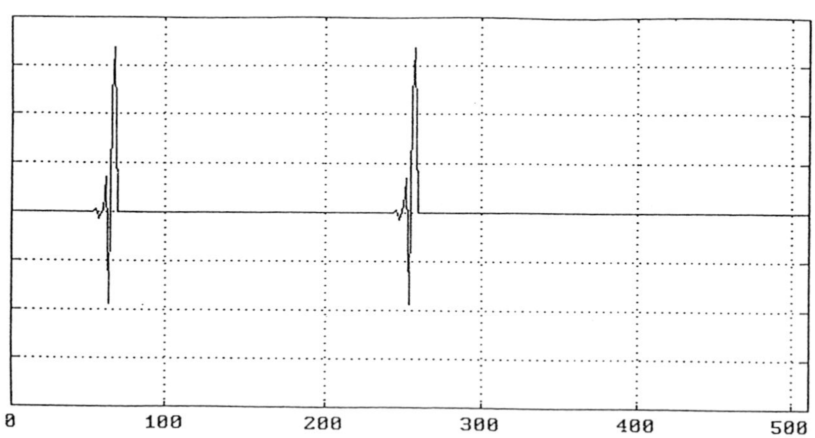

Figure 5. WT decomposition of the signal in Figure 3 (1st octave). D20 wavelet.

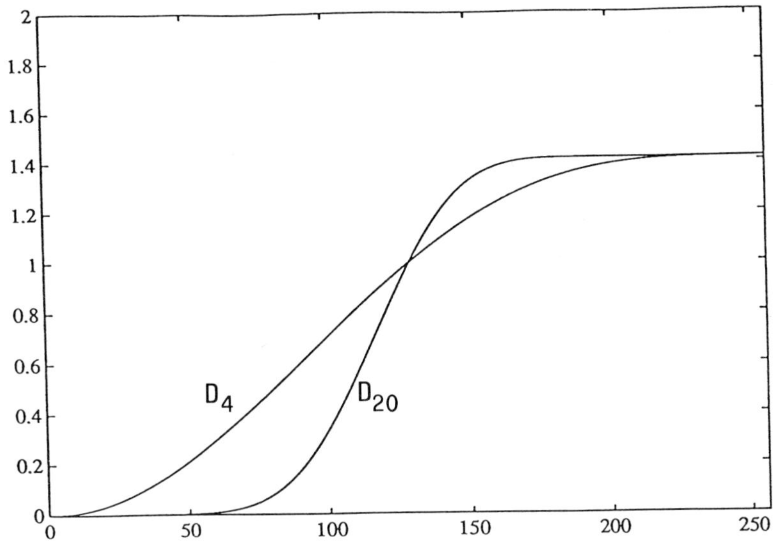

Figure 6. Frequency response of the HPF's corresponding to D4 and D20 wavelets.

§6 WT effectiveness for arrival times estimation: seismic data

As a second step of the analysis, we have applied the same subband coding scheme to the seismic trace of Figure 1. In Figure 7, the first octave of the decomposition is reported in order to evaluate its capability of identifying the onset of significant events, using the D20 wavelet.

It is noticed here that the various phases arrival times cannot be recognized at all from these coefficients. This rather disappointing result can be explained by the fact that the trace is affected by noise, whose spectrum spreads over all the significant frequency band and whose power is not negligeable with respect to the signal component. The presence of the noise completely cancels out any information which could be gained on the signal. This means that the trace de-noising procedure plays a very important role not only for achieving a SNR improvement, but also for detecting the arrival times by means of the WT.

In Figure 8, the WT coefficients of the *de-noised* trace are reported. The few residual coefficients carry information about the *onset of a signal*, as the coefficients due to the noise have been suppressed by the de-noising procedure.

If a trace is merely noisy, no WT coefficients survive the de-noising procedure. Therefore, the WT coefficients of the de-noised trace could be exploited as a signal detection method which is not based on a simple threshold decision on the received power.

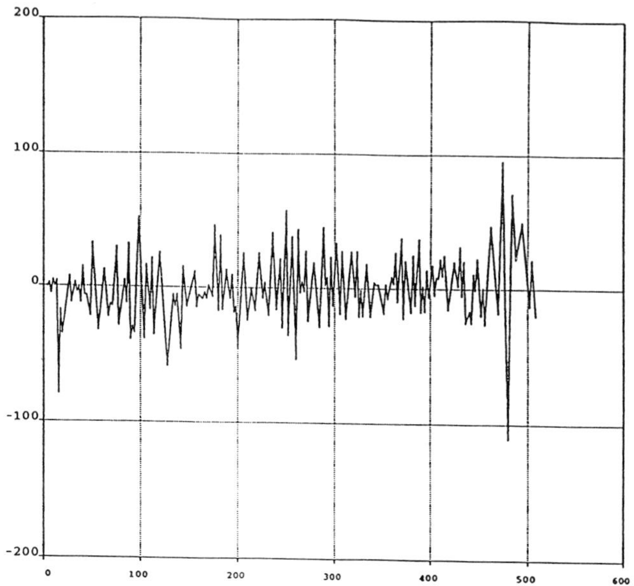

Figure 7. WT decomposition of the signal in Figure 1 (1st octave).

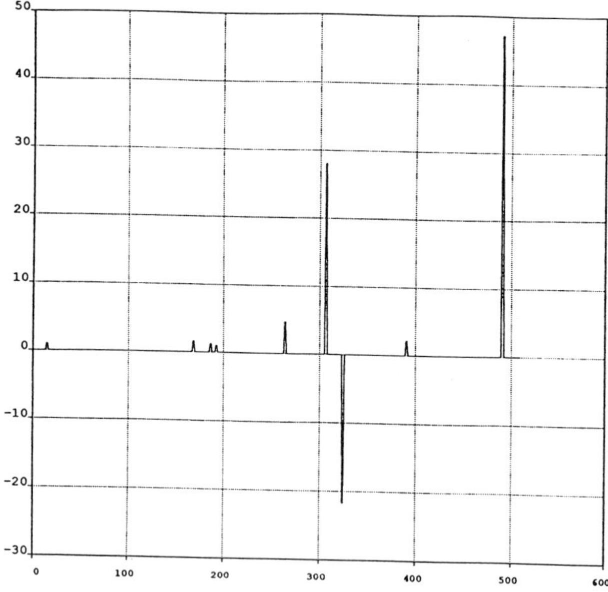

Figure 8. WT decomposition of the signal in Figure 2 (1st octave).

§7 Conclusion

The WT processing technique has been applied to seismic traces for exploiting its de-noising capability and its multiresolution properties. The following conclusion can be drawn.

The presence of noise makes the WT decomposition useless for the signal detection and the arrival times estimation. On the other hand, the WT coefficients are able to effectively localize the signal sharp variation points, (and in particular *the onset of a signal*) provided that the trace is previoulsy de-noised, as it has been proven by applying the technique on synthetic data. This property, in connection with the very effective signal de-noising capability offered by the WT, can be exploited for the seismic signal analysis, in particular for the arrival times estimation. Moreover, both the WT decomposition and the de-noising procedure have proved to be computationally efficient, so that these techniques can be profitably employed in a real-time system for an automatic event recognition.

Acknowledgments. The authors would like to thank Dr. Alessandrini of the National Institute of Geophysics for having supplied the seismic traces and for the interesting and helpful discussions.

References

1. Berman, Z., The uniqueness question of discrete wavelet maxima representation, Tech. Report TR 91–48, Syst. Res. Cent., Univ. of Maryland, College Park, April 1991.
2. Chui, C. K., *An Introduction to Wavelets*, Academic Press, 1992.
3. Chui, C. K., (ed.), *Wavelets: A Tutorial in Theory and Applications*, Academic Press, 1992.
4. Daubechies, I., Orthonormal bases of compactly supported wavelets, *Comm. Pure and Appl. Math.* **41** (7) (1988).
5. Goupillaud, P., A. Grossmann, and J. Morlet, Cycle-octave and related transforms in seismic signal analysis, *Geoexploration* **23**, Elsevier Science publisher B.V., Amsterdam, 1984/85, 85–102.
6. Lo Presti, L. and F. Neri, *L'analisi dei segnali*, C.L.U.T., Torino, 1992.
7. Mallat, S. and W. L. Hwang, Singularity detection and processing with wavelets, *IEEE Trans. Inform. Theory* **38** (2) (1992).
8. Mallat, S. and S. Zhong, Characterization of signals from multiscale edges, *IEEE Trans. Pattern Anal. and Machine Intell.* **14** (7) (1992).
9. Meyer, Y., *Ondelettes et Opérateurs, Tome I, Ondelettes*, Herrmann, Paris, 1990.
10. Rioul, O. and M. Vetterli, Wavelets and signal processing, *IEEE SP Magazine*, October 1991.
11. Roberto, V. and C. Chiaruttini, Seismic signal understanding: a knowledge-based recognition system, *IEEE Trans. Signal Processing* **40** (7) (1992).

12. Robinson, E. A. and M. T. Silvia, *Digital Foundations of Time Series Analysis; Wave Equation - Space-Time Processing*, Holden-Day Inc., San Francisco, 1981.

13. Vetterli, M. and C. Herley, Wavelets and filter banks: theory and design, *IEEE Trans. Signal Processing* **40** (9) (1992).

Gabriella Olmo
Department of Electronics, Politecnico,
Corso Duca degli Abruzzi 24,
I–10129 Torino
Italy
olmo@polito.it

Letizia Lo Presti
Department of Electronics, Politecnico,
Corso Duca degli Abruzzi 24,
I–10129 Torino
Italy
lopresti@polito.it

Mean Value Jump Detection: A Survey of Conventional and Wavelet Based Methods

Aline Denjean and Francis Castanié

Abstract. Many problems in signal processing are concerned with the detection of abrupt changes in the parameters of signals or systems. We focus our attention on the simple case of a mean value jump of a stationary random process, as it includes various other cases. The aim of the present study is to compare the performances of two classes of detectors, the first based on conventional segmentation algorithms and the second on the Continuous Wavelet Transform (CWT). The first section gives a brief overview of the methods, and highlights the 2-D wavelet detectors such as the sum of fixed scales slices or the 2-D correlation with the CWT jump signature. According to previous results [4,6], the optimal wavelet maximizes the signal to noise ratio in the (time, scale) plane. Unlike the usual representations, these detectors take advantage of information redundancy in the plane. The second section is devoted to the comparison between the different methods, by means of the detectors characteristics in terms of probability of false alarm and non-detection. A common criterion of detection delay is taken into account, as well as the accuracy of the detection date, or the possibility of detecting two close jumps. Finally, we give some indications relating to the performance of non-abrupt changes with the new optimal wavelet, and evaluate the possibility of those methods to deal with other jumps.

§1 Detection problems

The problem of detecting abrupt changes, also referred to as ruptures or non-stationarities, involves instantaneous changes or a few samples long changes in signals or dynamic systems [2]. It may be a complement of adaptive treatment, improving tracking capability of algorithms [8]. For many years, people in the industry have been interested in early detection of small changes out of a mix of information, both for financial and security reasons. Signal segmentation will lead to event detection, stationarity hypothesis testing, for example, for applications in fault detection, non-destructive testing, bio-medical, or space signal processing.

Wavelets: Theory, Algorithms, and Applications
Charles K. Chui, Laura Montefusco, and Luigia Puccio (eds.), pp. 573–584.

ISBN 0-12-174575-9

Most change detection problems can be classified into one of the following two types: a mean value jump (additive change) or a behavior modification around a mean level (non-additive change). The mean value jump studied in this survey constitutes a particularly important non-stationarity, due to the possibility of transforming numerous problems into a (non-optimal) mean value change. As an example, the state equation of a dynamic model may lead to a mean value jump in the innovation process (via Willsky's *Generalized Likelihood Ratio* principle [9]). We can also quote jumps in the signal moments (namely, power) or spectral jumps, as well as law testing like Gaussianity which results in a mean value jump in the order 4 moment.

The signal form is described by the simple expression:

$$y(t) = \Delta m.U(t - t_0) + x(t) = s(t) + x(t), \tag{1.1}$$

where $U(t)$ is a unit step, Δm the jump amplitude, $x(t)$ a stationary null mean random process with correlation function $R_x(\tau)$ and Power Spectral Density (PSD) $S_x(f)$. t_0 is the unknown jump location. $x(t)$ generally corresponds to some kind of innovation process of a Kalman filter [6,7]. The class of jumps embedded in such processes offers the advantageous property of almost whiteness, and the *Gaussian* assumption is reasonable for these processes.

The different problems can be divided into a general problem and sub-optimal problem formulations [2,8]. The general problem of an optimal estimator of the discontinuity date, as a result of a sequential hypothesis testing, has no known theoretical solution. Sub-optimal formulations consider the two different issues of discontinuity detection and date estimation separately, and lead to some restrictions among which is the possibility of multiple or composite hypotheses, or the possible use of change date and characteristic signal parameters estimation (before and after change) in the test formulation.

§2 Overview of the detection strategies

2.1 Conventional segmentation methods

We begin with a general presentation of these methods, the general parametric formulation of which results in the test of a mean value jump in the (Gaussian) innovation ϵ_n of a Kalman filter. This formulation is used for the processing of all problems relating to additive ruptures in linear multivariable systems excited by white noises. Unknown parameters such as the change date, or the post-change mean value are estimated via the basic tool of the Likelihood Maximum, according to the GLR-Willsky and modified GLR algorithms (GLR, GLRmod).

In addition, simpler methods are considered for a piecemeal constant signal embedded in a white noise with a priori knowledge of parameters. For example, the Filtered Derivative (FD) algorithm uses the derivation of a lowpass filter output and assumes the mean value before the jump μ_0 to be known. Furthermore, the Cumulative Sum of the Page-Hinkley algorithm (CUSUM-HK) represents an adaptive threshold integrator if we assume known mean values before and after the change μ_0 and μ_1 (via a *Likelihood Ratio* maximization upon change time); otherwise, it provides two parallel detectors of mean increase and decrease with the assumption of known μ_0 (possibly estimated by an elementary *sliding mean* filter) and minimum detection amplitude ν_m. For optimality, this algorithm needs a Likelihood Maximum jump amplitude estimator, which is given by the GLR method.

Thresholding is a detection strategy used in many algorithms such as FD (jointly to a counter), and HK (with an adaptive threshold). The GLR algorithm considers the correlation with an assumed change signature before thresholding, but the difficulty of determining the threshold leads us to use of the GLRmod algorithm. These methods produce different parameter estimations, in particular:

- the change date for HK, GLR, GLRmod, and also for FD where it is equal to the detection date;
- the jump amplitude for the GLR and GLRmod algorithms.

Another important element is the blind interval during which no detection can take place. The GLR and GLRmod formulations allow for an immediate updating depending on the jump amplitude estimation, whereas the FD and HK algorithms need a longer period of time for the new computation of the algorithms.

2.2 Wavelet based strategies

The Continuous Wavelet Transform (CWT) constitutes the basic tool of our new approach [5,7]. Indeed, in spite of the advantage of a white bi-dimensional noise for both a biorthogonal set like the Daubechies wavelet basis and a discrete white noise, the problem of a non- time-shift invariant change signature does not allow the use of a detection strategy based on the correlation with the change signature for the discrete case.

We recall the CWT definition of Equation (1.1) :

$$C_y(a,\tau) = a^{-1/2} \int_R y(t).\psi^* \left(\frac{t-\tau}{a}\right) dt, \tag{2.2.1}$$

where $\psi(t)$ is a L^2 normalized wavelet of Fourier Transform (FT) , $\phi(f)$.

The jump signature is easily computed, owing to the linearity of the transform:

$$C_y(a,\tau) = C_s(a,\tau) + C_x(a,\tau), \tag{2.2.2}$$

where $C_s(a,\tau)$ represents the deterministic conic jump signature, and $C_x(a,\tau)$ the transformed noise in the time-scale plane. We have

$$C_s(a,\tau) = -\Delta m.a^{1/2} I_\psi^* \left(\frac{t_0-\tau}{a}\right) \tag{2.2.3}$$

with $I_\psi(t) = \int_{-\infty}^{t} \psi(u)du$.

The detection issue here is to be able to recognize this signature which is included in the transformed noise in Equations (2.2.2) and (2.2.3). Consequently, we need to derive a statistical description of $C_x(a, \tau)$ and we have

$$E\Big[C_x(a,\tau)\Big] = 0 \qquad \forall a, \tau, \tag{2.2.4}$$

$$E\Big[C_x(a_1,\tau_1)C_x^*(a_2,\tau_2)\Big] = \mu(a_1, a_2, \tau_1 - \tau_2), \tag{2.2.5}$$

where $\mu(a_1, a_2, t)$ possesses the following FT:

$$M(a_1, a_2, f) = \sqrt{a_1 a_2} S_x(f)\phi^*(a_1 f)\phi(a_2 f). \tag{2.2.6}$$

From Equation (2.2.5), the variance can be derived as

$$Var\Big[C_x(a,\tau)\Big] = \int_R R_x(u) R_\psi\left(\frac{u}{a}\right) du \tag{2.2.7}$$

$$\sim S_x(0) \ when\ a \to \infty,$$

where $R_\psi(u) = \int_R \psi^*(t)\psi(t+u)dt$. These statistical results shed light on the interest of the transform, since the variance is asymptotically constant for large scales, whereas the signature amplitude increases to $a^{1/2}$ (see Equations (2.2.3) and (2.2.7)).

The expression of the Signal to Noise Ratio (SNR) in the transformed plane summarizes this last property as follows:

$$\gamma(a,\tau) = \frac{|\ C_s(a,\tau)\ |^2}{Var\Big[C_x(a,\tau)\Big]},$$

$$\gamma(a) \sim \Delta m^2 a \frac{|\ I_\psi(0)\ |^2}{S_x(0)} \qquad \forall\, \tau. \tag{2.2.8}$$

Indeed, the detection performance will increase with the scale, until it attains the limit set by the analysis window.

The choice of the wavelet basis takes into account both an optimality criterion and the possibility of dealing with an unbiased detection date estimator.

From an optimality point of view, based upon SNR maximization Equation (2.2.8), the class of solutions for a bounded support wavelet and an abrupt change is formed by the functions of constant module over the whole interval. Indeed, the Schwartz bound is reached for the colinearity of s and ψ. In particular, the Haar wavelet meets this condition. For other changes with error function erf, arctan, and polynomial shapes, the formulation of the new problem of optimization under constraint leads to a new optimal wavelet class.

In addition, some symmetry properties are desirable for the wavelet basis, yielding an unbiased estimate of the jump position. For example, Morlet and the symmetrical Haar wavelets are such that $|\ I_\psi(t)\ |$ is maximum for $t = 0$

(see Equation (2.2.3)). The symmetrical Haar wavelet basis introduced in this paragraph is nothing other than a version of the conventional Haar function definition rendered symmetrical. This symmetrical Haar basis will be given a special attention in our simulations since it meets the two previous conditions for an abrupt change detection problem.

The CWT detection strategies can be divided into three categories. A simple approach uses either CWT thresholding of a slice with a fixed scale, which should be large enough according to the above-mentioned SNR properties, or a CWT sum along the slices, with a distance between scales larger than the correlation radius of the stochastic process $C_x(a, \tau)$, for independence (see Equation (2.2.5)). However, it is worth noting that the information redundancy occurring with a smaller distance may improve detection.

The optimal approach concerns adapted filtering. However, this approach is hampered by a CWT noise with unknown $S_x(f)$ which is no longer white, and by an unrealizable filter. We are thus interested in sub-optimal strategies defined by the CWT correlation with the 1- or 2-D signature:

$$\Gamma_y(\tau) = \int_R C_y(a,t).M(a,\tau - t)dt,$$

or

$$\Gamma_y(\tau) = \int_{R^2} C_y(a,t).M(a,\tau - t)dadt, \tag{2.2.9}$$

where $M = C_s$. Among these various methods, the 2-D detectors take advantage of bi-dimensional information redundancy in the time-scale plane.

§3 A comparison of the general methods

3.1 Fundamental differences

Unlike the CWT representations, conventional detector formulations assume *a priori* knowledge of some parameters in the signal or the system. This characteristic of parametric or non-parametric formulations is the first difference between the methods. A second difference is the sequential or block processing. Conventional algorithms use sample by sample tests whereas the CWT detectors use signal windowing. These differences between the various representations increase the difficulty of defining common performance criteria for a theoretical comparison.

3.2 Performance evaluation

Let us begin with the conventional detection algorithms. The optimality test which is based on the *maximum likelihood* criterion either leads to the CUSUM algorithm of Page-Hinkley (HK) when μ_1 or ν_m are known, or to the GLR or GLRmod algorithms when ν_m and t_0 are estimated. For our problem of a piecemeal constant signal embedded in a white noise, a smoothed version of HK algorithm with a jump amplitude estimator is preferred.

A simple integrator is optimum for the same problem with the *SNR maximum* criterion (adapted filtering). Five intuitive performance criteria are given in [2]:

(1) Mean time between false alarm (fa),
(2) Probability of false alarm (pfa),
(3) Mean delay to detection,
(4) Probability of non-detection (*pnd*),
(5) Accuracy of the change time and magnitude estimates.

One possible test consists in minimizing (3) with fixed (1). Ideally, the desired properties are small fa, *nd*, plus small mean delay, but, in general, we are led to make a trade-off, depending on the relative importance of fa and *nd* which, in turn, depends on the particular type of problem. With these methods and an infinite duration, it is worth noting that we have available an estimator of the jump time from the detection time.

The wavelet based detectors that we have previously presented meet the *SNR maximum* optimality criterion. Performance has been proved to increase with scale, and *pnd* and pfa computation gives the characteristics of such detectors, via the application of the *Neyman-Pearson* test for the normal n-dimensional case. For its simplicity, the particular case of a CWT fixed scale slice is studied.

Let us give some notations. The scales are $(a_i)_{i=1\ldots n}$ and the time samples $(t_j)_{j=1\ldots m}$. C_y and M are, respectively, the plane samples of the CWT signal and the jump signature:

$$C_y(a) = \left(C_{y_j}(a)\right)_{j=1\ldots m},$$

$$M(a) = -\Delta m a^{1/2}\left(I_{\psi_j}(a)\right)_{j=1\ldots m},$$

$$C_y = \left(C_y(a_i)\right)_{i=1\ldots n}, \tag{3.2.1}$$

$$M = \left(M(a_i)\right)_{i=1\ldots n}. \tag{3.2.2}$$

The covariance matrix $\mu(a)$ and $\left(\mu(a_i, a_j)\right)_{i,j=1\ldots n}$ are defined as follows:

$$\mu_{ij}(a) = \left\{E\left[C_{y_i}(a).C^*_{y_j}(a)\right]\right\}_{i,j=1\ldots m},$$

$$\mu_{kl}(a_i, a_j) = \left\{E\left[C_{y_k}(a_i).C^*_{y_l}(a_j)\right]\right\}_{k,l=1\ldots m}. \tag{3.2.3}$$

The case of real CWT processes leads to the probability laws:

$$C_x \to N(0, \mu),$$

$$C_y \to N(M, \mu).$$

Given the general complex formulation:

$$\mu(a) = \mu_r(a) + i\mu_i(a),$$

$\mu(a)$ satisfies the hypotheses of a complex normal multivariate distribution, and the CWT noise and signal follow the respective probability laws:

$$\bar{C}_x \to N(0, \frac{1}{2}\bar{\mu}),$$

$$\bar{C}_y \to N(\bar{M}, \frac{1}{2}\bar{\mu}),$$

where $\bar{C}_y = \begin{pmatrix} C_{y_r} \\ C_{y_i} \end{pmatrix}$, $\bar{M} = \begin{pmatrix} M_r \\ M_i \end{pmatrix}$, and $\bar{\mu} = \begin{pmatrix} \mu_r & -\mu_i \\ \mu_i & \mu_r. \end{pmatrix}$

In the real case with twice the number of variables and according to the Neyman-Pearson test and Equations (3.2.1), (3.2.2) and (3.2.3), we have

$$\frac{L(\bar{C}_y/\bar{C}_s)}{L(\bar{C}_y/0)} > k(pfa),$$

$$L(\bar{C}_y/\bar{C}_s) = \frac{\sqrt{\det \bar{\mu}^{-1}}}{(2\pi)^{m*n}} e^{-1/2^t(\bar{C}_y - \bar{M})\bar{\mu}^{-1}(\bar{C}_y - \bar{M})},$$

$${}^t\bar{M}\bar{\mu}^{-1}\bar{C}_y > d = \ln k + \frac{1}{2}{}^t\bar{M}\bar{\mu}^{-1}\bar{M}. \tag{3.2.4}$$

For a fixed pfa, the threshold choice is increasingly difficult because of the multiplicity of variables

$$d = f\Big((a_i)_{i=1\dots n}, (t_j)_{j=1\dots m}, I_\psi, \Delta m\Big).$$

This leads to a one-dimensional problem. Denoting $\sigma^2 = {}^t\bar{M}\bar{\mu}^{-1}\bar{M}$, we have

$$pfa = \frac{1}{2}\left[1 - erf\left(\frac{d}{\sqrt{2}\sigma}\right)\right], \tag{3.2.5}$$

$$pnd = \frac{1}{2}\left[1 + erf\left(\frac{d - \sigma^2}{\sqrt{2}\sigma}\right)\right]. \tag{3.2.6}$$

Simulation gives preference to 2-D detectors due to the importance of redundant information, and the new expressions of pd and pfa can be derived from the application of the same Neyman-Pearson test.

In order to give some elements of comparison for the two types of detectors, let us consider an infinite length of signal. In this case, both types of representations offer a detection probability equal to 1 asymptotically. In addition, the CWT unitarity property is of great importance since it reveals the equivalence to signal adapted filtering.

In the more real situation of a finite duration, we point out pd variation in time in Figure 1. This stresses the need for a new mean delay definition [2] and indicates the influence of the wavelet support on the detection of two close jumps.

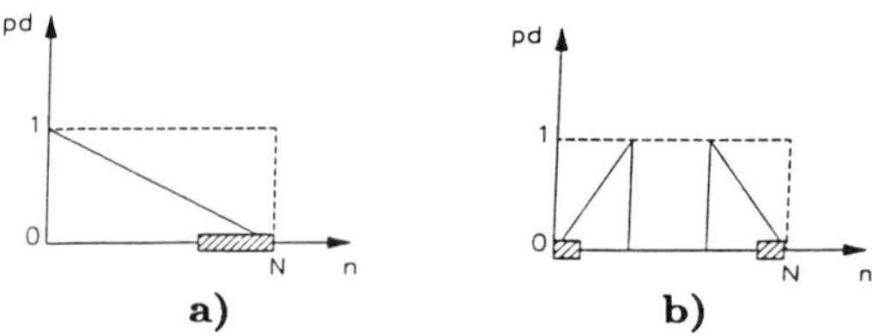

Figure 1. *pd* variation in time: a) Conventional methods; b) CWT slice.

§4 Simulation results

In this section, we give the example of the 2048 samples CWT of a mean value jump at date 800 and SNR $\Delta m^2/\sigma^2$ of -10 dB, with the symmetrical Haar wavelet of support 80. Figure 2 shows the bi-dimensional behavior of both the transform module $|CWT|$, and the adapted filtering $|FA2D|$ for each fixed scale slice, while Figure 3 describes the scale evolution of the maximum abscissa of these CWT slices.

The results of the 2-D detectors formed by the CWT module sum along fixed scale slices are presented along with the 2-D correlation in Figure 4. We can note that in the case of the Haar wavelet, more than for Morlet, for instance, it is more difficult to satisfy the condition of a scale step greater than the correlation radius given by the transformed noise covariance (Figures 5 and 6). However, the information redundancy actually seems favourable to the detection. The comparison with conventional methods is presented in Figure 7 obtained by applying the adapted filtering (*i.e.*, a simple integration) or the cumulative sum algorithm to the same signal.

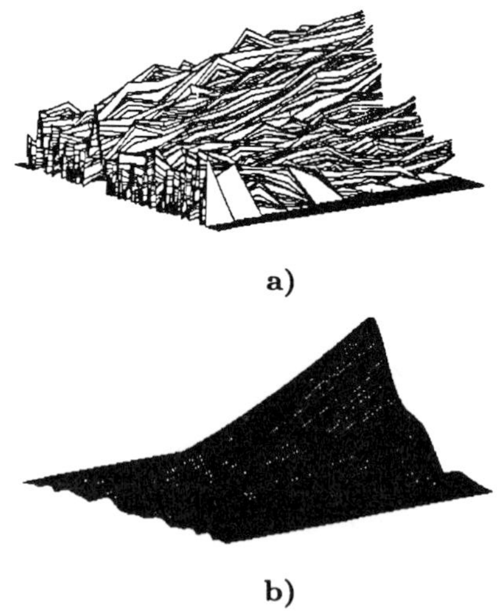

Figure 2. 3-D representations (scales 3 to 7.5, step 0.5): a) $|CWT|$; b) $|FA2D|$.

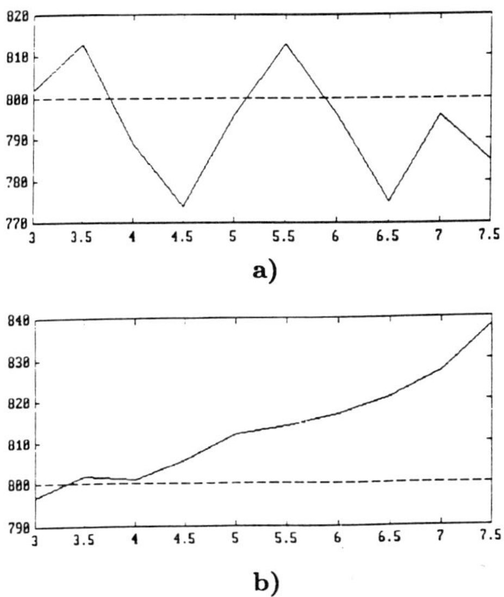

Figure 3. Maximum abscissa in τ (same scales): a) $|CWT|$; b) $|FA2D|$.

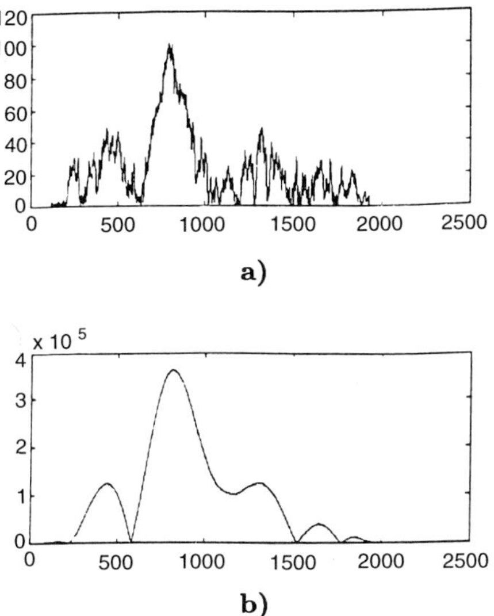

Figure 4. 2-D detectors (maximum abscissa in τ):

a) $|CWT|$ sum along fixed scale slices ($\tau = 796$);
b) 2-D correlation ($\tau = 815$).

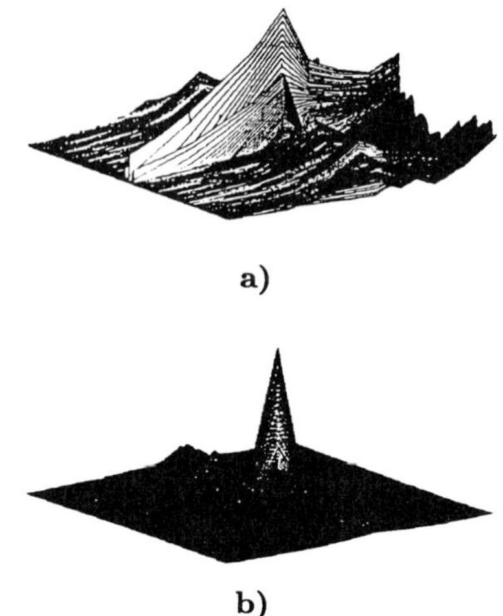

Figure 5. Covariance module (scales 2^{-5} to 2^{4}): a) Haar; b) Morlet.

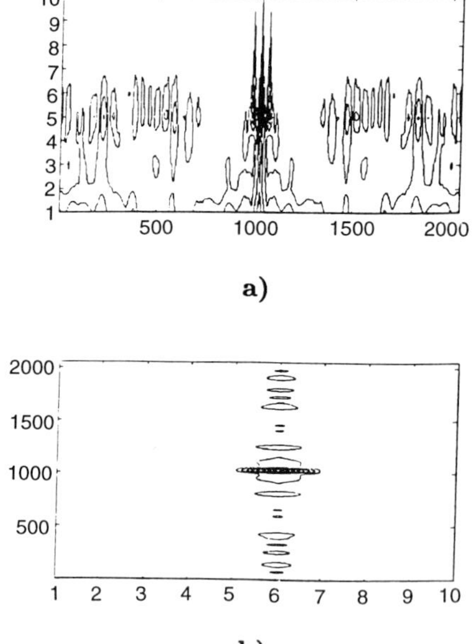

Figure 6. Covariance module contours (same scales): a) Haar; b) Morlet.

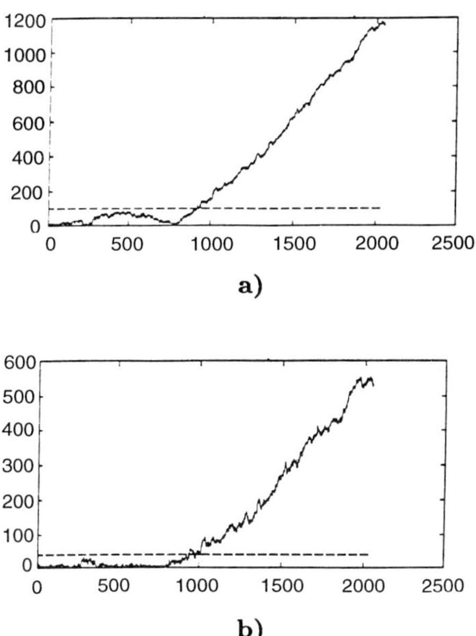

Figure 7. Conventional segmentation methods (detection at date τ):
a) Integrator (threshold$=100$, $\tau = 921$);
b) Cumulative sum (threshold$=40$, $\tau = 922$).

§5 Conclusion

In the present communication, faced with the absence of optimal rupture existence tests, we have presented new wavelet based detectors as an alternative to conventional methods for the particular non-stationarity of a mean value jump of a Gaussian process. The presentation of the two approaches has shed light on their fundamental implications and differences and has derived a first performance evaluation. Various simulations with small SNR have showed that the CWT is a powerful tool and have favored the 2-D wavelet detectors, even for less abrupt changes, where the new optimal wavelet class is to be calculated.

The possibility of extending these CWT detectors to other jumps such as power or model ones will be studied, based on a new signature involving order n moments. The particular case of a power jump is of great importance, since no adapted filtering is known.

References

1. Basseville, M. and A. Benveniste, Design and comparative study of some sequential jump detection algorithms for digital signals, *IEEE Trans. ASSP* **31** (1983), 521–534.
2. Basseville, M. and I. Nikiforov, *Detection of Abrupt Changes*, Prentice Hall, 1993.
3. Castanié, F. and P. Soulé, An improvement of non-stationarity detection by level-crossing analysis of linear prediction error, *Proc. of ICASSP'83*, Boston, 1983, 251–254.
4. Castanié, F. and A. Denjean, Mean value jumps detection using wavelet decomposition, *Proc. of IEEE SP Time-Frequency and Time-Scale Analysis*, Victoria, BC, Canada, 1992, 181–184.
5. Combes, J.M., A. Grossmann, and P. Tchamitchian, *Wavelets, Time-Frequency Methods and Phase Space*, Springer-Verlag, 1989.
6. Denjean A. and F. Castanié, Théorie 2-D de la détection de sauts de moyenne par décomposition en ondelettes, *GRETSI Conference*, Juan Les Pins (France), **1** (1993), 181–184.
7. Meyer, Y. and S. Roques, *Wavelet Analysis and Applications*, Editions Frontières, 1993.
8. Seminary, *Détection de ruptures dans les modèles dynamiques de signaux et systèmes*, C.N.R.S., Paris, 1984.
9. Willsky, A. S. and H. L. Jones, A generalized likelihood ratio approach to the detection and estimation of jumps in linear systems, *IEEE Trans. AC* **21** (1976), 108–112.

Aline Denjean
LEN7-GAPSE, ENSEEIHT,
2 rue Camichel,
31 071 Toulouse Cedex, France
denjean@len7.enseeiht.fr

Francis Castanié
LEN7-GAPSE, ENSEEIHT,
2 rue Camichel,
31 071 Toulouse Cedex, France
castanie@len7.enseeiht.fr

Comparison of Picture Compression Methods: Wavelet, Wavelet Packet, and Local Cosine Transform Coding

Mladen Victor Wickerhauser

Abstract. This is a survey of some new perspectives on transform coding image compression. We mentions some mathematical properties of sampling, entropy, and time-frequency localization of analysis-synthesis functions. Then we discuss several experiments in picture compression using wavelets, wavelet packets, local sines and cosines, and the adaptive "best-basis" method. We calculate the efficiency of several techniques to describe a fast transform from a library, with the aim of transforming each image into its own best-adapted coordinates for transmission with minimal overhead. Standard C algorithms for the local sine and cosine transforms are included in the appendix.

§1 Introduction

In this summary I will describe several experiments in picture compression using wavelets and the local cosine transform of Coifman and Meyer. I will also describe an adaptive wavelet transform coding method and a local cosine transform algorithm based on the idea of a "best basis," and provide Standard C algorithms for computing some of the described transforms.

§2 Relevant notions from mathematics

By a "picture" we will mean any function $S = S(x, y) \in l^2(\mathbf{Z}^2)$. In practice, of course, pictures will be nonzero only at finitely many points. Also, they must represent band-limited functions, since it is impossible to distinguish spatial frequencies above K cycles per unit distance from those below K if we use a sampling rate $2K$ samples per unit.

Wavelets: Theory, Algorithms, and Applications
Charles K. Chui, Laura Montefusco, and Luigia Puccio (eds.), pp. 585–621.

ISBN 0-12-174575-9

2.1 Sampled signals and approximation

Suppose $f \in L^2(\mathbf{R}^n)$ has d uniformly continuous derivatives, for $d \geq 0$, and that $\phi \in L^2(\mathbf{R})$ satisfies the additional conditions

$$\int_R \phi(x)\,dx = 0; \qquad \int_{\mathbf{R}} x^m \phi(x)\,dx = 0,\ 0 < m < d; \qquad \int_R x^d \phi(x)\,dx < \infty.$$

Then Taylor's theorem implies that the discrete values $f(k2^{-\nu})$ are very good approximations to the inner products $\langle f, \phi_{-\nu,k}\rangle$, for $k \in \mathbf{Z}^n$ and $\nu \in \mathbf{N}$. We obtain the estimate

$$\sup_{x \in I_{\nu,k}} |\langle f, \phi_{-\nu,k}\rangle - f(x)| < C2^{-\nu d},$$

where $I_{\nu,k} = \times \prod_{i=1}^{n} [2^{-\nu}k_i, 2^{-\nu}(k_i+1)[$, and $0 < C < \infty$ may be chosen independently of ν and k.

By a *signal* we shall mean a compactly supported function belonging to $C^d(\mathbf{R}^n)$, which thereby satisfies the hypothesis for f above. For simplicity, we will at first consider the 1-dimensional case. By rescaling we may assume that $\sup_{x\in[k,k+1[} |\langle f, \phi_{0,k}\rangle - f(x)| < \epsilon$ for some acceptably small $\epsilon > 0$. If $d > 1$, then this error decreases faster than the number of samples grows.

2.2 The Heisenberg uncertainty principle

The functions underlying transform coding techniques can be judged by their simultaneous localization in space and wavenumber. This localization cannot be better than a universal fixed amount, as shown by Heisenberg's inequality or the uncertainty principle.

A nonzero function of compact support cannot also have a compactly supported Fourier transform. The Fourier transform of a compactly supported function is entire analytic, and an entire function which vanishes on an open set must be zero everywhere. Various refinements of this result exist, posing difficulties for time-frequency analysis.

Heisenberg is credited with observation that in quantum mechanics, one cannot simultaneously specify both the location and momentum of a wave function to arbitrary accuracy. The product of the location and momentum uncertainties must exceed the quantization h. Without developing the whole of quantum mechanics, we can sketch the proof of this fact and relate it to the present discussion of time-frequency atoms.

Let $\psi \in L^2(\mathbf{R})$ be a wave function, normalized with $\|\psi\|^2 = 1$. The location and momentum operators are X and P, respectively, where $X\psi(x) = x\psi(x)$ and $P\psi(x) = i\hbar\frac{d\psi(x)}{dx}$. Assume that the wave function has a derivative in L^2 and that $x\psi(x)$ is in L^2. Then the location and momentum are calculated from the wave function by $\langle\psi, X\psi\rangle$ and $\langle\psi, P\psi\rangle$, where $\langle\cdot,\cdot\rangle >$ is the inner product in $L^2(\mathbf{R})$. Location evaluates to the integral $\int_{\mathbf{R}} x|\psi(x)|^2\,dx$, which may be interpreted as an expectation $E(x)$, since $|\psi(x)|^2$ is a probability density function. Momentum may be also be interpreted as an expectation by first applying the Fourier transform, a unitary change of variable:

$$\int_{\mathbf{R}} \bar{\psi}(x)\left(\frac{1}{ih}\frac{d\psi(x)}{dx}\right) dx = i\hbar \int_{\mathbf{R}} \xi|\hat{\psi}(\xi)|^2\, dx = E(\xi).$$

The uncertainties of location and momentum are $\Delta X = \sqrt{E(x^2) - E(x)^2}$ and $\Delta P = \sqrt{E(\xi^2) - E(\xi)^2}$, respectively. Now we ask the question, what is the minimum value of $\Delta X \Delta P$? Observe that we are asking how well we can localize a function in L^2 simultaneously in time and frequency.

Without loss, we may assume that the spatial coordinates and initial phase are chosen so that $E(x) = 0$ and $E(\xi) = 0$. Define $A(\psi) = E(x)^2 E(\xi^2)$ in this case; any minimum ψ for the functional A is also a minimum for $\Delta X \Delta P$.

Using the calculus of variations, we can compute critical points for A. We look for a stationary ψ, namely a function for which $\delta A(\psi) = 0$. The component of this variation in the direction of an arbitrary perturbation function $\eta(x)$ in L^2 is computed below, where we have adjusted the time units so that $\hbar = 1$:

$$\begin{aligned}
\langle \delta A(\psi), \eta\rangle &= \frac{d}{d\epsilon}\left(\int_{\mathbf{R}} x^2|\psi(x) + \epsilon\eta(x)|^2\, dx \times \int_{\hat{\mathbf{R}}} \xi^2|\hat{\psi}(\xi) + \epsilon\hat{\eta}(\xi)|^2\, d\xi\right)\Bigg|_{\epsilon=0} \\
&= \left(\int_{\mathbf{R}} x^2\left[\psi(x)\overline{\eta}(x) + \overline{\psi}(x)\eta(x)\right] dx\right) \times E(\xi^2) \\
&\quad + E(x^2) \times \left(\int_{\hat{\mathbf{R}}} \xi^2\left[\hat{\psi}(\xi)\overline{\hat{\eta}}(\xi) + \overline{\hat{\psi}}(\xi)\hat{\eta}(\xi)\right] d\xi\right) \\
&= 2\Re\left(E(\xi^2)\int_{\mathbf{R}} x^2\psi(x)\overline{\eta}(x)\, dx + E(x^2)\int_{\hat{\mathbf{R}}} \xi^2\hat{\psi}(\xi)\overline{\hat{\eta}}(\xi)\, d\xi\right) \\
&= \Re\langle \Delta P X^2\psi + \Delta X P^2\psi, \eta\rangle.
\end{aligned}$$

Now $\delta A = 0$ implies that the last expression vanishes for all η, which implies that

$$\Delta P X^2\psi + \Delta X P^2\psi = 0.$$

2.3 The entropy of a sequence

Let $p = \{p_i : i = 0, 1, 2, \ldots\}$ be a discrete probability distribution function (pdf), that is, $0 \leq p_i \leq 1$ for all i and $\sum_i p_i = 1$. The *entropy* [17] of p, $H(p)$, is defined to be the following sum:

$$H(p) = \sum_i p_i \log \frac{1}{p_i}.$$

In this sum, $0 \log 0$ is taken to be 0. Since all the terms in the sum are positive, $H(p) \geq 0$. Roughly speaking, entropy measures the logarithm of the number of meaningful coefficients in the signal. It is not hard to show (see, for example,

[20]) that if only N of the values p_i are nonzero, then $H(p) \le \log N$. Such a pdf is said to be concentrated into at most N values, but we can use entropy to define the theoretical dimension $d(p)$ of a pdf:

$$d(p) = e^{H(p)}.$$

This is a more general notion of concentration of a pdf, one that might be finite even if $p_i > 0$ for infinitely many values of i, so long as all but finitely many of those values are negligibly small.

Entropy and theoretical dimension are preserved by renumbering the probabilities p_i and, thus, we may assume that the pdf is strictly decreasing: $p_0 \ge p_1 \ge p_2 \ge \ldots$. We may measure the rate at which a pdf decreases in terms of its partial sum sequence $S_n = S_n(p) = \sum_{i=0}^{n} p_i$; since S_n increases and $\lim_{n\to\infty} S_n = 1$, we will say that the pdf p decreases faster than the pdf q if and only if $S_n(p) \ge S_n(q)$ for all $n = 0, 1, 2, \ldots$. Again, it is not hard to show (see [20]) that if p decreases faster than q by this definition, then $H(p) \le H(q)$ and thus $d(p) \le d(q)$.

Unfortunately, the converse is not true because rate of decrease gives only a partial order on pdfs. There are pairs of pdfs p and q for which $H(p)$ is finite and $H(q)$ is infinite, but the rates of decrease are not comparable because $S_n(q)$ starts out by dominating $S_n(p)$ only to fall behind as $n \to \infty$. However, theoretical dimension is still useful as an indicator of which pdf has the longer "tail" of nonnegligible probabilities.

Now let $x = \{x_k : k = 1, 2, \ldots\}$ be an arbitrary sequence of positive integers. This is what we obtain, for example, after approximating a sequence of positive real numbers by their integer parts. There are two distinct notions of entropy for such sequences, which happen to be roughly equivalent in the special cases we will be considering. Both require forming a pdf $p = p(x)$ from the sequence x, but this is done in different ways.

The first notion is the traditional entropy of a source. We consider x_k to an independent Bernoulli trial of some process taking the value $i \ge 0$ with probability p_i. By the law of large numbers, this may be approximated by

$$\lim_{N\to\infty} \frac{\#\{1 \le k \le N : x_k = i\}}{N}.$$

Then $H(p)$ gives the lower bound for the expected number of bits per trial required to describe the sequence x in the most efficient alphabet, one which uses short code sequences for the most common values and longer sequences for rare events. (To be completely honest here, when talking about "bits" we must use $\log_2$ rather than log in the definition of entropy and 2 rather than e for the base of the exponential in the definition of theoretical dimension). If $H(p)$ is small, then x is concentrated into a few values i, with all other values being rare. If we renumber p into decreasing order, then the common values of x_k will be labeled 0, 1, etc., the "small" values. The "large" values will be rare; we expect few of them in any finite subsequence of x.

The second notion is geometric. We form probabilities from the x_k's themselves by letting $\|x\| = \left(\sum_k |x_k|^2\right)^{1/2}$ and defining $p_i = |x_i|^2/\|x\|^2$ whenever $\|x\|$ is finite. This obviously gives a pdf, and if $H(p)$ is small then we may conclude that large values of x_k are rare since there can only be a small number of k for which $|x_k|^2$ has a significant portion of the total energy $\|x\|^2$. This notion of entropy can be generalized to arbitrary real or complex valued sequences. Furthermore, we can calculate $H(p)$ directly from x with the following formula:

$$H(p) = \log \|x\|^2 + \frac{\ell(x)}{\|x\|^2}; \qquad \ell(x) = \sum_k |x_k|^2 \log \frac{1}{|x_k|^2}.$$

Note that $H(p)$ is monotonic with $\ell(x)$, so that they are equally useful for comparison purposes.

§3 Transform coding methods

These work by performing a change of variable on the pixel values, with the expectation that most of the new coordinates can be discarded with minimal loss of image quality. The transforms must be invertible in exact arithmetic, but since some information will be discarded by rounding or quantization, they will not be perfectly invertible in practice.

3.1 Local cosine transform

Let $\{I_k\}$ be a collection of disjoint compact intervals of $\mathbf{R}$, indexed by the integers k. Coifman and Meyer showed how to construct an orthonormal basis of $L^2(\mathbf{R})$ subordinate to the partition $\mathbf{R} = \cup_k I_k$ in which the basis vectors are cosines multiplied by smooth bumps. For definiteness we will use a particular symmetric bump function

$$b(x) = \begin{cases} \sin \frac{\pi}{4}(1 + \sin \pi x), & \text{if } -\frac{1}{2} < x < \frac{3}{2}, \\ 0, & \text{otherwise.} \end{cases}$$

It is easy to see that this function is symmetric about the value $x = \frac{1}{2}$. It is smooth on $(-\frac{1}{2}, \frac{3}{2})$ with vanishing derivatives at the boundary points, so that it has a continuous derivative on $\mathbf{R}$. Notice that we can modify b to obtain more continuous derivatives by iterating the innermost $\sin \pi x$: n iterations will yield at least 2^{n-1} vanishing derivatives at $-\frac{1}{2}$ and $\frac{3}{2}$.

We may translate and dilate the bump b onto an interval I_k by the formula $b_k(x) = b(\frac{x-a_k}{|I_k|})$, where a_k is the left endpoint of I_k. Let $c(n,x) = \sqrt{2}\cos \pi n x$, and consider the set of functions $c(n+\frac{1}{2}, x)$ for $n \geq 0$, $x \in [0,1]$. These form an orthonormal basis for $L^2([0,1])$. They may also be dilated and translated to the interval I_k by the formula $c_k(n,x) = \frac{1}{\sqrt{|I_k|}} c(n, \frac{x-a_k}{|I_k|})$. Now define $\psi_{nk}(x) = b_k(x) c_k(n+\frac{1}{2}, x)$ for integers $n \geq 0$ and k. These form an orthonormal basis for $L^2(\mathbf{R})$: the proof is a direct calculation. Cosine may be replaced by sine, and there are some other modifications possible, but this set of functions is sufficiently general for our present purposes.

Observe that ψ_{nk} is well localized in both space and frequency. In space, it is compactly supported on a proper subset of $3I_k$, which is defined as that interval centered at the center of I_k but having three times the width. In frequency, $\hat{\psi}_{nk}$ consists of two modulated bumps centered at $n+\frac{1}{2}$ and $-n-\frac{1}{2}$, respectively, with spread equal to that of $\hat{b}_k$. But $\hat{b}_k$ is well localized because b_k is smooth.

The construction generalizes to higher dimensions. For multi-indices n and k, define $\Psi_{nk}(x) = \psi_{n_1k_1}(x_1)\ldots\psi_{n_dk_d}(x_d)$ to obtain an orthonormal basis for $\mathbf{R}^d$ made of tensor products. Of course, it is possible to use a different partition in each dimension, as well as different bump functions. For the present exposition, we will concentrate on the special case of 2 dimensions, all intervals I_k of equal width, and always the same $b(x)$, as defined above. At the end of the paper, we will consider dyadic intervals with a restricted range of widths, as well.

3.2 Discrete local cosine transform

The functions ψ_{nk} have discrete analogues which form a basis of $l^2(\mathbf{Z})$, or $l^2(\mathbf{T})$. For the former, let I_k be a finite interval of integers with least element a_k, and let $|I_k|$ be the number of elements in I_k. The following vectors form a basis of $l^2(\mathbf{Z})$:

$$\psi_{nk}(j) = b_k(j+\frac{1}{2})c_k(n+\frac{1}{2}, j+\frac{1}{2}), \qquad \text{for integer } k \text{ and } 0 \leq n < |I_k|.$$

Apart from the bells b_k, these are evidently the basis functions for the so-called DCT-IV transform. The particular bells chosen allow cosines on adjacent intervals to overlap while remaining orthogonal. A similar basis may be constructed over equally spaced points on the circle $\mathbf{T}$.

3.3 Implementation by folding

Rather than calculating inner products with the sequences ψ_{nk}, we can preprocess data so that standard fast DCT-IV algorithms may be used. This may be visualized as "folding" the overlapping parts of the bells back into the interval. This folding can be transposed onto the data, and the result will be disjoint intervals of samples which can be "unfolded" to produce smooth overlapping segments.

Suppose we wish to fold a function across 0, onto the intervals $[-1/2, 0)$ and $(0, 1/2]$, using the bell b defined above. Then folding replaces the function $f = f(x)$ with the left and right parts f_- and f_+:

$$\begin{aligned} f_-(x) &= b(-x)f(x) - b(x)f(-x), && \text{if } x \in [-\tfrac{1}{2}, 0); \\ f_+(x) &= b(x)f(x) + b(-x)f(-x), && \text{if } x \in (0, \tfrac{1}{2}]. \end{aligned}$$

The symmetry of b allows us to use $b(-x)$ instead of introducing the bell attached to the left interval.

Unfolding reconstructs f from f_- and f_+ by the following formulas:

$$f(x) = \begin{matrix} b(x)f_+(-x) + b(-x)f_-(x), & \text{if } x \in [-\frac{1}{2}, 0); \\ b(x)f_+(x) - b(-x)f_-(-x), & \text{if } x \in (0, \frac{1}{2}]. \end{matrix}$$

Composing these relations yields $f(x) = \big(b(x)^2 + b(-x)^2\big)f(x)$, which is verified by the bell b defined above, for which the sum of the squares is 1. We can translate and dilate these relations to all adjacent pairs of intervals, and of course it works for sequences as well. These details are best understood by reading the comments in the computer program in Section 7.

3.4 Block discrete local cosine transform coding

We propose a modification of the JPEG picture compression algorithm. We will replace the block discrete cosine transform (DCT) on 8x8 blocks with the block discrete local cosine transform (LCT). By the results of the previous section we expect to de-correlate adjacent samples about as well as with DCT, without introducing discontinuities at block boundaries. We can perform experiments of two types in order to evaluate this modification. First we calculate the rate distortion curves of some example pictures with this method and compare it to JPEG output. Results of one such calculation are attached to Figure 7. Second, we compare reconstructions from encodings at equal bit rates, but different methods, and judge them subjectively. This will verify the correlation between calculated and perceived distortion. One such comparison is available in Figures 8 through 24.

3.5 Adaptive block discrete local cosine transform coding

A further modification of the JPEG algorithm involves allowing the block size to grow if sections of the picture are regular on scales spanning many blocks. The main idea is that of the "best-basis" search algorithm, which compares the information cost of describing several adjacent blocks with that of describing their joint parent. In this case, the information cost is the entropy of the transformed coefficients, which can be computed for blocks of size 8x8, 16x16, 32x32, and so on, up to the dimensions of the picture itself.

The cost of adding adaptivity to the transform coding scheme is a header describing the decomposition of the picture into blocks. Suppose for definiteness that we will consider only dyadic blocks of sizes 8x8 or bigger in a picture of size 256x256. Then each block needs no more than 2.585 bits to describe which of the 6 levels it comes from, and there are 1024 blocks, so the overhead of the adaptive scheme is 2647 bits for 65536 pixels, or about 0.04 bits per pixel. It remains to be seen whether this will be regained by the improved fit of the adapted basis.

§4 Wavelet and wavelet packet methods

We introduce a generalization of subband coding which may be used to compress digitized pictures or sequences of pictures. The new method uses 2-dimensional perfect reconstruction quadrature mirror filters (QMFs) to recursively decompose an image into finer and finer subbands. This produces an overabundant set of transform coefficients. The complete set of coefficients may be found in $O(N \log N)$ operations.

4.1 Wavelets and scaling or sampling functions

Let ϕ be a scaling function for a multiresolution decomposition of $L^2(\mathbf{R})$. Such a function may be constructed with an arbitrary finite number of vanishing moments by iteration of a sufficiently long finite impulse response quadrature mirror filter. Let $h = \{h_j\}_{j=-R}^{R-1}$ be such a filter. Then ϕ satisfies

$$\phi(x) = \sum_{j=-R}^{R-1} h_j \phi(2x - j).$$

As in an earlier section, we construct a mother wavelet ψ from ϕ by taking

$$\psi(x) = \sum_{j=-R}^{R-1} g_j \phi(2x - j),$$

where $g = \{g_j\}_{j=-R}^{R-1}$ is determined from h by the formula $g_j = (-1)^j h_{-(j+1)}$.

The wavelet expansion $f = \sum_{\nu,k} \langle f, \psi_{\nu,k} \rangle \psi_{\nu,k}$ is calculated by the pyramid scheme of Mallat. Write $s_{\nu,k} = \langle f, \phi_{\nu,k} \rangle$, and $d_{\nu,k} = \langle f, \psi_{\nu,k} \rangle$. Denote by H and G the convolution-decimations with filters h and g, respectively. We calculate the following.

$$\begin{array}{ccccccccc}
f(k) \approx \langle f, \phi_{0k} \rangle & \xrightarrow{I} & s_{0,k} & \xrightarrow{H} & s_{1,k} & \xrightarrow{H} & s_{2,k} & \xrightarrow{H} & \cdots \\
 & & G\downarrow & & G\downarrow & & G\downarrow & & \\
 & & d_{1,k} & & d_{2,k} & & d_{3,k} & &
\end{array}$$

Each arrow costs $2R$ multiplications per coefficient.

If f is supported in $[1, N]$, then a sequence at scale ν has $N2^{-\nu}$ nonzero coefficients. We may assume for technical simplicity that N is a positive power of 2. Hence the pyramid reaches its top after $\log_2 N$ scales, requiring $4R(N-1)$ multiplications. Since R is independent of N (and is typically much smaller), this algorithm's complexity is linear. It yields N coefficients $\{d_{\nu,k} : 1 \le \nu \le \log_2 N, 1 \le k \le N2^{-\nu}\} \cup \{s_{(\log_2 N),1}\}$.

By the orthogonality of quadrature mirror filters, we have

$$\sum_{k=1}^{N} |\langle f, \phi_{0k}\rangle|^2 = |s_{(\log_2 N),1}|^2 + \sum_{\nu=1}^{\log_2 N} \sum_{k=1}^{N2^{-\nu}} |d_{\nu,k}|^2.$$

Denote by $s'_{\nu,k}$ and $d'_{\nu,k}$ the coefficients obtained from the pyramid scheme using $f(k)$ rather than $\langle f, \phi_{0k}\rangle$. If $|s'_{0k} - s_{0k}|^2 < \epsilon$ for $1 \le k \le N$, then $|d'_{\nu,k} - d_{\nu,k}|^2 < \epsilon N$ for $1 \le \nu \le \log_2 N$ and $1 \le k \le N2^{-\nu}$. If f has more than one derivative and we use wavelets with more than one vanishing moment, we can arrange that $\epsilon N \to 0$ as f is rescaled and $N \to \infty$.

Suppose that f is a signal with d continuous derivatives, and ψ is a mother wavelet with at least d vanishing moments. Then f is in the Sobolev space L^2_d, and its Littlewood-Paley characterization yields the formula:

$$\sum_{\nu=-\infty}^{\infty} \left(2^{-2\nu d} \sum_{k} |\langle f, \psi_{\nu k}\rangle|^2 \right) < \|f\|^2_{2,d}$$

Since f has compact support, there is some sufficiently large ν above which $\sum_k |\langle f, \psi_{\nu k}\rangle|^2 < C2^{\nu}$. This part of the series sums to a constant if $2d > 1$.

The sum converges, so that $|\langle f, \psi_{\nu k}\rangle|^2 < 2^{2\nu d}\|f\|^2_{2,d}$ as $\nu \to -\infty$. On the other hand, there are at most $C2^{-\nu}$ translates $\psi_{\nu k}$ for which $\langle f, \psi_{\nu k}\rangle$ is nonzero. We deduce that a decreasing rearrangement of the wavelet coefficients of f, written $\{c_j\}_{j=1}^{\infty}$ with corresponding wavelet ψ_j, will satisfy $|c_j|^2 < Cj^{-2d}$. This gives an estimate

$$\left\| f - \sum_{|c_j|^2 \ge \epsilon} c_j \psi_j \right\|^2 = \sum_{|c_j|^2 < \epsilon} |c_j|^2 \quad < \sum_{j > (\frac{\epsilon}{C})^{-1/2d}} j^{-2d} \quad < \quad C' \epsilon^{1-\frac{1}{2d}}.$$

Such an estimate is also valid for the finite sequences which arise when sampling a function supported on $[1, N]$. The N samples $\{f(j)\}_{j=1}^{N}$ may be compressed into a smaller number by replacing them with the largest few coefficients c_j. For a given error ϵ, we take the top $N' = (C/\epsilon)^{1/2d}$ coefficients. For very smooth signals where C is small and d large, we will obtain $N' \ll N$.

To compress a signal with d continuous derivatives, it is optimal to use an analyzing wavelet with d vanishing moments. With too few, the wavelet coefficients will not decrease as fast as possible. With too many, calculating the coefficients by filter convolution-decimation will become too expensive.

Figure 1 shows two wavelets and their scaling functions. The Haar wavelet has one vanishing moment, and has been used for a long time to represent nonsmooth (digital) signals. The "Coiflet" of 6 vanishing moments is a recent discovery, and is part of a family of almost symmetric smooth wavelets well suited to certain smooth signals.

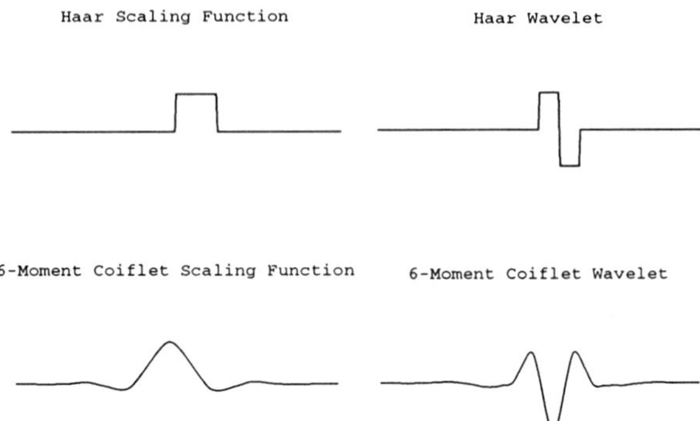

Figure 1. Haar and Coiflet wavelets and scaling functions.

4.2 Subband coding

A signal may be divided into frequency subbands by repeated application of convolution and decimation operators. For a 2-dimensional signal or "picture," these are usually tensor products of one-dimensional QMFs (separable filters), as defined below. There has also been considerable recent interest in nonseparable filters [2], for which the following discussion is equally valid, but for simplicity we used only separable filters in our fingerprint compression experiments.

Let $\{h_k\}, \{g_k\}$ belong to l^1, and define two decimating convolution operators $H : l^2 \to l^2$, $G : l^2 \to l^2$ as follows:

$$Hf_k = \sum_{j=-\infty}^{\infty} h_j f_{j+2k}, \qquad Gf_k = \sum_{j=-\infty}^{\infty} g_j f_{j+2k}.$$

H and G are called *quadrature mirror filters (or QMFs)* if they satisfy an orthogonality condition:

$$HG^* = GH^* = 0,$$

where H^* denotes the adjoint of H, and G^* the adjoint of G. They are further called *perfect reconstruction filters* if they satisfy the condition

$$H^*H + G^*G = I,$$

where I is the identity operator. These conditions translate to restrictions on the sequences $\{h_k\}, \{g_k\}$. Let m_0, m_1 be the bounded periodic functions defined by

$$m_0(\xi) = \sum_{k=-\infty}^{\infty} h_k e^{ik\xi}, \qquad m_1(\xi) = \sum_{k=-\infty}^{\infty} g_k e^{ik\xi}.$$

Then H, G are quadrature mirror filters if and only if the matrix below is unitary for all ξ:

$$\begin{pmatrix} m_0(\xi) & m_0(\xi+\pi) \\ m_1(\xi) & m_1(\xi+\pi) \end{pmatrix}.$$

This fact is proved in [8].

Figure 2 shows the traditional block diagram describing the action of a pair of quadrature mirror filters. On the left is convolution and down-sampling (by 2); on the right is up-sampling (by 2) and adjoint convolution, followed by summing of the components. The broken lines in the middle represent either transmission or storage.

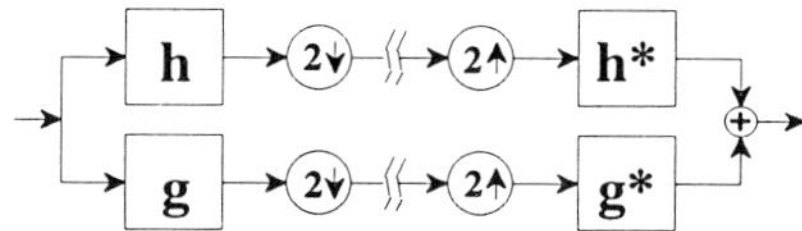

Figure 2. Block diagram of convolution-decimation and reconstruction.

Now we can define four two-dimensional convolution-decimation operators in terms of H and G, namely the tensor products of the pair of quadrature mirror filters:

$$F_0 \stackrel{\text{def}}{=} H \otimes H, \qquad F_0 v(x,y) = \sum_{i,j} v(i,j) h_{2x+i} h_{2y+j},$$

$$F_1 \stackrel{\text{def}}{=} H \otimes G, \qquad F_1 v(x,y) = \sum_{i,j} v(i,j) h_{2x+i} g_{2y+j},$$

$$F_2 \stackrel{\text{def}}{=} G \otimes H, \qquad F_2 v(x,y) = \sum_{i,j} v(i,j) g_{2x+i} h_{2y+j},$$

$$F_3 \stackrel{\text{def}}{=} G \otimes G, \qquad F_3 v(x,y) = \sum_{i,j} v(i,j) h_{2x+i} h_{2y+j}.$$

These convolution-decimations have the following adjoints:

$$F_0^* v(x,y) = \sum_{i,j} v(i,j) h_{2i+x} h_{2j+y},$$

$$F_1^* v(x,y) = \sum_{i,j} v(i,j) h_{2i+x} g_{2j+y},$$

$$F_2^* v(x,y) = \sum_{i,j} v(i,j) g_{2i+x} h_{2j+y},$$

$$F_3^* v(x,y) = \sum_{i,j} v(i,j) h_{2i+x} h_{2j+y}.$$

The orthogonality relations for this collection are as follows:

$$F_nF_m^* = \delta_{nm}I; \quad F_0^*F_0 \oplus F_1^*F_1 \oplus F_2^*F_2 \oplus F_3^*F_3 = I.$$

The space $l^2(\mathbf{Z}^2)$ of pictures may be decomposed into a partially ordered set $\mathbf{W}$ of subspaces $W(n,m)$, where $m \geq 0$, and $0 \leq n < 4^m$. These are the images of orthogonal projections composed of products of convolution-decimations. Put $W(0,0) = l^2$, and define recursively

$$W(4n+i, m+1) = F_{2i+j}^*F_{2i+j}W(n,m) \qquad \text{for } i = 0,1,2,3.$$

These subspaces may be partially ordered by a relation which we define recursively as well. We say W is a *precursor* of W' (write $W \prec W'$) if they are equal or if $W' = F^*FW$ for a convolution-decimation F in the set $\{F_0, F_1, F_2, F_3\}$. We also say that $W \prec W'$ if there is a finite sequence $V_1, \ldots, V_n$ of subspaces in $\mathbf{W}$ such that $W \prec V_1 \prec \ldots \prec V_n \prec W'$. This is well defined, since each application of F^*F increases the index m.

Subspaces of a single precursor $W \in \mathbf{W}$ will be called its *descendants*, while the first generation of descendants will naturally be called *children*. By the orthogonality condition,

$$W = F_0^*F_0W \oplus F_1^*F_1W \oplus F_2^*F_2W \oplus F_3^*F_3W.$$

The right hand side contains all the children of W.

As in [19], the coordinates with respect to the standard basis of $W(n,m)$ are in $F_{(1)} \ldots F_{(m)}W(0,0)$, where the particular filters $F_{(1)} \ldots F_{(m)}$ are determined uniquely by n. Therefore, we can express in standard coordinates the orthogonal projections of $W(0,0)$ onto the complete tree of subspaces $\mathbf{W}$ by recursively convolving and decimating with the filters.

We may think of the quadrature mirror filters H and G as low-pass and high-pass filters, respectively. Their tensor products form a partition of unity in the Fourier transform space. They can be described as nominally dividing the support set of the Fourier transform $\hat{S}$ of the picture into dyadic squares. If the filters were perfectly sharp, then this would be literally true, and the children of W would correspond to the 4 dyadic subsquares one scale smaller. We illustrate this in Figure 3.

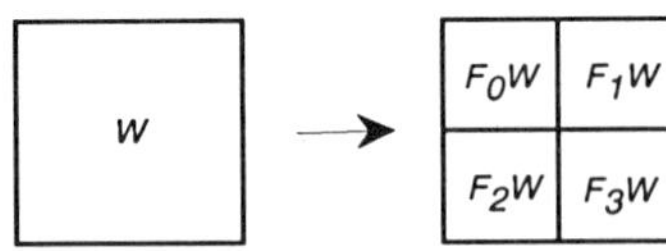

Figure 3. Four child subbands of a picture.

Figure 4 shows 2 generations of descendants, labeled as the complete decomposition of $\mathbf{R}^4 \times \mathbf{R}^4$. Within the dyadic squares are the n-indices of the corresponding subspaces at that level. If we had started with a picture of $N \times N$ pixels, then we could repeat this decomposition process $\log_2 N$ times.

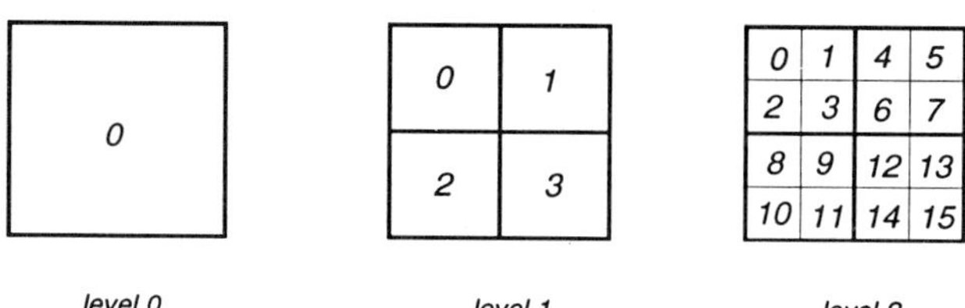

Figure 4. Ancestor subband and two generations of descendants.

4.3 Two-dimensional wavelet packets

An earlier paper [3] presented a method for generating a library of orthonormal basis by recursive QMF convolution-decimation. From an algorithmic standpoint, this is equivalent to recursive subband coding while retaining all intermediate subband decompositions. The coefficients produced at each stage are correlations of the signal with compactly-supported oscillatory functions called "wavelet packets." The analytic properties of these functions have been extensively studied [4,5]. If perfect reconstruction QMFs are used, then the wavelet packets satisfy some remarkable orthogonality properties.

From the tree $\mathbf{W}$ of subspaces we may choose a *basis subset*, defined as a collection of mutually orthogonal subspaces $W \in \mathbf{W}$, or lists of pairs (n, m), which together span the root. Basis subsets are in one-to-one correspondence with dyadic decomposition of the unit square. Classical subband coding takes coefficients from a fixed set of subbands, usually from a single level of the quadtree in Figure 6. Wavelet transform coding also extracts coefficients from a fixed collection of blocks, the octave subbands, which are schematically represented in the left part of Figure 5. The right hand part shows a typical dyadic decomposition down to level 4; its basis subset consists of the 16 pairs (0,1), (3,1), (4,2), (5,2), (6,2), (7,2), (8,2), (9, 2), (10,2), (45,3), (46,3), (47,3), (176,4), (177,4), (178,4), and (179,4). We may call such a basis subset B.

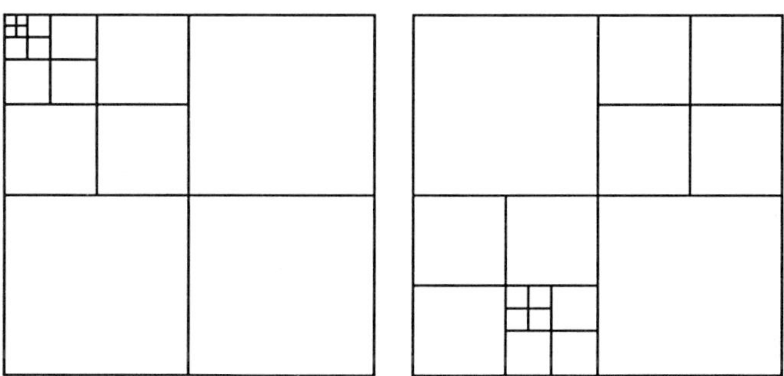

Figure 5. Octave and custom subbands.

Proposition 1. *The number of basis subsets grows exponentially with the size of the tree* $\mathbf{W}$, *which grows like* $O(N \log N)$ *with the number of pixels* N.

Proof: Let A_n be the number of bases in the library corresponding to a tree of $1+n$ levels, namely levels $0\dots n$. A decomposition to level n is only possible for a picture of size at least $N = 4^n$ pixels. Then $A_0 = 1$, and we can calculate $A_{n+1} = 1+A_n^4$, namely the root and combinations of the 4 independent children subtrees with A_n bases each. Simplifying this by discarding the 1 gives the estimate $A_{n+1} \geq 2^{4^n} = 2^N$.

To each subspace $W \in \mathbf{W}$ we may assign an information cost H_W. The quantity $H_W(S)$ measures the expense of including W in the decomposition used to represent the picture S. Define the *best basis* for representing S (with respect to H_W) to be the basis subset B_0 which minimizes

$$\sum_{W \in B} H_W(S)$$

over all basis subsets $B \subset \mathbf{W}$.

Some examples of information cost functions are listed in [6]. The simplest is the number of elements above a predetermined threshold ϵ, namely $H_W(S) = \#\{x \in S_W : |x| \geq \epsilon\}$, where S_W is the sequence of components of S in the direction of the standard basis vectors of W. This sequence is $F_{i_m} \dots F_{i_1} S$, where $W = F_{i_m}^* F_{i_m}^* \dots F_{i_1}^* F_{i_1} W(0,0)$. The following algorithm finds the basis subset with the fewest coefficients above the threshold.

Set a predetermined deepest level L. Label each subspace at level L as "kept," *i.e.*, the subspaces indexed by (n, L) for $0 \leq n < 4^L$. Next, set the level index m to $L-1$. Compare the information cost of the subspace $W(n,m)$ with the sum of the information costs of its children $W(4n, m+1)$, $W(4n+1, m+1)$, $W(4n+2, m+1)$, and $W(4n+3, m+1)$. If the parent is less than or equal to the sum of the children, then mark the parent as "kept." This means that by choosing the parent rather than the children, we will have fewer coefficients above the threshold in the representation of S. On the other hand, if the sum of the children is less than the parent, leave the parent unmarked but attribute to her the sum of the children's information costs. By passing this along, prior generations will always have their information costs compared to the least costly collection of descendants.

After all the subspaces at level $m = L-1$ have been compared to their children, decrement the level index and continue the comparison. We can proceed in this way until we have compared the root $W(0,0)$ to its 4 children. We claim that the topmost "kept" nodes in depth-first order constitute a best basis. That is, the collection of "kept" nodes W with no "kept" precursors is a basis subset which minimizes information cost. But this is easily proved by induction on the level index.

If we think of the coefficients below ϵ as negligible, we now have a basis in which the fewest coefficients are non-negligible. This cost function requires that we decide in advance what negligible means, which in some applications may not be feasible. This decision may be postponed by using a different measure of the concentration of energy into the coefficients. For example, there is an additive analog of Shannon entropy, namely,

$$H_W(S) = - \sum_{x \in S_W} x^2 \log x^2,$$

with S_W as above. This is related to the classical measure of the concentration of a probability distribution function.

4.4 Wavelet packets and the best basis

The "best-basis" method is to select a most efficient orthogonal representation of the picture from among these coefficients, using an efficiency functional chosen according to the application. Compression occurs when the coefficients are quantized and coded to remove redundancy, then stored together with an efficient representation of the chosen basis.

The overabundant set of coefficients is naturally organized into a quadtree of subspaces by frequency. Every connected subtree containing the root corresponds to a different orthonormal basis, and the totality of such bases forms a "library" of fast transforms. The most efficient of all the transforms in the library may be found by recursive comparison. The efficiency functionals are characterized by additivity across orthogonal decompositions. For such functionals, the choice algorithm will find the global minimum in $O(N)$ operations, where N is the number of pixels in the image.

The advantages of this method over traditional subband coding or fixed wavelet transform methods is that the basis is chosen automatically to best represent the particular image. Hence the name "best-basis." In this sense the transform is highly nonlinear. However, the transform is well-conditioned in the following sense: the basis choice has an exact representation and contributes no error, while the coefficient transform is orthonormal, *i.e.*, has condition number 1. Best-basis differs from statistical compression methods in that no statistical model of the ensemble of images is used. Compression can occur because pixel values are correlated by the smoothness of the sampled band-limited image. The exact nature of the correlation need not be known *a priori*, and deviant images will be automatically represented by deviant best bases.

Some related adaptive methods are adaptive quantization subband coding of images [16], and adaptive vector quantization of wavelet coefficients [12].

We present some results of using the method to compress high-resolution fingerprint images, using various notions of entropy as the efficiency functional and using mean-square deviation as the error criterion.

4.5 Compressing a best-basis representation

Suppose that B is a best-basis subset of W, chosen by counting coefficients above a predetermined threshold. We may then extract just these nonnegligible coefficients and transmit them, together with their locations in the tree. This number of coefficients is no greater than the number of pixels, since it is chosen after comparison with the original basis, among others. Define the *compression ratio* to be the ratio of retained coefficients to the original pixels. With thresholding and counting, the compression ratio measures how well a library represents a picture S at a fixed precision ϵ.

In practice it is sometimes necessary to fix the compression ratio, for example due to bandwidth limitations. In that case we may use the entropy cost function to obtain the most concentrated representation, and then take only as many of the largest coefficients as we can afford. This may be accomplished by first sorting into decreasing order by absolute value, then reading off the desired number of coefficients. Alternatively, since we know in advance how many coefficients we can use, it may be more efficient to bubble up the top few coefficients and discard the rest of the array. The second method is better if the number of retained coefficients is less than $\log N$, where N is the number of pixels.

For 2-dimensional signals (*i.e.*, pictures), we may use 3 analyzing wavelets and one scaling function formed from tensor products of 1-dimensional functions. Define

$$\begin{aligned}
ss_{\nu k} &= \int_{\mathbf{R}^2} f(x,y)\phi_{-\nu,k_x}(x)\phi_{-\nu,k_y}(y)\,dxdy,\\
sd_{\nu k} &= \int_{\mathbf{R}^2} f(x,y)\phi_{-\nu,k_x}(x)\psi_{-\nu,k_y}(y)\,dxdy,\\
ds_{\nu k} &= \int_{\mathbf{R}^2} f(x,y)\psi_{-\nu,k_x}(x)\phi_{-\nu,k_y}(y)\,dxdy,\\
dd_{\nu k} &= \int_{\mathbf{R}^2} f(x,y)\psi_{-\nu,k_x}(x)\psi_{-\nu,k_y}(y)\,dxdy.
\end{aligned}$$

The 2-dimensional pyramid scheme is then:

$$\begin{array}{ccccccc}
f(k) \approx \langle f\,,\,\phi_{0k_x}\otimes\phi_{0k_y}\rangle & \xrightarrow{I} & ss_{0k} & \xrightarrow{H\otimes H} & ss_{1k} & \xrightarrow{H\otimes H} \cdots \\
H\otimes G, H\otimes G, G\otimes G: & & \downarrow\downarrow\downarrow & & \downarrow\downarrow\downarrow & \\
 & & sd_{1k},\ldots & & sd_{2k},\ldots &
\end{array}$$

Each arrow costs $(2R)^2$ multiplications per coefficient. The total complexity is $(2R)^2(2N-2)^2$, which is linear in the number of samples. The algorithm may be readily applied to million-sample signals, or pictures of 1024×1024 pixels, using a contemporary desk top computer.

4.6 Relationships among the methods

The wavelet basis is one particular basis subset of $\mathbf{W}$, namely all elements in the standard bases of the subspaces $W(1,m)$, $W(2,m)$, and $W(3,m)$ for $m = 1,2,\ldots$, together with $W(0,L)$ if we fix a deepest level L. Numerous other authors have investigated representing signals in the wavelet basis; a small sampling of this work is the papers [1,3,10,11,12,13,18,21]. While this will obviously yield no better choice of basis than the "best-basis," the savings in computation and the absence of side information may warrant its consideration when the signals are of a particular class.

One may also consider the correlation of a picture with the complete set of tensor products of wavelet packets. These form a larger nonhomogeneous tree of subspaces which must be labeled with an x-scale and y-scale, rather than with a single scale as above. There is a more general notion of admissible subset, and a best-basis search algorithm to find extrema. This basis will produce higher compression ratios at a given threshold, at a cost of greater computational complexity and increased overhead describing the basis. The practical disadvantages rule out this generalization for image compression; further details may be found in [20].

4.7 Efficient coding of best-basis coefficients

The transformation from a 2-dimensional signal to its best-basis representation is nonlinear since the choice of basis depends upon the signal itself. Together with the coefficients we must include the extra information describing which basis was used, and we must somehow indicate which coefficient each quantized value represents.

There are at least 2 ways to include this latter information. Best-basis coefficients may be individually tagged with their coordinates in the best-basis tree. Suppose this tree begins with an $N \times N$ signal and decomposes it down to level L, where $L \leq \log_2 N$. Then there are LN^2 wavelet packet coefficients, and it takes $\log_2 LN^2$ bits to encode each individual one. This method is used and documented in the wavelet packet software programs available by anonymous ftp from the Yale Mathematics Department [14].

Alternatively, we may agree upon an ordering of the coefficients, write out the coefficients in this order after quantization, and then entropy code the entire list. If we were to use a single basis like wavelets or DCT, then we need never explicitly tag any coefficients. We obtain compression because the quantized stream of coefficients has a lower entropy that the original stream of bytes.

We shall use a variation of the second method to code best basis elements. Namely, we will include some side information which describes the chosen basis, and we shall then write all the (quantized) coefficients from that basis out into a stream for entropy coding. This method is substantially more efficient, and is essential for a competitive picture compression algorithm.

4.7.1 Describing the basis, quantizing all the coefficients

Imagine $L+1$ arrays of $N \times N$ numbers. The first array represents the original signal, which we may call Z. The second is a concatenation of the 4 subspaces obtained via separable filter convolution-decimation, *i.e.*, the spaces $F_0(X)F_0(Y)Z$, $F_1(X)F_0(Y)Z$, $F_0(X)F_1(Y)Z$, and $F_1(X)F_1(Y)Z$. Array m represents the concatenations of the 4^m subspaces that make up level m of the wavelet packet decomposition. Of course, we must have $0 \le m \le L \le \log_2 N$. Picture these arrays stacked one atop the other, as in Figure 6 below.

Suppose that from this collection of arrays we have chosen a best basis. This will be a subset of the coefficients having the property that if one element of a subspaces is in the basis, then that whole subspace is in the basis. Also, if a subspace is in the basis, then none of its descendent or ancestor subspaces are in the basis. Such a subset can be identified with a cover by dyadic subarrays. Looking down through the stack of arrays, this cover gives a tiling of the original $N \times N$ array by square subarrays of size $2^{-m}N \times 2^{-m}N$, where m is the level from which that particular subspace was chosen.

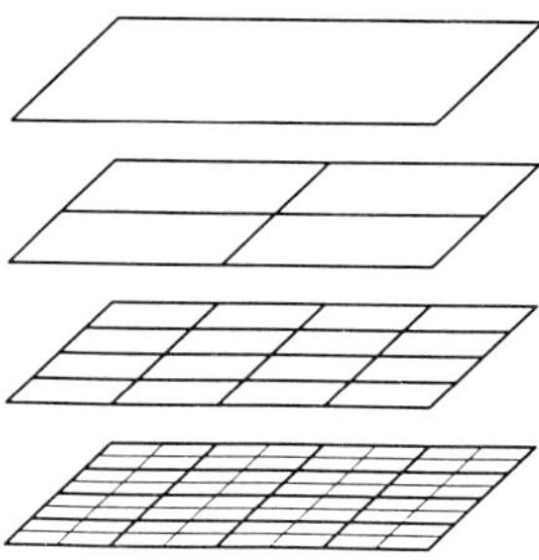

Figure 6. Quadtree of two-dimensional wavelet packet subbands.

4.7.2 Levels map

In this scheme, we use 2 arrays to describe the transformation, a "levels map" and a "coefficients list." The first has 2^{2L} integers of length $\log_2(1 + L)$ bits each, and describes the level from which a corresponding $N2^{-L} \times N2^{-L}$ block of coefficients in the next list was chosen. The second array contains the $N \times N$ coefficients from the best basis, scanned row-by-row or in some other agreed-upon pattern. We illustrate with 3 examples.

If the original signal turns out to be the best-basis representation, then every coefficient will be chosen from level 0, the levels map will consist of $2^L \times 2^L$ 0's, and the coefficients list will contain the original signal in the order it was scanned.

If the complete bottom level L turns out to give the best basis, then the levels map will contain $2^L \times 2^L$ L's and the coefficients list will contain all the coefficients from the bottom level L, scanned in some canonical order.

If the wavelet basis turns out to be the best, then the levels map will contain a description of the wavelet basis, and the coefficients list will contain the wavelet coefficients of the signal. To illustrate, suppose that $N = 16$, $L = 3$, and the signal is in fact the scaling function of amplitude 1, level 3 and position

(0, 0). Then the levels map and the coefficients list will look respectively like the following:

```
1 1 1 1 1 1 1 1    0 0 0 0 0 0 0 0 0 0 0 0 0 0 0 0
1 1 1 1 1 1 1 1    0 0 0 0 0 0 0 0 0 0 0 0 0 0 0 0
1 1 1 1 1 1 1 1    0 0 0 0 0 0 0 0 0 0 0 0 0 0 0 0
1 1 1 1 1 1 1 1    0 0 0 0 0 0 0 0 0 0 0 0 0 0 0 0
1 1 1 1 2 2 2 2    0 0 0 0 0 0 0 0 0 0 0 0 0 0 0 0
1 1 1 1 2 2 2 2    0 0 0 0 0 0 0 0 0 0 0 0 0 0 0 0
1 1 1 1 2 2 3 3    0 0 0 0 0 0 0 0 0 0 0 0 0 0 0 0
1 1 1 1 2 2 3 3    0 0 0 0 0 0 0 0 0 0 0 0 0 0 0 0
                   0 0 0 0 0 0 0 0 0 0 0 0 0 0 0 0
                   0 0 0 0 0 0 0 0 0 0 0 0 0 0 0 0
                   0 0 0 0 0 0 0 0 0 0 0 0 0 0 0 0
                   0 0 0 0 0 0 0 0 0 0 0 0 0 0 0 0
                   0 0 0 0 0 0 0 0 0 0 0 0 0 0 0 0
                   0 0 0 0 0 0 0 0 0 0 0 0 0 0 0 0
                   0 0 0 0 0 0 0 0 0 0 0 0 0 0 1 0
                   0 0 0 0 0 0 0 0 0 0 0 0 0 0 0 0
```

The number of coefficients and side data is large, but the information content is low, and entropy coding of this list will greatly shorten it.

We can obtain an estimate of the overhead cost of this method prior to entropy coding. There are 4^L additional integers each of length $\log_2(1+L)$ bits. This gives a total of $4^L \log_2(1+L)/N^2$ bits per pixel, which we can control by making $L < \log_2 N$.

4.7.3 Subspace lists

A basis may also be described by listing the subspaces it contains. One method is to list subspaces by level. We use 3 arrays:

1. an array `num[`L`]` of integers giving the number of subspaces chosen at each level,
2. an array of arrays `subspace[][]` listing the subspaces chosen from each level, and
3. an array containing the complete set of chosen coefficients.

The array `num[]` in item 1 contains L entries of varying length: 1 bit for entry 0, which describes whether level 0 is chosen; $2m$ bits for entry m which tells how many of the 4^m subspaces on level m are in the best-basis; and so on up to $2L$ bits for level L. Together the subspaces must account for all of the coefficients, so we have the relation:

$$\texttt{num[0]} * N^2 + \texttt{num[1]} * N^2/4 + \texttt{num[2]} * N^2/4^2 + \ldots + \texttt{num[}L\texttt{]} * N^2/4^L = N^2,$$

which implies that $\texttt{num[}L\texttt{]} = 4^L(1 - \texttt{num[0]} - \ldots - \texttt{num[}L-1\texttt{]}/4^{L-1})$, so it is not necessary to transmit this value to describe the basis. The total number of bits in the array `num[]` is, thus, $1 + 2\sum_{m=1}^{L-1} m = 1 + (L-1)L = L^2 - L + 1$.

The array `subspace[]` of arrays in item 2 contains L entries of length `num[0]`, `num[1]`, ..., `num[L-1]`, respectively. Array `subspace[m][]` contains integers of length $2m$ bits; array `subspace[0]` need not be allocated, since there is a unique subspace at level 0. Also, array `subspace[L][]` need not be allocated. Suppose that the subspace numbering scheme assigns the indices $4k$, $4k+1$, $4k+2$, and $4k+3$ to the subspaces descended from k. The subspaces in level L will be labeled by the integers $0, \ldots, 4^L$, and the ones which are actually present are the survivors after the indices $4^{L-m}k, \ldots, 4^{L-m}(k+1)-1$ are deleted for each subspace k at level $0 \leq m < L$.

With this numbering scheme for the subspaces, the wavelet basis of the example above would be represented by `num[]`$=\{0,3,3\}$, and the following subspaces list:

$$\begin{aligned} \text{subspace}[1][] &= 0,\ 1,\ 2 \\ \text{subspace}[2][] &= 12,\ 13,\ 14 \end{aligned}$$

The coefficients list will look the same as the above.

Hence the total number of bits in `subspace[][]` is

$$\sum_{m=1}^{L-1} 2m * \texttt{num}[m] \leq 2(L-1)4^{L-1}.$$

This coding method requires an extra $(L^2 - L + 1 + 2(L-1)4^{L-1})/N^2$ bits per pixel, which we again control by limiting L.

4.7.4 Coding the tree

The subspaces in the best basis are encountered in depth-first order as they are selected, and this order can be used to code the quadtree. The side information consists of an array `next[]` of integers which describe at which level the next node in the best basis resides. The nodes themselves are traversed in depth-first order, and the level of the next node is determined by a very short integer of only $log_2 L$ bits.

Some extra economy is possible, since the presence in the best basis of a node at level m sometimes implies that the following nodes can only be in levels $m, \ldots, L$. In an extreme case of this phenomenon, the first subspace at the deepest level L implies that all of its siblings are also in the best basis. Hence whenever the deepest level appears, it is not necessary to follow it with 3 "L" symbols. Further, since level 0 (the root of the tree, or original signal) is essentially meaningless we can use that value instead of L to represent the deepest level, thus using only the range $0, \ldots, L-1$ in the coding of the tree.

In the wavelet basis case above, the depth-first-search encounter order list would be $\{1,1,1,2,1,1,0\}$. We can either write the coefficients list in the same manner as before, or we can dump the coefficients from each subspace as it is encountered in depth-first order. In the example case, that would produce a stream of small square arrays:

```
0 0 0 0 0 0 0 0        0 0 0 0 0 0 0 0        0 0 0 0 0 0 0 0
0 0 0 0 0 0 0 0        0 0 0 0 0 0 0 0        0 0 0 0 0 0 0 0
0 0 0 0 0 0 0 0        0 0 0 0 0 0 0 0        0 0 0 0 0 0 0 0
0 0 0 0 0 0 0 0        0 0 0 0 0 0 0 0        0 0 0 0 0 0 0 0
0 0 0 0 0 0 0 0        0 0 0 0 0 0 0 0        0 0 0 0 0 0 0 0
0 0 0 0 0 0 0 0        0 0 0 0 0 0 0 0        0 0 0 0 0 0 0 0
0 0 0 0 0 0 0 0        0 0 0 0 0 0 0 0        0 0 0 0 0 0 0 0
0 0 0 0 0 0 0 0        0 0 0 0 0 0 0 0        0 0 0 0 0 0 0 0
0 0 0 0    0 0 0 0        0 0 0 0    0 0      0 0  0 0
0 0 0 0    0 0 0 0        0 0 0 0    0 0      0 0  0 0
0 0 0 0    0 0 0 0        0 0 0 0                         1 0
0 0 0 0    0 0 0 0        0 0 0 0                         0 0
```

This side information costs at most $4^{L-1}\log_2 L$ bits, or $4^{L-1}\log_2 L/N^2$ bits per pixel. The worst cases occur when every subspace in the bottom row is chosen and the description is `next[]` $= \{0, 0, \ldots, 0\}$ (4^{L-1} 0's), or when the next-to-bottom row is chosen and we have `next[]`$=\{L-1, \ldots, L-1\}$ (4^{L-1} $L-1$'s). The side information will also be compressed losslessly as part of the compression algorithm, and because of the redundancy in both of these cases the lossless compression will be extremely efficient.

4.8 Reconstruction

A picture represented as coefficients may be reconstructed by calculating the value at each point of the appropriate linear combination of wavelet packets. We cascade this computation for efficiency, as follows. First we allocate enough memory for the deepest level of the tree of subspaces that contains retained coefficients, and insert the coefficients into their appropriate locations. Then we reconstruct the parent subspaces by applying the adjoints of the convolution and decimation operators, which produces part of the next deepest level. Into this we add the retained coefficients which belong in that level, at their respective locations, and reconstruct the parents of this level. We continue in this manner until we have reconstructed the root, which now contains the picture.

4.9 Operation counts

Suppose that S is an N-element picture. Applying convolution-decimations to generate the tree of coefficient sequences requires $O(N \log N)$ operations. Calculating information costs has complexity $O(N \log N)$. Labeling "kept" subspaces is equivalent to a breadth-first search through the tree, which has complexity $O(N)$. Locating topmost "kept" subspaces is equivalent to a depth-first search, with complexity $O(N)$, and filling an output register with coefficients from the best basis takes an additional $O(N)$ operations.

We can perform a radix sort to determine the largest coefficients in the output register: this has complexity $O(N \log N)$. The alternative, extracting the top t coefficients, requires $O(tN)$ operations. We choose the more efficient method: in either case the total complexity of the compression algorithm is $O(N \log N)$.

Reconstruction from the retained coefficients has the same complexity as generating all the coefficients, since we must in general reproduce the entire tree, and the convolution-decimations have the same complexity as their adjoints. The constant in $O(N \log N)$ is smaller because no additional steps (such as searching for a best basis) are needed during reconstruction.

4.10 Tricks and optimizations

There are several enhancements commonly used in practice. Prior to sorting, wavelet packet coefficients can be weighted by their visibility, a psychophysiological observable measuring the human eye's spatial frequency response. They must then be unweighted prior to the reconstruction of the signal. This permutes the decreasing rearrangement so as to minimize the weighted error. Note that such bounded weighting amounts to conjugating the compression projection by a Calderón–Zygmund operator.

It is also possible to optimize the storage of the selected wavelet packet coefficients. For example, not all coefficients need to be quantized at the same precision. The relative quantizations are determined by experiment. Finally, the resulting bit stream can be entropy-coded to minimize the number of bits per picture.

§5 How to compare coding methods

All "lossy" transform coding methods introduce errors and distortion, but different methods produce radically different kinds and quantities of artifacts. Even when the errors can be quantified, their magnitude may not adequately describe the "visibility" of the artifacts they represent, so it is always necessary to judge the quality of a lossy compression through experiments involving its ultimate consumer. In the case of images to be examined by humans, this might be done with subjective evaluations by a large number of observers. In the case of machine analyzed images, a coding method may be judged by its transparency: how little it distorts subsequent computations.

5.1 Transform coding gain

One way to judge a compression method is by its effectiveness at concentrating pixel energy into a small number of codewords. This concentration is called "coding gain" and it can be quantified using yet another notion of entropy.

Suppose $\{f_n\}_{n=1}^N$ is a finite time series with autocovariance matrix $m_{ij} = \rho^{|i-j|}$, where ρ is close to 1. This corresponds to a first-order Gauss-Markov process with adjacent correlation coefficient ρ; such time series have been used to model pictures and speech signals for purposes of comparing transform coding schemes. An ideal scheme would de-correlate the samples, or equivalently diagonalize m; this is done by the Karhunen-Loéve transformation, which is impractically slow for our purposes. Instead, we shall use fast orthogonal transformations and measure the quality of approximate diagonalization by the so-called energy compaction or maximum transform coding gain, which is defined as

$$E(m) = \frac{\frac{1}{N}\sum_{i=1}^{N} m_{ii}}{(\prod_{i=1}^{N} m_{ii})^{1/N}}.$$

If m is diagonal, then the time series coefficients are independent random variables with variances m_{ii}. The entropy of the source of the time series is then $H(m) = \sum_i \log m_{ii}$, and in this case $E(m)$ is seen to be a decreasing function of $H(m)$.

Now let U and V be orthogonal matrices, whose columns may be considered an orthonormal basis. We call U a better basis than V if $E(UmU^*) > E(VmV^*)$, or equivalently, if $H(UmU^*) < H(VmV^*)$. For $\rho = 0.95$ and a matrix of order 32, the energy compaction for the local cosine transform is 9.47, while that for the Karhunen-Loéve basis is 9.54. By contrast, the discrete cosine transform gives 9.43.

5.2 Rate-distortion curves

Since the coefficients produced by the transform coding method will ultimately be quantized or approximated by a small number of chosen values, methods can be compared by the signal to noise ratio (distortion) produced by quantization and entropy coding giving a particular compression ratio (rate). Each coding method thereby yields a rate-distortion curve for each image, and one method is better than another if its rate-distortion curves consistently or on average have lower distortion at the desired rates.

Figure 7 shows the sample picture and rate-distortion curves for various coding methods applied to a typical picture intended for human consumption. Alternatively, Table 1 shows how mean square error increases as fewer and fewer transform coefficients are retained.

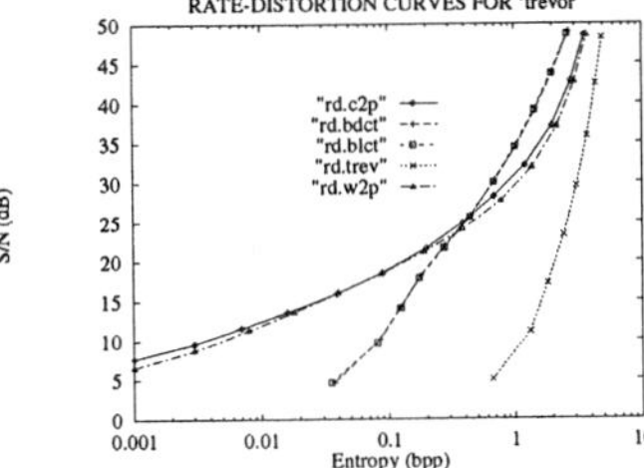

Figure 7. Original picture and rate-distortion curves for 5 compression methods.

5.3 Reduction in entropy

Write S for the original picture sequence and S' for the sequence written in the best-basis. We may ask what is the expected reduction in entropy obtained by going from S to S'. This is equivalent to evaluating

$$\int_{|x|^2=1} [H(x') - H(x)]\, d\omega(x),$$

where x' is x written in best-basis coordinates, and $\omega(x)$ is normalized surface measure on the $(N-1)$-sphere.

The simplest example is a 2-point signal $s = (s_0, s_1)$, periodized, so that the only available filters are the Haar filters $\sqrt{2}p = \{1,1\}$, $\sqrt{2}q = \{1,-1\}$. If we suppose that $\|s\| = 1$, then the original entropy is just $H(s) = -|s_0|^2 \log|s_0|^2 - |s_1|^2 \log|s_1|^2$. The only other basis gives the coefficients $s_0' = (s_0 + s_1)/2$, $s_1' = (s_0 - s_1)/2$, with corresponding entropy $H(s') = -|s_0'|^2 \log|s_0'|^2 - |s_1'|^2 \log|s_1'|^2$. We parameterize $s_0 = \cos x$, $s_1 = \sin x$ and evaluate the integral numerically to get

$$\begin{aligned}\frac{1}{2\pi}\int_0^{2\pi} [H(s) - H(s')]\, dx &= \frac{2}{\pi}\int_{\pi/8}^{3\pi/8} [H(s) - H(s')]\, dx \\ &= -0.210393231149915.\end{aligned}$$

Now the expected entropy of a 2-point distribution is

$$\frac{1}{2\pi}\int_0^{2\pi} H(s)\, dx = 0.3862942192715892$$

so that the expected entropy in the best basis is 0.1759009881216742.

Roughly speaking, entropy measures the logarithm of the number of meaningful coefficients in the signal. The above entropies correspond to 1.47 and 1.19 coefficients by this analogy, a reduction of 19%.

We have computed the entropy of the picture, and its entropy in the best basis obtainable with filters of 1, 2, 3, 5, and 10 vanishing moments. Hence we have also computed the exponential function of the entropy. These results are listed in Table 2 below.

Table 1. Rate versus distortion for compressions with Daubechies filters.

Filter Moments	Compression Ratio	MS (%)	MSE (dB)
10	100	6.58048	11
10	50	3.85890	14
10	20	1.54101	18
10	10	0.61173	22
10	5	0.16766	27
10	2	0.00755	41
5	100	5.87643	12
5	50	3.39416	14
5	20	1.34303	18
5	10	0.50403	22
5	5	0.12961	28
5	2	0.00376	44
3	100	6.18733	12
3	50	3.51632	14
3	20	1.36315	18
3	10	0.50358	22
3	5	0.12848	28
3	2	0.00332	44
2	100	6.32141	11
2	50	3.70738	14
2	20	1.46627	18
2	10	0.54608	22
2	5	0.14227	28
2	2	0.00370	44
1	100	7.36173	11
1	50	4.38014	13
1	20	1.83089	17
1	10	0.73331	21
1	5	0.18761	27
1	2	0.00317	44

Table 2. Entropy reduction by wavelet transformation.

Filter Moments	Entropy	Reduction	Theoretical Dimension
(original)	9.449	-	12700
10	4.371	-5.078	79
5	4.265	-5.184	71
3	4.397	-5.052	81
2	4.313	-5.136	75
1	4.345	-5.104	77

Figure 8. DCT versusLCT, 8 bits per coefficient.

5.4 Subjective evaluation

For the subjective evaluation, Figure 8 compare the perceived distortion for two methods which have the same distortion energy.

Figure 9 is a 256×256-pixel computer-rendered picture, created by Craig Kolb with a ray-tracing program. Pixels are stored as 8-bit integers.

The remaining pictures in Figures 10 to 24 are reconstructed from compressions performed with the author's implementation of the 2-dimensional pyramid algorithm. Coefficients are sorted then selected in decreasing order of absolute value. The definition of compression ratio as it is used in the captions is the number of pixels divided by the number of coefficients used to represent the picture.

§6 Transforming compressed pictures

As described in [7], the wavelet packet coefficients of a picture contain a mixture of spatial and spectral information. A list of the most energetic subspaces used in a compressed picture conveys a signature for the picture. Certain operators are very efficiently represented by their action on the wavelet packet coefficients. Some examples include spatial filtering and local image enhancement, edge and texture detection, and local rescaling.

For purposes of explanation we will assume that the picture is a function of 2 real variables supported in the region $[0,1] \times [0,1]$, with a resolution of 2^{-L}. Let (n, m, k) be the index of a coefficient in the complete wavelet packet expansion of a picture S. Here $m = 0, 1, \ldots, L$. We may divide $0 \leq n < 4^m$ into n_x and n_y by taking the odd and even bits in its binary expansion, respectively. These can be arranged in "sequency" order (by Gray-encoding; see [20]) and then will approximately correspond to x and y frequencies. Likewise, k may be divided into its x and y components k_x and k_y. Then the transforms described above may be defined by their action on (n_x, n_y, m, k_x, k_y) as follows:

Figure 9. Original ray-traced picture.

Spatial filtering: to remove (or attenuate) high frequency components in a particular direction $\alpha = \tan \frac{y}{x}$, simply discard any coefficient for which $|\alpha - \tan \frac{n_y}{n_x}| < \epsilon$ with $n_x > C$, $n_y > C$, where the cutoff frequency C and the directionality ϵ are parameters of the filter.

Local image enhancement: to remove high frequency noise from a particular region of the picture, employ spatial filtering as described above, but only on those coefficients for which 2^m is less than the diameter of the region, and for which $2^m(k_x, k_y)$ is a point in the region.

Edge detection: suppose we wish to find an edge of scale m_0 in a picture of resolution L, *i.e.*, a white region which darkens to black in a distance $2^{m_0 - L}$. Such an edge will contribute large coefficients to scales $1, 2, \ldots, m_0$ at high frequencies. We may graph it by selecting only coefficients $c(n_x, n_y, m, k_x, k_y)$ (above a (large) threshold, with $m < m_0$ and n greater than an appropriate monotone function of m, and then plot points at $2^m(k_x, k_y)$.

Texture detection: textures may be characterized by constant ratios between wavelet packet coefficients at nearby translations. Suppose for example that we wish to detect a texture in which $c(n_0, m_0, k+1) = -c(n_0, m_0, k+1)$ for all k in some region. An operator which added a coefficient at k to its neighbor at $k+1$ would have 0's in its range at k, indicating where that texture was located.

Local rescaling: we simply replace $c(n, m, k)$ with $c(n', m', k')$ for a restricted range of k's. The map $n \mapsto n'$, etc., is determined by the rescaling. For example, if $n' = 2n$ and $m' = m+1$, then we will increase the magnification of the picture locally with little change in the frequency content.

The survey article [18] describes these and other operations and their fast implementations in greater detail.

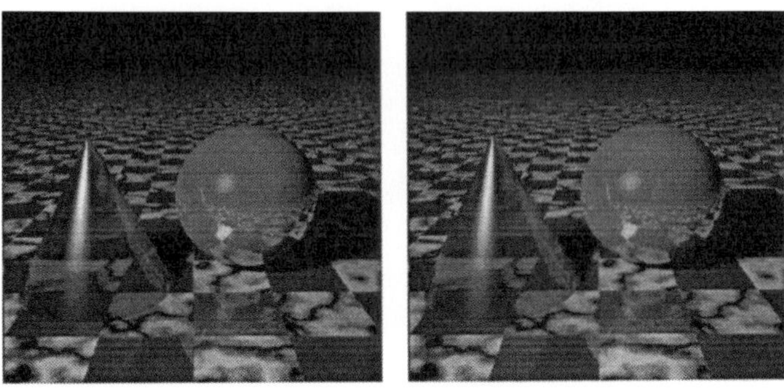

Figure 10. D 2 filter, compression ratios 2 and 5.

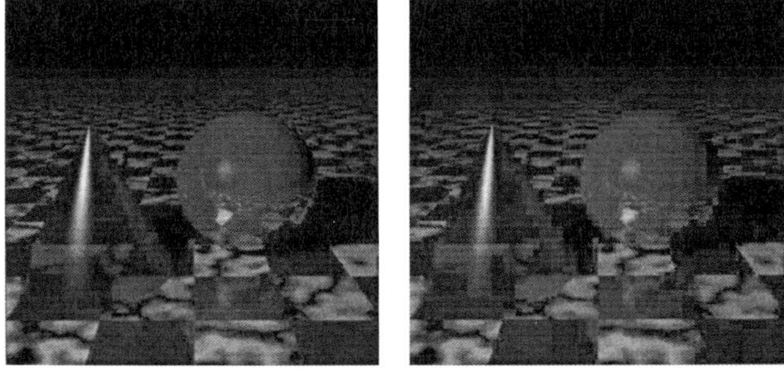

Figure 11. D 2 filter, compression ratios 10 and 20.

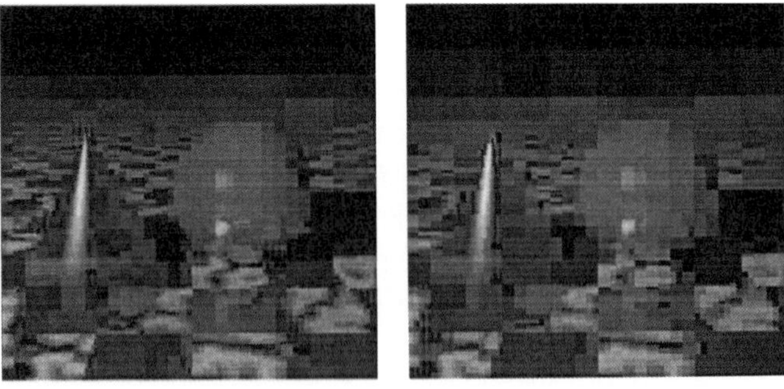

Figure 12. D 2 filter, compression ratios 50 and 100.

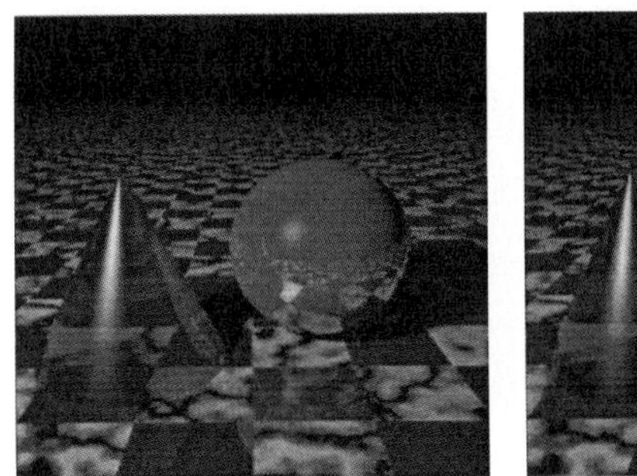

Figure 13. D 4 filter, compression ratios 2 and 5.

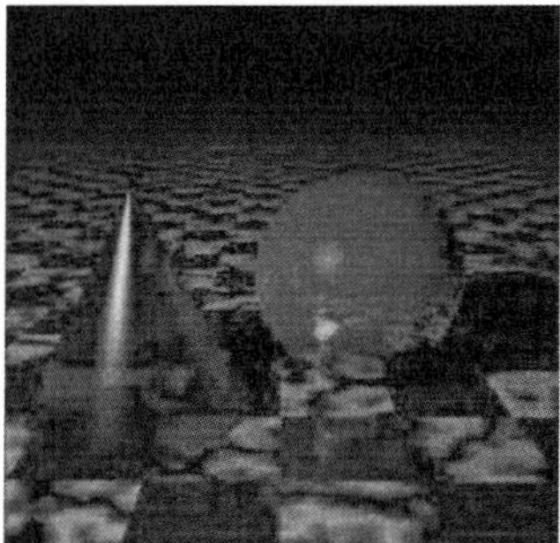

Figure 14. D 4 filter, compression ratios 10 and 20.

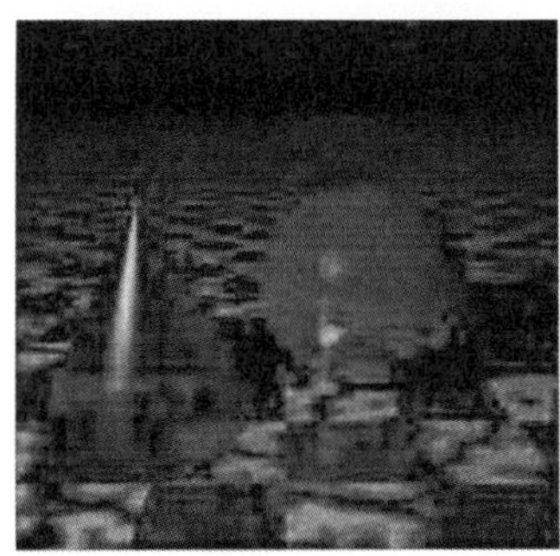
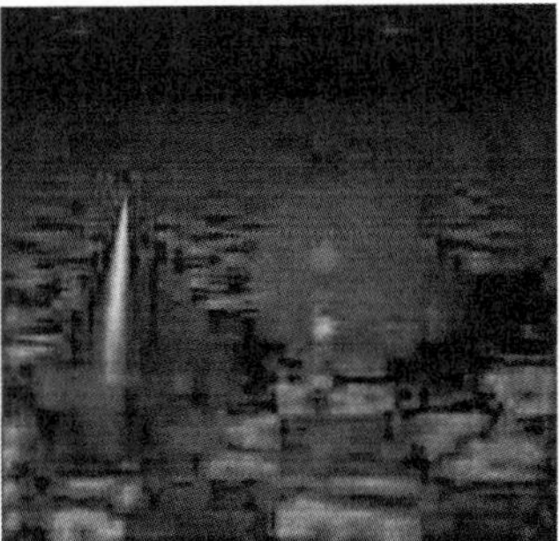

Figure 15. D 4 filter, compression ratios 50 and 100.

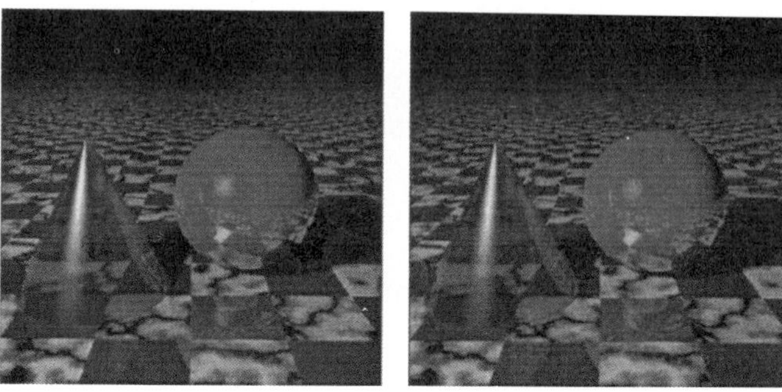

Figure 16. D 6 filter, compression ratios 2 and 5.

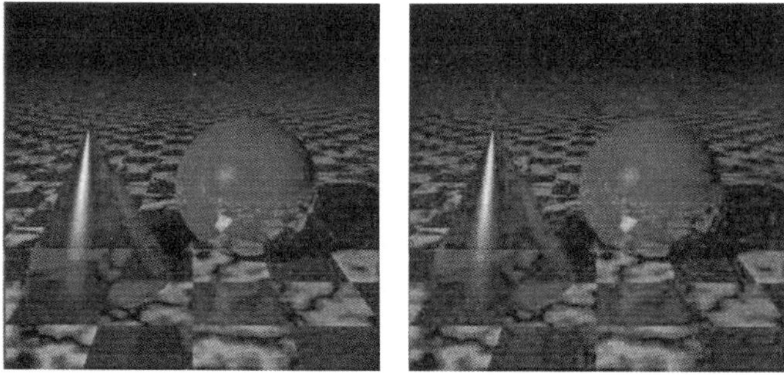

Figure 17. D 6 filter, compression ratios 10 and 20.

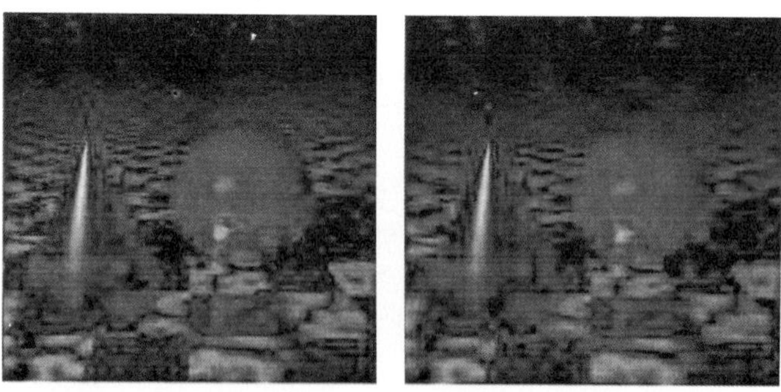

Figure 18. D 6 filter, compression ratios 50 and 100.

Figure 19. D 10 filter, compression ratios 2 and 5.

 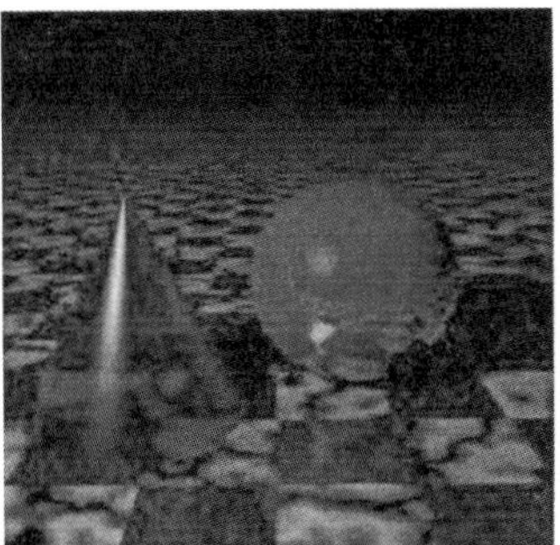

Figure 20. D 10 filter, compression ratios 10 and 20.

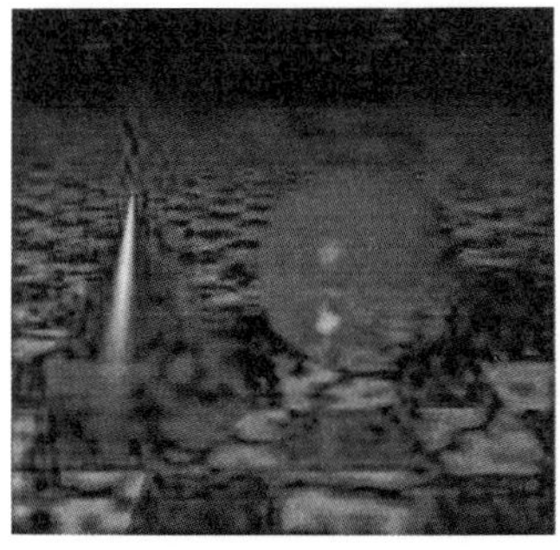 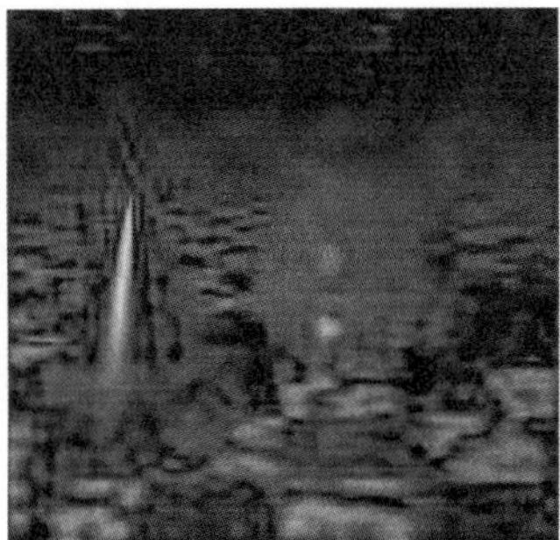

Figure 21. D 10 filter, compression ratios 50 and 100.

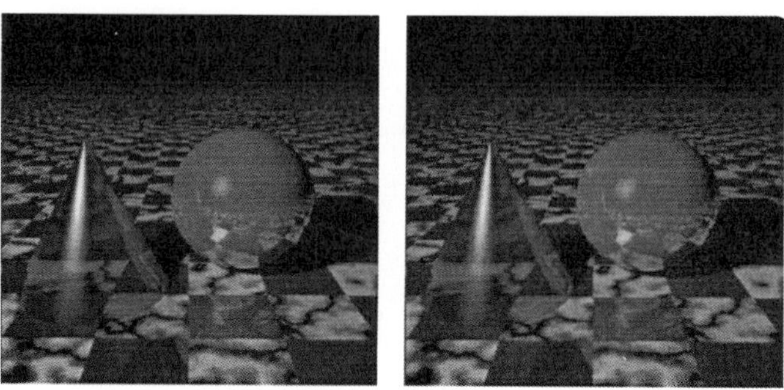

Figure 22. D 20 filter, compression ratios 2 and 5.

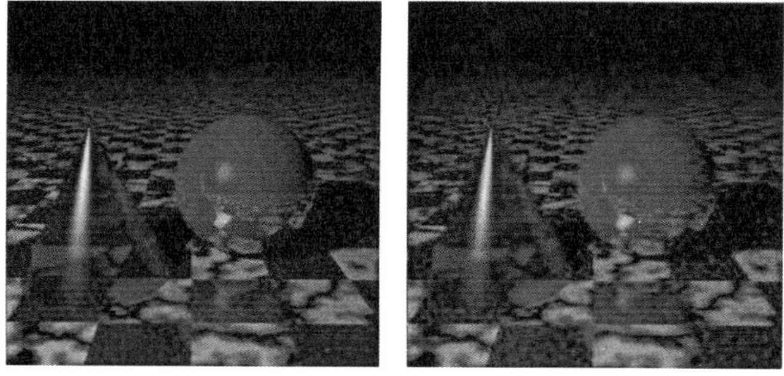

Figure 23. D 20 filter, compression ratios 10 and 20.

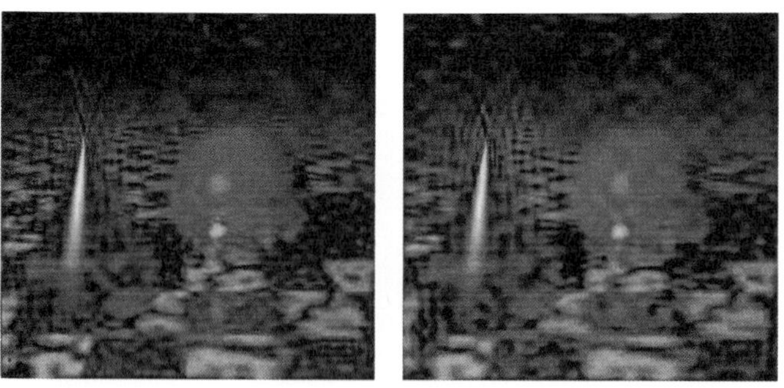

Figure 24. D 20 filter, compression ratios 50 and 100.

§7 Source programs

Below is a printout of three C programs which perform block discrete LCT on a 256x256 pixel grey scale image. The first is a header file "fold.h" which contains most of the subroutines. A standard library package "interface.o" (not listed here) contains standard procedures for reading file names from the command line and so on. Its procedures are declared in "interface.h", but only `open_file()` and `open_different_file()` are used in these programs, and their purpose is self-evident:

```
/*** fold.h Header file for 2-D LCT and iLCT  programs. ***/
#include "interface.h"
char *inputname, *outputname;
FILE *inputfile, *outputfile;

/*** Trivial functions to read and write single floats: */
void getv( float *value ) { fscanf(inputfile, "%f", value); }
void putv( float value )  { fprintf(outputfile, "%f\n", value); }

/*** Take a sequence X[] and put B[]-determined combinations into it: */
void fold_in_place(float *X, float *B, int bellReach) {
  int i, j;  float temp;
  for(i=0, j= -1; i<bellReach; i++, j--) {
    temp = B[i]*X[i] + B[j]*X[j];
    X[j] = B[i]*X[j] - B[j]*X[i];
    X[i] = temp;
  }
}

/*** Inverse of fold_in_place(): */
void unfold_in_place( float *X, float *B, int bellReach ) {
  int i, j;  float temp;
  for(i=0, j=-1; i<bellReach; i++, j--) {
    temp = B[i]*X[i] - B[j]*X[j];
    X[j] = B[i]*X[j] + B[j]*X[i];
    X[i] = temp;
  }
}

/*** Fill an array of arrays with cosines: */
float **make_cos_tab( int n ) {
  float **pcosine;  int i, j;  double freq, norm;
  pcosine = (float **) malloc(n*sizeof(float *));
  norm = sqrt(2.0/(double)n);    /* ...to get unit vector */
  for (i=0; i<n; i++)    {
    pcosine[i] = (float *)calloc(n, sizeof(float));
    freq = (1.570796326794897)*(double)(2*i + 1)/(double)n;
    for(j= 0; j<n; j++)
      pcosine[i][j] = (float)(norm*cos(freq*(0.5+(double)j)));
```

```
  }
  return(pcosine);
}

/*** Find DCT-IV in place by inner products: */
void dctiv( float *x, float *y, int n, float **pcosine ) {
  int k, j;  float sum;
  for(k=0; k<n; k++) {  /* Loop over the frequency parameter */
    sum = 0.0;
    for(j= 0; j<n; j++) sum += pcosine[k][j]*x[j];
    y[k] = sum;
  }
  for(k=0; k<n; k++) x[k] = y[k];
}

/*** Allocate and assign an array with the left edge of a bell */
float *make_bell( int bellReach, int zeroContact ) {
  double x;  int i, j;  float *bell;
  bell = (float *)calloc((unsigned)(2*bellReach),sizeof(float));
  bell = &bell[bellReach]; /* Fill bell[] with cutoff function */
  for(j= -bellReach; j<bellReach; j++) { /* iterate  sin() */
    x =    ((double)j+0.5)/(double)bellReach; /* real no. in [-1, 1] */
    for(i=0; i<zeroContact; i++) x = sin( HALFPI*x );
    bell[j] = (float)sin( (1.570796326794897)*(0.5 + 0.5*x) );
  }
  return(bell);
}
```

The program “fold.c” performs the transform to BDLCT coefficients, listing them in order in the output file specified on the command line or interactively:

```
/*** fold.c
    Perform a 2-D local cosine transform on an 256x256 picture divided into
square blocks of dimension 8x8 which overlap (orthogonally) by 4 samples.
Read samples, row by row, from specified 'inputfile' via the function
getv(), which is defined in fold.h. Output the coefficients column by
column into 'outputfile'. */
#include "fold.h"

main( int argc, char **argv ) {
  float *row, *col[256], *bell, y[8], **pcosine;  int i, j;
  open_file("input","rb",&inputname,&inputfile,argc,argv);
  open_different_file(inputname, "output","wb",
                    &outputname,&outputfile,argc,argv);
  row = (float *)calloc((unsigned)(256+2*4), sizeof(float));
  row = &row[4];/* Set array range to [-4, 256+4) */
  for(j=0; j<256; j++) {
    col[j] = (float *)calloc((unsigned)(256+2*4), sizeof(float));
    col[j] = &col[j][4]; /* Set array range to [-4, 256+4) */
```

```
  }
  bell = make_bell(4, 1); /* Make and store the bell[] */
  pcosine = make_cos_tab(8); /* Make the table of cosines */
  for(i=0; i<256; i++) {/* Loop over rows */
    for(j=0; j<256; j++) getv(&row[j]);  /* Read a row */
    for(j= -4; j<0; j++) row[j] = row[256+j];/* Periodize */
    for(j= 0; j< 4; j++) row[256+j] = row[j];
    for(j=0; j≤256; j+= 8) fold_in_place(&row[j], bell, 4);
    for(j=0; j<256; j+= 8) dctiv(&row[j], y, 8, pcosine);
    for(j=0; j<256; j++) col[j][i] = row[j]; /* Transpose */
  }
  for(j=0; j<256; j++) {/* Loop over columns */
    for(i= -4; i<0; i++) col[j][i] = col[j][256+i]; /* Periodize */
    for(i= 0; i< 4; i++) col[j][256+i] = col[j][i];
    for(i=0; i≤256; i+= 8) fold_in_place(&col[j][i], bell, 4);
    for(i=0; i<256; i+= 8) dctiv(&col[j][i], y, 8, pcosine);
  }
  for(j=0; j<256; j++) for(i=0; i<256; i++)  putv(col[j][i]);
  exit(0);
}
```

The program "unfold.c" inverts the BDLCT and dumps the reconstructed picture into a specified output file:

```
/*** unfold.c
    Invert the 2-D local cosine transform on the encoding of a picture  of size
256x256 pixels divided into square blocks of dimension 8x8 which overlap
(orthogonally) by 4 samples.  Assume that the coefficients are presented
column-by-column in 'inputfile'. Output the recontructed samples
row-by-row to 'outputfile'. */
#include "fold.h"

main(int argc, char **argv) {
  float *row[256], *col, *bell, y[8], **pcosine;  int i, j;
  open_file("input","rb",&inputname,&inputfile,argc,argv);
  open_different_file(inputname, "output","wb",
                  &outputname,&outputfile,argc,argv);
  col = (float *)calloc((unsigned)(256+2*4), sizeof(float));
  col = &col[4];     /* Set array range to [-4, 256+4) */
  for(j=0; j<256; j++) {
    row[j] = (float *)calloc((unsigned)(256+2*4), sizeof(float));
    row[j] = &row[j][4]; /* Set array range to [-4, 256+4) */
  }
  bell = make_bell(4, 1); /* Make and store the bell[] */
  pcosine = make_cos_tab(8); /* Make the table of cosines */
  for(i=0; i<256; i++) {     /* Loop over colums */
    for(j=0; j<256; j++) getv(&col[j]); /* Read column */
    for(j=0; j<256; j+= 8) dctiv(&col[j], y, 8, pcosine);
    for(j= -4; j<0; j++) col[j] = col[256+j];     /* Periodize */
    for(j= 0; j< 4; j++) col[256+j] = col[j];
```

```
    for(j=0; j≤256; j+= 8) unfold_in_place(&col[j], bell, 4);
    for(j=0; j<256; j++) row[j][i] = col[j]; /* Transpose*/
  }
  for(j=0; j<256; j++) {   /* Loop over rows */
    for(i=0; i<256; i+= 8) dctiv(&row[j][i], y, 8, pcosine);
    for(i= -4; i<0; i++) row[j][i] = row[j][256+i]; /* Periodize */
    for(i= 0; i< 4; i++) row[j][256+i] = row[j][i];
    for(i=0; i≤256; i+= 8) unfold_in_place(&row[j][i], bell, 4);
  }
  for(j=0; j<256; j++) for(i=0; i<256; i++) putv(row[j][i]);
  exit(0);
}
```

Acknowledgments. Research supported in part by NFS Grant number DMS-9302828 and AFOSR contract number F49620-92-J-0106.

References

1. Bradley, J. N., C. M. Brislawn, and T. E. Hopper, The FBI wavelet/scalar quantization standard for gray-scale fingerprint image compression, in *Visual Information Processing II* **1961**, Orlando, Florida, SPIE, April 1993.
2. Cohen, I. and I. Daubechies, Non-separable bidimensional wavelet bases, *Revista Matemática Iberoamericana*, 1992, to appear.
3. Coifman, R. R., Y. Meyer, S. R. Quake, and M. V. Wickerhauser, Signal processing and compression with wavelet packets, in *Progress in Wavelet Analysis and Applications, Proceedings of the International Conference "Wavelets and Applications"*, Y. Meyer and S. Roques (eds.), 77–93 Editions Frontieres, Toulouse, France, June 1990, 8–13.
4. Coifman, R. R. Y. Meyer, and M. V. Wickerhauser, Size properties of wavelet packets, in *Different Perspectives on Wavelets*, I. Daubechies (ed.), Am. Math. Soc., 453–470.
5. Coifman, R.R. Y. Meyer, and M.V. Wickerhauser, Wavelet analysis and signal processing, in *Wavelets and Their Applications*, G. Beylkin, R. R. Coifman, I. Daubechies, S. Mallat, Y. Meyer, L. Raphael, and M. Ruskai, (eds.) Jones and Bartlett, Boston, 1992, 153–178.
6. Coifman, R.R. and M.V. Wickerhauser, Entropy based algorithms for best basis selection, *IEEE Trans. Inform. Theory* **32** (1992), 712–718.
7. Coifman, R..R. and M..V Wickerhauser, Wavelets and adapted waveform analysis: A toolkit for signal processing and numerical analysis, in *Different Perspectives on Wavelets*, I. Daubechies (ed.), Minicourse Lecture Notes, Am. Math. Soc., 119–153.
8. Daubechies, I., Orthonormal bases of compactly supported wavelets, *Comm. Pure and Appl. Math.* **61** (1988), 909–996.
9. Daubechies, I., (ed.), *Different Perspectives on Wavelets*, No. 47, in Proceedings of Symposia in Applied Mathematics, Am. Math. Soc., San Antonio, Texas, 11-12 January, 1993.

10. Donoho, D. L., Unconditional bases are optimal bases for data compression and for statistical estimation, *Appl. and Comp. Harmonic Anal.* **1** (1) (1993), 100–115.
11. Devore, R., B. Jawerth, and B..J. Lucier, Image compression through wavelet transform coding, *IEEE Trans. Inform. Theory* **38** (1992), 719–746.
12. Mathieu, P., M. Barlaud, and M. Antonini, Compression d'images par transformée en ondelette et quantification vectorielle, *Traitment du Signal* **7** (2) (1990), 101–115.
13. Mallat, S. G. and J. Froment, Compression d'images et ondelettes: une double approche contours et textures, in *Problémes Non-Linéaires Appliqués, Ondelettes et Paquets D'Ondes*, Pierre-Louis Lyons (ed.), 227–247, INRIA, Roquencourt, France, Minicourse Lecture Notes, 17–21 June 1991.
14. Pascal, A., Yale University Department of Mathematics, InterNet Anonymous File Transfer (ftp) Site pascal.math.yale.edu [128.36.23.1], 1989–present.
15. Ruskai, M. B., et al., (eds.), *Wavelets and Their Applications*, Jones and Bartlett, Boston, 1992.
16. Ramchandran, K. and M. Vetterli, Efficient quantization for adaptive wave-packet tree structures, Columbia University, 1 June 1991, preprint.
17. Shannon, C. E. and W. Weaver, *The Mathematical Theory of Communication*, The University of Illinois Press, Urbana, 1964.
18. Wickerhauser, M. V., High-resolution still picture compression, *Digital Signal Processing: a Review Journal* **2** (4) (1992), 204–226.
19. Wickerhauser, M. W., Best-adapted wavelet packet bases, in *Different Perspectives on Wavelets*, I. Daubechies (ed.), Am. Math. Sóc., 1993, 155–171.
20. Wickerhauser, M. V., *Adapted Wavelet Analysis from Theory to Software*, AK Peters, Ltd., Wellesley, Massachusetts, 9 April 1994. 486+xii pages, to appear.
21. Zubair, L., K. R. Sreenivasan, and M. V. Wickerhauser, Compression of turbulence data and images using wavelet packets, in *Studies in Turbulence*, T. Gadsky, S. Sirkar, and C. Speziale (eds.), Springer Verlag, New York, 1991.

Mladen Victor Wickerhauser
Department of Mathematics
Washington University
St. Louis, Missouri 63130
victor@math.wustl.edu

Subject Index

WAVELET ANALYSIS AND ITS APPLICATIONS

CHARLES K. CHUI, SERIES EDITOR

1. Charles K. Chui, *An Introduction to Wavelets*
2. Charles K. Chui, ed., *Wavelets: A Tutorial in Theory and Applications*
3. Larry L. Schumaker and Glenn Webb, eds., *Recent Advances in Wavelet Analysis*
4. Efi Foufoula-Georgiou and Praveen Kumar, eds., *Wavelets in Geophysics*
5. Charles K. Chui, Laura Montefusco, and Luigia Puccio, eds., *Wavelets: Theory, Algorithms, and Applications*